Lecture Notes in Computer Science 2248
Edited by G. Goos, J. Hartmanis, and J. van Leeuwen

Springer-Verlag Berlin Heidelberg GmbH

Colin Boyd (Ed.)

Advances in Cryptology – ASIACRYPT 2001

7th International Conference on the Theory
and Application of Cryptology and Information Security
Gold Coast, Australia, December 9-13, 2001
Proceedings

Series Editors

Gerhard Goos, Karlsruhe University, Germany
Juris Hartmanis, Cornell University, NY, USA
Jan van Leeuwen, Utrecht University, The Netherlands

Volume Editor

Colin Boyd
Queensland University of Technology
School of Data Communications
Information Security Research Centre
GPO Box 2434, Brisbane Q4001, Australia
E-mail: boyd@isrc.qut.edu.au

Cataloging-in-Publication Data applied for

Die Deutsche Bibliothek - CIP-Einheitsaufnahme

Advances in cryptology : proceedings / ASIACRYPT 2001, 7th International Conference on the Theory and Application of Cryptology and Information Security, Gold Coast, Australia, December 9 - 13, 2001. Colin Boyd (ed.). - Berlin ; Heidelberg ; New York ; Barcelona ; Hong Kong ; London ; Milan ; Paris ; Tokyo : Springer, 2001
(Lecture notes in computer science ; Vol. 2248)

CR Subject Classification (1998): E.3, G.2.2, D.4.6, K.6.5, F.2.1-2, C.2, J.1

ISSN 0302-9743
ISBN 978-3-540-42987-6 ISBN 978-3-540-45682-7 (eBook)
DOI 10.1007/978-3-540-45682-7

This work is subject to copyright. All rights are reserved, whether the whole or part of the material is concerned, specifically the rights of translation, reprinting, re-use of illustrations, recitation, broadcasting, reproduction on microfilms or in any other way, and storage in data banks. Duplication of this publication or parts thereof is permitted only under the provisions of the German Copyright Law of September 9, 1965, in its current version, and permission for use must always be obtained from Springer-Verlag. Violations are liable for prosecution under the German Copyright Law.

http://www.springer.de

© Springer-Verlag Berlin Heidelberg 2001

Originally published by Springer-Verlag Berlin Heidelberg New York in 2001.

Typesetting: Camera-ready by author, data conversion by PTP-Berlin, Stefan Sossna
Printed on acid-free paper SPIN: 10845949 06/3142 5 4 3 2 1 0

Preface

The origins of the Asiacrypt series of conferences can be traced back to 1990, when the first Auscrypt conference was held, although the name Asiacrypt was first used for the 1991 conference in Japan. Starting with Asiacrypt 2000, the conference is now one of three annual conferences organized by the International Association for Cryptologic Research (IACR). The continuing success of Asiacrypt is in no small part due to the efforts of the Asiacrypt Steering Committee (ASC) and the strong support of the IACR Board of Directors.

There were 153 papers submitted to Asiacrypt 2001 and 33 of these were accepted for inclusion in these proceedings. The authors of every paper, whether accepted or not, made a valued contribution to the success of the conference. Sending out rejection notifications to so many hard working authors is one of the most unpleasant tasks of the Program Chair.

The review process lasted some 10 weeks and consisted of an initial refereeing phase followed by an extensive discussion period. My heartfelt thanks go to all members of the Program Committee who put in extreme amounts of time to give their expert analysis and opinions on the submissions. All papers were reviewed by at least three committee members; in many cases, particularly for those papers submitted by committee members, additional reviews were obtained. Specialist reviews were provided by an army of external reviewers without whom our decisions would have been much more difficult. A list of their names is included overleaf; I hope this is complete, but if there are omissions please be assured this was not intentional. My thanks go to all of them.

In addition to the contributed papers, I was delighted to be able to secure two eminent and engaging speakers for the invited talks at the conference. Arjen K. Lenstra talked on “Impossible Security: Matching AES Security Using Public Key Systems” and Brendan McKay talked on “Debunking the Bible Codes”. As is traditional at all IACR conferences, a rump session was held to give the opportunity to hear latest results and work in progress on a wide variety of topics. I would like to thank Bill Caelli for agreeing to take charge of this event with his usual flair.

The smooth running of Asiacrypt 2001 was engineered by an Organizing Committee led by the General Chair, Ed Dawson, and his deputy Mark Looi. Other members of the committee were Andrew Clark, Ernest Foo, Betty Hansford, Lauren May, Christine Orme and Jason Thomas.

I received sterling advice from many experienced people at all stages of the program preparation. Members of the ASC and the IACR board were all very supportive. Special mention must go to Tatsuaki Okamoto who acted as Advisory Member of the committee and provided advice based on his considerable experience. I would also like to particularly thank previous chairs of IACR conferences, Mihir Bellare, Bart Preneel, and Joe Kilian, who got very used to being bothered by me with requests for advice on all kinds of problems I had not encountered before and were always prepared to give their insightful opinions.

Any conference today relies heavily on technology to ease the administrative burden. All paper submissions to Asiacrypt 2001 were received electronically using the web based submission software which has been provided by Chanathip Namprempre. Papers were then seamlessly imported into the review software which was kindly provided by COSIC, Katholieke Universiteit Leuven, courtesy of Bart Preneel. The submission software was supported by COSIC's Wim Moreau, who was extremely helpful in providing advice and bug fixes, and went to great lengths to provide extra features in the software at very short notice. Installing and maintaining the software at ISRC was Andrew Clark. Andrew worked tirelessly to ensure that the web server and review server were (almost) always available, provided several additional features to the software, and generally worked miracles to solve all the problems I came up with.

I was assisted in many different ways by numerous other ISRC members. Ed Dawson, as General Chair and also as Director of ISRC, provided his support throughout. Greg Maitland and Kapali Viswanathan came to my rescue on numerous occasions.

Having seen all the people who contributed to the process of preparing the program, it may be deduced that I did very little myself. Nevertheless all those late nights and weekends went somewhere and I would like to acknowledge the forbearance of my family $D+C^3$, and my colleagues and research students, who experienced severe denial of service at many times over the months leading up to the conference.

October 2001 Colin Boyd

ASIACRYPT 2001

December 9-13, 2001, Gold Coast, Australia

Sponsored by the
International Association for Cryptologic Research (IACR)

General Chair
Ed Dawson, Queensland University of Technology, Australia

Program Chair
Colin Boyd, Queensland University of Technology, Australia

Program Committee

Masayuki Abe	NTT Laboratories, Japan
Ronald Cramer	BRICS & University of Aarhus, Denmark
ZongDuo Dai	Univ. of Science and Technology of China
Rosario Gennaro	IBM TJ Watson Research Centre, USA
Jovan Golić	Gemplus, Italy
Chi-Sung Laih	National Cheng Kung Univ., Taiwan
Kwok Yan Lam	PrivyLink International Ltd, Singapore
Pil Joong Lee	POSTECH, Korea
Arjen K Lenstra	Citibank, USA; TU Eindhoven, The Netherlands
Wenbo Mao	HP Laboratories, UK
pascal Paillier	Gemplus, France
Vincent Rijmen	Cryptomathic, Belgium
Bimal Roy	Indian Statistical Institute
Rei Safavi-Naini	University of Wollongong, Australia
Kouichi Sakurai	Kyushu University, Japan
Nigel Smart	University of Bristol, UK
Stefan Wolf	University of Waterloo, Canada
Moti Yung	CertCo, USA
Yuliang Zheng	Monash University, Australia

Advisory Member:
Tatsuaki Okamoto (Asiacrypt 2000 Program Chair) NTT Laboratories, Japan

External Reviewers

Joonsang Baek
Li Bao
Rana Barua
Alexandre Benoit
Simon Blackburn
Ian Blake
Daniel Bleichenbacher
Wieb Bosma
Eric Brier
Jan Camenisch
Ran Canetti
Denis Carabin
Sandeepan Choudhury
Andrew Clark
Christophe Clavier
Jean-Sebastien Coron
Robert Coulter
Ed Dawson
Jean-Francois Dhem
Ye Ding Feng
Matthias Fitzi
Matt Franklin
Atushi Fujioka
Eiichiro Fujisaki
Steven Galbraith
Shuhong Gao
Juan Garay
Pierrick Gaudry
Dieter Gollmann
Juanma González Nieto
Kishan C. Gupta
Stuart Haber
Ou Haiwen
Shai Halevi
Marie Henderson
Florian Hess
Nick Howgrave-Graham
D.J. Guan
Markus Jakobsson
Thomas Johansson
Marc Joye
Ari Juels
Meng Chow Kang
Jonathan Katz
Kazukuni Kobara
Reto Kohlas
Hartono Kurino
Peter Landrock
Pierre-Yvan Liardet
Yehuda Lindell
Christoph Ludwig
Anna Lysyanskaya
Greg Maitland
Subhamoy Maitra
Alfred Menezes
Bernd Meyer
Bill Millan
Shingo Miyazaki
Sean Murphy
David Naccache
Phong Nguyen
Jesper Buus Nielsen
Wakaha Ogata
Katsu Okeya
Sarbani Palit
Kenny Paterson
Beatrice Peirani
Fabien Petitcolas
Wang Ping
Florence Ques
Alon Rosen
Atri Rudra
Yasu Sakai
Palash Sarkar
Claus Schnorr
Gadiel Seroussi
Nickolas Sheppard
Igor Shparlinski
Leonie Simpson
David Soldera
Martijn Stam
Ron Steinfeld
Hung-Min Sun
Willy Susilo
Koutarou Suzuki
S.C. Tai
Tsuyoshi Takagi
Chik How Tan
Christophe Tymen
Wen-Guey Tzeng
Shigenori Uchiyama
Salil Vadhan
Eric Verheul
Kapali Viswanathan
Huaxiong Wang
Yejing Wang
Michael Wiener
Yi Xun
Jeff Jianxin Yan

Table of Contents

Block Ciphers

Provable Security

Threshold Cryptography

Two-Party Protocols

Zero Knowledge

Cryptographic Building Blocks

Elliptic Curve Cryptography

Anonymity

Author Index

Cryptanalysis of the NTRU Signature Scheme (NSS) from Eurocrypt 2001

Craig Gentry[1], Jakob Jonsson[2], Jacques Stern[3*], and Michael Szydlo[2]

[1] DoCoMo Communications Laboratories USA, Inc.
cgentry@dcl.docomo-usa.com
[2] RSA Laboratories, 20 Crosby Drive, Bedford, MA 01730, USA
{jjonsson,mszydlo}@rsasecurity.com
[3] Dépt d'Informatique, Ecole normale Supérieure, Paris, France
Jacques.Stern@ens.fr

Abstract. In 1996, a new cryptosystem called NTRU was introduced, related to the hardness of finding short vectors in specific lattices. At Eurocrypt 2001, the NTRU Signature Scheme (NSS), a signature scheme apparently related to the same hard problem, was proposed. In this paper, we show that the problem on which NSS relies is much easier than anticipated, and we describe an attack that allows efficient forgery of a signature on any message. Additionally, we demonstrate that a transcript of signatures leaks information about the secret key: using a correlation attack, it is possible to recover the key from a few tens of thousands of signatures. The attacks apply to the recently proposed parameter sets NSS251-3-SHA1-1, NSS347-3-SHA1-1, and NSS503-3-SHA1-1 in [2]. Following the attacks, NTRU researchers have investigated enhanced encoding/verification methods in [11].

Keywords: NSS, NTRU, Signature Scheme, Forgery, Transcript Analysis, Lattice, Cryptanalysis, Key Recovery, Cyclotomic Integer.

1 Introduction

Recently, Hoffstein, Pipher, and Silverman introduced a public-key signature scheme called NSS (the "NTRU Signature Scheme") [9]. This scheme is related to the NTRU cryptosystem, which was first introduced at the CRYPTO '96 rump session. An attack on NTRU was quickly found by Coppersmith and Shamir (see [4]), which led the authors to adopt larger parameters, and reformulate the underlying hard problem as a lattice problem. The current version of NTRU, as published in [6], remains unbroken. NSS is also related to the problem of finding short vectors in certain lattices, and is an improvement over an early version [7] presented at the CRYPTO 2000 rump session. This version proved to be insecure, which the designers observed at an early stage. Ilya Mironov [14] made the same observation independently a few months later. Basically, it

* This work has been partially supported by the French Ministry of Research under the RNRT Project "Turbo-Signatures"

C. Boyd (Ed.): ASIACRYPT 2001, LNCS 2248, pp. 1–20, 2001.
© Springer-Verlag Berlin Heidelberg 2001

appeared that signatures leaked information about the private key, which allowed for statistical attacks.

To eliminate the disclosed weaknesses, certain adaptations were made, yielding the scheme described in [9] and [8]. Unfortunately, these signatures *still* leak information about the private key. More precisely, it turns out that correlations between certain coefficients in the signature and the private key are sufficient to recover the entire public key.

Moreover, and even more dramatic, is a direct forgery attack which enables an adversary to sign arbitrary messages without any knowledge of the private key. While the flaw does not rule out potentially secure future revisions, our analysis shows that the scheme as presented in [9], [8] and [2] is completely insecure.

This paper is organized as follows. In section 2 we provide background and describe NSS in more detail. In section 3 we describe the efficient forgery procedure. Next, in section 4 we explain how to recover the key by examining valid signatures. In section 5 we discuss some revisions suggested by the authors of NSS to repair the signature scheme.

2 Description of NSS

Here we review some mathematics that underlie NSS, and give a brief description of the scheme. We refer readers to [9] and [8] for more detailed information.

2.1 Background Mathematics

The key underlying mathematical structure of the scheme is the polynomial ring

$$R = \mathbb{Z}_q[X]/(X^N - 1) \tag{1}$$

where N and q are integers. In practice, N is prime (e.g., 251) and q is a power of 2 (e.g., 128). Elements in R are polynomials of degree (at most) $N-1$ and with coefficients in the range $(-q/2, q/2]$.

Multiplication in this ring is like ordinary polynomial multiplication, but subject to the relations $X^{N+k} = X^k$ for any $k \geq 0$. This means that the coefficient of X^k in the product $a * b$ of $a = a_0 + a_1 X + \ldots + a_{N-1}X^{N-1}$ and $b = b_0 + b_1 X + \ldots + b_{N-1}X^{N-1}$ is

$$(a * b)_k = \sum_{i+j=k \bmod N} a_i b_j. \tag{2}$$

The multiplication of two polynomials in R is also called the *convolution product* of the two polynomials. For any polynomial $a \in R$, it is also convenient to introduce the *convolution matrix* of a as follows: Let M_a be the $N \times N$ matrix indexed by $\{0, \ldots, N-1\}$, where the element on position (i, j) is equal to $a_{(j-i) \bmod N}$. With this representation, the product of a and b can be also expressed as the product of the row vector $(a_0, \ldots, a_{N-1})$ with the matrix M_b. From now on, we will freely identify any polynomial with its corresponding row vector.

While the ring (1) may seem unnatural at first, it is directly related to the ring of integers in the cyclotomic field

$$\mathbb{Q}(\zeta_N) = \mathbb{Q}[X]/(X^{N-1} + ... + X + 1). \quad (3)$$

This field $\mathbb{Q}(\zeta_N)$ is a field extention of the rational numbers $\mathbb{Q}$, and has a subring of *Algebraic Integers*, $\mathbb{Z}(\zeta_N)$, analogous to the ordinary integers $\mathbb{Z} \subset \mathbb{Q}$. In fact, the set of polynomials $p \in R$ with $p(1) = 1$ is isomorphic to the integers $\mathbb{Z}(\zeta_N) \subset \mathbb{Q}(\zeta_N)$, and the convolution product described above in (2) is simply the ordinary multiplication operation in this field.

This field has been extensively studied and has been proposed for use in other cryptographic applications such as factoring and as basis for a public key cryptosystem (e.g. [17]), and is likely to appear in further analysis of NTRU related cryptosystems. However, further familiarity with this field is not required for the rest of this paper.

2.2 The NSS Signature Scheme

The public key of NSS consists of a polynomial h of degree $N-1$, and the private key of the scheme consists of two polynomials f and g with "small coefficients" such that $f * h = g$, where the polynomials are elements of $R = \mathbb{Z}_q[X]/(X^N - 1)$, and q and N are typically 128 and 251.

In order to describe the scheme further, additional parameters are needed. These parameters include the integer p, which is typically chosen to be 3, and the integers d_f, d_g and d_m, whose suggested values are respectively 70, 40 and 32. The latter parameters are used to define several families of polynomials denoted by $\mathcal{L}(d_1, d_2)$, a notation that refers to the set of polynomials of degree at most $N-1$ with d_1 coefficients 1, d_2 coefficients -1 and all other coefficients 0.

Key generation: Two polynomials f and g are defined as

$$f = f_0 + pf_1$$

$$g = g_0 + pg_1$$

where f_0 and g_0 are publicly known small polynomials (typically $f_0 = 1$ and $g_0 = 1 - 2X$). The polynomial f_1 is randomly chosen from $\mathcal{L}(d_f, d_f)$ and similarly g_1 is randomly chosen from $\mathcal{L}(d_g, d_g)$. It is required that f be invertible (i.e., there exists some f^{-1} with $f * f^{-1} = 1 \bmod q$). This is true with very high probability; in any case the preceding step may be repeated by choosing a different polynomial f_1.

Signature generation: To sign a message, one transforms the message to be signed into a message representative according to a hash function-based procedure such as that described in [2]. This message representative is a polynomial in $\mathcal{L}(d_m, d_m)$. The signer first computes

$$w = m + w_1 + pw_2$$

where w_1, w_2 are two polynomials with small coefficients generated at random in a rather complex manner that is described in Appendix B. The signer next computes the convolution

$$s = f * w \bmod q$$

and outputs the pair (m, s) as the signature of m.

Signature verification: A signature (m, s) consists of the message m together with the polynomial s of degree $N-1$, with coefficients reduced modulo q. Signature verification depends on two further parameters $D_{\min}$ and $D_{\max}$ (paper [9] suggests $D_{\min} = 55$ and $D_{\max} = 87$, together with the parameters suggested above), and upon the concept of *Deviation.* Given two polynomials, A and B of degree $N-1$, the deviation $\mathrm{Dev}(A, B)$ is the function that counts the number of coefficients where $(A \bmod q) \bmod p$ and $(B \bmod q) \bmod p$ differ. Here, modular reduction computes the coefficients in the interval $(-q/2, q/2]$ (resp. $(-p/2, p/2]$). If A and B are two random polynomials in the ring $\mathbb{Z}_q[X]/(X^N-1)$ and p equals 3, we would expect $\mathrm{Dev}(A, B)$ to be about $\frac{2}{3}N \approx 167$, since the probability that A_i and B_i differ modulo 3 is about $\frac{2}{3}$.

To verify a signature, first it is checked that $s \neq 0$. Then the polynomial $t = s * h \pmod q$ is computed, and the two conditions

$$D_{\min} \leq \mathrm{Dev}(s, f_0 * m) \leq D_{\max}$$

$$D_{\min} \leq \mathrm{Dev}(t, g_0 * m) \leq D_{\max}$$

are checked. If both conditions hold, the signature is accepted as valid.

The soundness of the scheme follows from technical estimates, which we omit. It should be noted that signature generation does not necessarily produce valid signatures. With the above parameters, signature verification fails in twenty percent of the cases and, when this happens, the signer has to create another signature.

3 Forgery Attacks

Paper [9] claims that a signature essentially proves possession of the secret trapdoor. Further, it envisions several potential attacks and concludes that the security of the system, with the above parameters, is comparable to RSA with 1024 bit moduli. We show that an attacker can generate forgeries (with slightly fewer than $D_{\max} = 87$ deviations) almost as quickly as the signer can generate signatures, without any knowledge of the private key. Furthermore, the attacker can generate forgeries with substantially fewer than $D_{\max}$ deviations by using lattice reduction.

3.1 Basic Forgery Attack: The Principle

In [9] and [8], NSS and NTRU are described as being based on essentially the same hard lattice problem. In fact, the problem underlying NSS is more of an error correction problem and, as demonstrated in many papers (see e.g. [16]), such problems take much larger dimensions to become hard.

The attack is very simple, once the perspective has been changed, as just indicated. The attacker's task is to find a pair of polynomials (s,t) that satisfy $t = s * h \pmod q$, as well as the deviation requirements:

$$55 \leq \mathrm{Dev}(s, f_0 * m) \leq 87;$$

$$55 \leq \mathrm{Dev}(t, g_0 * m) \leq 87.$$

Since s and t have $2N$ coefficients altogether, and the equation $t = s * h \pmod q$ imposes N linear constraints, the attacker has N degrees of freedom remaining in s and t with which he can try to satisfy the deviation requirements. With these N degrees of freedom, he sets

$$s_i \equiv (f_0 * m)_i \bmod p$$

and

$$t_j \equiv (g_0 * m)_j \bmod p$$

for $\lfloor N/2 \rfloor$ coefficients of s and $\lceil N/2 \rceil$ coefficients of t — i.e., he chooses about half the coefficients of s and half of t to be non-deviating. The remaining halves of s and t are left to chance. Since the chosen half of s (resp. t) has no deviations, and the remaining half will probabilistically deviate in about $\frac{2}{3}$ of the positions, overall about $\frac{1}{3}$ of the coefficients of s (resp. t) will deviate. Since $\frac{1}{3}N \approx 84 \leq D_{\max}$ for $(N, D_{\max}) = (251, 87)$, this process will usually generate a valid forgery after only a few iterations. In general, if $p = 3$ and $D_{\max} \geq \frac{1}{3}N$, then this attack will generate forgeries regardless of the size of N.

3.2 Basic Forgery Attack: The Details

In practice, the attack is slightly more complicated than the above, because it is possible that the constraints on s and t are incompatible. In this case, we say the attacker is *unlucky*. To avoid being unlucky, the attacker constrains only $k < N/2$ coefficients each of s and t. By setting up linear equations based on the constraints on t, we obtain a system of k linear equations modulo q over the $(N-k)$ free unknowns. The coefficients of the unknowns in this system form a $k \times (N-k)$ submatrix M of M_h whose coefficients are modulo q integers. We make the heuristic assumption that these coefficients are independent random bits, when reduced modulo 2.

Lemma 1. *Based on the heuristic assumption, the attacker is unlucky with probability at most* $\epsilon = \frac{1}{2^{N-2k}}$

Proof of lemma: We show that, with probability at least $1-\epsilon$, the columns of M mod 2 generate the entire k-dimensional space over the two-element field. If this holds, the system has rank k and it has solutions modulo 2 and modulo q as well, since q is a power of 2. Now, for every k-bit vector x, a column vector v is such that the inner product (v,x) is zero with probability 1/2. Since there are $N-k$ independent columns, x is orthogonal to all column vectors with probability $\frac{1}{2^{N-k}}$. Since there are 2^k possible values for x, we get that, with probability at least $1-\frac{1}{2^{N-2k}}$, there is no vector orthogonal to all column vectors of M mod 2. This means that these column vectors span the entire space. □

Setting $k=121$, the attacker will be lucky with probability at least $1-2^{-9}$. Assuming he is lucky, the attack now amounts to solving a system of 121 equations with 130 unknowns. However, a closer look shows that the matrix corresponding to this system does not depend on m, provided the attacker keeps the same selection of coordinates for his constraints; only the "righthand side" of the linear system does. This makes possible standard preprocessing of the linear system. To keep things simple, assume that, by suitably reindexing coefficients, one has brought the constrained coefficients of s in front and made the constrained coefficients of t the trailing block. Then, the matrix M of the system that the attacker has to solve is at the right bottom corner of M_h, defined by the last k rows of M_h and its last $N-k$ columns. Further relabeling makes the last k columns of M an invertible submatrix U. Again U lies at the bottom right corner of M_h. Thus, once the attacker precomputes the inverse U^{-1} of U, he may thereafter generate solutions to the linear system by choosing the $N-2k=9$ middle coordinates of s arbitrarily and obtaining the k last ones by a single multiplication by U^{-1}. For $k=121$, one readily checks that the obtained solution will satisfy the deviations requirement with probability $\geq 1/4$, so the attacker can expect to obtain the desired forgery after only 4 such multiplications. This makes forgery almost as fast as regular signature generation.

Alternatively, the attacker may search for a solution whose number of deviations lies closer to the middle of the interval $(D_{\min}, D_{\max})$, simply by searching through the 128^9 solutions to his linear equations. In a relatively short time, he can expect to find a solution (s,t) for which s and t have, for example, only 75 deviations.

A computer program written in C confirms the above analysis. Specifically, we have carried out the following two experiments:

1. The first with a public key that we manufactured, corresponding to the parameters from [9].
2. The second with a public key coming from one challenge from the NTRU web site. This challenge is for the encryption scheme. Unfortunately, there is no challenge for the signature scheme, but we wished to make it clear that we were working without the secret key. The challenge uses $N=263$ instead of $N=251$. We left the other parameters unchanged and observe that raising N only makes the forgery slightly more difficult.

We found forgeries whose distance pairs are respectively $(75,74)$ and $(79,79)$, close to the middle of the interval $(D_{\min}, D_{\max})$.

3.3 Forgery Attack with Lattice Reduction

In this section we make use of *lattice reduction*, a technique to find useful $\mathbb{Z}$ bases of lattices (discrete subgroups of $\mathbb{R}^n$). The celebrated LLL algorithm [13] is one of a family of algorithms that find bases containing short vectors in a lattice, and has found many uses in cryptology. The contemporary survey [15] provides an overview of lattice techniques and [1] provides detailed descriptions of many forms of the LLL reduction algorithm. In this paper, we use LLL as a black box algorithm to find a vector of short Euclidean norm in a lattice defined by the $\mathbb{Z}$ span of the rows of a matrix.

We can strengthen the basic forgery attack described above by supplementing it with a lattice reduction technique. We exploit the fact that we have considerable freedom when choosing the constrained coefficients of s and t and make the observation that all possible 'simple forgeries' differ from a given one by a $2N$-dimensional vector from an easily defined lattice. In other words, the idea here is (1) to generate an initial (s'', t'') using the basic forgery attack, and then (2) to correct some of the initial signature's deviations using lattice reduction. This hybrid approach allows us to generate forgeries averaging about 56 deviations in a few minutes.

Let (s'', t'') be the initial signature obtained using the basic forgery attack. Since $t'' = s'' * h \pmod q$, the vector (s'', t'') is in the lattice generated by the rows of the following matrix [1]:

$$L_{CS} = \begin{bmatrix} I_{(N)} & M_h \\ 0 & qI_{(N)} \end{bmatrix},$$

where $I_{(N)}$ denotes the N-dimensional identity matrix. In the basic forgery attack, to describe it in a slightly different fashion than previously, we found an invertible $k \times k$ submatrix U of M_h and then reordered the rows and columns of L_{CS} to obtain

$$L_{CS,2} = \begin{bmatrix} I_{(N-k)} & 0 & R & S \\ 0 & I_{(k)} & T & U \\ 0 & 0 & qI_{(N-k)} & 0 \\ 0 & 0 & 0 & qI_{(k)} \end{bmatrix},$$

where the invertibility of U made it easy to set the first k (actually, first $N-k$) and last k columns to whatever values we desired, modulo q. So, without loss of generality, we assume that in our initial signature (s'', t''), the first k coefficients of s'' and last k coefficients of t'' are chosen to be non-deviating (understanding that since the rows and columns of L_{CS} were reordered, s'' and t'' have been relabeled).

The attacker now would like to find some way of correcting the $N-k$ deviating coefficients of s'' (resp. t'') without touching the k non-deviating coefficients of s'' (resp. t''). To this end, the attacker would like to find a set of *harmless* row

[1] Coppersmith and Shamir introduced this lattice in their attack on NTRU [4]. Since that time, the inventors of NTRU have hypothesized that the security of NTRU and NSS is related to the apparently hard problem of finding short vectors in this lattice.

vectors in the lattice generated by $L_{CS,2}$ that contain zeros in the first k and last k positions, so that, for any vector (v_s, v_t) in this set, the pair $(s'' + v_s, t'' + v_t)$ will still be non-deviating in its first k and last k coefficients, while possibly having fewer deviations in its other positions.

We obtain the set of harmless vectors by making a slight modification to $L_{CS,2}$, obtaining a different lattice basis for the same lattice:

$$L_{CS,3} = \begin{bmatrix} I_{(N-k)} & -V & R-VT & 0 \\ 0 & I_{(k)} & T & U \\ 0 & qI_{(k)} & 0 & 0 \\ 0 & 0 & qI_{(N-k)} & 0 \\ 0 & 0 & 0 & qI_{(k)} \end{bmatrix},$$

where $V = SU^{-1} \pmod q$. To check that both generated lattices are indeed the same, one simply considers a linear combination of the first N rows of $L_{CS,2}$, corresponding to the sequence of coefficients $(\alpha_1, \cdots, \alpha_N)$. Writing the coefficients blockwise as (A_1, A_2), we see that exactly the same vector modulo q is obtained from the rows of $L_{CS,3}$ by a linear combination corresponding to $(A_1, A_2 + VA_1)$. The result follows. Notice that the rows $k+1$ to $N-k$ and rows $N+1$ to $2N$ of $L_{CS,3}$ have no nonzero coefficients in the first k or last k positions. We let $L_{harmless}$ be the lattice generated by these $(2N-2k)$ harmless vectors. These vectors are clearly linearly independent, so we conclude that the dimension of $L_{harmless}$ is exactly $(2N-2k)$.

Now, how do we use the lattice of harmless vectors to improve upon (s'', t'')? We will construct a lattice in which short vectors correspond to vectors with small deviations. Then we can search for a harmless vector, which, when added to (s'', t'') is a very short vector. This problem is an example of a closest vector lattice problem (CVP), related to the shortest vector lattice problem (SVP). See [15] for some comments on the relationship of the CVP to the SVP. To this end, we consider the lattice

$$L_{pq} = \begin{bmatrix} pL_{harmless} \\ (s', t') \end{bmatrix},$$

where (s', t') is the row vector with coefficients modulo pq satisfying $s' \equiv s'' \bmod q$ and $t' \equiv t'' \bmod q$, as well as $s' \equiv (f_0 * m) \bmod p$ and $t' \equiv (g_0 * m) \bmod p$ (again, and hereafter, keeping the relabeling in mind). For any row vector (v_s, v_t) in this lattice, $v_s * h = v_t \pmod q$. Moreover, v_s and v_t will satisfy one of three equations modulo p, depending on the value of the scalar coefficient of (s', t'):

$$v_s \equiv v_t \equiv 0 \bmod p, \text{ or}$$

$$v_s \equiv (f_0 * m) \bmod p \text{ and } v_t \equiv (g_0 * m) \bmod p, \text{ or}$$

$$-v_s \equiv (f_0 * m) \bmod p \text{ and } -v_t \equiv (g_0 * m) \bmod p\,.$$

If we could find a (v_s, v_t) with small coefficients — for example, in the range $(-q/2, q/2]$ — that does not satisfy the first condition[2] $v_s \equiv v_t \equiv 0 \bmod p$, then

[2] We observe in practice that this first condition may be avoided empirically with high probability via a small modifications of the lattice L_{pq}.

either (v_s, v_t) or $(-v_s, -v_t)$ would be a valid forgery having zero deviations. Unfortunately, finding a short (v_s, v_t) appears to be a hard lattice problem that cannot be solved in any reasonable time for lattices as large as L_{pq}.

So, instead of attempting to reduce L_{pq}, we select c columns of L_{pq}, corresponding to unchosen coefficients of (s'', t''), and define L_{final} to be the submatrix of L_{pq} consisting of these c columns. The lattice generated by L_{final} is only c-dimensional. We then apply lattice reduction to L_{final}, obtaining a c-dimensional output vector. Every coefficient of the output vector that falls in the interval $(-q/2, q/2]$ is now non-deviating. In general, the expected number of deviations for s (resp. t) after this process is $(2N - 2k - c)/3 + \Delta/2$, where Δ is the expected number of coefficients of the c-dimensional output vector that are outside the interval $(-q/2, q/2]$.

For concreteness, when attacking the "practical implementation of NSS," the attacker might set k to be 95 and c to be 150 and reduce the resulting lattice using a blocksize of 20. The lattice reduction algorithm is completed a few minutes, and empirically, the resulting s and t typically each deviate in about 56 positions. For NSS to be secure, $D_{\max}$ would, of course, have to be set much lower than 56 to ensure that the hybrid forgery attack fails with high probability.

4 Transcript Attacks

4.1 Description of the Attack

In this section we show how to recover the private keys f and g by examining a transcript of signatures. A transcript consists of some number of pairs (m, s) of messages with valid signatures created by the NSS signature algorithm. We also obtain t for each message via the relation $t = s * h \pmod q$. The basis of the attack is to examine the distributions of the s or t coefficients for a subset of messages m. By setting one coefficient of m to a fixed value, the distributions of the coefficients of s and t converge to a limiting distribution which depends on a chosen coefficient of the secret key f or g. Thus we compare sample distributions of s or t to precomputed estimations of the limiting distribution for each possible value of f or g's coefficient.

As mentioned above, both the NTRU corporation research team and Mironov observed that if the *averages* of these distributions were dependent on the key coefficients, the private keys would be extremely rapidly recovered by essentially averaging the signatures. This problem was quickly corrected in the following version of NSS [8], by altering the signature algorithm to guarantee that the average of these distributions would be indeed independent of the private key coefficients. However certain s and t distributions do depend on the possible f and g coefficient values and are still quite distinct from one another. Comparing these distributions to one another or to a precomputed distribution leads to an exposure of the private key. One interpretation of the attack is that it is an exploitation of information leaked through the higher moments of the signatures.

The signature of a message s is obtained via an algorithm which chooses w_1 and w_2 according to an intricate algorithm (see [8]), and sets

$$s = f * (m + w_1 + pw_2).$$

This algorithm to choose w_1 and w_2 is described in Appendix B and it is easily observed to be constructed so as to avoid the simple averaging attack.

All of our experiments have used the suggested parameters $q = 128$, $p = 3$, and $N = 251$, although the technique is generally applicable. For this parameter set, the polynomials w_1 and w_2 have approximately 25 and 64 nonzero entries each, and m is set to have 32 coefficients equal to 1 and 32 equal to -1. The coefficients of s thus depend on the private key f, the message m and the randomly generated polynomials w_1 and w_2. The situation is entirely similar with g and t since

$$t = g * (m + w_1 + pw_2).$$

In order to obtain the coefficient f_k, we fix indices i_0 and j_0 with $i_0 = j_0 + k \bmod N$, and examine the distribution of s_{i_0} over a transcript of messages with m_{j_0}=1. Unraveling the convolution arithmetic, we have

$$s_{i_0} = \sum_{j+k=i_0} f_k(m_j + w_{1,j} + pw_{2,j}).$$

We note that the quantity $W_j = m_j + w_{1,j} + pw_{2,j}$ is nearly (but not exactly, due to a quirk of the w_1 generation) identically distributed for each index j, when the distribution is taken over random values of m. We consider s_{i_0} to be the sum of the random variables W_j, and because f has exactly 140 nonzero entries, s_{i_0} is nearly a sum of 140 identically distributed random variables drawn from a fixed distribution. However, requiring that $m_{j_0} = 1$ (or 0 or -1) distinguishes the random variable W_{j_0} from the others. Our observation is that the term $f_k W_{j_0}$ in the sum defining s_{i_0} will contribute differently depending on the value of f_k.

Since an explicit calculation of the distribution of s_i would necessarily rely on the complex formulas for w_1 and w_2, we tested the heuristic reasoning above with several numerical experiments. There are many possible variants of this approach. For example, one could also set $m_j = 0$ or $m_j = -1$ for the appropriate coefficient, and thereby extract additional information from a given size transcript. We mention here only one key optimization. Although we fixed the index i_0 above, in fact *every* coefficient of m may be potentially used to obtain information about each coefficient of f. Namely, for a single message-signature pair, examining s_i for all indices i such that $m_j = 1$ and $i + j = k$ speeds up the convergence by a factor of 32, since m has 32 coefficients equal to 1. Thus we essentially examine the distribution

$$s_{i,j} = \sum_{j+k=i, m_j=1} f_k(m_j + w_{1,j} + pw_{2,j})$$

over a large set of transcripts. We performed several computer experiments which implemented the above optimized statistical analysis. Our programs, written in C, were able to recover the private key with a very high degree of accuracy.

4.2 Efficiency of the Attack

To create the estimated background limiting distribution, we simply created several million messages, each signed by a different private key, and calculated the distributions of s_k conditional on $m_j = 1$, and f_k assuming a particular value in the set $\{-3, 0, 3\}$. These statistics were gathered individually for each coefficient of f_k, but for simplicity of exposition we combine them, and define the three probability distributions $\mathbf{F_0}$, $\mathbf{F_3}$, $\mathbf{F_{-3}}$, to be the limiting distributions of s_i given $m_j = 1$, and the prescribed f_k value.

Given a valid transcript of signed messages, for each coefficient index i, the sample distribution of s_i is formed, and denoted $\mathbf{S_i}$. Next $\mathbf{S_i}$ is compared to each of the distributions $\mathbf{F_0}$, $\mathbf{F_3}$, $\mathbf{F_{-3}}$, according to some distribution comparison method. To do this, we define $S_i(x)$ be the probability that $s_i = x$ for some $x \bmod q$. Similarly define $F_{0,i}(x)$, $F_{3,i}(x)$, and $F_{-3,i}(x)$ to be the respective probabilities that $s_i = x$ (conditional on the prescribed value of f_k and $m_j = 1$ for $i = j + k$). One simple, effective measure useful for distinguishing these distributions is defined as

$$\Delta_i(v) = \sum_x (F_{v,i}(x) - A_i(x))(S_i(x) - A_i(x)),$$

where $A_i(x)$ is the average of the frequencies ($F_{0,i}(x)$, $F_{3,i}(x)$, and $F_{-3,i}(x)$). Thus for each coefficient i of f, we calculate $\Delta_i(v)$ for $v \epsilon \{-3, 0, 3\}$. Next, we ordered the values $\Delta_i(3)$ and $\Delta_i(-3)$ and select the smallest 70 values to identify the coefficients with $f = 3$, and $f = -3$ respectively.

There are clearly many other ways in which the distributions could be compared, for example with the L_2 norm. The convergence obtained with the above metric efficiently recovered the key coefficients, and alternative measures were only used in subsequent confirming experiments. We briefly note that the first coefficient of f has a slightly different distribution than the other indices, but this may be easily adjusted for, and is of minimal importance as it is just a single index.

After predicting the private key, we compared it to the actual private key, and checked our results. Here we summarize the number of mistakes made for several applications of this technique to transcripts of different lengths.

Signatures	Trials	Average Errors
100,000	31	7.3
300,000	16	2.6
400,000	5	1.2

The incorrectly predicted coefficients all correspond to indices which were near the end of the 70 minimal values in the orderings of $\Delta_i(3)$ and $\Delta_i(-3)$. In fact, in each trial, we identified a subset of 40 such 'dubious' indices before comparing to the private key, and verified that all of the errors were located at such indices. Given this localization of the errors, we conclude that it is feasible via direct search to obtain the *exact* private key given our estimated private key.

Depending upon the size of the index subset to examine, we estimate that it is possible to obtain the exact key via direct search, even if the guess has up to 10 errors [3] Thus with our method of examining the s distribution, the key

[3] Assuming the 10 errors are so localized, an upper bound on the number of potential corrections to f is equal to the binomial coefficient $(40, 10)$, or less than 2^{29}.

f may be completely deduced with as little as 100,000 signatures. We note also that significant partial information about a key's values may be used to greatly speed up certain lattice attacks, and in particular lattice reduction techniques may also be used to correct the estimated keys with a larger error tolerance than the brute force search method described above. These optimization techniques are not described further in this paper.

We note that it is likely that examining t rather than s would yield improved convergence rates. This conjecture is based on the fact that g is defined to have 80 nonzero entries rather than 140. We did not test this hypothesis directly in the above situation, but rather in the subsequent statistical attack on an NSS variant which we now describe.

4.3 An NSS Variant

Although the NSS version published in [8] was the subject of our first analysis, several variants proposed for the recent EESS standard [2] use a different private key structure. These key structures were proposed to increase the signing efficiency. Recall that the key space notation $\mathcal{L}(d, d)$ indicates a polynomial with d coefficients equal to 1 and d coefficients equal to -1. In the original version f was chosen to be $f = 1 + 3f_1$ where $f_1 \in \mathcal{L}(70, 70)$, and $g = 1 - 2x + 3g_1$ where $g_1 \in \mathcal{L}(40, 40)$.

The optimized key space is formed as follows. $f = 1 + 3f_1 * f_2$ and $g = 1 + 2x + 3g_1 * g_2$, where $f_1 \in \mathcal{L}(7, 7)$, $f_2 \in \mathcal{L}(5, 5)$, $g_1 \in \mathcal{L}(5, 5)$, and $g_2 \in \mathcal{L}(4, 4)$.

Because of cancelation or correlation in the product, f and g typically contain fewer nonzero elements and contain several coefficients equal to 6 or -6. Thus while the original scheme has private keys with a known number of coefficients that assume values in the set $\{3, 0, -3\}$, the new key have differing numbers of coefficients which typically assume values in the set $\{6, 3, 0, -3, -6\}$. (We ignore the first few indices of f and g for simplicity).

At first glance this appears to make the creation of the precomputed limiting distributions difficult. However, there are actually very few possible cases to consider. For example, a typical g has 62 coefficients equal to 3 or -3 and 5 equal to 6 or -6. The various other possibilities may be tried sequentially, in order of probability. Alternatively, we note that it is also true that the limiting distributions of s and t distinguish between the key structures with fewer or greater numbers of 6 and -6 coefficients very rapidly, without a need to fix values of m_j.

We found that the new private key structures led to even faster convergence. Several factors were changed simultaneously in the following experiment. First, we analyzed the distribution of t instead of that of s. Secondly, we assumed the number of coefficients in 6,-6 and 3,-3 was known, and did not attempt to deduce it. Thirdly, we used the L_2 norm to compare the distributions. Finally, a two-stage algorithm first found the 6 and -6 coefficients (very easily), and the remaining indices were ordered by the L_2 distances to the precomputed distributions. The values of f_k were predicted according to this order. We found few errors in these predictions, with a smaller number of signatures.

Signatures	Trials	Average Errors
30,000	10	5.6
50,000	10	4.8
100,000	5	1.8
200,000	5	1.0

As with the standard keys, it is possible to identify a subset of questionable indices for which the guess may be in error. Therefore even a direct search is feasible to obtain the exact private key. Thus we conclude that this last technique would find the exact private key with a transcript of size 30,000.

Further optimizations are possible. For example, for a hybrid attack one may estimate both keys f and g via a method described above, and then assign confidence measures to each index. We then assume that the $N/2$ coefficients of f and $N/2$ coefficients of g that have the highest confidence measures are in fact correctly chosen. The remaining coefficients are determined by the relation $g = f * h$ as in section 3, and finally we check that the deduced key pair (f, g) is correct. Only enough signatures needed to provide half of each of f and g would be needed to obtain the exact key. Another promising optimization would be to use the value of the message coefficients m_j to make an educated guess to the values of m_j before they were reduced modulo q, and compare these distributions. Refinements of this strategy might reduce the number of signatures to ten or twenty thousand. However, in light of our efficient forgery and the fact that the NSS scheme has recently been replaced with a revised version, such optimizations are not pursued further in this paper.

5 Countermeasures

Subsequent to the discovery of these attacks, the authors of NSS began searching for a secure revision of the NSS signature scheme. Jeffrey Hoffstein outlined several techniques to alter the scheme at Eurocrypt 2001. These modifications were formalized shortly thereafter in a technical note on the NTRU web site [11], with further improvements in the second draft [12].

Shortly after this paper was initially submitted, the authors of NSS settled on a revision of NSS, complete with suggested parameter choices. The precise definition of the revised scheme may be found in a preliminary standards document [2]. Currently, the third draft of this standard is available at the Consortium for Efficient Embedded Security web page [3].

The revised scheme does indeed appear to resist the attacks described in this paper. We do not rigorously define the new scheme here, but only mention the revised scheme's salient features and how they obviate the above attacks. Further details may currently be found in technical notes, a preprint, and a standards document [11,12,10,3].

The following is a partial list of the modifications.

1. **Private Key Generation:** In the version of NSS attacked in this paper, $f = f_0 + pf_1$ and $g = g_0 + pg_1$ where f_0 and g_0 are public parameters. In the revised scheme, $f = u + pf_1$ and $g = u + pg_1$ where u is kept private.

2. **Verification Criteria:** Verification is no longer based on the single criterion of deviations, but on multiple tests.
   – Norm Conditions: Verify that $|p^{-1}(s-m) \bmod q| < B$ and $|p^{-1}(t-m) \bmod q| < B$, where B is some bound on the *centered norms* [11].
   – Coefficient Distribution Checks: Perform a battery of specific checks (in [3]) on the distributions of the coefficients of s and t.
   – Moment Balancing: Optionally, use an alternate method of w_1 and w_2 creation, which alters the coefficients to include higher moment balancing.

These alterations were made to avoid the attacks presented in this paper, and therefore seem rather *ad hoc.* In particular, the verification protocol is strikingly lengthy [3], consisting of 17 steps! However the new key component u, norm conditions, and distributional criteria do appear to improve the security.

First, we discuss the new key component u. This is a very clever method of masking the combination of m coefficients which determine the distribution of w_0 values. Without the tool of controlling this distribution via selecting subsets of the messages, (say with $m_j = 1$) our transcript analysis can not effectively directly obtain distributions which are sensitive to the private key coefficient values. Adding u appears to make the distributions very close, even given millions of signatures. This renders the key recovery attack much less effective. Alternatively, the moment balancing techniques may also be used to make the distributions very close to one another.

Although the new verification protocol is a much less elegant revision than the use of u, it appears to serve its purpose of making forgery more difficult. The norm conditions relate the forgery problem of revised NSS to a (presumably hard) closest vector problem; the deviations criterion did not accomplish this. Also, the distribution checks appear to screen out forgeries generated by the forgery attacks above. However, it is unclear whether these new verification criteria are sufficient. It is likely that an attacker could already satisfy the norm conditions by simply using our (unmodified) forgery attack with the lattice reduction. Further cryptanalysis may show that it is possible to refine our attack to satisfy the distribution checks, as well.

The authors of NSS give some interesting analysis on how well the new scheme resists the attacks presented here [10]. They include a description of the new verification checks, a careful distributional analysis of the coefficients of the signatures in the new scheme, and a heuristic argument that signature forgery is as hard as a closest vector problem, assuming the adversary is given no transcript of previous signatures.

The new scheme is expected to receive renewed scrutiny, and since the key generation, signing and verification processes differ substantially, both forgery and key recovery techniques should be re-evaluated.

6 Conclusion

We wish to mention that our attack does not endanger the NTRU encryption scheme. On the other hand, we think that it shows the benefits of the *provable security* approach taken by cryptographic research in the last few years.

NSS had no security proof at all, not even relative to a precisely described lattice problem of some form. Lacking such proof, one could not easily argue that NSS was immune to potential simple attacks, as demonstrated by the present work. Following the attack, NTRU researchers have investigated enhanced encoding/verification methods in [11]. It appears that such methods can offer a form of provable security by reducing forgery to solving a well defined lattice attack. This rules out the method of section 3. However, such a reduction would not apply to an attacker who takes advantage of transcripts of previously obtained signatures, as in section 4. We believe that the heuristic approach taken by NSS designers makes it extremely difficult to prevent such transcript attacks.

Acknowledgments. The authors would like to thank Julien P. Stern for help with a C-program and discussions, Philip Hirschhorn for providing real signature transcripts, Burt Kaliski, Phong Nguyen and Yiqun Lisa Yin for helpful discussions, and lastly Jeffrey Hoffstein, Jill Pipher, and Joseph Silverman who, after their conception of NSS, were also supportive of cryptanalysis research efforts.

References

1. H. Cohen. A Course in Computational Algebraic Number Theory. Graduate Texts in Mathematics, 138. Springer, 1993.
2. Consortium for Efficient Embedded Security. Efficient Embedded Security Standard (EESS) # 1: Draft 1.0. Previously posted on `http://www.ceesstandards.org`.
3. Consortium for Efficient Embedded Security. Efficient Embedded Security Standard (EESS) # 1: Draft 3.0. Available from `http://www.ceesstandards.org`.
4. D. Coppersmith and A. Shamir. Lattice Attacks on NTRU. In Proc. of Eurocrypt '97, LNCS 1233, pages 52–61. Springer-Verlag, 1997.
5. G. H. Hardy, E. M. Wright. An Introduction to the Theory of Numbers, 5th edition. Oxford University Press, 1979.
6. J. Hoffstein, J. Pipher and J.H. Silverman. NTRU: A New High Speed Public Key Cryptosystem. In Proc. of Algorithm Number Theory (ANTS III), LNCS 1423, pages 267–288. Springer-Verlag, 1998.
7. J. Hoffstein, J.H. Silverman. NSS: The NTRU Signature Scheme. Preliminary version, August 2000.
8. J. Hoffstein, J. Pipher, J.H. Silverman. NSS: The NTRU Signature Scheme. Preprint, November 2000. Available from `http://www.ntru.com`.
9. J. Hoffstein, J. Pipher, J.H. Silverman. NSS: The NTRU Signature Scheme. In Proc. of Eurocrypt '01, LNCS 2045, pages 211–228. Springer-Verlag, 2001.
10. J. Hoffstein, J. Pipher, J.H. Silverman. NSS: The NTRU Signature Scheme: Theory and Practice. Preprint, 2001. Available from `http://www.ntru.com`.
11. J. Hoffstein, J. Pipher, J.H. Silverman. Enhanced encoding and verification methods for the NTRU signature scheme. Previously posted on `http://www.ntru.com/technology/tech.technical.htm`.
12. J. Hoffstein, J. Pipher, J.H. Silverman. Enhanced encoding and verification methods for the NTRU signature scheme (ver. 2). May 30, 2001. Available from `http://www.ntru.com/technology/tech.technical.htm`.
13. A. Lenstra, H. Lenstra, and L. Lovasz. Factoring polynomials with rational coefficients. Math. Ann. 261, pages 515–534, 1982.

14. I. Mironov. A Note on Cryptanalysis of the Preliminary Version of the NTRU Signature Scheme. Preprint, January 2001. Available at `http://eprint.iacr.org/2001/005/`.
15. P. Nguyen and J. Stern. Lattice Reduction in Cryptology: An Update. In Proc. of Algorithm Number Theory (ANTS IV), LNCS 1838, pages 85–112. Springer-Verlag, 2000.
16. J. Stern. A method for finding codewords of small weight. Coding Theory and applications, LNCS 388, pages 106–113. Springer-Verlag, 1989.
17. R. Scheidler and H. C. Williams. A public-key cryptosystem utilizing cyclotomic fields. Designs, Codes and Cryptography 6, pages 117–131, 1995.

A An Example of Signature Forgery

Here we give an example of how to forge signatures using the public key. Let parameters be as defined in NSS251-3-SHA1-1 [2]; $N = 251$, $p = 3$, $q = 128$, $V_m = 32$, $\text{Dev}^i_{\text{min}} = 55$, $\text{Dev}^i_{\text{max}} = 87$, $f_0 = 1$, $g_0 = 1 - 2X$. Let the public key $h = f^{-1} * g \pmod q$ be

	0	1	2	3	4	5	6	7	8	9	a	b	c	d	e	f
0 :	1	21	-59	-54	1	-33	-13	-11	-21	11	-30	31	-7	18	-61	85
1 :	3	41	52	-39	-30	4	-36	41	-11	56	46	-7	-7	7	-8	16
2 :	-58	-5	32	-3	-29	59	54	-25	53	48	47	32	-5	28	-9	-9
3 :	37	24	-50	17	-26	-58	10	39	4	-23	-55	-63	-29	-19	0	31
4 :	10	16	-25	28	29	-62	24	27	57	31	62	-61	35	39	-27	5
5 :	17	-22	22	28	32	41	14	-62	-18	-58	15	61	25	9	63	-9
6 :	47	30	0	58	58	-60	13	55	4	9	-62	11	58	-34	-39	13
7 :	40	27	36	-15	24	-31	37	23	31	55	-12	-20	43	-61	1	27
8 :	-44	-10	11	58	-63	-51	-46	-21	-6	-28	-17	-58	-28	6	21	-58
9 :	58	-3	10	-8	-26	48	12	64	2	14	-55	-20	-33	-24	-40	6
a :	-13	42	56	-23	-63	26	-52	-29	-4	35	12	-19	-24	47	-21	60
b :	-15	-17	63	62	55	17	-61	5	30	24	-32	-44	17	29	-63	57
c :	60	-25	-47	-51	2	11	-35	-44	-15	-5	7	-9	43	36	-18	-60
d :	-53	-2	-44	33	-27	-35	-17	5	-17	14	0	2	-6	49	29	-48
e :	-31	64	-8	64	-46	12	36	-57	23	-9	39	45	19	54	-21	49
f :	-7	-43	40	60	-45	20	-50	5	54	-13	-45					

Let the message to be signed be

	0	1	2	3	4	5	6	7	8	9	a	b	c	d	e	f	0	1	2	3	4	5	6	7	8	9	a	b	c	d	e	f
0:	0	0	0	0	0	0	0	0	1	0	0	0	-	0	0	1	0	0	0	0	0	0	0	-	1	1	0	0	0	0	-	-
2:	0	0	0	0	0	0	0	-	0	0	0	1	0	1	0	0	0	0	0	0	-	0	0	0	0	1	0	-	-	0	0	-
4:	0	0	0	-	-	-	0	0	-	0	0	0	0	1	0	0	0	0	-	-	1	0	1	0	0	0	0	0	1	0	0	0
6:	0	1	0	0	0	0	0	0	1	0	1	0	0	0	1	-	1	0	0	0	0	-	0	0	0	1	0	0	-	0	0	0
8:	0	0	1	0	-	0	0	-	0	0	0	0	0	0	0	0	1	0	0	0	0	0	0	0	0	0	0	0	0	0	-	0
a:	0	0	1	-	-	0	0	0	0	0	0	-	-	-	0	0	0	1	0	0	0	1	0	0	0	0	1	1	0	-	0	0
c:	0	0	0	0	0	-	0	0	0	0	0	0	0	0	1	1	0	1	-	0	-	0	0	0	-	0	0	1	0	0	0	0
e:	0	1	0	0	0	0	0	0	0	0	0	0	0	0	0	1	-	0	0	0	1	0	0	0	0	0	1					

(– denotes the integer -1). We now find an initial signature (s", t") by imposing $k = 95$ constraints on both s and t. For clarity in this example, we impose these constraints on the first 95 coefficients of s" and last 95 coefficients of t". Then, from the many possible (s", t"), we may get s" equal to

	0	1	2	3	4	5	6	7	8	9	a	b	c	d	e	f
0 :	0	0	0	0	0	0	0	0	1	0	0	0	-1	0	0	1
1 :	0	0	0	0	0	0	0	-1	1	1	0	0	0	0	-1	-1
2 :	0	0	0	0	0	0	0	-1	0	0	0	1	0	1	0	0
3 :	0	0	0	0	-1	0	0	0	0	1	0	-1	-1	0	0	-1
4 :	0	0	0	-1	-1	-1	0	0	-1	0	0	0	0	1	0	0
5 :	0	0	-1	-1	1	0	1	0	0	0	0	0	1	0	56	0
6 :	-2	38	32	-41	-32	38	-4	-21	-4	8	-47	-57	-40	27	3	39
7 :	-44	14	33	52	-5	34	57	4	16	-4	-45	-18	-23	-58	-22	6
8 :	56	59	5	-57	-33	-55	19	-41	52	26	50	-54	2	57	-27	-30
9 :	47	9	36	-42	-17	-50	-7	-44	-55	-47	-30	-45	-39	34	36	7
a :	-32	-19	4	23	-43	-40	-3	59	22	-52	46	42	24	-12	-19	7
b :	24	-43	64	-41	54	-31	-13	-31	-49	-55	57	-54	-56	-60	-48	-20
c :	-36	26	4	18	16	-61	33	45	-16	53	59	64	-60	-13	35	-47
d :	-23	50	45	44	-52	53	49	-29	-52	35	54	53	-15	50	-18	26
e :	-7	-1	30	-50	-17	-14	-54	31	-59	35	-21	-44	-14	62	-15	-5
f :	36	27	-6	6	36	29	-12	1	58	19	21					

and t" equal to

	0	1	2	3	4	5	6	7	8	9	a	b	c	d	e	f
0 :	25	-30	15	62	49	-24	-24	-12	15	-17	33	24	-61	64	-16	-57
1 :	-31	18	23	-29	27	39	-20	-35	-13	2	-54	39	36	-33	16	-13
2 :	-20	-45	-20	-3	25	10	54	-37	-33	41	-41	-47	-31	-15	31	-14
3 :	-52	16	-45	-10	-56	-22	-42	52	8	-20	55	13	30	32	-28	41
4 :	-57	25	49	-14	52	-38	-41	-35	22	-36	-27	-13	36	35	45	-10
5 :	54	-31	-9	3	-57	-37	9	-9	-16	-60	-59	14	18	26	-45	25
6 :	12	-40	11	31	41	5	-37	9	12	-21	-45	4	42	-18	-2	-29
7 :	-52	4	19	54	57	52	-23	-34	-31	-63	-60	-51	-14	42	2	13
8 :	56	-16	30	44	14	-37	-8	51	33	26	9	-12	-62	47	14	3
9 :	-50	18	-10	-33	24	-48	-4	60	-50	26	60	26	0	0	-1	-1
a :	0	0	1	0	1	-1	0	0	0	0	0	-1	1	1	-1	0
b :	0	1	1	0	0	1	1	0	0	0	1	-1	1	-1	-1	0
c :	0	0	0	0	0	-1	-1	0	0	0	0	0	0	0	1	-1
d :	1	1	0	-1	-1	-1	0	0	-1	-1	0	1	1	0	0	0
e :	0	1	1	0	0	0	0	0	0	0	0	0	0	0	0	1
f :	0	-1	0	0	1	1	0	0	0	0	1					

The pattern of deviations between s" and $(f_0 * m)$ looks as follows (each star denotes a deviation):

```
................................................................
..............................*.*******.**..*.**.*.***.***..***.
***.*******.*...**..******...*****...*.*********.***.*********.*
.**.*...****.**.****.***.*.*.*.***.***.***.***..*...**.****
```

For t" and $(g_0 * m)$ the pattern is:

```
...**...*....****.**..***.*...***.*.**..***.**.***.**..**..**.*.
****..****.*.***.***.****.****.*.******.***....**.*..******.****
****..****..***.***...*.**.*....................................
...........................................................
```

At this point s" and t" have 108 and 98 deviations, respectively. We now apply lattice reduction to coefficient positions 95 through 169 in s" and 81 through 155 in t" (the 75 leftmost coefficients in s" and 75 rightmost coefficients in t" that have not yet been constrained, for a total of 150 columns). For s, we get:

	0	1	2	3	4	5	6	7	8	9	a	b	c	d	e	f
0 :	0	0	0	0	0	0	0	0	4	0	0	0	-4	0	0	4
1 :	0	0	0	0	0	0	0	-4	4	4	0	0	0	0	-4	-4
2 :	0	0	0	0	0	0	0	-4	0	0	0	4	0	4	0	0
3 :	0	0	0	0	-4	0	0	0	0	4	0	-4	-4	0	0	-4
4 :	0	0	0	-4	-4	-4	0	0	-4	0	0	0	0	4	0	0
5 :	0	0	-4	-4	4	0	4	0	0	0	0	0	4	0	-30	0
6 :	9	-26	21	-54	39	33	21	-6	-47	9	43	-42	12	33	-50	-49
7 :	13	-24	3	6	-63	-19	12	33	6	7	-30	-36	-28	-12	-12	0
8 :	9	57	28	24	-52	12	18	20	6	-33	15	51	9	-33	-3	-9
9 :	-30	-17	-27	-21	6	9	-27	0	-36	-18	-42	-9	57	3	38	36
a :	36	-18	-47	35	47	-15	27	63	3	-12	-30	-22	-56	-40	10	57
b :	-49	-16	58	20	-53	-26	-19	29	-46	51	0	-49	-36	-29	15	-47
c :	-42	12	-52	51	40	47	42	-30	51	-53	-11	6	-39	51	13	-9
d :	-40	-51	-29	26	-32	44	3	27	-35	-9	55	-58	-60	0	-62	-17
e :	51	-26	30	-43	-50	7	-10	-8	-29	5	36	18	-30	-46	-21	42
f :	61	25	39	56	5	27	56	29	51	19	59					

For t we get:

	0	1	2	3	4	5	6	7	8	9	a	b	c	d	e	f
0 :	-37	41	-52	21	57	-57	-56	-63	-53	-2	5	40	-38	57	-62	18
1 :	7	57	-61	-32	18	-6	32	33	37	-30	36	62	-27	-15	54	-4
2 :	-34	-31	-51	24	-25	-8	62	57	38	28	-1	25	-50	-63	-63	-12
3 :	18	-10	-6	2	-39	-29	54	-13	-62	55	34	-35	-28	-60	-26	39
4 :	-2	-17	-44	-53	-38	-63	1	-19	-54	52	53	-61	50	10	-36	33
5 :	-27	-21	53	19	3	40	25	31	-33	-12	-54	-27	58	4	36	-15
6 :	21	-20	-5	-48	36	21	-30	42	4	-5	16	-56	0	-33	-41	-21
7 :	-39	1	30	18	-6	11	-43	6	-27	64	4	-6	-10	2	59	-3
8 :	30	-6	-8	-20	-31	20	3	17	-43	-15	-6	-15	15	-9	30	-3
9 :	-36	52	19	-3	-12	-9	-48	-48	27	-18	12	15	0	0	-4	-4
a :	0	0	4	0	4	-4	0	0	0	0	0	-4	4	4	-4	0
b :	0	4	4	0	0	4	4	0	0	0	4	-4	4	-4	-4	0
c :	0	0	0	0	0	-4	-4	0	0	0	0	0	0	0	4	-4
d :	4	4	0	-4	-4	-4	0	0	-4	-4	0	4	4	0	0	0
e :	0	4	4	0	0	0	0	0	0	0	0	0	0	0	0	4
f :	0	-4	0	0	4	4	0	0	0	0	4					

The deviation pattern for s is:

```
...............................................................
...............................................................
...........................................*.*.*****.***.**.*.*
..*.*....**....*******..*.**..**...*******...*.***.**.**.**
```

The deviation pattern for t is:

```
***...*...******..**..******..**.*..****.**..**..*.***.**..*****
****.********.*..............................................*.
...............................................................
...........................................................
```

Thus, we have produced an s and t that have 47 and 54 deviations from $(f_0 * m)$ and $(g_0 * m)$ respectively. These values are indeed even below the suggested parameter value of $\mathrm{Dev}^i_{\min} = 55$, which shows that our forgeries would pass even stricter deviation requirements.

Obviously the s and t of this example have highly unusual coefficient distributions modulo q, which the verifier could easily detect, but this need not be the case in general. We can make the coefficient distribution of s and t more ordinary by 1) constraining random coefficient positions and 2) distributing the values of the constrained coefficients of s" and t" more randomly modulo q, rather than setting them all equal to −1, 0 or 1.

B Determination of w_1 and w_2

The following pseudocode may also be found the appendix of [8]

```
let w2 have 32 +1's and 32 -1's
set w1[] to 0

compute s = f * (mes + 3 w2)
compute t = g * (mes + 3 w2)

reduce s and t modulo q
reduce s and t modulo p

//create w1, first try

for(i=0;i<N;i++)
  if(s[i] != mes[i] AND  t[i] != mes[i] AND  s[i] == t[i])
    w1[i] = (mes[i] - s[i]) mod p

  if(s[i] != mes[i] AND  t[i] != mes[i] AND  s[i] != t[i])
    w1[i] = 1 or -1 with 50% probability
loop
```

```
//create w1, second try

for(i=0;i<N;i++)
  if(s[i] != mes[i] AND  t[i] == mes[i])
     w1[i] = (mes[i] - s[i]) mod p with 1/4 probability

  if(s[i] == mes[i] AND  t[i] != mes[i])
     w1[i] = (mes[i] - t[i]) mod p with 1/4 probability

  if(w1 has more than 25 nonzero coefficients)
     break out of the loop
loop

// modify w2 to prevent averaging attack

for(i=0;i<N;i++)
    with probability 1/p, w2[i] = w2[i] - (mes[i] + w1[i])

w = w1 + 3 w2
```

On the Insecurity of a Server-Aided RSA Protocol

Phong Q. Nguyen[1*] and Igor E. Shparlinski[2**]

[1] CNRS/Département d'Informatique, École normale supérieure
45 rue d'Ulm, 75005 Paris, France
pnguyen@ens.fr http://www.di.ens.fr/~pnguyen/
[2] Department of Computing, Macquarie University
Sydney, NSW 2109, Australia
igor@ics.mq.edu.au http://www.comp.mq.edu.au/~igor/

Abstract. At Crypto '88, Matsumoto, Kato and Imai proposed a protocol, known as RSA-S1, in which a smart card computes an RSA signature, with the help of an untrusted powerful server. There exist two kinds of attacks against such protocols: passive attacks (where the server does not deviate from the protocol) and active attacks (where the server may return false values). Pfitzmann and Waidner presented at Eurocrypt '92 a passive meet-in-the-middle attack and a few active attacks on RSA-S1. They discussed two simple countermeasures to thwart such attacks: renewing the decomposition of the RSA private exponent, and checking the signature (in which case a small public exponent must be used). We present a new lattice-based provable passive attack on RSA-S1 which recovers the factorization of the RSA modulus when a very small public exponent is used, for many choices of the parameters. The first countermeasure does not prevent this attack because the attack is a one-round attack, that is, only a single execution of the protocol is required. Interestingly, Merkle and Werchner recently provided a security proof of RSA-S1 against one-round passive attacks in some generic model, even for parameters to which our attack provably applies. Thus, our result throws doubt on the real significance of security proofs in the generic model, at least for server-aided RSA protocols. We also present a simple analysis of a multi-round lattice-based passive attack proposed last year by Merkle.

Keywords: Cryptanalysis, RSA signature, Server-aided protocol, Lattices.

1 Introduction

Small units like chip cards or smart cards have the possibility of computing, storing and protecting data. Today, many of these cards include fast and secure coprocessors allowing to quickly perform the expensive operations needed

* Work supported in part by the RNRT "Turbo-signatures" project of the French Ministry of Research.

** Work supported in part by the Australian Research Council.

C. Boyd (Ed.): ASIACRYPT 2001, LNCS 2248, pp. 21–35, 2001.
© Springer-Verlag Berlin Heidelberg 2001

by public key cryptosystems. However, a large proportion of the cards consists of cheap cards with too limited computing power for such tasks. To overcome this problem, extensive research has been conducted under the generic name "server-aided secret computations" (SASC). In the SASC protocol, the client (the smart card) wants to perform a secret computation (for example, RSA signature generation) by borrowing the computing power of an untrusted powerful server without revealing its secret information. One distinguishes two kinds of attacks against such protocols: attacks where the server follows rigorously the protocol are called *passive attacks*, while attacks where the server may return false computations are called *active attacks*. Attacks are called multi-round when they require several executions of the protocol between the same parties.

Most of the SASC protocols proposed for RSA signatures have been shown to be either inefficient or insecure (see for instance the two recent examples [13,10]), which explains why, to our knowledge, none of these protocols has ever been used in practice. Many of these protocols are variants of the protocols RSA-S1 and RSA-S2 proposed by Matsumoto, Kato and Imai [8] at Crypto '88, which use a random linear decomposition of the RSA private exponent. At Eurocrypt '92, Pfitzmann and Waidner [15] presented several natural meet-in-the-middle passive attacks and some efficient active attacks against RSA-S1 and RSA-S2. To prevent such attacks, they discussed two countermeasures which should be used together: one is to renew the decomposition of the private exponent at each signature, the other is to check the signature before the end of the protocol, which is a well-known countermeasure but requires a very small public exponent since the check is performed by the card.

The first countermeasure was effective against the original active attacks of [15], but Merkle [10] showed last year at ACM CCS '00 that the resulting scheme was still insecure. Indeed, he presented an efficient lattice-based multi-round passive attack, which was successful (in practice) against many choices of the parameters. Merkle's paper [10] included an analysis of the attack, inspired by well-known lattice-based methods [5] to solve the subset sum problem. However, the analysis was rather technical and not exactly correct (it assumed a distribution of the parameters which was not the one induced by the protocol). We present a simple analysis of a slight variant of Merkle's attack, which enables to explain experimental results, and to provide provable results for certain choices of the parameters.

The main contribution of this paper is a new lattice-based passive attack which recovers the private exponent (like Merkle's attack), but only in the case a very small public exponent is used (which is the second countermeasure). Interestingly, this attack is only one-round in the sense that a single execution of the protocol is sufficient, whereas Merkle's attack is multi-round, requiring many signatures produced by the card with the help of the same server. Consequently, the first countermeasure has no impact on this new attack. And these results point out the limits of the *generic model*, as applied to the security analysis of server-aided RSA protocols. Indeed, Merkle and Werchner [11] proved at PKC '98 that the RSA-S1 protocol was secure against one-round passive attacks in the generic model, in the sense that all generic attacks have complexity at least that of a square-root attack (better than the meet-in-the-middle attack presented by Pfitzmann and Waidner [15]). Roughly speaking, in this context, generic attacks (see [11] for a precise definition) do not take advantage of special properties of the group used. However, our attack shows that the RSA-S1 scheme is not even secure against one-round passive attacks in the standard model of computation. In particular, the attack provably works against certain choices of the parameters to which the square-root attack cannot apply. Thus, contrary to what Merkle and Werchner claimed in [11], the generic model is not appropriate for investigating the security of server-aided RSA protocols.

The rest of the paper is organized as follows. In Section 2, we make a short description of the RSA-S1 server-aided protocol and review some useful background. We refer to [8,15] for more details. In Section 3, we present our variant of Merkle's lattice-based attack, together with an analysis. In Section 4, we present our new lattice-based attack on low-exponent RSA-S1.

2 Background

2.1 The RSA-S1 Server-Aided Protocol

Let N be an RSA-modulus and let φ denote the Euler function. Let e and d be respectively the RSA public and private exponents:

$$ed \equiv 1 \pmod{\varphi(N)}.$$

For an integer s we denote by $[s]$ the set of integers of the interval $[0, s-1]$ and by $[s]_\pm$ the set of integers of the interval $[-s+1, s-1]$.

Let k, ℓ and m be positive integers and let $\mathcal{B}_{k,\ell,m}$ be the set of vectors

$$\mathbf{f} = (f_1, \dots, f_m) \in \left[2^\ell\right]^m$$

with $\gcd(f_1, \dots, f_m, \varphi(N)) = 1$ and with

$$\sum_{i=1}^{m} \mathrm{wt}(f_i) = k, \tag{1}$$

where $\mathrm{wt}(f)$ denotes the Hamming weight, that is, the sum of binary digits of an integer $f \geq 0$.

The RSA-S1 server-aided protocol from [8] computes an RSA signature $x^d \pmod{N}$ with the help of an (untrusted) server in the following way:

The RSA-S1 Protocol.

Step 1 The card selects a vector $\mathbf{f} = (f_1, \dots, f_m) \in \mathcal{B}_{k,\ell,m}$ at random accordingly to any fixed probability distribution.

Step 2 The card sends a vector $\mathbf{d} = (d_1, \dots, d_m) \in [\varphi(N)]^m$ chosen uniformly at random from the set of vectors satisfying the congruence

$$\sum_{i=1}^{m} f_i d_i \equiv d \pmod{\varphi(N)}, \tag{2}$$

if possible. Otherwise the card returns to Step 1.

Step 3 The card asks the server to compute and return $z_i \equiv x^{d_i} \pmod{N}$, $i = 1, \dots, m$.

Step 4 The card computes

$$x^d \equiv \prod_{i=1}^{m} z_i^{f_i} \pmod{N}.$$

Our description follows the presentation of [10] rather than the one of the original paper [8]. For instance, [8] asks that $\sum_{i=1}^{m} \mathrm{wt}(f_i) \le k$ instead of (1) but this difference is marginal as all our results can easily be adapted to this case.

For Step 4, the card mainly has two possibilities, due to memory restrictions. One is the square-and-multiply method, which requires at most $k\ell$ modular multiplications and very little memory. The other is the algorithm of [4], which enables to compute $\prod_{i=1}^{m} z_i^{f_i} \pmod{N}$ efficiently but requires more memory than the square-and-multiply method. When using this algorithm, to optimize the choice of the parameters, one should remove the restriction (1) and replace the choice $f_i \in [2^\ell]$ by $f_i \in [h]$ where h is some small integer, not necessarily a power of 2. The algorithm then requires at most $m+h-3$ modular multiplications, and the temporary storage of either m or $h-1$ elements, according to whether the card stores all the m elements $z_1, \dots, z_m$, or the $h-1$ elements $t_j = \prod_{f_i=j} z_i$, $1 \le j < h$ (which must be computed upon reception of the z_i's). Other known tricks to speed-up the computation of products of exponentiations (see [6] and [9, Sect. 14.6]) do not seem to be useful in this context.

The protocol requires the transfer of approximately $2m \log N$ bits. Since the bandwidth of a cheap smartcard is typically 9600 bauds, this means that m must be restricted to low values. For instance, with a 1024-bit modulus, the value $m = 50$ already represents 10.7 seconds.

2.2 Passive Attacks on RSA-S1

Notice that the protocol is broken as soon as the f_i's are disclosed. Indeed, the integer $\sum_{i=1}^{m} f_i d_i$ is congruent to the RSA private exponent modulo $\varphi(N)$, and therefore enables to sign any message (and this can be checked thanks to the public exponent e). And, of course, one may further recover the factorization of N in randomized polynomial time, from $e\sum_{i=1}^{m} f_i d_i - 1$ which is a non-zero multiple of $\varphi(N)$ (see for instance [9, Section 8.2.2]).

The authors of [8] claimed that the only possible passive attack was to exhaustive search the f_i's, which requires roughly C operations where:

$$C = \binom{m\ell}{k}.$$

But obviously, one can devise simple meet-in-the-middle passive attacks. Pfitzmann and Waidner [15] noticed that one could split $(f_1, \ldots, f_m)$ as $(g_1, \ldots, g_m) + (h_1, \ldots, h_m)$ where $\sum \mathrm{wt}(g_i) \leq \sum \mathrm{wt}(h_i) = \lceil k/2 \rceil$, and deduced an attack with time and space complexity roughly:

$$\binom{m\ell}{\lceil k/2 \rceil}.$$

The attack of [15] is however not optimal: the complexity can easily be improved using a trick used by Coppersmith [18] in a meet-in-the-middle attack against the discrete logarithm problem with low Hamming weight. By choosing random subsets of cardinality $\lceil m\ell/2 \rceil$ inside $\{1, \ldots, m\ell\}$, one obtains a randomized meet-in-middle-attack with time and space complexity roughly:

$$\sqrt{k}\binom{\lceil m\ell/2 \rceil}{\lceil k/2 \rceil}.$$

Thus, we obtain an attack of complexity roughly the square root $\sqrt{C}$ of that of exhaustive search. Therefore in our numerical experiments we mainly consider sets of parameters for which $C \geq 2^{120}$. Note however that even with $C \approx 2^{100}$, the square-root attack is not much practical, due to memory constraints.

In [11], Merkle and Werchner proposed an adaptation of generic algorithms (see [17]) to server-aided RSA protocols, and showed that any one-round passive generic attack on RSA-S1 had complexity at least $\Omega(\sqrt{C})$.

In [15], Pfitzmann and Waidner also presented a few active attacks which cannot be avoided by increasing the parameters contrary to the passive attacks mentioned previously. They discussed two countermeasures to prevent their own active attacks:

- Renewing the decomposition of the private exponent d at each execution of the protocol, as described in Steps 1 and 2.
- Verifying the signature $x^d \pmod N$ before releasing it, by computing $(x^d)^e \pmod N$ and checking that it is equal to x. This countermeasure is well-known and requires a very small public exponent e (otherwise there is no computational advantage in using the server to compute $x^d \pmod N$).

The second countermeasure seems necessary but is not sufficient to prevent one of the active attacks of [15], and it creates the attack of Section 4. The first countermeasure prevents all the active attacks of [15], but creates the passive attack of Merkle [10], which we analyze in Section 3. Interestingly, it seems that the attacks of Section 3 and 4 do not apply to the RSA-S2 protocol, which is a CRT variant of RSA-S1 (see [8,15]). The situation is reminiscent of that of RSA with small private exponent, in which the best attack known [3] fails if the private exponent is small modulo both $p-1$ and $q-1$.

2.3 Lattices

Our attacks are based on lattice basis reduction, a familiar tool in public-key cryptanalysis. We give a brief overview of lattice theory (see the survey [14] for a list of references). In this paper, we call a *lattice* any subgroup of $(\mathbb{Z}^n, +)$: in the literature, these are called integer lattices. For any set of vectors $\mathbf{b}_1, \ldots, \mathbf{b}_d \in \mathbb{Z}^n$, we define the set of all integral linear combinations:

$$L(\mathbf{b}_1, \ldots, \mathbf{b}_d) = \left\{ \sum_{i=1}^{d} n_i \mathbf{b}_i : n_i \in \mathbb{Z} \right\}.$$

By definition, $L(\mathbf{b}_1, \ldots, \mathbf{b}_d)$ is a lattice, called the lattice spanned by the vectors $\mathbf{b}_1, \ldots, \mathbf{b}_d$. A *basis* of a lattice L is a set of linearly independent vectors $\mathbf{b}_1, \ldots, \mathbf{b}_d$ such that:

$$L = L(\mathbf{b}_1, \ldots, \mathbf{b}_d).$$

In any lattice, there is always at least one basis, and in general, there are in fact infinitely many lattice bases. But all the bases of a lattice L have the same number of elements, called the *rank* or *dimension* of the lattice. All the bases also have the same d-dimensional volume, which is by definition the square root of the determinant $\det_{1 \le i,j \le d} \langle \mathbf{b}_i, \mathbf{b}_j \rangle$, where $\langle , \rangle$ denotes the Euclidean inner product. This volume $\mathrm{vol}(L)$ is called the volume or determinant of the lattice. When the lattice dimension d is equal to the space dimension n, this volume is simply the absolute value of the determinant of any lattice basis.

For a vector $\mathbf{a}$, we denote by $\|\mathbf{a}\|$ its Euclidean norm. A basic problem in lattice theory is the shortest vector problem (SVP): given a basis of a lattice L, find a non-zero vector $\mathbf{v} \in L$ such that $\|\mathbf{v}\|$ is minimal among all non-zero lattice vectors. Any such vector is called a shortest lattice vector. It is well-known that the Euclidean norm of a shortest lattice vector is always less than $\sqrt{d}\mathrm{vol}(L)^{1/d}$, d denoting the lattice dimension. In "usual" lattices, one does not expect the norm of a shortest lattice vector to be much less than this upper bound.

Many attacks in public-key cryptanalysis work by reduction to SVP, or to approximating SVP (see the survey [14]). The shortest vector problem was recently shown to be NP-hard under randomized reductions [1], and therefore, it is now widely believed that there is no polynomial-time algorithm to solve SVP. However, there exist polynomial-time algorithms which can provably approximate SVP. The first algorithm of that kind was the celebrated LLL lattice basis reduction algorithm of Lenstra, Lenstra and Lovász [7]. We use the best deterministic polynomial-time algorithm currently known to approximate SVP, which is due to Schnorr [16] and is based on LLL:

Lemma 1. *There exists a deterministic polynomial time algorithm which, given as input a basis of an s-dimensional lattice L, outputs a non-zero lattice vector* $\mathbf{u} \in L$ *such that:*

$$\|\mathbf{u}\| \le 2^{O\left(s \log^2 \log s / \log s\right)} \min \left\{ \|\mathbf{z}\| : \quad \mathbf{z} \in L, \mathbf{z} \ne 0 \right\}.$$

Recently, Ajtai *et al.* [2] discovered a randomized algorithm which slightly improves the approximation factor $2^{O\left(s \log^2 \log s / \log s\right)}$ to $2^{O(s \log \log s / \log s)}$. In practice, the best algorithm to approximate SVP is a heuristic variant of Schnorr's algorithm [16]. Interestingly, these algorithms typically perform much better than theoretically expected: they often return a shortest lattice vector, provided that the lattice dimension is not too large. Hence, it is useful to predict what can be achieved efficiently if an SVP-oracle (that is, an algorithm which solves SVP) is available. For instance, this was done for the subset sum problem [5]. However, unless the lattice dimension is extremely small, it is hard to predict beforehand whether an SVP-instance is solvable in practice, which means that experiments are always necessary in this case.

3 An Analysis of Merkle's Multi-round Attack

3.1 Merkle's Attack

The attack of Merkle [10] is based on the following observation: Because for each $\mathbf{f} = (f_1, \ldots, f_m) \in \mathcal{B}_{k,\ell,m}$ and $\mathbf{d} = (d_1, \ldots, d_m) \in [\varphi(N)]^m$

$$0 < \sum_{i=1}^{m} f_i d_i < k 2^\ell \varphi(N)$$

we have

$$\sum_{i=1}^{m} f_i d_i \equiv d + j\varphi(N)$$

with $j \in [k2^\ell]$, that is, j cannot take too many distinct values.

It is shown in [10] that regardless of the distribution of the vectors $\mathbf{f} \in \mathcal{B}_{k,\ell,m}$ with probability at least $1/k2^\ell$ for two pairs $\mathbf{f}_1 = (f_1, \ldots, f_m)$, $\mathbf{d}_1 = (d_1, \ldots, d_m)$, and $\mathbf{f}_2 = (f_{m+1}, \ldots, f_{2m})$, $\mathbf{d}_2 = (d_{m+1}, \ldots, d_{2m})$ of vectors produced by the above protocol we have the following equation (over the integers rather than modulo N):

$$\sum_{i=1}^{m} f_i d_i = \sum_{i=m+1}^{2m} f_i d_i. \tag{3}$$

In fact, any rule to select the above vectors gives rise to a collision after at most $k2^\ell$ executions of the protocol. Besides, the "birthday paradox" suggests that a collision is likely to happen after roughly $k^{1/2}2^{\ell/2}$ executions of the protocol.

The linear equation (3) is unusual because each f_i is small (compared to the d_i's), and this can be interpreted in terms of lattices. More precisely, it is argued in [10] that $(\mathbf{f}_1, \mathbf{f}_2)$ is the shortest vector in a particular lattice related to the homogeneous equation (3) and the congruences

$$\sum_{i=1}^{m} f_i d_i \equiv \sum_{i=m+1}^{2m} f_i d_i \equiv d \pmod{\varphi(N)}. \tag{4}$$

However, the analysis presented by Merkle is not sufficient, because it assumes a distribution of the parameters which is not the one of the protocol (see [10,

Theorem 2.1]). And no result is proposed without SVP-oracles. Hence, Merkle's attack, as presented in [10], is not a proved attack, even under the assumption of an SVP-oracle, which is not so unusual for a lattice-based attack. Nevertheless, the experiments conducted by Merkle (see [10]) showed that the attack was successful in practice against many choices of the parameters. Thus, it was interesting to see whether Merkle's attack could be proved, with or without SVP-oracles. Here, we provide a proof, for a slight variant of Merkle's attack. The analysis we present can in fact be extended to the original attack, but our variant is slightly simpler to describe and to analyze, while the difference of efficiency between the two attacks is marginal.

3.2 A Variant of Merkle's Attack

We work directly with the lattice corresponding to (3): Let $\mathcal{L}(\mathbf{d}_1, \mathbf{d}_2)$ be the $(2m-1)$-dimensional lattice formed by all vectors $\mathbf{z} \in \mathbb{Z}^{2m}$ with

$$\sum_{i=1}^{m} z_i d_i = \sum_{i=m+1}^{2m} z_i d_i.$$

This lattice is the simplest case of an orthogonal lattice (as introduced in [12]), and one can compute a basis of such lattices in polynomial time. It can easily be showed that the volume of the lattice is given by:

$$\mathrm{vol}(\mathcal{L}(\mathbf{d}_1, \mathbf{d}_2)) = \frac{\left(d_1^2 + \ldots + d_{2m}^2\right)^{1/2}}{\gcd(d_1, \ldots, d_{2m})}.$$

Thus, one would expect its shortest non-zero vector to have a norm around:

$$(2m-1)^{1/2}\mathrm{vol}\left(\mathcal{L}(\mathbf{d}_1, \mathbf{d}_2)\right)^{1/(2m-1)} \approx (2m-1)^{1/2}\varphi(N)^{1/(2m-1)}.$$

On the other hand, the vector $\mathbf{f} = (f_1, \ldots, f_{2m})$ belongs to this lattice, and has a norm of at most $k^{1/2}2^{\ell}$. Hence, if $k^{1/2}2^{\ell}$ is much smaller than $(2m-1)^{1/2}\varphi(N)^{1/(2m-1)}$, we expect $\mathbf{f}$ to be the shortest vector of $\mathcal{L}(\mathbf{d}_1, \mathbf{d}_2)$, and if it is smaller enough, then the gap between $\mathbf{f}$ and the other lattice vectors guarantees that the algorithm of Lemma 1 will find it. Once $\mathbf{f}$ is known, one can derive the value $\sum_{i=1}^{m} f_i d_i$, which is congruent to the RSA private exponent modulo $\varphi(N)$, and therefore enables to sign any message. And one may further recover the factorization of N in randomized polynomial time, from $e\sum_{i=1}^{m} f_i d_i - 1$ which is a non-zero multiple of $\varphi(N)$ (see for instance [9, Section 8.2.2]).

In [10], the original attack of Merkle worked with a slight variant of the lattice $\mathcal{L}(\mathbf{d}_1, \mathbf{d}_2)$, to take advantage of the fact that $f_i \in [2^{\ell}]$ and not $f_i \in [2^{\ell}]_{\pm}$. Such a trick was used for the subset sum problem [5]. However, this trick is not as useful here, because the distributions are different. This means that the difference between our variant and the original attack is marginal.

3.3 Theoretical Results

The previous reasoning can in fact be made rigorous by a tight analysis, which gives rise to the following result:

Theorem 1. *There is a deterministic algorithm $\mathcal{A}$ which, given as input an RSA modulus N, together with a public exponent e, and a set $\mathcal{D}$ of $k2^\ell$ vectors $\mathbf{d} \in [\varphi(N)]^m$ corresponding to a certain set $\mathcal{F}$ of vectors $\mathbf{f} \in \mathcal{B}_{k,\ell,m}$ generated by $k2^\ell$ independent executions of* RSA-S1, *outputs a value $\mathcal{A}(\mathcal{D})$ in time polynomial in $k, 2^\ell, m, \log N$ such that:*

$$\Pr_{\mathcal{D}}\left[\mathcal{A}(\mathcal{D}) \equiv d \pmod{\varphi(N)}\right] \geq 1 - \frac{k^{m+2}2^{2\ell(m+2)+O(m^2\log^2\log m/\log m)}}{\varphi(N)}$$

where the probability is taken over all random choices of $\mathcal{D}$ for the given $\mathcal{F}$.

Proof. Given a set $\mathcal{D}$ of $k2^\ell$ vectors $\mathbf{d}$ associated with the protocol RSA-S1, which corresponds to a certain set $\mathcal{F}$ of $k2^\ell$ (unknown) vectors $\mathbf{f} \in \mathcal{B}_{k,\ell,m}$, the algorithm $\mathcal{A}$ selects all possible pairs of such vectors $\mathbf{d}_1$ and $\mathbf{d}_2$ and uses the algorithm of Lemma 1 to find a short vector $\mathbf{u}$ in the $(2m-1)$-dimensional lattice $\mathcal{L}(\mathbf{d}_1, \mathbf{d}_2)$ formed by all vectors $\mathbf{z} \in \mathbb{Z}^{2m}$ such that

$$\sum_{i=1}^{m} z_i d_i = \sum_{i=m+1}^{2m} z_i d_i.$$

We know that there is at least one pair $(\mathbf{d}_1, \mathbf{d}_2)$ such that the equation (3) holds. Notice that for any $\mathbf{f} \in \mathcal{B}_{k,\ell,m}$, we have

$$\|f\|^2 = \sum_{i=1}^{m} f_i^2 < 2^\ell \sum_{i=1}^{m} f_i \leq k2^{2\ell}. \tag{5}$$

Thus, if we apply the algorithm of Lemma 1 to $\mathcal{L}(\mathbf{d}_1, \mathbf{d}_2)$, we obtain a vector $\mathbf{u} = (u_1, \ldots, u_{2m})$ such that:

$$\begin{aligned}\|\mathbf{u}\|^2 &\leq 2^{O(m\log^2\log m/\log m)} \min\left\{\|\mathbf{z}\|^2, \quad \mathbf{z} \in \mathcal{L}(\mathbf{d}_1, \mathbf{d}_2)\right\} \\ &\leq 2^{O(m\log^2\log m/\log m)} \left(\|\mathbf{f}_1\|^2 + \|\mathbf{f}_2\|^2\right) \\ &\leq k2^{2\ell+O(m\log^2\log m/\log m)}.\end{aligned}$$

Therefore, there exists some integer $U = k^{1/2}2^{\ell+O(m\log^2\log m/\log m)}$ such that $|u_i| < U$ for $i = 1, \ldots, 2m$, that is, $\mathbf{u} \in [U]_\pm^{2m}$.

We write $\mathbf{u} = (\mathbf{u}_1, \mathbf{u}_2)$ where $\mathbf{u}_1, \mathbf{u}_2 \in [U]_\pm^m$ and say that $\mathbf{u}$ is *similar* to the concatenation $(\mathbf{f}_1, \mathbf{f}_2)$ if either $\mathbf{u}_1$ is non-zero and parallel to $\mathbf{f}_1$, or $\mathbf{u}_2$ is non-zero and parallel to $\mathbf{f}_2$. Notice that if one knows a vector $\mathbf{u} \neq 0$ similar to $\mathbf{f}_1, \mathbf{f}_2$, one obtains at most 2^ℓ possible values for either $\mathbf{f}_1$ or $\mathbf{f}_2$. And if $\mathbf{f}_1$ or $\mathbf{f}_2$ is correct, then $\langle \mathbf{f}_1, \mathbf{d}_1 \rangle$ or $\langle \mathbf{f}_2, \mathbf{d}_2 \rangle$ is congruent to d modulo $\varphi(N)$, which can be checked by signing a message. Hence it is enough to show that with probability at least

$1 - k^{m+2}2^{2\ell(m+2)+O(m^2 \log^2 \log m/ \log m)}\varphi(N)^{-1}$ the vector $\mathbf{u} = (\mathbf{u}_1, \mathbf{u}_2)$ returned by the algorithm of Lemma 1 is similar to $(\mathbf{f}_1, \mathbf{f}_2)$.

First for $\mathbf{f}_1, \mathbf{f}_2 \in \mathcal{B}_{k,\ell,m}$ we estimate the size of the set $\mathcal{E}(\mathbf{f}_1, \mathbf{f}_2)$ of pairs of vectors $\mathbf{d}_1, \mathbf{d}_2 \in [\varphi(N)]^m$ such that for some $\mathbf{u} = (\mathbf{u}_1, \mathbf{u}_2) \in [U]_{\pm}^{2m}$ which is not similar to $(\mathbf{f}_1, \mathbf{f}_2)$ we have the equation

$$\sum_{i=1}^{m} u_i d_i = \sum_{i=m+1}^{2m} u_i d_i. \tag{6}$$

Let us fix a nonzero vector $\mathbf{u} = (\mathbf{u}_1, \mathbf{u}_2) \in [U]_{\pm}^{2m}$ and a vector $(\mathbf{f}_1, \mathbf{f}_2) \in \mathcal{B}_{k,\ell,m}$ which are not similar. Without loss of generality we may assume that $\mathbf{u}_2 \neq 0$ and is not parallel to $\mathbf{f}_2$ and that $f_{2m} \neq 0$. Then excluding d_{2m} from (6) using (3), we obtain an equation

$$\sum_{i=1}^{m} c_i d_i = \sum_{i=m+1}^{2m-1} c_i d_i \tag{7}$$

with $c_i = u_i - f_i u_{2m}/f_{2m}$, $i = 1, \ldots, 2m-1$. By our assumption, for at least one $i \geq m+1$, the coefficient $c_i \neq 0$. Without loss of generality we may assume that $c_{2m-1} \neq 0$. Then the first congruence in (4) gives us at most $2^\ell \varphi(N)^{m-1}$ possible values for $\mathbf{d}_1 = (d_1, \ldots, d_m)$. Indeed, assuming that $f_m \neq 0$ and selecting the integers $d_1, \ldots, d_{m-1} \in [\varphi(N)]$ arbitrarily, we obtain a congruence of the form $f_m d_m \equiv D \pmod{\varphi(N)}$ which has at most $\gcd(f_m, \varphi(N)) \leq f_m < 2^\ell$ solutions $d_m \in [\varphi(N)]$. Finally, for any of $\varphi(N)^{m-2}$ possible choices of $d_{m+1}, \ldots, d_{2m-2} \in [\varphi(N)]^{m-2}$ the equation (7) gives at most one value for d_{m-1} and then the second congruence in (4) gives us at most $\gcd(f_{2m}, \varphi(N)) \leq f_{2m} < 2^\ell$ possible values for d_{2m}. So the total number of solutions for such $\mathbf{u}$ is at most $2^{2\ell}\varphi(N)^{2m-3}$. The total number of such vectors is at most U^{2m}. Thus we finally derive

$$\begin{aligned}\#\mathcal{E}(\mathbf{f}_1, \mathbf{f}_2) &\leq (2U)^{2m} 2^{2\ell} \varphi(N)^{2m-3} \\ &\leq k^m 2^{2\ell(m+1)+O(m^2 \log^2 \log m/ \log m)} \varphi(N)^{2m-3}.\end{aligned}$$

For each vector $\mathbf{f} \in \mathcal{B}_{k,\ell,m}$ there are exactly $\varphi(N)^{m-1}$ vectors $\mathbf{d} \in [\varphi(N)]^m$ satisfying the congruence (2). Therefore, the probability that there is a pair of vectors $\mathbf{f}_1, \mathbf{f}_2 \in \mathcal{F}$ such that the corresponding vectors $\mathbf{d}_1, \mathbf{d}_2 \in \mathcal{D}$ satisfy $\mathbf{d}_1, \mathbf{d}_2 \in \mathcal{E}(\mathbf{f}_1, \mathbf{f}_2)$ is at most

$$\begin{aligned}&\frac{(\#\mathcal{F})^2 k^m 2^{2\ell(m+1)+O(m^2 \log^2 \log m/ \log m)} \varphi(N)^{2m-3}}{\varphi(N)^{2m-2}} \\ &\qquad = k^{m+2} 2^{2\ell(m+2)+O(m^2 \log^2 \log m/ \log m)} \varphi(N)^{-1},\end{aligned}$$

and the result follows. □

Assuming that an SVP-oracle is available, we derive much stronger estimates.

Theorem 2. *There is a deterministic algorithm $\mathcal{A}$ which, given an access to an SVP-oracle and as input an RSA modulus N, together with a public exponent e, a set $\mathcal{D}$ of $k2^\ell$ vectors $\mathbf{d} \in [\varphi(N)]^m$ corresponding to a certain set $\mathcal{F}$ of vectors $\mathbf{f} \in \mathcal{B}_{k,\ell,m}$ generated by $k2^\ell$ independent executions of* RSA-S1, *outputs a value $\mathcal{A}(\mathcal{D})$ in time polynomial in $k, 2^\ell, m, \log N$ such that:*

$$\Pr_{\mathcal{D}}\left[\mathcal{A}(\mathcal{D}) \equiv d \pmod{\varphi(N)}\right] \geq 1 - \frac{k^{m+2}2^{2(\ell m+2\ell+m)}}{\varphi(N)}$$

where the probability is taken over all random choices of $\mathcal{D}$ for the given $\mathcal{F}$.

As in [10], instead of waiting for $k2^\ell$ executions of RSA-S1 one may also restrict to only two executions, which yields the following version of Theorems 1 and 2:

Theorem 3. *There is a deterministic algorithm $\mathcal{A}$ which, given as input an RSA modulus N, together with a public exponent e, a pair of vectors $\mathbf{d}_1, \mathbf{d}_2 \in [\varphi(N)]^m$ corresponding to a pair of vectors $\mathbf{f}_1, \mathbf{f}_2 \in \mathcal{B}_{k,\ell,m}$ generated by two independent executions of* RSA-S1, *outputs a value $\mathcal{A}(\mathbf{d}_1, \mathbf{d}_2)$ in time polynomial in $k, 2^\ell, m, \log N$ such that:*

$$\Pr_{\mathbf{d}_1,\mathbf{d}_2}\left[\mathcal{A}(\mathbf{d}_1, \mathbf{d}_2) \equiv d \pmod{\varphi(N)}\right] \geq \frac{1}{k2^\ell} - \frac{k^m 2^{2\ell(m+1)+O(m^2 \log^2 \log m/\log m)}}{\varphi(N)}$$

where the probability is taken over all random choices of $\mathbf{d}_1, \mathbf{d}_2$ for the given $\mathbf{f}_1, \mathbf{f}_2$.

Theorem 4. *There is a deterministic algorithm $\mathcal{A}$ which, given access to an SVP-oracle and as input an RSA modulus N, together with a public exponent e, a pair of vectors $\mathbf{d}_1, \mathbf{d}_2 \in [\varphi(N)]^m$ corresponding to a pair of vectors $\mathbf{f}_1, \mathbf{f}_2 \in \mathcal{B}_{k,\ell,m}$ generated by two independent executions of* RSA-S1, *makes a single call to the SVP-oracle with the lattice $\mathcal{L}(\mathbf{d}_1, \mathbf{d}_2)$ and outputs a value $\mathcal{A}(\mathbf{d}_1, \mathbf{d}_2)$ in time polynomial in $k, 2^\ell, m, \log N$ such that:*

$$\Pr_{\mathbf{d}_1,\mathbf{d}_2}\left[\mathcal{A}(\mathbf{d}_1, \mathbf{d}_2) \equiv d \pmod{\varphi(N)}\right] \geq \frac{1}{k2^\ell} - \frac{k^m 2^{2(\ell m+\ell+m)}}{\varphi(N)}$$

where the probability is taken over all random choices of $\mathbf{d}_1, \mathbf{d}_2$ for the given $\mathbf{f}_1, \mathbf{f}_2$.

Notice that unless k (and thus $\ell \geq k/m$) is exponentially large compared to m, which is completely impractical, the terms k^{m+2} and k^m in the bounds of Theorems 1 and 3 respectively, can be included in the term $2^{O(m^2 \log^2 \log m/\log m)}$.

3.4 Experiments

In practice, the attack is as efficient as Merkle's original attack, due to the fact that strong lattice basis reduction algorithms behave like oracles for the shortest vector problem up to moderate dimension. In [10], Merkle reported the experimental results presented in Table 1. Notice however that none of the sets of parameters of Table 1 leads to an efficient protocol (for the card).

Table 1. Experiments with Merkle's attack

m	k	ℓ	Success (%)	Complexity of the sqrt attack
25	28	11	100	2^{62}
32	26	10	100	2^{62}
38	26	9	100	2^{63}
42	26	8	100	2^{63}
48	26	7	70	2^{63}
56	26	6	10	2^{63}

4 A New One-Round Attack on Low Exponent RSA-S1

4.1 Description of the Attack

We now assume that a very small public exponent e is used. We also assume that the secret primes p and q defining $N = pq$ have approximately the same length. Let $s = p+q = O(N^{1/2})$. We have $\varphi(N) = N - s + 1$. When the RSA-S1 protocol is performed once, we have:

$$\sum_{i=1}^{m} f_i d_i \equiv d \pmod{\varphi(N)},$$

and therefore,

$$\sum_{i=1}^{m} f_i e d_i \equiv 1 \pmod{\varphi(N)}.$$

From (5) we see that there exists $r \in \left[k2^{\ell}e\right]$ such that

$$\sum_{i=1}^{m} f_i e d_i = 1 + r\varphi(N) = 1 + r(N - s + 1).$$

Hence

$$\sum_{i=1}^{m} f_i e d_i = 1 + r - rs \pmod{N}, \tag{8}$$

where $|1 + r - rs| = O(k2^{\ell}eN^{1/2})$. We thus obtain a linear equation modulo N where the unknown coefficients f_i and $1 + r - rs$ are all relatively small. This suggests to define the $(m+1)$-dimensional lattice $\mathcal{L}_{e,N}(\mathbf{d})$ spanned by the rows of the following matrix:

$$\begin{pmatrix} N & 0 & 0 & \dots & 0 \\ ed_1 & eR & 0 & \dots & 0 \\ ed_2 & 0 & eR & \ddots & \vdots \\ \vdots & \vdots & \ddots & \ddots & 0 \\ ed_m & 0 & \dots & 0 & eR \end{pmatrix}$$

where $R = \lfloor N^{1/2} \rfloor$. Obviously, the volume of this lattice is $\mathrm{vol}(\mathcal{L}_{e,N}(\mathbf{d})) = e^m N R^{m/2}$. Therefore, one would expect its shortest vector to be of norm roughly $(m+1)^{1/2} e^{m/(m+1)} N^{(m+2)/(2m+2)}$. On the other hand, the lattice contains the target vector

$$\mathbf{t} = (1 + r - rs, f_1 eR, \ldots, f_m eR),$$

whose norm is $\|\mathbf{t}\| = O\left(k 2^\ell e N^{1/2}\right)$ because of (5). Hence, the target vector is likely to be the shortest vector in this lattice if $k e^{1/(m+1)} 2^\ell$ is much smaller than $m^{1/2} N^{1/(2m+2)}$. Note that this condition is satisfied for sufficiently large N and that it is very similar to the heuristic condition we obtained in Section 3.2, which suggests that the efficiency of the attacks of Section 4 and 3 should be comparable. In case the target vector is really much smaller than the other lattice vectors, then the algorithm of Lemma 1 finds it. Once the target vector is known, we can recover a private exponent equivalent to d thanks to $\sum_{i=1}^m f_i d_i$, which enables to sign any message, as in Merkle's attack. Again, one may further derive a not too large multiple of $\varphi(N)$, which yields the factorization of N in randomized polynomial time.

4.2 Theoretical Results

The previous attack can be proved, using the same counting arguments of the proof of Theorem 1:

Theorem 5. *There is a deterministic algorithm $\mathcal{A}$ which, given as input an RSA modulus $N = pq$ such that $p + q = O(N^{1/2})$, together with a public exponent e, and a vector $\mathbf{d} \in [\varphi(N)]^m$ corresponding to a certain vector $\mathbf{f} \in \mathcal{B}_{k,\ell,m}$ generated by* RSA-S1, *outputs a value $\mathcal{A}(\mathbf{d})$ in time polynomial in $k, 2^\ell, m, \log N$ such that:*

$$\Pr_{\mathbf{d}}\left[\mathcal{A}(\mathbf{d}) \equiv d \pmod{\varphi(N)}\right] \geq 1 - \frac{k^{m+1} e^{m+1} 2^{\ell(m+1) + O(m^2 \log^2 \log m / \log m)}}{N^{1/2}}$$

where the probability is taken over all random choices of $\mathbf{d}$ for the given $\mathbf{f}$.

Proof. The algorithm $\mathcal{A}$ starts by applying the algorithm of Lemma 1 to find a short vector $\mathbf{w} \neq 0$ in the $(m+1)$-dimensional lattice $\mathcal{L}_{e,N}(\mathbf{d})$. Since $\mathbf{t}$ is a lattice vector and because $p + q = O(N^{1/2})$, we have:

$$\|\mathbf{w}\| \leq 2^{O(m \log^2 \log m / \log m)} \|\mathbf{t}\| = k e 2^{\ell + O(m \log^2 \log m / \log m)} N^{1/2}.$$

By definition of the lattice, $\mathbf{w}$ is of the form:

$$\mathbf{w} = (u_0 N + \sum_{i=1}^m e d_i u_i, u_1 eR, \ldots, u_m eR),$$

where each u_i is an integer.

Therefore, there exists some integer $U = k e 2^{\ell + O(m \log^2 \log m / \log m)}$ such that $|u_i| < U$ for $i = 1, \ldots, 2m$. Thus $\mathbf{u} = (u_1, \ldots, u_m) \in [U]_{\pm}^m$. We may assume that $\|\mathbf{w}\| < N$ otherwise the right hand side of the inequality of the theorem is negative, making the bound trivial. Then necessarily $\mathbf{u} \neq 0$. We also have

$$\sum_{i=1}^{m} ed_i u_i \equiv w_0 \pmod{N} \tag{9}$$

for some $w_0 \in [W]_\pm$ where $W = O\left(ke2^{\ell+O(m\log^2\log m/\log m)}N^{1/2}\right)$.

Clearly, we may assume that $2^\ell \leq \min\{p,q\}$ otherwise the result is trivial. Thus for any $i = 1,\dots,m$ with $f_i \neq 0$ we have $\gcd(f_i, N) = 1$. As before we see that for each w_0 and for each $\mathbf{u} \in [U]_\pm^m$ not parallel to $\mathbf{f}$ there are at most $\varphi(N)^{m-2}$ vectors $\mathbf{d} \in [\varphi(N)]^m$ satisfying both (8) and (9). Therefore the total number of vectors $\mathbf{d} \in [\varphi(N)]^m$ which satisfy (8) and at least one congruence (9), for some $w_0 \in [W]_\pm$ and some nonzero vector $\mathbf{u} \in [U]_\pm^m$ not parallel to $\mathbf{f}$, is at most

$$2^{m+1}WU^m\varphi(N)^{m-2} = k^{m+1}e^{m+1}2^{\ell(m+1)+O(m^2\log^2\log m/\log m)}N^{1/2}\varphi(N)^{m-2}.$$

Taking into account that $\varphi(N) \geq N/2$ we obtain the desired result. □

Of course, the same proof provides a stronger result if an SVP-oracle is available:

Theorem 6. *There is a deterministic algorithm $\mathcal{A}$ which, given access to an SVP-oracle and as input an RSA modulus $N = pq$ such that $p + q = O(N^{1/2})$, together with a public exponent e, vector $\mathbf{d} \in [\varphi(N)]^m$ corresponding to a certain vector $\mathbf{f} \in \mathcal{B}_{k,\ell,m}$ generated by* RSA-S1, *makes a single call to the SVP-oracle with the lattice $\mathcal{L}_{e,N}(\mathbf{d})$ and outputs a value $\mathcal{A}(\mathbf{d})$ in time polynomial in $k, 2^\ell, m, \log N$ such that:*

$$\Pr_{\mathbf{d}}\left[\mathcal{A}(\mathbf{d}) \equiv d \pmod{\varphi(N)}\right] \geq 1 - \frac{k^{m+1}e^{m+1}2^{\ell(m+1)+O(m)}}{N^{1/2}}$$

where the probability is taken over all random choices of $\mathbf{d}$ *for the given* $\mathbf{f}$.

Certainly one can obtain similar results when the primes p and q are not balanced, although the probability of success decreases.

4.3 Experiments

We made a few experiments with a (balanced) 1024-bit RSA modulus and a public exponent $e = 3$, using Victor Shoup's NTL library [19]. The experiments have confirmed the heuristic condition. By applying standard floating point LLL reduction, and improved reduction if necessary, we have been able to recover the private exponent for all the parameters considered by Merkle in his own experiments [10] (see Table 1). The success rate has been 100%, except with the case $(m,k,\ell) = (56,26,6)$ where it is 65% (for this case, Merkle only achieved a 10% success rate). We also made some experiments on other (more realistic) sets of parameters. For instance, over 100 samples, we have always been able to recover the factorization with $(m,k,\ell) = (60,30,3)$, $(70,30,2)$ and $(80,40,1)$. The attack takes at most a couple of minutes, as the lattice dimension is only $m+1$. These results show that no set of parameters for RSA-S1 provides sufficient security without being impractical for the card.

References

1. M. Ajtai, 'The shortest vector problem in L_2 is NP-hard for randomized reductions', *Proc. 30th ACM Symp. on Theory of Comput.*, ACM, 1998, 10-19.
2. M. Ajtai, R. Kumar and D. Sivakumar, 'A sieve algorithm for the shortest lattice vector problem' *Proc. 33rd ACM Symp. on Theory of Comput.*, ACM, 2001, 601–610.
3. D. Boneh and G. Durfee, 'Cryptanalysis of RSA with private key d less than $N^{0.292}$', *Proc. of Eurocrypt '99*, Lect. Notes in Comp. Sci., Vol. 1592, Springer-Verlag, Berlin, 1999, 1–11.
4. E. Brickell, D.M. Gordon, K.S. McCurley, and D. Wilson, 'Fast exponentiation with precomputation', *Proc. Eurocrypt '92*, Lect. Notes in Comp. Sci., Vol. 658, Springer-Verlag, Berlin, 1993, 200–207.
5. M.J. Coster, A. Joux, B.A. LaMacchia, A.M. Odlyzko, C.-P. Schnorr, and J. Stern, 'Improved low-density subset sum algorithms', *Comput. Complexity*, **2** (1992), 111–128.
6. D. M. Gordon, 'A survey of fast exponentiation methods', *J. of Algorithms*, **27** (1998), 129–146.
7. A. K. Lenstra, H. W. Lenstra and L. Lovász, 'Factoring polynomials with rational coefficients', *Mathematische Annalen*, **261** (1982), 515–534.
8. T. Matsumoto, K. Kato, and H. Imai, 'Speeding up secret computations with insecure auxiliary devices', *Proc. Crypto '88*, Lect. Notes in Comp. Sci., Vol. 403, Springer-Verlag, Berlin, 1990, 497–506.
9. A. J. Menezes, P. C. van Oorschott and S. A. Vanstone, *Handbook of applied cryptography*, CRC Press, Boca Raton, FL, 1996.
10. J. Merkle, 'Multi-round passive attacks on server-aided RSA protocols', *Proc. 7th ACM Conf. on Computer and Commun. Security*, ACM, 2000, 102–107.
11. J. Merkle and R. Werchner, 'On the security of server-aided RSA protocols', *Proc. PKC '98*, Lect. Notes in Comp. Sci., Vol.1431, Springer-Verlag, Berlin, 1998, 99–116.
12. P. Q. Nguyen and J. Stern, 'Merkle-Hellman revisited: A cryptanalysis of the Qu–Vanstone cryptosystem based on group factorizations', *Proc. Crypto '97*, Lect. Notes in Comp. Sci., Vol.1294, Springer-Verlag, Berlin, 1997, 198–212.
13. P. Q. Nguyen and J. Stern, 'The Béguin–Quisquater server-aided RSA protocol from Crypto'95 is not secure', *Proc. Asiacrypt '98*, Lect. Notes in Comp. Sci., Vol.1514, Springer-Verlag, Berlin, 1998, 372–379.
14. P. Q. Nguyen and J. Stern, 'The two faces of lattices in cryptology', *Proc. CALC '01*, Lect. Notes in Comp. Sci., Vol.2146, Springer-Verlag, Berlin, 2001, 146–180.
15. B. Pfitzmann and M. Waidner, 'Attacks on protocols for server-aided RSA computation', *Proc. Eurocrypt '92*, Lect. Notes in Comp. Sci., Vol.658, Springer-Verlag, Berlin, 1993, 153–162.
16. C. P. Schnorr, 'A hierarchy of polynomial time basis reduction algorithms', *Theor. Comp. Sci.*, **53** (1987), 201–224.
17. V. Shoup, 'Lower bounds for discrete logarithms and related problems', *Proc. Eurocrypt '97*, Lect. Notes in Comp. Sci., Vol.1233, Springer-Verlag, Berlin, 1997, 256–266.
18. D. Stinson, 'Some baby-step giant-step algorithms for the low Hamming weight discrete logarithm problem', To appear in *Mathematics of Computation*.
19. V. Shoup, 'NTL computer package version 5.0', *Available from* `http://www.shoup.net/`.

The Modular Inversion Hidden Number Problem

Dan Boneh[1], Shai Halevi[2], and Nick Howgrave-Graham[2]

[1] Department of Computer Science, Stanford University, CA, USA
dabo@cs.stanford.edu
[2] IBM T.J. Watson Research Center, NY, USA
{shaih,nahg}@watson.ibm.com

Abstract. We study a class of problems called Modular Inverse Hidden Number Problems (MIHNPs). The basic problem in this class is the following: Given many pairs $\langle x_i, \text{MSB}_k\left((\alpha+x_i)^{-1} \bmod p\right)\rangle$ for random $x_i \in \mathbb{Z}_p$ the problem is to find $\alpha \in \mathbb{Z}_p$ (here $\text{MSB}_k(x)$ refers to the k most significant bits of x). We describe an algorithm for this problem when $k > (\log_2 p)/3$ and conjecture that the problem is hard whenever $k < (\log_2 p)/3$. We show that assuming hardness of some variants of this MIHNP problem leads to very efficient algebraic PRNGs and MACs.

Keywords: Hidden number problems, PRNG, MAC, Approximations, Modular inversion, Lattices, Coppersmith's attack

1 Introduction

In recent years several new complexity assumptions were used to construct efficient cryptosystems. The Decision Diffie-Hellman assumption (DDH) was used to construct chosen ciphertext secure encryption [7] and number theoretic pseudo random functions [15]. The Strong RSA assumption was used to construct efficient signature schemes [10,8]. In this paper we introduce a new class of algebraic complexity assumptions which we call the Modular Inverse Hidden Number Problem (MIHNP). Using MIHNP we construct an efficient number theoretic Pseudo Random Number Generator (PRNG) and an efficient MAC. The basic step in evaluating the MAC and the PRNG is one modular inversion modulo a moderate size prime. No expensive exponentiations are needed.

To describe the basic MIHNP we introduce the following notation that will be used throughout the paper: For an m-bit prime p and $y \in \mathbb{Z}_p$ we use $\text{MSB}_k(y \bmod p)$ to denote any integer $Y \in \mathbb{Z}_p$ satisfying $|Y - y| < p/2^k$. In other words, Y is an approximation to y that (usually) matches y on the k most significant bits. We write $\text{MSB}_k(y)$ where there is no ambiguity about the modulus p. In addition, throughout the paper we define the inverse of $0 \in \mathbb{Z}_p$ to be 0. We consistently use Greek characters to denote hidden values.

MIHNP. An instance of the basic MIHNP problem is as follows: let p be a fixed m-bit prime and k, n be positive integers. Let α be a random hidden element of $\mathbb{Z}_p$. We are given p, k, and $\left\langle x_i, \text{MSB}_k(\frac{1}{\alpha+x_i})\right\rangle$ for *random* values $x_1, \ldots, x_n$.

C. Boyd (Ed.): ASIACRYPT 2001, LNCS 2248, pp. 36–51, 2001.
© Springer-Verlag Berlin Heidelberg 2001

The problem is to find α. The δ-MIHNP assumption states that there is no polynomial time algorithm for the Basic-MIHNP problem whenever $k < \delta m$.

In other words, given *many* approximations to $(\alpha + x_i)^{-1} \bmod p$ for random $x_i \in \mathbb{Z}_p$ the problem is to find α. The parameters m, n, k are security parameters for the problem. Note that when $n > 2(m/k)$ the hidden number α is uniquely defined with high probability and consequently there is a unique answer to this problem. We show a lattice-based algorithm, that solves this problem when $k > m/3$. We also explain why this algorithm does not extend to solve it for $k < m/3$. As our algorithm represents the current state-of-the-art in lattice reduction techniques, we conjecture that such techniques cannot be used beyond the $m/3$ bound. More generally, we conjecture that the δ-MIHNP assumption holds for any $\delta < 1/3$. In the next section we introduce several variants of MIHNP that are useful for cryptographic constructions. We also show that the MIHNP problem has a simple limited random self reduction.

MIHNP is closely related to several other Hidden Number Problems (HNPs). Hidden number problems were introduced in [4] where they were used to prove the bit security of the Diffie-Hellman secret in $\mathbb{Z}_p$. The standard HNP is as follows: let $\alpha \in \mathbb{Z}_p$ be a hidden random number. Given $\text{MSB}_k(\alpha \cdot x_i \bmod p)$ for random $x_1, \ldots, x_n \in \mathbb{Z}_p$ the problem is to find α. The standard, HNP can be efficiently solved when $k = O(\sqrt{|p|})$, and this solution forms the basis of the bit-security result in [4] (as well as an attack on weak versions of the Digital Signature Algorithm (DSA), see [13]). This is in contrast to MIHNP which appears to be hard even when k is a constant fraction of $|p|$.

2 Approximate Modular Inversion Problems

We introduce several variants of the basic MIHNP and study their properties. The first variant of MIHNP, which we call the *Computational-MIHNP*, is useful for constructing a MAC.

Computational-MIHNP: An instance of the C-MIHNP problem is as follows: let p be a fixed m-bit prime and k, n be positive integers. Let α be a random hidden element of $\mathbb{Z}_p$. We are given p, k, and $\left\langle x_i, \text{MSB}_k(\frac{1}{\alpha+x_i})\right\rangle$ for *random* values $x_1, \ldots, x_n$. The problem is to construct another pair $\left\langle x, \text{MSB}_k(\frac{1}{\alpha+x})\right\rangle$ for some $x \neq x_i$. The δ-CMIHNP assumption states that there is no polynomial time algorithm for this problem whenever $k < \delta m$.

Although we cannot prove the equivalence of this problem to the basic MIHNP, we do not know of an algorithm for solving it without first discovering the secret α from the given input. The second variant, which we call the *Decisional-MIHNP* is useful for constructing PRNGs.

Decisional-MIHNP: An instance of the D-MIHNP problem is as follows: let p be a fixed m-bit prime and k, n be positive integers. Let α be a random hidden element of $\mathbb{Z}_p$. We are given p and k. The problem is to distinguish the following two ensembles:

$$\left\{ x_1, \text{MSB}_k(\tfrac{1}{x_1+\alpha}), \ldots, x_n, \text{MSB}_k(\tfrac{1}{x_n+\alpha}) \right\} \quad \text{and}$$

$$\left\{ \quad x_1, \text{MSB}_k(r_1), \quad \ldots, \quad x_n, \text{MSB}_k(r_n) \quad \right\}$$

where $\alpha, x_1, \ldots, x_n, r_1, \ldots, r_n$ are chosen uniformly at random in $\mathbb{Z}_p$. The δ-DMIHNP assumption states that no polynomial time algorithm can distinguish these two ensembles with non-negligible advantage whenever $k < \delta m$.

As before, we cannot reduce this problem to either of the previous problems, but we know of no algorithms for D-MIHNP, other than first finding the hidden element α. In a sense, it seems that the tools that we have for designing algorithms for these problems are too crude to distinguish between these variants.

This situation is somewhat analogous to the situation with the various discrete-logarithm assumptions. The basic MIHNP can be viewed as an analog of the Discrete-Log Problem (DLP): given $g^\alpha \bmod p$ find the hidden number α. Just as DLP is often insufficient for cryptographic constructions, we need stronger assumptions that the basic MIHNP for the constructions in this paper. The C-MIHNP can be viewed as an analog of the Computational Diffie-Hellman assumption (CDH), and D-MIHNP is the analog of the Decision Diffie-Hellman assumption (DDH). As is the case with the various MIHNP problems, we also do not have reductions between the various discrete-log problems, yet the only algorithms that we know for solving any of them involve solving discrete-log.

2.1 Random Self Reduction for MIHNP

The MIHNP problem has a simple limited random self reduction among instances modulo the same prime p. The reduction shows that for a prime p if finding $\alpha \in \mathbb{Z}_p$ is hard for a worst case α then it is also hard for a random $\alpha \in \mathbb{Z}_p$.

Suppose there is an algorithm $\mathcal{A}$ that solves the Basic-MIHNP problem with probability ϵ, where the probability is taken over the choice of the x_i's and also over the choice of α. We show that this implies an algorithm $\mathcal{B}$ for solving Basic-MIHNP that works for any fixed α with probability ϵ, where this time the probability is over the choice of the x_i's only.

Given an instance of Basic-MIHNP, $\langle x_i, y_i \rangle$, $i = 1, \ldots, n$, algorithm $\mathcal{B}$ picks a random $r \in \mathbb{Z}_p$, and runs algorithm $\mathcal{A}$ on the Basic-MIHNP problem defined by the tuples $\langle x_i + r, y_i \rangle$, $i = 1, \ldots, n$.

Note that if the original MIHNP instance corresponds to the hidden number α, then the new instance will correspond to the hidden number $\alpha' = \alpha - r$, which is random and independent of the x_i's. It follows that with probability ϵ, the algorithm $\mathcal{A}$ indeed returns α', and then $\mathcal{B}$ can add back r to recover α.

We call this a limited random self reduction, since we only randomize the solution α, and not the elements $x_1, \ldots, x_n$. The computational MIHNP and decisional MIHNP have similar limited random self reductions.

3 Security Analysis of the MIHNP

In this section we analyze the security of MIHNP. We show how to apply the currently known technology in algebraic cryptanalysis to MIHNP, and demonstrate the limitations of that technology when applied to this problem. We know of no better way to distinguish the pairs $\langle x_i, \text{MSB}_k((\alpha + x_i)^{-1})\rangle$ from random, other than to actually recover the secret α (and use the knowledge of this to verify the bits), and so this is the problem we address. That is, we assume that we have a system of equations

$$(\alpha + x_i)(b_i + \epsilon_i) = 1 \pmod{p} \qquad i = 0, \ldots, n, \tag{1}$$

where $\alpha \in Z_p$ is the (large, secret) variable we aim to discover, the x_i's are known, but randomly chosen elements of Z_p, the b_i's are the known most significant bits, and the ϵ_i are variables that correspond to the unknown low order bits, so we have $|\epsilon_i| \leq 2^{m-k}$ for all i. Observe that once we find any of the ϵ_i's we can discover the secret α immediately, from the fact that $\alpha = 1/(b_i + \epsilon_i) - x_i \pmod{p}$. However, as we shall see, typically we find all the ϵ_i simultaneously, or none at all.

We attempt to solve MIHNP using lattice techniques. We set up a lattice that incorporates the relations from Eq. (1), so that the bound on the size of the ϵ_i's will correspond to some small vector in the lattice. If we can make the argument that this vector is by far smaller than any other vector in this lattice, then we could use the LLL lattice reduction algorithm [14] to find it, thereby recovering the ϵ_i's. This framework was used in [4] to solve the original HNP.

Looking at Eq. (1), however, we find that these relations cannot be used directly to set up a lattice. The reason is that each of these relations has a term of the form $\alpha \cdot \epsilon_i$, where α is unbounded (i.e., it can be as large as p), and the ϵ_i's change from one relation to the next. To use in a lattice, one must first "linearize" these relations, and doing so would introduce a new unbounded variable for each of the products $\alpha\epsilon_i$. (We stress that current technology has no problem handling either changing small unknowns such as the ϵ_i, or fixed large unknowns such as α. It is *the product of the two* that makes this problem hard.)

We are therefore forced to eliminate the unknown α from the relations of Eq. (1), before we can use them to set up a lattice. Given the $n+1$ relations from Eq. (1), we eliminate the unknown α, and produce n relations of the form:

$$(x_i - x_0)(b_0 + \epsilon_0)(b_i + \epsilon_i) - (b_0 + \epsilon_0) + (b_i + \epsilon_i) \;=\; 0 \pmod{p} \tag{2}$$

These relations are already in a form that is amenable for use in a lattice, and we can apply to them (an extension of) the techniques from [6,12], as we now explain. We start by re-writing the left hand side of Eq. (2) as a polynomial in the unknowns ϵ_0 and ϵ_i, namely:

$$f_i(\epsilon_0, \epsilon_i) \stackrel{\text{def}}{=} (x_i - x_0)\epsilon_0\epsilon_i + (b_0(x_i - x_0) + 1)\epsilon_i + (b_i(x_i - x_0) - 1)\epsilon_0 + (b_0 b_i(x_1 - x_0))$$

Notice that the coefficients of this polynomial are known to us (since we know all the b_i's and x_i's), and therefore we can set up a lattice based on their values. To simplify notation, we denote below $f_i(\epsilon_0, \epsilon_i) = A_i\epsilon_0\epsilon_i + B_i\epsilon_i + C_i\epsilon_0 + D_i$.

3.1 First Attempt: A Linear Approach

As a first attempt at a solution, we set up a lattice of dimension $3n+2$ as follows. The lattice is spanned by the rows of a real matrix M that has the following general structure:

$$M = \begin{pmatrix} E & R \\ 0 & P \end{pmatrix}$$

where E and P are diagonal matrices of dimensions $(2n+2) \times (2n+2)$ and $n \times n$, respectively, and R is a $(2n+2) \times n$ matrix. Each of the first $2n+2$ rows of M is associated with one of the terms in relations from Eq. (2) (i.e., the constant term, the terms ϵ_i, and the terms $\epsilon_0\epsilon_i$), and each of the last n columns is associated with one of the n relations.

The matrix R incorporates the relations themselves. The (i,j) entry in this matrix is just the coefficient in the j'th relation of the term corresponding to row i. The diagonal entries of the matrix P are all equal to p, and the diagonal entries of the matrix E correspond to the bounds on the terms associated with each row. Specifically, if the term which is associated with row i is bounded by B, then entry (i,i) in E is equal to $1/B$. That is, the row corresponding to the constant term has diagonal entry 1, rows corresponding to ϵ_i have diagonal entries $1/2^{m-k}$, and rows corresponding to $\epsilon_0\epsilon_j$ have diagonal entries $1/2^{2(m-k)}$. An example for the matrix M for $n=2$ is given in Figure 1.

$$M = \begin{pmatrix} 1 & 0 & 0 & 0 & 0 & 0 & D_1 & D_2 \\ 0 & 2^{k-m} & 0 & 0 & 0 & 0 & C_1 & C_2 \\ 0 & 0 & 2^{k-m} & 0 & 0 & 0 & B_1 & 0 \\ 0 & 0 & 0 & 2^{k-m} & 0 & 0 & 0 & B_2 \\ 0 & 0 & 0 & 0 & 2^{2(k-m)} & 0 & A_1 & 0 \\ 0 & 0 & 0 & 0 & 0 & 2^{2(k-m)} & 0 & A_2 \\ 0 & 0 & 0 & 0 & 0 & 0 & p & 0 \\ 0 & 0 & 0 & 0 & 0 & 0 & 0 & p \end{pmatrix} \begin{matrix} \text{row corresponds to} \\ \dots\dots 1 \\ \dots\dots \epsilon_0 \\ \dots\dots \epsilon_1 \\ \dots\dots \epsilon_2 \\ \dots\dots \epsilon_0\epsilon_1 \\ \dots\dots \epsilon_0\epsilon_2 \\ \\ \\ \end{matrix}$$

Fig. 1. The matrix M for the case $n=2$.

We can now view each one of the relations of Eq. (2) as holding over the integers, by explicitly introducing the appropriate multiple of p. Namely, we have:

$$A_i\epsilon_0\epsilon_i + B_i\epsilon_i + C_i\epsilon_0 + D_i + p \cdot \kappa_i = 0 \tag{3}$$

From the way we constructed this system of n polynomial relations, we know that it has an integer solution $\epsilon_i = e_i, \kappa_i = k_i$ in which all the e_i's are bounded below 2^{m-k}. Let $\boldsymbol{v}$ be a $(3n+2)$ integer vector containing the values of all the terms in our system of equations, according to this solution. Namely, we set

$$\boldsymbol{v} \stackrel{\text{def}}{=} \langle 1,\ e_0,\ \ldots,\ e_n,\ e_0 e_1,\ \ldots,\ e_0 e_n,\ k_1,\ \ldots,\ k_n \rangle$$

It follows that for this integer vector $\boldsymbol{v}$ we get:

$$\boldsymbol{v} \cdot M = \left\langle 1,\ \frac{e_0}{2^{m-k}},\ \ldots,\ \frac{e_n}{2^{m-k}},\ \frac{e_0 e_1}{2^{2(m-k)}},\ \ldots,\ \frac{e_0 e_n}{2^{2(m-k)}},\ 0,\ \ldots,\ 0 \right\rangle$$

Thus, the lattice point $\boldsymbol{v} \cdot M$ has only $2n+2$ non-zero entries, and each of these is less than 1, so its Euclidean norm is less than $\sqrt{2n+2}$.

On the other hand, it is easy to see that the determinant of the lattice $L(M)$ equals $p^n/2^{(m-k)(3n+1)}$, so making use of the Gaussian heuristic[1] for short lattices vectors, we expect that our vector is the shortest point in $L(M)$ as long as

$$\sqrt{2n+2} \ll \sqrt{3n+2} \left(2^{(k-m)(3n+1)} \cdot p^n \right)^{1/(3n+2)}. \tag{4}$$

Whenever this condition is met we will assume that an adversary can recover the vector $\boldsymbol{v}$ using lattice reduction methods such as LLL (although we note that in practice, the adversary may not find this that easy unless $\boldsymbol{v} \cdot M$ is the shortest vector by a substantial margin).

Substituting $p \approx 2^m$ into Eq. (4) and ignoring low-order terms, this condition is simplified to $2^k \gg 2^{2m/3}$. Therefore, this method can only be used when the number of bits of $1/(\alpha + x_i)$ that we see is more than $2m/3$ (alternatively, when the number of bits that we are missing is less than $m/3$). This gives an algorithm for Basic-MIHNP when $\delta \geq 2/3$.

The dimension of the lattice. We remark that the same bounds (but no better) could also be achieved from a lattice of smaller dimension that utilizes the fact that $\boldsymbol{v} \cdot M$ has n trailing zeros. However in this analysis we are only interested in showing that there exist bounds on m even for lattices of arbitrary dimension (i.e. we effectively allow the adversary the power to reduce lattices of arbitrary dimension). This means we can ignore efficiency issues regarding the dimension of the lattice, and opt for the easier way to describe and extend the lattices (as above). This assumption is particularly important to note in the subsequent section where the dimension grows exponentially in n.

3.2 Making Use of Multiples

To improve upon the bound of $m/3$, we apply a technique due to Coppersmith [6] to make better use of the relations in Eq. (2). Namely, instead of using only these relations in our lattice, we can use also relations that are derived by taking

[1] In fact, it is possible to prove rigorously, that when the x_i's are chosen at random, the vector $\boldsymbol{v} \cdot M$ is (with high probability) the shortest vector in $L(M)$. This proof will appear in the full version of this paper.

products of them. For example, since we have $f_1(\epsilon_0, \epsilon_1) = 0$ and $f_2(\epsilon_0, \epsilon_2) = 0$, we also know that $\epsilon_2 f_1(\epsilon_0, \epsilon_1) = 0$, and also $f_1(\epsilon_0, \epsilon_1) \cdot f_2(\epsilon_0, \epsilon_2) = 0$. Moreover, since the original relations hold modulo p, then the last relation holds also modulo p^2.

Of course, these additional relations introduce new terms that were not present in the original one. (For example, the relation $f_1 f_2 = 0$ from above has a term $\epsilon_0\epsilon_1\epsilon_2$, which we did not have in the original system.) Nonetheless, we hope that weighing the additional relations against the additional terms, we would be able to get a better result. Hence, our goal here is to add as many relations as possible, while keeping the number of additional terms as small as possible.

Once we decide on a set of relations to use, we construct the lattice in exactly the same way as above. Namely, if we have r relations and t terms, we construct a $(r + t) \times (r + t)$ matrix M with the same structure as above. That is, the top left $t \times t$ sub-matrix E is diagonal with entries that correspond to the (bounds on the) different terms, to its right we put a $t \times r$ matrix R that corresponds to our relations, and at the bottom left we put a diagonal matrix P that would take care of the modular reductions. One difference is that now, if the i'th relation holds modulo p^i, then the corresponding diagonal entry of P will be p^i (rather than just p).

Constructing a lattice. The key aspect of this approach is to choose which relations to put in the lattice, and to analyze the parameters achieved by this lattice. Below we think of the process of adding relations to the lattice as happening in phases. In phase d, we add to the lattice relations that are obtained by multiplying up to d of the original relations. These new relations look like $f_{i_1} \cdots f_{i_d} = 0 \bmod p^d$ for some $0 < i_1, \ldots, i_d \leq n$.

We note that once we have in the lattice some relations (and all their terms), we might as well add other relations that use only terms that already appear in the lattice. For example, if we have the relation $f_1 f_2 = 0$ in the lattice, we might as well also add the relation $\epsilon_1 f_2 = 0$, since every term that appears in $\epsilon_1 f_2$ must already appear in $f_1 f_2$ (because f_1 includes the term ϵ_1). Therefore, once we have $f_1 f_2$ and all its terms, we can add $\epsilon_1 f_2$ "for free". The only exception is that we have to make sure that the relations in the lattice are linearly independent. For example, once we have in the lattice the relations $f_2, \epsilon_0 f_2, \epsilon_1 f_2$ and $f_1 f_2$, we cannot add also the relation $\epsilon_0\epsilon_1 f_2$, as it is linearly dependent on the other relation, by the equality $f_1 f_2 = A_1\epsilon_0\epsilon_1 f_2 + B_1\epsilon_1 f_2 + C_1\epsilon_0 f_2 + D_1 f_2$.

Notations and conventions. In the analysis below we talk about the "weight" of relations or terms. The weight of a relation is the number of original relations that are multiplied. For example, the relation $\epsilon_1 f_2 = 0$ has weight 1, and $f_1 f_2 = 0$ has weight 2. We note that if a relation has weight i, then this relation holds modulo p^i (but not necessarily modulo p^{i+1}). The weight of a term is just its degree. For example, the weight of $\epsilon_0^2\epsilon_1$ is 3. With this notation, the determinant of the lattice is proportional to the total weight of all the relations, and inversely proportional to the total weight of all the terms that are used in these relations.

More precisely, when p is an m-bit prime, and the bound on the ϵ_i's is 2^{m-k}, (i.e., the number of bits of $1/(\alpha+x)$ that we see is k), the determinant of the lattice is roughly $2^{m\cdot\mathrm{weight(relations)}-(m-k)\cdot\mathrm{weight(terms)}}$.

Recall that our goal is to maximize the determinant (since this would imply that the lattice is unlikely to have short vectors, other than the one corresponding to our solution). Below we show, however, that we always have weight(relations) $\le$ 2weight(terms)/3. Therefore, to get $det(L) > 1$, we must have $m - k < 2m/3$.

The relations. In this analysis we assume that the number n of the original relations can be made as large as we want. Since we aim to show that the approach is bound to fail beyond $2m/3$, we can make this assumption without loss of generality (as adding relations can only help the algorithm). When we analyze phase d, we assume that $n \gg d$, so that we get a good approximation of the sum $\sum_{i=1}^{d}\binom{n}{i}$ by taking just the last term, $\binom{n}{d}$.

The relations that we add in phase d are all the $\binom{n}{d}$ relations of weight d, that are obtained by multiplying d distinct relations (from the original n), and then adding all the relations that are now "for free". We again note that since $n \gg d$, then a vast majority of the relations in phase d are of this form.

Analysis. We start by analyzing the weight of the terms. Since each f_i has all the possible terms for a multi-linear function in ϵ_0, ϵ_i, it follows that a product of d distinct f_i's have all the possible terms with degree at most d in ϵ_0, and at most 1 in all the other ϵ_i's.

We group the terms according to the number of ϵ_i's *other than* ϵ_0 in them. Clearly, we have exactly $(d+1)\binom{n}{j}$ terms with exactly j ϵ_i's other than ϵ_0. (We have $\binom{n}{j}$ ways to choose the ϵ_i's, and then ϵ_0 can have any degree between 0 and d.) The weight of these terms ranges from j (if the degree of ϵ_0 is 0) to $j+d$ (if the degree of ϵ_0 is d). Therefore, the total weight of all the terms is

$$\mathrm{weight(terms)} = \sum_{j=0}^{d}\left(\binom{n}{j}\cdot(j+(j+1)+\cdots+(j+d))\right)$$

$$= \sum_{j=0}^{d}\left(\binom{n}{j}\cdot(d+1)(j+\frac{d}{2})\right)$$

Recall now that we assume that n is large enough with respect to d, so that $\sum_{j=0}^{d}\binom{n}{j} = \binom{n}{d}(1+o(1))$. This implies that also

$$\mathrm{weight(terms)} = \binom{n}{d}(d+1)(3d/2)(1+o(1))$$

By the same argument, the number of terms is $(d+1)\binom{n}{d}(1+o(1))$.

We now proceed to analyze the weight of the relations. First, observe that we cannot have more relations than terms in the lattice, since otherwise we get linear dependencies. Thus, there are at most $(d+1)\binom{n}{d}(1+o(1))$ relations. Moreover, the weight of each of these relations cannot be more than d, since in phase d we only multiply up to d of the f_i's. Therefore, the total weight of all the relations is bounded by

$$\text{weight(relations)} \leq d \cdot (d+1)\binom{n}{d}(1+o(1))$$

In fact, it is possible to show that this bound is tight, and the total weight of the relations that we get is at least $d^2\binom{n}{d}$. We conclude that in our lattice we must have

$$\frac{\text{weight(relations)}}{\text{weight(terms)}} \leq \frac{d \cdot (d+1)\binom{n}{d}(1+o(1))}{\binom{n}{d}(d+1)(3d/2)(1+o(1))} \leq \frac{2}{3} + o(1)$$

(We remark that a more careful analysis can even show a bound of $2/3 - o(1)$.)

3.3 Conclusions from the Analysis of MIHNP

We showed that the Basic-MIHNP problem can be efficiently solved whenever we are given more than 1/3 of the bits of $(\alpha+x_i)^{-1} \bmod p$. The analysis does not extend beyond 1/3. (Moreover, near 1/3 the dimension of the lattice makes it completely infeasible to reduce.) For this reason, we conjecture that this problem is hard when we are given less than 1/3 of the bits even if a large number of random samples x_i are given. That is, we conjecture that the δ-MIHNP assumption holds whenever $\delta < 1/3$.

3.4 Other Variants of MIHNP

The tools that we devised to analyze the MIHNP can be used also to analyze similar problems. For example, in Section 4 we will be interested in a problem where we are given pairs $(x_i, \beta/(\alpha+x_i) \bmod p)$, $i = 1 \dots n$, and we need to recover both α and β. The corresponding relations that we get are

$$(\alpha + x_i)(b_i + \epsilon_i) = \beta \pmod{p} \qquad\qquad i = 0, \dots, n,$$

Again, we have a problem with the terms $\alpha\epsilon_i$, but when we eliminate α as before, we would get terms $\beta\epsilon_i$. Hence, to be able to set up a lattice we need to eliminate both α and β. More generally we may consider relations of the form

$$R_i: \ (x_{i0} + y_{i0}\epsilon_i) + \sum_{j=1}^{r}(x_{ij} + y_{ij}\epsilon_i)\alpha_j = 0$$

where the x_{ij}'s and y_{ij}'s are random and known, the α_i's are unknown and unbounded, but common to all these relations, and each ϵ_i is an unknown unique to relation i, but for which we have some bound. As before, the terms $\epsilon_i\alpha_j$ cannot be handled by standard lattice reduction techniques, so we need to first eliminate the α_j's. We now show that if we need to eliminate r such "unbounded variables", then the lattice-reduction techniques from above can only be used when $k/m > r/(r+2)$ (i.e., the number of hidden bits is less than $m \cdot \frac{2}{r+2}$. Note that for MIHNP we have $r = 1$, and indeed we got the bound $k/m > 1/3$. For the case above of two variables, we get $k/m > 1/2$. Hence, this problem is harder than MIHNP in the sense that it can only be solved when $\delta = k/m > 1/2$.

Assume that we are given given $n+r$ relations. We can set a linear system in the r unknowns $\alpha_1 \ldots \alpha_r$ and r relations $R_{n+1}, \ldots, R_{n+r}$, solve for the unknowns, and then substitute the solution in all the other n relations $R_1 \ldots R_n$ (each time multiplying by the common denominator, to get a polynomial relation rather than a rational one).[2]

Using Cramer's rule for the solution of a linear system, it is easy to verify that the terms that we substitute for the α_j's are multi-linear in $\epsilon_{n+1} \cdots \epsilon_{n+r}$. Hence, after eliminating the "unbounded variables", we are left with n relations $f_i = 0, i = 1 \ldots n$, where f_i is a multi-linear relation in $\epsilon_i, \epsilon_{n+1} \cdots \epsilon_{n+r}$. These relations are the ones we use to set-up a lattice.

As we did for MIHNP, we set-up a lattice not only using the f_i's themselves, but also using products of them. As before, we use relations that we obtain by multiplying d distinct f_i's (for some parameter d, and under the assumption that $n \gg d$). A product of d such f_i's is a relations

$$p(\epsilon_{i_1}, \epsilon_{i_2}, ..., \epsilon_{i_d}, \epsilon_{n+1}, \ldots, \epsilon_{n+r}) = 0$$

where p is multi-linear in the ϵ_{i_j}'s, and has degree d in $\epsilon_{n+1} \cdots \epsilon_{n+r}$. We want to count the total weight of the terms and relations in this lattice. As we know, if $n \gg d$, then it is sufficient to consider only these terms that include exactly d distinct e_i's, other than $\epsilon_{n+1} \cdots \epsilon_{n+r}$. So there are $\binom{n}{d}$ ways of choosing the ϵ_{i_j}'s, and for each choice we have $(d+1)^r$ possible combinations of the degrees of $\epsilon_{n+1} \cdots \epsilon_{n+r}$. Namely, for a specific choice of $\epsilon_{i_1}, \epsilon_{i_2}, ..., \epsilon_{i_d}$, the terms that we get are exactly all the terms in the expression

$$(\epsilon_{i_1} \cdot \epsilon_{i_2} \ \cdots \ \epsilon_{i_d}) \cdot ((1 + \epsilon_{n+1} + \cdots + \epsilon_{n+1}^d) \ \cdots \ (1 + \epsilon_{n+r} + \cdots + \epsilon_{n+r}^d))$$

This means that for this choice of ϵ_{i_j}'s, we have $(d+1)^r$ terms, and the weight of these terms vary between d and $d + rd$. The total weight of all these terms is

$$\sum_{k_1=0}^{d} \sum_{k_2=0}^{d} \cdots \sum_{k_r=0}^{d} (d + k_1 + k_2 + \ldots k_r) \ = \ (d+1)^r \cdot (d + rd/2)$$

Therefore, we have $\binom{n}{d}(d+1)^r$ terms, of total weight $\binom{n}{d}(d+1)^r \cdot d(1 + r/2)$. On the other hand, we cannot have more relations than terms, and the weight of a relation cannot be more than d, so the total weight of the relations is

[2] Clearly, this is not the only way to eliminate the unbounded variables. For example, we can solve different sets of relations for these unknowns, depending on the relation to which we want to substitute. However, tracing through the arguments below, the method we use here seems to give the smallest number of terms.

at most $\binom{n}{d}(d+1)^r \cdot d$. (This bound is tight, since it can be shown that for random relations, the total weight is at least $\binom{n}{d}(d+1)^r \cdot (d-r)$.) Recall that the determinant of our lattice is roughly $2^{m \cdot \text{weight(relations)} - (m-k) \cdot \text{weight(terms)}}$. To get the determinant above 1, we therefore must have

$$m \cdot \binom{n}{d}(d+1)^r d > (m-k) \cdot \binom{n}{d}(d+1)^r d(1+r/2)$$

which means that $m > (m-k)(1+r/2)$, or $k/m > r/(r+2)$.

4 Cryptographic Applications

The apparent intractability of MIHNP, suggests that it may be useful as the basis for cryptographic applications. Indeed, we show below how to use the decision-MIHNP assumption and the computational-MIHNP assumptions, respectively, to get an efficient pseudorandom generator and a MAC.

4.1 Pseudorandom Generator

The decision-MIHNP immediately suggests a construction of a PRNG. The input to this generator would be "the secret" a, and n random points $x_1..x_n \in \mathbb{Z}_p$. The output would be the points $x_1..x_n$ together with (say) 1/4 of the bits of $1/(a+x_i) \bmod p$ for all i. More precisely, we have the following system:

Parameters. The parameters of the system include an m-bit prime p, and two other parameters, n and k, where k specifies how many bits of $1/(a+x)$ we output, and n specifies how many x'es we have in the input of the generator. These parameters are discussed in more details below.

The generator. On parameters p, n and k, the generator input is a sequence $(x_1, .., x_n, a)$ of $n+1$ elements in Z_p. The output is the sequence

$$G(a, x_1, .., x_n) \stackrel{\text{def}}{=} \left(x_1, ..., x_n, \text{MSB}_k(\frac{1}{a+x_1}), ..., \text{MSB}_k(\frac{1}{a+x_n})\right)$$

The security of this generator follows immediately from the decision-MIHNP assumption. We note that this is a pseudorandom number generator, but not pseudorandom bit generator, since the output distribution is not the uniform one. There are standard techniques for transforming this to a pseudorandom bit generator. Any of a number of standard extractors could be used for this purpose [16,11].

Proposition 1. *Under the D-MIHNP assumption, G is a secure pseudorandom generator.*

One point worth mentioning is the re-keying of the generator from the previous output. It is well known, see [3], that it is secure to do this, if the underlying generator is itself secure. In our case this means that we may fix the $x_1, ..., x_n$ once at the start of the whole procedure, and then use just the $\text{MSB}_k(1/(a+x_i))$ part of the output to re-key a and form the output bits of the PRNG.

Parameters and performance. The parameters m (the size of the prime p) and k (the number of bits to output from each $1/(a+x_i)$) must be chosen such that solving the MIHNP with k output bits modulo a prime of size $|p| = m$ is infeasible. More precisely, if we assume that the threshold for feasible solution is when the adversary sees $\geq m/3$ of the bits of $1/(a+x_i)$, and we want security level of 2^r, we need to make sure that our generator outputs at most $m/3 - r$ of the bits. This means that we have a tradeoff between the number of the bits that we output (which is related to the expansion of the generator) and the size of the prime that we work with.

A reasonable setting is to set $r = k$ (i.e., output as many bits of $1/(a+x_i)$ as our security parameter). With this setting, we should choose m so that $k \leq m/3 - k$, namely $m \geq 6k$. (Another constraint is that to get security level 2^r, we must hide at least $2r$ bits of $1/(a+x_i)$, to avoid birthday-type attacks. In the current setting, however, this constraint is subsumed by the previous one.) An invocation of the generator G stretches a random input of length $(n+1)m$ bits, into a pseudorandom output of length $n(m+k)$ bits. Hence, each invocation generates $nk - m$ pseudorandom bits.

For a numerical example, assume that we want to get security level of 2^{80}. We then set $m = 6 \cdot 80 = 480$ and $k = 80$ (i.e., we work with a 480-bit prime, and output 80 of the bits of $1/(a+x_i)$). With these parameters, each invocation of G generates $80n - 480$ pseudorandom bits (so we must choose $n > 6$ to get any expansion). In our example below we use $n = 10$.

A naive implementation of this generator would require n modular inversions to compute $\text{MSB}_k(\frac{1}{a+x_i}), i = 1...n$. Therefore, the cost of this implementation is roughly k bits per inversion (for a sufficiently large n). Keeping with the numerical example above, choosing, for example, $n = 10$, the size of the seed (which is the amount of state we keep) is 4800 bits (= 600 bytes), and we get $nk - m = 320$ pseudorandom bits at the cost of 10 inversions, or 32 bits per modular inversion. Keeping a larger state results in more bits per inversion. For example, setting $n = 20$, we have 9600 bits (= 1200 bytes) of state, and we get 1120 bits at the cost of 20 inversions, which is 56 bits per inversion.

Even this naive implementation is already quite fast. With a careful implementation, the cost of modular inversion can be as small as only a few multiplications [1]. Moreover, since we work in a relatively small field, the operations can be quite fast. Finally, we note that the modular inversions are independent of each other, so it is trivial to parallelize this computation.

Speedup via batching. One way to speed up the computation, is to trade modular inversions for multiplications by using batching. The idea, first discovered by Peter Montgomery, is as follows: To compute $1/(a+x_i), i = 1...n$, we first compute the product $\pi = \prod_i(a+x_i)$, then invert only this product to get $\pi^{-1} \bmod p$, and finally compute $1/(a+x_j) = \pi^{-1} \cdot \prod_{i \neq j}(a+x_i)$. It is not hard to see that one can compute all the values $1/(a+x_i)$ using only $3(n-1)$ multiplications and one modular inversion. (For this, one needs to keep in memory up to n intermediate values during the computation.) If inversion is more expensive than three multiplications (as is the case for all the multi-precision software libraries that we know), then this implementation will be more efficient than the naïve one.

Back to our numeric example, with $n = 10$ we get 320 bits for one inversion and 28 multiplications, which is about 11 bits per multiplication. With $n = 20$ we get 1120 bits for one inversion and 58 multiplications, which is roughly 19 bits per multiplication. Hence, our generator is more efficient than other algebraic generators, e.g. the pseudorandom generator due to Gennaro [9] which is based on the problem of discrete-log with small exponent. The generator of [9] generates approximately one pseudorandom bit per multiplication. Furthermore, Gennaro's generator uses a much larger prime field. Other algebraic generators, such as the Blum-Blum-Shub generator [2], generate a small number of pseudorandom bits per multiplication modulo a much larger modulus than the one we use. The exact comparison of our generator to BBS depends on the number of bits per round output by the BBS generator.

Even faster variants. We can increase the speed even further by slightly modifying the generator itself. Below we describe two such modifications.

Re-defining the output. To speed the batching implementation, we change the output of the generator, so that it would be easier to compute this output from the intermediate value π^{-1} that we get during the computation. Specifically, we set

$$G'(a, x_1, .., x_n) \stackrel{\text{def}}{=} \left(x_1, ..., x_n, \text{MSB}_k(\frac{1}{\pi_1}), ..., \text{MSB}_k(\frac{1}{\pi_n}) \right)$$

where the π_j's are defined by $\pi_j = \prod_{i \neq j}(a + x_i)$.

We stress that the security of G' does not seem to be equivalent to our original D-MIHNP problem. Rather, this generator defines yet another variant of D-MIHNP. Still, the analysis from Section 3 applies in exactly the same manner to this variant too.

Implementing G' using the batching technique takes only $2m - 1$ multiplications and one inversion (and can also be parallelized easier than with G). Hence, in our numerical example we get 10 bits per multiplication for $n = 8$, or 28 bits per multiplication for $n = 20$.

Using a harder MIHNP problem. Another possibility is to use harder variants of MIHNP. For example, instead of using $f_\alpha(x) = 1/(\alpha + x)$ as our underlying function, we can use $f_{\alpha,\beta}(x) = \beta/(\alpha + x)$ where both α and β are secret.

(We mention in passing that just like the original MIHNP, this variant too has limited random self-reducibility (in α and β). This is because we can set $y_i = (x_i + s) \cdot r^{-1}$ and then we have $\frac{\beta}{\alpha+y} = \frac{\beta r}{(\alpha r+s)+x}$. For any fixed α, β (with $\beta \neq 0$), if we choose r, s uniformly at random (with $r \neq 0$) then $\alpha r + s, \beta r$ are uniformly random and independent.)

From the analysis in Section 3.4, it follows that this problem is infeasible to solve when the number of "missing bits" is more than $m/2$ (as opposed to $2m/3$ for the original MIHNP). This means that we may be able to output as many as $m/2 - r$ bits for security level of 2^r. Assuming that we still set $r = k$, this argument suggests that we must set m (the size of our prime) so that $k \leq m/2-k$, or $m \geq 4k$.

Each invocation of the generator now stretches $m(n+2)$ random bits into $n(m+k)$ pseudorandom bits, so we get $nk - 2m$ pseudorandom bits per invocation. For a numerical example, to get security level of 2^{80}, we choose $m = 320$, and $k = 80$ (i.e., work with a 320-bit prime and output 80 bits of $\beta/(\alpha + x_i)$). Working with $n = 10$, we have 3840 bits of state and 160 bits per invocation (same as for the $n = 8$ example from above). However, this generator does roughly 25% more operations, but in a smaller field ($|p| = 320$ instead of $|p| = 480$), so we expect it to be nearly twice as fast. Similarly, using $n = 22$ we get 1120 bits per application with state of 7040 bits, which is the same number of bits per invocation, and somewhat smaller state than the $n = 20$ example above. Again, we do 10% more operation over a smaller field, so we expect the overall running time to be roughly twice as fast.

4.2 Message Authentication Code

The computational-MIHNP directly implies an efficient "weak MAC", secure under known (random) message attacks. The parameters p and k are chosen just as for the generator, and the secret MAC key is an element $\alpha \in Z_p$. To authenticate a message χ, one adds the authentication tag

$$\mathrm{MAC}_{\alpha}(\chi) = \mathrm{MSB}_k\left(\frac{1}{\alpha + x}\right)$$

Proposition 2. *Suppose the δ-Computational MIHNP assumption holds. Then when $k < \delta(\log_2 p)$ the above MAC is secure under known (random) message attacks.*

The proof is immediate. The cost per MAC computation is thus just one modular inversion. Moreover, if we need to compute MAC for many messages $x_1 ... x_n$, we can use the same batching trick from the previous section to speed up this computation.

The "weak MAC" above can be converted to a MAC secure against chosen message attack, using standard techniques. For example, one could apply the MAC to a random string r and then use a one-time signature based on r to sign the message x. However, these generic conversion techniques (from security against a known message attack to security against a chosen message attack) make the MAC much less efficient. We do not know whether the MAC from above is by itself secure against chosen message attack. This would require a version of the computational MIHNP assumption, where the x_i's can be chosen by the attacker. Currently we cannot tell whether this chosen message MIHNP problem is intractable.

5 Conclusions and Open Problems

In this paper we proposed the MIHNP, and variants thereof, as new and potentially hard mathematical problems. We presented a few efficient cryptographic constructions based on these problems. To justify the hardness of these MIHNP problems we used the most up-to-date lattice analysis techniques to solve MIHNP and even allowed the attacker the power to reduce infeasibly large lattices. Our best algorithm works whenever the fraction of given bits is greater than one third of the length of the modulus. However, the lattice based approach does not extend to solve the MIHNP when less than a third of the bits is given. We therefore conjectured the MIHNP is hard in this case. MIHNP is an interesting and efficient building block for cryptographic systems. It clearly deserves further study.

One particularly interesting question to answer is how much easier the MIHNP problem becomes if the x_i are not randomly chosen, but adversarially chosen. If the Computational-MIHNP remains hard when the x_i's are chosen adversarially then we obtain an efficient MAC from MIHNP. Also, it is very interesting to see whether any non-lattice approaches shed any light on the hardness of these MIHNP problems.

Lastly we mention that the analysis we have used for MIHNP, can be heuristically applied to certain modular polynomials arising from using the Diffie–Hellman protocol with elliptic curves (ECDH). As was recently done in [5], we may apply our results to proving statements on the bit security of ECDH. Specifically, to prove the bit security of ECDH on a specific curve E, it is sufficient to solve the following hidden-number problem (called ECHNP): We are given

$$\left\langle (x_i, y_i),\ \mathrm{MSB}_k\Big(\Big(\frac{\psi - y_i}{\chi - x_i}\Big)^2 - x_i - \chi\Big) \right\rangle$$

for many *random* points $(x_i, y_i) \in E$, and we need to find the hidden point $(\chi, \psi) \in E$. The (heuristic) analysis from Section 3.4 can be applied to this problem too, and it suggests that ECHNP can be solved for $\delta > 3/5$. This would mean that given an algorithm that computes the top $3/5$ fraction of bits in (the x-coordinate of) the ECDH secret, one can devise an algorithm to compute all the bits. However, since the analysis in Section 3.4 is only a heuristic, one does not immediately get a proof of bit-security for ECDH.

We leave further details to a subsequent paper, but mention that while we were able to convert the heuristic analysis into a formal proof in some cases, the result that we get is very weak: We can only prove that for some small constant $\epsilon \approx 0.02$, computing a $(1-\epsilon)$ fraction of the bits in the ECDH secret is as hard as computing them all. This is related to a recent result of Boneh and Shparlinksi [5] which shows that if ECDH is hard on some curve E then there is no single efficient algorithm that predicts one bit of the ECDH secret for many curves isomorphic to E. Our result applies to blocks of bits (rather than a single bit), but is stronger than [5] in the sense that it applies to a specific curve rather than a family of curves. We show that if ECDH is hard on a specific curve then the top $(1 - \epsilon)$ fraction of the bits of the ECDH secret on that curve cannot be efficiently computed.

References

1. E. Bach, J. Shallit, "Algorithmic number theory, Volume I: efficient algorithms", MIT press, 1996.
2. L. Blum, M. Blum, M. Shub, "A simple unpredictable pseudo-random number generator", SIAM J. Comput. 15, 2 (1986) 364–383.
3. M. Blum and S. Micali. *How to Generate Cryptographically Strong Sequences of Pseudo-Random Bits.* SIAM J.Computing, 13(4):850–864, November 1984.
4. D. Boneh, Venkatesan R., "Hardness of Computing the Most Significant Bits of Secret Keys in Diffie-Hellman and Related Schemes", Proc. of Crypto, 1996, pp. 129–142, 1996.
5. D. Boneh, I. Shparlinksi, "On the unpredictability of bits of the elliptic curve Diffie–Hellman scheme", In *Advances in Cryptology - CRYPTO 2001*, volume 2139 of *Lecture Notes in Computer Science*, pp. 201–212. Springer-Verlag, 2001.
6. D.Coppersmith, "Small solutions to polynomial equations, and low exponent RSA vulnerabilities", J. of Cryptology, Vol. 10, pp. 233–260, 1997.
7. R. Cramer and V. Shoup, "A practical public key cryptosystem provably secure against adaptive chosen ciphertext attack", in proc. Crypto '98, pp. 13–25, 1998.
8. R. Cramer and V. Shoup, "Signature schemes based on the Strong RSA Assumption", Proc. 6th ACM Conf. on Computer and Communications Security, 1999.
9. R. Gennaro. An improved pseudo-random generator based on discrete log. In *Advances in Cryptology - CRYPTO 2000*, volume 1880 of *Lecture Notes in Computer Science*, pp. 469–481. Springer-Verlag, 2000.
10. R. Gennaro, S. Halevi, T. Rabin, "Secure hash-and-sign signature without random oracles", Proc. Eurocrypt '99, pp. 123–139, 1999.
11. R. Impagliazzo, D. Zuckerman, "How to Recycle Random Bits", FOCS, 1989.
12. N. Howgrave-Graham. Finding small roots of univariate modular equations revisited. In proceedings *Cryptography and Coding*, Lecture Notes in Computer Science, vol. 1355, Springer-Verlag, pp. 131–142, 1997.
13. N. Howgrave-Graham, N. Smart. Lattice attacks on digital signature schemes. manuscript.
14. A. Lenstra, H. Lenstra, and L. Lovász. Factoring polynomials with rational coefficients. *Mathematische Annalen*, vol. 261, pp. 515–534, 1982.
15. M. Naor, O. Reingold, "Number theoretic constructions of efficient pseudo random functions", Proc. FOCS '97. pp. 458–467.
16. A. Ta-Shma, D. Zuckerman, and S. Safra, "Extractors from Reed-Muller Codes", FOCS, 2001.

Secure Human Identification Protocols

Nicholas J. Hopper and Manuel Blum

Computer Science Department, Carnegie Mellon University, 5000 Forbes Ave.
Pittsburgh PA 15213, USA
{hopper,mblum}@cs.cmu.edu

Abstract. One interesting and important challenge for the cryptologic community is that of providing secure authentication and identification for unassisted humans. There are a range of protocols for secure identification which require various forms of trusted hardware or software, aimed at protecting privacy and financial assets. But how do we verify our identity, securely, when we don't have or don't trust our smart card, palmtop, or laptop?
In this paper, we provide definitions of what we believe to be reasonable goals for secure human identification. We demonstrate that existing solutions do not meet these reasonable definitions. Finally, we provide solutions which demonstrate the feasibility of the security conditions attached to our definitions, but which are impractical for use by humans.

1 Introduction

Consider the problem of human identification. A human H wishes to prove his identity to a computational device C. The channel over which H and C will communicate is insecure and possibly controlled by an adversary. The protocol which accomplishes this task must satisfy the property that no adversary, even one who has witnessed past identifications, may successfully impersonate H except with negligible probability. Complicating matters further, H and C would like to reuse the secret they share for many identifications.

This problem arises on a daily basis in our society, yet the solutions to date are inadequate for several reasons. The traditional password approach is unacceptable, since a network snoop can record the password and will then be able to falsely authenticate as the user at will. Schemes which build a cryptographically strong key from some initial weak secret, such as SRP and EKE, require trusted hardware and software, since the computations involved are far beyond the abilities of most humans. Zero-knowledge schemes such as Fiat-Shamir [1] require trusted hardware which can be stolen or compromised. One-time passwords [2] are just that – good for only a single authentication; pads of such passwords are vulnerable to theft and still require a large ratio of "key material" to authentications.

These schemes all require the human to have some computational or memory aid to securely authenticate himself. In this paper we seek a solution that is viable

C. Boyd (Ed.): ASIACRYPT 2001, LNCS 2248, pp. 52–66, 2001.
© Springer-Verlag Berlin Heidelberg 2001

for the traveler who lost his luggage, or the purchaser who forgot his wallet. We believe that practical scenarios such as these justify the need for such a solution.

An alternative to the above schemes (SRP, EKE, Fiat-Shamir, one-time passwords) is a challenge-response protocol:

- The user and computer share a secret.
- The computer randomly challenges the user
- The user responds in such a way that an adversary cannot easily learn the secret.

Papers by Matsumoto and Imai [3], Wang *et al* [4], and Matsumoto [5] provide schemes which are sufficient for a small number of authentications. In their case, the secret can be recovered in polynomial time once a linear (in the size of the secret) number of authentications have been witnessed by an eavesdropper. (In our case, the number of authentications that must be witnessed to recover a secret in polynomial time is quadratic in the size of the challenge, which in turn is superpolynomial in the size of the secret.) Naor and Pinkas [6] give an identification protocol which is secure for a number of identifications which is linear in the size of the challenge and which requires a low-tech hardware item: a transparency. If stolen, the transparency can be copied and used to masquerade successfully as the legitimate user.

It is the goal of this paper to suggest that protocols which allow unaided humans to identify themselves securely and repeatedly may be feasible and should be a goal of the cryptographic community. In Section 2 we provide security definitions which we contend should be the goal of human identification protocols. In Section 3, we give examples of some cryptographic primitives which humans can execute without assistance. Section 4 gives a protocol which is provably secure against eavesdropping adversaries, based on these primitives; Section 5 outlines a protocol which is heuristically secure against arbitrary adversaries. This protocol is composed of a small number of steps that are individually feasible for humans. As a whole, however, the protocol requires too much computation (and possibly too much memory) to be practical for most humans.

2 Definitions

We begin by formally defining the notion of an identification protocol, and what we will mean for a protocol to be human executable. We then define two notions of security, in terms of passive and active adversaries. Finally we show how some traditional solutions to this problem either fail to satisfy the conditions of human execution or security.

2.1 Human Identification Protocols

We follow [7] in defining a protocol as a pair of (public, probabilistic) interacting programs (H, C) with auxiliary inputs; we denote the result of interaction between H and C with inputs x and y as $\langle H(x), C(y)\rangle$ and we denote the transcript of bits exchanged during their interaction by $T(H(x), C(y))$. A protocol yields some form of identification if H and C accept with high probability when run with the same auxiliary input and reject with high probability when run with different auxiliary input.

Definition 1. *An* identification protocol *is a pair of probabilistic interactive programs* (H, C) *with shared auxiliary input* z*, such that the following conditions hold:*

- *For all auxiliary inputs* z*,* $Pr[\langle H(z), C(z)\rangle = \texttt{accept}] > 0.9$
- *For each pair* $x \neq y$*,* $Pr[\langle H(x), C(y)\rangle = \texttt{accept}] < 0.1$

When $\langle H, C\rangle = \texttt{accept}$*, we say that* H verifies his identity to C*,* C authenticates H*, or* H authenticates to C.

In this paper we are interested in the case where H can be executed by a human. For the reasons outlined in Section 1, we rule out any form of computational aid. Additionally, we allow for occasional human error and varying abilities of the human population:

Definition 2. *An identification protocol* (H, C) *is said to be* (α, β, t) - human executable *if at least a* $(1 - \alpha)$ *portion of the human population can perform the computations* H *unaided and without errors in at most* t *seconds, with probability greater than* $1 - \beta$.

An ultimate goal might be to design a $(.1, .1, 10)$-human executable identification protocol that also meets the security definitions defined subsequently; the protocols we give here are on the order of $(.9, .2, 300)$-human executable, which is clearly not practical as a replacement for traditional solutions to the problem. Still, since they meet our security conditions, we believe they provide evidence that such a protocol is feasible.

A practical issue concerns whether the claim "(H, C) is (α, β, t)-human executable" can be demonstrated. Since we lack a well-defined model of human computation, establishing the claim rigorously seems infeasible in most cases. However, we believe that for the present, in many cases such claims can be evaluated intuitively. In cases where they cannot, empirical evidence should suffice.

2.2 Security Definitions

We give both a weak characterization of security, in terms of *passive* adversaries, and a strong characterization of security, in terms of *active* adversaries. Both characterizations are parameterized by a pair (p, k) where p gives the probability that a computationally bounded attacker can successfully simulate H to C after k interactions with H and/or C.

Definition 3. *An identification protocol* (H, C) *is* (p, k)-secure against passive adversaries *if for all computationally bounded adversaries* $\mathcal{A}$*,*

$$Pr[\langle \mathcal{A}(T^k(H(z), C(z))), C(z)\rangle = \texttt{accept}] \leq p \ ,$$

where $T^k(H(z), C(z))$ *is a random variable sampled from* k *independent transcripts* $T(H(z), C(z))$.

That is, even after a passive adversary has witnessed k identification sessions between H and C, he still cannot successfully masquerade as H with probability greater than p. A passive adversary models the eavesdropper or "shoulder-surfer" who is willing to watch H identify himself but does not control the communication channel between H and C. On the other hand, an active adversary is permitted to control the channel between H and C, which leads to a much stronger definition of security.

Definition 4. *An identification protocol* (H, C) *is* (p, k)-secure against active adversaries *if for all computationally bounded adversaries* $\mathcal{A}$*,*

$$Pr[\langle \mathcal{A}(T^k(\mathcal{A}, H(z), C(z))), C(z)\rangle = \texttt{accept}] < p \ ,$$

where $T^k(\mathcal{A}, H(z), C(z))$ *denotes a random variable sampled from* k *sessions where* $\mathcal{A}$ *is allowed to observe and make arbitrary changes to the communications between* H *and* C.

This last definition is a theoretical goal which in practice is not achieved by any existing solution to this problem, except for the case $k = 1$. For example, most password-based protocols may be compromised in one authentication by a trojan horse which records the user's password before performing (or failing to perform) the computational steps involved. Therefore, we will relax this condition as follows. We will allow a third outcome for the interaction of H and C (and any third parties): we will allow H to reject C. This will be denoted by $\langle H(\cdot), C(\cdot)\rangle = \perp$. Our relaxed security requirement is that after eavesdropping on k identification sessions, $\mathcal{A}$ still has probability at most q of interacting with H and C without being detected:

Definition 5. *An identification protocol* (H, C) *is* (p, q, k)-detecting against active adversaries *if for all computationally bounded adversaries* $\mathcal{A}$*,*

- $Pr[\langle H(z), \mathcal{A}(T^k(H(z), C(z)))\rangle \neq \perp] < q$
- $Pr[\langle \mathcal{A}(T^k(H(z), C(z))), C(z)\rangle = \texttt{accept}] < p$.

In this setting, we deprive the adversary $\mathcal{A}$ of the opportunity to interfere with communication between H and C. For a protocol satisfying this security condition, H should consider his communications with C to be compromised once H rejects C, and should not respond to any further authentication requests until the parties may securely exchange a new secret z'.

We note that in the human-executable setting some parameters may be relaxed when compared with computationally intensive protocols for identification. For example, a "standard" cryptographic goal for an identification protocol might be a protocol which is $(2^{-m}, 2^{-m}, 2^{100})$-detecting against active adversaries. But when a human is providing the transcripts for $T^k(C(z), H(z))$ it is quite reasonable to expect that security for 10^6 authentications will be sufficient, since a human would take decades to provide so many. Further, many applications which require human authentication are apparently more tolerant to false positives; for example, most automated teller machines have a confidence level of only 10^{-4}. Thus a $(10^{-6}, 10^{-6}, 10^7)$-detecting protocol may be acceptable for humans.

3 Plausible Hard Problems

In this section we introduce two computational problems as candidates for constructing secure human executable authentication protocols, along with some evidence that these computational problems are hard. Both problems can be characterized as loosely based on the sparse subset sum problem, taken over vectors of digits, with some twists intended to allow more authentications.

3.1 Learning Parity in the Presence of Noise

Suppose the secret shared between the human and the computer is a vector $\mathbf{x}$ of length n over $GF(2)$. Authentication proceeds as follows: The computer, C, generates a random n-vector $\mathbf{c}$ over $GF(2)$ and sends it to the human, H, as a challenge. H responds with the bit $r = \mathbf{c} \cdot \mathbf{x}$, the inner product over $GF(2)$. C accepts if $r = \mathbf{c} \cdot \mathbf{x}$. Clearly on a single authentication, C accepts a legitimate user H with probability 1, and an impostor with probability $\frac{1}{2}$; iteration k times results in accepting an impostor with probability 2^{-k}. Unfortunately, after observing $O(n)$ challenge-response pairs between C and H, the adversary M can use Gaussian elimination to discover the secret $\mathbf{x}$ and masquerade as H.

Suppose we introduce a parameter $\eta \in (0, \frac{1}{2})$ and allow H to respond incorrectly with probability η; in that case the adversary can no longer simply use Gaussian elimination to learn the secret $\mathbf{x}$. This is an instance of the problem of *learning parity with noise* (LPN). In fact the problem of learning $\mathbf{x}$ becomes NP-Hard in the presence of errors; it is NP-Hard to even find an $\mathbf{x}$ satisfying more than half of the challenge-response pairs collected by M [8]. Of course, the hardness results of Håstad [8] simply imply that there exist instances of this problem which cannot be solved in polynomial time unless P=NP; it is still possible that the problem is tractable in the random case. However, Kearns[9] has shown that in the random case, parity is not efficiently learnable in the statistical query model; and all known efficient learning algorithms for noisy concepts can be cast in this model. Additionally, Blum *et al* [10] show that for the case of uniformly distributed challenges, weak prediction is equivalent to strong prediction – that is, any algorithm to predict the next response bit with probability

$\frac{1}{2} + \frac{1}{n^c}$ can be used to recover the underlying parity function; and any algorithm which can learn LPN when the parity function is chosen uniformly can be used to learn arbitrary parity functions. Further, the best known algorithm for the general random problem, due to Blum, Kalai and Wasserman, requires $2^{\Omega(n/\log n)}$ challenge-response pairs and works in time $2^{\Omega(n/\log n)}$; here we will give some evidence that this problem is, in fact, uniformly hard and cannot be solved in time and sample size $poly(n, 1/(\frac{1}{2} - \eta))$.

In the following, we will refer to an instance of LPN as a $m \times n$ matrix $\mathbf{A}$ (where $m = poly(n)$); a m-vector $\mathbf{b}$, and a noise parameter η; the problem is to find a n-vector $\mathbf{x}$ such that $|\mathbf{Ax} - \mathbf{b}| \leq \eta m$, where $|\mathbf{x}|$ denotes the Hamming weight of the vector $\mathbf{x}$.

Lemma 1. *(Pseudo-randomizability)*
Any instance of LPN can be transformed in polynomial time into an instance chosen uniformly at random from a space of 2^{n^2} *possibilities.*

PROOF: Choose the $n \times n$ matrix $\mathbf{R} \in_U \{0,1\}^{n^2}$; Then if there is a solution to the instance $(\mathbf{AR}, \mathbf{b}, \eta)$, say $\mathbf{y}$, then we have:

$$|(\mathbf{AR})\mathbf{y} - \mathbf{b}| \leq \eta m \ ,$$

and if we let $\mathbf{x} := \mathbf{Ry}$ we find that $\mathbf{Ax} = \mathbf{A}(\mathbf{Ry}) = (\mathbf{AR})\mathbf{y}$, which yields the desired $\mathbf{x}$, since:

$$|\mathbf{Ax} - \mathbf{b}| = |(\mathbf{AR})\mathbf{y} - \mathbf{b}| \leq \eta m \ .$$

Thus there is a polynomial-time transformation between adversarial instances and and a large class of random instances, such that a solution to the randomly chosen instance can be transformed into a solution to the adversarial instance. Phrased differently, each instance of LPN belongs to a space of $O(2^{n^2})$ instances such that either all of the instances are easy or only a negligible fraction are easy. This is similar to the situation with discrete logarithms, where either all of the instances modulo a given prime are easy, or only a negligible fraction are easy.

Lemma 2. *(Log-Uniformity)*
If there exists an algorithm $\mathcal{A}$ capable of solving a $1/poly(n)$ fraction of the instances $(\mathbf{A}, \mathbf{b}, \eta)$ of LPN in time $poly(n, \log(1/(\frac{1}{2} - \eta)))$, then with high probability, any instance can be solved in time $poly(n, \log(1/(\frac{1}{2} - \eta)))$.

PROOF: Let $\epsilon(\eta) = \frac{1}{2} - \eta$, and let $\mathcal{A}$ be an algorithm which solves random instances in time $poly(n, \log(1/\epsilon(\eta)))$. Let $(\mathbf{A}, \mathbf{b}, \eta)$ be an adversarial instance of LPN. Create the new instance $(\mathbf{A}', \mathbf{b}', \eta')$ as follows:

- For each row of $\mathbf{A}$, randomly choose n other rows of A and use the sum of these rows as the corresponding entry in $\mathbf{A}'$
- Fill in the corresponding entry in $\mathbf{b}'$ by adding the corresponding rows of $\mathbf{b}$.
- Set $\eta' := \frac{1}{2} - \frac{1}{2}(1 - 2\eta)^{n+1}$

Given the error rate η in the initial instance, the error rate η' is correct, by the following lemma (due to Blum, Kalai, and Wasserman):

Lemma 3. *Let* $(a_1, b_1), \ldots, (a_s, b_s)$ *be samples from* $(\mathbf{A}, \mathbf{b}, \eta)$*; then* $b_1 + \ldots + b_s$ *is the correct label for* $a_1 + \ldots + a_s$ *with probability* $\frac{1}{2} + \frac{1}{2}(1 - 2\eta)^s$.

The proof follows by induction on s [11]. The resulting instance is distributed uniformly; so with probability $1/poly(n)$, $\mathcal{A}$ solves it in time $poly(n, \log(1/\epsilon(\eta')))$. But note that:

$$\begin{aligned} \epsilon(\eta') &= \frac{1}{2}(1 - 2\eta')^{n+1} \\ &= \frac{1}{2}(1 - 2(\frac{1}{2} - \epsilon(\eta)))^{n+1} \\ &= \frac{1}{2}(2\epsilon(\eta))^{n+1} \end{aligned}$$

so that $poly(n, \log(1/\epsilon(\eta'))) = poly(n, \log(1/\epsilon(\eta)))$; since the expected number of attempts to find an instance soluble by $\mathcal{A}$ is $poly(n)$, $\mathcal{A}$ solves adversarial instances in time $poly(n, \log(1/\epsilon(\eta)))$.

Conjecture 1. (Hardness of LPN)
LPN is uniformly hard in n and η: there is no algorithm to solve a uniformly chosen instance $(\mathbf{A}, \mathbf{b}, \eta)$ in time $poly(n, 1/(\frac{1}{2} - \eta))$ with non-negligible probability.

EVIDENCE:

- (LPN) is not efficiently learnable in the statistical query model; combined with the uniformity results of Blum *et al* this suggests that uniformly chosen inputs are hard.
- The best known algorithm for the random case, given by Blum, Kalai, and Wasserman, has superpolynomial complexity.
- Lemmas 1 and 2.

This assumption is not unprecedented: the McEliece public-key cryptosystem [12] relies on a related assumption, and the pseudo-random generator proposed by Blum, Furst, Kearns and Lipton [10] is secure under a very similar assumption.

In adapting this problem to use by humans, we restrict the hamming weight of the secret vector x to be k, where k is roughly logarithmic in n, the length of the challenge. Rather than taking the inner product of vectors over $GF(2)$, challenges are vectors of decimal digits, and responses are the sum without carries (i.e., modulo 10) of the digits in the positions corresponding to the non-zero entries of x. Our best algorithm for solving instances of this related problem has complexity $\binom{n}{k/2}$. The algorithm proceeds by evaluating all possible hamming-weight $k/2$ vectors on the challenges, and applying hashing to find pairs of vectors which sum to the correct response on roughly a fraction $1 - \eta$ of challenges. Note that while this attack is better than the brute force approach of guessing all weight-k vectors – which has complexity $\binom{n}{k}$ – the complexity is still superpolynomial when k is logarithmic in n.

3.2 Sum of k Mins

Let $z = \langle (x_1, y_1), (x_2, y_2), \ldots, (x_k, y_k) \rangle$ be a set of pairs (x_i, y_i) of integers mod n. Let $\mathbf{v} \in \{0, \ldots, 9\}^n$, and define $f(\mathbf{v}, z)$ by:

$$f(\mathbf{v}, z) = \sum_{i=1}^{k} \min\{\mathbf{v}[x_i], \mathbf{v}[y_i]\} \bmod 10 \ .$$

Then the sum of k mins problem is: given m pairs $(\mathbf{v_1}, u_1), \ldots (\mathbf{v_m}, u_m)$, where $\mathbf{v_i} \in \{0, \ldots, 9\}^n$, $u_i \in \{0, \ldots, 9\}$, and $k \log_{10} n \leq m \leq \binom{n}{2}$, find a set z such that $u_i = f(\mathbf{v_i}, z)$ for all $i = 1, \ldots, m$.

An algebraic approach to this problem is to form the system of equations given by:

$$\begin{bmatrix} v_{1,1,2} & v_{1,1,3} & \cdots & & v_{1,n-1,n} \\ v_{2,1,2} & \cdots & v_{2,i,j} & \cdots & \\ \vdots & & & \ddots & \\ v_{m,1,2} & \cdots & v_{m,i,j} & \cdots & v_{m,n-1,n} \end{bmatrix} \begin{bmatrix} z_{1,2} \\ z\vdots \\ z_{i,j} \\ \vdots \\ z_{n-1,n} \end{bmatrix} = \begin{bmatrix} u_1 \\ u_2 \\ \vdots \\ u_m \end{bmatrix} \pmod{10} \ ,$$

where $v_{k,i,j} = \min\{\mathbf{v_k}[i], \mathbf{v_k}[j]\}$, $z_{i,j} = 1$ if $(i,j) \in z$, and $1 \leq i < j \leq n$. If $m \geq \binom{n}{2}$ we expect to solve this system uniquely by Gaussian elimination. When $m < \binom{n}{2}$, on the other hand, this approach leads to a sparse subset sum problem. The best known algorithms for these instances have complexity roughly $\binom{n(n-1)/2}{k/2}$ (which is greater than $\binom{n}{k}$ when $k > 3$).

Another approach to the problem is a form of maximum-likelihood estimation (MLE): for some subset of the locations in z, try all possible values, while modeling the remaining inputs to $f(\cdot, z)$ as uniform random variables (an accurate model when the $\mathbf{v_i}$ are chosen at random). Choose the subset of locations which gives the best chance of observing the output values u_i. If the subset of z we are guessing has l locations, this algorithm has complexity $\binom{n}{l}$. However, to succeed in selecting correct locations, the algorithm may require many samples (perhaps more than $\binom{n}{2}$).

For any distribution $\mathcal{D}$, the maximum probability of distinguishing between $\mathcal{D}$ and the uniform distribution on the same range U is $\Delta(\mathcal{D}, U)$, where

$$\Delta(A, B) = \frac{1}{2} \sum_{e \in E} |Pr_A[e] - Pr_B[e]| \ ,$$

i.e., the statistical distance between A and B. Thus the expected minimum number of samples required to distinguish between $\mathcal{D}$ and U is $1/\Delta(\mathcal{D}, U)$. Therefore calculating this distance for the distribution of the modulo 10 sum of k mins will help us develop lower bounds on the required sample complexity for MLE.

To calculate the statistical distance between k mins and uniformly random digits, we derive an expression which will allow us to calculate the probability of obtaining a digit as the sum of k mins. Let P_d^k denote the probability of obtaining the digit d as a sum of k mins. Then the P_d^1 are easily obtainable by enumerating all pairs of digits. For $k > 1$, we note that for each d, there are 10 ways to obtain d as the sum of k mins: for each digit d', we obtain d' from one min and $d - d' \bmod 10$ from the other $k - 1$ mins. In other words, we can write the recurrence $P_d^k = \sum_{0 \leq d' \leq 9} P_{d-d'}^{k-1} P_{d'}^1$, which leads to the observation that dynamic programming is sufficient for obtaining the distribution over k mins. Table 1 yields the result of applying this procedure to calculate the expected minimum sample complexity to distinguish between the uniform distribution on $\{0, \ldots, 9\}$ and a sum of k mins, for $k \leq 12$.

Table 1. Distribution of sum of k mins, and expected minimum number of samples required to distinguish from uniform

k	1	2	3	4	5	6	7	8	9	10	11	12
#S	4	14	44	140	532	1346	4154	12848	39696	122682	379100	1171498

Note that the essential meaning of this table is that without guessing more than 12 locations from a challenge, an adversary cannot expect to use statistical procedures to learn a sum of 12 mins password with fewer than 1,171,498 challenge-response pairs. In general, we can protect against this attack by choosing k such that the number of required samples is greater than $\binom{n}{2}$, since $\binom{n}{2}$ samples are sufficient for Gaussian elimination.

4 Security against Passive Adversaries

In this section, we will give a protocol which is (p, k)-secure against passive adversaries but not against arbitrary adversaries. We also give some empirical evidence that it is $(0.9, 0.25, 160)$-human executable. Intuitively, C generates the coefficient matrix of some LPN instance while H generates the output vector and some errors. Thus after a number of repetitions C can be reasonably sure that H knows the shared secret vector $\mathbf{x}$.

Protocol 1

Shared Secrets: H and C share a secret 0-1 vector $\mathbf{x}$ with $|\mathbf{x}| = k$.

Authentication:

(C1) C sets $i := 0$
- Repeat m times:
  - (C2) C selects a random challenge $\mathbf{c} \in_R \{0, 1\}^n$ and sends it to H
  - (H1) With probability $1 - \eta$, H responds with $r := \mathbf{c} \cdot \mathbf{x}$, otherwise H responds with $r := 1 - \mathbf{c} \cdot \mathbf{x}$.
  - (C3) if $r = \mathbf{c} \cdot \mathbf{x}$, C increments i.

(C4) if $i \geq (1 - \eta)m$, C accepts H.

Theorem 1. *If H guesses random responses r, C will accept H with probability at most*

$$\left(\frac{1}{2}\right)^m \sum_{i=(1-\eta)m}^{m} \binom{m}{i} \leq e^{-c_0 m} ,$$

where $c_0 \geq \frac{2}{3}$ is a constant depending only on η.

PROOF: Let X be the random variable denoting the number of times H guesses correctly; since this probability is at most $\frac{1}{2}$, the probability of guessing correctly exactly i times out of m is $\binom{m}{i}\left(\frac{1}{2}\right)^m$; the first result follows from summing the probabilities of guessing correctly $(1-\eta)m$ or more times; the second result follows by a Chernoff bound with $c_0 = (3-2\eta)^2/6 \geq (3-1)^2/6 = \frac{2}{3}$.

Theorem 2. *If LPN is hard, then Protocol 1 is $(e^{-\frac{2}{3}m}, poly(n))$ - secure against a passive adversary.*

PROOF: Obvious. Since a passive adversary can only observe challenge-response pairs $(\mathbf{c}, r)$, obtaining the secret $\mathbf{x}$ can only be accomplished via solving the LPN problem.

Unfortunately, as previously mentioned, this protocol is not secure against an active adversary: suppose M can insert arbitrary challenges into the interaction; then M can record $n/m(1-\eta)^2$ successful authentications and replay them back to H, discarding $(\mathbf{c}, r)$ pairs which do not match; the remaining pairs will have no errors and can be solved by Gaussian elimination. Additionally, this protocol must be iterated many times in order to achieve any sort of security.

As an additional consideration for the human user, the challenges $\mathbf{c}$ could be selected from $\{0, \ldots, 9\}^n$ and the arithmetic done modulo 10, a natural base for many humans. This reduces the number of iterations necessary for a given security level by a constant factor. It also requires modifying the method of making an error: in cases when an error is to be made, the response should be chosen uniformly from $\{0, \ldots, 9\}$.

Note that assuming the best known attack complexity is optimal, we can choose parameters which will provide ample security in this setting. For example, when $n = 1000$ and $k = 19$, the best known attack's complexity of $\binom{n}{\lceil k/2 \rceil}$ is roughly 2^{78}. This compares favorably with common minimum strength guidelines for choosing cryptographic parameters.

To assess the property of human executability, we conducted the following experiment. A computer implementing this authentication system with $m = 7$, $\eta = \frac{1}{7}$, $n = 200$ and $k = 15$ was attached to a Coke machine in our department's lounge. The system was also implemented as a web page, which provided a tutorial in its use. Students and faculty were permitted to access the web page as often as they wished, and a free Coke was given to anyone who could successfully authenticate himself to the computer attached to the coke machine. In a one week period, 54 users attempted 195 authentications and successfully completed 155. The average time per successful authentication was 166 seconds, and the average time per unsuccessful authentication was 171 seconds. Thus it is empirically clear that there is some value α for which this is a $(\alpha, .25, 160)$-human executable identification protocol which is secure against computationally bounded eavesdropping adversaries.

5 Security against Arbitrary Adversaries

The protocol of the previous section is quite insecure against an adversary who is capable of modifying the communications between H and C. For example, by simply replaying the same challenge $\mathbf{r}$ back to H several times, $\mathcal{A}$ can compute the true value of $\mathbf{r} \cdot \mathbf{x}$ and thus after collecting n such error-free values, can learn the secret $\mathbf{x}$ by Gaussian elimination. Even if we simply replace weight-k LPN by sum of $\lceil k/2 \rceil$ mins, the problem persists. That is, while simply replaying the same challenge to H will no longer allow $\mathcal{A}$ to learn the secret z, replaying the same challenge with a slight change — for example, changing a single '9' to a '0' — will still allow $\mathcal{A}$ to learn the secret z with $O(n)$ well-chosen challenges.

Thus we seek to make it difficult for $\mathcal{A}$ to submit arbitrary challenges to H in place of those sent by C. To do so, we will introduce two mechanisms. First, *Error-Correcting Challenges* have the property that it with very high probability a challenge cannot be modified in a small number of locations. Second, we require the challenges to satisfy some concept which is hard to learn without membership queries, such as satisfying $f(\mathbf{r}, z) = 0 \bmod 10$ for an independent k-mins password z.

5.1 Error-Correcting Challenges

Blum *et al.* [13] show how a function which is linear with probability $1 - \delta$ can be self-corrected to a linear function which matches the given function with probability $1 - 2\delta$. Self-correction of this form is used in many Probabilistically Checkable Proof (PCP) arguments. We propose that a similar error-detecting/self-correcting approach can be applied to the challenges in our system, resulting in a system which has the property that with high probability an adversary cannot make local changes to a challenge.

The protocol proposed in this document will use the self-correction algorithm of [13] to achieve this goal. A legitimate challenge will consist of $w \times h$ 10×10 squares of digits, or $n = 100wh$ digits. Each square will be generated by choosing 3 digits (a, b, c) uniformly at random; then the digit at location x, y will have the value $L(x, y) = ax + by + c \bmod 10$. Linearity can be tested by choosing a random point $\boldsymbol{x}$ (mod 10) and random offset mod 10, $\boldsymbol{r}$, and testing whether $L(\boldsymbol{x}) = L(\boldsymbol{x}+\boldsymbol{r}) - L(\boldsymbol{r}) + L(0)$. If a challenge square passes this test several times then we say that it is close to linear, and in the subsequent phase we will access the value of a location $\boldsymbol{x}$ by accessing its "self-corrected" value at the randomly chosen offset $\boldsymbol{r}$, which is given by $L(\boldsymbol{x} + \boldsymbol{r}) - L(\boldsymbol{r}) + L(0)$. Thus if we reject a challenge which contains a highly non-linear square and self-correct otherwise, with high probability an adversary will be unable to effect a local change to a challenge.

5.2 The Protocol

Coupled with a deterministic response protocol to prevent replay attacks, we obtain the protocol outlined below.

Protocol 2

Shared Secrets: H and C share two sum of k mins secrets p_1 and p_2, and a secret digit d. As in Section 3, we denote by $f(c, p_i)$ the result of taking the sum of the self-corrected min of each pair in p_i for the challenge c.

Authentication: Repeat m times for confidence 10^{-m}:

(C1) Uniformly pick wh sets of parameters (a_i, b_i, c_i) and form the error-correcting challenge for these parameters, $c = ECC(\boldsymbol{a}, \boldsymbol{b}, \boldsymbol{c})$. If $f(c, p_1) \neq d$, repeat until the condition holds. Send the resulting challenge to H:

$$C \to H \ : \ c = ECC(\boldsymbol{a}, \boldsymbol{b}, \boldsymbol{c}) \ .$$

(H1) Test each square for linearity. Reject if any square is not close to linear. (Report a network infiltration to system administrator and choose a new password)

(H2) Check that $f(c, p_1) = d$. If not, reject and report a network infiltration to system administrator.

(H3) Respond with the self-corrected sum of mins for the password p_2:

$$H \to C \ : \ r = f(c, p_2) \ .$$

(C2) Reject if $r \neq f(c, p_2)$.

C accepts H if it has not rejected after m rounds.

Intuitively, we use self-correction on error-correcting challenges to make it infeasible for an adversary to make local changes to a challenge. Thus, to make a membership query, the adversary must make global changes to the challenge, yet since $f(\cdot, p_1)$ is distributed essentially uniformly any global change will be caught with probability at least 0.9.

Thus heuristically, we have a protocol which is $(0.1, 0.1, \binom{n}{2})$-detecting against computationally bounded adversaries. With the challenge size $n = 900$, $k = 12$ and $m = 6$, the best known attack on sum of k mins given fewer than $\binom{n}{2}$ samples has complexity greater than $\binom{900}{12}$, which is roughly 2^{89}. Thus the security of the system appears to be quite high.

It seems reasonable that a human can learn to do linearity testing on sight, since error-correcting challenges form distinctive patterns of digits; thus the human computational load in this protocol may be as low as 96 base 10 sums and 24 mins to compute the response to a single challenge. For confidence 10^{-6}, this translates to a protocol which requires a minimum of 576 base 10 sums plus considerable search effort. Therefore, while this protocol offers a great deal of security against arbitrary computationally bounded adversaries, it seems unlikely to be of practical significance on its own.

6 Some Inherent Limitations

The approach to human-executable primitives taken here has some inherent limitations which may, unfortunately, make it difficult to improve on these protocols without a new approach. We now consider a large class of "similar" protocols in which the shared secret is a set of k of relevant locations in a n-digit challenge. We will show that, assuming the "meet in the middle" attack of time complexity $O(\binom{n}{k/2})$ is optimal, there is no significantly harder function in this class, computationally speaking, than parity with noise.

We model the human as a finite automaton which sequentially processes k inputs by transitions between states in the range $\{1, \ldots, Q\}$ and which gives an output in the range $\{1, \ldots, d\}$. Since humans have highly bounded memory, this model seems fitting for human computation in this application. We assume that the transition table for this automaton may change between inputs but is publicly known.

Now consider an attack which uses m challenge-response pairs and processes all sequences of locations of length $k/2$. For each location sequence, the automaton is run forward $k/2$ steps, producing a string in the range $\{1, \ldots, d\}^m$. This string and its location are inserted in a hash table with Q^m spaces. Also, for each sequence of $k/2$ locations, for each challenge, the automaton is started from each intermediate state $\{1, \ldots, Q\}$ and the list of intermediate states which produced the correct response is retained. For each challenge, the expected number of intermediate states retained will be Q/d. Thus we will expect approximately $(Q/d)^m$ sequences of intermediate states to match the correct responses for each sequence of $k/2$ locations; each of these intermediate state sequences can be inserted into the same hash table of size Q^m. Any match in the hash table between a "first-half" sequence and a "second-half" sequence suggests a length k sequence of locations which matches on the m challenge-response pairs under consideration; such a sequence can be tested against the $O(k \log_d n)$ challenge-response pairs required to uniquely determine the secret k locations.

Now we assess the total computational work factor for this attack. First, for each sequence of $k/2$ locations, $(Q/d)^m$ length-m sequences must be inserted into the hash table, for a total of $O((Q/d)^m n^{k/2})$ work. Also, each collision between a "first-half" sequence and a "second-half" sequence will require some work to check against the full set of challenge-response pairs. For an appropriate family of universal hash functions, the expected number of collisions will be

$$\frac{n^{k/2} \times (Q/d)^m n^{k/2}}{Q^m} = \frac{n^k}{d^m} .$$

Choosing m to minimize the sum $(Q/d)^m n^{k/2} + \frac{n^k}{d^m}$ results in the choice

$$m = \frac{\log \frac{n^{k/2} \log d}{\log(Q/d)}}{\log Q} ,$$

and gives the total work factor

$$O(n^{k(1-\frac{\log d}{2 \log Q})}) .$$

Thus if d and Q are close, or equal as in our protocols, an attacker can always break such a protocol by guessing only about half of the shared secret. On the other hand, decreasing d relative to Q increases the number of challenges a user must respond to for a given confidence level, while increasing Q adds to the cognitive load on the human. Thus while some incremental improvement in the computational security of our protocols may be possible, overall our choice of primitives represent a close to optimal tradeoff between computational difficulty and human cognitive load for this class of protocols.

7 Conclusions

We believe that the search for protocols providing secure, reusable authentication to unaided humans is an interesting and important pursuit for the cryptographic community. In this paper, we have shown that no current solutions to this problem exist. We have provided definitions that we believe are reasonable goals for such protocols, and we have given protocols which achieve the security conditions attached to these goals. While we do not argue that the protocols we present are practical solutions to this problem – executing the protocols and remembering the secrets seem too hard – we believe that they are surprisingly close to practical while offering a good deal of security. Thus we believe that they suggest that more practical solutions may exist, which can match or even exceed their security conditions. We invite the reader to surpass them.

Acknowledgements. The authors wish to thank Avrim Blum for several discussions concerning the difficulty of parity with noise, Moni Naor for pointing out that the subset sum meet-in-the-middle attack can still be applied in the presence of noise, and Preston Tollinger for conducting the Coke Machine experiment. This material is based upon work supported under a National Science Foundation Graduate Research Fellowship.

References

1. Fiat, A., Shamir, A.: How to prove yourself: Practical solutions to identification and signature problems. In Odlyzko, A.M., ed.: Advances in Cryptology—CRYPTO '86. Volume 263 of Lecture Notes in Computer Science., Springer-Verlag, 1987 (1986) 186–194
2. Lamport, L.: Password authentication with insecure communication. Communications of the ACM **24** (1981)
3. Matsumoto, T., Imai, H.: Human identification through insecure channel. In Davies, D.W., ed.: Advances in Cryptology—EUROCRYPT 91. Volume 547 of Lecture Notes in Computer Science., Springer-Verlag (1991) 409–421
4. Wang, C.H., Hwang, T., Tsai, J.J.: On the Matsumoto and Imai's human identification scheme. In Guillou, L.C., Quisquater, J.J., eds.: Advances in Cryptology—EUROCRYPT 95. Volume 921 of Lecture Notes in Computer Science., Springer-Verlag (1995) 382–392

5. Matsumoto, T.: Human-computer cryptography: An attempt. In Neuman, C., ed.: 3rd ACM Conference on Computer and Communications Security, New Delhi, India, ACM Press (1996) 68–75
6. Naor, M., Pinkas, B.: Visual authentication and identification. In Kaliski Jr., B.S., ed.: Advances in Cryptology—CRYPTO '97. Volume 1294 of Lecture Notes in Computer Science., Springer-Verlag (1997) 322–336
7. Goldreich, O.: Foundations of cryptography (fragments of a book). Available electronically at `http://theory.lcs.mit.edu/~oded/frag.html` (1998)
8. Håstad, J.: Some optimal inapproximability results. In: Proceedings of the Twenty-Ninth Annual ACM Symposium on Theory of Computing, El Paso, Texas (1997) 1–10
9. Kearns, M.: Efficient noise-tolerant learning from statistical queries. In: Proceedings of the Twenty-Fifth Annual ACM Symposium on the Theory of Computing, San Diego, California (1993) 392–401
10. Blum, A., Furst, M., Kearns, M., Lipton, R.J.: Cryptographic primitives based on hard learning problems. In Stinson, D.R., ed.: Advances in Cryptology—CRYPTO '93. Volume 773 of Lecture Notes in Computer Science., Springer-Verlag (1993) 278–291
11. Blum, A., Kalai, A., Wasserman, H.: Noise-tolerant learning, the parity problem, and the statistical query model. In: Proceedings of the Thirty-Second Annual ACM Symposium on Theory of Computing, Portland, Oregon (2000)
12. McEliece, R.J.: A public-key cryptosystem based on algebraic coding theory. Technical report, Jet Propulsion Laboratory (1978) Deep Space Network Progress Report.
13. Blum, M., Luby, M., Rubinfeld, R.: Self-testing/correcting with applications to numerical problems. In: Proceedings of the Twenty Second Annual ACM Symposium on Theory of Computing, Baltimore, Maryland (1990) 73–83

Unbelievable Security

Matching AES Security Using Public Key Systems

Arjen K. Lenstra

Citibank, N.A. and Technische Universiteit Eindhoven
1 North Gate Road, Mendham, NJ 07945-3104, U.S.A.
arjen.lenstra@citicorp.com

Abstract. The Advanced Encryption Standard (AES) provides three levels of security: 128, 192, and 256 bits. Given a desired level of security for the AES, this paper discusses matching public key sizes for RSA and the ElGamal family of protocols. For the latter both traditional multiplicative groups of finite fields and elliptic curve groups are considered. The practicality of the resulting systems is commented upon. Despite the conclusions, this paper should not be interpreted as an endorsement of any particular public key system in favor of any other.

1 Introduction

The forthcoming introduction [12] of AES-128, AES-192, and AES-256 creates an interesting new problem. In theory, AES-128 provides a very high level of security that is without doubt good enough for any type of commercial application. Levels of security higher than AES-128, and certainly those higher than AES-192, are beyond anything required by ordinary applications. Suppose, nevertheless, that one is not satisfied with the level of security provided by AES-128 and insists on using AES-192 or AES-256. This paper considers the question what key sizes of corresponding security one should then be using for the following public key cryptosystems:

- RSA and RSA multiprime (RSA-MP; the earliest reference is [14]).
- Diffie-Hellman and ElGamal-like systems [10,15] based on the discrete logarithm problem in prime order subgroups of
  - multiplicative groups of prime fields.
  - multiplicative groups of extension fields: fields of fixed small characteristic and compressed representation methods (LUC [17] and XTR [8]).
  - groups of elliptic curves over prime fields (ECC, [1]).

These are the most popular systems and the only ones that are widely accepted. Systems that have recently been introduced and that are still under scrutiny are not included, with the exception of XTR – it is included because this paper sheds new light on its alleged performance equivalence to ECC. Also discussed are performance issues related to the usage of keys of the resulting sizes.

C. Boyd (Ed.): ASIACRYPT 2001, LNCS 2248, pp. 67–86, 2001.
© Springer-Verlag Berlin Heidelberg 2001

The introduction of the AES will soon bring along the introduction of cryptographic hash functions of matching security levels [13], namely SHA-256, SHA-384, and SHA-512. Because many common subgroup based cryptographic protocols use subgroup orders and hashes of the same sizes, the decision what subgroup size to use with AES-ℓ becomes easy: use subgroups of prime order q with $\lceil \log_2 q \rceil = 2\ell$. For ECC that settles the issue, from a practical point of view at least. This is reflected in the revised standard FIPS 186-2 [11]. For the other subgroup systems the finite field size remains to be decided upon. It may be assumed that both for properly chosen finite fields and for ECC the resulting subgroup operation is slower than a single application of the AES or SHA. It follows that, with respect to the familiar exhaustive search, collision, and square-root attacks against AES-ℓ, SHA-2ℓ, and properly chosen subgroups, respectively, the weakest links will be the AES and the SHA, not the subgroup based system.

It may be argued that the question addressed in this paper is of academic interest only. Indeed, it remains to be seen if the security obtained by actual realization and application of 'unbelievably secure' systems such as AES-192, AES-256, or matching public key systems, will live up to the intended theoretical bounds. That issue is beyond the scope of this article. Even under the far-fetched assumption that implementations are perfect, it is conceivable that the actual security achieved by the AES is less than the intended one. Thus, even though one may be happy with the (intended) security provided by AES-128, one may cautiously decide to use AES-256 and match it with a public key system of 'only' 128-bit security [21]. Therefore, and to give the theme of this paper somewhat wider applicability, not only public key sizes matching AES-192 and AES-256 are presented, but also the possibly more realistic sizes matching DES, 2K3DES, 3K3DES, and AES-128. Here iK3DES refers to triple DES with i keys.

This paper is organized as follows. Issues concerning security levels of the cryptosystems under consideration are discussed in Section 2. RSA moduli sizes of security equivalent to the symmetric systems, now and in the not too distant future, are presented in Section 3. The security of RSA-MP, i.e., the minimal factor size of (matching) RSA moduli, is discussed in Section 4. Section 5 discusses matching finite field sizes for a variety of finite fields as applied in systems based on subgroups of multiplicative groups (i.e., not ECC): prime fields, extension fields with constant extension degree, and fields with constant (small) characteristic. Section 6 discusses various performance related issues, such as total key lengths and relative runtimes of cryptographic operations. A summary of the findings is presented in Section 7.

2 Security Levels

2.1 Breaking Cryptosystems

Throughout this paper breaking a symmetric cryptosystem means retrieving the symmetric key. Breaking RSA means factoring the public modulus, and breaking a subgroup based public key system means computing the discrete logarithm of a public subgroup element with respect to a known generator. Attacks based on protocol specific properties or the size of public or secret exponents are not considered. Thus, this paper lives in an idealized world where only key search and number theoretic attacks count. For any real life situation this is a gross oversimplification. But real life security cannot be obtained without resistance against these basic attacks.

2.2 Equivalence of Security

Under the above attack model, two cryptosystems provide the same level of security if the expected effort to break either system is the same. This way of comparing security levels sounds simpler than it is, because 'effort' can be interpreted in several ways. In [7] two possible ways are distinguished to compare security levels:

- Two cryptosystems are *computationally equivalent* if breaking them takes, on average, the same computational effort.
- Two cryptosystems are *cost equivalent* if acquiring the hardware to break them in the same expected amount of time costs the same.

Both types of equivalence have their pros and cons. The computational effort to break a cryptosystem can, under certain assumptions, be estimated fairly accurately. If the assumptions are acceptable, then the outcome should be acceptable as well. Computational effort does not take into account that it may be possible to attack one systems using much simpler and cheaper hardware than required for the other. The notion of cost equivalence attempts to include this issue as well. But it is an inherently much less precise measure, because cost of hardware can impossibly be pinpointed.

2.3 Symmetric Key Security Levels

A symmetric cryptosystem provides d-bit security if breaking it requires on average 2^{d-1} applications of the cryptosystem. Throughout this paper the following assumptions are made:

1. Single DES provides 56-bit security.
2. 2K3DES provides 95-bit security.
3. 3K3DES provides 112-bit security [15, page 360].
4. AES-ℓ provides ℓ-bit security, for $\ell = 128, 192, 256$.

The single DES estimate is based on the effort spent by recent successful attacks on single DES, such as described in [5]. The 2K3DES estimate is based on the approximately 100-bit security estimate from [20] combined with the observation that since 1990 the price of memory has come down relative to the price of processors. It may thus be regarded as an estimate that is good only for cost equivalence purposes. However, the computationally equivalent estimate may not be much different. The commonly used 112-bit estimate for 3K3DES is of a computational nature and ignores memory costs that far exceed processor costs. The best realistic attack uses parallel collision search on a machine with about a million terabytes of memory, and would lead to a security level of 116 bits[1]. This is more conservative than the classic meet-in-the-middle attack, which would lead to 128-bit (cost-equivalent) security. These comments on 2K3DES and 3K3DES security levels are due to Mike Wiener [21].

As far as the AES estimates are concerned, there is no a priori reason to exclude the possibility of substantial cryptanalytic progress affecting the security of the AES, in particular given how new the AES is. It is assumed, however, that if the AES estimates turn out to be wrong, then the AES will either be patched (cf. the replacement of SHA by SHA-1), or that it will be replaced by a new version of the proper and intended security levels.

The security provided by a symmetric cryptosystem is not necessarily the same as its key length. The above assumptions hold only if all keys are full-length. Systems of intermediate strength can be obtained by fixing part of the keys. This possibility is not further discussed in this paper (but see Figure 1).

It is assumed that symmetric keys are used for a limited amount of time and a limited encryption volume. Issues related to the limited block length of the DES and its variants are therefore of no concern in this paper.

2.4 Public Key Security Levels

Security levels of public key systems are determined by comparing them to symmetric key security levels. This means that computational and cost equivalence have to be distinguished.

In [7] it is argued that computational and cost equivalence are equivalent measures for the comparison of the security of symmetric systems and ECC. Not explicitly mentioned in [7], and therefore worth mentioning here, is the related fact that the amount of storage needed by the most efficient known attack on ECC (parallelized Pollard rho) does not depend on the subgroup order, but only on the relative cost of processors and storage [21]. In any case, if AES-128 and a certain variant of ECC are computationally equivalent, then they may be considered to be cost equivalent as well.

For the other public key systems, however, there is a gap between computational and cost equivalence. For example, it follows from [7] that AES-128 and about 3200-bit RSA are currently computationally equivalent. With respect to cost equivalence, AES-128 is currently more or less equivalent to 2650-bit RSA. This last estimate depends on an assumption about hardware prices and increases with cheaper hardware. See Section 3 for details. In Sections 3 to 5 both types of equivalence are used to determine public key parameters that provide security equivalent to the symmetric systems. The approach used is based on [7], but entirely geared towards the current application. The results from [7] have been criticized as being conservative [16] – prospective users of AES-192 or AES-256 may be even more conservative as far as security related choices are concerned. The non-ECC entries of most tables consist of two numbers, referring to the cost and computationally equivalent figures, respectively.

[1] Each 4-fold memory reduction doubles the runtime.

3 RSA Modulus Sizes of Matching Security

3.1 Current Equivalence

Let

$$L[n] = e^{1.923(\log n)^{1/3}(\log\log n)^{2/3}}$$

be the approximate asymptotic growth rate of the expected time required for a factoring attack against an RSA modulus n using the fastest currently known factoring algorithm, the number field sieve (NFS). This runtime does not depend on the size of the factors of n. It depends only on the size of the number n being factored.

As in [7] actual factoring runtimes are extrapolated to obtain runtime estimates for larger factoring problems. The basis for the extrapolation is the fact that the computational effort required to factor a 512-bit RSA modulus is about 50 times smaller than required to break single DES. With the asymptotic runtime given above it follows that a k-bit RSA modulus currently offers security computationally equivalent to a symmetric cryptosystem of d-bit security and speed comparable to single DES if

$$L[2^k] \approx 50 * 2^{d-56} * L[2^{512}].$$

Furthermore, according to the estimates given in [7], a k'-bit RSA modulus currently offers security cost equivalent to the same symmetric cryptosystem if

$$L[2^{k'}] \approx \frac{50 * 2^{d-56} * L[2^{512}]}{26 * P}.$$

In the latter formula P indicates the (wholesale) price of a stripped down PC of average performance and with reasonable memory. In [7] the default choice $P = 100$ is made. Any other price within a reasonable range of the default choice will have little effect on the sizes of the resulting RSA moduli. See [7, Section 3.2.5] for a more detailed discussion of this issue.

Unlike [7], the relatively speed of the different symmetric cryptosystems under consideration is ignored. The differences observed – comparable implementations of 3DES may be three times slower than single DES, but the AES may be three times faster – are so small that they have hardly any effect on the sizes of the resulting RSA moduli. If desired the right hand sides of the formulas above may be multiplied by v if the symmetric system under consideration is per application v times slower than single DES (using comparable implementations).

3.2 Expected Future Equivalence

Improved hardware may be expected to have the same effect on the security of symmetric and asymmetric cryptosystems. It may therefore be assumed that over time the relative security of symmetric cryptosystems and RSA is affected only by new cryptanalytic insights that affect one system but not the other.

As far as cryptanalytic progress against symmetric cryptosystems is concerned, it is assumed that they are patched or replaced if a major weakness is found, cf. 2.2.

Progress in factoring, i.e., cryptanalytic progress against RSA, is common. The past effects of improved factoring methods closely follow a Moore-type law [7]. Extrapolation of this observed behavior implies the following. In year $y \geq 2001$ a k-bit RSA modulus may be expected to offer security computationally equivalent to a symmetric cryptosystem of d-bit security if

$$L[2^k] \approx 50 * 2^{d-56+2(y-2001)/3} * L[2^{512}].$$

Cost equivalence is achieved in year y for a k'-bit RSA modulus if

$$L[2^{k'}] \approx \frac{50 * 2^{d-56+2(y-2001)/3} * L[2^{512}]}{26 * P},$$

with P as in 3. As in 3 effects of the symmetric cryptosystem speed are ignored, and $P = 100$ is a reasonable default choice. For $y = 2001$ the formulas are the same as in 3, even though, compared to [7], two years of factoring progress should have been taken into account. Such progress has not been reported in the literature. If progress had been obtained according to Moore's law, its effect on RSA moduli sizes matching the AES would have been between one and two percent, which is negligible.

3.3 Resulting RSA Modulus Sizes

The formulas from 3 and 3.1 with $P = 100$ lead to the RSA modulus sizes in Table 1. The first (lower) number corresponds to the bit-length of a cost equivalent RSA modulus, the second (higher) number is the more conservative bit length of a computationally equivalent RSA modulus. Currently equivalent sizes are given in the row for year 2001, and sizes that can be expected to be equivalent in the years 2010, 2020, and 2030, are given in the rows for those years. It is assumed that factoring progress until 2030 behaves as it behaved since about 1970, i.e., that it follows a Moore-type law. If new factoring progress is found to be unlikely, the numbers given in the row for year 2001 should be used for all other years instead. If factoring progress is expected, but at a slower rate than in the past, one may for instance use the 2010 data for 2020. The data as presented in the table, however, and in particular the computationally equivalent sizes, may be interpreted as 'conservative'. It should be understood that, even for the conservative choices, there is no guarantee that surprises will not occur.

The numbers in Table 1 are not rounded or manipulated in any other way. That is left to the user, cf. [7, Remark 4.1.1]. For the 416-bit RSA modulus cost equivalent in 2001 to single DES, see also Table 2. As an example suppose an RSA modulus size has to be determined for an application that uses AES-192 and that is supposed to be in operation until 2020. It follows from Table 1 that RSA moduli should be used of eight to nine thousand bits long. Using RSA moduli of only three to four thousand bits length would undermine the apparently desired

Table 1. Matching RSA modulus sizes.

Year	DES		2K3DES		3K3DES		AES-128		AES-192		AES-256	
2001	416	620	1333	1723	1941	2426	2644	3224	6897	7918	13840	15387
2010	518	747	1532	1955	2189	2709	2942	3560	7426	8493	14645	16246
2020	647	906	1773	2233	2487	3046	3296	3956	8042	9160	15574	17235
2030	793	1084	2035	2534	2807	3408	3675	4379	8689	9860	16538	18260

security level (namely, higher than AES-128). Five to seven thousand bit RSA moduli would make the public system stronger than AES-128, as desired, but would also make RSA the weakest link if AES-192 lives up to the expectations.

4 RSA Factor Sizes of Matching Security

Let

$$E[n,p] = (\log_2 n)^2 e^{\sqrt{2 \log p \log \log p}}$$

be the approximate asymptotic growth rate of the expected time required by the elliptic curve method (ECM) to find a factor p of a composite number n (assuming that such a factor exists). This runtime depends mostly on the size of the factor p, and only polynomially on the size of the number n being factored. It follows that smaller factors can be found faster. A regular RSA modulus n has two prime factors of about $(\log_2 n)/2$ bits. In that case the ECM can in general be expected to be slower than the NFS, so the ECM runtime does not have to be taken into account in Section 3. In RSA-MP the RSA modulus has more than two prime factors. This implies that the factors should be chosen in such a way that they cannot be found faster using the ECM than using the NFS. In this section it is analysed how many factors an RSA-MP modulus may have so that the overall security is not affected. It is assumed that the modulus size is chosen according to Table 1, so that the moduli offer security equivalent to the selected symmetric cryptosystem with respect to NFS attacks. It is also assumed that all factors have approximately the same size.

From the definitions of $L[n]$ and $E[n,p]$ it follows that, roughly, the factors p of an RSA-MP modulus n should grow proportionally to

$$n^{(\log n)^{-\frac{1}{3}}}.$$

The size $\log_2 p$ should therefore grow as $(\log_2 n)^{2/3}$, and an RSA-MP modulus n may, asymptotically, have approximately $O((\log_2 n)^{1/3})$ factors. Such asymptotic results are, however, of hardly any interest for this paper.

Instead, given an RSA modulus (chosen according to Table 1) an explicit bound is needed for the number of factors that may be allowed. To derive such a bound the approach from [7] cited in 2.3 is used of extrapolating actual runtimes to derive expected runtimes for larger problem instances. The basis for the extrapolation is the observation that finding a 167-bit factor of a 768-bit number can be expected to require an about 80 times smaller computational effort than breaking single DES ([7, Section 5.9] and [22]). Let n' be an RSA modulus that offers security (computationally or cost) equivalent to a symmetric cryptosystem of d-bit security and speed comparable to single DES (i.e., n' is chosen according to Table 1). An RSA-MP modulus n with smallest prime factor p and with $\log n \approx \log n'$ offers security equivalent to the same symmetric cryptosystem if

$$E[n, p] \geq 80 * 2^{d-56} * E[2^{768}, 2^{167}].$$

Here it is assumed that it is reasonable not to expect substantial improvements of the ECM, and that for application of the ECM itself computational and cost equivalence are the same [16]. Given the least p satisfying the above formula, the recommended number of factors of an RSA-MP modulus n equals $m = \lfloor \log n / \log p \rfloor$. The resulting numbers of factors are given in Table 2, along with the bit lengths $\lceil (\log_2 n)/m \rceil$ of the factors, with the computationally equivalent result below the cost equivalent one. Note that $\log_2 p \leq \lceil (\log_2 n)/m \rceil$. For single DES and a cost equivalent RSA modulus in 2001 this approach would lead to a single 416-bit factor, since factoring a composite 416-bit RSA modulus using the ECM can be expected to be easier than breaking single DES. For that reason, that entry is replaced by 'two 217-bit factors'.

Table 2. Number of factors and factor size for matching RSA-MP moduli.

Year	DES	2K3DES	3K3DES	AES-128	AES-192	AES-256
2001	2 : 217 2 : 310	2 : 667 3 : 575	2 : 971 3 : 809	3 : 882 3 : 1075	4 : 1725 4 : 1980	4 : 3460 5 : 3078
2010	2 : 259 3 : 249	3 : 511 4 : 489	3 : 730 4 : 678	3 : 981 4 : 890	4 : 1857 5 : 1699	5 : 2929 5 : 3250
2020	3 : 216 4 : 227	3 : 591 4 : 559	3 : 829 4 : 762	4 : 824 4 : 989	4 : 2011 5 : 1832	5 : 3115 6 : 2873
2030	3 : 265 5 : 217	4 : 509 5 : 507	4 : 702 5 : 682	4 : 919 5 : 876	5 : 1738 5 : 1972	5 : 3308 6 : 3044

It can be seen that for a fixed symmetric cryptosystem the number of factors allowed in RSA-MP increases over time. This is mostly due to the fact that the growing moduli sizes 'allow' more primes of the same size, and to a much smaller degree due to the fact that larger moduli make application of the ECM slower.

Almost the same numbers as in Table 2 are obtained if the factor 80 is replaced by any other number in the range $[80/5, 80 * 5]$. Uncertainty about the precise expected behavior of the ECM is therefore not important, as long as the estimate is in an acceptable range.

It may be argued that $E[n, p]$ should include a factor $\log p$. It would make finding larger factors harder compared to the definition used above, and thus would lead to more factors per RSA-MP modulus. For Table 2 it hardly matters. Similarly, the factor $(\log_2 n)^2$ in $E[n, p]$ may be replaced by $(\log_2 n)^{\log_2 3}$ (or something even smaller) if faster multiplication techniques such as Karatsuba (or an even faster method) are used. The effect of these changes on Table 2 is small: for computational equivalence to 2K3DES in 2010 and for cost equivalence to AES-128 in 2020 it would result in three instead of four factors.

4.1 Remark

Although strictly speaking besides the scope of this paper, Table 3 gives the number of factors that may be allowed in RSA-MP moduli of bit lengths 1024, 2048, 4096, and 8192 with the cost equivalent number followed by the computationally equivalent one. It follows, for example, that in the conservative computationally equivalent model one would currently allow three factors in a 1024-bit RSA-MP modulus. But, using less conservative cost equivalence one would, more conservatively, allow only two factors in a 1024-bit modulus (see also Figure 1). This is consistent with the fact that for cost equivalence 1024-bit moduli are considered to be more secure than for computational equivalence: currently just 74 bits for the latter but 85 bits for the former.

Table 3. Number of factors for RSA-MP popular modulus sizes.

Year	1024		2048		4096		8192	
2001	2	3	3	3	3	4	4	4
2010	2	3	3	4	3	4	4	5
2020	3	4	3	4	4	4	4	5
2030	3	5	4	5	4	5	5	5

5 Finite Field Sizes of Matching Security

In this section subgroups refer to prime order subgroups of multiplicative groups of finite fields. Public key systems based on the use of subgroups can either be broken by directly attacking the subgroup or by attacking the finite field.

As mentioned in Section 1 the subgroup size will in practice be determined by the hash size. The latter follows immediately from the symmetric cryptosystem choice if the AES is used. Because the subgroup order is prime, the subgroup offers security equivalent to the symmetric cryptosystem as far as direct subgroup attacks are concerned. It remains to select the finite field in such a way that it provides equivalent security as well. That is the subject of this section.

5.1 Fixed Degree Extension Fields

Let p be a prime number and let $k > 0$ be a fixed small integer. The approximate asymptotic growth rate of the expected time to compute discrete logarithms in $\mathbf{F}_{p^k}^*$ is $L[p^k]$, where L is as in 3. An RSA modulus n and a finite field $\mathbf{F}_{p^k}$ therefore offer about the same level of security if n and p^k are of the same order of magnitude (disregarding the possibility of subgroup attacks in $\mathbf{F}_{p^k}^*$). It is generally accepted that for such n, p, and k factoring n is somewhat easier than computing discrete logarithms in $\mathbf{F}_{p^k}^*$. For the present purposes the distinction is negligible. Furthermore, it is reasonable to assume the same rate of cryptanalytic progress for factoring and computing discrete logarithms. It follows that Table 1 can be used to obtain matching fixed degree extension field sizes: to find $\log_2 p$ divide the numbers given in Table 1 by the fixed extension degree k.

5.2 Prime Fields

It follows from 5.1 that if prime fields are used (i.e., $k = 1$), then conservative field sizes (i.e., $[\log_2 p]$) are given by the numbers in Table 1.

As an example suppose a subgroup and prime field size have to be determined for an application that uses AES-256 and that is supposed to be in operation until 2010. Since SHA-512 will be used in combination with AES-256, the most practical subgroup order is a 512-bit prime. Furthermore, it follows from Table 1 that the prime determining the prime field should be about fifteen thousand bits long. Using eight thousand bits or less would undermine the apparently desired security level (namely, higher than AES-192). A nine to fourteen thousand bit prime would make the public system stronger than AES-192, as desired, but would also make the prime field discrete logarithm the weakest link.

5.3 Extension Fields of Degrees 2 and 6

LUC and XTR reduce the representation size of subgroup elements by using their trace over a certain subfield so that the representation belongs to the subfield as well. This does not affect the security and increases the computational efficiency [8,17].

LUC. LUC uses a subgroup of $\mathbf{F}_{p^2}^*$ of order dividing $p+1$ and traces over $\mathbf{F}_p$. It follows from 5.1 that the size of the prime field $\mathbf{F}_p$ can be found by dividing the numbers from Table 1 by $k = 2$. Table 4 contains the resulting values of $[\log_2 p]$.

XTR. XTR uses a subgroup of $\mathbf{F}_{p^6}^*$ of order dividing $p^2 - p + 1$ and traces over $\mathbf{F}_{p^2}$. The size of the underlying prime field $\mathbf{F}_p$ can be found by dividing the numbers from Table 1 by $k = 6$, resulting in the $[\log_2 p]$-values in Table 5.

5.4 Remark

For many of the LUC and XTR key sizes in Tables 4 and 5 there is an integer $e > 1$ such that $(\log_2 p)/e \geq \log_2 q$. This implies that the fields $\mathbf{F}_p$ in LUC and $\mathbf{F}_{p^2}$ in XTR can be replaced by $\mathbf{F}_{\bar{p}^e}$ (LUC) and $\mathbf{F}_{\bar{p}^{2e}}$ (XTR), where $\log_2 \bar{p}^e \approx \log_2 p$ (see [8, Section 6]). Because as a result $\log_2 \bar{p} \geq \log_2 q$, proper $\bar{p}$ and q can still be found efficiently, in ways similar to the ones suggested in [8]. In XTR care must taken that q and $\bar{p}$ are chosen so that q is a prime divisor of $\phi_{6e}(\bar{p})$, the $6e$-th cyclotomic polynomial evaluated at $\bar{p}$, which divides $\bar{p}^{2e} - \bar{p}^e + 1$. In LUC q must divide $\phi_{2e}(\bar{p})$, a divisor of $\bar{p}^e + 1$. With a proper choice of minimal polynomial for the representation of the elements of $\mathbf{F}_{\bar{p}^e}$ (LUC) or $\mathbf{F}_{\bar{p}^{2e}}$ (XTR), this leads to smaller public keys and potentially a substantial speedup (also of the parameter selection). The numbers in Section 6 do not take this possibility into account.

5.5 Small Characteristic Fields

Let p be a small fixed prime (such as 2), and let $k > 0$ be an extension degree. The approximate asymptotic growth rate of the time to compute discrete logarithms in $\mathbf{F}_{p^k}^*$ for small fixed p is

$$e^{c(\log p^k)^{1/3}(\log\log p^k)^{2/3}}$$

Table 4. $\lceil \log_2 p \rceil$ for matching LUC prime fields.

Year	DES	2K3DES	3K3DES	AES-128	AES-192	AES-256
2001	208 310	667 862	971 1213	1322 1612	3449 3959	6920 7694
2010	259 374	766 978	1095 1355	1471 1780	3713 4247	7323 8123
2020	324 453	887 1117	1244 1523	1648 1978	4021 4580	7787 8618
2030	397 542	1018 1267	1404 1704	1838 2190	4345 4930	8269 9130

Table 5. $\lceil \log_2 p \rceil$ for matching XTR prime fields.

Year	DES	2K3DES	3K3DES	AES-128	AES-192	AES-256
2001	70 104	223 288	324 405	441 538	1150 1320	2307 2565
2010	87 125	256 326	365 452	491 594	1238 1416	2441 2708
2020	108 151	296 373	415 508	550 660	1341 1527	2596 2873
2030	133 181	340 423	468 568	613 730	1449 1644	2757 3044

Table 6. $\lceil \log_2 p^k \rceil$ for matching small characteristic fields.

Year	DES	2K3DES	3K3DES	AES-128	AES-192	AES-256
2001	455 732	1767 2357	2690 3440	3781 4695	10637 12318	22210 24823
2010	592 912	2066 2711	3073 3883	4249 5227	11508 13269	23570 26277
2020	770 1140	2432 3140	3535 4414	4809 5861	12524 14377	25139 27954
2030	977 1398	2835 3608	4037 4986	5412 6539	13594 15539	26771 29694

for c oscillating in the interval $[1.526, 1.588]$ (cf. [3]). Since the smallest c leads to the more conservative field sizes, let

$$L'[p^k] = e^{1.526(\log p^k)^{1/3}(\log\log p^k)^{2/3}}.$$

This function is similar to L as defined in 3, but has a smaller constant in the exponent. This has serious implications for the choice of the field size p^k for small fixed p, compared to the case where k is fixed (as in 5.1). Computing discrete logarithms in $\mathbf{F}_{2^{607}}$ requires an about 25 times smaller computational effort than breaking single DES [19]. It follows that a small fixed characteristic field $\mathbf{F}_{p^k}$ currently offers security computationally equivalent to a symmetric cryptosystem of d-bit security and speed comparable to single DES if

$$L'[p^k] \approx 25 * 2^{d-56} * L'[2^{607}].$$

With respect to cost equivalence and expected future equivalence the same approach as in 3 and 3.1 is used: divide the right hand side by $26 * P$ for cost equivalence, and multiply it by $2^{2(y-2001)/3}$ for future equivalence. The resulting values of $\lceil \log_2 p^k \rceil$ for small characteristic fields are given in Table 6 (for $P = 100$); for $p = 2$ the numbers indicate the recommended value for k. Historically, subgroups of multiplicative groups of characteristic two finite fields were mostly of interest because of their computational advantages. Comparing the numbers in Table 1 and Table 6, however, it is questionable if the computational advantages outweigh the disadvantage of the relatively large field size.

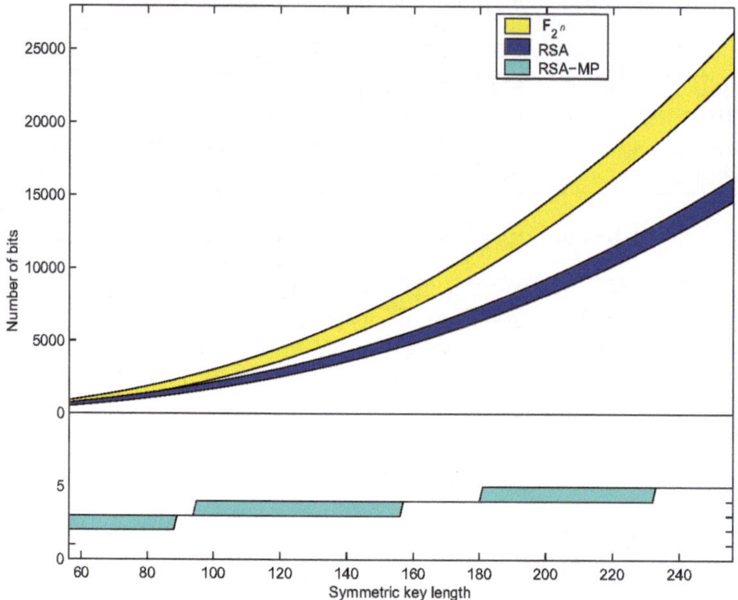

Fig. 1. The sizes from Tables 1 and 6 and the numbers of factors from Table 2 for the year 2010. The shaded areas are bounded from above by the computationally equivalent curves and step function and from below by the cost equivalent ones.

6 Performance Issues

Assume that public key sizes are chosen according to Tables 1 to 6 to match a symmetric cryptosystem of d-bit security. In this section the impact on public key size overhead and computational requirements is discussed.

6.1 Public Key Sizes

In Table 7 public key sizes are given for three scenarios. The regular public key refers to all bits contained in the public key. In an ID-based set-up the public key is reconstructed based on the user's identity and an additional number of overhead bits. Refer to [6] for ID-based public key compression for RSA. For subgroup based systems ID-based methods can trivially be designed in almost any number of ways. In a shared public key environment users share a large part of the public key data. In that case only the part that is unique for each user has to be counted.

For subgroup based systems the public key consists of a description of the subgroup, the generator g, its prime order q, and the public point $h = g^s$ (or its trace), where s is the secret key. The generator itself can usually be derived at the cost of an exponentiation of an element with a small representation, and is not counted. In an ID-based system the description of the subgroup and q can be reconstructed from the user's identity and, say, 64 additional bits, which leads to a total public key overhead of 64 bits plus the bits required to describe h. In a shared environment all users use the same g and q, so h is the only part of the public key that is unique for each user.

Fixed degree extension fields are not considered in Table 7, because in that case one may as well use LUC or XTR. The choice of subgroups of multiplicative groups of small characteristic fields is limited. Using such subgroups therefore makes sense only in a context where the public key data, with the exception of the public point h, are shared.

Table 7. Number of bits required for public key data.

PKC	regular	ID-based	shared
RSA, public exponent e $\log_2 n$ from Table 1 $2\log_2 p = \log_2 n$	$\log_2 e + \log_2 n$	$\log(\frac{1}{2}\log_2 n) + \frac{1}{2}\log_2 n$	n/a
RSA-MP, public exponent e $\log_2 n$ from Table 1 m, $\log_2 p$ from Table 2	$\log_2 e + \log_2 n$	$\log\log_2 p + \frac{m-1}{m}\log_2 n$	n/a
$\mathbf{F}_p$, $\log_2 p$ from Table 1	$2\log_2 p$	$64 + \log_2 p$	$\log_2 p$
$\mathbf{F}_{p^k}$, small p, $\log_2 p^k$ from Table 6	n/a	n/a	$\log_2 p^k$
LUC, $\log_2 p$ from Table 4	$2d + 2\log_2 p$	$64 + \log_2 p$	$\log_2 p$
XTR, $\log_2 p$ from Table 5	$2d + 3\log_2 p$	$64 + 2\log_2 p$	$2\log_2 p$
ECC, $\log_2 p = 2d$	$9d + 1$	$3d + 65$	$2d + 1$

For LUC and XTR the public key sizes follow from [17] and [8]. For ECC the description of the subgroup requires a finite field and an elliptic curve over the field. With d as above, the field and curve take at most $2d$ and $4d$ bits, respectively. About d and $2d + 1$ bits are required for the subgroup order q and the public point h. This leads to $9d + 1$ bits for ordinary ECC, $3d + 65$ bits for ID-based ECC (since the information about q must be present), and $2d + 1$ for shared ECC. For ECC the sizes do not depend on the year. To illustrate the public key size formulas, public key sizes for the year 2010 are given in Table 8, rounded to two significant digits.

6.2 Communication Overhead for Subgroup Based Systems

Each message in the Diffie-Hellman key agreement protocol consists of the representation of a subgroup element. The communication overhead per message is given in the last column of Table 7. ElGamal encryption has the same overhead (on top of the length of the message itself). The communication overhead of ElGamal-based message recovery signature schemes is equal to $2d$.

Table 8. Number of bits of public key data.

PKC	DES	2K3DES	3K3DES	AES-128	AES-192	AES-256
			regular			
RSA(-MP)	550 780	1600 2000	2200 2700	3000 3600	7500 8500	15000 16000
$\mathbf{F}_p$	1000 1500	3100 3900	4400 5400	5900 7100	15000 17000	29000 32000
LUC	630 860	1700 2100	2400 2900	3200 3800	7800 8900	15000 17000
XTR	370 490	960 1200	1300 1600	1700 2000	4100 4600	7800 8600
ECC	510	860	1000	1200	1700	2300
			ID-based			
RSA	270 380	770 990	1100 1400	1500 1800	3700 4300	7300 8100
RSA-MP	270 500	1000 1500	1500 2000	2000 2700	5600 6800	12000 13000
$\mathbf{F}_p$	580 810	1600 2000	2300 2800	3000 3600	7500 8600	15000 16000
LUC	320 440	830 1000	1200 1400	1500 1800	3800 4300	7400 8200
XTR	240 310	580 720	790 970	1000 1300	2500 2900	4900 5500
ECC	230	350	400	450	640	830
			shared			
$\mathbf{F}_p$	520 750	1500 2000	2200 2700	2900 3600	7400 8500	15000 16000
$\mathbf{F}_{p^k}$, small p	590 910	2100 2700	3100 3900	4200 5200	12000 13000	24000 26000
LUC	260 370	770 980	1100 1400	1500 1800	3700 4200	7300 8100
XTR	170 250	510 650	730 900	980 1200	2500 2800	4900 5400
ECC	110	190	230	260	390	510

6.3 Computational Requirements

In this section the relative theoretical computational requirements are estimated for the most common cryptographic applications of the public key cryptosystems discussed above: encryption, decryption, signature generation, and signature verification. No actual runtimes are given. For software implementations the theoretical estimates should give a reasonable prediction of the actual relative performance. For implementations using dedicated hardware, such as special-purpose exponentiators, all predictions concerning RSA and prime field subgroups are most likely too pessimistic. However, as soon as special-purpose hardware is available for ECC, LUC, or XTR, the relative performance numbers should again be closer to reality.

For subgroup based systems common ElGamal-like schemes are used where decryption and signing each require a single subgroup exponentiation, encryption requires two separate subgroup exponentiations, and signature verification requires the product of two subgroup exponentiations (a 'double exponentiation'). The Diffie-Hellman key agreement protocol has, per party, the same cost as encryption, i.e., two separate subgroup exponentiations.

It is assumed that squaring and multiplication in the finite field $\mathbf{F}_p$ and the ring $\mathbf{Z}/n\mathbf{Z}$ of integers modulo n take the same amount of time if $\log_2 p \approx \log_2 n$. A squaring in $\mathbf{Z}/n\mathbf{Z}$ is assumed to take 80% of the time of a multiplication in $\mathbf{Z}/n\mathbf{Z}$. Basic exponentiation methods are used, i.e., no window tricks. This hardly affects the relative performance. Precomputation of the value g^t with $\log_2 t \approx (\log_2 q)/2$ combined with double exponentiation is used for subgroup based signature generation. For XTR the methods from [18] are used. The LUC and ECC estimates follow from [18, Section 7]. For ECC the time to recover the y-coordinates of subgroup elements is not counted.

The resulting runtime expressions for the four basic cryptographic functions are given in Table 9. Small characteristic fields are not included because the

Table 9. Number of multiplications in $\mathbf{F}_p$ (unless noted otherwise).

PKC matching symmetric system of d-bit security	encryption	signature verification	decryption	signature generation
RSA, public exponent e $\log_2 n$ from Table 1 $2\log_2 p = \log_2 n$	$1.3\log_2 e$ in $\mathbf{Z}/n\mathbf{Z}$		sequential: $2.6\log_2 p$ 2 in parallel: $1.3\log_2 p$	
RSA-MP, public exponent e $\log_2 n$ from Table 1 m, $\log_2 p$ from Table 2	$1.3\log_2 e$ in $\mathbf{Z}/n\mathbf{Z}$		sequential: $1.3m\log_2 p$ m in parallel: $1.3\log_2 p$	
$\mathbf{F}_p$, $\log_2 p$ from Table 1	$5.2d$	$3.1d$	$2.6d$	$1.6d$
LUC, $\log_2 p$ from Table 4	$6.4d$	$3.5d$	$3.2d$	$1.8d$
XTR, $\log_2 p$ from Table 5	$21d$	$12d$	$10d$	$6d$
ECC, $\log_2 p = 2d$	$36d$	$20d$	$18d$	$10d$

relative speed of $\mathbf{F}_{2^k}$ and $\mathbf{F}_p$ arithmetic is too platform dependent. Despite potential advantages of hardware $\mathbf{F}_{2^k}$-arithmetic, the large value that is required for k may make these fields unattractive for very high security non-ECC cryptographic applications.

As an illustration of the data in Table 9, the relative performance of the cryptographic operations is given in Table 10 for the year 2010, rounded to two significant digits. For Table 10 the time $M(L)$ for modular multiplication of L-bit integers is proportional to L^2. This corresponds to regular hardware implementations. The unit of time is the time required for a single multiplication in $\mathbf{Z}/n\mathbf{Z}$ for a 1024-bit integer n. This arbitrary choice has no influence on the relative performance. For RSA and RSA-MP the sequential ('S') and parallel ('P') performance is given, with the number of parallel processors and the relative parallel runtime separated by a semicolon. RSA encryption and signature verification for $e = 3$ or $e = 2^{17}+1$ goes about 20 or 3 times faster, respectively, than for a random 32-bit public exponent as in Table 10.

For higher security public key systems other than ECC the finite field and ring sizes get so large that implementation using Karatsuba-like multiplication techniques should be worthwhile. In software implementations this can easily be realized. In Table 11 the relative performance for the year 2010 is given using Karatsuba-like modular multiplication. This implies that $M(L)$ is proportional to $L^{\log_2 3}$, as opposed to L^2 as in Table 10. The unit of time in Table 11 is the time required for a single Karatsuba-like multiplication in $\mathbf{Z}/n\mathbf{Z}$ for a 1024-bit integer n. Since this may be different from the time required for a regular 1024-bit modular multiplication (as in Table 10), the numbers in Tables 10 and 11 are not comparable.

Table 10. Relative performance using regular arithmetic for the year 2010.

PKC	DES	2K3DES	3K3DES	AES-128	AES-192	AES-256
			$\log_2 n$			
RSA(-MP)	520 750	1500 2000	2200 2700	2900 3600	7400 8500	15000 16000
			$\log_2 p$			
$\mathbf{F}_p$	520 750	1500 2000	2200 2700	2900 3600	7400 8500	15000 16000
LUC	260 370	770 980	1100 1400	1500 1800	3700 4200	7300 8100
XTR	90 130	260 330	370 450	490 590	1200 1400	2400 2700
ECC	112	190	224	256	384	512
		encryption (with $\log_2 e = 32$ for RSA and RSA-MP)				
RSA(-MP)	11 22	93 150	190 290	340 500	2200 2900	8500 10000
$\mathbf{F}_p$	75 160	1100 1800	2700 4100	5500 8000	53000 69000	270000 340000
LUC	23 48	340 550	820 1300	1700 2500	16000 21000	84000 100000
XTR	8 17	120 200	290 450	610 890	5800 7600	30000 37000
ECC	24	120	190	290	970	2300
			decryption			
RSA (S)	43 130	1100 2300	3300 6200	7900 14000	130000 190000	970000 1300000
RSA (P)	{2 : 22 2 : 65	{2 : 560 2 : 1200	{2 : 1600 2 : 3100	{2 : 3900 2 : 7000	{2 : 63000 2 : 95000	{2 : 490000 2 : 660000
RSA-MP (S)	43 57	500 580	1400 1500	3500 3500	32000 30000	160000 210000
RSA-MP (P)	{2 : 22 3 : 19	{3 : 170 4 : 150	{3 : 480 4 : 390	{3 : 1200 4 : 870	{4 : 7900 5 : 6100	{5 : 31000 5 : 43000
$\mathbf{F}_p$	37 77	550 900	1300 2000	2700 4000	26000 34000	140000 170000
LUC	11 24	170 280	410 630	840 1200	8100 11000	42000 51000
XTR	4 9	61 99	150 230	300 450	2900 3800	15000 18000
ECC	12	59	96	140	490	1100
			signature generation			
RSA (S)	43 130	1100 2300	3300 6200	7900 14000	130000 190000	970000 1300000
RSA (P)	{2 : 22 2 : 65	{2 : 560 2 : 1200	{2 : 1600 2 : 3100	{2 : 3900 2 : 7000	{2 : 63000 2 : 95000	{2 : 490000 2 : 660000
RSA-MP (S)	43 57	500 580	1400 1500	3500 3500	32000 30000	160000 210000
RSA-MP (P)	{2 : 22 3 : 19	{3 : 170 4 : 150	{3 : 480 4 : 390	{3 : 1200 4 : 870	{4 : 7900 5 : 6100	{5 : 31000 5 : 43000
$\mathbf{F}_p$	23 48	340 550	820 1300	1700 2500	16000 21000	84000 100000
LUC	6 13	93 150	230 340	460 680	4400 5800	23000 28000
XTR	2 5	36 58	85 130	180 260	1700 2200	8700 11000
ECC	7	32	53	79	270	630
		signature verification (with $\log_2 e = 32$ for RSA and RSA-MP)				
RSA(-MP)	11 22	93 150	190 290	340 500	2200 2900	8500 10000
$\mathbf{F}_p$	44 92	660 1100	1600 2400	3300 4800	31000 41000	160000 200000
LUC	13 26	190 300	450 690	930 1400	8900 12000	46000 57000
XTR	5 10	71 120	170 260	350 520	3400 4400	17000 21000
ECC	13	65	110	160	530	1300

As an example of an application of Tables 10 and 11, suppose AES-192 is used in 2010 along with a cost equivalent public key system. With regular (quadratic growth) modular arithmetic, ECC encryption takes time equivalent to about 970 regular multiplications modulo a 1024-bit modulus. This can be expected to be about twice faster than RSA encryption (with a 32-bit public exponent), and about six times faster than XTR encryption. But with Karatsuba-like arithmetic, RSA encryption takes time equivalent to about 960 Karatsuba multiplications modulo a 1024-bit modulus (but using a 7400-bit modulus). This can be expected to be about 1.5 times faster than ECC encryption, and about six times faster than XTR. For decryption, however, RSA is substantially slower than both ECC and XTR for either type of arithmetic, even if RSA-MP is used on four parallel processors.

6.4 Parameter Selection

For all public key systems except ECC, parameter selection is dominated by the generation of the primes defining the moduli, finite fields, and subgroup orders. For each L-bit prime to be generated, the generation time is proportional to $M(L)L^2$. A more precise runtime function depends on a wide variety of implementation choices that are not discussed here. Obviously, parameter selection for high security RSA, prime field, or LUC based systems will be slow compared to RSA-MP and, in particular, XTR.

For systems based on a subgroup of $\mathbf{F}_{p^k}$ for fixed small p public key data are usually shared (except for the public point h). For such systems the speed of parameter selection is therefore not an important issue.

ECC parameters can be found in expected polynomial time. Nevertheless, even for security equivalent to 2K3DES the solution is not yet considered to be sufficiently practical for systems with non-shared keys. The slow growth of the parameter sizes implies, however, that if a satisfactory solution is found for current (relatively low) security levels, then the solution will most likely also work fast enough for very high security levels. For ECC over fields of characteristic two this goal is close to being achieved [4].

7 Summary of Findings

Matching AES-192 or AES-256 security levels with public key systems requires public key sizes far beyond anything in regular use today. For instance, to match the security of AES-192 with RSA, it would be prudent to use moduli of about 7000 bits. But given current resources, the overall practicality of RSA with such moduli is questionable. Encryption and signature verification are faster than for any other system if the public exponent is small, but the modulus itself may be prohibitively large. RSA-MP fares a little better. But even if fully parallelized it is still relatively unattractive. An interesting observation is that computationally equivalent RSA-MP moduli often allow more factors than the (smaller) cost equivalent ones, and may thus attain greater decryption and signature generation speed (at the cost of a higher level of parallelism).

Table 11. Relative performance using Karatsuba arithmetic for the year 2010.

PKC	DES	2K3DES	3K3DES	AES-128	AES-192	AES-256
			$\log_2 n$			
RSA(-MP)	520 750	1500 2000	2200 2700	2900 3600	7400 8500	15000 16000
			$\log_2 p$			
$\mathbf{F}_p$	520 750	1500 2000	2200 2700	2900 3600	7400 8500	15000 16000
LUC	260 370	770 980	1100 1400	1500 1800	3700 4200	7300 8100
XTR	90 130	260 330	370 450	490 590	1200 1400	2400 2700
ECC	112	190	224	256	384	512
		encryption (with $\log_2 e = 32$ for RSA and RSA-MP)				
RSA(-MP)	14 25	79 120	140 190	220 300	960 1200	2800 3300
$\mathbf{F}_p$	99 180	940 1400	1900 2700	3500 4800	23000 29000	90000 110000
LUC	40 72	380 560	800 1100	1500 2000	9400 12000	37000 44000
XTR	23 41	220 320	450 630	820 1100	5400 6600	21000 25000
ECC	60	240	360	510	1500	3100
			decryption			
RSA (S)	76 200	1300 2400	3200 5500	6800 11000	74000 110000	430000 560000
RSA (P)	2 : 38 2 : 99	2 : 630 2 : 1200	2 : 1600 2 : 2700	2 : 3400 2 : 5600	2 : 37000 2 : 53000	2 : 220000 2 : 280000
RSA-MP (S)	76 100	660 790	1700 1800	3600 3700	25000 25000	100000 130000
RSA-MP (P)	2 : 38 3 : 34	3 : 220 4 : 200	3 : 560 4 : 460	3 : 1200 4 : 930	4 : 6200 5 : 4900	5 : 20000 5 : 26000
$\mathbf{F}_p$	49 88	470 690	970 1400	1800 2400	12000 14000	45000 53000
LUC	20 36	190 280	400 560	730 980	4700 5800	18000 22000
XTR	12 21	110 160	230 320	410 560	2700 3300	10000 12000
ECC	30	120	180	260	730	1500
			signature generation			
RSA (S)	76 200	1300 2400	3200 5500	6800 11000	74000 110000	430000 560000
RSA (P)	2 : 38 2 : 99	2 : 630 2 : 1200	2 : 1600 2 : 2700	2 : 3400 2 : 5600	2 : 37000 2 : 53000	2 : 220000 2 : 280000
RSA-MP (S)	76 100	660 790	1700 1800	3600 3700	25000 25000	100000 130000
RSA-MP (P)	2 : 38 3 : 34	3 : 220 4 : 200	3 : 560 4 : 460	3 : 1200 4 : 930	4 : 6200 5 : 4900	5 : 20000 5 : 26000
$\mathbf{F}_p$	30 54	290 420	600 840	1100 1500	7100 8800	28000 33000
LUC	11 20	110 160	220 310	400 540	2600 3200	10000 12000
XTR	7 12	63 93	130 180	240 320	1600 1900	6100 7200
ECC	17	65	100	140	400	840
		signature verification (with $\log_2 e = 32$ for RSA and RSA-MP)				
RSA(-MP)	14 25	79 120	140 190	220 300	960 1200	2800 3300
$\mathbf{F}_p$	59 110	560 820	1200 1600	2100 2900	14000 17000	54000 63000
LUC	22 40	210 310	440 610	800 1100	5200 6400	20000 24000
XTR	13 24	130 190	260 370	480 650	3100 3900	12000 14000
ECC	33	130	200	280	800	1700

The unattractive sizes of RSA moduli of high security levels is entirely due to the number field sieve. If it had not been invented, and the asymptotically slower quadratic sieve factoring algorithm would still be the fastest factoring algorithm, then at least until 2030 RSA moduli of 2048, 4096, and 8192 bits would be good matches for AES-128, AES-192, and AES-256, respectively. But, it could have been worse too: if the special number field sieve would apply to RSA moduli, then RSA moduli would have to be chosen according to Table 6 instead of Table 1, i.e., considerably larger.

Compared to RSA and RSA-MP, subgroups of prime fields have the same size problem. They are much slower for encryption and signature verification. Decryption and signature generation is competitive only in environments where RSA and RSA-MP cannot be parallelized. Furthermore, subgroups of prime fields are consistently outperformed by LUC and XTR. So, unless second and sixth degree extension fields turn out to be less secure than currently believed, subgroups of prime fields are not competitive.

Similarly, LUC is consistently outperformed by XTR[2]. Unless a dramatic breakthrough occurs in the fixed degree extension field discrete logarithm problem, XTR is a good choice if one insists on using a non-ECC subgroup public key system. It has the additional advantages that parameter selection is easy and that current special purpose RSA modular multipliers (that can handle public moduli up to, say, 1024 bits) may be used even for very high security applications (possibly using Remark 5.4). The latter is also possible for LUC (if Remark 5.4 is used), may be possible for RSA-MP, but is out of the question for RSA or prime field subgroups.

Overall, ECC suffers the smallest performance degradation when moving to very high security levels. Generation of ECC public keys in a non-shared set-up remains problematic, for all security levels. If that is not a concern, and barring cryptanalytic progress affecting the elliptic curve discrete logarithm problem, the choice is obvious.

For current security levels, i.e., comparable to 1024-bit RSA, the choice is between RSA, RSA-MP, XTR, and ECC and will mostly depend on the application. For current higher security levels, comparable to 2048-bit RSA, the theoretical performance gap between ECC and the other public key systems already becomes noticeable, with only XTR still within range of ECC. However, hardware accelerators are currently available for 2048-bit RSA and RSA-MP, but not for other security equivalent public key systems. So, for the next few years RSA and RSA-MP will still be the methods of choice in many practical circumstances where security equivalent to 2048-bit RSA is required. This may change radically if new types of hardware accelerators are developed. And even if that does not happen, it will change eventually, i.e., for higher security levels, because special purpose hardware cannot beat the asymptotics.

Disclaimer. The contents of this paper are the sole responsibility of the author and not of his employer. The author does not accept any responsibility for the use of the material presented in this paper. Despite his academic involvement

[2] However, for LUC it is in general faster to test if a value is correctly formatted, i.e., if it is the trace of a proper subgroup element. Refer to [9] for details.

with XTR, the author does not have any financial or other material interests in any of the cryptosystems discussed in this paper.

Acknowledgments. The author thanks Eric Verheul and Mike Wiener for their many insightful comments on earlier versions of this paper and Martijn Stam for his assistance with Figure 1.

References

1. I. Blake, G. Seroussi, N. Smart, *Elliptic curves in cryptography*, Cambridge University Press, 1999.
2. H. Cohen, A. Miyaji, T. Ono, *Efficient elliptic curve exponentiation using mixed coordinates*, Proceedings Asiacrypt'98, LNCS 1514, Springer-Verlag 1998, 51-65.
3. D. Coppersmith, *Fast evaluation of logarithms in fields of characteristic two*, IEEE Trans. Inform. Theory 30 (1984) 587-594.
4. R. Harley, Rump session presentations at Eurocrypt 2001 and Crypto 2001; data available from argote.ch/Research.html.
5. P.C. Kocher, *Breaking DES*, RSA Laboratories' Cryptobytes, v. 4, no 2 (1999), 1-5; also at www.rsasecurity.com/rsalabs/pubs/cryptobytes.
6. A.K. Lenstra, *Generating RSA moduli with a predetermined portion*, Proceedings Asiacrypt'98, LNCS 1514, Springer-Verlag 1998, 1-10.
7. A.K. Lenstra, E.R. Verheul, *Selecting cryptographic key sizes*, to appear in the Journal of Cryptology; available from www.cryptosavvy.com.
8. A.K. Lenstra, E.R. Verheul, *The XTR public key system*, Proceedings of Crypto 2000, LNCS 1880, Springer-Verlag 2000, 1-19; available from www.ecstr.com.
9. A.K. Lenstra, E.R. Verheul, *Fast irreducibility and subgroup membership testing in XTR*, Proceedings PKC 2001, LNCS 1992, Springer-Verlag 2001, 73-86; available from www.ecstr.com.
10. A.J. Menezes, P.C. van Oorschot, S.A. Vanstone, *Handbook of applied cryptography*, CRC Press, 1997.
11. National institute of standards and technology, *Digital signature standard*, FIPS Publication 186-2, February 2000.
12. National institute of standards and technology, //csrc.nist.gov/encryption/aes/.
13. National institute of standards and technology, //csrc.nist.gov/cryptval/shs.html.
14. R.L. Rivest, A. Shamir, L.M. Adleman, *Cryptographic communications system and method*, U.S. Patent 4,405,829, 1983.
15. B. Schneier, *Applied cryptography*, second edition, Wiley, New York, 1996.
16. R.D. Silverman, *A cost-based security analysis of symmetric and asymmetric key lengths*, RSA Laboratories Bulletin 13, April 2000.
17. P. Smith, C. Skinner, *A public-key cryptosystem and a digital signature system based on the Lucas function analogue to discrete logarithms*, Proceedings of Asiacrypt '94, LNCS 917, Springer-Verlag 1995, 357-364.
18. M. Stam, A.K. Lenstra, Speeding up XTR, Proceedings Asiacrypt 2001, Springer-Verlag 2001, this volume; available from www.ecstr.com.
19. E. Thome, *Computation of discrete logarithms in* $\mathbf{F}_{2^{607}}$, Proceedings Asiacrypt 2001, Springer-Verlag 2001, this volume.
20. P.C. van Oorschot, M.J. Wiener, *A known-plaintext attack on two-key triple encryption*, Proceedings Eurocrypt'90, LNCS 473, Springer-Verlag 1991, 318-325.
21. M.J. Wiener, personal communication, August 2001.
22. P. Zimmermann, personal communication, 1999.

A Probable Prime Test with Very High Confidence for $n \equiv 1 \bmod 4$

Siguna Müller*

University of Klagenfurt, Department of Mathematics, 9020 Klagenfurt, Austria
siguna.mueller@uni-klu.ac.at

Abstract. Although the Miller-Rabin test is very fast in practice, there exist composite integers n for which this test fails for 1/4 of all bases coprime to n. In 1998 Grantham developed a probable prime test with failure probability of only 1/7710 and asymptotic running time 3 times that of the Miller-Rabin test. For the case that $n \equiv 1 \bmod 4$, by S. Müller a test with failure rate of 1/8190 and comparable running time as for the Grantham test was established. Very recently, with running time always at most 3 Miller-Rabin tests, this was improved to 1/131040, for the other case, $n \equiv 3 \bmod 4$. Unfortunately the underlying techniques cannot be generalized to $n \equiv 1 \bmod 4$. Also, the main ideas for proving this result do not extend to $n \equiv 1 \bmod 4$.
Here, we explicitly deal with $n \equiv 1 \bmod 4$ and propose a new probable prime test that is extremely efficient. For the first round, our test has average running time $(4 + o(1)) \log_2 n$ multiplications or squarings mod n, which is about 4 times as many as for the Miller-Rabin test. But the failure rate is much smaller than $1/4^4 = 1/256$. Indeed, for our test we prove a worst case failure probability less than 1/1048350. Moreover, each iteration of the test runs in time equivalent to only 3 Miller-Rabin tests. But for each iteration, the error is less than 1/131040.

Keywords: Probable Prime Testing, Error Probability, Worst Case Analysis, Quadratic-Field Based Methods, Combined Tests

1 Introduction

1.1 Motivation

Large prime numbers are essential for most cryptographic applications. Perhaps the most common probabilistic prime test is the *Strong Fermat Test* (Miller-Rabin Test), which consists of testing that $a^s \equiv 1$, resp. $a^{2^j s} \equiv -1 \bmod n$ for some $0 \leq j \leq r-1$ where $n-1 = 2^r s$ with s odd. Although exponentiation modulo n can be performed extremely fast, the catch with this, as with any probable prime test, is the existence of pseudoprimes. This means that certain composite integers are identified as primes by the test.

* Research supported by the Austrian Science Fund (FWF), FWF-Project no. P 14472-MAT

C. Boyd (Ed.): ASIACRYPT 2001, LNCS 2248, pp. 87–106, 2001.
© Springer-Verlag Berlin Heidelberg 2001

In a typical cryptographic scenario, some of the involved parties may be malicious. If an adversary manages to sell composites as primes, this usually compromises the security of the corresponding protocol. As strong pseudoprimes can easily be *constructed* this often allows fooling a pseudoprimality testing device that utilizes Miller-Rabin only. As an example, strong pseudoprimes are known with respect to all forty-six prime bases a up to 200 [3]. While a composite number can be a strong pseudoprime for at most $1/4$ of all bases coprime to n, there exist composites that actually do pass for this largest possible bound of the $1/4$ bases. Moreover, such numbers can efficiently be characterized and constructed [9]. Although it is known that $a = 2$ is a *witness* (for the compositeness of n) for most odd composites, it was shown in [2] that there are infinitely many Carmichael numbers whose least witness is larger than $(\log n)^{1/3 \log\log\log n}$. Also, it is conjectured [2] that there are $x^{1/5}$ Carmichael numbers $n \leq x$ for which there is *no base* a in any given set of $\frac{1}{11}\log x$ distinct integers $\leq x$ that proves n composite by the Miller-Rabin test.

While Miller-Rabin works well for any average number n on input a random base a, due to the fact that pseudoprimes can be constructed, the cautious might want to minimize the chance of being sold a composite instead of a prime.

There exist a number of deterministic algorithms for primality testing (see e.g., [7,10,12,20,32]), which however require rather involved theory and implementation. The advantage with pseudoprimality testing still is, that these approaches are a lot faster and can much more easily be realized in practice.

The result of this paper is a new probable prime test which is considerably more reliable than the previous proposals, but which still is much easier to describe and implement than the deterministic tests.

1.2 The Proposed Test

The main ideas for the pseudoprimality tests [16,26,28] consist of a combined Miller-Rabin test by utilizing both, the original $\mathbb{F}_p$-based algorithm, as well as the quadratic field (QF)- based analogue. An additional testing criterion in [26, 28] is based on the underlying (Cipolla related) square root finding algorithm modulo primes p (Lemma 1 below). If the result is not a correct root modulo n, n is disclosed as composite. Otherwise, this gives an additional testing condition.

Here, we incorporate yet another root-finding algorithm when $n \equiv 1 \bmod 4$. This constitutes a counterpart to the very recent results in [28] for $n \equiv 3 \bmod 4$ (the easier case). Via some efficient algorithm we test for what should be a square root of some Q mod n, $\left(\frac{Q}{n}\right) = 1$. This automatically constitutes a strengthened version of a Miller-Rabin Test. Consecutively, we test for the square root of 1 in the quadratic extension. We show how the root finding part can be obtained with low cost, with simultaneously obtaining a speed-up for the evaluation of the QF-part, as well as a reduction of the failure rate.

In essence, the test for any $n \equiv 1 \bmod 4$ runs as follows. We incorporate the same trivial testing conditions in our precomputation as does Grantham, [16]. Also, as in [16], we assume that n is not a perfect square.

0. (Precomputation)
   - If n is divisible by a prime up to $\min\{B, \sqrt{n}\}$, where $B = 50000$, declare n to be composite and stop.
   - If $\sqrt{n} \in \mathbb{Z}$ declare n to be composite and stop.
1. (Parameter Selection)
   Select randomly $P \in \mathbb{Z}_n$, Q in $\mathbb{Z}_n^*$ such that $\left(\frac{Q}{n}\right) = 1$, $\left(\frac{D}{n}\right) = -1$.
2. (Square Root Part)
   - Run one of the square root finding algorithms of sect. 2.3 for the root of Q modulo n.
     If the root finding algorithm declares n composite, stop.
   - Let a be the root of Q obtained, and let $P' \leftarrow P/a \bmod n$.
3. (QF-Based Part)
   - Let $\alpha(P',1)$, $\overline{\alpha}(P',1)$ be the roots of $x^2 - P'x + 1$.
   - Test, if $\alpha(P',1)^{(n+1)/2} \equiv \overline{\alpha}(P',1)^{(n+1)/2} \bmod n$.
     // For efficient practical realization see sect. 3.1.
     If not, n is composite and stop.
   - Compute $\gcd(\alpha(P',1)^{(n+1)/2} \pm 1, n)$. If one of these reveals a proper factor of n, output the factor. Otherwise declare n to be a probable prime.

The above describes the first round of the test. When being iterated, some of the calculations can be done more efficiently (see sect. 3.3).

1.3 The Results of This Paper

The main result of this paper is the following theorem. As in [16], one selfridge is equivalent to the time required for one round of Miller-Rabin.

Theorem 1. *A composite integer $n \equiv 1 \bmod 4$ passes k iterations of the proposed test with worst case failure probability less than $1/1048350 \cdot 1/131040^{k-1}$, which is approximately $1/2^{17k+3}$.*

For k iterations, the above test has average running time $3k+1$ selfridges.

In detail, the result can be stated as follows.

- For one round of the proposed test, the exact failure is less than $1/2^{20} + 1/(2 \cdot B^2) < 1/1048350$ and the average running time is 4 selfridges.
- For each additional iteration, the proposed test has worst case failure probability $1/2^{17} + 4/B^2 < 1/131040$ and average running time 3 selfridges.

The first round failure rate should be contrasted to the worst case error probability $1/256$ of four iterations of the Miller-Rabin test. For two iterations of the proposed test this is $1/(1.37 \cdot 10^{11})$, opposed to $1/16384$, for three iterations $1/(1.8 \cdot 10^{16})$ opposed to $1/1048576$, etc.

The estimate is based on worst case analysis and on the assumption of the existence of special (bad) composites. Otherwise, the result would even be better.

The number of pairs that pass the proposed test (so-called 'liars') can explicitly be determined. This number of liars is largest for integers n of the form like $p-1|n-1$ and $p+1|n+1$ for all primes $p|n$. Such special types of numbers must be very rare and it is not even known whether they exist at all. This shows the difficulty for composites to pass the test with respect to varied parameters. Thus, the average case error rate is expected to be much smaller (see [14,37]).

Below, we describe one method how the underlying algorithms can *easily and efficiently be evaluated.* This is based on a naive multiply/add arithmetic and can easily be implemented with low effort. Alternatively, this could be achieved via the computation of elements in a quadratic extension field [22], the evaluation of second-order recurrences and Lucas chains [9,16,23,34,40], or of powers of 2×2 matrices [35].

For modular exponentiation, many improvements to the conventional powering ladder have been designed. We hope that analogously to the many tools for speeding up exponentiation in the prime field, similar devices for the QF- part will further improve on the practicality of the proposed test.

1.4 Related Work

A number of probable prime tests have been proposed which are based on various testing functions [1,6,8,19,33]. It turns out that the methods based on different underlying techniques are the most reliable ones, whereas those based on one technique only, allow the generation of pseudoprimes, even with respect to varied testing parameters. From a practical viewpoint however, the suggestions based on third and higher-order recurrences seem to be too expensive.

Pomerance, Selfridge, Wagstaff [33] and *Baillie, Wagstaff* [8] proposed a test based on both the Fermat test and on second-order (Lucas) sequences, which is very powerful. Although the underlying criteria can be evaluated extremely fast, no composite number is known for which this probable prime test fails. Indeed, nobody has yet claimed the $620 that is offered for such an example. While it is not known whether this test does allow any pseudoprimes at all, some heuristics indicate that such composites actually might exist [31]. Although the specific choice of the parameters makes the routine easy to describe, it might increase the chance of generating any pseudoprimes with respect to these parameters. Some related tests based on different parameters have been implemented in several computer-algebra systems which however turned out to be quite weak [30]. It is not known how reliable other parameters to this test are. Also, there is no quantifying measure to determine how reliable it actually is.

Several **probabilistic tests** have been published, for which an *explicit estimate on the worst case failure probability* is known.

- The Miller-Rabin test is usually taken as a unit measure with running time 1 selfridge [16] and worst case failure 1/4.
- J. Grantham [16] proposed an extremely efficient test with worst case failure rate 1/7710 and asymptotic running time 3 selfridges. Unfortunately, the practical implementation is rather involved and it seems that on average

$4.5 \log_2(n)$ multiplications (instead of the asymptotic $(3+o(1))\log_2(n)$) are necessary.
- By S. Müller [26] a probable prime test for the case that $n \equiv 1 \bmod 4$ was developed. The test has running time similar to the Grantham test, but with worst case error probability 1/8290 per round. This bound, however, can only be achieved for at least two iterations of the test.
- Recently, a proposal has been made [28] for $n \equiv 3 \bmod 4$ with failure rate 1/131040 but only 3 selfridges running time.

Jaeschke's tables [17] of strong pseudoprimes show that these occur very frequently for $n \equiv 1 \bmod 4$. Unfortunately, the techniques for the most effective test above, [28], are exclusive for the case $n \equiv 3 \bmod 4$. As the condition $4|n+1$ constitutes a critical requirement for both the methods employed, as well as for the failure estimate, this cannot be extended to $n \equiv 1 \bmod 4$. Our results will be improvements and extensions of the methods of [26]. Indeed, for integers $n \equiv 1 \bmod 4$ essentially new techniques will be developed in this paper.

Relevance to Cryptography: For cryptographic applications, it is often necessary to generate pseudoprimes which are primes except for arbitrary small error rate. E.g., if a probability $1/2^{100}$ is to be achieved, one needs

- 50 iterations of Miller-Rabin, which is 50 selfridges,
- 8 iterations of the Grantham test, which is (asymptotically) 24 selfridges,
- 6 iterations of the proposed test, which is only 19 selfridges.

Due to the simple evaluation method of the proposed test via a naive powering ladder (sect. 3.1), we hope that this theoretical improvement will have some practical significance as well.

2 The New Idea

2.1 Some Fundamental Properties

Unless stated otherwise, let p, p_i be an odd prime, respectively an odd prime divisor of an integer $n \equiv 1 \bmod 4$ that is to be tested for primality. For simplicity we use the abbreviations of [36], *psp*(a), *epsp*(a), *spsp*(a), to denote, respectively, a pseudoprime, an Euler pseudoprime, and a strong pseudoprime, to base a.

Let $\epsilon(p) = \left(\frac{D}{p}\right)$ and $\epsilon(n) = \left(\frac{D}{n}\right)$, for $D = P^2 - 4Q$ the discriminant of $x^2 - Px + Q$ with characteristic roots $\alpha = \alpha(P,Q)$, $\overline{\alpha} = \overline{\alpha}(P,Q)$. We will assume that $\gcd(2QD, n) = 1$.

A number of probable prime tests are based on suitable properties in $\mathbb{F}_{p^2}$. As with the Miller-Rabin test in $\mathbb{F}_n$, when $n = p$ is prime, for both roots $y \in \mathbb{F}_{n^2}$ of $x^2 - Px + Q$ with $\epsilon(n) = -1$, one has, $y^u \equiv 1 \bmod n$, or $y^{2^k u} \equiv -1 \bmod n$ for some $0 \le k \le t-1$, where $n^2 - 1 = 2^t u$ with u odd. The exponent $2^k u = \frac{n^2-1}{2^j}$ is still too large for obtaining strong testing conditions. More restrictive ones are being obtained via $y^{n-\epsilon(n)} \equiv 1$, respectively $Q \bmod n$, according as $\epsilon(n) = 1$ or

-1. As the former case constitutes an ordinary Fermat condition, in combination with a Fermat test, it only makes sense to test for the latter one. Thus, unless stated otherwise, we will throughout assume $\epsilon(n) = -1$.

Composite integers n fulfilling $y^{n-\epsilon(n)} \equiv Q \bmod n$ for $\epsilon(n) = -1$ are known as quadratic field based pseudoprimes w.r.t. (P, Q), abbrev. *QFpsp*(P, Q).

If $\left(\frac{Q}{n}\right) = 1$ and if $\alpha, \overline{\alpha}$ denote the two roots y, then, for n prime, the two roots need to evaluate to the same value, even with the smaller exponent $(n-\epsilon(n))/2$ in place of $n - \epsilon(n)$, i.e., we must have $\alpha^{(n-\epsilon(n))/2} \equiv \overline{\alpha}^{(n-\epsilon(n))/2} \bmod n$. Composite integers fulfilling this criterion are denoted *elpsp*(P, Q). In our case, for $n \equiv 1 \bmod 4$ and $\epsilon(n) = -1$, the value $(n+1)/2$ is odd which already constitutes the strong Lucas test and the pseudoprimes are denoted *slpsp*(P, Q).

Lemma 1. *Let* $\epsilon(n) = -1$ *and let* $n \equiv 1 \bmod 4$ *be a composite integer that fulfills* $\alpha^{n-\epsilon(n)} \equiv \overline{\alpha}^{n-\epsilon(n)} \equiv Q \bmod n$ *for* $\left(\frac{Q}{n}\right) = 1$. *Then n is both psp*(Q) *and QFpsp*(Q). *If* $\alpha^{(n-\epsilon(n))/2} \equiv \overline{\alpha}^{(n-\epsilon(n))/2} \bmod n$ *then n is slpsp*(P, Q) *for* $\left(\frac{D}{n}\right) = -1$ *and, moreover,* $\left(\frac{Q}{p}\right) = 1$ *for all prime divisors p of n.*

Proof. This follows directly from the proof of Theorem 3, [26], because for $n \equiv 1 \bmod 4$, $(n - \epsilon(n))/2 = (n + 1)/2$ is odd. □

The above conditions are tested in [16], however, Grantham does not consider the nature of the value $\alpha^{(n-\epsilon(n))/2}$ modulo n. In [26], a formula was obtained when n is a prime, and this was used to establish a new pseudoprimality test.

Proposition 1. *If α is any root of* $x^2 - Px + Q$, *and if* $a^2 \equiv Q \bmod n$ *for n prime, then* $\alpha^{\frac{n-\epsilon(n)}{2}} \equiv \overline{\alpha}^{\frac{n-\epsilon(n)}{2}} \bmod n$, *and this is equivalent to* $\left(\frac{P+2a}{n}\right) \bmod n$, *if* $\epsilon(n) = 1$, *and equivalent to* $\left(\frac{P+2a}{n}\right) a \bmod p$, *if* $\epsilon(n) = -1$.

Often a composite n fulfills the condition $\alpha^{n-\epsilon(n)} \equiv \overline{\alpha}^{n-\epsilon(n)} \equiv Q \bmod n$, but not the stronger one of Proposition 1. In that case $\gcd(\alpha^{(n-\epsilon(n))/2} \pm a, n)$ is a proper factor of n. This is the final condition being tested in Step 3 of the test.

2.2 The Main Problem

While the values $\alpha(P, Q)^k$, $\overline{\alpha}(P, Q)^k$, and $Q^k \bmod n$ theoretically can be evaluated with less than $(3+o(1)) \log_2 n$ multiplications [16], the practical application of the techniques in [16] is rather involved. For general Q, the fastest algorithm is given in [16]. Unfortunately, this requires special representation of k in terms of shortest addition chains. Brauer's Theorem [18] guarantees that asymptotically the number of multiplications in such shortest addition chains is $o(log(n))$, that is, it is vanishingly small compared to the number of squarings needed. This gives the *asymptotically* small running time of the Grantham test, but in practice, the required number of multiplications seems to be more like $4.5 \log_2(n)$.

For $\alpha = \alpha(P, Q)$, $\overline{\alpha} = \overline{\alpha}(P, Q)$, define the Lucas functions by $U_m(P, Q) = \frac{\alpha^m - \overline{\alpha}^m}{\alpha - \overline{\alpha}}$ and $V_m(P, Q) = \alpha^m + \overline{\alpha}^m$. It can be shown that these are always integers (see, e.g., [41]).

Thus, for the QF-based tests (with, as usual, $\epsilon(n) = -1$), the condition $\alpha^{(n+1)/2} \equiv \overline{\alpha}^{(n+1)/2} \bmod n$, is equivalent to the vanishing of $U_k(P,Q) \bmod n$ for $k = (n+1)/2$. This, in turn can easily be checked via the condition

$$DU_k(P,Q) = 2V_{k+1}(P,Q) - PV_k(P,Q) \tag{1}$$

by means of two V- values, which is much easier than evaluating the U- function.

Moreover, the computation of $V_k(P,Q)$ for $Q = 1$ is much easier and faster than for general Q. Thus, it is natural to ask, how easily the required $V_k(P,Q)$, $V_{k+1}(P,Q)$ can be computed via some shifted parameters (P',Q') with $Q' = 1$.

A transformation between $V_k(P,Q)$ and $V_{2k}(\hat{P},1)$ is given in [13]. Unfortunately this induces a shift of the degree from k to $2k$ and cannot be applied in our scenario, which requires $k = (n-\epsilon(n))/2 = (n+1)/2$ to remain odd.

As in our case Q is a square, we apply the following well-known identities,

$$V_k(ca,a^2) = a^k V_k(c,1), \quad aU_k(ca,a^2) = a^k U_k(c,1). \tag{2}$$

Hence, if $\alpha(P/a,1)^{(n+1)/2} \equiv \overline{\alpha}(P/a,1)^{(n+1)/2} \equiv \pm 1 \bmod n$ and $a^2 \equiv Q \bmod n$, then also $\alpha(P,Q)^{(n+1)/2} \equiv \overline{\alpha}(P,Q)^{(n+1)/2} \equiv \pm a^{(n+1)/2} \bmod n$.

Our *main goal* is a method for the separate computation of a root a of Q modulo n and for the evaluation of $\alpha(P/a,1)^k$, which in total is faster than the evaluation of $\alpha(P,Q)^k$, and which also induces a smaller failure rate. In detail, for the former,

- Find a practical root-finding algorithm that returns the root a of Q, $\left(\frac{Q}{n}\right) = 1$ for n prime, but with high probability discloses n as composite, otherwise.
- If the value a returned is a correct root of Q modulo n, then this should impose restrictive pseudoprimality conditions on n.

Remark 1. 1. If a is indeed a correct root of $Q \bmod n$, then the QF- part of the proposed test implies $\alpha(P,Q)^{(n+1)/2} \equiv \overline{\alpha}(P,Q)^{(n+1)/2} \equiv \pm a^{(n+1)/2} \bmod n$. If the root-finding algorithm imposes the condition $a^{(n-1)/2} \equiv \pm 1 \bmod n$ on n, then the above quantity is congruent to $\pm a \bmod n$ (see Proposition 1) and in that case n is also *spsp*(Q).

2. This shows why the case $n \equiv 3 \bmod 4$ in [28] is easier to deal with. Not only can the root be efficiently computed via $Q^{(n+1)/4} \bmod n$, but also, even when n is composite, this implies that $a^{(n-1)/2} \equiv \pm 1 \bmod n$.
3. While the root-finding algorithms for $n \equiv 1 \bmod 4$ are more expensive, they will be used in a way so as to induce some additional testing conditions.

2.3 Square Roots Modulo n and Conditions on the Pseudoprimes

The case that $\boxed{n \equiv 1 \bmod 4}$.

Let $n = 2^r s + 1$, with s odd, and call r the order of n. Suppose $\left(\frac{u}{n}\right) = -1$ and $(u^s)^{2^{r-1}} \equiv -1 \bmod n$. Then the 2-Sylow subgroup S_r of $\mathbb{Z}_n^*$ is cyclic of order 2^r. Shanks' root-finding algorithm [38] is based on the relation $a^2 \equiv bQ \bmod n$

for some b in some S_k. When n is prime, there exist new k, b, a such that this condition still holds and the index k decreases. Subsequently b gets pushed down into smaller subgroups of S_k until finally $b \in S_0 = \{1\}$, and the solution is found.

Note that the algorithm hinges on the existence of some u as above. But that criterion is not limited to n being prime. Modulo n, that condition on the u will either fail, or often the result of the algorithm will not be a root of Q. Indeed, the algorithm of Shanks not only efficiently performs Step 2 of the proposed test, but also works as an efficient probable prime test (see also [29]).

// *Detailed Description of Step 2 of the Proposed Test.*
INPUT: $n = 2^r s + 1$, $2 \nmid s$, $\left(\frac{Q}{n}\right) = 1$.
OUTPUT: a, a square root of $Q \bmod n$, or 'n is composite'.

1. (Precomputation)
   Choose randomly $u \in \mathbb{Z}_n^*$ with $\left(\frac{u}{n}\right) = -1$. Let $z \leftarrow u^s \bmod n$. If not $z^{2^{r-1}} \equiv -1 \bmod n$, declare n to be composite.
2. (Initialization)
   Let $k \leftarrow r - 1$, $t \leftarrow Q^{(s-1)/2} \bmod n$, $a \leftarrow Qt \bmod n$, $b \leftarrow at \bmod n$.
3. (Body of the Algorithm)
   While $b \not\equiv 1 \bmod n$ (*)
   $m \leftarrow 1$, $B \leftarrow b$, *found* $\leftarrow$ false;
   While $m < k$ and *found* $=$ false (**)
   if $B = 1$ then OUTPUT $g \leftarrow \gcd(B_0 - B, n)$;
   // proper factor of n found
   if $B = -1$ then *found* $\leftarrow$ true;
   else $m \leftarrow m + 1$, $B_0 \leftarrow B$, $B \leftarrow B^2 \bmod n$.
   If *found* $=$ false then OUTPUT 'n is composite'.
   // otherwise we have $B \equiv b^{2^{m-1}} \equiv -1 \bmod n$
   Update $t \leftarrow z^{2^{k-m-1}}$, $z \leftarrow t^2$, $b \leftarrow bz$, $a \leftarrow at \bmod n$, $k \leftarrow m$.
4. OUTPUT $\pm a \bmod n$.

The algorithm always returns a root of Q when n is prime. This also holds for $n \equiv 3 \bmod 4$. Note the more restrictive condition (**), $b^{2^{m-1}} \equiv -1 \bmod n$ for $m \geq 1$, as opposed to the original one by Shanks, $b^{2^m} \equiv 1 \bmod n$. This introduces an additional pseudoprimality testing condition.

Lemma 2. *If a composite n passes the precomputation, then n is spsp(u). If the original b is congruent to 1 modulo n, or if n fulfills condition (**) at least for the first loop (*), then n is spsp(Q) and $a^{(n-1)/2} \equiv \pm 1 \bmod n$.*

Moreover, n passes at most $r-1$ iterations of the loop (*), where $r = \nu_2(n-1)$. Additionally, for $k \geq 2$ and random input Q, n passes k iterations of (*) with probability at most $1/3^k$.

Proof. The first assertions are obvious. Now suppose $n - 1$ is at least divisible by 2^3 and that n enters the loop (*) at least twice.

Note that after each iteration (*) the relation $a^2 \equiv Qb \bmod n$ holds. Once $b \equiv 1 \bmod n$, the desired solution is found.

From the previous iteration we have $b^{2^{m-1}} \equiv -1 \bmod n$. Let $h = z^{2^{k-m}}$. From the latter condition and the fact that $u^{(n-1)/2} \equiv -1 \bmod n$ it follows exactly as when n is a prime, that h has order 2^m modulo n.

If firstly $h^{2^{m-1}} \equiv -1 \bmod n$ (what would happen if n were prime), then $hb = t^2 b$ has order dividing 2^{m-1} and the new b enters the next loop. But this means that each new b has an order which is at least by one factor in 2 smaller than the previous b. This explains the condition that each new m has to be less than k (which was the previous m). Equivalently, the sequence of the k in the loop are strictly decreasing, so that altogether there are less than r iterations of (*) (unless n is already previously disclosed as composite).

On the other hand, if $h^{2^{m-1}} \not\equiv -1 \bmod n$, but $b^{2^{m-1}} \equiv -1 \bmod n$, then $(hb)^{2^{m-1}} \not\equiv 1 \bmod n$ and (hb) (which is the new b) has order 2^m, as does the previous b. In this case, the new b does not fulfill (**).

It follows from above that unless the algorithm already terminated, we have $b^{2^M} \equiv 1 \bmod n$ for some M. If $M = 0$, we are done. Otherwise, we are seeking the smallest m with $b^{2^{m-1}} \equiv -1 \bmod n$, when $b \neq 1$, i.e., when $m \geq 1$. In that case, in analogy to the Miller-Rabin test, the first such power of b before 1 has to be -1. When we first arrive at 1, without encountering -1, n is immediately disclosed as composite, and the gcd above obviously yields a proper factor of n. In exactly such a case the algorithm terminates at a point where it would not if n were prime. Thus, the above algorithm terminates much faster for composites. Precisely, it terminates for each case where $b^{2^m} \equiv 1 \bmod n$, but $b^{2^{m-1}} \equiv 1 \bmod p$ for one prime p dividing n, and $b^{2^{m-1}} \equiv -1 \bmod q$ for another prime $q|n$. It does not terminate when $b^{2^{m-1}} \equiv -1 \bmod p$ for all $p|n$. If n is the product of two primes, the latter only happens in one out of three cases, while if n has more factors, the probability not to terminate is even smaller. Thus, in at most 1 out of 3 cases each additional iteration of (*) does not terminate. The desired assertion follows from the hypothesis that the Q are randomly chosen (subject only to the condition $\left(\frac{Q}{n}\right) = -1$), which implies that all the b values are random. □

For the special case $\boxed{n \equiv 5 \bmod 8}$ the above can be achieved even simpler.

// *Alternative Case of Step 2 of the Proposed Test.*

1. Select randomly $d \in \mathbb{Z}_n^*$.
   If n is not $spsp(2d^2)$, declare n to be composite.
2. Let $z \leftarrow (2d^2Q)^{(n-5)/8} \bmod n$ and $i \leftarrow z^2 \cdot 2d^2Q \bmod n$.
3. If not $i^2 \equiv -1 \bmod n$, declare n to be composite, otherwise $a \equiv zdQ(i-1) \bmod n$ is a square root of Q modulo n.

When n is known to be prime, this always gives is a square-root of Q via *one* exponentiation only (then clearly the first step can be omitted).

Lemma 3. *If a composite $n \equiv 5 \bmod 8$ passes the above algorithm, then a and i are correct roots of Q and $-1 \bmod n$, respectively. Moreover, n is $spsp(2d^2)$, as well as $spsp(2d^2Q)$. As a consequence, n is also $epsp(Q)$.*

Proof. This follows since any *epsp*(a) for $\left(\frac{a}{n}\right) = -1$ is already *spsp*(a). □

Remark 2. For $d = 1$ the above algorithm was proposed by Atkin [5], and actually constitutes a *deterministic* root-finding method for primes $n \equiv 5 \bmod 8$.

Step 1 is necessary to have n *epsp*(Q), which will be required below. We incorporate the random value d to minimize the failure probability by means of Miller-Rabin with respect to the *random* base $2d^2$.

Corollary 1. *Suppose a composite integer n passes the proposed test. Then, in the case of the square root finding algorithm for $n \equiv 5 \bmod 8$, this implies $\alpha(P,Q)^{(n+1)/2} \equiv \overline{\alpha}(P,Q)^{(n+1)/2} \equiv \pm a^{(n+1)/2} \bmod n$, and in the case of the Shanks-based root finding algorithm, the latter value is congruent to $\pm a \bmod n$. In both cases, $\alpha(P,Q)^{n+1} \equiv \overline{\alpha}(P,Q)^{n+1} \equiv Q \bmod n$.*

Proof. The first part follows from above. Note that if n passes the root-finding algorithm then it is *epsp*(Q). But if n is *epsp*(Q) and *elpsp*(P,Q), then by well-known results [24], this implies, $\alpha(P,Q)^{n+1} \equiv \overline{\alpha}(P,Q)^{n+1} \equiv Q \bmod n$. □

3 Performance

3.1 Evaluation of the QF-Based Part

By property (1), the QF-part can be evaluated via the V- functions only. Using the identities, $V_{2k}(P,1) = V_k(P,1)^2 - 2$ and $V_{2k+1}(P,1) = V_k(P,1)V_{k+1}(P,1) - P$, this can be done via a simple powering ladder analogously as for exponentiation.

The algorithm in [34] can easily be modified to obtain two consecutive V-values, as required. The operations are done modulo n.

```
INPUT: m = ∑_{j=0}^{l} b_j 2^j, the binary representation of m, and P.
OUTPUT: The pair V_m(P,1) and V_{m+1}(P,1).

1. (Initialization) Set d_1 ← P, d_2 ← P^2 − 2.
2. (Iterate on j) For j from l − 1 down to 1 do
        If b_j = 1, set d_1 ← d_1 d_2 − P, d_2 ← d_2^2 − 2.
        If b_j = 0, set d_2 ← d_1 d_2 − P, d_1 ← d_1^2 − 2.
3. (Evaluate) Let w_1 ← d_1 d_2 − P, w_2 ← d_1^2 − 2.
   If b_0 = 1 return (w_1, P w_1 − w_2), else return (w_2, w_1).
```

Thus, the pair $V_{(n+1)/2}(P,1)$, $V_{(n+1)/2+1}(P,1)$ may be computed modulo n using fewer than $2\log_2(n)$ multiplications mod n and $\log_2 n$ additions mod n. Half of the multiplications mod n are squarings mod n.

// Detailed Description of Step 3 of the Proposed Test.

- Let $k = (n+1)/2$ and evaluate $(V_k(P',1), V_{k+1}(P',1))$ modulo n.
- Test, if $2V_{k+1}(P',1) \equiv P'V_k(P',1) \bmod n$. If not, declare n to be composite.
- Compute $\gcd(V_k(P',1) \pm 2, n)$. If this reveals a factor of n, output the factor. Otherwise declare n to be a probable prime.

3.2 Runtime-Analysis

In [16], J. Grantham suggested a unit measure for a probable prime test based on the running time of the Miller-Rabin test. An algorithm with input n is said to have running time of k *selfridges* if it can be computed in $(k + o(1)) \log_2 n$ multiplications mod n. For simplicity, squarings are counted as multiplications.

As exponentiation to the tth power can be done in $(1 + o(1)) \log_2 t$ multiplications by using easily constructed addition chains [18], the Miller-Rabin test has running time of at most 1 selfridge.

Theorem 2. – *For random input Q and u, the proposed test, via the general root finding algorithm, has average running time* 4 *selfridges.*
– *For the $n \equiv 5 \bmod 8$ based root finding algorithm, the proposed test always has running time less than* 4 *selfridges.*

Proof. By the above, Step 3 of the proposed test requires at most two selfridges.

The Atkin-based method always requires two exponentiations, so we only need to consider the general Shanks-based root finding algorithm. Precomputation and initialization require one exponentiation each. It follows from [21] that the number of multiplications averaged over primes $n \equiv 1 \bmod 4$ is $o(1) \log_2(n)$. The additional squarings that we require in Step 1 for the Miller-Rabin test base u can be comprised in the $o(1) \log_2(n)$ multiplications above.

We upper bound the number of multiplications in the worst case required by the loop (*). If n is $spsp(u)$ then $z = u^s$ generates S_r, the 2-Sylow subgroup of $\mathbb{Z}_n^*$. So S_r has order 2^r, S_{r-1} has order 2^{r-1}, is generated by z^2, and in general, S_{r-i} has order 2^{r-i} and is generated by z^{2^i} for $i = 0, 1, ..., r$.

The condition (**) indicates in which of the 2-subgroups b is in. Alternatively, we can consider the values that k takes in the algorithm, which also (except for the first k), specifies the subgroup where b is in. Namely, $b \in S_k \setminus S_{k-1}$.

Recall that the sequences of the k-values have to be strictly decreasing. E.g., for order $r = 4$, the possible k-sequences are, $(4,1)$, $(4,2)$, $(4,2,1)$, $(4,3)$, $(4,3,1)$, $(4,3,2)$, $(4,3,2,1)$. Generally there are 2^{r-1} such k-sequences.

For random Q's and u's the values b are random as well and it can be shown (see [21, p. 235]) that every k-sequence has the same probability. Lindhurst determined the total number of multiplications C_r over all the possible k-sequences and then divided by the number of sequences, 2^{r-1}, to get the average. Then, the average number of multiplications (after the initialization), is $C_r/2^{r-1} = (r^2 + 7r - 12)/4 + 1/2^{r-1}$ (see [21, p. 236]).

Although all sequences are equally like, they can be grouped into those with the same length. The 2^{r-1} sequences of order r are obtained by fixing the r as first value of the sequence, and by determining the $\binom{r-1}{1}, \binom{r-1}{2}, \ldots, \binom{r-1}{r-1}$ subsequences $(k_1, .., k_l)$ of respective lengths $1, 2, \ldots, r-1$. This shows that an average sequence is expected to have length about $r/2$. Equivalently, on average, the loop (*) is iterated $r/2$ times. Additionally, for $r \geq 8$, more than 99 % of all sequences have length between $\lfloor \frac{r}{4} \rfloor$ and $\lceil \frac{3r}{4} \rceil$.

But then Lemma 2 and Lemma 6 below implies that on average at most $\frac{(r^2+7r-12)/4+1/2^{r-1}}{3^{\lfloor r/4 \rfloor}}$ multiplications are to be expected before the algorithm terminates, when n is composite. Comparing numerator and denominator, we see that this is much less than for primes, as repeated iterations of (*) are much less likely. □

3.3 The Iterated Test

After the first round it is more efficient to shorten each of the following iterations, instead of re-running the entire procedure. For entire iteration, we would achieve a failure probability of about $1/2^{20k}$ and $4k$ selfridges for k rounds.

Below, the failure rate of the QF-based part will be shown to be much smaller than the one based on the root finding algorithm. Yet, each of those parts requires about two selfridges. When being iterated, it is more efficient to repeat only a part of the root finding algorithm, whilst obtaining the full QF-part. In fact, for both of the above root finding algorithms, the first step is only required at the first round. This motivates the following shortened version of the proposed test for any iterations after the first.

// Iterations After the First Round of the Test.

1. (Parameter Selection) As above.
2. (Square Root Part)
   - Let u and d, accordingly, be the values of the first round of the proposed test in Step 1 of the root finding part.
   - Run one of the above root finding algorithms by skipping the corresponding Step 1.
     If the algorithms declares n composite, stop.
   - Let a and P' be as above.
3. (QF-Based Part) As above.

4 The Probability Estimate

The proof of Theorem 1 will be given in a sequence of auxiliary results. The general idea is to determine an upper bound on the number of the liars (i.e., pairs that pass) and to upper bound the ratio of these to the number of all pairs possible as input to the test. It was shown in [16] that for n an odd composite, not a perfect square, the number of pairs (P, Q), such that $\left(\frac{P^2-4Q}{n}\right) = -1$ and $\left(\frac{Q}{n}\right) = 1$, $1 < \gcd(P^2 - 4Q, n) < n$, or $1 < \gcd(Q, n) < n$, is more than $n^2/4$.

4.1 The QF-Based Part

Underlying all the pseudoprimality tests based on quadratic fields is the investigation of the powers of the characteristic roots α, $\overline{\alpha}$. It is well known that if n is any integer with $\gcd(Q, n) = 1$ then there is a positive integer m such that $\alpha(P,Q)^m \equiv \overline{\alpha}(P,Q)^m \bmod n$. Let $\rho = \rho(n,P,Q)$ be the least such positive integer. This is usually called the *rank of appearance* (apparition) [36,41].

The rank of appearance has the following properties (see [11,36,39]).

$$\alpha(P,Q)^m \equiv \overline{\alpha}(P,Q)^m \bmod k \text{ if and only if } \rho(k,P,Q)|m, \tag{3}$$

$$\rho(p,P,Q) \mid p - \epsilon(p), \text{ and } \rho(p,P,Q) \mid (p-\epsilon(p))/2 \text{ iff } (Q/p) = 1, \tag{4}$$

$$\rho(lcm(m_1,...,m_k)) = lcm(\rho(m_1),...,\rho(m_k)), \tag{5}$$

$$\text{If } p^c || \ \alpha(P,Q)^{\rho(p,P,Q)} - \overline{\alpha}(P,Q)^{\rho(p,P,Q)} \text{ then } \rho(p^e,P,Q) = p^{\max(e,c)-c}\rho(p,P,Q). \tag{6}$$

A necessary condition for the test to pass is $\alpha(P,Q)^{n+1} \equiv \overline{\alpha}(P,Q)^{n+1} \bmod n$. Since $p \not| \, n+1$ for $p|n$, we need not consider the pairs (P,Q) modulo p^α whose rank is a multiple of p (compare (6)). Thus, it suffices to investigate the parameters whose rank is an odd divisor of $p - \epsilon(p)$, since $(n+1)/2$ is odd for $n \equiv 1 \bmod 4$.

Given n, the task is to count the number of the liars (P,Q), which is determined by the rank of appearance of each of these pairs. But this requires knowledge of the individual quadratic residue symbols $\left(\frac{Q}{p_i}\right)$ and $\epsilon(p_i) = \left(\frac{P^2-4Q}{p_i}\right)$ for all primes $p_i|n$.

Generally, these values are not known for the number n to be tested for primality. However, certain conditions on these symbols are automatically satisfied when a composite n indeed passes the test. Specifically, by Lemma 1 it suffices to consider the case that $\left(\frac{Q}{p}\right) = 1$ for any prime p dividing n. We separately consider the values $\epsilon(p_i)$.

Definition 1. *Let* $n = \prod_{i=1}^{\omega} p_i^{\alpha_i}$*, where* $\omega = \omega(n)$ *is the number of different prime factors of* n*. For* $1 \le i \le \omega$ *let* $\epsilon = \epsilon(p_i) \in \{1,-1\}$*, and call* $(\epsilon) = (\epsilon(p_1),...,\epsilon(p_\omega))$ *the signature modulo* n *with respect to* P *and* Q*, when* $\left(\frac{P^2-4Q}{p_i}\right) = \left(\frac{D}{p_i}\right) = \epsilon(p_i)$ *for all* i*. Similarly, we call each* $\epsilon(p_i)$ *the signature modulo* $p_i|n$*, and* $\epsilon(p)$ *the signature modulo any prime* p*.*

Throughout, P is assumed to be different from 0, since otherwise the rank of appearance modulo n is always equal to 2. (This is no restriction as for $P = 0$ always $(D/n) = 1$ in our case.) Proposition 2 was proved in [28] and Proposition 3 was proved in [25].

Proposition 2. *Let* k*,* $p \not| k$*, be a positive integer and* $\epsilon \in \{-1,1\}$ *a constant. For a fixed value of* P_0*,* $P_0 \neq 0$*, the number of* $Q \bmod p^\alpha$ *such that* $\left(\frac{Q}{p}\right) = 1$*,* (P_0,Q) *has signature* $\epsilon \bmod p$*, and* $\alpha(P_0,Q)^k \equiv \overline{\alpha}(P_0,Q)^k \bmod p^\alpha$*, equals* $\frac{1}{2}\left(\gcd(k,\frac{p-\epsilon}{2}) - 2\right)$ *if* $2|k$ *and* $2|\frac{p-\epsilon}{2}$*, and* $\frac{1}{2}\left(\gcd(k,\frac{p-\epsilon}{2}) - 1\right)$*, otherwise.*

Proposition 3. *Let* k *be a positive integer with* $p \not| k$ *and* $\epsilon \in \{-1,1\}$ *a constant. For a fixed value of* Q_0*,* $\left(\frac{Q_0}{p}\right) = 1$*, the number of* $P \bmod p^\alpha$ *such that* (P,Q_0) *has signature* ϵ *and* $\alpha(P,Q_0)^k \equiv \overline{\alpha}(P,Q_0)^k \bmod p^\alpha$ *is,* $\frac{1}{2}\gcd(k,p-\epsilon) - 1$*, when* $\nu_2(k) \ge \nu_2(p-\epsilon)$*, and* $\gcd(k,\frac{p-\epsilon}{2}) - 1$*, otherwise.*

Corollary 2. *Let a signature* (ϵ) *be fixed. Then the number of pairs* (P, Q) *that fulfill* $\alpha(P,Q)^{(n+1)/2} \equiv \overline{\alpha}(P,Q)^{(n+1)/2} \bmod n$ *with respect to this signature is at most* $\frac{1}{2^\omega}\prod_{i=1}^{\omega}\left(\gcd(\frac{n+1}{2}, \frac{p_i-\epsilon(p_i)}{2}) - 1\right)^2 \cdot \prod_{i=1}^{\omega} p_i^{\alpha_i - 1}$.

Lemma 4. *Let* $n \equiv 1 \bmod 4$ *be an odd integer, not a perfect square. If* p_j *is a prime such that* p_j^2 *divides* n*, then* n *is slpsp*(P,Q) *for* $\left(\frac{D}{n}\right) = -1$ *with probability less than* $1/(8p_j)$.

Proof. Let (ϵ) be a fixed signature. By Corollary 2, the number of liars (P, Q) with respect to this signature is at most $(1/2^{3\omega})\prod_{i=1}^{\omega}(p_i - 1)^2 \cdot \prod_{i=1}^{\omega} p_i^{\alpha_i-1}$.

If $\omega = 2$, then there are two possible signatures with $\left(\frac{D}{n}\right) = -1$ and so the number of the liars is at most $(1/2^5)\cdot\prod_{i=1}^{\omega}(p_i-1)^2\cdot\prod_{i=1}^{\omega} p_i^{\alpha_i-1}$. This gives a failure probability of less than $(1/2^3)\cdot\prod_{i=1}^{\omega} p_i^{\alpha_i+1}/\prod_{i=1}^{\omega} p_i^{2\alpha_i} = 1/(2^3\cdot\prod_{i=1}^{\omega} p_i^{\alpha_i-1}) \leq 1/(8p_j)$. For $\omega \geq 3$ there are always less than 2^ω different signatures with $\left(\frac{D}{n}\right) = -1$ and the number of liars is less then $(1/2^{2\omega})\cdot\prod_{i=1}^{\omega} p_i^{\alpha_i+1}$ which gives a probability of at most $1/(2^{2\omega-2}p_j) \leq 1/(2^4 p_j)$. Finally, if $\omega = 1$, so that $n = p_j^{\alpha_j}$, then necessarily $\alpha_j > 2$ by hypothesis and the probability in this case is at most $1/(2p_j^2)$. □

Typical for pseudoprimality testing based on the Fermat/QF-based combinations is the fact that $\left(\frac{D}{p}\right) = 1$ becomes rather unlikely for $p|n$ when $\left(\frac{D}{n}\right) = -1$.

Proposition 4. *The number of pairs* $(P, Q) \bmod n$ *for which a squarefree integer* n *with* ω *prime factors fulfills* $\alpha(P,Q)^{(n+1)/2} = \overline{\alpha}(P,Q)^{(n+1)/2} \bmod n$ *such that* $\left(\frac{P^2-4Q}{p_i}\right) = 1$ *for some* $p_i|n$*, is given as follows. It is less than* $\frac{5n\phi(n)}{2^6 B}$ *if* $\omega = 2$*, less than* $n\phi(n)\left(\frac{\omega}{2^{3\omega-2}B} + \frac{1}{2^{2\omega-4}B^2}\right)$ *if* $\omega \geq 4$ *is even, and less than* $\frac{n\phi(n)}{B^2}$ *if* ω *is odd.*

Proof. See the proof to Proposition 5 in [28], where exactly the number of such pairs is being established. □

Remark 3. For $\omega = 2$ the proof in [28] shows that the above quantities are only obtained for strongest divisor properties, like $\mathrm{odd}(p_i+1)|n+1$ for one $p_i|n$, and $\mathrm{odd}(p_j+1)|t(n+1)$ for $t = 3$ and the other $p_j|n$. Otherwise, the results would be much smaller.

When the test passes for some fixed $Q = Q_0$, then we have for each parameter P, $\alpha^{(n+1)/2} \equiv \overline{\alpha}^{(n+1)/2} \bmod n$, and this is either equivalent to $a^{(n+1)/2} \bmod n$, or to $-a^{(n+1)/2} \bmod n$, where a is independent of P, and by the root finding algorithms is uniquely determined by the Q_0. For all P that pass, this determines a specific general 'multiplier' $S \equiv a^{(n+1)/2}$, resp. $S \equiv -a^{(n+1)/2}$ modulo n. The proof to the next result is analogous to Lemma 5, [26] (see Proposition 4, [28]).

Lemma 5. *Let $n \equiv 1 \bmod 4$ be any composite integer, and $Q = Q_0$, as well as some 'multiplier' S be fixed. If p is any prime dividing n, then there are at most $\frac{1}{2}\left(\gcd(\frac{n+1}{2}, p-\epsilon(p)) - 1\right)$ elements P with $\left(\frac{P^2-4Q_0}{p}\right) = \epsilon(p)$ for which $\alpha(P,Q)^{\frac{n+1}{2}} \equiv \overline{\alpha}(P,Q)^{\frac{n+1}{2}} \equiv S \bmod p$.*

Corollary 3. *For a squarefree $n \equiv 1 \bmod 4$ let (ϵ) be a fixed signature. Then the number of pairs (P,Q) with $\alpha(P,Q)^{(n+1)/2} \equiv \overline{\alpha}(P,Q)^{(n+1)/2} \equiv \pm a^{(n+1)/2} \bmod n$ w.r.t. this signature is at most $\frac{1}{2^{2\omega-1}} \prod_{i=1}^{\omega} (\gcd(\frac{n+1}{2}, \frac{p_i-\epsilon(p_i)}{2}) - 1)^2$.*

Remark 4. It is essential that $n \equiv 1 \bmod 4$ to have $(n+1)/2$ odd. For $n \equiv 3 \bmod 4$ analogous, but more involved results can be obtained, [28].

This gives the error rate for each iteration of the test (after the first round).

Theorem 3. *Let P and Q be randomly chosen in Step 1 of the proposed test. Let $n \equiv 1 \bmod 4$ be a composite integer which is not a perfect square and not divisible by primes up to B. Then the probability that n fulfills $\alpha(P,Q)^{(n+1)/2} \equiv \overline{\alpha}(P,Q)^{(n+1)/2} \equiv \pm a^{(n+1)/2} \bmod n$ for $a^2 \equiv Q \bmod n$, is given as follows.*

- *If n is not a product of exactly three prime factors, it is less than $1/2^{17} + 4/B^2 < 1/131040$.*
- *If n is the product of three different primes, and if n is further epsp(Q), then it is less then $4/B^2 + 3(B^2+1)/2(B^4 - 3B^2)$.*

Proof. When n is not squarefree, Lemma 4 gives the result. If a squarefree n has an even number of prime factors we apply Proposition 4, where the probability becomes largest for $\omega = 2$ in which case it is less than $5/(2^4 B) < 1/160000$.

Further, if $n = p_1p_2p_3$ is squarefree and has exactly 3 prime factors, we can use Lemma 2.11 of [16]. In this Lemma, Grantham separately considers the cases, $\left(\frac{P^2-4Q}{p_i}\right) = 1$ for some i, and $\left(\frac{P^2-4Q}{p_i}\right) = -1$ for all i. By Proposition 4 (which corresponds to Lemma 2.9 of [16] when ω is odd), the former case yields a probability of $4/B^2$. In the latter case, necessarily $\alpha^{n+1} \equiv Q \bmod p_i$ and $\alpha^{p_i+1} \equiv Q \bmod p_i$ so that $\alpha^{n-p_i} \equiv 1 \bmod p_i$, since n is *epsp*(Q) by hypothesis (see Corollary 1). This congruence holds for exactly $\gcd(n - p_i, p_i^2 - 1)$ elements. Since n has only three factors, these gcd$'s$ cannot all be equal to its maximal value, $p_i^2 - 1$. Indeed, Grantham gives an upper limit for these quantities. From this, he obtains the probability for such pairs which pass the test. By adding both cases, the probability can be bounded by $4/B^2 + 3(B^2+1)/2(B^4 - 3B^2)$.

It remains to consider the case where n is squarefree and divisible by an odd number ω of at least 5 prime factors. The number of pairs with $\left(\frac{D}{p}\right) = 1$ for at least one $p|n$ is again by Proposition 4 less than n^2/B^2. So it suffices to consider the pairs with $\left(\frac{D}{p}\right) = -1$ for all primes $p|n$. In this case the number of pairs is by Corollary 3 at most $(1/2^{4\omega-1}) \prod (p_i - 1)^2$. When adding these two cases, the probability is upper bounded by $1/2^{17} + 4/B^2$ which is less than $1/131040$. □

4.2 The Square Root Finding Based Part

It is well-known that when a is taken randomly from $\mathbb{Z}_n^*$, the probability for a to be a Miller-Rabin liar is at most 1/4. If $\left(\frac{a}{n}\right)$ is fixed to some special value (e.g., -1), then in that case there are only $\phi(n)/2$ such a as possible input values to the Miller-Rabin test. Yet, even for fixed jacobi symbol, we show below that in our case the failure rate is smaller than the expected 2/4.

The following result can immediately be verified.

Proposition 5. *Let n be spsp(u), where $\left(\frac{u}{n}\right) = -1$, and let p be any prime divisor of n. Then, if $\left(\frac{u}{p}\right) = -1$, we have $\nu_2(p-1) = \nu_2(n-1)$, and if $\left(\frac{u}{p}\right) = 1$, we have $\nu_2(p-1) > \nu_2(n-1)$.*

Notation: Let $\nu(n)$ denote the largest integer such that $2^{\nu(n)}$ divides $p-1$ for each prime p dividing n. As above, write $n-1 = 2^r s$ with s odd.

Proposition 6. *Suppose n is spsp(a) for $\left(\frac{a}{n}\right) = -1$. Then $a \in \mathcal{S}_{-1}(n)$ where $\mathcal{S}_{-1}(n) = \{a \bmod n : a^{2^{\nu(n)-1}s} \equiv -1 \bmod n\}$. Moreover, we have $\#\mathcal{S}_{-1}(n) = 2^{(\nu(n)-1)\omega(n)} \prod_{p|n} \gcd(s, p-1)$.*

Proof. If $\left(\frac{a}{n}\right) = -1$ then there exists $p|n$ with $\left(\frac{a}{p}\right) = -1$ and by Proposition 5, $\nu_2(p-1) = r = \nu(n)$. Moreover, in that case, $\nu_2(p-1) = \nu_2(\mathrm{ord}_p(a))$. By a standard result for n being *spsp*(a) (see e.g., [2]), we also have $\nu_2(\mathrm{ord}_p(a)) = \nu_2(\mathrm{ord}_q(a))$, so that $\nu_2(\mathrm{ord}_q(a)) = \nu(n)$ for any $q|n$. In particular, if $a^{2^i s} \equiv -1 \bmod n$ for some $0 \le i \le r-1$ (the first case for n being *spsp*(a)), then $a^{2^{\nu(n)-1}s} \equiv -1 \bmod q$ for any $q|n$. Note also that the case $a^s \equiv 1 \bmod n$ (the second case for n being *spsp*(a)), is impossible, since $a^{(n-1)/2} \equiv -1 \bmod n$ by hypothesis.

The cardinality $\#\mathcal{S}_{-1}(n)$ follows from [13, p. 128]. □

Lemma 6. *Suppose an odd composite integer n, not a perfect square, is not the product of exactly three prime factors. Let $a \in \mathbb{Z}_n^*$ be chosen randomly from the set of all b with $\left(\frac{b}{n}\right) = -1$. Then the probability that n is spsp(a) is given as follows. If $n = p_1 p_2$ where $p_1 = 2^k t + 1$ and $p_2 = 2^{k+1} t + 1$, $2 \nmid t$, it is at most 1/4. Otherwise, it is at most 1/8.*

Proof. We follow the proof of Lemma 3.4.8. in [13]. Then the desired probability can be determined via

$$\frac{\phi(n)}{2\#\mathcal{S}_{-1}(n)} = \frac{1}{2} \prod_{p^\alpha || n} p^{\alpha-1} \frac{p-1}{2^{\nu(n)-1} \gcd(s, p-1)}.$$

Note that each factor $(p-1)/(2^{\nu(n)-1} \gcd(s, p-1))$ is an even integer. Then, if $\omega(n) \ge 4$, we have $\phi(n)/(2\#\mathcal{S}_{-1}(n)) \ge 1/2 \cdot (2^4) = 8$.

If $\omega(n) = 2$, we distinguish the following cases. Suppose $2^{\nu(n)+2}|p-1$ for one $p|n$. Then $2^{\nu(n)-1}\gcd(s, p-1) \le (p-1)/8$ and therefore $\phi(n)/(2\#\mathcal{S}_{-1}(n)) \ge 1/2 \cdot (2 \cdot 8) = 8$.

Now, let $2^{\nu(n)+\delta}|p-1$ for one $p|n$, where δ equals 0 or 1. Write the two primes in the form $p_1 = 2^{\nu(n)+\delta}t_1 + 1$ and $p_2 = 2^{\nu(n)}t_2 + 1$.

For the case that $t_1 \ne t_2$, Arnault [4, p. 877] showed that $t_1|s$ and $t_2|s$ is simultaneously impossible. This means that for at least one p_i, $\gcd(s, p_i - 1) \le t_i/3$. If $\delta = 0$, then $\phi(n)/(2\#\mathcal{S}_{-1}(n)) \ge 1/2 \cdot (2 \cdot 6) = 6$, while if $\delta = 1$, this introduces an additional factor of 2, and $\phi(n)/(2\#\mathcal{S}_{-1}(n)) \ge 12$.

The result of the Lemma follows, since for $\left(\frac{Q}{n}\right) = -1$ and $\omega(n) = 2$ there is one prime factor p_i with $\left(\frac{Q}{p_i}\right) = 1$, so that by Proposition 5, $\nu_2(p_i - 1) > \nu_2(n-1) = \nu_2(p_j - 1) = \nu(n)$. This means we do have $\delta = 1$, as required.

Finally, the special case $p_1 = 2^k t + 1$ and $p_2 = 2^{k+1}t + 1$ implies $t|s$, in which case $\phi(n)/(2\#\mathcal{S}_j(n)) \ge 1/2 \cdot (2 \cdot 4)$, since $2^{\nu(n)+1}|p_2 - 1$. □

4.3 Proof of the Main Result

Proof of Theorem 1. Suppose firstly that n is a product of three different prime factors. Then Theorem 3 and Lemma 2, respectively Lemma 3, give the result.

For the Atkin-based root finding method Lemma 3 asserts that n is *spsp*$(2d^2)$. Since $\left(\frac{2d^2}{n}\right) = -1$ as $\left(\frac{2}{n}\right) = -1$ for $n \equiv 5 \bmod 8$, we can apply Lemma 6. By assumption, d is chosen randomly in the square root finding algorithm. For random selection of this basis, the condition on n to be *spsp*$(2d^2)$ is independent of the QF-based test.

If n is not such a special two-factor integer as described in Lemma 6, this introduces a factor of 1/8 (for each random d) in addition to the failure probability obtained above in Theorem 3 for the test that checks the QF-condition.

If n passes the Shanks-based method, it firstly is *spsp*(u) for u with $\left(\frac{u}{n}\right) = -1$. For randomly chosen u this again introduces a factor of 1/8 in the failure probability.

Finally, for both types of the root finding algorithms, if n does have the special two-factor form, then it follows easily that Proposition 4 introduces a much smaller failure rate than above (the corresponding number of the QF-liars, which is based on the quantities $\gcd(n+1, p \pm 1)$, becomes much smaller when the odd part of $p-1$ divides $n-1$). In total, for $\omega = 2$ the largest failure rate applies to the general type of two factor numbers.

Thus, we have the failure rate, for the first round, $F_1 = 1/2^{20} + 1/(2B^2)$, and for $k-1$ additional iterations, $F_1 \cdot (1/2^{17} + 4/B^2)^{k-1}$. For larger k the B proportion is negligible, so that for a total of k rounds we have failure approximately $1/(2^{20} \cdot 2^{17(k-1)}) = 1/2^{17k+3}$. □

5 Open Problems and Further Remarks

While with the much smaller failure rate of 1/1048350, our test has running time 4 times that of the Miller-Rabin test. We do not know how effectively the failure rate still can be reduced, when allowing more time for evaluation (for each round). On the other hand, the question is, how to optimally tackle the tradeoff between the reliability and the running time, and what the limits for a test with much larger running time are, so that it practically still makes sense.

Strong pseudoprimes with respect to at least 4 random bases exist very often. Below $n = 1000$ there are 54 such composites with at least four non-trivial bases as liars. Our test, without trial division and, for simplicity $d = 1$ (see Lemma 3) for $n \equiv 5 \bmod 8$, would for any possible pairs of parameters detect these.

Similarly, it is extremely easy to construct strong pseudoprimes with respect to at least 4^2, 4^3, ..., random bases. We do not know, computationally, how much more effort is required for the generation of pseudoprimes for the iterated proposed test (say, for the $n \equiv 5 \bmod 8$ algorithm with $d = 1$). Here, the typical Fermat/Lucas restrictions come into play and considerably limits the effectiveness of the Fermat- based generation methods for pseudoprimes.

On the other hand, sometimes it seems that many repeated iterations would not be necessary, if the input parameters have certain advantageous values. For the Miller-Rabin test, it is known that the bases $2, 3, 5, 7$ seem to work better, as they are primitive roots for most primes.

Even more effectively, the special choice of the parameters in the Baillie-PSW test essentially improves on its reliability.

If the proposed tests were run for one pair of parameters only, it is not known to what extent, and for which parameters it is most reliable.

Note added in proof: I. Damgård and G. Frandsen recently established a QF-based test with average case error estimates [15].

Acknowledgements. I am grateful to the following individuals, who were most helpful to me in writing this paper. In alphabetical order, they are: W. Bosma, H. Dobbertin, J. Grantham, A. Lenstra, W.B. Müller, and H.C. Williams. I would also like to thank the anonymous referees for their insightful remarks.

References

1. Adams, W., Shanks, D., Strong primality tests that are not sufficient. *Math. Comp.* **39**, 255-300 (1982).
2. Alford, W.R., Granville, A., Pomerance, C., On the difficulty of finding reliable witnesses. *Algorithmic Number Theory*, LNCS **877**, 1-16 (1994).
3. Arnault, F., Rabin-Miller primality test: Composite numbers which pass it. *Math. Comp.* **64**, no. 209, 355 - 361 (1995).
4. Arnault, F., The Rabin-Monier theorem for Lucas pseudoprimes. *Math. Comp.* **66**, 869 - 881 (1997).
5. Atkin, A.O.L., Probabilistic Primality Testing. *INRIA Res. Rep. 1779*, 159-163 (1992).

6. Atkin, A.O.L., Intelligent primality test offer. *Computational Perspectives on Number Theory* (D. A. Buell, J.T. Teitelbaum, eds.), Proceedings of a Conference in Honor of A.O.L. Atkin, International Press, 1-11 (1998).
7. Atkin, A.O.L., Morain, F., Elliptic curves and primality proving. *Math. Comp.* **61**, 29-68 (1993).
8. Baillie, R., Wagstaff, S.S., Lucas pseudoprimes. *Math. Comp.* **35**, 1391-1417 (1980).
9. Bleichenbacher, D., *Efficiency and Security of Cryptosystems based on Number Theory.* Dissertation ETH Zürich (1996).
10. Bosma, W., Van der Hulst, M.-P., Faster primality testing. *EUROCRYPT' 89*, LNCS **434**, 652-656 (1990).
11. Carmichael R.D., On sequences of integers defined by recurrence relations. *Quart. J. Pure Appl. Math.* **48**, 343-372 (1920).
12. Cohen, H., Lenstra H. W., Primality testing and Jacobi sums. *Math. Comp.* **42**, 297-330 (1984).
13. Crandall, R., Pomerance, C., Prime Numbers. A Computational Perspective. Springer-Verlag (2001).
14. Damgård, I., Landrock, P., Pomerance, C., Average case error estimates for the strong probable prime test. *Math. Comp.* **61**, no. 203, 177-194 (1993).
15. Damgård, I., Frandsen, G. S., An extended quadratic Frobenius primality test with average case error estimates. Draft, University of Aarhus, Denmark, August 31 (2001).
16. Grantham, J., A probable prime test with high confidence. *J. Number Theory* **72**, 32-47 (1998).
17. Jaeschke, G., On strong pseudoprimes to several bases. *Math. Comp.* **61**, 915-926, (1993).
18. Knuth, D., The Art of Computer Programming. Vol. 2/Seminumerical Algorithms. Addison-Wesley, 1997.
19. Kurtz G., Shanks, D., Williams, H.C., Fast primality tests for numbers less than $50 \cdot 10^9$. *Math. Comp.* **46**, 691-701 (1986).
20. A. K. Lenstra, H. W. Lenstra Jr., The Development of the Number Field Sieve. Springer-Verlag, Berlin, 1993.
21. Lindhurst, S., An analysis of Shank's Algorithm for computing square roots in finite fields. *CRM* Proceedings and Lecture Notes, Vol. 19, 231-242 (1999).
22. Menezes, A., van Oorschot, P.C., Vanstone, S., Handbook of Applied Cryptography. CRC (1997).
23. Montgomery, P., Evaluating recurrences of form $X_{m+n} = f(X_m, X_n, X_{m-n})$ via Lucas chains. Preprint.
24. More, W., The LD probable prime test. *Contemporary Mathematics*, **225**, 185-191 (1999).
25. Müller, S., On the combined Fermat/Lucas probable prime test. In: Walker, M. (ed.) *Cryptography and Coding*, LNCS **1746**, Springer - Verlag, 222-235 (1999).
26. Müller, S., On probable prime testing and the computation of square roots mod n. *Algorithmic Number Thory*, ANTS IV, Proceedings, Wieb Bosma (ed.), LNCS **1838**, 423-437 (2000).
27. Müller, S., On the rank of appearance and the number of zeros of the Lucas sequences over F_q. *Finite Fields and Applications*, H. Niederreiter, A. Enge (eds.), 390-408, Springer (2001).
28. Müller, S., A probable prime test with very high confidence for $n \equiv 3 \bmod 4$. Submitted.

29. Müller, S., On probable prime testing and Shanks' root finding algorithm. Preprint, University of Klagenfurt 2001.
30. Pinch, R. G. E., Some primality testing algorithms. Preprint (1993).
31. Pomerance, C., Are there counter-examples to the Baillie-PSW primality test? In: Lenstra, A.K. (ed.) Dopo Le Parole aangeboden aan Dr. A. K. Lenstra. Privately published Amsterdam (1984).
32. C. Pomerance, The number field sieve. *Proceedings of Symposia in Applied Mathematics,* Vol. **48**, pp. 465–480 (1994).
33. Pomerance, C., Selfridge, J. L., Wagstaff, S.S., Jr., The pseudoprimes to $25 \cdot 10^9$. *Math. Comp.* **35**, no. 151, 1003–1026 (1980).
34. Postl, H., Fast evaluation of Dickson Polynomials. *Contrib. to General Algebra* **6**, 223-225 (1988).
35. Riesel, H., Prime Numbers and Computer Methods for Factorization. Birkhäuser (1994).
36. Ribenboim, P., *The New Book of Prime Number Records.* Berlin, Springer (1996).
37. Shoup, V., Primality testing with fewer random bits. *Computational Complexity* **3**, 355–367 (1993).
38. Shanks, D., Five number-theoretic algorithms. *Proc. of the second Manitoba Conf. on numerical mathematics*, Thomas, R.S.D., Williams, H.C. (eds.), 51-70 (1972).
39. Somer, L., On Lucas d-Pseudoprimes. In: *Applications of Fibonacci Numbers, Volume 7*, Bergum G.E., Philippou, A.N., Horadam, A.F. (eds.), Kluwer, 369-375, (1998).
40. Williams, H.C., A $p+1$ method of factoring. *Math. Comp.* **39**, no. 159, 225–234 (1982).
41. Williams, H.C., *Éduard Lucas and Primality Testing.* John Wiley & Sons (1998).

Computation of Discrete Logarithms in $\mathbb{F}_{2^{607}}$

Emmanuel Thomé

Laboratoire d'Informatique (LIX)
École polytechnique
91128 Palaiseau Cedex
FRANCE
Emmanuel.Thome@polytechnique.fr

Abstract. We describe in this article how we have been able to extend the record for computations of discrete logarithms in characteristic 2 from the previous record over $\mathbb{F}_{2^{503}}$ to a newer mark of $\mathbb{F}_{2^{607}}$, using Coppersmith's algorithm. This has been made possible by several practical improvements to the algorithm. Although the computations have been carried out on fairly standard hardware, our opinion is that we are nearing the current limits of the manageable sizes for this algorithm, and that going substantially further will require deeper improvements to the method.

1 Introduction

Among the most common paradigms upon which public key cryptographic schemes rely are the difficulty of the factorization of large integers (for the RSA cryptosystem), and the difficulty of computing discrete logarithms in appropriate groups (for the Diffie-Hellman key exchange protocol [14], ElGamal cryptosystem [16], and ElGamal and Schnorr [38] signature schemes). Appropriate groups for discrete logarithm cryptosystems are multiplicative groups of finite fields, the group of points of elliptic curves [26,33], and also the jacobians of curves of higher genus [27,4,18]. The level of security reached by the use of these different groups varies a lot. Both the factorization of large numbers [29] and the computation of discrete logarithms in finite fields [11,19,3] can be addressed in subexponential time. This in turn has implications on the security of some elliptic curves cryptosystems, where the discrete logarithm problem on the curve reduces to the discrete logarithm problem on (an extension of) the curve's definition field [32,17]. This applies in particular to supersingular elliptic curves, where the MOV reduction [32] makes the discrete logarithm problem subexponential.

This being said, the existence of a subexponential attack does not automatically rule out a cryptosystem. A thorough account on which computations a cryptanalist can do with the current technology is necessary. While a tremendous amount of work (and CPU time) has been put towards the factorization of larger and larger numbers (S. Cavallar et al. used the Number Field Sieve to factor numbers as big as 512 bits [6,9], and even up to 774 bits numbers of a special form [7]), the computation of discrete logarithms in finite fields does

C. Boyd (Ed.): ASIACRYPT 2001, LNCS 2248, pp. 107–124, 2001.
© Springer-Verlag Berlin Heidelberg 2001

not seem to looked at so frequently. For prime fields, a recent work by Joux and Lercier [22] computed logarithms in $\mathbb{F}_p$ with p having 120 decimal digits, i.e. 399 bits. For fields of characteristic 2, Gordon and McCurley [20] almost* computed logarithms in $\mathbb{F}_{2^{503}}$, but that was back in 1993. This makes it hard, today, to make a reasonable guess on how difficult a characteristic 2 finite field discrete logarithm problem actually is. Subsequently, when the discrete logarithm on an elliptic curve reduces to some finite field of characteristic 2, it is not easy to tell how big this field should be for the cryptosystem to be secure.

In this context, our goal was to investigate how far we could go today in computing discrete logarithms in $\mathbb{F}_{2^n}$. The fastest algorithm for this purpose is due to Coppersmith [11] and has complexity $O(\exp((c+o(1))n^{\frac{1}{3}}(\log n)^{\frac{2}{3}}))$, for a small constant $c \approx 1.4$. This complexity makes it comparable to the Number Field Sieve [29], when addressing the factorization of an n-bit number. The 503-bit discrete logarithm record of Gordon and McCurley [20] was done using massively parallel supercomputers at Sandia National Laboratories. As far as we know, no recent state-of-the-art computations have been achieved. For our computations, we used *standard* hardware: the typical computers we used were much like everybody's desktop PC. Nonetheless, we have been able to carry the record to a few digits higher than before by computing discrete logarithms in $\mathbb{F}_{2^{607}}$.

Section 2 of this article outlines Coppersmith's algorithm. Section 3 reviews the rationales that drive the choice of each individual parameter in the algorithm. Sections 4 to 8 detail how we addressed the difficulties showing up in several parts of the algorithm. Section 9 shows the technical data on how the computations went along.

At the time of this writing, the computations over $\mathbb{F}_{2^{607}}$ are not finished. The sieving part is completed, and the linear algebra is underway. The computation of the solution to the linear system is expected to be finished by the beginning of the autumn 2001. As a very last-minute news, Joux and Lercier [23] appear to have computed logarithms in $\mathbb{F}_{2^{521}}$, using the general function field sieve approach [2]. This approach is fairly different from the one adopted here, and is not addressed in this paper. However, the result presented by [23] is highly encouraging.

2 Coppersmith's Algorithm

Throughout this article, we will let K denote the field $\mathbb{F}_{2^n}$, which will be represented as the quotient $\mathbb{F}_2[X]/(f(X))$, where f is a monic irreducible polynomial of degree n over $\mathbb{F}_2$. We will often talk of the elements of K merely as polynomials. It will be understood that what we actually mean is a class of polynomials inside this quotient. Likewise, the *degree* of a non-zero element of K will be the minimum degree of the polynomials representing it (always between 0 and $n-1$).

* The computations had not been fully carried out, since the resulting linear system was never solved

It will sometimes be convenient to write f as $X^n + f_1$, where f_1 is a polynomial. For the purposes of the algorithm, f_1 will be chosen so as to have the smallest possible degree. It is believed, but not proven, that such an f_1 exists whenever we allow its degree to grow as $O(\log n)$.

Coppersmith's algorithm belongs to the family of *index-calculus* algorithms. This means that we first select a factor base $\mathcal{B}$, and aim at computing the logarithms of its elements. For this, we gather a collection of relations among them. The relations will be of the form $\prod_{i=1}^{l} \pi_i^{e_i} = 1$, where the π_i's are the elements of the factor base. For reasons that will become clear later, this is referred to as the sieving part. This part can easily be distributed. Once we have enough relations involving the elements of the factor base, we obtain their logarithms as the solution of a (usually huge) linear system (we take the log of each relation). This is the linear algebra part. Implementations can be done efficiently on multiprocessor shared-memory machines, but such computers are expensive. Distribution of the computation across a network of not-so-expensive computers is very hard. The knowledge of all these logarithms, if the factor base is big enough, enables us to compute any logarithm in K easily. We will not detail that third part here since it is far easier than the two others. The interested reader might consult Coppersmith's original article [11] for reference.

The factor base $\mathcal{B}$ consists of all irreducible polynomials with degree less than a chosen bound b. It is known that $\mathcal{B}$ has roughly $\frac{2^{b+1}}{b}$ elements (see for instance [31]). Up to now, Coppersmith's algorithm is very resemblant to Adleman's [1, 5,3], which computes discrete logarithms in any Galois field, no matter the characteristic (but with poorer complexity than Coppersmith's). The key difference is in the production of linear relations. To build relations among the elements of $\mathcal{B}$, we choose random relatively prime polynomials A and B of degrees d_A and d_B, respectively. Let k be a power of 2 near $\sqrt{n/d_A}$, and $h = \lceil \frac{n}{k} \rceil$. Then we write:

$$C = AX^h + B,$$
$$D = C^k = A^k X^{hk} + B^k \equiv A^k X^{hk-n} f_1 + B^k \ [f].$$

An appropriate choice of the parameters keeps the degrees of C and D balanced, around $\sqrt{nd_A}$. For each such produced pair, we want to know whether it is *smooth* or not. The pair (C, D) is *smooth* when both polynomials have their irreducible factors inside $\mathcal{B}$. Of course, the bigger the factor base, the more likely this is. A smooth pair will give us a linear relation among the logarithms of the elements of $\mathcal{B}$, since if we denote them π_i, $1 \le i \le \#\mathcal{B}$, we can find integers α_i and β_i such that:

$$C = \prod_i \pi_i^{\alpha_i}, D = \prod_i \pi_i^{\beta_i}, \Rightarrow DC^{-k} = \prod_i \pi_i^{\beta_i - k\alpha_i} = 1,$$
$$\Rightarrow \sum_i (\beta_i - k\alpha_i) \log \pi_i \equiv 0 \ [2^n - 1]$$

Once we have gathered enough relations, we are facing a (fairly big) linear system that has to be solved, the unknowns being the logarithms of the elements of $\mathcal{B}$.

3 Choice of the Parameters

Coppersmith's algorithm introduces many parameters that may seem arbitrary at first glance. In [11], Coppersmith computed the asymptotical optimum value for each of them. We will not redo this analysis here, but rather briefly discuss the practical importance of each of the parameters, especially taking care of implementation realities like available hardware.

The choice of b. This main parameter, whose asymptotical optimum value is $n^{1/3}(\log n)^{2/3}$ controls the ratio between the work amounts in the first and second stages. The bigger b, the easier the first stage (even if we have twice as many relations to produce, the probabilities of smoothness increase drastically with b). On the other hand, increasing b by 1 almost doubles the size of the linear system in the second stage. Since the linear algebra is hardly distributable, the available hardware enforces a strong limit on the size of this system (otherwise the matrix would not fit into memory).

The choice of d_A *and* d_B. Originally, Coppersmith grouped them as a single parameter chosen asymptotically "near b"*. These parameters account for the number of pairs to test. Taking into account the probability of smoothness, we have to make sure that the $2^{d_A+d_B+1}$ available coprime (A, B) pairs will be enough to produce the required number of relations among the elements of $\mathcal{B}$. Of course, the sad news is that increasing d_A and d_B raises the degrees of C and D, and hence lowers the probability of smoothness. We have split Coppersmith's single parameter in two because it is usually possible to choose d_B a little bit above d_A without increasing the degrees of C and D (the optimum difference between the two is $\frac{hk-n+\deg f_1}{k}$). Therefore, we can maximize the number of pairs which are available.

The choice of k. Ideally, C and D have almost the same degree, their optimal value being $n^{2/3}(\log n)^{1/3}$. In fact, these can be somewhat unbalanced from the practical point of view. The parameter k is there to keep these polynomials in the same range, but unfortunately the requirement that k be a power of 2 gives us little control over it. The asymptotical best value for k is $\sqrt{\frac{n}{d_A}} = \left(\frac{n}{\log n}\right)^{\frac{1}{3}}$. For the problems we are concerned about, $k = 4$ appeared to be the correct choice. It might be that, at $n = 607$, we are nearing the cross-over point between $k = 4$ and $k = 8$, but $k = 8$ is still inadequate. One other aspect about the choice of k is that half of the coefficients in the linear system are $-k$ (the other

* An asymptotic ratio is computed in [11], depending on the algorithm used for linear algebra

ones being 1's). This brings a complication to the linear algebra (the structured gaussian elimination, namely), which could only be worsened by the choice of a bigger k.

The choice of f_1. Another hidden parameter lies in the choice of f_1. Usually, one can choose among a couple of candidates for f_1. The ones of low degree have a clear advantage due to the influence on $\deg D$, but [20] shows that polynomials with small factors are also worth investigating. The reader is referred to [20] for a thorough discussion on the choice of f_1.

In our computations, for $n = 607$, the following parameters were chosen: $b = 23$ (hence $\#\mathcal{B} = 766,150$), $d_A = 21, d_B = 28, k = 4, h = 152$. As for the choice of f_1, it turned out that $X^9+X^7+X^6+X^3+X+1$ had an overwhelming advantage, being simultaneously the candidate of smallest degree and with only small factors: f_1 factorizes as $(X+1)^2(X^2+X+1)^2(X^3+X+1)$. Given these parameters, the respective degrees of C and D were 173 and 112.

4 Description of the Polynomial Sieve

In Coppersmith's original version of the algorithm, the smooth pairs were located by repeatedly applying a smoothness test to all pairs of the allowed range. Gordon and McCurley [20], as an alternative, designed an efficient polynomial sieve, which helped to reduce the time spent on each pair (smooth or not). The idea is as follows. For A fixed, we maintain a big array of integers (initially 0) associated to the different pairs to be tested, that is, all the possible B's. Let g be an irreducible polynomial. We want to add $\deg g$ to the values associated to the B's* satisfying:

$$B \equiv AX^h \ [g]. \tag{E}$$

Doing this sieve efficiently implies being able to step quickly through all multiples of g. This can be done without awkward polynomial multiplications using *Gray codes.* For any non-zero positive integer x, let $l(x)$ denote the index of the least significant bit set in the binary representation of x (starting at $l(1) = 0, l(2) = 1$). Then the congruence class of $AX^h \mod g$ among the polynomials of degree less than or equal to d_B is given by the set of values of the sequence defined by: $B_0 = AX^h \mod g$, $B_i = B_{i-1} + X^{l(i)}g$, for $0 < i < 2^{1+d_B-d_g}$. Of course, it is worthwhile to precompute the $X^j g$'s, since these differ from each other only by arithmetic shifts.

This sieve is done for a certain collection of irreducible polynomials. One can also take into account the contribution of powers of irreducible polynomials, adding $\deg g$ to all $B's$ satisfying $B \equiv AX^h \ [g^j]$. If the sieve is done for all irreducible polynomials g, and also their powers, the value in each table cell is precisely the degree of the smooth part in the factorization of the associated quantity $C = AX^h + B$ (an entry for which the congruence holds modulo

* Or, equivalently, the *pairs*, since A remains fixed.

g^j accumulates a total contribution of $j \deg g$ from the consecutive sieves with $g, g^2, \ldots, g^j$). Therefore, an entry in the table which has a value of $\deg C$ automatically corresponds to a pair with C smooth.

In real life, one does not use all the relevant irreducible polynomials for the sieve, and an important improvement comes from the use of incomplete sieves. Two parts of the sieve are actually very expensive: the sieve over small irreducibles on the one hand, because there are many cells to update for each small irreducible, and the sieve over big irreducibles on the other hand because the initialization cost is high (and the number of irreducibles of a given degree raises with the degree). Therefore, we considered skipping these parts. Doing so, we lose accuracy, because the smoothness of C is only evaluated from the contribution of medium-size polynomials. Instead of $\deg C$, we use as *qualification bound* the average contribution from medium-size polynomials to a smooth C. If the standard deviation of this quantity is high, it will be hard to recognize pairs yielding smooth C's among the set of all pairs to be considered. Since the subsequent distinction between useful and useless pairs is done on a per-pair basis (a factorization job, in fact) their number should not grow too much. We found interest in skipping the sieve over irreducibles of degree 1 to 9, because their total contribution to smooth polynomials did not deviate too much from its average value, whereas we only skipped high degree 23, because otherwise we would have had to lower drastically the qualification bound to catch sufficiently many of the pairs yielding a smooth C, which in turn would have made the factorization cost too high.

Based on the same ideas, it is not always worthwhile to sieve over powers of an irreducible polynomial. Locating cells corresponding to pairs divisible by g^j for an irreducible polynomial g and an integer i is practically pointless if the expected number of cells to update is too small (this number is $2^{d_B+1-j \deg g}$). In fact, the only powers that we found worthwhile to sieve with were squares of polynomials of degrees 10 and 11.

It could be tempting to try to also do a sieve with D, but the situation is quite different. The initialization of the sieve must be done with $B_0 = A(X^{hk-n} f_1)^{1/k} \bmod g$, for g an irreducible polynomial. This computation is more complicated than previously. Also, this only works when g is an irreducible polynomial, and not when it is a power of an irreducible, because a k-th root might not exist modulo g^j. This difficulty is due to the same particularity of D that Gordon and McCurley already noticed in [20]: this polynomial is more likely to be square-free than it would be if it were random (and therefore it is less likely to be smooth). As we have just seen, this last point is not too disturbing since one hardly uses powers of irreducibles for the sieve.

Sieving over D turned out not to be useful in our case, since the first sieve (over C) already eliminated most of the pairs, and eventually testing the smoothness of D on a per-pair basis was more efficient. Nonetheless, sieving over D only instead of C could be useful in different settings, depending on how $\deg C$ and $\deg D$ compare to each other. In $\mathbb{F}_{2^{607}}$, the parameter k seems to be better around 4, and as a consequence, the degrees of C and D are not really balanced:

$\deg C$ is much higher. If we were about to carry out computations in, say, $\mathbb{F}_{2^{997}}$, $k = 8$ would probably be a better choice. And $\deg D$ would become automatically bigger than $\deg C$. A sieve over D in this situation would therefore enable us to discard much more pairs than its counterpart (because there would be very few smooth D's), and the benefit in the factorization part would probably compensate for the sieve's relative drawback.

5 Using Large Primes

One well-known improvement to the sieving part of index calculus algorithms is the so-called *large prime variation*. The idea is that aside plain, *full* relations, we allow *partial* relations, corresponding to pairs which are smooth up to a certain number of big irreducible cofactors (above the factor base bound) called *large primes*. Afterwards, these partial relations are matched together when this is possible in order to eliminate the cofactors. The partial relations come almost for free in the sieving stage, since they would otherwise have been discarded at the end of the factorization stage and not earlier. The degree of large primes must of course be kept under a certain bound: allowing for too large "large primes" eventually brings no benefit. From our point of view, this approach fits well here. We merely have to lower the qualification bound from $\deg C$ to $\deg C - \mathcal{L}$, where $\mathcal{L}$ is the maximum allowed degree of large primes.

When we allow only one *large prime*, matching partial relations together involves only a hashing process in order to be able to spot partial relations containing an already met large prime. The number of full relations reconstructed this way grows quadratically vs. the number of partial relations. When up to two large primes are used (see [30]), an algorithm resembling "union-find" helps to find *cycles*: relation after relation, we build a graph whose vertices are the large primes. An edge connects two vertices if a partial relations exists involving them. There is also a special vertex named "1", to which all primes involved alone in a partial relation are connected. Under certain conditions*, a cycle in this graph will give us a free full relation. The overhead is small, but this cycle detection has to be implemented with care because managing a graph with more than 10^8 edges among 2.10^9 vertices can turn out to be quite awkward. More elaborate schemes allow the processing of partial relations with more large primes, see for instance [15]. Recently, in the course of the record-breaking factorization of RSA-155, S. Cavallar proposed in [8] an efficient scheme for this large prime matching task, inspired by *structured gaussian elimination* like in [37]. We lacked the required time to investigate the respective efficiency of all of these different strategies when applied to our case. This is a real concern here, because while the multi-large-prime schemes have proven to be very efficient in the factorization context, this is not completely clear for discrete logarithms. Factorization algorithms use relations that are defined up to squares, that is, with exponents defined over $\mathbb{F}_2$. For discrete logarithms, exponents are defined in a big finite ring, here

* Slight complications are brought by the fact that our coefficients are not defined over $\mathbb{F}_2$.

$\mathbb{Z}/(2^n - 1)\mathbb{Z}$. When combining partial relations with large primes in common, one can only cancel one large prime at a time. For this reason, the landscape is quite different.

Our computations have been carried out using the *double* large prime variation, that works well even with regard to the coefficient issue. Two large primes were allowed. For efficiency reasons (discussed in section 7), only the factorization of C could have two large primes, while D was restricted to only a single large prime. 10% of the relations had actually only one large prime, and among the remaining relations (that had two large primes), 30% had both large primes on the same side (the C side, actually), the rest of the relations having their large primes balanced on each side. We did the cycle detection using a straightforward union-find algorithm. Figures about the cycle detection can be found in section 9.

6 Grouping Sieves

As it is described above, the sieve algorithm uses an array of fixed size, namely 2^{d_B+1} bytes (assuming one byte per sieve location). Our setup had $d_B = 28$, so this makes a sieve area of 512MB, far above what is acceptable. Furthermore, it was not certain by the beginning of the sieve whether the outcome of pairs with a polynomial B of big degree would eventually be used or not. We decided to have a first look at the pairs from which we knew that the outcome would be better, that is, the pairs with smaller B's, and defer the analysis of less promising pairs to a later time. Our strategy was to decompose the whole sieving job in *chunks* indexed by fixed parts A_f and B_f of the polynomials A and B. The chunks consisted of areas of the form:

$$\begin{aligned}\text{chunk}(A_f, B_f) =& \{(A, B) = (A_f X^{\delta_A+1} + A_v, B_f X^{\delta_B+1} + B_v),\\ & \deg A_v \leq \delta_A, \deg B_v \leq \delta_B\}, \text{with } \delta_A = 6, \delta_B = 24.\end{aligned}$$

Each chunk could be sieved by the machine handling it in any suitable way. The most straightforward approach is to do $2^7 = 128$ sieves, each of them addressing 2^{25} bytes, that is 32MB, for the sieve area.

Since we ran the job using idle time on many not-so-powerful machines, this was still too much memory to be used for some of them. A further possibility is to divide the 32MB sieve area into yet more (say 2^γ, with γ a small integer), smaller sieve areas (of size $2^{-\gamma} \times$ 32MB). But when the sieve area becomes so small, the initialization cost becomes too important. The expensive task is the modular reduction $AX^h \bmod g$, which is performed for each g. One can precompute the initialization data for the 2^γ sieves, but even after doing that, we were unsatisfied with the cost of the initialization, and tried to trim it down even more.

We wanted to achieve this without letting additional bits of B vary, but rather sieving over several A's at a time. This is possible because for reasonably close A's, the initialization for a given g is almost the same. In the following paragraphs, g will denote either an irreducible polynomial, or a power of an irreducible polynomial. Inside a given sieve with A completely fixed, we want to find the solutions to:

$$B_f X^{\delta_B+1} + B_v \equiv AX^h \ [g].$$

If we allow some of the lowest bits of A, say ϵ of them (with $\epsilon \leq \delta_A + 1$, of course) to vary, the equation becomes:

$$B_v + \alpha X^h \equiv AX^h + B_f X^{\delta_B+1} \ [g], \text{ with } \deg \alpha < \epsilon. \qquad \text{(E')}$$

The solutions to this equation form an affine subspace $\mathbb{S}$ of the $\mathbb{F}_2$-vector space $\mathbb{V} = F \oplus G$, with $F = \langle 1, X, X^2, \dots X^{\delta_B} \rangle$ and $G = \langle X^h, \dots X^{h+\epsilon-1} \rangle$. The expected dimension of $\mathbb{S}$ is $\dim \mathbb{S} = \delta_B + 1 + \epsilon - d_g$. We will try to find $\mathbb{S}$ using linear algebra over $\mathbb{F}_2$. The idea behind this is that arithmetic shifts and logical operations take almost no time compared to a polynomial division or multiplication. We will consider two situations.

The easy case is when $d_g \leq \delta_B + 1$. $\mathbb{S}$ writes down as $s_0 + \mathbb{S}'$, with a point $s_0 = AX^h + B_f X^{\delta_B+1} \mod g$, and an underlying vector space $\mathbb{S}'$ spanned by the $X^i g$ for $0 \leq i \leq \delta_B - d_g$, and the $X^{h+i} + (X^{h+i} \mod g)$ for $0 \leq i < \epsilon$. We claim that the computation of these generators costs very little above 2 modular reductions since once $X^h \mod g$ has been computed, inferring the $X^{h+i} \mod g$ inductively is easy (one bit test and one exclusive-or if needed). If we did independent sieves we would have needed 2^ϵ modular reductions (which can be anything but cheaper).

If $d_g > \delta_B + 1$, we extend $\mathbb{V}$ to $\bar{\mathbb{V}} = \bar{F} \oplus G$, $\bar{F} = F \oplus \langle X^{\delta_B+1}, \dots X^{d_g-1} \rangle$. Let $\bar{\mathbb{S}}$ be the set of solutions of E' in $\bar{\mathbb{V}}$. A point $\bar{s}_0$ in $\bar{\mathbb{S}}$ is obtained as in the previous case, and generators of the underlying vector space $\bar{\mathbb{S}}'$ are the $u + \phi(u)$ for $u \in G$, ϕ being the linear map from G to $\bar{F}$ that reduces a polynomial mod g. Using gaussian elimination, we can find a point $s_0 \in \mathbb{S}$ deduced from $\bar{s}_0$ and $\bar{\mathbb{S}}'$ if such a point exists, and the generators of $\mathbb{S}'$ (the vector space underlying $\mathbb{S}$) are the $u + \phi(u)$ for $u \in \phi^{-1}(F)$. This involves finding the kernel of a $(\dim \bar{F} - \dim F) \times \epsilon$ matrix, which is expected to be quite easy (perhaps a dozen CPU cycles). Although the case where $d_g > \delta_B + 1$ is unlikely to be met often in practice (we don't want to sieve when $\deg g$ is too big), we will augment this quick description with an example. Suppose we have the following setup:

$$\begin{aligned} g &= (X^{14} + X^{13} + X^{12} + X^{10} + X^8 + X^5 + 1)^2, \\ \bar{s}_0 &= X^{27} + X^{26} + X^3 + 1, \\ h &= 152, \quad \delta_B = 24, \quad \epsilon = 3. \end{aligned}$$

The first three columns of the following matrix are the $\dim \bar{F} - \dim F$ most significant coefficients of the polynomials $u + \phi(u)$ for $u = X^{h+i}$ and $0 \leq i < \epsilon$. The last one contains the leading coefficients of $\bar{s}_0$:

$$T = \begin{pmatrix} 0\,1\,0\,1 \\ 1\,0\,1\,1 \\ 0\,0\,0\,0 \end{pmatrix}$$

Doing a gaussian elimination on the columns of T, one easily obtains:

$$s_0 = \bar{s}_0 + X^h + \phi(X^h) + X^{h+1} + \phi(X^{h+1}) \in \mathbb{S},$$
$$\phi^{-1}(F) = \langle X^h + X^{h+2} \rangle.$$

Of course, this example is a bit particular in that the last row of T is zero, therefore the dimension of $\mathbb{S}'$ is one, instead of the expected value 0. Other cases can occur: for instance, if $\bar{s}_0$ had had a non-zero coefficient in X^{25}, then we would have had no solutions to the equation E' inside $\mathbb{V}$.

We have shown two ways to play with the memory available to the siever. These can actually be mixed together. Using parameters γ and ϵ together, a chunk is divided in $2^{\delta_A+1+\gamma-\epsilon}$ sieves, each of them using $2^{\delta_B+1+\epsilon-\gamma}$ bytes of memory. The influence of the two parameters γ and ϵ is shown on figure 1. Timings are in seconds runtime on 450MHz Pentium II's. The percentages show the timing difference versus the standard sieve (which has $\gamma = \epsilon = 0$). Three figures are present in each table cell. The figures are always normalized to reflect the time needed to sieve a (fictitious) 128MB sieve area. The first one, on which the effect of both γ and ϵ is the most striking, shows the time spent in initializing the sieve (or, in re-reading again and again the precomputed initialization data when $\gamma > 0$. Precomputation time in this case is also included). The second figure is the time spent in the sieve itself, that is, adding $\deg g$ to each table cell corresponding to a pair divisible by g, for all possible g's. The effects of γ and ϵ on this sieving time are hardly noticeable (the variations are likely to be due to operating system overhead). The third figure is the total time spent including allocation overhead and final pair detection (but not the factorization, which comes afterwards, and is irrelevant here). On the right and the bottom, the actual memory sizes used by the sieve area are given. Since our jobs have been running in background on otherwise used machines, we preferred not to use too much memory. Using the setting $\gamma = 4$, $\epsilon = 3$ was a satisfying compromise, with a mere 16MB sieving area.

7 Factorization of the Pairs

Once good pairs have been located, the actual production of the relations (or partial relations) requires the factorization of the pairs (C, D). Efficient algorithms exist for polynomial factorization, but our actual problem here is not the usual one. Instead of the factorization of one huge polynomial (of degree several thousands for instance), we have to deal with the factorization of a huge number of relatively small polynomials (in our case, their degree is less than 200). Therefore, asymptotically better behaving algorithms might not be worthwhile. Furthermore, we are willing to give up as soon as we suspect the polynomial might not be smooth after all. In a few words, merely applying some classical distinct degree factorization algorithm can turn out to be a considerable waste of time. We built a factorization scheme based on several specific improvements that turned out to be worthwhile.

	$\epsilon = 0$	$\epsilon = 1$	$\epsilon = 2$	$\epsilon = 3$	
$\gamma = 0$	33.28 (0%) 105.68 (0%) 144.24 (0%)	22.76 (-31%) 109.96 (+4%) 138.20 (-4%)	12.26 (-63%) 107.75 (+1%) 125.35 (-13%)	7.01 (-78%) 110.62 (+4%) 122.98 (-14%)	
$\gamma = 1$	38.84 (+16%) 109.48 (+3%) 153.80 (+6%)	24.46 (-26%) 108.94 (+3%) 138.88 (-3%)	12.94 (-61%) 110.84 (+4%) 129.29 (-10%)	6.89 (-79%) 107.73 (+1%) 119.95 (-16%)	256MB
$\gamma = 2$	46.16 (+38%) 107.12 (+1%) 158.60 (+9%)	28.66 (-13%) 109.38 (+3%) 143.50 (0%)	14.87 (-55%) 105.92 (0%) 126.12 (-12%)	7.99 (-75%) 106.29 (0%) 119.63 (-17%)	128MB
$\gamma = 3$	63.04 (+89%) 108.92 (+3%) 179.08 (+24%)	35.22 (+5%) 109.98 (+4%) 152.66 (+5%)	19.37 (-41%) 109.46 (+3%) 134.31 (-6%)	10.18 (-69%) 105.58 (0%) 121.11 (-16%)	64MB
$\gamma = 4$	96.56 (+190%) 108.68 (+2%) 210.72 (+46%)	54.56 (+63%) 111.26 (+5%) 171.28 (+18%)	28.27 (-15%) 110.42 (+4%) 144.14 (0%)	14.69 (-55%) 106.84 (+1%) 126.87 (-12%)	32MB
		2MB	4MB	8MB	16MB

Fig. 1. Influence of γ and ϵ on the sieving time.

The pairs that constitute the input to the factorization step are such that C has a reasonable probability to be smooth: it has been selected for this purpose. D, however, has no reason to be smooth. Therefore, the first thing to try out is a smoothness test on D, in order to avoid useless computations on all pairs with non-smooth D's. The smoothness test applied is the same as in [11], except that we want to allow large primes. The b-smooth part of D is computed as

$$D_{\text{smooth}} = \gcd\left(D, D' \prod_{j=1+\lfloor\frac{b}{2}\rfloor}^{b} (X^{2^j} + X) \mod D\right).$$

In some cases (if D has a very big square factor), D_{smooth} might not actually be b-smooth, but that's exceptional. Concerning the cofactor $\frac{D}{D_{\text{smooth}}}$, we are facing a design choice, since we can either allow only one large prime, that is, allow a cofactor of degree at most $\mathcal{L}$, or permit several large primes, setting for instance the cofactor bound to the looser $2\mathcal{L}$. However, in the latter case, the cofactor needs not factor kindly into two large primes of degree less than $\mathcal{L}$ (actually, it is most likely not to). The best choice depends on what we want to do with partial relations. In our experiments, less than 1% of the D's passed the former test, while around 12% passed the latter (hence there were more pairs to be factorized afterwards, resulting in a 25% increase of the factorization cost). Since we used only two large primes in total, the yield was better when using the first, more restrictive test. If we were to allow three large primes or more (cf [15]), the second, looser, test would probably be more adequate.

Once D has passed the test, and has therefore an acceptable probability to be smooth (save the cofactors), we need to factorize C and D. Originally, when running our program on smaller examples like $\mathbb{F}_{2^{313}}$, we found it useful to track down small factors either by explicit trial division or by precomputations and table lookup. The idea is to quickly compute the valuation of a given polynomial with respect to some irreducible. Of course, this is trivial for the valuation with respect to X. Let us explain briefly how this can be done for the valuation with respect to $X+1$. We notice that $(X+1)^{16} = X^{16}+1$. Since our implementation represents the polynomials over $\mathbb{F}_2$ using one bit per coefficient, computing a remainder modulo $X^{16}+1$ is fast. Assuming we have a 32-bit machine, this requires less than $\frac{\deg P}{32} + 3$ operations (exclusive "OR"s, one shift and one "AND"). If we have a precomputed table holding the values of $\nu(Q)$ (ν is the valuation) for all polynomials Q of degree 0 to 15 (this requires 32KB), we can obtain $\nu_g(P)$ with high probability. Indeed, we have $\nu_g(P \mod X^{16}+1) = \nu_g(P)$ unless $P \equiv 0\ [X^{16}+1]$ (in which case we have an inequality $\leq$). Once we have this value, we merely have to do one division by the appropriate (precomputed) power of $X+1$. If the valuation is at least 16, we repeat the operation on the cofactor. From the basic observation that a remainder modulo a cyclotomic polynomial is easily computable, we could extend this approach for irreducibles of degree up to 4. Alas, the improvement obtained from this method was not significant for the case of $n = 607$, probably because the average degree of C and D made the contribution of little factors too small for this to be useful. We also tried to factor the relations by sieving with all or part of the possible irreducibles that could appear in the factorization, but this brought no significant improvement either.

Since it turned out that our attempts towards removing some of the factors by hand were not worthwhile, the whole factorization job was achieved by a general-purpose factorization algorithm (in any case, if we did remove some of the factors by trial division, the cofactor would have still had to be factorized via such an algorithm). We used Niederreiter's algorithm [35], which proved four times faster than a classical distinct degree factorization procedure. The explanation of this lies of course in the small degree of our polynomials, and in the fact that we work over $\mathbb{F}_2$, for which Niederreiter's algorithm is well suited.

8 Improvements to the Linear Algebra Stage

The sparse matrix emerging from the sieving has roughly $\frac{2^{b+1}}{b}$ columns, and a bigger number of lines (we had a 40% excess). This matrix is extremely sparse: the number of non-zero terms (called the weight) of a given line corresponding to a smooth pair (C, D) is actually the number of distinct factors in the factorization of DC^{-k}. Most relations are also obtained from recombinations of partial relations, so the weight for a recombination of s relations is s times the average number of factors in a factorization like DC^{-k}. In our case, this amounts to an average weight of the lines for the whole matrix of 67.7. Handling such systems requires well-suited algorithms, designed to take advantage of the sparsity as much as possible. Actually, this is a well studied subject, since sparse matrices arise in many domains. For the literature about sparse matrices coming from discrete logarithm or factorization problems, one can consult [37,43,12,34, 25,28]. Two particularly annoying points are relevant to our case. Unlike linear system that arise from factorization problems, ours is defined over a big field, $\mathbb{Z}/(2^{607}-1)\mathbb{Z}$. Second, unlike what happens with Adleman's algorithm [1], or with the number field sieve when applied to the discrete logarithm problem [19], our coefficients are not always ± 1. As explained earlier in this article, half of them are $\pm k$.

In order to solve our system, we first apply the well-known *structured gaussian elimination* as described in [37]. This algorithm takes advantage of both the sparsity of the matrix, and also of the "unbalanced" shape of its lines: each line in the matrix corresponds to a relation, and the coefficients on the left correspond to small factors, while those on the right correspond to big factors. The probability of a given polynomial to be divisible by a given factor of degree d being $\frac{1}{2^d}$, the density of the matrix is much higher on the left part (small factors) than on the right part (big factors). The structured gaussian elimination starts from the right end of the matrix (which is extremely sparse) and tries to remove lines and columns without increasing (if at all) the matrix density.

We modified the original process described in [37] in the spirit of what is done in [42]: we evaluate, at each step, the influence of each possible operation to the cost of the linear system solving algorithm that follows the SGE. The better steps towards the reduction of the linear algebra cost are taken, until nothing interesting can be done anymore. This process is able to shrink down the matrix to a fraction of its original size. Here, having many coefficients equal to $\pm k$ on input causes lines to be multiplied quite often while pivoting is done. Since a given line cannot be multiplied too many times (otherwise we would have to allow the coefficient to grow above one machine word), this makes the elimination less efficient.

Afterwards, we found it enlightening to use the *block Wiedemann* algorithm. This algorithm has been proposed by Coppersmith in [12], extending a previous algorithm by Wiedemann [43]. Another algorithm, the block Lanczos algorithm [34], is often preferred to the block Wiedemann algorithm. We used the latter because it gave us an opportunity to successfully experiment the accelerating procedure described in [39]: the crux of the block Wiedemann algorithm is the computation of a linear generator for a matrix sequence (a matrix analogue to the Berlekamp-Massey algorithm), and [39] uses FFT to reduce the complexity of this task from $O(N^2)$ to $O(N\log^2 N)$, achieving a 50 times speedup for the computation undertaken here. The block Wiedemann algorithm performs well both theoretically and in practice. See [40,41,24,25,39] for several insights on the algorithm. The block Wiedemann algorithm is interesting in the fact that at least for one part of the algorithm, several machines holding a private copy of the matrix (for which they need to have the proper amount of memory) can each do a part of the work without communication between them. Therefore, one can regard this as a partial distribution. We found that the optimal number of machines to be used simultaneously in this computation was 4 (luckily, we had that number of machines able to hold the 400MB matrix in RAM).

9 Computations over $\mathbb{F}_{2^{607}}$

The comprehensive sieving part took about 19,000 MIPS years. As a comparison, the factorization of RSA-155 required 8,000 MIPS years. The outcome of the sieving processes, in terms of relations per hour, dropped from 1000 relations (full or partial) per hour with the very first chunks (the degrees were still small) to 400 afterwards, and eventually 100 for the very last ranges of data. Almost all the sieving area up to $d_B = 28$ has been needed (a more thorough usage of this area could have been achieved if we did not use incomplete sieves, but the trade off was clear in their favor). Of these relations, of course, most were partial ones. The total amount of data produced by these sub-processes nears 10GB. The cycle detection algorithm ran approximately for one day and produced the biggest part of the relations at the end: 815,726 relations were reconstructed using cycles of length going from 2 to 40. All of these cycles were linked to the special vertex "1", which is not surprising given the size of the corresponding connected component. More than 650,000 relations were obtained from cycles of length 3 or more, which shows that using the *double* large prime variation was a winning choice. Meanwhile, we only produced 217,867 genuine full relations. Additional data can be found in table 1. The average weight of these 1,033,593 relations in total was 67.7, the maximum weight being 524. We discarded the relations whose weight was above 120, since these were definitely too heavy to be useful. We were left with 904,004 relations, involving 765,427 columns (the average weight dropped to 64.3).

We ran a structured gaussian elimination algorithm (SGE) on this matrix. The schedule time for SGE was approximately one day. We were able to divide by two the cost of the subsequent block Wiedemann algorithm. The matrix obtained after the SGE had size $484,603 \times 484,603$ with an average line weight of 106.7. One can find this reduction ratio quite disappointing compared to ratios typically achieved in other contexts. This could be a consequence of the fact that most relations were recombined ones. These were therefore denser, and had coefficients somewhat bigger than other lines, which impairs the reduction ratio of the SGE.

The block Wiedemann algorithm is currently underway, in the process of finding an element of the kernel of this matrix. We expect it to be finished by the beginning of the autumn. It should be noted that since $2^{607} - 1$ is prime, the linear algebra task cannot be eased anyhow by the chinese remainder theorem.

Table 1. Data from computations in $\mathbb{F}_{2^{607}}$

Size of the factor base	766,150 polynomials
Total number of relations	1,033,593 relations
Full relations	217,867 relations
Cycles obtained	815,726 cycles
Partial relations used (≤ 2 large primes)	60,128,419 relations
Large primes involved	85,944,405 polynomials
Relations with only one large prime	5,992,928 relations
Cycles of length 2	150,566 cycles
Cycles of length 3	142,031 cycles
Cycles of length 4	123,900 cycles
Cycles of length 5	101,865 cycles
Cycles of length 6 or more	297,364 cycles
Size of the biggest cycle	40 edges
Size of the biggest connected component	22,483,158 edges
Size of the second biggest connected component	167 edges
Number of connected components with 1 edge	22,025,908 components
Number of connected components with 2 edges	2,726,940 components
Number of connected components with 3 edges	848,691 components

10 Conclusion

Computation of discrete logarithms in $\mathbb{F}_{2^{607}}$ is now a matter of weeks (linear algebra is in its last phase). As was predicted by Gordon and McCurley in the conclusion of their article [20], this was far from an easy task, and the computation took enormous proportions. Today's supercomputers might achieve the work we did in quite a reasonable time, but going further will necessarily imply more advanced techniques, including, but probably not limited to, the use of four large primes (taking into account the remark on the coefficient issue in section 5). The conclusion of our computation is that one can not seriously claim that discrete logarithms in, say, $\mathbb{F}_{2^{997}}$, are within the reach of a computation of the type we have undertaken. A very well-funded institution (e.g. governmental) could perhaps go that far, but this is much likely to involve a tremendous (and highly expensive) computational effort. An implication of our work to how we should regard the security of an elliptic curve cryptosystem with a MOV reduction [32] of the discrete logarithm problem to the discrete logarithm problem in a field $\mathbb{F}_{2^n}$, is that if n is around $1,000$, attacking such a problem is very hard, and if n is around $1,200$, this size is twice above the computational mark that we have just set. Therefore, the security of such a cryptosystem in the latter case can be seen as no lower than the security of an RSA-1024 cryptosystem, given that RSA-512 schemes have been successfully attacked using computational means comparable to ours.

Acknowledgements. Our program has been written in `C`, using the `ZEN` computer algebra package [10] and the `GMP` package [21] for multiprecision integer arithmetic. CPU time has been (and is being) provided by several institutions. Three units at École polytechnique, Palaiseau, France, provided most of the sieving time: the student computer clusters, UMS MEDICIS, and LIX (computer science research group). On a smaller scale, CPU time has been used for the sieving at École normale supérieure, Paris, France, and also at the department of Mathematics of the University of Illinois at Chicago, while the author was visiting this institution in 1999/2000. Large prime matching has been entirely done at LIX. Linear algebra involved mostly resources from LIX, and also from UMS MEDICIS to the extent possible given the distribution constraints for this task. We would like to express our grateful thanks to all the users at these places, and to the IT staff who have always been very helpful.

A special thank goes to François Morain, who has been helping the author with questions and comments since the beginning of this work, and throughout the preparation and achievement of this record.

References

1. L. M. Adleman. A subexponential algorithm for the discrete logarithm problem with applications to cryptography. In *Proc. 20th FOCS*, pp. 55–60. IEEE, 1979.
2. L. M. Adleman. The function field sieve. In L. Adleman and M.-D. Huang, eds., *ANTS-I*, vol. 877 of *Lecture Notes in Comput. Sci.*, pp. 108–121. Springer-Verlag, 1994. Proc. 1st Algorithmic Number Theory Symposium, Cornell University, May 6–9, 1994.
3. L. M. Adleman and J. DeMarrais. A subexponential algorithm for discrete logarithms over all finite fields. *Math. Comp.*, 61(203):1–15, July 1993.
4. S. Arita. Algorithms for computations in Jacobians of C_{ab} curve and their application to discrete-log-based public key cryptosystems. In *Proceedings of Conference on The Mathematics of Public Key Cryptography, Toronto, June 12–17*, 1999.
5. I. F. Blake, R. Fuji-Hara, R. C. Mullin, and S. A. Vanstone. Computing logarithms in finite fields of characteristic two. *SIAM J. Alg. Disc. Meth.*, 5(2):276–285, June 1984.
6. CABAL. Factorization of RSA-140 using the number field sieve. Available online at `ftp://ftp.cwi.nl/pub/herman/NFSrecords/RSA-140`, Feb. 1999.
7. CABAL. 233-digit SNFS factorization. Available online at `ftp://ftp.cwi.nl/pub/herman/SNFSrecords/SNFS-233`, Nov. 2000.
8. S. Cavallar. Strategies in filtering in the number field sieve. In W. Bosma, ed., *Proc. ANTS-IV*, vol. 1838 of *Lecture Notes in Comput. Sci.*, pp. 209–231. Springer-Verlag, 2000.
9. S. Cavallar et al. Factorization of a 512-bit RSA modulus. In B. Preneel, ed., *Proc. EUROCRYPT 2000*, vol. 1807 of *Lecture Notes in Comput. Sci.*, pp. 1–18. Springer-Verlag, 2000.
10. F. Chabaud and R. Lercier. `ZEN`, a toolbox for fast computation in finite extensions over finite rings. Homepage at `http://www.di.ens.fr/~zen`.
11. D. Coppersmith. Fast evaluation of logarithms in fields of characteristic two. *IEEE Trans. Inform. Theory*, IT–30(4):587–594, July 1984.

12. D. Coppersmith. Solving linear equations over GF(2) via block Wiedemann algorithm. *Math. Comp.*, 62(205):333–350, Jan. 1994.
13. T. Denny and V. Müller. On the reduction of composed relations from the number field sieve. In H. Cohen, ed., *Proc. ANTS-II*, vol. 1122 of *Lecture Notes in Comput. Sci.*, pp. 75–90. Springer-Verlag, 1996.
14. W. Diffie and M. E. Hellman. New directions in cryptography. *IEEE Trans. Inform. Theory*, IT–22(6):644–654, Nov. 1976.
15. B. Dodson and A. K. Lenstra. NFS with four large primes: an explosive experiment. In D. Coppersmith, ed., *Proc. CRYPTO '95*, vol. 963 of *Lecture Notes in Comput. Sci.*, pp. 372–385. Springer-Verlag, 1995.
16. T. ElGamal. A public-key cryptosystem and a signature scheme based on discrete logarithms. *IEEE Trans. Inform. Theory*, IT–31(4):469–472, July 1985.
17. G. Frey and H.-G. Rück. A remark concerning m-divisibility and the discrete logarithm in the divisor class group of curves. *Math. Comp.*, 62(206):865–874, Apr. 1994.
18. S. D. Galbraith, S. M. Paulus, and N. P. Smart. Arithmetic on superelliptic curves. To appear in *Mathematics of Computation*, 2001.
19. D. M. Gordon. Discrete logarithms in GF(p) using the number field sieve. *SIAM J. Discrete Math.*, 6(1):124–138, Feb. 1993.
20. D. M. Gordon and K. S. McCurley. Massively parallel computation of discrete logarithms. In E. F. Brickell, ed., *Proc. CRYPTO '92*, vol. 740 of *Lecture Notes in Comput. Sci.*, pp. 312–323. Springer-Verlag, 1993.
21. T. Granlund. `GMP`, the GNU multiple precision arithmetic library. Homepage at `http://www.swox.se/gmp`.
22. A. Joux and R. Lercier. Discrete logarithms in GF(p) (120 decimal digits). Email to the `NMBRTHRY` mailing list; available at `http://listserv.nodak.edu/archives/nmbrthry.html`, Apr. 2001.
23. A. Joux and R. Lercier. Discrete logarithms in GF(2^n) (521 bits). Email to the `NMBRTHRY` mailing list; available at `http://listserv.nodak.edu/archives/nmbrthry.html`, Sept. 2001.
24. E. Kaltofen. Analysis of Coppersmith's block Wiedemann algorithm for the parallel solution of sparse linear systems. *Math. Comp.*, 64(210):777–806, July 1995.
25. E. Kaltofen and A. Lobo. Distributed matrix-free solution of large sparse linear systems over finite fields. *Algorithmica*, 24:331–348, 1999.
26. N. Koblitz. Elliptic curve cryptosystems. *Math. Comp.*, 48(177):203–209, Jan. 1987.
27. N. Koblitz. Hyperelliptic cryptosystems. *J. of Cryptology*, 1:139–150, 1989.
28. B. A. LaMacchia and A. M. Odlyzko. Solving large sparse linear systems over finite fields. In A. J. Menezes and S. A. Vanstone, eds., *Proc. CRYPTO '90*, vol. 537 of *Lecture Notes in Comput. Sci.*, pp. 109–133. Springer-Verlag, 1990.
29. A. K. Lenstra and H. W. Lenstra, Jr., eds. *The development of the number field sieve*, vol. 1554 of *Lecture Notes in Math.* Springer, 1993.
30. A. K. Lenstra and M. S. Manasse. Factoring with two large primes. *Math. Comp.*, 63(208):785–798, Oct. 1994.
31. R. Lidl and H. Niederreiter. *Finite fields.* Number 20 in Encyclopedia of mathematics and its applications. Addison–Wesley, Reading, MA, 1983.
32. A. Menezes, T. Okamoto, and S. A. Vanstone. Reducing elliptic curves logarithms to logarithms in a finite field. *IEEE Trans. Inform. Theory*, IT–39(5):1639–1646, Sept. 1993.
33. A. J. Menezes. *Elliptic curve public key cryptosystems.* Kluwer Academic Publishers, 1993.

34. P. L. Montgomery. A block Lanczos algorithm for finding dependencies over GF(2). In L. C. Guillou and J.-J. Quisquater, eds., *Proc. EUROCRYPT '95*, vol. 921 of *Lecture Notes in Comput. Sci.*, pp. 106–120, 1995.
35. H. Niederreiter. A new efficient factorization algorithm for polynomials over small finite fields. *Appl. Algebra Engrg. Comm. Comput.*, 4:81–87, 1993.
36. A. M. Odlyzko. Discrete logarithms in finite fields and their cryptographic significance. In T. Beth, N. Cot, and I. Ingemarsson, eds., *Proc. EUROCRYPT '84*, vol. 209 of *Lecture Notes in Comput. Sci.*, pp. 224–314. Springer-Verlag, 1985.
37. C. Pomerance and J. W. Smith. Reduction of huge, sparse matrices over finite fields via created catastrophes. *Experiment. Math.*, 1(2):89–94, 1992.
38. C. P. Schnorr. Efficient signature generation by smart cards. *J. of Cryptology*, 4(3):161–174, 1991.
39. E. Thomé. Fast computation of linear generators for matrix sequences and application to the block wiedemann algorithm. In B. Mourrain, ed., *Proc. ISSAC '2001*, pp. 323–331. ACM Press, 2001.
40. G. Villard. A study of Coppersmith's block Wiedemann algorithm using matrix polynomials. Research Report 975, LMC-IMAG, Grenoble, France, Apr. 1997.
41. G. Villard. Further analysis of Coppersmith's block Wiedemann algorithm for the solution of sparse linear systems. In W. W. Küchlin, ed., *Proc. ISSAC '97*, pp. 32–39. ACM Press, 1997.
42. D. Weber and T. Denny. The solution of McCurley's discrete log challenge. In H. Krawczyk, ed., *Proc. CRYPTO '98*, vol. 1462 of *Lecture Notes in Comput. Sci.*, pp. 458–471. Springer-Verlag, 1998.
43. D. H. Wiedemann. Solving sparse linear equations over finite fields. *IEEE Trans. Inform. Theory*, IT–32(1):54–62, Jan. 1986.

Speeding Up XTR

Martijn Stam[1,*] and Arjen K. Lenstra[2]

[1] Technische Universiteit Eindhoven
P.O.Box 513, 5600 MB Eindhoven, The Netherlands
stam@win.tue.nl

[2] Citibank, N.A. and Technische Universiteit Eindhoven
1 North Gate Road, Mendham, NJ 07945-3104, U.S.A.
arjen.lenstra@citicorp.com

Abstract. This paper describes several speedups and simplifications for XTR. The most important results are new XTR double and single exponentiation methods where the latter requires a cheap precomputation. Both methods are on average more than 60% faster than the old methods, thus more than doubling the speed of the already fast XTR signature applications. An additional advantage of the new double exponentiation method is that it no longer requires matrices, thereby making XTR easier to implement. Another XTR single exponentiation method is presented that does not require precomputation and that is on average more than 35% faster than the old method. Existing applications of similar methods to LUC and elliptic curve cryptosystems are reviewed.

Keywords: XTR, addition chains, Fibonacci sequences, binary Euclidean algorithm, LUC, ECC.

1 Introduction

The XTR public key system was introduced at Crypto 2000 [10]. From a security point of view XTR is a traditional subgroup discrete logarithm system, as was proved in [10]. It uses a non-standard way to represent and compute subgroup elements to achieve substantial computational and communication advantages over traditional representations. XTR of security equivalent to 1024-bit RSA achieves speed comparable to cryptosystems based on random elliptic curves over random prime fields (ECC) of equivalent security. The corresponding XTR public keys are only about twice as large as ECC keys, assuming global system parameters – without the last requirement the sizes of XTR and ECC public keys are about the same. Furthermore, parameter initialization from scratch for XTR takes a negligible amount of computing time, unlike RSA and ECC.

This paper describes several important speedups for XTR, while at the same time simplifying its implementation. In the first place the field arithmetic as described in [10] is improved by combining the modular reduction steps. More importantly, a new application of a method from [15] is presented that results in

* The first author is sponsored by STW project EWI.4536

C. Boyd (Ed.): ASIACRYPT 2001, LNCS 2248, pp. 125–143, 2001.
© Springer-Verlag Berlin Heidelberg 2001

an XTR exponentiation iteration that can be used for three different purposes. In the first place these improvements result in an XTR double exponentiation method that is on average more than 60% faster than the double exponentiation from [10]. Such exponentiations are used in XTR ElGamal-like signature verifications. Furthermore, they result in two new XTR single exponentiation methods, one that is on average about 60% faster than the method from [10] but that requires a one-time precomputation, and a generic one without precomputation that is on average 35% faster than the old method.

Examples where precomputation can typically be used are the 'first' of the two exponentiations (per party) in XTR Diffie-Hellman key agreement, XTR ElGamal-like signature generation, and, to a lesser extent, XTR-ElGamal encryption. The new generic XTR single exponentiation can be used in the 'second' XTR Diffie-Hellman exponentiation and in XTR-ElGamal decryption. As a result the runtime of XTR signature applications is more than halved, the time required for XTR Diffie-Hellman is almost halved, and XTR-ElGamal encryption and decryption can both be expected to run at least 35% faster (with encryption running more than 60% faster after precomputation).

The method from [15] was developed to compute Lucas sequences. It can thus immediately be applied to the LUC cryptosystem [18]. It was shown [16] that it can also be applied to ECC. The resulting methods compare favorably to methods that have been reported in the literature [5]. Because they are not generally known their runtimes are reviewed at the end of this paper.

The double exponentiation method from [10] uses matrices. The new method does away with the matrices, thereby removing the esthetically least pleasing aspect of XTR. For completeness, another double exponentiation method is shown that does not require matrices. It is directly based on the iteration from [10] and does not achieve a noticeable speedup over the double exponentiation from [10], since the matrix steps that are no longer needed, though cumbersome, are cheap.

This paper is organized as follows. Section 2 reviews the results from [10] needed for this paper. It includes a description of the faster field arithmetic and matrix-less XTR double exponentiation based on the iteration from [10]. The 60% faster (and also matrix-less) XTR double exponentiation is presented in Section 3. Applications of the method from Section 3 to XTR single exponentiation with precomputation and to generic XTR single exponentiation are described in Sections 4 and 5, respectively. In Section 6 the runtime claims are substantiated by direct comparison with the timings from [10]. Section 7 reviews the related LUC and ECC results.

2 XTR Background

For background and proofs of the statements in this section, see [10]. Let p and q be primes with $p \equiv 2 \bmod 3$ and q dividing $p^2 - p + 1$, and let g be a generator of the order q subgroup of $\mathbf{F}_{p^6}^*$. For $h \in \mathbf{F}_{p^6}^*$ its trace $Tr(h)$ over $\mathbf{F}_{p^2}$ is defined as the sum of the conjugates over $\mathbf{F}_{p^2}$ of h:

$$Tr(h) = h + h^{p^2} + h^{p^4} \in \mathbf{F}_{p^2}.$$

Because the order of h divides $p^6 - 1$ the trace over $\mathbf{F}_{p^2}$ of h equals the trace of the conjugates over $\mathbf{F}_{p^2}$ of h:

$$(1) \qquad Tr(h) = Tr(h^{p^2}) = Tr(h^{p^4}).$$

If $h \in \langle g \rangle$ then its order divides $p^2 - p + 1$, so that

$$Tr(h) = h + h^{p-1} + h^{-p}$$

since $p^2 \equiv p - 1 \bmod (p^2 - p + 1)$ and $p^4 \equiv -p \bmod (p^2 - p + 1)$. In XTR elements of $\langle g \rangle$ are represented by their trace over $\mathbf{F}_{p^2}$. It follows from (1) that XTR makes no distinction between an element of $\langle g \rangle$ and its conjugates over $\mathbf{F}_{p^2}$.

The discrete logarithm (DL) problem in $\langle g \rangle$ is to compute for a given $h \in \langle g \rangle$ the unique $y \in \{0, 1, \ldots, q - 1\}$ such that $g^y = h$. The XTR-DL problem is to compute for a given $Tr(h)$ with $h \in \langle g \rangle$ an integer $y \in \{0, 1, \ldots, q-1\}$ such that $Tr(g^y) = Tr(h)$. If y solves an XTR-DL problem then $(p - 1)y$ and $-py$ (both taken modulo q) are solutions too. It is proved in [10, Theorem 5.2.1] that the XTR-DL problem is equivalent to the DL problem in $\langle g \rangle$, with similar equivalences with respect to the Diffie-Hellman and Decision Diffie-Hellman problems. Furthermore, it is argued in [10] that if q is sufficiently large (which will be the case), then the DL problem in $\langle g \rangle$ is as hard as it is in $\mathbf{F}^*_{p^6}$. This argument is the most commonly misunderstood aspect of XTR and therefore rephrased here.

Because of the Pohlig-Hellman algorithm [17] and the fact that $p^6 - 1 = (p-1)(p+1)(p^2+p+1)(p^2-p+1)$, the general DL problem in $\mathbf{F}^*_{p^6}$ reduces to the DL problems in the following four subgroups of $\mathbf{F}^*_{p^6}$:

- The subgroup of order $p - 1$, which can efficiently be embedded in $\mathbf{F}_p$.
- The subgroup of order $p+1$ dividing p^2-1, which can efficiently be embedded in $\mathbf{F}_{p^2}$ but not in $\mathbf{F}_p$.
- The subgroup of order $p^2 + p + 1$ dividing $p^3 - 1$, which can efficiently be embedded in $\mathbf{F}_{p^3}$ but not in $\mathbf{F}_p$.
- The subgroup of order $p^2 - p + 1$, which cannot be embedded in any true subfield of $\mathbf{F}_{p^6}$.

So, to solve the DL problem in $\mathbf{F}^*_{p^6}$ in the most general case, four DL problems must be solved. Three of these DL problems can efficiently be reformulated as DL problems in multiplicative groups of the true subfields $\mathbf{F}_p$, $\mathbf{F}_{p^2}$, and $\mathbf{F}_{p^3}$ of $\mathbf{F}_{p^6}$. With the current state of the art of the DL problem in extension fields, these latter three problems are believed to be strictly (and substantially) easier than the DL problem in $\mathbf{F}^*_{p^6}$. But that means that the subgroup of order $p^2 - p + 1$ is, so to speak, the subgroup that is responsible for the difficulty of the DL problem in $\mathbf{F}^*_{p^6}$. With a proper choice of q dividing $p^2 - p + 1$, this subgroup DL problem is equivalent to the problem in $\langle g \rangle$. This implies that the DL problem in $\langle g \rangle$ is as hard as it is in $\mathbf{F}^*_{p^6}$, unless the latter problem is not as hard as it is currently believed to be. It also follows that, if the DL problem in $\langle g \rangle$ is easier than it is in $\mathbf{F}^*_{p^6}$, then the problem in $\mathbf{F}^*_{p^6}$ can be at most as hard as it is in $\mathbf{F}^*_p$, $\mathbf{F}^*_{p^2}$, or $\mathbf{F}^*_{p^3}$. Proving such a result would require a major breakthrough.

Thus, for cryptographic purposes and given the current state of knowledge regarding the DL problem in extension fields, XTR and $\mathbf{F}_{p^6}$ give the same security. For p and q of about 170 bits the security is at least equivalent to 1024-bit RSA and approximately equivalent to 170-bit ECC.

XTR has two main advantages compared to ordinary representation of elements of $\langle g \rangle$:

- It is shorter, since $Tr(h) \in \mathbf{F}_{p^2}$, whereas representing an element of $\langle g \rangle$ requires in general an element of $\mathbf{F}_{p^6}$, i.e., three times more bits;
- It allows faster arithmetic, because given $Tr(g)$ and u the value $Tr(g^u)$ can be computed substantially faster than g^u can be computed given g and u.

In this paper it is shown that $Tr(g^u)$ can be computed even faster than shown in [10].

Throughout this paper, c_u denotes $Tr(g^u) \in \mathbf{F}_{p^2}$, for some fixed p and g of order q as above. Note that $c_0 = 3$. In [10,11,12] it is shown how p, q, and c_1 can be found quickly. In particular there is no need to find an explicit representation of $g \in \mathbf{F}_{p^6}$.

2.1 Improved $\mathbf{F}_{p^2}$ Arithmetic. Because $p \equiv 2 \bmod 3$, the zeros α and α^p of the polynomial $(X^3-1)/(X-1) = X^2+X+1$ form an optimal normal basis for $\mathbf{F}_{p^2}$ over $\mathbf{F}_p$. An element $x \in \mathbf{F}_{p^2}$ is represented as $x_1\alpha + x_2\alpha^2$ with $x_1, x_2 \in \mathbf{F}_p$. From $\alpha^2 = \alpha^p$ if follows that $x^p = x_2\alpha + x_1\alpha^2$, so that p-th powering in $\mathbf{F}_{p^2}$ is free. In [10] the product $(x_1\alpha + x_2\alpha^2)(y_1\alpha + y_2\alpha^2)$ is computed by computing x_1y_1, x_2y_2, $(x_1+x_2)(y_1+y_2) \in \mathbf{F}_p$, so that $x_1y_2 + x_2y_1 \in \mathbf{F}_p$ and the product

$$(x_2y_2 - x_1y_2 - x_2y_1)\alpha + (x_1y_1 - x_1y_2 - x_2y_1)\alpha^2 \in \mathbf{F}_{p^2}$$

follow using four subtractions. This implies that products in $\mathbf{F}_{p^2}$ can be computed at the cost of three multiplications in $\mathbf{F}_p$ (as usual, the small number of additions and subtractions is not counted).

For a regular multiplication of $u, v \in \mathbf{F}_p$ the field elements u and v are mapped to integers $\bar{u}, \bar{v} \in \{0, 1, \ldots, p-1\}$, the integer product $\bar{w} = \bar{u}\bar{v} \in \mathbf{Z}$ is computed (the 'multiplication step'), the remainder $\bar{w} \bmod p \in \{0, 1, \ldots, p-1\}$ is computed (the 'reduction step'), and finally the resulting integer $\bar{w} \bmod p$ is mapped to $\mathbf{F}_p$. The reduction step is somewhat costlier than the multiplication step; the mappings between $\mathbf{F}_p$ and $\mathbf{Z}$ are negligible. The same applies if Montgomery arithmetic [13] is used, but then the reduction and multiplication step are about equally costly.

It follows that the computation of $(x_1\alpha + x_2\alpha^2)(y_1\alpha + y_2\alpha^2)$ can be made faster by computing, in the above notation, $\bar{w}_1 = \bar{x}_2\bar{y}_2 - \bar{x}_1\bar{y}_2 - \bar{x}_2\bar{y}_1 \in \mathbf{Z}$ and $\bar{w}_2 = \bar{x}_1\bar{y}_1 - \bar{x}_1\bar{y}_2 - \bar{x}_2\bar{y}_1 \in \mathbf{Z}$ using four integer multiplications, followed by two reductions $\bar{w}_1 \bmod p$ and $\bar{w}_2 \bmod p$. This works both for regular and Montgomery arithmetic. Because the intermediate results are at most $3p^2$ in absolute value the resulting final reductions are of the same cost as the original reductions (with additional subtraction correction in Montgomery arithmetic, at negligible extra cost). As a result, products in $\mathbf{F}_{p^2}$ can be computed at the cost of just two and a half multiplications in $\mathbf{F}_p$, namely the usual three multiplication steps and just two reduction steps. If regular arithmetic is used the speedup can be expected to be somewhat larger. It follows in a similar way that the computation of $xz - yz^p \in \mathbf{F}_{p^2}$ for $x, y, z \in \mathbf{F}_{p^2}$ can be reduced from four multiplications in $\mathbf{F}_p$ to the same cost as three multiplications in $\mathbf{F}_p$; refer to [10, Section 2.1] for the details of that computation. Combining, or postponing, the reduction steps in this way is not at all new. See for instance [4] for a much earlier application.

This results in the following improved version of [10, Lemma 2.1.1].

Lemma 2.2 *Let $x, y, z \in \mathbf{F}_{p^2}$ with $p \equiv 2 \bmod 3$.*
i. *Computing x^p is free.*
ii. *Computing x^2 takes two multiplications in $\mathbf{F}_p$.*
iii. *Computing xy costs the same as two and a half multiplications in $\mathbf{F}_p$.*
iv. *Computing $xz - yz^p$ costs the same as three multiplications in $\mathbf{F}_p$.*

Efficient computation of c_u given p, q, and c_1 is based on the following facts.

2.3 Facts. Fact 2b follows from Lemma 2.2 and Facts 1b and 2a. The other facts are derived as in [10].

1. Identities involving traces of powers, with $u, v \in \mathbf{Z}$:
   a) $c_{-u} = c_{up} = c_u^p$. It follows from Lemma 2.2.*i* that negations and p-th powers can be computed for free.
   b) $c_{u+v} = c_u c_v - c_v^p c_{u-v} + c_{u-2v}$. It follows from Lemma 2.2.*i* and *iv* that c_{u+v} can be computed at the cost of three multiplications in $\mathbf{F}_p$ if c_u, c_v, c_{u-v}, and c_{u-2v} are given.
   c) If $c_u = \tilde{c}_1$, then $\tilde{c}_v$ denotes the trace of the v-th power g^{uv} of g^u, so that $c_{uv} = \tilde{c}_v$.
2. Computing traces of powers, with $u \in \mathbf{Z}$:
   a) $c_{2u} = c_u^2 - 2c_u^p$ takes two multiplications in $\mathbf{F}_p$.
   b) $c_{3u} = c_u^3 - 3c_u^{p+1} + 3$ costs four and a half multiplications in $\mathbf{F}_p$, and produces c_{2u} as a side-result.
   c) $c_{u+2} = c_1 c_{u+1} - c_1^p c_u + c_{u-1}$ costs three multiplications in $\mathbf{F}_p$.
   d) $c_{2u-1} = c_{u-1} c_u - c_1^p c_u^p + c_{u+1}^p$ costs three multiplications in $\mathbf{F}_p$.
   e) $c_{2u+1} = c_{u+1} c_u - c_1 c_u^p + c_{u-1}^p$ costs three multiplications in $\mathbf{F}_p$.

Let S_u denote the triple (c_{u-1}, c_u, c_{u+1}); thus $S_1 = (3, c_1, c_1^2 - 2c_1^p)$. The triple $S_{2u-1} = (c_{2(u-1)}, c_{2u-1}, c_{2u})$ can be computed from S_u and c_1 by applying Fact 2a twice to compute $c_{2(u-1)}$ and c_{2u} based on c_{u-1} and c_u, respectively, and by applying Fact 2d to compute c_{2u-1} based on $S_u = (c_{u-1}, c_u, c_{u+1})$ and c_1. This takes seven multiplications in $\mathbf{F}_p$. The triple S_{2u+1} can be computed in a similar fashion from S_u and c_1 at the cost of seven multiplications in $\mathbf{F}_p$ (using Fact 2e to compute c_{2u+1}).

Let v be a non-negative integer, and let $v = \sum_{i=0}^{r-1} v_i 2^i$ be the binary representation of v, where $v_i \in \{0, 1\}$, $r > 0$, and $v_{r-1} = 1$. It is well known that the v-th power of an element of, say, a finite field can be computed using the ordinary square and multiply method based on the binary representation of v. A similar iteration can be used to compute S_{2v+1}, given S_1.

2.4 XTR Single Exponentiation (cf. [10, Algorithm 2.3.7]). Let S_1, c_1, and $v_{r-1}, v_{r-2}, \ldots, v_0 \in \{0,1\}$ be given, let $y = 1$ and $e = 0$ (so that $2e+1 = y$; the values y and e are included for expository purposes only). To compute S_{2v+1} with $v = \sum_{i=0}^{r-1} v_i 2^i$, do the following for $i = r-1, r-2, \ldots, 0$ in succession:

Bit off If $v_i = 0$, then compute S_{2y-1} based on S_y and c_1, replace S_y by S_{2y-1} (and thus S_{2e+1} by $S_{2(2e)+1}$ because it follows from $2e+1 = y$ that $2(2e)+1 = 4e+1 = 2y-1$), replace y by $2y-1$, and e by $2e$ (so that the invariant $2e+1 = y$ is maintained).

Bit on Else if $v_i = 1$, then compute S_{2y+1} based on S_y and c_1, replace S_y by S_{2y+1} (and thus S_{2e+1} by $S_{2(2e+1)+1}$ because it follows from $2e+1 = y$ that $2(2e+1)+1 = 4e+3 = 2y+1$), replace y by $2y+1$, and e by $2e+1$ (so that the invariant $2e+1 = y$ is maintained).

As a result $e = v$. Because $2e+1 = y$ the final S_y equals S_{2v+1}. Note that v_{r-1}, or any other v_i, does not have to be non-zero.

Both the 'bit off' and the 'bit on' step of Algorithm 2.4 take seven multiplications in $\mathbf{F}_p$. Thus, given an odd positive integer $t < q$ and S_1, the triple $S_t = (c_{t-1}, c_t, c_{t+1})$ can be computed in $7\log_2 t$ multiplications in $\mathbf{F}_p$. In [10] this was $8\log_2 t$ because of the slower field arithmetic used there. The restriction that t is odd and positive is easily removed: if t is even, then first compute S_{t-1} and next apply Fact 2c, and if t is negative, then use Fact 1a.

In Algorithm 2.4, the trace c_1 of g in $S_1 = (c_0, c_1, c_2) = (3, c_1, c_1^2 - 2c_1^p)$ can be replaced by the trace c_t of the t-th power g^t of g (cf. Fact 1c): with $\tilde{c}_1 = c_t$, $\tilde{S}_1 = (\tilde{c}_0, \tilde{c}_1, \tilde{c}_2) = (3, c_t, c_{2t}) = (3, c_t, c_t^2 - 2c_t^p)$, and the previous paragraph, the triple $\tilde{S}_v = (\tilde{c}_{v-1}, \tilde{c}_v, \tilde{c}_{v+1}) = (c_{(v-1)t}, c_{vt}, c_{(v+1)t})$ can be computed in $7\log_2 v$ multiplications in $\mathbf{F}_p$, for any positive integer $v < q$.

Now let $v = \sum_{i=0}^{r-1} v_i 2^i$ as above and let

$$v' = 2^r k + v = \sum_{i=0}^{s+r-1} v_i 2^i$$

for some integer $k \geq 1$. After the first s iterations of the application of Algorithm 2.4 to S_1, c_1, and $v_{s+r-1}, v_{s+r-2}, \ldots, v_0$ the value for e equals k and $S_y = S_{2k+1}$. The remaining r iterations result in $S_{2v'+1} = S_{2^{r+1}k+2v+1}$, and are the same as if Algorithm 2.4 was applied to S_y (as opposed to S_1) and $v_{r-1}, v_{r-2}, \ldots, v_0$. It follows that if Algorithm 2.4 is applied to S_{2k+1}, c_1, and $v_{r-1}, v_{r-2}, \ldots, v_0$, then the resulting value is $S_{2^{r+1}k+2v+1}$. Note that the v_i's do not have to be non-zero. Thus, given any (odd or even) $t < 2^{r+1}$, S_k, and c_1, the triple $S_{2^{r+1}k+t}$ can be computed in $7\log_2 t$ multiplications in $\mathbf{F}_p$. This leads to the following double exponentiation method for XTR.

2.5 Matrix-Less XTR Double Exponentiation. Let a and b be integers with $0 < a, b < q$, and let S_k and c_1 be given. To compute c_{bk+a} do the following.

1. Let r be such that $2^r < q < 2^{r+1}$.
2. Compute $d = b/2^{r+1} \bmod q$ and $t = a/d \bmod q$.
3. Compute $S_{2^{r+1}k+t}$:
   - Use Facts 2a and 2e to compute S_{2k+1} based on S_k.
   - If t is odd let $t' = t$, else let $t' = t - 1$.
   - Let $t' = 2v + 1$.
   - Let $v = \sum_{i=0}^{r-1} v_i 2^i$ with $v_i \in \{0,1\}$ (and $v_{r-1}, v_{r-2}, \ldots$ possibly zero).
   - Apply Algorithm 2.4 to S_{2k+1}, c_1, and $v_{r-1}, v_{r-2}, \ldots, v_0$, resulting in $S_{2^{r+1}k+t'}$.
   - If t is odd then $S_{2^{r+1}k+t} = S_{2^{r+1}k+t'}$, else use Fact 2c to compute $S_{2^{r+1}k+t} = S_{2^{r+1}k+t'+1}$ based on $S_{2^{r+1}k+t'}$.
4. Let $\tilde{c}_1 = c_{2^{r+1}k+t}$.
5. Compute $\tilde{S}_1 = (\tilde{c}_0, \tilde{c}_1, \tilde{c}_2) = (3, \tilde{c}_1, \tilde{c}_1^2 - 2\tilde{c}_1^p)$ (cf. Fact 1c).
6. Apply Algorithm 2.4 to $\tilde{S}_1$, $\tilde{c}_1$ and the bits containing the binary representation of d, resulting in $\tilde{S}_d = (\tilde{c}_{d-1}, \tilde{c}_d, \tilde{c}_{d+1})$.
7. The resulting $\tilde{c}_d$ equals $c_{d(2^{r+1}k+t) \bmod q} = c_{bk+a}$.

Algorithm 2.5 takes about $14 \log_2 q$ multiplications in $\mathbf{F}_p$. This is a small constant number of multiplications in $\mathbf{F}_p$ better than [10, Algorithm 2.4.8] (assuming the faster field arithmetic is used there too). For realistic choices of q the speedup achieved using Algorithm 2.5 is thus barely noticeable. Nevertheless, it is a significant result because the fact that the matrices as required for [10, Algorithm 2.4.8] are no longer needed, facilitates implementation of XTR. In Section 3 of this paper a more substantial improvement over the double exponentiation method from [10] is described that does not require matrices either.

3 Improved Double Exponentiation

In this section it is shown how c_{bk+a} can be computed based on S_k and c_1 (or, equivalently, based on $S_{k-1} = (c_{k-2}, c_{k-1}, c_k)$ and c_1, cf. Fact 2.3.1b) in a single iteration, as opposed to the two iterations in Algorithm 2.5. For greater generality, it is shown how $c_{bk+a\ell}$ is computed, based on c_k, c_ℓ, $c_{k-\ell}$, and $c_{k-2\ell}$.

A rough outline of the new XTR double exponentiation method is as follows. Let $u = k$, $v = \ell$, $d = b$, and $e = a$. It follows that $ud + ve = bk + a\ell$ and that c_u, v_v, c_{u-v}, and c_{u-2v} are known. The values of d and e are decreased, while at the same time u and v (and thereby c_u, c_v, c_{u-v}, and c_{u-2v}) are updated, in order to maintain the invariant $ud + ve = bk + a\ell$. The changes in d and e are effected in such a way that at a given point $d = e$. But if $d = e$, then $bk + a\ell = ud + ve = d(u + v)$, so that $c_{bk+a\ell}$ follows by computing c_{u+v} and next $c_{d(u+v)}$ (cf. Fact 2.3.1c).

There are various ways in which d and e can be changed. The most efficient method to date was proposed by P.L. Montgomery in [15], for the computation of second degree recurrent sequences. The method below is an adaptation of [15, Table 4] to the present case of third degree sequences.

3.1 Simultaneous XTR Double Exponentiation. Let a, b, c_k, c_ℓ, $c_{k-\ell}$, and $c_{k-2\ell}$ be given, with $0 < a, b < q$. To compute $c_{bk+a\ell}$ do the following.

1. Let $u = k$, $v = \ell$, $d = b$, $e = a$, $c_u = c_k$, $c_v = c_\ell$, $c_{u-v} = c_{k-\ell}$, $c_{u-2v} = c_{k-2\ell}$, $f_2 = 0$, and $f_3 = 0$ (u and v are carried along for expository purposes only).
2. As long as d and e are both even, replace (d, e) by $(d/2, e/2)$ and f_2 by f_2+1.
3. As long as d and e are both divisible by 3, replace (d, e) by $(d/3, e/3)$ and f_3 by $f_3 + 1$.
4. As long as $d \neq e$ replace $(d, e, u, v, c_u, c_v, c_{u-v}, c_{u-2v})$ by the 8-tuple given below.
   a) If $d > e$ then
      i. if $d \leq 4e$, then $(e, d-e, u+v, u, c_{u+v}, c_u, c_v, c_{v-u})$.
      ii. else if d is even, then $(\frac{d}{2}, e, 2u, v, c_{2u}, c_v, c_{2u-v}, c_{2(u-v)})$.
      iii. else if e is odd, then $(\frac{d-e}{2}, e, 2u, u+v, c_{2u}, c_{u+v}, c_{u-v}, c_{-2v})$.
      iv. **optional:**
      else if $d \equiv e \bmod 3$, then $(\frac{d-e}{3}, e, 3u, u+v, c_{3u}, c_{u+v}, c_{2u-v}, c_{u-2v})$.
      v. else (e is even), then $(\frac{e}{2}, d, 2v, u, c_{2v}, c_u, c_{2v-u}, c_{2(v-u)})$.
   b) Else (if $e > d$)
      i. if $e \leq 4d$, then $(d, e-d, u+v, v, c_{u+v}, c_v, c_u, c_{u-v})$.
      ii. else if e is even, then $(\frac{e}{2}, d, 2v, u, c_{2v}, c_u, c_{2v-u}, c_{2(v-u)})$.
      iii. else if d is odd, then $(\frac{e-d}{2}, d, 2v, u+v, c_{2v}, c_{u+v}, c_{v-u}, c_{-2u})$.
      iv. **optional:**
      else if $e \equiv 0 \bmod 3$, then $(\frac{e}{3}, d, 3v, u, c_{3v}, c_u, c_{3v-u}, c_{3v-2u})$.
      v. **optional:**
      else if $e \equiv d \bmod 3$, then $(\frac{e-d}{3}, d, 3v, u+v, c_{3v}, c_{u+v}, c_{2v-u}, c_{v-2u})$.
      vi. else (d is even), then $(\frac{d}{2}, e, 2u, v, c_{2u}, c_v, c_{2u-v}, c_{2(u-v)})$.
5. Apply Fact 2.3.1b to c_u, c_v, c_{u-v}, and c_{u-2v}, to compute $\tilde{c}_1 = c_{u+v}$.
6. Apply Algorithm 2.4 to $\tilde{S}_1 = (3, \tilde{c}_1, \tilde{c}_1 - 2\tilde{c}_1^p)$, $\tilde{c}_1$, and the binary representation of d, resulting in $\tilde{c}_d = c_{d(u+v)}$ (cf. Fact 2.3.1c). Alternatively, and on average faster, apply Algorithm 5.1 described below to compute $\tilde{c}_d = c_{d(u+v)}$ based on $\tilde{c}_1$ (note that this results in a recursive call to Algorithm 3.1).
7. Compute $c_{2^{f_2}d(u+v)}$ based on $c_{d(u+v)}$ by applying Fact 2.3.2a f_2 times.
8. Compute $c_{3^{f_3}2^{f_2}d(u+v)}$ based on $c_{2^{f_2}d(u+v)}$ by applying Fact 2.3.2b f_3 times.

The asymmetry between Steps 4a and 4b is caused by the asymmetry between u and v, i.e., c_{u-2v} is available but c_{v-2u} is not. As a consequence, the case '$d \equiv 0 \bmod 3$' is slower than the case '$e \equiv 0 \bmod 3$' (Step 4(b)iv), and its inclusion would slow down Algorithm 3.1.

Steps 4(a)i and 4(b)i each require a single application of Fact 2.3.1b at the cost of three multiplications in $\mathbf{F}_p$. Steps 4(a)v and 4(b)ii each require two applications of Fact 2.3.2a at the cost of $2+2 = 4$ multiplications in $\mathbf{F}_p$. Steps 4(a)ii, 4(a)iii, 4(b)iii, and 4(b)vi each require an application of Fact 2.3.1b and two applications of Fact 2.3.2a at the cost of $3+2+2 = 7$ multiplications in $\mathbf{F}_p$. The three optional steps 4(a)iv, 4(b)iv, and 4(b)v each require two applications of Fact 2.3.1b and one application of Fact 2.3.2b for a total cost of $3+3+4.5 = 10.5$ multiplications in $\mathbf{F}_p$.

In Table 1 the number of multiplications in $\mathbf{F}_p$ required by Algorithm 3.1 is given, both with and without optional steps 4(a)iv, 4(b)iv, and 4(b)v. Each set of entries is averaged over the same collection of 2^{20} randomly selected t's, a's, and b's, with t of the size specified in Table 1 and a and b randomly selected from $\{1, 2, \ldots, t-1\}$. For regular double exponentiation $t \approx q$, but $t \approx \sqrt{q}$ for the application in Section 4. It follows from Table 1 that inclusion of the optional steps leads to an overall reduction of more than 6% in the expected number of multiplications in $\mathbf{F}_p$. For the optional steps it is convenient to keep track of the residue classes of d and e modulo 3. These are easily updated if any of the other steps applies, but require a division by 3 if either one of the optional steps is carried out. It depends on the implementation and the platform whether or not an overall saving is obtained by including the optional steps. In most software implementations it will most likely be worthwhile.

Table 1. Empirical performance of Algorithm 3.1, with $0 < a, b < t$.

multiplications in $\mathbf{F}_p$

$\lceil \log_2 t \rceil = T$	including steps 4(a)iv, 4(b)iv, and 4(b)v: average	standard deviation σ	$\sigma/\sqrt{T}$	without steps 4(a)iv, 4(b)iv, and 4(b)v: average	standard deviation σ	$\sigma/\sqrt{T}$
60	$350.01 = 5.83T$	$20.5 = 0.34T$	2.65	$372.89 = 6.21T$	$30.0 = 0.50T$	3.88
70	$410.42 = 5.86T$	$22.2 = 0.32T$	2.65	$437.41 = 6.25T$	$32.6 = 0.47T$	3.89
80	$470.84 = 5.89T$	$23.7 = 0.30T$	2.65	$501.94 = 6.27T$	$34.8 = 0.44T$	3.90
90	$531.21 = 5.90T$	$25.2 = 0.28T$	2.66	$566.36 = 6.29T$	$37.0 = 0.41T$	3.90
100	$591.63 = 5.92T$	$26.5 = 0.27T$	2.65	$630.85 = 6.31T$	$39.1 = 0.39T$	3.91
110	$652.03 = 5.93T$	$27.8 = 0.25T$	2.65	$695.40 = 6.32T$	$41.1 = 0.37T$	3.92
120	$712.39 = 5.94T$	$29.1 = 0.24T$	2.66	$759.87 = 6.33T$	$43.0 = 0.36T$	3.93
130	$772.78 = 5.94T$	$30.2 = 0.23T$	2.65	$824.31 = 6.34T$	$44.6 = 0.34T$	3.92
140	$833.19 = 5.95T$	$31.5 = 0.22T$	2.66	$888.91 = 6.35T$	$46.4 = 0.33T$	3.92
150	$893.66 = 5.96T$	$32.5 = 0.22T$	2.65	$953.34 = 6.36T$	$48.1 = 0.32T$	3.93
160	$953.98 = 5.96T$	$33.6 = 0.21T$	2.66	$1017.79 = 6.36T$	$49.7 = 0.31T$	3.93
170	$1014.42 = 5.97T$	$34.7 = 0.20T$	2.66	$1082.36 = 6.37T$	$51.3 = 0.30T$	3.93
180	$1074.84 = 5.97T$	$35.7 = 0.20T$	2.66	$1146.88 = 6.37T$	$52.7 = 0.29T$	3.93
190	$1135.19 = 5.97T$	$36.6 = 0.19T$	2.66	$1211.34 = 6.38T$	$54.3 = 0.29T$	3.94
200	$1195.58 = 5.98T$	$37.6 = 0.19T$	2.66	$1275.82 = 6.38T$	$55.7 = 0.28T$	3.94
210	$1256.05 = 5.98T$	$38.5 = 0.18T$	2.66	$1340.23 = 6.38T$	$57.1 = 0.27T$	3.94
220	$1316.42 = 5.98T$	$39.5 = 0.18T$	2.66	$1404.75 = 6.39T$	$58.5 = 0.27T$	3.94
230	$1376.87 = 5.99T$	$40.3 = 0.18T$	2.66	$1469.36 = 6.39T$	$59.7 = 0.26T$	3.94
240	$1437.25 = 5.99T$	$41.2 = 0.17T$	2.66	$1533.89 = 6.39T$	$61.1 = 0.25T$	3.94
250	$1497.61 = 5.99T$	$42.0 = 0.17T$	2.66	$1598.22 = 6.39T$	$62.3 = 0.25T$	3.94
260	$1558.00 = 5.99T$	$42.9 = 0.17T$	2.66	$1662.80 = 6.40T$	$63.7 = 0.24T$	3.95
270	$1618.47 = 5.99T$	$43.8 = 0.16T$	2.66	$1727.31 = 6.40T$	$64.9 = 0.24T$	3.95
280	$1678.74 = 6.00T$	$44.5 = 0.16T$	2.66	$1791.85 = 6.40T$	$66.1 = 0.24T$	3.95
290	$1739.17 = 6.00T$	$45.3 = 0.16T$	2.66	$1856.32 = 6.40T$	$67.2 = 0.23T$	3.94
300	$1799.57 = 6.00T$	$46.1 = 0.15T$	2.66	$1920.88 = 6.40T$	$68.4 = 0.23T$	3.95

Conjecture 3.2 *Given integers a and b with $0 < a, b < q$ and trace values c_k, c_ℓ, $c_{k-\ell}$, and $c_{k-2\ell}$, the trace value $c_{bk+a\ell}$ can on average be computed in about $6\log_2(\max(a,b))$ multiplications in $\mathbf{F}_p$ using Algorithm 3.1.*

It follows that XTR double exponentiation using Algorithm 3.1 is on average faster than the XTR single exponentiation from [10] (given in Algorithm 2.4), and more than twice as fast as the previous methods to compute $c_{bk+a\ell}$ ([10, Algorithm 2.4.8 and Theorem 2.4.9] and Algorithm 2.5). An additional advantage of Algorithm 3.1 is that, like Algorithm 2.5, it does not require matrices.

These advantages have considerable practical consequences, not only for the performance of XTR signature verification (Section 6), but also for the accessibility and ease of implementation of XTR. In Sections 4 and 5 consequences of Algorithm 3.1 for XTR single exponentiation are given.

Based on Table 1 the expected practical behavior of Algorithm 3.1 is well understood, and the practical merits of the method are beyond doubt. However, a satisfactory theoretical analysis of Algorithm 3.1, or the second degree original from [15], is still lacking. The iteration in Algorithm 3.1 is reminiscent of the binary and subtractive Euclidean greatest common divisor algorithms. Iterations of that sort typically exhibit an unpredictable behavior with a wide gap between worst and average case performance; see for instance [1,7,19] and the analysis attempts and open problems in [15].

This is further illustrated in Figure 1. There the average number of multiplications for $\lceil \log_2 t \rceil = 170$ is given as a function of the value of the constant in Steps 4(a)i and 4(b)i of Algorithm 3.1. The value 4 is close to optimal and convenient for implementation. However, it can be seen from Figure 1 that a value close to 4.8 is somewhat better, if one's sole objective is to minimize the number of multiplications in $\mathbf{F}_p$, as opposed to minimizing the overall runtime. The curves in Figure 1 were generated for constants ranging from 2 to 8 with stepsize 1/16, per constant averaged over the same collection of 2^{20} randomly selected t's, a's, and b's. The remarkable shape of the curves – both with at least four local minima – is a clear indication that the exact behavior of Algorithm 3.1 will be hard to analyse. It is of no immediate importance for the present paper and left as a subject for further study.

Remark 3.3 As shown in Appendix A other small improvements can be obtained by distinguishing more different cases than in Algorithm 3.1. The version presented above represents a good compromise that combines reasonable overhead with decent performance. In practical circumstances the performance of Algorithm 3.1 is on average close to optimal.

Remark 3.4 If Algorithm 3.1 is implemented using the slower field arithmetic from [10, Lemma 2.1.1], as opposed to the improved arithmetic from 2.1, it can on average be expected to require $7.4 \log_2(\max(a, b))$ multiplications in $\mathbf{F}_p$. This is still more than twice as fast as the method from [10] (using the slower arithmetic), but more than 20% slower than Conjecture 3.2.

Remark 3.5 Unlike the XTR exponentiation methods from [10], different instructions are carried out by Algorithm 3.1 for different input values. This makes Algorithm 3.1 inherently more vulnerable to environmental attacks than the methods from [10] (cf. [10, Remark 2.3.9]). If the possibility of such attacks is a concern, then utmost care should be taken while implementing Algorithm 3.1.

4 Single Exponentiation with Precomputation

Suppose that for a fixed c_1 several c_u's for different u's, with $0 < u < q$, have to be computed. In this section it is shown that, after a small amount of precom-

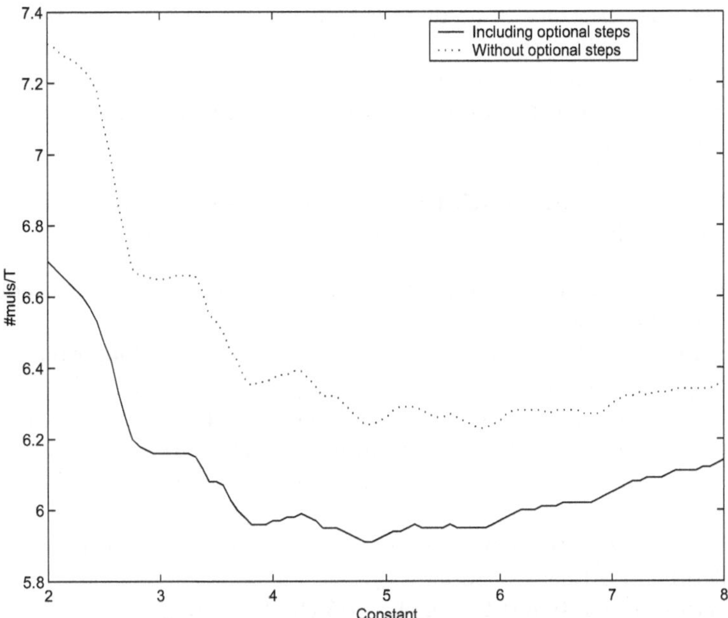

Fig. 1. Dependence on the value of the constant.

putation, this can be done using Algorithm 3.1 in less than half the number of multiplications in $\mathbf{F}_p$ that would be required by Algorithm 2.4.

Let $t = 2^{\lceil (\log_2 q)/2 \rceil}$, and suppose that $S_{t-1} = (c_{t-2}, c_{t-1}, c_t)$ has been precomputed based on c_1. For any $u \in \{0, 1, \ldots, q-1\}$ non-negative integers a and b of at most $1 + (\log_2 q)/2$ bits can simply be computed such that $u = bt + a$. Given S_{t-1} and c_1, the value c_u can then be computed using Algorithm 3.1 with $k = t$ and $\ell = 1$. This leads to the following precomputation and XTR single exponentiation with precomputation.

4.1 Precomputation. Let c_1 be given. To precompute values t and $S_{t-1} = (c_{t-2}, c_{t-1}, c_t)$ do the following.

1. Let $t = 2^{\lceil (\log_2 q)/2 \rceil}$, $v = (t-2)/2$, and let $v_{r-1}, v_{r-2}, \ldots, v_0$ be the binary representation of v (so $v_i = 1$ for $0 \le i < r$ for $t = 2^{[\log_2 q)/2]}$).
2. Apply Algorithm 2.4 to $S_1 = (3, c_1, c_1^2 - 2c_1^p)$, c_1, and $v_{r-1}, v_{r-2}, \ldots, v_0$ to compute $S_{2v+1} = S_{t-1}$.

The value S_{t-1} computed by Algorithm 4.1 consists of the traces of three consecutive powers of the subgroup generator corresponding to c_1. Algorithm 4.1 takes essentially a single application of Algorithm 2.4, and thus about $3.5 \log_2 q$ multiplications in $\mathbf{F}_p$, since $\log_2 t \approx (\log_2 q)/2$. Improved XTR single exponentiation Algorithm 5.1 given below would require more than a single application, because it produces just the trace of a single power, and not its two 'nearest neighbors' as well. With [11, Theorem 5.1], which for most t's allows fast computation of c_{t+1} given c_1, c_{t-1}, and c_t, two applications of Algorithm 5.1 would suffice. But that is still expected to be slower than a single application of Algorithm 2.4, as follows from Corollary 5.3.

4.2 XTR Single Exponentiation with Precomputation. Let u, c_1, t, and S_{t-1} be given, with $0 < u < q$. To compute c_u, do the following.

1. Compute non-negative integers a and b such that $u = bt + a \bmod q$ and a and b are at most about $\sqrt{q}$:
   - If $\log_2(t \bmod q) \approx (\log_2 q)/2$ (as in 4.1), then use long division to compute a and b such that $u = b(t \bmod q) + a$.
   - Otherwise, use the lattice-based method described in 4.4. With the proper choice of t this results in a and b that are small enough.
2. If $b = 0$, then compute $c_a = c_u$ using either Algorithm 2.4 or Algorithm 5.1, based on c_1.
3. Otherwise, if $a = 0$, then compute $\tilde{c}_b = c_{tb} = c_u$ using either Algorithm 2.4 or Algorithm 5.1, based on $\tilde{c}_1 = c_t$.
4. Otherwise, if $a \neq 0$ and $b \neq 0$, then do the following:
   - Let $k = t$, $\ell = 1$, so that $S_{t-1} = (c_{k-2\ell}, c_{k-\ell}, c_k)$ and $c_\ell = c_1$.
   - Use Algorithm 3.1 to compute $c_{bk+a\ell} = c_u$ based on a, b, c_k, c_ℓ, $c_{k-\ell}$, and $c_{k-2\ell}$.

Obviously, any t of about the same size as $\sqrt{q}$ will do. A power of 2, however, facilitates the computation of a and b in Step 1 of Algorithm 4.2. Algorithm 4.2 allows easy implementation and, apart from the precomputation, the performance overhead on top of the call to Algorithm 2.4, 5.1, or 3.1 is negligible. The expected runtime of Algorithm 4.2 follows from Conjecture 3.2.

Corollary 4.3 *Given integers u and t with $0 < u < q$ and $\log_2 t \approx (\log_2 q)/2$ and trace values c_1, c_t, c_{t-1}, and c_{t-2}, the trace value c_u can on average be computed in about $3\log_2 u$ multiplications in $\mathbf{F}_p$ using Algorithm 4.2.*

This is more than 60% faster than Algorithm 2.4 as described in [10] using the slower field arithmetic. It can be used in the first place by the owner of the XTR key containing c_1. Thus, XTR signature generation can on average be done more than 60% faster than before [10, Section 4.3]. It can also be used by shared users of an XTR key, such as in Diffie-Hellman key agreement. However, it only affects the first exponentiation to be carried out by each party: party A's computation of c_a given c_1 and a random a can be done on average more than 60% faster, but the computation of c_{ab} based on the value c_b received from party B is not affected by this method. See Section 5 how to speedup the computation of c_{ab} as well.

The precomputation scheme may also be useful for XTR-ElGamal encryption [10, Section 4.2]. In XTR-ElGamal encryption the public key contains two trace values, c_1 and c_k, where k is the secret key. The sender (who does not know k)

picks a random integer b, computes c_b based on c_1, computes c_{bk} based on c_k, uses c_{bk} to (symmetrically) encrypt the message, and sends the resulting encryption and c_b to the owner of k. If the sender uses XTR-ElGamal encryption more than once with the same c_1 and c_k, then it is advantageous to use precomputation. In this application *two* precomputations have to be carried out, once for c_1 and once for c_k. The recipient has to compute c_{bk} based on the value c_b received (and its secret k). Because c_b will not occur again, precomputation based on c_b does not make sense for the party performing XTR-ElGamal decryption.

4.4 Fast Precomputation. It is shown that the choice $t = p$ leads to a faster precomputation, while only marginally slowing down Step 1 of Algorithm 4.2. The triple $S_{p-1} = (c_{p-2}, c_{p-1}, c_p)$ follows from $c_p = c_1^p$ (Fact 2.3.1a), $c_{p-1} = c_1$ (because if g is a root with trace c_1, then $g^{p^2} = g^{p-1}$ is one of its conjugates and has the same trace), and from the fact that, according to [12, Proposition 5.7], c_{p-2} can be computed at the cost of a square-root computation in $\mathbf{F}_p$. Here it is assumed that the public key containing p, q, and c_1 contains an additional single bit of information to resolve the square-root ambiguity[1]. Thus, if $p \equiv 3 \bmod 4$ recipients of XTR public key data with p and q of the above form can do the precomputation of S_{p-1} at a cost of at most $\approx 1.3 \log_2 p$ multiplications in $\mathbf{F}_p$, assuming the owner of the key sends the required bit along. The storage overhead (on top of c_1) for S_{p-1} is just a single element of $\mathbf{F}_{p^2}$, as opposed to three elements for S_{t-1} as in 4.1.

If $p \bmod q \approx \sqrt{q}$, then non-negative a and b of order about $\sqrt{q}$ in Step 1 of Algorithm 4.2 can be found at the cost of a division with remainder. This is, for instance, the case if p and q are chosen as $r^2 + 1$ and $r^2 - r + 1$, respectively, as suggested in [10, Section 3.1]. However, usage of such primes p and q is not encouraged in [10] because of potential security hazards related to the use of primes p of a 'special form'.

Interestingly, and perhaps more surprisingly, sufficiently small a and b exist and can be found quickly in the general case as well. Let L be the two-dimensional integral lattice $\{(e_1, e_2)^T \in \mathbf{Z}^2 : e_1 + e_2 p \equiv 0 \bmod q\}$. If $(e_1, e_2)^T \in L$, then

$$(e_1 + e_2) - e_1 p \equiv -e_2 p + e_2 + e_2 p^2 = e_2(p^2 - p + 1) \equiv 0 \bmod q$$

so that $(e_1 + e_2, -e_1)^T \in L$. Let $v_1 = (e_1, e_2)^T$ be the shortest non-zero vector of L (using the L_2-norm). It may be assumed that $e_1 \geq 0$. It follows that $e_2 \geq 0$, because otherwise $(e_1+e_2, -e_1)^T$ or $(-e_2, e_1+e_2)^T \in L$ would be shorter than v_1. If v_2 is the shortest of $(e_1 + e_2, -e_1)^T, (-e_2, e_1 + e_2)^T \in L$, then $|v_2| < 2|v_1|$ and $\{v_1, v_2\}$ is easily seen to be a shortest basis for L, with $e_1^2 + e_1 e_2 + e_2^2 = q$ and $e_1, e_2 \leq \sqrt{q}$. This implies that given $\{v_1, v_2\}$ and any integer vector $(-u, 0)^T$, there is a vector $(a, b)^T$ with $0 \leq a, b \leq 2\sqrt{q}$ such that $(-u + a, b)^T \in L$. It follows that $-u + a + bp \equiv 0 \bmod q$, i.e., $u \equiv bp + a \bmod q$ as desired. Using the initial basis $\{(q, 0)^T, (-p, 1)^T\}$, the vector v_1 can be found quickly [3, Algorithm

[1] The statement in [12, Proposition 5.7] that this requires a square-root computation in $\mathbf{F}_{p^2}$, as opposed to $\mathbf{F}_p$, is incorrect. This follows immediately from the proof of [12, Proposition 5.7].

1.3.14], and for any u the vector $(a, b)^T$ can easily be computed. In [6, Section 4] a similar construction was independently developed for ECC scalar multiplication.

Corollary 4.5 *Given an integer u with $0 < u < q$ and trace values c_1 and c_{p-2}, the trace value c_u can on average be computed in about $3\log_2 u$ multiplications in $\mathbf{F}_p$ using Algorithm 4.2.*

The owner of the key must explicitly compute c_{p-2} in order to compute the ambiguity-resolving bit. Thus, the owner cannot take advantage of fast precomputation. This adds a minor cost to the key creation.

5 Improved Single Exponentiation

In this section it is shown how Algorithm 3.1 can be used to obtain an XTR single exponentiation method that is on average more than 25% faster than Algorithm 2.4. That is 35% faster than the single exponentiation from [10] based on the slower field arithmetic. Using Algorithm 3.1 to obtain an on average faster XTR single exponentiation is straightforward: to compute c_u with $0 < u < q$ based on c_1 just apply Algorithm 3.1 to $k = \ell = 1$ and any positive a, b with $a+b = u$, then a speedup of more than 14% over Algorithm 2.4 can be expected according to Table 1.

The 25% faster method uses this same approach, but exploits the freedom of choice of a and b: if a and b, i.e., d and e in Algorithm 3.1, can be selected in such a way that the iteration in Step 4 of Algorithm 3.1 favors the 'cheap' steps, while still quickly decreasing d and e, then Algorithm 3.1 should run faster than for randomly selected a and b. Given the various substeps of Step 4 of Algorithm 3.1 and the associated costs, a good way to split up u in the sum of positive a and b seems to be such that b/a is close to the golden ratio $\phi = \frac{1+\sqrt{5}}{2}$, i.e., the asymptotic ratio between two consecutive Fibonacci numbers. This can be seen as follows. If the initial ratio between d and e is close to ϕ, then Step 4(a)i applies and d, e is replaced by $e, d-e$. This corresponds to a 'Fibonacci-step back' so that the ratio between the new d and e (i.e., e and $d-e$) can again be expected to be close to ϕ. Furthermore, the sum of d and e is reduced by a factor ϕ, which is a relatively good drop compared to the low cost of Step 4(a)i (namely, three multiplications in $\mathbf{F}_p$). This leads to the following improved XTR single exponentiation.

5.1 Improved XTR Single Exponentiation. Let u and c_1 be given, with $0 < u < q$. To compute c_u, do the following.

1. Let $a = \text{round}(\frac{3-\sqrt{5}}{2}u)$ and $b = u - a$ (where round(x) is the integer closest to x). As a result $b/a \approx \phi$ as above.
2. Let $k = \ell = 1$, $c_k = c_\ell = c_1$, $c_{k-\ell} = c_0 = 3$, $c_{k-2\ell} = c_{-1} = c_1^p$ (cf. Fact 2.3.1a).
3. Apply Algorithm 3.1 to a, b, c_k, c_ℓ, $c_{k-\ell}$, and $c_{k-2\ell}$, resulting in $c_{bk+a\ell} = c_u$.

Proposition 5.2 *In the call to Algorithm 3.1 in Step 3 of Algorithm 5.1, the values of d and e in Step 4 of Algorithm 3.1 are reduced to approximately half their original sizes using a sequence of approximately $\log_\phi \sqrt{u}$ iterations using just Step 4(a)i.*

Proof. Let $m = \text{round}(\log_\phi u)$. Asymptotically for $m \to \infty$ the values a and b in Algorithm 5.1 satisfy $b/a = \phi + \epsilon_1$ with $|\epsilon_1| = O(2^{-m})$. Furthermore, for $n \to \infty$, the n-th Fibonacci number F_n satisfies $\frac{F_n}{F_{n-1}} = \phi + \epsilon_2$ with $|\epsilon_2| = O(2^{-n})$. It follows that $a = \frac{F_{m-1}}{F_m} b + \epsilon_3$, where $|\epsilon_3|$ is bounded by a small positive constant.

Define $(d_0, e_0) = (b, a)$ and $(d_i, e_i) = (e_{i-1}, d_{i-1} - e_{i-1})$ for $i > 0$. With induction it follows from $a = \frac{F_{m-1}}{F_m} b + \epsilon_3$ that

$$d_i = \frac{F_{m-i}}{F_m} b - (-1)^i F_i \epsilon_3 \tag{2}$$

for $0 \leq i < m$. Algorithm 3.1 as called from Algorithm 5.1 will perform Fibonacci steps as long as $e_i < d_i < 2e_i$. But as soon as $d_i > 2e_i$ this nice behavior will be lost. From $e_i = d_{i+1}$ and (2) it follows that $d_i > 2e_i$ is equivalent to

$$\frac{F_{m-i-3}}{F_m} b < (-1)^{i-1} F_{i+3} \epsilon_3.$$

Because F_m/b and $|\epsilon_3|$ are both bounded by small positive constants, the first time this condition will hold is when F_{m-i-3} and F_{i+3} are of the same order of magnitude, i.e., $m - i - 3 \approx i + 3$. Thus, the Fibonacci behavior is lost after about $m/2 = \log_\phi \sqrt{u}$ iterations, at which point $d_i \approx \sqrt{u}$ (this follows from (2)). This completes the proof of Proposition 5.2.

Based on Proposition 5.2, a heuristic average runtime analysis of Algorithm 5.1 follows easily. The Fibonacci part consists of about $\log_\phi \sqrt{u}$ iterations consisting of just Step 4(a)i of Algorithm 3.1, at a total cost of $3 \log_\phi \sqrt{u} \approx 2.2 \log_2 u$ multiplications in $\mathbf{F}_p$. Once the Fibonacci behavior is lost, the remaining d and e are assumed to behave as random integers of about the same order of magnitude as $\sqrt{u}$, so that, according to Conjecture 3.2, the remainder can on average be expected to take about $6 \log_2 \sqrt{u} = 3 \log_2 u$ multiplications in $\mathbf{F}_p$.

Corollary 5.3 *Given an integer u with $0 < u < q$ and a trace value c_1, the trace value c_u can on average be computed in about $5.2 \log_2 u$ multiplications in $\mathbf{F}_p$ using Algorithm 5.1.*

This corresponds closely to the actual practical runtimes. It is more than 25% better than Algorithm 2.4. Without the optional steps in Algorithm 3.1 the speedup is reduced to about 22%.

Remark 5.4 If insufficient precision is used in the computation of a and b in Step 1 of Algorithm 5.1, then ϵ_3 in the proof of Proposition 5.2 is no longer bounded by a small constant. It follows that $d_i > 2e_i$ already holds for a smaller value of i, implying that the Fibonacci behavior is lost earlier. A precise analysis of the expected performance degradation as a function of the lack of precision is straightforward. In practice this effect is very noticeable.

If a and b happen to be such that all steps are Fibonacci steps, then the cost would be $4.3 \log_2 u$. This is fewer than $\log_2 u$ multiplications in $\mathbf{F}_p$ better than the average behavior obtained.

6 Timings

To make sure that the methods introduced in this paper actually work, and to discover their runtime characteristics, all new methods were implemented and tested. In this section the results are reported, in such a way that the results can easily and meaningfully be compared to the timings reported in [10].

Algorithm 2.5 was implemented, tested for correctness, and it was confirmed that the speedup over the double exponentiation from [10] is negligible. However, implementing Algorithm 2.5 was shown to be significantly easier than it was for the matrix-based method from [10]. Thus, Algorithm 2.5 may still turn out to be valuable if Algorithm 3.1 cannot be used (Remark 3.5).

The methods from Sections 3, 4, and 5 were implemented as well, and incorporated in cryptographic XTR applications along with the old methods from [10]. The resulting runtimes are reported in Table 2. Each runtime is averaged over 100 random keys and 100 cryptographic applications (on randomly selected data) per key. The timings for the XTR single exponentiations with precomputation do not include the time needed for the precomputations. The latter are given in the last two rows. All times are in milliseconds on a 600 MHz Pentium III NT laptop, and are based on the use of a generic and not particularly fast software package for extended precision integer arithmetic [8]. More careful implementation should result in much faster timings. The point of Table 2 is however not the absolute speed, but the relative speedup over the methods from [10].

The RSA timings are included to allow a meaningful interpretation of the timings: if the RSA signing operation runs x times faster using one's own software and platform, then most likely XTR will also run x times faster compared to the figures in Table 2. For each key an odd 32-bit RSA public exponent was randomly selected. 'CRT' stands for 'Chinese Remainder Theorem'. For a theoretical comparison of the runtimes of RSA, XTR, ECC, and various other public key systems at several security levels, refer to [9].

Table 2. RSA, old XTR, and new XTR runtimes.

method		key selection	signing	verifying	encrypting	decrypting
1020-bit RSA	with CRT	908 ms	40 ms	5 ms	5 ms	40 ms
	without CRT		123 ms			123 ms
170-bit XTR	old	64 ms	10 ms	21 ms	21 ms	10 ms
	new, no precomputation	62 ms	7.3 ms	8.6 ms	15 ms	7.3 ms
	new, with precomputation		4.3 ms		8.6 ms	
	precomputation 4.1		4.4 ms		8.8 ms	
	fast precomputation 4.4		1.6 ms		6.0 ms	

7 Application to LUC and ECC

The exponentiations in LUC [18] and ECC when using the curve parameterization proposed in [14] can be evaluated using second degree recurrences. For LUC this is described in detail in [15]. For ECC it is described in [16] and follows by combining [14] and [15]. For ease of reference the resulting runtimes are summarized in this section.

7.1 LUC. Let p and q be primes such that q divides $p+1$, and let g be a generator of the order q subgroup of $\mathbf{F}_{p^2}^*$. In LUC elements of $\langle g \rangle$ are represented by their trace over $\mathbf{F}_p$. Let $v_n \in \mathbf{F}_p$ denote the trace over $\mathbf{F}_p$ of g^n.

Conjecture 7.2 *(cf. Conjecture 3.2) Given integers a and b with $0 < a, b < q$ and trace values v_k, v_ℓ, and $v_{k-\ell}$, the trace value $v_{bk+a\ell}$ can on average be computed in about $1.49 \log_2(\max(a,b))$ multiplications and $0.33 \log_2(\max(a,b))$ squarings in $\mathbf{F}_p$, using the method implied by [15, Table 4].*

Corollary 7.3 *(cf. Corollary 4.3) Given integers u and t with $0 < u < q$ and $\log_2 t \approx (\log_2 q)/2$ and trace values v_1, v_t, and v_{t-1}, the trace value v_u can on average be computed in about $0.75 \log_2 u$ multiplications and $0.17 \log_2 u$ squarings in $\mathbf{F}_p$ using a generalization of Algorithm 4.2.*

Corollary 7.4 *(cf. Corollary 5.3) Given an integer u with $0 < u < q$ and a trace value v_1, the trace value v_u can on average be computed in about $1.47 \log_2 u$ multiplications and $0.17 \log_2 u$ squarings in $\mathbf{F}_p$ using a generalization of Algorithm 5.1.*

7.5 ECC. Let E be an elliptic curve over a prime field $\mathbf{F}_p$, let $E(\mathbf{F}_p)$ be the group of points of E over $\mathbf{F}_p$, and let $G \in E(\mathbf{F}_p)$ be a point of prime order q. As usual, the group operation in $E(\mathbf{F}_p)$ is written additively.

Conjecture 7.6 *(cf. Conjecture 3.2) Given integers a and b with $0 < a, b < q$ and points kG, ℓG, and $(k-\ell)G$, the x-coordinate of the point $(bk+a\ell)G$ can on average be computed in approximately $7 \log_2(\max(a,b))$ multiplications and $3.7 \log_2(\max(a,b))$ squarings in $\mathbf{F}_p$, using the method implied by [15, Table 4] combined with the elliptic curve parameterization from [14].*

Corollary 7.7 *(cf. Corollary 4.3) Given integers u and t with $0 < u < q$ and $\log_2 t \approx (\log_2 q)/2$ and points G, tG, and $(t-1)G$, the x-coordinate of the point uG can on average be computed in about $3.5 \log_2 u$ multiplications and $1.8 \log_2 u$ squarings in $\mathbf{F}_p$ using a generalization of Algorithm 4.2.*

Corollary 7.8 *(cf. Corollary 5.3) Given an integer u with $0 < u < q$ and a point G, the x-coordinate of the point uG can on average be computed in about $6.4 \log_2 u$ multiplications and $3.3 \log_2 u$ squarings in $\mathbf{F}_p$ using a generalization of Algorithm 5.1.*

The single scalar multiplication algorithms are competitive with the ones described in the literature [5]. The double scalar multiplication algorithm from [16] (and as slightly adapted to obtain Conjecture 7.6) is substantially better than other ECC double scalar multiplication methods reported in the literature [2]. For appropriate elliptic curves Corollary 7.7 can be combined with the method proposed in [6], so that the runtime of Corollary 7.7 would hold for Corollary 7.8.

8 Conclusion

The XTR public key system as published in [10] is one of the fastest, most compact, and easiest to implement public key systems. In this paper it is shown that it is even faster and easier to implement than originally believed. The matrices from [10] can be replaced by the more general iteration from Section 3. This results in 60% faster XTR signature applications, substantially faster encryption, decryption, and key agreement applications, and more compact implementations.

Acknowledgment. The authors thank Peter Montgomery from Microsoft Research whose remarks [16] stimulated this research.

References

1. E. Bach, J. Shallit, *Algorithmic Number Theory*, The MIT Press, 1996.
2. M. Brown, D. Hankerson, J. López, A. Menezes, *Software implementation of the NIST elliptic curves over prime fields*, Proceedings RSA Conference 2001, LNCS 2020, Springer-Verlag 2001, 250-265.
3. H. Cohen, *A course in computational algebraic number theory*, GTM 138, Springer-Verlag 1993.
4. H. Cohen, A.K. Lenstra, *Implementation of a new primality test*, Math. Comp. 48 (1987) 103-121.
5. H. Cohen, A. Miyaji, T. Ono, *Efficient elliptic curve exponentiation using mixed coordinates*, Proceedings Asiacrypt'98, LNCS 1514, Springer-Verlag 1998, 51-65.
6. R.P. Gallant, R.J. Lambert, S.A. Vanstone, *Faster point multiplication on elliptic curves with efficient endomorphisms*, Proceedings Crypto 2001, LNCS 2139, Springer-Verlag 2001, 190-200.
7. D.E. Knuth, *The art of computer programming, Volume 2, Seminumerical Algorithms*, third edition, Addison-Wesley, 1998.
8. A.K. Lenstra, *The long integer package FREELIP*, available from www.ecstr.com.
9. A.K. Lenstra, *Unbelievable security: matching AES security using public key systems*, Proceedings Asiacrypt 2001, Springer-Verlag 2001, this volume.
10. A.K. Lenstra, E.R. Verheul, *The XTR public key system*, Proceedings of Crypto 2000, LNCS 1880, Springer-Verlag 2000, 1-19; available from www.ecstr.com.
11. A.K. Lenstra, E.R. Verheul, *Key improvements to XTR*, Proceedings of Asiacrypt 2000, LNCS 1976, Springer-Verlag 2000, 220-233; available from www.ecstr.com.
12. A.K. Lenstra, E.R. Verheul, *Fast irreducibility and subgroup membership testing in XTR*, Proceedings PKC 2001, LNCS 1992, Springer-Verlag 2001, 73-86; available from www.ecstr.com.

13. P.L. Montgomery, *Modular multiplication without trial division*, Math. Comp. 44 (1985) 519-521.
14. P.L. Montgomery, *Speeding the Pollard and elliptic curve methods of factorization*, Math. Comp. 48 (1987) 243-264.
15. P.L. Montgomery, *Evaluating recurrences of form $X_{m+n} = f(X_m, X_n, X_{m-n})$ via Lucas chains,* January 1992; ftp.cwi.nl: /pub/pmontgom/Lucas.pz.gz.
16. P.L. Montgomery, Private communication: *expon2.txt, Dual elliptic curve exponentiation*, manuscript, Microsoft Research, August 2000.
17. S.C. Pohlig, M.E. Hellman, *An improved algorithm for computing logarithms over $GF(p)$ and its cryptographic significance*, IEEE Trans. on IT, 24 (1978), 106-110.
18. P. Smith, C. Skinner, *A public-key cryptosystem and a digital signature system based on the Lucas function analogue to discrete logarithms*, Proceedings of Asiacrypt '94, LNCS 917, Springer-Verlag 1995, 357-364.
19. B. Vallée, *Dynamics of the binary Euclidean algorithm: functional analysis and operators*, Algorithmica 22 (1998), 660-685; and other related papers available from www.users.info-unicaen.fr/~brigitte/Publications/.

A Further Improved Double Exponentiation

Almost 2% can be saved compared to Algorithm 3.1 by distinguishing more cases in Step 4. This is done by replacing Step 4 of Algorithm 3.1 by the following:

4. As long as $d \neq e$ replace $(d, e, u, v, c_u, c_v, c_{u-v}, c_{u-2v})$ by the 8-tuple given below.
   a) If $d > e$ then
      i. if $d \leq 5.5e$, then $(e, d-e, u+v, u, c_{u+v}, c_u, c_v, c_{v-u})$.
      ii. else if d and e are odd, then $(\frac{d-e}{2}, e, 2u, u+v, c_{2u}, c_{u+v}, c_{u-v}, c_{-2v})$.
      iii. else if $d \leq 6.4e$, then $(e, d-e, u+v, u, c_{u+v}, c_u, c_v, c_{v-u})$.
      iv. else if $d \equiv e \bmod 3$, then $(\frac{d-e}{3}, e, 3u, u+v, c_{3u}, c_{u+v}, c_{2u-v}, c_{u-2v})$.
      v. else if d is even, then $(\frac{d}{2}, e, 2u, v, c_{2u}, c_v, c_{2u-v}, c_{2(u-v)})$.
      vi. else if $d \leq 7.5e$, then $(e, d-e, u+v, u, c_{u+v}, c_u, c_v, c_{v-u})$.
      vii. else if $de \equiv 2 \bmod 3$, then $(\frac{d-2e}{3}, e, 3u, 2u+v, c_{3u}, c_{2u+v}, c_{u-v}, c_{-u-2v})$.
      viii. else (e is even), then $(\frac{e}{2}, d, 2v, u, c_{2v}, c_u, c_{2v-u}, c_{2(v-u)})$.
   b) Else (if $e > d$)
      i. if $e \leq 5.5d$, then $(d, e-d, u+v, v, c_{u+v}, c_v, c_u, c_{u-v})$.
      ii. else if e is even, then $(\frac{e}{2}, d, 2v, u, c_{2v}, c_u, c_{2v-u}, c_{2(v-u)})$.
      iii. else if $e \equiv d \bmod 3$, then $(\frac{e-d}{3}, d, 3v, u+v, c_{3v}, c_{u+v}, c_{2v-u}, c_{v-2u})$.
      iv. else if $de \equiv 2 \bmod 3$, then $(d, \frac{e-2d}{3}, u+2v, 3v, c_{u+2v}, c_{3v}, c_{u-v}, c_{u-4v})$.
      v. else if $e \leq 7.4d$, then $(d, e-d, u+v, v, c_{u+v}, c_v, c_u, c_{u-v})$.
      vi. else if d is odd, then $(\frac{e-d}{2}, d, 2v, u+v, c_{2v}, c_{u+v}, c_{v-u}, c_{-2u})$.
      vii. else if $e \equiv 0 \bmod 3$, then $(\frac{e}{3}, d, 3v, u, c_{3v}, c_u, c_{3v-u}, c_{3v-2u})$.
      viii. else (d is even), then $(\frac{d}{2}, e, 2u, v, c_{2u}, c_v, c_{2u-v}, c_{2(u-v)})$.

Steps 4(a)vii and 4(b)iv require 13.5 and 12.5 multiplications in $\mathbf{F}_p$, respectively. The cost of the other steps is as in Section 3. The average cost to compute $c_{bk+a\ell}$ turns out to be about $5.9 \log_2(\max(a, b))$ multiplications in $\mathbf{F}_p$. Omission of Steps 4(a)iii, 4(a)vi, and 4(b)v, combined with a constant 4 instead of 5.5 in Steps 4(a)i and 4(b)i leads to an almost 1% speedup over Algorithm 3.1.

An Efficient Implementation of Braid Groups

Jae Choon Cha[1], Ki Hyoung Ko[1], Sang Jin Lee[1], Jae Woo Han[2], and Jung Hee Cheon[3]

[1] Department of Mathematics
Korea Advanced Institute of Science and Technology, Taejon, 305–701, Korea.
{jccha,knot,sjlee}@knot.kaist.ac.kr

[2] National Security Research Institute, Taejon, 305–335, Korea.
jwhan@etri.re.kr

[3] International Research center for Information Security
Information and Communications University, Taejon, 305–732, Korea.
jhcheon@icu.ac.kr

Abstract. We implement various computations in the braid groups via practically efficient and theoretically optimized algorithms whose pseudo-codes are provided. The performance of an actual implementation under various choices of parameters is listed.

1 Introduction

A new cryptosystem using the braid groups was proposed in [5] at Crypto 2000. Since then, there has been no serious attempt to analyze the system besides one given by inventors [7]. We think that this is because the braid group is not familiar to most of cryptographers and cryptanalysts. The primary purpose to announce our implementation is to encourage people to attack the braid cryptosystem. In [7], a necessary condition for the instances of the mathematical problem which the braid cryptosystem is based on is found so that it makes the mathematical problem intractable. This means that a key selection is crucial to maintain the theoretical security of the braid cryptosystem. Thus the key generation is one of the areas where much research is required and we think that the search for strong keys should be eventually aided by computers. This is the secondary purpose of our implementation.

In this paper we discuss implementation issues of the braid group given by either the Artin presentation [2] or the band-generator presentation [1]. Due to the analogy between the two presentations, our implementations on the two presentations are basically identical, except the low-level layer consisting of data structures and algorithms for canonical factors, which play the role of the building blocks for braids. Even though the algorithms of the present implementation in the braid groups are our initial work, they are theoretically optimized so that all of single operations can be executed at most in $\mathcal{O}(n \log n)$ where n is the braid index n that is the security parameter corresponding to the block sizes in other cryptosystems. This excellent speed is achieved because the canonical factors are expressed as permutations that can be efficiently and naturally handled by computers. The efficiency of the implementation shows that the braid group

C. Boyd (Ed.): ASIACRYPT 2001, LNCS 2248, pp. 144–156, 2001.
© Springer-Verlag Berlin Heidelberg 2001

is a good source of cryptographic primitives [5,6]. It is hard to think of any other non-commutative groups that can be digitized as efficiently as the braid group. Matrix groups are typical examples of non-commutative groups and in fact any group can be considered as a matrix group via representations. But the group multiplication in the braid group of index n is faster than the multiplication of $(n \times n)$ matrices.

This paper is organized as follows. Section 2 is a quick review of the minimal necessary background on braid groups. In Section 3 and 4, we develop data structures and algorithms for canonical factors and braids, respectively. In Section 5, we show how to generate random braids. In Section 6, we discuss the performance of our implementation, through the braid cryptosystems in [7]. Section 7 is our conclusion.

2 A Quick Review of the Braid Groups

A *braid* is obtained by laying down a number of parallel strands and intertwining them so that they run in the same direction. In our convention, this direction is horizontally toward the right. The number of strands is called the *braid index*. The set B_n of isotopy classes of braids of index n has a group structure, called the *n-braid group*, where the product of two braids x and y is nothing more than laying down the two braids in a row and then matching the end of x to the beginning of y.

Any braid can be decomposed as a product of simple braids. One type of simple braids is the *Artin generators* σ_i that have a single crossing between i-th and $(i+1)$-st strand as in Figure 1 (a), and the other type is the *band-generators* a_{ts} that have a single half-twist band between t-th and s-th strand running over all intermediate strands as in Figure 1 (b).

The n-braid group B_n is presented by the Artin generators $\sigma_1, \ldots, \sigma_{n-1}$ and relations $\sigma_i\sigma_j = \sigma_j\sigma_i$ for $|i-j| > 1$ and $\sigma_i\sigma_j\sigma_i = \sigma_j\sigma_i\sigma_j$ for $|i-j| = 1$. On the other hand, B_n is also presented by the band-generators a_{ts} for $n \geq t > s \geq 1$ and relations $a_{ts}a_{rq} = a_{rq}a_{ts}$ for $(t-r)(t-q)(s-r)(s-q) > 0$ and $a_{ts}a_{sr} = a_{tr}a_{ts} = a_{sr}a_{tr}$ for $n \geq t > s > r \geq 1$.

These will be called the *Artin presentation* and the *band-generator presentation*, respectively. There are theoretically similar solutions to the word and conjugacy problems in B_n for both presentations [1,2,3]. The band-generator presentation has a computational advantage over the Artin as far as the word problem is concerned. Since almost all the machineries are identical in the two theories, it will be convenient to introduce unified notation so that we may review both theories at the same time.

1. Let B_n^+ be the monoid defined by the same generators and relations in a given presentation. Elements in B_n^+ are called *positive braids* or *positive words*. The relations in the Artin and band-generator presentations preserve word-length of positive braids and so the word-length is easy to compute for positive braids. The natural map $B_n^+ \to B_n$ is injective. [1,4]. There are no known presentations of B_n except these two that enjoy this injection property needed for a fast solution to the word problem.

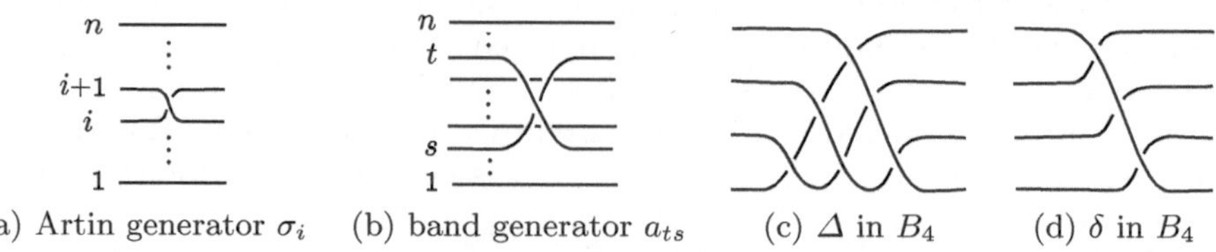

(a) Artin generator σ_i (b) band generator a_{ts} (c) Δ in B_4 (d) δ in B_4

Fig. 1. Generators and fundamental braids

2. There is a *fundamental braid* $\mathbf{D}$. It is $\Delta = (\sigma_1 \cdots \sigma_{n-1})(\sigma_1 \cdots \sigma_{n-2}) \cdots \sigma_1$ in the Artin presentation and $\delta = a_{n(n-1)}a_{(n-1)(n-2)} \cdots a_{21}$ in the band-generator presentation as shown in Figure 1 (c), (d). The fundamental braid $\mathbf{D}$ can be written in many distinct ways as a positive word in both presentations. Due to this flexibility, it has two important properties:
   (i) For each generator a, $\mathbf{D} = aA = Ba$ for some $A, B \in B_n^+$.
   (ii) For each generator a, $a\mathbf{D} = \mathbf{D}\tau(a)$ and $\mathbf{D}a = \tau^{-1}(a)\mathbf{D}$ where τ is the automorphism of B_n defined by $\tau(\sigma_i) = \sigma_{n-i}$ for the Artin presentation and $\tau(a_{ts}) = a_{(t+1)(s+1)}$ for the band-generator presentation.
3. There are partial orders '$\leq$', '$\leq_L$' and '$\leq_R$' in B_n. For two words V and W in B_n, we say that $V \geq W$ (resp. $V \geq_L W$, $V \geq_R W$) if $V = PWQ$ (resp. $V = WP$, $V = PW$) for some $P, Q \in B_n^+$. If a word is compared against either the empty word e or a power of $\mathbf{D}$, all three orders are equivalent due to the property (ii) above. Note that the partial orders depend on a presentation of B_n and W is a positive word if and only if $W \geq e$.
4. For two elements V and W in a partial order set, the *meet* $V \wedge W$(resp. *join* $V \vee W$) denotes the largest (resp. smallest) element among all elements smaller (resp. larger) than V and W. If both the meet and join always exist for any pair of elements in a partially order set, the set is said to have a *combinatorial lattice structure*. The braid group B_n has a combinatorial lattice structure for '$\leq_L$' and '$\leq_R'$ in any of both presentations [3,1]. When we want to distinguish the meet and join for left and right versions, we will use '$\wedge_L$', '$\wedge_R$', '$\vee_L$' and '$\vee_L$'.
5. A braid satisfying $e \leq A \leq \mathbf{D}$ is called a *canonical factor* and $[0,1]_n$ denotes the set of all canonical factors in B_n. The cardinality of $[0,1]_n$ is $n!$ for the Artin presentation, and the nth Catalan number $C_n = \dfrac{(2n)!}{n!(n+1)!}$ for the band-generator presentation. Note that C_n is much smaller than $n!$ and this is one of main reasons why it is sometimes computationally easier to work with the band-generator presentation than the Artin presentation.
6. For a positive braid P, a decomposition $P = A_0P_0$ is *left-weighted* if $A_0 \in [0,1]_n$, $P_0 \geq e$, and A_0 has the maximal length (or maximal in '$\leq_L$') among all such decompositions. A left-weighted decomposition $P = A_0P_0$ is unique. A_0 is called the *maximal head* of P. The notion 'right-weighted' can be also defined similarly.
7. Any braid W given as a word can be decomposed uniquely into

$$W = \mathbf{D}^u A_1 A_2 \cdots A_k, \qquad e < A_i < \mathbf{D},\ u \in \mathbb{Z}, \tag{1}$$

where the decomposition A_iA_{i+1} is left-weighted for each $1 \le i \le k-1$. This decomposition, called the *left canonical form* of W, is unique and so it solves the word problem. The integer u (resp. $u+k$) is called the *infimum* (resp. *supremum*) of W and denoted by $\inf(W)$ (resp. $\sup(W)$). The infimum (resp. supremum) of W is the smallest (resp. largest) integer m such that $\mathbf{D}^{-m}W \ge e$ (resp. $\le e$). The *canonical length* of W, denoted by $\mathrm{len}(W)$, is given by $k = \sup(W) - \inf(W)$ and will be used as an important parameter together with the braid index n. The right canonical form of W can be also defined similarly.

3 Canonical Factors

3.1 Data Structures

A canonical factor in the Artin presentation of B_n can be identified with the associated n-permutation, which is obtained by replacing the i-th Artin generator σ_i by the transposition of i and $i+1$. We represent an n-permutation as an array A of n integers, where $A[i]$ is equal to the image of i under the permutation. A is called a *permutation table.*

A canonical factor in the band-generator presentation is also uniquely determined by the associated permutation. Thus a canonical factor can be represented by a permutation table as before, but a permutation is associated to a canonical factor in the band-generator presentation only if it is a product of "disjoint parallel descending cycles" [1]. Two descending cycles $(s_i s_{i-1} \cdots s_1)$ and $(t_j t_{j-1} \cdots t_1)$, where $s_i > \cdots > s_1$ and $t_j > \cdots > t_1$, are called *parallel* if s_a and s_b do not separate t_c and t_d (i.e. $(s_a - t_c)(s_a - t_d)(s_b - t_c)(s_b - t_d)$ is positive) for all $1 \le a < b \le i$ and $1 \le c < d \le j$. Thus a canonical factor can also be represented by an array X of n integers where $X[i]$ is the maximum in the descending cycle containing i. X is called a *descending cycle decomposition table.* The permutation table is useful for products and inverses, and the descending cycle decomposition table is useful for the meet operation discussed later. The two tables can be converted in $\mathcal{O}(n)$ time. Thus any one of them can be chosen to implement the braid groups without affecting the complexities of algorithms. We describe concrete algorithms in Algorithm 1 and 2.

Algorithm 1 *Convert a permutation table to a descending cycle decomposition table.*

Input: *permutation table A of length n.*
Output: *descending cycle decomposition table X.*

```
for i ← 1 to n do X[i] ← 0;
for i ← n to 1 step −1 do begin
    if (X[i] = 0) then X[i] ← i;
    if (A[i] < i) then X[A[i]] ← X[i];
end
```

Algorithm 2 *Convert a descending cycle decomposition table to a permutation table.*

> ***Input:*** *Descending cycle decomposition table X.*
> ***Output:*** *Permutation table A.*
> *(We need an array Z of size n.)*

```
for i ← 1 to n do Z[i] = 0;
for i ← 1 to n do begin
    if (Z[X[i]] = 0) then A[i] ← X[i] else A[i] ← Z[X[i]];
    Z[X[i]] ← i;
end
```

3.2 Operations

Comparison. Two given canonical factors are identical if and only if their representations given by either permutation tables or descending cycle decomposition tables are identical. Thus the comparison is an $\mathcal{O}(n)$ operation.

Product and Inverse. The product and inverse operations in permutation groups are done in $\mathcal{O}(n)$. If the product of two canonical factors is again a canonical factor, the composition of associated permutations is the permutation associated to the product in both presentations. Hence in this case the product of canonical factors is computed in $\mathcal{O}(n)$.

The Automorphism τ^u. The automorphism τ defined by $\tau(a) = \mathbf{D}^{-1}a\mathbf{D}$ sends canonical factors to canonical factors. An arbitrary power $\tau^u(a)$ for a canonical factor a can also be computed in $\mathcal{O}(n)$, independent of u, since the permutation table of $\mathbf{D}^u$ can be obtained immediately from the parity (resp. the modulo n residue class) of u in the Artin (resp. band-generator) presentation.

Meet. In the Artin presentation, an algorithm computing the meet of two canonical factors with $\mathcal{O}(n \log n)$ running time and $\mathcal{O}(n)$ space is known [3, Chapter 9]. We explain the idea of the algorithm briefly. Suppose that A and B are canonical factors and $C = A \wedge_L B$ be the left meet. We view A, B and C as permutation tables. The algorithm sorts the integers $1, \dots, n$ according to the order "$\prec$" defined by $x \prec y$ if and only $C[x] < C[y]$. The final result is the permutation table of the inverse of C, and by inverting it the permutation table of C is obtained. Using the standard divide-conquer trick, we divide the sequence to be sorted into two parts, to say X and Y, sort each of X and Y recursively, and merge them according to $\prec$. In the merging step, we need to compare integers $x \in X$ and $y \in Y$ according to $\prec$. The essential point is that $y \prec x$ if and only if the infimum of $A[i]$ over all $i \in X$ lying in the right-hand side of x is greater than the supremum of $A[j]$ over all $j \in Y$ lying in the left-hand side of y, and the analogous condition holds for B. This can be checked in constant time using tables of infimums and supremums, which can be constructed before the merge step in linear time proportional to the sum of the sizes of X and Y. Hence the total timing is equal to that of standard divide-conquer sorting, $\mathcal{O}(n \log n)$. We describe the left meet algorithm explicitly in Algorithm 3.

Algorithm 3 *Compute the meet of two canonical factors in the Artin presentation.*

> ***Input:*** *Permutation tables A, B*
> ***Output:*** *The permutation table C of the meet $A \wedge_L B$.*
> *(We need arrays U, V, W of size n.)*

```
Initialize C as the identity permutation;
Sort C[1]···C[n] according to A and B (see the subalgorithm below);
C ← inverse permutation of C;
```

Subalgorithm: *Sort $C[s] \cdots C[t]$ according to A and B.*

```
if t ≤ s then return;
m ← ⌊(s + t)/2⌋;
Sort C[s]···C[m] according to A and B;
Sort C[m + 1]···C[t] according to A and B;
U[m] ← A[C[m]];
V[m] ← B[C[m]];
if s < m then
  for i ← m − 1 to s step − 1 do begin
     U[i] ← min(A[C[i]], U[i + 1]);
     V[i] ← min(B[C[i]], V[i + 1]);
  end
U[m + 1] ← A[C[m + 1]];
V[m + 1] ← B[C[m + 1]];
if t > m + 1 then
  for i ← m + 2 to t do begin
     U[i] ← max(A[C[i]], U[i − 1]);
     V[i] ← max(B[C[i]], V[i − 1]);
  end
l ← s;
r ← m + 1;
for i ← s to t do begin
   if (l > m) ∨ ((r ≤ t) ∧ (U[l] > U[r]) ∧ (V[l] > V[r]))
     then W[i] ← C[r]; r ← r + 1;
      else W[i] ← C[l]; l ← l + 1;
end
for i ← s to t do C[i] ← W[i];
```

The right meet is computed in a similar way, or alternatively by the identity $A \wedge_R B = (A^{-1} \wedge_L B^{-1})^{-1}$, where the inverse notations denote the inverses in the permutation group.

In the band-generator presentation, it is known that the meet of two canonical factors can be computed in $\mathcal{O}(n)$ time [1]. Basically, the meet is obtained by computing the refinement of the two partitions of $\{1, \ldots, n\}$ that corresponds to the parallel descending cycle decompositions. We describe below an algorithm to compute the meet, which is an improved version of one in [1]. We remark that the left meet and the right meet are the same in the band-generator presentation.

Algorithm 4 *Compute the meet of two canonical factors in band-generator presentation.*

> ***Input:*** *Descending cycle decomposition tables A and B.*
> ***Output:*** *The descending cycle decomposition table C of the meet $A \wedge B$.*

```
for i ← 1 to n do U[i] ← n − i + 1;
Sort U[1]···U[n] such that
     (A[U[i]], B[U[i]], U[i]) is descending in the dictionary order;
j ← U[n]; C[j] ← j;
for i ← n − 1 to 1 step − 1 do begin
   if (A[j] ≠ A[U[i]]) ∨ (B[j] ≠ B[U[i]]) then j ← U[i];
   C[U[i]] ← j;
end
```

The complexity is determined by the sorting step since all the other parts are done in linear time. In braid cryptosystems, it is expected that n is not so large (perhaps less than 500) and hence it is practically reasonable to apply the bucket sort algorithm. The bucket sort algorithm can be applied twice to sort pairs $(A[U[i]], B[U[i]], U[i])$ lexicographically. (Recall that the original order is preserved as much as possible by the bucket sort.) Since we have at most n possibilities for the values of $A[U[i]]$ and $B[U[i]]$, both space and execution time are linear in n. In some situations, the following trade-off of space and execution time is useful. We may sort the pairs $(A[U[i]], B[U[i]])$ using the bucket sort algorithm once, where $\mathcal{O}(n^2)$-space is required but the practical execution speed is improved. To save space (e.g. on small platforms), usual sorting algorithms by comparisons (e.g. divide-conquer sort) can be applied to get an $\mathcal{O}(n \log n)$ algorithm that requires no additional space.

4 Braids

4.1 Data Structures

Writing a given braid as $\beta = \mathbf{D}^q A_1 A_2 \cdots A_\ell$, where q is an integer and each A_i is a canonical factor, we represent the braid as a pair $\beta = (q, (A_i))$ of an integer q and a list of ℓ canonical factors (A_i) in both presentations. We note that this representation is not necessarily the left canonical form of β, and hence ℓ may be greater than the canonical length of β.

A braid given as a word in generators is easily converted into the above form, in both presentations, by rewriting each negative power σ^{-1} of generators as a product of $\mathbf{D}^{-1}$ and a canonical factor $\mathbf{D}\sigma^{-1}$ and collecting every power of $\mathbf{D}$ at the left end using the fact $(\prod A_i)\mathbf{D}^{\pm 1} = \mathbf{D}^{\pm 1}(\prod \tau^{\pm 1}(A_i))$ for any sequence of canonical factors A_i. This is done in $\mathcal{O}(n\ell)$, where n is the braid index and ℓ is the length of the given word.

4.2 Operations

Group Operations. Basic group operations are easily implemented. From the identity

$$(\mathbf{D}^p A_1 \cdots A_\ell)(\mathbf{D}^q B_1 \cdots B_{\ell'}) = \mathbf{D}^{p+q}\tau^q(A_1)\cdots\tau^q(A_\ell)B_1\cdots B_{\ell'} \quad (2)$$

the multiplication of two braids is just the juxtaposition of two lists of permutation and applying τ. The inverse of a braid can be computed using the formula

$$(\mathbf{D}^q A_1 \cdots A_\ell)^{-1} = \mathbf{D}^{-q-\ell}\tau^{-q-\ell}(B_\ell)\cdots\tau^{-q-1}(B_1) \quad (3)$$

where $B_i = A_i^{-1}\mathbf{D}$, viewing A_i and $\mathbf{D}$ as permutations. Since a power of τ is computed in linear time in n, braid multiplication and inversion have complexity $\mathcal{O}(\ell n)$. A conjugation consists of two multiplications and one inversion, and hence also has the complexity $\mathcal{O}(\ell n)$.

Left Canonical Form. A representation of a braid can be converted into the left canonical form by the algorithms in [3, Chapter 9] and [1]. Given a positive braid $P = A_1 \cdots A_\ell$, where A_i is a canonical factor, the algorithm computes the maximal heads of $A_{\ell-1}A_\ell$, $A_{\ell-2}A_{\ell-1}A_\ell$, $\ldots$, $A_1 \cdots A_\ell = P$ sequentially using the following facts [3, Chaper 9] [2] [1].

1. For any positive braid A and P, the maximal head of AP is the maximal head of the product of A and the maximal head of P.
2. For two canonical factors A and B, the maximal head of AB is $A((\mathbf{D}A^{-1}) \wedge_L B)$, where the inverse is taken in the permutation group.

From these facts, the i-th maximal head is the maximal head of the product of $A_{\ell-i}$ and the $(i-1)$-st maximal head, and it can be computed using meet operation once. At the last step, we obtain the left weighted decomposition $P = B_1P_1$. Doing it again for P_1, we obtain the left weighted decomposition $P_1 = B_2P_2$, and repeating this, finally we obtain the left canonical form of P. Note that this process is very similar to the bubble sort, where the maximum (or minimum) of given elements is found at the first stage, and repeat it for the remaining elements. The complexity of left canonical form algorithm is the same as that of the bubble sort: complexities are $\mathcal{O}(\ell^2 n \log n)$ and $\mathcal{O}(\ell^2 n)$ in the Artin presentation and the band-generator presentations, respectively. The difference comes from the complexity of the meet operation. We describe the left canonical form algorithm in a concrete form.

Algorithm 5 *Convert a braid into the left canonical form.*

Input: *A braid representation* $\beta = (p, (A_i))$.
Output: *The left canonical form of* β.

$\ell \leftarrow \ell(\beta)$;
$i \leftarrow 1$;

while $(i < \ell)$ do begin
    $t \leftarrow \ell;$
    for $j \leftarrow \ell - 1$ to i step -1 do begin
        $B \leftarrow (\mathbf{D}A_j^{-1}) \wedge_L A_{j+1};$
        if (B is nontrivial) then begin
            $t \leftarrow j;\ A_j \leftarrow A_j B;\ A_{j+1} \leftarrow B^{-1} A_{j+1};$
        end
    end
    $i \leftarrow t + 1;$
end
while $(\ell > 0) \wedge (A_1 = \mathbf{D})$ do begin
    Remove A_1 from β; $\ell \leftarrow \ell - 1;\ p \leftarrow p + 1;$
end
while $(\ell > 0) \wedge$ (A_ℓ is trivial) do begin
    Remove A_ℓ from β; $\ell \leftarrow \ell - 1;$
end

The multiplications and inversions in lines 6 and 8 are performed viewing $\mathbf{D}$, B and A_k as permutations.

We remark that Algorithm 5 can be modified for parallel processing. For convenience, we denote the job of lines 6–9 for (i,j) by $S(i,j)$. Then $S(i,j)$ can be processed after $S(i-1,j-1)$ is finished. Thus the jobs $S(1,k), S(2,k+2), \ldots, S(\ell-1, k+2(\ell-2))$ can be processed simultaneously for $k = \ell-1, \ell-2, \ldots, 1, 0, -1, \ldots, -\ell+3$. ($S(i,j)$ for invalid (i,j) is ignored here.) This method offers algorithms with $\mathcal{O}(\ell n \log n)$ and $\mathcal{O}(\ell n)$ execution time in the Artin and the band-generator presentation, using $\mathcal{O}(\ell)$ processors.

Comparison. In order to compare two braids β_1 and β_2 with ℓ_1 and ℓ_2 canonical factors, we need to convert them into their canonical forms since the same braid can be represented in different forms. Assuming β_1 and β_2 are in left canonical form, the comparison is done by comparing the exponents of $\mathbf{D}$ and the lists of canonical factors, and so has complexity $\mathcal{O}(\min\{\ell_1, \ell_2\} \cdot n)$. Without the assumption, the total complexity of comparison is equal to that of the conversion into left canonical form, $\mathcal{O}(\min\{\ell_1, \ell_2\} \cdot n \log n)$ and $\mathcal{O}(\min\{\ell_1, \ell_2\} \cdot n)$ for the Artin presentation and band-generator presentation, respectively. (Note that for comparison, Algorithm 5 can be executed simultaneously for β_1 and β_2 to extract the canonical factors in the left canonical forms, and stopped if either different canonical factors are found or nothing is left for any one of β_1 and β_2.)

5 Random Braids

Random braids play an important role in braid cryptosystems [5,7]. Since the braid group B_n is discrete and infinite, a probability distribution on B_n makes no sense. But there are finitely many positive n-braids with ℓ canonical factors, we may consider randomness for these braids. Since such a braid can be generated by concatenating ℓ random canonical factors, the problem is reduced to how to choose a random canonical factors in both presentations.

5.1 Artin Presentation

In the Artin presentation of B_n, a canonical factor can be chosen randomly by generating a random n-permutation. It is well known that this is done by using a random number oracle $(n-1)$ times; we start with the identity permutation table A, and for $i = 1, 2, \ldots, n-1$, pick a random number j between i and n and swap $A[i]$ and $A[j]$.

5.2 Band-Generator Presentation

In the band-generator presentation, we need more complicated arguments. Parallel descending cycle decompositions can be identified with non-crossing partitions of the set $\{1, \ldots, n\}$. It is known that they are again naturally bijective to the set BS_n of *ballot sequences* $s_1 s_2 \cdots s_{2n}$ of length $2n$, which are defined to be sequences satisfying $s_1 + \cdots + s_k \geq 0$ for all k and $s_1 + \cdots + s_{2n} = 0$ (e.g. see [8]). Of course, $|BS_n|$ is equal to the n-th Catalan number C_n. The recurrence relation

$$C_n = C_0 C_{n-1} + C_1 C_{n-2} + \cdots + C_{n-1} C_0 \tag{4}$$

can be naturally interpreted by means of ballot sequences as follows. For a given ballot sequence $s_1 \cdots s_{2n}$, choose the minimal i such that $s_1 + \cdots + s_i = 0$. Then $s_1 = 1$, $s_i = -1$ and the subsequences $s_2 \cdots s_{i-1}$ and $s_{i+1} \cdots s_{2n}$ are again ballot sequences of length $2(i-1)$ and $2(n-i)$, respectively. This establishes a bijection between BS_n and the disjoint union $\bigcup_{i=1}^{n-1} BS_{i-1} \times BS_{n-i}$. We inductively define a linear order on BS_n via the bijection, by the following rules: elements in $BS_{i-1} \times BS_{n-i}$ are smaller than elements in $BS_{j-1} \times BS_{n-j}$ if and only if $i < j$, and elements in $BS_{i-1} \times BS_{n-i}$ are lexicographically ordered. Then a random ballot sequence can be generated as follows. Choose a random number k between 1 and C_n, and take the k-th ballot sequence. Algorithm 6 does the second step, by tracing the above bijection recursively. By an induction, it can be shown that the running time of Algorithms 6 is $\mathcal{O}(n \log n)$.

Algorithm 6 *Construct the k-th ballot sequence of length $2n$.*

> ***Input:*** *An integer k between 1 and C_n.*
> ***Output:*** *The k-th ballot sequence $s_1 \cdots s_{2n}$.*
>
> if $k \leq C_0 C_{n-1}$ then $i \leftarrow 1$;
> elseif $k > C_n - C_{n-1} C_0$ then begin $i \leftarrow n$; $k \leftarrow k - C_n + C_{n-1} C_0$; end
> else for $i \leftarrow 1$ to n do
>   if $(k \leq C_{i-1} C_{n-i})$ then break;
>   else $k \leftarrow k - C_{i-1} C_{n-i}$;
> $x \leftarrow \lfloor k / C_{n-i} \rfloor$; $y \leftarrow k - x C_{n-i}$;
> $s_1 \leftarrow 1$; $s_{2i-1} \leftarrow -1$;
> if $i > 1$ then $s_2 \cdots s_{2i-2} \leftarrow$ the $(x+1)$-st ballot sequence of length $2(i-1)$;
> if $i < n$ then $s_{2i} \cdots s_{2n} \leftarrow$ the $(y+1)$-st ballot sequence of length $2(n-i)$;

A ballot sequence can be transformed to a permutation table associated to a canonical factor in the band generator presentation, via the correspondence between ballot sequences and non-crossing partitions of $\{1,\ldots,n\}$ [8]. We describe an $\mathcal{O}(n)$ algorithm.

Algorithm 7 *Convert a ballot sequence to a disjoint cycle decomposition table.*

> ***Input:*** *A ballot sequence* $s_1 \cdots s_{2n}$.
> ***Output:*** *A permutation table* A.
> *(We need a stack* S *of maximal size* n.*)*

```
for i ← 1 to 2n do begin
    if s_i = 1 then push i into S;
                else begin
                        Pop j from S;
                        if i is odd then A[(i+1)/2] = j/2
                                     else A[j/2] = (i+1)/2;
                    end
end
```

In the above discussion, we assume that the Catalan numbers C_n is known. It is not a severe problem, since a table of C_n can be computed very quickly using the recurrence relation $C_{n+1} = (4n+2)C_n/(n+2)$. If you want to avoid division of big integers, the recurence relation (4) is useful.

We finish this section with a remark on the distribution generated by out algorithm. Since the same braid can be represented in different ways in our implementation, the distribution is not uniform on the set of positive n-braids of canonical length ℓ. However, the distribution has a property that more complex braids, which can be represented in more different ways, are generated with higher probability. It seems to be a nice property for braid cryptosystems.

6 Performance

In this section we consider the braid cryptosystem proposed in [7], which is a revised version of one in [5]. Let LB_n and UB_n be the subgroups of B_n generated by the Artin generators $\sigma_1,\ldots,\sigma_{\lfloor n/2\rfloor-1}$ and $\sigma_{\lfloor n/2\rfloor},\ldots,\sigma_n$, respectively. A secret key is given as a pair (a_1,a_2), where a_1 and a_2 are in LB_n, and the associated public key is a pair (x,y) such that $y = a_1xa_2$. The encryption and decryption scheme is as follows.

Encryption Given a message $m \in \{0,1\}^M$,
1. Choose $b_1, b_2 \in UB_n$.
2. Ciphertext is $(c_1,c_2) = (b_1xb_2, H(b_1yb_2) \oplus m)$.

Decryption Given a ciphertext (c_1,c_2), $m = H(a_1c_1a_2) \oplus c_2$.

In the above scheme, $H\colon B_n \to \{0,1\}^M$ is a collision-free hash function. H can be obtained by composing a collision free hash function of bitstrings into $\{0,1\}^M$ with a conversion function of braids into bitstrings. A braid given as its left canonical form $\mathbf{D}^u A_1 \cdots A_\ell$ can be converted into a bitstring by dumping the integer u and the permutation tables of A_i as binary digits for $i = 1, \ldots, \ell$ sequentially. Since different braids are converted into different bitstrings, this conversion can be used as a part of the hash H.

We remark that if the secret key is of the form (a, a^{-1}) and b_1^{-1} is taken as b_2 in the above encryption procedure, the cryptosystem in [5] is obtained. Hence in performance issues, there is no difference between the cryptosystems in [7] and [5].

The above scheme is easily implemented based on our works. In the encryption, two random braid generations, four multiplications and two left canonical form operations are involved. In the decryption, two multiplications and one left canonical form operation are involved. Thus both operations have running time $\mathcal{O}(\ell^2 n \log n)$ and $\mathcal{O}(\ell^2 n)$ in the Artin and the band-generator presentation, respectively. In Table 1, we show the performance of an implementation of the cryptosystem using the Artin presentation, at various security parameters suggested in [5]. The security levels are estimated using the results of [7]. In order to focus on the performance of braid operations, the execution time of the hash function is ignored. This experiment is performed on a computer with a Pentium III 866MHz processor.

Table 1. Performance of the braid cryptosystem at various parameters

n	ℓ	Block Size (Kbyte)	Encryption Speed (Block/sec)	Encryption Speed (Kbyte/sec)	Decryption Speed (Block/sec)	Decryption Speed (Kbyte/sec)	Security Level
100	15	1.97	74.46	146.53	95.60	188.13	2^{85}
150	20	4.36	37.44	163.40	47.42	206.94	2^{125}
200	30	9.34	17.21	160.71	22.30	208.26	2^{199}
250	40	16.36	10.61	173.66	13.62	222.78	2^{280}

7 Conclusion

Table 2 summaries braid algorithms discussed and their complexities. In Input and Output columns, PT, DT, AB and BB mean a permutation table, a descending cycle decomposition table, a braid given by the Artin presentation and a braid given by the band-generator presentations, respectively. As usual n is the braid index and ℓ the maximum of canonical lengths (or numbers of canonical factors) of input braids, except for the comparison algorithm, where ℓ denotes the minimum of canonical lengths of two given braids. The complexities of the algorithms are measured by the number of steps required. The space complexities of the algorithms are easily seen to be either constant or linear.

Table 2. Complexities of braid algorithms

Operation	Input	Output	Complexity	Reference
PT → DT	PT	DT	$\mathcal{O}(n)$	Alg. 1
DT → PT	DT	PT	$\mathcal{O}(n)$	Alg. 2
Product	PT	PT	$\mathcal{O}(n)$	3.2
Inverse	PT	PT	$\mathcal{O}(n)$	3.2
τ^k	PT	PT	$\mathcal{O}(n)$	3.2
Meet (Artin)	PT	PT	$\mathcal{O}(n \log n)$	Alg. 3
Meet (Band)	DT	DT	$\mathcal{O}(n)$	Alg. 4
Comparison	PT (or DT)	True/False	$\mathcal{O}(n)$	3.2
Random (Artin)		PT	$\mathcal{O}(n)$	5.1
Random (Band)		PT	$\mathcal{O}(n \log n)$	5.2, Alg. 6, 7
Product	AB (or BB)	AB (or BB)	$\mathcal{O}(\ell n)$	4.2
Inverse	AB (or BB)	AB (or BB)	$\mathcal{O}(\ell n)$	4.2
Left Canonical	AB	AB	$\mathcal{O}(\ell^2 n \log n)$	Alg. 5
Form	BB	BB	$\mathcal{O}(\ell^2 n)$	Alg. 5
Comparison	AB	True/False	$\mathcal{O}(\ell^2 n \log n)$	4.2
	BB	True/False	$\mathcal{O}(\ell^2 n)$	4.2
Random		AB	$\mathcal{O}(\ell n)$	5
		BB	$\mathcal{O}(\ell n \log n)$	5

Acknowledgements. The first three authors were supported in part by the Ministry of Science and Technology under the National Research Laboratory Grant 2000–2001 program.

References

1. J. S. Birman, K. H. Ko and S. J. Lee, *A new approach to the word and conjugacy problem in the braid groups*, Advances in Mathematics 139 (1998), 322-353.
2. E. A. Elrifai and H. R. Morton, *Algorithms for positive braids*, Quart. J. Math. Oxford 45 (1994), 479–497.
3. D. Epstein, J. Cannon, D. Holt, S. Levy, M. Paterson and W. Thurston, *Word processing in groups*, Jones & Bartlett, 1992.
4. F. A. Garside, *The braid group and other groups*, Quart. J. Math. Oxford 20 (1969), no. 78, 235–254.
5. K. H. Ko, S. J. Lee, J. H. Cheon, J. H. Han, J. S. Kang and C. Park, *New public key cryptosystem using braid groups*, Advances in Cryptology, Proceedings of Crypto 2000, Lecture Notes in Computer Science 1880, ed. M. Bellare, Springer-Verlag (2000), 166–183.
6. E. Lee, S. J. Lee and S. G. Hahn, *Pseudorandomness from braid groups*, Advances in Cryptology, Proceedings of Crypto 2001, Lecture notes in Computer Science 2139, ed. J. Kilian, Springer-Verlag (2001), 486–502.
7. K. H. Ko, et al., *Mathematical security analysis of braid cryptosystems*, preprint.
8. R. P. Stanley, *Enumerative combinatorics*, Wadsworth and Brooks/Cole, 1986.

How to Achieve a McEliece-Based Digital Signature Scheme

Nicolas T. Courtois[1,2], Matthieu Finiasz[1,3], and Nicolas Sendrier[1]

[1] Projet Codes, INRIA Rocquencourt
BP 105, 78153 Le Chesnay - Cedex, France
Nicolas.Sendrier@inria.fr

[2] Systèmes Information Signal (SIS), Toulon University
BP 132, F-83957 La Garde Cedex, France
courtois@minrank.org
http://www.minrank.org/

[3] École Normale Supérieure, 45, rue d'Ulm, 75005 Paris.
finiasz@ens.fr

Abstract. McEliece is one of the oldest known public key cryptosystems. Though it was less widely studied than RSA, it is remarkable that all known attacks are still exponential. It is widely believed that code-based cryptosystems like McEliece do not allow practical digital signatures. In the present paper we disprove this belief and show a way to build a practical signature scheme based on coding theory. Its security can be reduced in the random oracle model to the well-known *syndrome decoding problem* and the distinguishability of permuted binary Goppa codes from a random code. For example we propose a scheme with signatures of 81-bits and a binary security workfactor of 2^{83}.

Keywords: digital signature, McEliece cryptosystem, Niederreiter cryptosystem, Goppa codes, syndrome decoding, short signatures.

1 Introduction

The RSA and the McEliece [11] public key cryptosystems, have been proposed back in the 70s. They are based on intractability of respectively *factorization* and *syndrome decoding problem* and both have successfully resisted more than 20 years of cryptanalysis effort.

RSA became the most widely used public key cryptosystem and McEliece was not quite as successful. Partly because it has a large public key, which is less a problem today, with huge memory capacities available at very low prices. However the main handicap was the belief that McEliece could not be used in signature. In the present paper we show that it is indeed possible to construct a signature scheme based on Niederreiter's variant [12] on the McEliece cryptosystem.

The cracking problem of RSA is the problem of extracting e-th roots modulo N called the RSA problem. All the general purpose attacks for it are structural

C. Boyd (Ed.): ASIACRYPT 2001, LNCS 2248, pp. 157–174, 2001.
© Springer-Verlag Berlin Heidelberg 2001

attacks that factor the modulus N. It is a hard problem but sub-exponential. The cracking problem for McEliece is the problem of decoding an error correcting code called Syndrome Decoding (SD). There is no efficient structural attacks that might distinguish between a permuted Goppa code used by McEliece and a random code. The problem SD is known to be NP-hard since the seminal paper of Berlekamp, McEliece and van Tilborg [3], in which authors show that complete decoding of a random code is NP-hard.

All among several known attacks for SD are fully exponential (though faster than the exhaustive search [4]), and nobody has ever proposed an algorithm that behaves differently for *complete decoding* and the *bounded decoding* problems within a (slightly smaller) distance accessible to the owner of the trapdoor. In [6] Kobara and Imai review the overall security of McEliece and claim that

> *[...] without any decryption oracles and any partial knowledge on the corresponding plaintext of the challenge ciphertext, no polynomial-time algorithm is known for inverting the McEliece PKC whose parameters are carefully chosen.*

Thus it would be very interesting to dispose of signature schemes based on such hard decoding problems. The only solution available up to date was to use zero-knowledge schemes based on codes such as the SD scheme by Stern [19]. It gives excellent security but the signatures are very long. All tentatives to build practical schemes failed, see for example [20].

Any trapdoor function allows digital signatures by using the unique capacity of the owner of the public key to invert the function. However it can only be used to sign messages the hash value of which lies in the ciphertext space. Therefore a signature scheme based on trapdoor codes must achieve complete decoding. In the present paper we show how to achieve complete decoding of Goppa codes for some parameter choices.

The paper is organized as follows. First we explain in §2 and §3 how and for which parameters to achieve complete decoding of Goppa codes. In §4 we present a practical and secure signature scheme we derive from this technique. Implementation issues are discussed in §5, and in particular, we present several tradeoffs to achieve either extremely short signatures (81 bits) or extremely fast verification. In §6 we present an asymptotic analysis of all the parameters of the system, proving that it will remain practical and secure with the evolution of computers. Finally in §7 we prove that the security of the system relies on the syndrome decoding problem and the distinguishability of Goppa codes from random codes.

2 Signature with McEliece

The McEliece cryptographic scheme is based on error correcting codes. It consists in randomly adding errors to a codeword (as it would happen in a noisy channel) and uses this as a cipher. The decryption is done exactly as it would be done to correct natural transmission errors. The security of this scheme simply relies on the difficulty of decoding a word without any knowledge of the structure of the code. Only the legal user can decode easily using the trap. The Niederreiter variant - equivalent on a security point of view [8] - uses a syndrome (see below) as ciphertext, and the message is an error pattern instead of a codeword (see Table 1).

2.1 A Brief Description of McEliece's and Niederreiter's Schemes

Let $\mathbf{F}_2$ be the field with two elements $\{0,1\}$. In the present paper, C will systematically denote a binary linear code of length n and dimension k, that is a subspace of dimension k of the vector space $\mathbf{F}_2^n$. Elements of $\mathbf{F}_2^n$ are called words, and elements of C are codewords. A code is usually given in the form of a generating matrix G, lines of which form a basis of the code. The parity check matrix H is a dual form of this generating matrix: it is the $n \times (n-k)$ matrix of the application of kernel C. When you multiply a word (a codeword with an error for example) by the parity check matrix you obtain what is called a syndrome: it has a length of $n-k$ bits and is characteristic of the error added to the codeword. It is the sum of the columns of H corresponding to the non-zero coordinates of the error pattern. Having a zero syndrome characterizes the codeword and we have $G \times H = 0$.

Let C be a binary linear code of length n and dimension k correcting t errors (i.e. minimum distance is at least $2t+1$). Let G and H denote respectively a generator and a parity check matrix of C. Table 1 briefly describes the two main encryption schemes based on code. In both case the *trap* is a t-error correcting procedure for C. It enables decryption (*i.e.* finding the closest codeword to a given word or equivalently the word of smallest Hamming weight with a prescribed syndrome).

Table 1. McEliece and Niederreiter code-based cryptosystems

	McEliece	**Niederreiter**
public key:	G	H
cleartext:	$x \in \mathbf{F}_2^k$	$x \in \mathbf{F}_2^n,\ w_H(x) = t$
ciphertext:	$y = xG + e,\ w_H(e) = t$	$y = Hx^T$
ciphertext space:	$\mathbf{F}_2^n$	$\mathbf{F}_2^{n-k}$

The *secret key* is a code C_0 (usually a Goppa code) whose algebraic structure provides a fast decoder. The public code is obtained by randomly permuting the coordinates of C_0 and then choosing a random generator or parity check matrix:

$$G = UG_0P \quad \text{or} \quad H = VH_0P$$

where G_0 and H_0 are a generator and a parity check matrix of C_0, U and V are non-singular matrices ($k \times k$ and $(n-k) \times (n-k)$ respectively) and P is a $n \times n$ permutation matrix.

The security of these two systems is proven to be equivalent [8] and is based on two assumptions:

- solving an instance of the decoding problem is difficult,
- recovering the underlying structure of the code is difficult.

The first assumption is enforced by complexity theory results [3,2,16], and by extensive research on general purpose decoders [7,18,4]. The second assumption received less attention. Still the Goppa codes used in McEliece are known by coding theorists for thirty years and so far no polynomially computable property is known to distinguish a permuted Goppa code from a random linear code.

2.2 How to Make a Signature

In order to obtain an efficient digital signature we need two things: an algorithm able to compute a signature for any document such that they identify their author uniquely, and a fast verification algorithm available to everyone.

A public-key encryption function can be used as a signature scheme as follows:

1. hash (with a public hash algorithm) the document to be signed,
2. decrypt this hash value as if it were an instance of ciphertext,
3. append the decrypted message to the document as a signature.

Verification just applies the public encryption function to the signature and verifies that the result is indeed the hash value of the document. In the case of Niederreiter or any other cryptosystem based on error correcting codes the point 2 fails. The reason is that if one considers a random syndrome it usually corresponds to an error pattern of weight greater than t. In other word, it is difficult to generate a random ciphertext unless it is explicitly produced as an output of the encryption algorithm.

One solution to the problem is to obtain for our code an algorithm to decode any syndrome, or at least a good proportion of them. It is the object of the next section.

2.3 Complete Decoding

Complete decoding consists of finding a nearest codeword to any given word of the space. In a syndrome language that is being able to find an error pattern corresponding to any given syndrome. This means decoding syndromes corresponding to errors of weight greater than t.

An approach to try to perform complete decoding would be to try to correct a fixed additional number of errors (say δ). To decode a syndrome corresponding to an error of weight $t + \delta$ one should then add δ random columns from the parity check matrix to the syndrome and try to decode it. If all of the δ columns correspond to some error positions then the new syndrome obtained will correspond to a word of weight t and can be decoded by our trapdoor function. Else we will just have to try again with δ other columns, and so on until we can

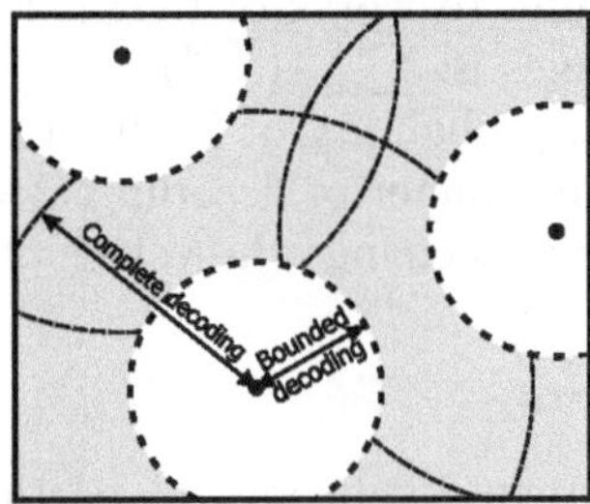

Fig. 1. From bounded decoding to complete decoding

decode one syndrome. Like this we can decode any syndrome corresponding to an error of weight less than or equal to $t + \delta$. If δ is large enough we should be able to decode any possible syndrome. However, a large δ will lead to a small probability of success for each choice of δ columns. This means that we will have to adapt the parameters of our code to obtain a δ small enough and in the same time keep a good security for our system.

This can be viewed from an different angle. Adding a random column of the parity check matrix to a syndrome really looks like choosing another random syndrome and trying to decode it. Choosing parameters for the code such that δ is small enough simply consists of increasing the density of the decodable syndromes in the space of all the syndromes, this is increasing the probability for a random syndrome to be decodable. This method will therefore take a first random syndrome (given by the hash function) and try to decode it, then modify the document and hash it again until a decodable syndrome is obtained.

The object of the next section will be to choose parameters such that the number of necessary attempts is small enough for this method to work in a reasonable time.

3 Finding the Proper Parameters

The parameters of the problem are the dimension k of the code, its length n and the maximum number t of errors the code can correct. These parameters affect all aspects of the signature scheme: its security, the algorithmic complexity for computing a signature, the length of the signature... We will start by exploring the reasons why the classical McEliece parameters are not acceptable and continue with what we wish to obtain.

3.1 Need for New Parameters

With the classical McEliece parameters ($n = 1024$, $k = 524$, $t = 50$) we have syndromes of length $n - k = 500$. This makes a total of 2^{500} syndromes. Among these only those corresponding to words of weight less than 50 are decodable. The number of such syndromes is: $\sum_{i=1}^{50}\binom{1024}{i} \simeq 2^{284}$

Therefore there is only a probability of 2^{-216} of success for each syndrome. This would mean an average number of decoding attempts of 2^{216} which is far too much. We will hence have to change the values of n, k and t.

3.2 Choosing Parameters

Binary Goppa codes are subfield subcodes of particular alternant codes [10, Ch. 12]. For a given integer m, there are many (about $2^{tm}/t$) t-error correcting Goppa codes of dimension $n - tm$ and length $n = 2^m$.

We are looking for parameters which lead to a good probability of success for each random syndrome. The probability of success will be the ratio between the number of decodable syndromes $\mathcal{N}_{dec}$ and the total number of syndromes $\mathcal{N}_{tot}$. As n is large compared with t we have:

$$\mathcal{N}_{dec} = \sum_{i=1}^{t}\binom{n}{i} \simeq \binom{n}{t} \simeq \frac{n^t}{t!}$$

and for Goppa codes $\mathcal{N}_{tot} = 2^{n-k} = 2^{mt} = n^t$. Therefore the probability of success is:

$$\mathcal{P} = \frac{\mathcal{N}_{dec}}{\mathcal{N}_{tot}} \simeq \frac{1}{t!}$$

This probability doesn't depend of n and the decoding algorithm has a polynomial complexity in m ($= \log_2 n$) and t. Therefore the signature time won't change a lot with n. As the security of the Goppa code used increases rapidly with n we will then be sure to find suitable parameters, both for the signature time and the security.

3.3 Secure Parameters

A fast bounded decoding algorithm can perform about one million decoding in a few minutes[1]. From the previous section, the number of decoding attempt to get one signature will be around $t!$, so get a reasonable signature scheme, t should not be more than 10. However for the codes correcting such a little number of errors we need to have very long codewords in order to achieve good security.

The Table 2 shows the binary workfactors for the Canteaut-Chabaud attack [4] on the McEliece cryptosystem (see section 6 for more details on the complexity of these attacks). We assume that an acceptable security level is of 2^{80} CPU operations, corresponding roughly to a binary workfactor of 2^{86}. Therefore, in our signature scheme, we need a length of at least 2^{15} with 10 errors or 2^{16} with 9 errors.

Though it is slightly below or security requirement, the choice $(2^{16}, 9)$ is better as it runs about 10 times faster.

[1] our implementation performs one million decodings in 5 minutes, but it can still be improved

Table 2. Cost for decoding

	n						
	2^{11}	2^{12}	2^{13}	2^{14}	2^{15}	2^{16}	2^{17}
$t = 8$	$2^{52.0}$	$2^{56.8}$	$2^{61.4}$	$2^{65.3}$	$2^{67.8}$	$2^{70.5}$	$2^{73.3}$
$t = 9$	$2^{54.6}$	$2^{59.9}$	$2^{69.3}$	$2^{74.0}$	$2^{78.8}$	$2^{83.7}$	$2^{88.2}$
$t = 10$	$2^{60.9}$	$2^{66.8}$	$2^{72.3}$	$2^{77.4}$	$2^{87.4}$	$2^{90.9}$	$2^{94.6}$

4 The Signature Scheme

With the chosen parameters we have a probability of 1/9! to decode each syndrome. We will therefore have to try to decode about 9! random syndromes. To do so we will simply use a counter i and hash it with the document: the hashed syndrome obtained will then depend of i, and by changing i we can have as many as we need. The signature scheme works as follows.

Let h be a hash function returning a binary word of length $n - k$ (the length of a syndrome). Let D be our document and $s = h(D)$. We denote $[\cdots s \cdots \mid \cdot i \cdot]$ the concatenation of s and i and $s_i = h([\cdots s \cdots \mid \cdot i \cdot])$.

The signature algorithm will compute the s_i for i starting at 0 and increasing by 1 at each try, until one of the syndromes s_i is decodable. We will note i_0 the first index for which s_i is decodable, and we will use this syndrome for the signature. As explained in section 2.2 the signature will then be the decrypted message, which is in our case the word z of length n and weight 9, such that $Hz^T = s_{i_0}$. However the signature will also have to include the value of i_0 for the verification. The signature will therefore be $[\cdots z \cdots \mid \cdot i_0 \cdot]$.

Signature length: the length of the signature will mainly depend of the way used to store z. It is a word of length $n = 2^{16}$ so the dumb method would be to use it directly to sign. However its weight is only 9 so we should be able to compress it a little. There are $\binom{2^{16}}{9} \simeq 2^{125.5}$ word of weight 9 so they could be indexed with a 126 bit counter. Let $i_1 < \ldots < i_9$ denote the positions of the non-zero bits of z. We define the index I_z of z by:

$$I_z = 1 + \binom{i_1}{1} + \binom{i_2}{2} + \ldots + \binom{i_9}{9}$$

The number of bits used to store i_0 isn't reducible: in average it's length is $\log_2(9!) \simeq 18.4$ bits. So the signature will be $[\cdots I_z \cdots \mid \cdot i_0 \cdot]$ with an average total length of $125.5 + 18.4 \simeq 144$ bits.

Note that using McEliece encryption scheme instead of Niederreiter's would not be satisfactory here. The signature would have a size larger than k bits (the size of a plaintext). And it would grow radiply with m if t is small. With the parameters above, the signature would have a length of 65411 bits!

Signature algorithm

- hash the document D into $s = h(D)$
- compute $s_i = h([\cdots s \cdots \mid \cdot i \cdot])$ for $i = 0, 1, 2 \ldots$
- find i_0 the smallest value of i such that s_i is decodable
- use our trapdoor function to compute z such that $Hz^T = s_{i_0}$
- compute the index I_z of z in the space of words of weight 9
- use $[\cdots I_z \cdots \mid \cdot i_0 \cdot]$ as a signature for D

Verification algorithm is much simpler (and faster)

- recover z from its index I_z
- compute $s_1 = Hz^T$ with the public key H
- compute $s_2 = h([\cdots h(D) \cdots \mid \cdot i_0 \cdot])$ with the public hash function
- compare s_1 and s_2: if they are equal the signature is valid

4.1 Attacks on the Signature Length

Having such short signatures enables attacks independent on the strength of the trapdoor function used, which are inherent to the commonly used method of computing a signature by inversion of the function. This generic attack runs in the square root of the exhaustive search. Let F be any trapdoor function with an output space of cardinality 2^r. The well known birthday paradox forgery attack computes $2^{r/2}$ hash[2] values $\mathrm{MD}(m_i)$ for some chosen messages, and picks at random $2^{r/2}$ possible signatures. One of these signatures is expected to correspond to one of the messages.

With our parameters the syndromes have a length of 144 bits and the complexity of the attack is the complexity of sorting the $2^{144/2} = 2^{72}$ values which is $2^{72} \times 72 \times 144 \simeq 2^{85}$ binary operations. This attack is not more threatening than the decoding attack, and in addition it requires a memory of about $2^{72} \times 72$ bits. Note also that the above attack depends on the syndrome length and not on the signature length, this will remain true later, even in the variants with shorter signature length.

5 Implementation Aspects

For any signature scheme there is an easy security preserving tradeoff between signature length and verification time. One may remove any h bits from the signature if one accepts exhaustive verification in 2^h for each possible value of the h missing bits. In the case of syndrome-based signature, one can do much better. As the signature consists of an error pattern of weight t, one may send

[2] MD denotes a cryptographic hash function with output of r bits

only $t-1$ out of the t errors. The verifier needs to decode the remaining error and this is much faster that the exhaustive search. More generally we are going to show that concealing a few errors (between 1 and 3) remains an excellent compromise as summarized in Table 3.

5.1 Cost of a Verification

Let s denote the hash value of the message and z denote the error pattern of weight t such that $Hz^T = s$. As z is the signature, we can compute $y = Hz^T$ by adding the t corresponding columns. The signature is accepted if y is the equal to s. The total cost of this verification is t column operations[3].

If u is a word of weight $t-1$ whose support is included in the support of z, we compute $y = s + Hu^T$, which costs $t-1$ column operations, and we check that y is a column of H, which does not cost more than one column operation if the matrix H is properly stored in a hash table.

Omitting two errors. Let us assume now that the word u transmitted as signature has weight $t-2$. There exists a word x of weight 2 such that $Hx^T = y = Hu^T$. We are looking for two columns of H whose sum is equal to y. All we have to do is to add y to any column of H and look for a match in H. Again if the columns of H are properly stored, the cost is at most $2n$ column operations.

This can be improved as the signer can choose which 2 errors are left to verifier to correct and omits in priority the positions which will be tested first, this divides the complexity in average by t (*i.e.* the match will be found in average after n/t tries).

Omitting more errors. In general, if u has weight $t-w$, we put $y = s + Hu^T$ and we need to compute the sum of y plus any $w-1$ columns of H and check for a match among the columns of H. Proper implementation will cost at most $3\binom{n}{w-1}$ column operations (yes, it is always 3, don't ask why!).

Again, if the signer omits the set of w errors which are tested first, the average cost can be divided by $\binom{t}{w-1}$.

Note that if more than 2 errors are not transmitted, the advantage is not better than the straightforward time/length tradeoff.

5.2 Partitioning the Support

Punctured code. Puncturing a code in p positions consist in removing the corresponding coordinates from the codewords. The resulting code has length $n-p$ and, in general, the same dimension[4] k. Without loss of generality we can

[3] In this section we will count all complexities in terms of column operations, one column operation is typically one access to a table and one operation like an addition or a comparison

[4] the actual dimension is the rank of a matrix derived from a generating matrix by removing the p columns

Table 3. Tradeoffs for the 9-error correcting Goppa code of length 2^{16}

remaining errors	cost [(a)] of verification		signature length
0	t	9	144 bits
1	t	9	132 bits
2	$2n/t$	2^{14}	119 bits
3	$3\binom{n}{2}/\binom{t}{2}$	2^{27}	105 bits
4	$3\binom{n}{3}/\binom{t}{3}$	2^{40}	92 bits

[(a)] in column operations (≈ 4 to 8 CPU clocks).

assume that the punctured positions come first. A parity check matrix H' of C' can be derived from any parity check matrix H of C by a Gaussian elimination: for some non-singular $(n-k) \times (n-k)$ matrix U we have

$$UH = \left(\begin{array}{c|c} I & R \\ \hline 0 & H' \end{array} \right),$$

where I denotes the $p \times p$ identity matrix.

Given a syndrome s we wish to find $z \in \mathbf{F}_2^n$ of weight t such that $s = Hz^T$. We denote s'' and s' respectively the p first and the $n-p-k$ last bits of Us and z' the last $n-p$ bits of z. Let $w \leq t$ denote the weight of z'.

$$\begin{cases} s = Hz^T \\ w_H(z) \leq t \end{cases} \Leftrightarrow \begin{cases} s' = H'z'^T \\ w_H(z') + w_H(Rz'^T + s'') \leq t \end{cases}$$

Shorter signatures. We keep the notations of the previous section. We partition the support of C into n/l sets of size l. Instead of giving the $t-w$ positions, we give the $t-w$ sets containing these positions. These $p = l(t-w)$ positions are punctured to produce the code C'. To verify the signature s we now have to correct w errors in C', *i.e.* find z' of weight w such that $s' = Hz'^T$. The signature is valid if there exists a word z' such that

$$w_H(z') \leq w \tag{1}$$

$$w_H(Rz'^T + s'') \leq t - w \tag{2}$$

We may find several values of z' verifying (1), but only one of them will also verify (2). If l is large, we have to check equation (2) often. On the other hand, large values of l produce shorter signatures. The best compromise is $l = m$ or a few units more.

The cost for computing H' is around $tm2^{m-1}$ column operations (independently of l and w). The number of column operations for decoding errors in C' is the same as in C but columns are smaller.

The signature size will be $\log_2 \left(\binom{n/l}{t-w} t! \right)$. If more than 3 errors are not transmitted, the length gain is not advantageous.

Table 4. Tradeoffs for $m = 16$, $t = 9$ and $l = m$

remaining errors (w)	cost $^{(a)}$ of verification	signature length
1	2^{22}	100 bits
2	2^{22}	91 bits
3	2^{27}	81 bits
4	2^{40}	72 bits

$^{(a)}$ in column operations (≈ 2 to 6 CPU clocks).

5.3 New Short Signature Schemes

With parameters $m = 16$ and $t = 9$, there are three interesting trade-offs between verification time and signature length. All three of them have the same complexity for computing the signature (in our our implementation the order of magnitude is one minute) and the same security level of about 2^{80} CPU operations.

Fast verification (CFS1). We transmit 8 out of the 9 error positions, the verification is extremely fast and the average signature length is $\log_2\left(t!\binom{n}{t-1}\right) = 131.1 < 132$ bits.

Short signature (CFS3). We partition the support in 2^{12} cells of 16 bits and we transmit 6 of the 9 cells. The verification time is relatively long, around one second and the average signature length is $\log_2\left(\binom{n/l}{t-w}t!\right) = 80.9 < 81$ bits.

Half & half (CFS2). We transmit the rightmost 7 error positions (out of 9). The verification algorithm starting with the left positions will be relatively fast in average, less than one millisecond. The average signature length is $\log_2\left(t!\binom{n}{t-2}\right) = 118.1 < 119$ bits.

In all three cases, to obtain a constant length signature one should be able to upper bound the number of decoding attempts. This is not possible, however by adding 5 bits to the signature the probability of failing to sign a message is less than 2^{-46}, and with 6 bits it drops to 2^{-92}.

5.4 Related Work

It seems that up till now the only signature scheme that allowed such short signatures was Quartz [14] based on HFE cryptosystem [13]. It is enabled by a specific construction that involves several decryptions in order to avoid the birthday paradox forgery described in 4.1 that runs in the square root of the exhaustive search. This method is apparently unique to multivariate quadratic cryptosystems such as HFE and works only if the best attack on the underlying trapdoor is well above the square root of the exhaustive search [13,14]. Such is not the case for the syndrome decoding problems.

6 Asymptotic Behavior

In order to measure the scalability of the system, we will examine here how the complexity for computing a signature and the cost of the best known attack evolve asymptotically. We consider a family of binary t-error correcting Goppa codes of length $n = 2^m$. These codes have dimension $k = n - tm$.

6.1 Signature Cost

We need to make $t!$ decoding attempts, for each of these attempts we need the following.

1. *Compute the syndrome.* As we are using Niederreiter's scheme we already have the syndrome, we only need to expand it into something usable by the decoder for alternant codes, the vector needed has a size of $2tm$ bits and is obtained from the syndrome by a linear operation, this costs $O(t^2m^2)$ operations in $\mathbf{F}_2$.
2. *Solve the key equation.* In this part, we apply Berlekamp-Massey algorithm to obtain the locator polynomial $\sigma(z)$, this costs $O(t^2)$ operations in $\mathbf{F}_{2^m}$.
3. *Find the roots of the locator polynomial.* If the syndrome is decodable, the polynomial $\sigma(z)$ splits in $\mathbf{F}_{2^m}[z]$ and its roots give the error positions. Actually we only need to check that the polynomial splits: that is $\gcd(\sigma(z), z^{2^m} - z) = \sigma(z)$. This requires t^2m operations in $\mathbf{F}_{2^m}$.

We will assume that one operation in $\mathbf{F}_{2^m}$ requires m^2 operations in $\mathbf{F}_2$, the total number of operations in $\mathbf{F}_2$ to achieve a signature is thus proportional to $t!t^2m^3$.

6.2 Best Attacks Complexity

Decoding attacks. The best known (and implemented) attack by decoding is by Canteaut and Chabaud [4] and its asymptotic time complexity is (empirically) around $(n/\log_2 n)^{f(t)}$ where $f(t) = \lambda t - c$ is an affine function with λ not much smaller than 1 and c is a small constant between 1 and 2.

Good estimates of the asymptotic behavior of the complexity of the best known general decoding techniques are given by Barg in [2]. In fact, when the rate $R = k/n$ of the code tends to 1, the time *and space* complexity becomes $2^{n(1-R)/2(1+o(1))}$, which, for Goppa codes, gives $n^{t(1/2+o(1))}$.

Structural attack. Very little is know about the distinguishability of Goppa codes. In practice, the only structural attack [9] consists in enumerating all Goppa codes and then testing equivalence with the public key. The code equivalence problem is difficult in theory [15] but easy in practice [17]. There are $2^{tm}/t$ binary t-error correcting Goppa codes of length $n = 2^m$, because of the properties of extended Goppa codes [10, Ch. 12, §4] only one out of mn^3 must be tested and, finally, the cost for equivalence testing cannot be lower than $n(tm)^2$ (a Gaussian elimination). Putting everything together leads to a structural attack whose cost is not less than tmn^{t-2} elementary operations.

Table 5. Characteristics of the signature scheme based on a $(n = 2^m, k = n - tm, d \geq 2t + 1)$ binary Goppa code

signature cost	$t!t^2m^3$
signature length[1]	$(t-1)m + \log_2 t$
verification cost[1]	t^2m
public key size	$tm2^m$
cost of best decoding attack	$2^{tm(1/2+o(1))}$
cost of best structural attack	$tm2^{m(t-2)}$

[1]One error position omitted

6.3 Intrinsic Strengths and Limitations

In Table 5 all complexities are expressed in terms of t and $m = \log_2 n$ and we may state the following facts:

- the signature cost depends exponentially of t,
- the public-key size depends exponentially of m,
- the security depends exponentially of the product tm.

From this we can draw the conclusion that if the system is safe today it can only be better tomorrow, as its security will depend exponentially of the signature size. On the other hand the signature cost and the key size will always remain high, as we will need to increase t or m or both to maintain a good security level. However, relatively to the technology, this handicap will never be as important as it is today and will even decrease rapidly.

7 Security Arguments

In this section we reduce the security of the proposed scheme in the random oracle model to two basic assumptions concerning hardness of general purpose decoding and pseudo-randomness of Goppa codes. We have already measured the security in terms of the work factor of the best known decoding and structural attacks. We have seen how the algorithmic complexity of these attacks will evolve asymptotically. The purpose of the present section is to give a formal proof that breaking the CFS signature scheme implies a breakthrough in one of two well identified problems. This reduction gives an important indication on where the cryptanalytic efforts should be directed.

One of these problem is decoding, it has been widely studied and a major improvement is unlikely in the near future. The other problem is connected to the classification of Goppa codes or linear codes in general. Classification issues are in the core of coding theory since its emergence in the 50's. So far nothing significant is known about Goppa codes, more precisely there is no known property invariant by permutation and computable in polynomial time which characterizes Goppa codes. Finding such a property or proving that none exists would be an important breakthrough in coding theory and would also probably seal the fate, for good or ill, of Goppa code-based cryptosystems.

7.1 Indistinguishability of Permuted Goppa Codes

Definition 1 (Distinguishers). *A T-time distinguisher is a probabilistic Turing machine running in time T or less such that it takes a given F as an input and outputs $\mathcal{A}^F$ equal to 0 or 1. The probability it outputs 1 on F with respect to some probability distribution $\mathcal{F}$ is denoted as:*

$$Pr[F \leftarrow \mathcal{F} : \mathcal{A}^F = 1]$$

Definition 2 ((T, ε)-PRC). *Let $\mathcal{A}$ be a T-time distinguisher. Let* $\text{RND}(n, k)$ *be the uniform probability distribution of all binary linear (n, k)-code. Let $\mathcal{F}(n, k)$ be any other probability distribution. We define the distinguisher's advantage as:*

$$Adv_{\mathcal{F}}^{PRC}(\mathcal{A}) \stackrel{def}{=} \Big| Pr[F \leftarrow \mathcal{F}(n, k) : \mathcal{A}^F = 1] - Pr[F \leftarrow \text{RND}(n, k) : \mathcal{A}^F = 1] \Big| .$$

*We say that $\mathcal{F}(n, k)$ is a (T, ε)-**PRC** (Pseudo-Random Code) if we have:*

$$\max_{T-\text{time } \mathcal{A}} Adv_{\mathcal{F}}^{PRC}(\mathcal{A}) \quad \leq \varepsilon .$$

7.2 Hardness of Decoding

In this section we examine the relationships between signature forging and two well-known problems, the *syndrome decoding problem* and the *bounded-distance decoding problem*. The first is NP-complete and the second is conjectured NP-hard.

Definition 3 (Syndrome Decoding - SD).

Instance: *A binary $r \times n$ matrix H, a word s of $\mathbf{F}_2^r$, and an integer $w > 0$.*
Problem: *Is there a word x in $\mathbf{F}_2^n$ of weight $\leq w$ such that $Hx^T = s$?*

This decision problem was proven NP-complete [3]. Achieving complete decoding for any code can be done by a polynomial (in n) number of calls to SD. Actually the instances of SD involved in breaking code-based systems are in a particular subclass of SD where the weight w is bounded by the half of the minimum distance of the code of parity check matrix H. Is has been stated by Vardy in [16] as:

Definition 4 (Bounded-Distance Decoding - BD).

Instance: *An integer d, a binary $r \times n$ matrix H such that every $d-1$ columns of H are linearly independent, a word s of $\mathbf{F}_2^r$, and an integer $w \leq (d-1)/2$.*
Problem: *Is there a word x in $\mathbf{F}_2^n$ of weight $\leq w$ such that $Hx^T = s$?*

It is probably not NP because the condition on H is NP-hard to check. However several prominent authors [1,16] conjecture that BD is NP-hard.

Relating signature forging and BD. An attacker who wishes to forge for a message M a signature of weight t with the public key H, has to find a word of weight t whose syndrome lies in the set $\{h(M,i) \mid i \in \mathbf{N}\}$ where $h()$ is a proper cryptographic hash function (see §4). Under the random oracle model, the only possibility for the forger is to generate any number of syndromes of the form $h(M,i)$ and to decode one of them this cannot be easier than $\mathrm{BD}(2t+1, H, h(M,i), t)$ for some integer i.

Relating signature forging and SD. Let us consider the following problem:

Definition 5 (List Bounded-Distance Decoding - LBD).

Instance: *An integer d, a binary $r \times n$ matrix H such that every $d-1$ columns of H are linearly independent, a subset S of $\mathbf{F}_2^r$, and an integer $w \leq \lfloor (d-1)/2 \rfloor$.*
Problem: *Is there a word x in $\mathbf{F}_2^n$ of weight $\leq w$ such that $Hx^T \in S$?*

Using this problem we will show how we may relate the forging of a signature to an instance of SD:

- In practice the forger must at least solve $\mathrm{LBD}(2t+1, H, S, t)$ where $S \subset \{h(M,i) \mid i \in \mathbf{N}\}$. The probability for the set S to contain at least one correctable syndrome is greater than $1 - \mathbf{e}^{-\lambda}$ where $\lambda = |S|\binom{n}{t}/2^r$. This probability can be made arbitrarily close to one if the forger can handle a set S big enough.
- Similarely, from any syndrome $s \in \mathbf{F}_2^r$, one can derive a set $R_{s,\delta} \subset \{s + Hu^T \mid u \in \mathbf{F}_2^n, w_H(u) \leq \delta\}$ where $\delta = d_{vg} - t$ and d_{vg} is an integer such that $\binom{n}{d_{vg}} > 2^r$. With probability close to $1 - \mathbf{e}^{-\mu}$ where $\mu = |R_{s,\delta}|\binom{n}{t}/2^r$, we have $\mathrm{LBD}(2t+1, H, R_{s,\delta}, t) = \mathrm{SD}(H, s, d_{vg})$. Thus solving $\mathrm{LBD}(2t+1, H, R_{s,\delta}, t)$ is at least as hard as solving $\mathrm{SD}(H, s, d_{vg})$.
- We would like to conclude now that forging a signature is at least as hard as solving $\mathrm{SD}(H, s, d_{vg})$ for some s. This would be true if solving $\mathrm{LBD}(2t+1, H, S, t)$ was harder than solving $\mathrm{LBD}(2t+1, H, R_{s,\delta}, t)$ for some s, which seems difficult to state. Nevertheless, with sets S and $R_{s,\delta}$ of same size, it seems possible to believe that the random set (S) will not be the easiest to deal with.

Though the security claims for our signature scheme will rely on the difficulty of $\mathrm{BD}(2t+1, H, s, t)$, it is our belief that it can reduced to the hardness of $\mathrm{SD}(H, s, d_{vg})$ (note that d_{vg} depends only of n and r, not of t). If we assume the pseudo-randomness of the hash function $h()$ and of Goppa codes these instances are very generic.

7.3 Security Reduction

We assume that the permuted Goppa code used in our signature scheme is a $(T_{Goppa}, 1/2)$-**PRC**, i.e. it cannot be distinguished from a random code with an advantage greater than 1/2 for all adversaries running in time $< T_{Goppa}$.

We assume that an instance of $\mathrm{BD}(2t+1, H, s, t)$ where H and s are chosen randomly cannot be solved with probability greater than 1/2 by an adversary running in time $< T_{BD}$.

Theorem 1 (Security of CFS). *Under the random oracle assumption, a T-time algorithm that is able to compute a valid pair message+signature for CFS with a probability $\geq 1/2$ satisfies:*

$$T \geq \min\left(T_{Goppa}, T_{BD}\right).$$

Proof (sketch): Forging a signature is at least as hard as solving $\mathrm{BD}(2t+1, H, s, t)$ where $s = h(M, i)$ (see §7.2) and H is the public key. Under the random oracle assumption, the syndrome $h(M, i)$ can be considered as random. If someone is able to forge a signature in time $T < T_{BD}$, then with probability 1/2 the matrix H has been distinguished from a random one and we have $T \geq T_{Goppa}$. □

8 Conclusion

We demonstrated how to achieve digital signatures with the McEliece public key cryptosystem. We propose 3 schemes that have tight security proofs in random oracle model. They are based on the well known hard *syndrome decoding problem* that after some 30 years of research is still exponential. The Table 6 summarizes the concrete security of our schemes compared to some other known signature schemes.

Table 6. McEliece compared to some known signature schemes

base cryptosystem	RSA	ElGamal	EC	HFE	McEliece/Niederreiter		
signature scheme	RSA	DSA	ECDSA	Quartz	CFS1	CFS2	CFS3
data size(s)	1024	160/1024	160	100	144		

security							
structural problem	factoring	DL(p)	Nechaev group?	HFEv-	Goppa $\stackrel{?}{=}$ PRCode		
best structural attack	$\mathbf{2^{102}}$	2^{102}	∞	$> 2^{97}$	2^{119}		
inversion problem	RSAP	DL(q)	EC DL	MQ	SD		
best inversion attack	2^{102}	$\mathbf{2^{80}}$	$\mathbf{2^{80}}$	2^{100}	$\mathbf{2^{83}}$		

efficiency							
signature length	1024	320	321	128	132	119	**81**
public key [kbytes]	0.2	0.1	0.1	71	1152		
signature time 1 GHz	9 ms	1.5 ms	5 ms	15 s	10 – 30 s		
verification time 1 GHz	9 ms	2 ms	6 ms	40 ms	$< 1\ \mu$s	< 1 ms	$\approx$ 1s

The proposed McEliece-based signature schemes have unique features that will make it an exclusive choice for some applications while excluding other. On one hand, we have seen that both key size and signing cost will remain high, but will evolve favorably with technology. On the other hand the signature length and verification cost will always remain extremely small. Therefore if there is no major breakthrough in decoding algorithms, it should be easy to keep up with the Moore's law.

References

1. A. Barg. Some new NP-complete coding problems. *Problemy Peredachi Informatsii*, 30:23–28, 1994 (in Russian).
2. A. Barg. *Handbook of Coding theory*, chapter 7 – Complexity issues in coding theory. North-Holland, 1999.
3. E. R. Berlekamp, R. J. McEliece, and H. C. van Tilborg. On the inherent intractability of certain coding problems. *IEEE Transactions on Information Theory*, 24(3), May 1978.
4. A. Canteaut and F. Chabaud. A new algorithm for finding minimum-weight words in a linear code: Application to McEliece's cryptosystem and to narrow-sense BCH codes of length 511. *IEEE Transactions on Information Theory*, 44(1):367–378, January 1998.
5. N. Courtois, M. Finiasz, and N. Sendrier. How to achieve a McEliece-based digital signature scheme. Cryptology ePrint Archive, Report 2001/010, February 2001. `http://eprint.iacr.org/` et RR-INRIA 4118.
6. K. Kobara and H. Imai. Semantically secure McEliece public-key cryptosystems -Conversions for McEliece PKC-. In *PKC'2001*, LNCS, Cheju Island, Korea, 2001. Springer-Verlag.
7. P. J. Lee and E. F. Brickell. An observation on the security of McEliece's public-key cryptosystem. In C. G. Günther, editor, *Advances in Cryptology – EUROCRYPT'88*, number 330 in LNCS, pages 275–280. Springer-Verlag, 1988.
8. Y. X. Li, R. H. Deng, and X. M. Wang. On the equivalence of McEliece's and Niederreiter's public-key cryptosystems. *IEEE Transactions on Information Theory*, 40(1):271–273, January 1994.
9. P. Loidreau and N. Sendrier. Weak keys in McEliece public-key cryptosystem. *IEEE Transactions on Information Theory*, 47(3):1207–1212, April 2001.
10. F. J. MacWilliams and N. J. A. Sloane. *The Theory of Error-Correcting Codes*. North-Holland, 1977.
11. R. J. McEliece. A public-key cryptosystem based on algebraic coding theory. *DSN Prog. Rep.*, Jet Prop. Lab., California Inst. Technol., Pasadena, CA, pages 114–116, January 1978.
12. H. Niederreiter. Knapsack-type crytosystems and algebraic coding theory. *Prob. Contr. Inform. Theory*, 15(2):157–166, 1986.
13. J. Patarin. Hidden fields equations (HFE) and isomorphisms of polynomials (IP): two new families of asymmetric algorithms. In *Eurocrypt'96*, LNCS, pages 33–48, 1996.
14. J. Patarin, L. Goubin, and N. Courtois. 128-bit long digital signatures. In *Cryptographers' Track Rsa Conference 2001*, San Francisco, April 2001. Springer-Verlag. to appear.
15. E. Petrank and R. M. Roth. Is code equivalence easy to decide? *IEEE Transactions on Information Theory*, 43(5):1602–1604, September 1997.

16. A. Vardy. The Intractability of Computing the Minimum Distance of a Code. *IEEE Transactions on Information Theory*, 43(6):1757–1766, November 1997.
17. N. Sendrier. Finding the permutation between equivalent codes: the support splitting algorithm. *IEEE Transactions on Information Theory*, 46(4):1193–1203, July 2000.
18. J. Stern. A method for finding codewords of small weight. In G. Cohen and J. Wolfmann, editors, *Coding theory and applications*, number 388 in LNCS, pages 106–113. Springer-Verlag, 1989.
19. J. Stern. A new identification scheme based on syndrome decoding. In D. R. Stinson, editor, *Advances in Cryptology - CRYPTO'93*, number 773 in LNCS, pages 13–21. Springer-Verlag, 1993.
20. J. Stern. Can one design a signature scheme based on error-correcting codes ? In *Asiacrypt 1994*, number 917 in LNCS, pages 424–426. Springer-Verlag, 1994. Rump session.

Efficient Traitor Tracing Algorithms Using List Decoding

Alice Silverberg[1*], Jessica Staddon[2**], and Judy L. Walker[3***]

[1] Department of Mathematics
Ohio State University
Columbus, OH, USA
silver@math.ohio-state.edu

[2] Xerox PARC
Palo Alto, CA, USA
jstaddon@parc.xerox.com

[3] Department of Mathematics and Statistics
University of Nebraska
Lincoln, NE, USA
jwalker@math.unl.edu

Abstract. We use powerful new techniques for list decoding error-correcting codes to efficiently trace traitors. Although much work has focused on constructing traceability schemes, the complexity of the tracing algorithm has received little attention. Because the TA tracing algorithm has a runtime of $O(N)$ in general, where N is the number of users, it is inefficient for large populations. We produce schemes for which the TA algorithm is very fast. The IPP tracing algorithm, though less efficient, can list all coalitions capable of constructing a given pirate. We give evidence that when using an algebraic structure, the ability to trace with the IPP algorithm implies the ability to trace with the TA algorithm. We also construct schemes with an algorithm that finds all possible traitor coalitions faster than the IPP algorithm. Finally, we suggest uses for other decoding techniques in the presence of additional information about traitor behavior.

1 Introduction

Traceability schemes are introduced in [9] and have been extensively studied in the intervening years for use as a piracy deterrent. We focus on one of the few aspects of this area of work that has received little attention: the complexity

* Silverberg would like to thank MSRI, Bell Labs Research Silicon Valley, NSA, and NSF.

** Much of this work was completed while Staddon was employed by Bell Labs Research Silicon Valley.

*** Walker is partially supported by NSF grants DMS-0071008 and DMS-0071011.

C. Boyd (Ed.): ASIACRYPT 2001, LNCS 2248, pp. 175–192, 2001.
© Springer-Verlag Berlin Heidelberg 2001

of the traitor tracing algorithms. We show that powerful new techniques for the list decoding of error-correcting codes enable us to construct traceability schemes with very fast traitor tracing algorithms. Further, we use list decoding to give new algorithms for producing a list of all coalitions capable of creating a given pirate. In addition, we discuss potential applications of other decoding methods to the problem of tracing traitors, suggest alternative approaches when additional information is known about the way the traitors are operating, and examine the relationship between two important tracing algorithms.

In a popular model for traceability schemes a unique set (possibly ordered) of r symbols is associated with each user. For example, the set may be associated with a user's software CD, or contained in a smartcard the user has for the purpose of viewing encrypted pay-TV programs (in the latter case, the set corresponds to a set of keys). When a coalition forms to commit piracy, it must construct a set to associate with the pirate object. In the case of unordered sets, this pirate set consists of r symbols, each of which belongs to at least one coalition member's set. If the sets are ordered, the coalition members must form an ordered pirate set in which the symbol in each position is identical to the symbol in the same position in the ordered set of some coalition member. In either scenario a traitor tracing algorithm is applied to the pirate, and identifies an actual traitor or traitors. The approach we take here is to use error-correcting codes to construct traceability schemes in which the sets are ordered. The ordered (as opposed to the unordered) set scenario yields naturally to coding theoretic techniques and has many practical applications ([10,7]).

We first focus on the TA traitor tracing algorithm (following the terminology in [40]), that identifies as traitors all users who share the most with the pirate. In general the TA algorithm runs in $O(N)$ time, where N is the number of users. However, this paper shows that for suitable constructions based on error-correcting codes, tracing can be accomplished in time polynomial in $c \log N$, where c is the maximum coalition size. This is a significant improvement, as we expect c to be much smaller than N. The constructions in this paper match the best previously known schemes in this model in terms of the alphabet size that is required to achieve a certain level of traceability for a given codeword length, and exceed all earlier schemes in the speed with which they trace (at least) one traitor.

We also consider the IPP tracing algorithm (following the terminology in [23]). The IPP algorithm identifies all coalitions capable of making a pirate and looks for a common member(s) amongst these coalitions. Hence, the IPP property seems to be a more fundamental traceability property. In general this algorithm runs in time $O(crN^c)$, where r is the length of each codeword, and hence is even less efficient than the TA algorithm. However, there are two good reasons to be interested in IPP codes. First, the extra computational burden of the IPP algorithm has led to the question (see [37]) of whether IPP schemes may beat TA schemes in other respects, namely, in terms of the number of codewords for a fixed set of parameters. We provide evidence that for schemes with enough structure to enable efficient tracing algorithms, increasing the number of codewords causes tracing to fail with *both* the TA and IPP algorithms. Hence, IPP codes do not appear to yield efficiency improvements in this respect. Secondly, as part of the IPP tracing process, additional valuable piracy information is amassed, namely, a list of all coalitions capable of creating the pirate in question. Such a list is not a by-product of the TA algorithm, but is a useful part of a security audit. We show that when error-correcting codes are used to construct TA traceability codes (which are also IPP codes, by a result in [37]), list decoding techniques can be used to construct new algorithms for finding all such coalitions. We give an algorithm that is more efficient than the brute force approach of the IPP algorithm of evaluating each coalition for its ability to create the pirate, thereby answering an open question in [37].

This paper gives the first applications of list decoding to the traitor tracing problem in the above model, although Zane [48] uses such techniques to address the related problem of watermark detection. (See Section 1.1 below for a discussion of this, and other, related work.) These list decoding techniques are receiving wide attention in the coding theory community, and improvements and generalizations are being rapidly produced. We believe that in this paper we have merely scratched the surface of the potential applications of decoding techniques to traceability. In the last section we discuss the use of other decoding methods when additional information is known about the traitors or how they operate, giving directions for future work in this area.

Overview. Section 1.1 covers related work on traceability and broadcast encryption and Section 2 covers the necessary background on traceability and coding theory. Section 3 describes how to construct efficient traceability schemes. Section 4 considers the relationship between TA and IPP traceability schemes, providing justification for our restriction to the TA case, and raising some questions concerning the relationship between TA and IPP for linear codes. Section 5 shows how codes of sufficiently large minimum distance enable a more efficient algorithm for finding all coalitions of traitors. A discussion of other potential applications of coding theoretic ideas and techniques to traceability questions is given in Section 6.

1.1 Related Work

The phrase *traitor tracing* is coined in [9] (see also the extended version [10]). In traceability schemes, users are each given an ordered (as in [9,7,15,37], for example), or unordered (as in [40], for example) set of keys.

In [6] (see also the revised version [7]), methods for creating TA traceability codes are given for the purpose of fingerprinting digital data. Lower bounds and additional constructions of TA traceability schemes are given in [40], while lower bounds are also proven in [27,26]. In addition, [26] provides a tracing algorithm for schemes in [27].

The problem of combining broadcast encryption and traceability is studied in [41,16,29,46].

Variations on the models of [10,7] have been studied in recent years. *Dynamic* models (here we study a static model), in which it is possible to get additional evidence of piracy in order to "test" traitor guesses, are studied in [15,3,33]. A public-key traitor tracing scheme is given in [5]. One of the nice properties of the scheme in [5] is that it is possible to identify *all* traitors. We note that although our algorithms in Sect. 3 can only guarantee the identification of one traitor, they do so in significantly faster time (polynomial in $c \log N$, versus $O(N\log^2 N\log\log N)$ in [5], with N the number of codewords and c the maximum coalition size).

In [31,11], ways in which accountability can be added to the model are discussed. For example, to improve upon the strength of the deterrent, in [11] committing piracy efficiently necessitates revealing sensitive information. In [17], a system in which pirate pay-TV decoders can only work for short periods of time is presented. As noted in [17], traceability can be a useful addition to a long-lived broadcast encryption scheme. If keys are allocated to smartcards in such a way as to ensure some traceability, it is possible to keep a list of traitor smartcards over time. If the smartcard of one particular user appears on the list frequently despite many smartcard refreshments (i.e., key changes) this mounting evidence makes it increasingly likely that the user is actually guilty, and not simply a victim of smartcard theft. Hence, as long as traceability schemes are efficient, they can quickly yield useful information during system audits.

Recently, the identifiable parent property (IPP) tracing algorithm has garnered attention [23,2,37] (also, very similar ideas are studied in [39]). In [23], a combinatorial characterization of 2-IPP schemes is presented. Additional constructions of and bounds for IPP schemes appear in [2,37].

A coding theoretic approach is taken in [25] to study the related problem of blacklisting users in a broadcast encryption scheme, but that paper does not address the question of tracing.

Our approach takes advantage of recent powerful list decoding methods, which originated with the work of Sudan [42]. In list decoding the input is a received word and the output is the list of all codewords within a given Hamming distance of the received word. Sudan's results by themselves are not strong enough to be applicable in the setting in which the TA algorithm succeeds in finding traitors (as opposed to identifying probable traitors), since the decoding procedure in [42] is not capable of correcting enough errors in the code. However, Sudan's work has recently been extended to enable it to efficiently correct more errors; i.e., it extends the radius of the Hamming ball around the received word in which it can find all the codewords in time polynomial in the length of the codewords. The improvements in [19] are precisely sufficient to be applicable to the setting where the TA algorithm succeeds. An additional advantage of this method is that it gives a list containing one or more traitors, rather than only one. Efficient list decoding algorithms now exist for Reed-Solomon codes, more general algebraic geometry codes, and some concatenated codes.

List decoding techniques are applied to the problem of watermarking in [48]. Whereas in traceability schemes each user has a unique codeword, in the watermarking scenario each user needs to be given the same "document" V, taken in [48] to be a vector of real numbers between 0 and 1. To prevent users from distributing pirated copies of V, each user is given a distinct, slightly modified "watermarked" version of V. The CKLS media watermarking scheme [8] is modified in [48] so that the watermarks are chosen from a set of randomly generated CKLS codes according to a Reed-Solomon code. Given a suspected pirate copy of V, the results of [42] on list decoding can then be used to identify one or more traitors.

Here, we consider the related question of traceability schemes, and we apply list decoding results for algebraic geometry codes and certain concatenated codes in addition to Reed-Solomon codes. In [48], Reed-Solomon codes are used to obtain vectors of real numbers between 0 and 1 to serve as a watermark, while here the error-correcting codes themselves are the traceability schemes.

We note that algebraic geometry codes appear to have been under-utilized in cryptological applications. For example, the results of [34] can be used to give better explicit examples of c-frameproof codes than those obtained in [7]. The codes constructed in [34] are concatenated codes (see below) where the outer code is an algebraic geometry code coming from a Hermitian curve, while those used in [7] come from pseudo-random graphs (see [1]).

2 Background on Codes and Traceability

In this section we give definitions, notation, and background on codes, traceability, and the decoding techniques that form the basis for our tracing algorithms.

2.1 Definitions and Notation

A *code* C of *length* r is a subset of Q^r, where Q is a finite alphabet. The elements of C are called *codewords*; each codeword has the form $x = (x_1, \cdots, x_r)$, where $x_i \in Q$ for $1 \leq i \leq r$. Subsets of C will be called *coalitions*.

For any coalition $C_0 \subseteq C$, we define the set of *descendants* of C_0, denoted $\mathsf{desc}(C_0)$ by

$$\mathsf{desc}(C_0) = \{w \in Q^r : w_i \in \{x_i : x \in C_0\}, \text{ for all } 1 \leq i \leq r\} \ .$$

The set $\mathsf{desc}(C_0)$ consists of the r-tuples that could be produced by the coalition C_0.

We define $\mathsf{desc}_c(C)$ to be the set of all $x \in Q^r$ for which there exists a coalition C_0 of size at most c such that $x \in \mathsf{desc}(C_0)$. In other words, $\mathsf{desc}_c(C)$ consists of the r-tuples that could be produced by a coalition of size at most c.

For $x, y \in Q^r$, let $I(x,y) = \{i : x_i = y_i\}$.

Definition 1. *A code C is a c-TA (traceability) code if for all coalitions C_i of size at most c, if $w \in \mathsf{desc}(C_i)$ then there exists $x \in C_i$ such that $|I(x,w)| > |I(z,w)|$ for all $z \in C - C_i$.*

In other words, C is a c-TA code if, whenever a coalition of size at most c produces a pirate word w, there is an element of the coalition which is closer to w than any codeword not in the coalition.

Codes with the identifiable parent property (IPP) are another type of traceability code.

Definition 2. *A code C is a c-IPP code if for all $w \in \mathsf{desc}_c(C)$, the intersection of the coalitions C_i of size at most c such that $w \in \mathsf{desc}(C_i)$ is nonempty.*

Suppose C is a code of length r. The *(Hamming) distance* between two elements x and y of Q^r is $r - |I(x, y)|$. The *minimum distance* of the code C is the smallest distance between distinct codewords of C.

If C is a c-IPP code and $w \in \mathsf{desc}_c(C)$, then the *traitors* that can produce the *pirate* w are the codewords that lie in all coalitions C_i of size at most c such that $w \in \mathsf{desc}(C_i)$.

When implementing one of the traceability codes just described, one randomly chooses a set of symbols $\{s_{(i,y)}\}$ with $i \in \{1, \dots, r\}$ and y in the alphabet Q, and the collection of symbols corresponding to a given user is determined by the codeword associated with that user. For example, if the codeword $x = (x_1, \dots, x_r)$ is associated with user u, then the set of symbols associated with user u is $S_u = \{s_{(1,x_1)}, \dots, s_{(r,x_r)}\}$. It is S_u, not x, that the user stores (e.g., S_u is embedded in the user's CD or smartcard). The encryption step makes the model of pirate behavior that we consider reasonable. Since the symbols are generated randomly it is essentially impossible to guess a symbol, and hence a coalition is only able to form a pirate out of its pooled collection of symbols. In other words, moving from codewords to symbols thwarts algebraic attacks (such as, for example, the attack on [27] found in [41,5]). Although a coalition may be able to write down any codeword (this information may be public), it can only generate the symbol associated with an entry in the codeword if there is a coalition member that agrees with the codeword in that position.

2.2 Background Traceability Results

The following result, which is Lemma 1.3 of [37], is very useful for showing that a code is c-IPP.

Lemma 1. ([37], Lemma 1.3) *Every c-TA code is a c-IPP code.*

As shown in [37], there are c-IPP codes that are not c-TA. We give a simple example of a 2-IPP code that is not 2-TA.

Example 1. Let $u_1 = (0, 0, 1)$, $u_2 = (1, 0, 0)$, and $u_3 = (2, 0, 0)$. The code $\{u_1, u_2, u_3\}$ is clearly 2-IPP, since the first entry of a pirate determines a traitor. The coalition $\{u_1, u_2\}$ can produce the pirate $w = (0, 0, 0)$. However, $|I(u_1, w)| = |I(u_2, w)| = |I(u_3, w)| = 2$, so the code is not 2-TA.

Note that for c-IPP codes, traitor tracing is roughly an $O(\binom{N}{c})$ process, where N is the total number of codewords in the code. A traitor tracing algorithm for a c-TA code takes as input a $w \in \mathsf{desc}_c(C)$ and outputs a codeword x such that $|I(x,w)|$ is largest. Hence for c-TA codes, tracing is an $O(N)$ process, in general.

The next result, which is proved in [37] (see Theorem 4.4 of that paper; see also [9] and [10]), shows that for codes with large enough minimum distance the TA algorithm suffices, and consists of finding codewords within distance $r - \frac{r}{c}$ from the pirate. Further, all codewords within this distance will be traitors.

Theorem 1. ([37], Theorem 4.4) *Suppose C is a code of length r, c is a positive integer, and the minimum distance d of C satisfies $d > r - \frac{r}{c^2}$. Then*

(i) C is a c-TA code;
(ii) if C_0 is a coalition of size at most c, and $w \in \mathsf{desc}(C_0)$, then:
  (a) there exists an element of C_0 within distance $r - \frac{r}{c}$ of w, and
  (b) every codeword within distance $r - \frac{r}{c}$ of w is in the coalition C_0.

2.3 Linear Codes

Linear codes are a very important class of codes. We will say that a code of length r is *linear*, or linear over F_q, if the alphabet is a finite field F_q and the code is a linear subspace of the vector space F_q^r. The *dimension* of the code is its dimension as a vector space. If C is a linear code over F_q of dimension k, then $|C| = q^k$.

Reed-Solomon codes are among the most widely-used linear codes, with many useful applications (e.g., compact disks). To obtain a Reed-Solomon code of length r and dimension k over the finite field F_q, fix r distinct elements $\alpha_1, \ldots, \alpha_r$ of F_q. The codewords are exactly the r-tuples $(f(\alpha_1), \ldots, f(\alpha_r))$ as f runs over (the zero polynomial and) all polynomials of degree less than k in $F_q[x]$. Note that a basis for the code over F_q is

$$\{(1, \ldots, 1), (\alpha_1, \ldots, \alpha_r), (\alpha_1^2, \ldots, \alpha_r^2), \ldots, (\alpha_1^{k-1}, \ldots, \alpha_r^{k-1})\} .$$

Since two distinct polynomials of degree less than k agree on at most $k-1$ points, the minimum distance of the code is $r - k + 1$.

A useful generalization of Reed-Solomon codes are *algebraic geometry (AG) codes* (see, for example, [18,38,44]). The linear codes with the "best" known parameters asymptotically are AG codes [45]. One advantage of AG codes is that they are not, in general, bound by the restriction that $r \leq q$, as was the case for the Reed-Solomon codes above. Being freed of this constraint allows us to have a smaller alphabet (and in applications, fewer keys), for given choices of the other parameters. Hermitian codes, coming from Hermitian curves, are examples of AG codes that have nice properties and can be defined explicitly. For those familiar with the below terminology (such knowledge is not essential for appreciating the results of this paper), we note that for our purposes it suffices to consider the one-point codes $C_X(\mathcal{P}, \ell P_0)$ which can be defined as follows. Start with a smooth, absolutely irreducible curve X of genus g defined over a finite field F_q, a set $\mathcal{P} = \{P_1, \ldots, P_r\}$ of r distinct F_q-rational points on X, another F_q-rational point P_0 on X which is not in the set $\mathcal{P}$, and an integer ℓ. The codewords are then the r-tuples $(f(P_1), \ldots, f(P_r))$, where f runs over the rational functions on X whose only pole is P_0, where the multiplicity is at most ℓ. If $2g - 2 < \ell < r$, this code has dimension $\ell + 1 - g$ and minimum distance at least $r - \ell$. Reed-Solomon codes can be viewed as algebraic geometry codes by taking X to be the projective line, $\mathcal{P}$ to be the set of points corresponding to the r chosen field elements, P_0 to be the point at infinity, and $\ell = k - 1$.

Concatenated codes are codes which are "concatenated" from two other codes. When two linear codes are concatenated, the product of their lengths (resp., dimensions, resp., minimum distances) is the length (resp., dimension, resp., minimum distance) of the (linear) concatenated code. There are linear concatenated codes for small alphabets which have good list decoding capabilities, i.e., a small list of possible codewords can be recovered even when a large percentage of the symbols are in error or have been erased [20].

We refer the reader to [18,28,38,44] for more information on coding theory.

2.4 Decoding

In the theory of error-correcting codes, a codeword is transmitted through a noisy channel and an element of Q^r (i.e., a *word*) is received. The receiver (or *decoder*) then tries to determine as accurately as possible which codeword was transmitted.

If d is the minimum distance of the code, then the receiver can "correct" $t = \lfloor \frac{d-1}{2} \rfloor$ errors; i.e., there is at most one codeword within distance t of the received word. The radius t is called the *error-correction bound* or the *packing radius*. *Minimum-distance* (or *nearest-neighbor*) *decoding* finds the closest codeword to the received word. In practice, minimum-distance decoding is very slow. In *bounded-distance* decoding, the decoder finds a codeword within a specified distance of the received word, if one exists. In the bounded-distance decoding decision problem, the inputs are a linear code over a given finite field, a received word, and a specified distance t, and the output is a yes or no answer to the question of whether there is a codeword within distance t of the received word. This decision problem is known to be NP-complete [4].

In *list decoding*, the goal is to output the list of all codewords within a specified distance of the received word. In [42] and [43], Sudan gave the first efficient methods for list decoding that run in time polynomial in the length of the codewords. Since then, Sudan's list decoding technique has been improved, generalized, and refined [35,36,19,20,21,22,24,30,32,47,12,13]. The runtimes for the steps of the algorithm have been improved, the number of errors that can be "corrected" has been increased, and the technique has been shown to be applicable to a larger class of codes. Sudan's original algorithm is for Reed-Solomon codes. Other codes for which the techniques have been shown to apply include AG codes (for which the focus has been on Hermitian codes) and certain concatenated codes (see [20], where the "outer code" is a Reed-Solomon or AG code and the "inner code" is a Hadamard code).

In *erasure decoding*, some positions of the received word are garbled or "erased", and cannot be identified. In this case the decoder knows that errors occurred in those positions. In *erasure-and-error decoding*, the decoder receives a word with some erasures and some errors, and determines the transmitted word, or a list of possible transmitted words (given some appropriate bounds on the numbers of errors and erasures).

In *soft-decision decoding*, instead of receiving a (*hard-decision*) word, the decoder receives a reliability matrix that states the probability that any given element of the alphabet was sent in any given position. Using this "soft" information, a soft-decision decoder outputs the most likely transmitted codeword(s).

3 Efficient Tracing Algorithms via List Decoding

In this section we show how the efficiency of the TA tracing algorithm can be greatly improved when the traceability scheme is based on certain error-correcting codes, and the tracing algorithm uses fast list decoding methods. What is an $O(N)$ process in general becomes a process that runs in time polynomial in $c \log N$. These constructions match the best previously known traceability schemes in this model in terms of the alphabet size that is required to support a given level of traceability and codeword length (roughly speaking, the alphabet size is $O(N^{\frac{c^2}{r}})$). The following theorem describes constructions based on Reed-Solomon, algebraic geometry, and concatenated codes. One advantage of considering all three types of codes is that the appropriate code choice for the traceability scheme depends on the desired parameters.

Theorem 2. *(i) Let C be a Reed-Solomon code of length r and dimension k over a finite field F_q of size at most 2^r. If c is an integer, $c \geq 2$, and $r > c^2(k-1)$, then C is a c-TA code and there is a traitor tracing algorithm that runs in time $O(r^{15})$. If $r = (1+\delta)c^2(k-1)$ then the algorithm runs in time $O(\frac{r^3}{\delta^6})$. For $r = \Theta(c^2 k)$, the runtime is $O(c^{30} \log_q^{15} N)$.*

(ii) Let X be a nonsingular plane curve of genus g defined over a finite field F_q, $\mathcal{P}$ a set of r distinct F_q-rational points on X, P_0 an F_q-rational point on X which is not in $\mathcal{P}$, and k an integer such that $k > g-1$. Let c be an integer such that $c \geq 2$ and $r > c^2(k+g-1)$, assume that $q \leq 2^r$, and assume the pre-processing described in [19] has occurred. Then the one-point AG code $C_X(\mathcal{P}, (k+g-1)P_0)$ is a c-TA code with a traitor tracing algorithm that runs in time polynomial in r.

(iii) If k and c are positive integers, q is a prime power, $q > c^2 \geq 4$, and δ is a real number such that $0 < \delta \leq \frac{q/c^2-1}{q-1}$, then there exists an explicit linear c-TA code over the field F_q of length $r = O(\frac{k^2}{\delta^3 \log(1/\delta)})$ (or length $r = O(\frac{k}{\delta^2 \log^2(1/\delta)})$) and dimension k with a polynomial (in r) traitor tracing algorithm.

Proof. (i) Since C is a Reed-Solomon code, the minimum distance d satisfies $d = r - k + 1$. The condition $r > c^2(k-1)$ is then equivalent to the condition $d > r - r/c^2$. By Theorem 1, C is a c-TA code and traitor tracing amounts to finding a codeword within distance $r - r/c$ of the pirate. Theorem 12 and Corollary 13 of [19] imply that if $t > \sqrt{(k-1)r}$ then all codewords within distance $r - t$ of a given word can be listed in time $O(r^{15})$, and if $t^2 = (1+\delta)(k-1)r$ then the runtime is $O(\frac{r^3}{\delta^6})$. Taking $t = r/c$ gives the desired results. (Note that $k = \log_q N$.)

(ii) The minimum distance d of the code satisfies $d \geq r - k - g + 1$ (see, for example, Theorem 10.6.3 of [28]). By our choice of c we have $d \geq r - k - g + 1 > r - r/c^2$ and $r - r/c < r - \sqrt{r(k+g-1)}$. By Theorem 27 of [19], there exists an algorithm that runs in time polynomial in r that outputs the list of codewords of distance less than $r - \sqrt{r(k+g-1)}$ from a given word. Now apply Theorem 1.

(iii) Theorems 7 and 8 and Corollaries 2 and 3 of [20] imply that there exists an explicit concatenated code over F_q of the correct length r and dimension k, with minimum distance $d \geq (1 - \frac{1}{q})(1-\delta)r$, with a polynomial time list decoding algorithm for e errors, as long as $e < (1 - \sqrt{\delta})(q-1)r/q$. The condition $\delta \leq \frac{q/c^2 - 1}{q-1}$ implies that $d > r - r/c^2$ and that the upper bound on the number of errors is satisfied when $e \leq r - r/c$. The result now follows from Theorem 1. □

We emphasize that further improvements in the runtime of list decoding algorithms are being rapidly produced. It seems that some of these results will bring the runtime down to $O(r \log^3 r)$ for Reed-Solomon codes, at least in certain cases (see [12]). The list decoding algorithm in [19] for AG codes was improved in [47] (see Theorems 3.4 and 4.1), where an explicit runtime was also given.

4 Comparative Analysis of TA and IPP Traceability

The results in this section justify a focus on TA (as opposed to IPP) schemes. In this paper we have been using the additional structure provided by linear codes to construct schemes for which the TA tracing algorithm is efficient. We know by Lemma 1 that c-TA codes are also c-IPP codes. However the converse fails ([37]; see also Example 1 above). If constructions of schemes for which the IPP tracing algorithm is efficient (i.e., significantly reduced from $O(\binom{N}{c})$ time) are possible, it is reasonable to expect this to be accomplished by introducing an algebraic structure. Here we give evidence that doing so may enable the inherently more efficient TA algorithm to be used to identify traitors. Hence, it is unclear that c-IPP schemes yield any advantage over c-TA schemes in finding a traitor.

First, we prove a necessary condition on Reed-Solomon codes, under which they yield c-TA set systems. This condition is that the minimum distance is greater than $r - r/c^2$, where r is the length of the codewords. This result suggests a potential method for generating examples of schemes that are c-IPP but not c-TA, namely, decreasing the minimum distance. Next we demonstrate through a family of counterexamples that in fact this approach does not work in general; when the minimum distance is $r - r/c^2$ it is possible to find Reed-Solomon codes for which both the IPP and TA tracing algorithms fail.

We recall that there is a natural way to produce unordered sets from the ordered sets that constitute the code: to a codeword $x = (x_1, \ldots, x_r)$, associate the set $x' = \{(1, x_1), \ldots, (r, x_r)\}$. We define TA and IPP set systems (as opposed to TA and IPP codes) in the natural way, with the noteworthy difference that a pirate *unordered set* consists of r elements such that each element is a member of some coalition member's set. This is a generalization of our earlier definition because it is not necessary to have one element of the form (i, y_i) for each $i = 1, \ldots, r$.

The following theorem is a partial converse of Theorem 1.

Theorem 3. *If $c \geq 2$ is an integer and C is a Reed-Solomon code of length r with minimum distance $d \leq r - \frac{r}{c^2}$, then the set system corresponding to C is not a c-TA set system.*

Proof. As above, if $u \in C$, write $u' = \{(1, u_1), \ldots, (r, u_r)\}$ for the associated element of the set system. Choose a codeword $v = (v_1, \ldots, v_r)$ in C. We will show that a coalition of size at most c exists which does not contain v', but which can implicate v'. In other words, we will construct a pirate set w which can be created by a coalition $\{u'_1, \ldots, u'_b\}$ with $b \leq c$ that does not contain v', but which satisfies $|v' \cap w| \geq |u'_i \cap w|$ for every i. Let $\delta = r - d = k - 1$, where k is the dimension of the code C. By assumption, $\delta \geq r/c^2$.

First, assume $c\delta \leq r$. For $i = 1, \ldots, c$, choose $u_i \in C$, distinct from v, which agrees with v on the δ positions $(i-1)\delta + 1, \ldots, i\delta$. (To do this, simply find a polynomial h_i of degree δ which vanishes on the δ field elements corresponding to these δ positions, and let u_i be the codeword corresponding to the polynomial $f - h_i$, where f is the polynomial corresponding to v.) Notice that, since two distinct codewords can agree on at most δ positions, each u'_i contains at least $r - c\delta$ elements which are not in v' or in u'_j for any $j \neq i$. Since $r - c\delta \geq 0$ and $c \geq 2$, we have $r - c\delta \geq \lceil \frac{r - c\delta}{c} \rceil = \lceil \frac{r}{c} \rceil - \delta$. We can therefore form a pirate set w so that for every i, $|u_i \cap w| \leq \delta + (\lceil \frac{r}{c} \rceil - \delta) = \lceil \frac{r}{c} \rceil$ and $|v' \cap w| = c\delta \geq \lceil \frac{r}{c} \rceil$. Thus the TA algorithm will mark v' as a traitor.

If on the other hand $c\delta > r$, simply choose $u_1, \ldots, u_j$ as above, where $j = \lfloor \frac{r}{\delta} \rfloor < c$, and choose $u_{j+1} \neq v$ to agree with v on the last $r - j\delta$ positions. The coalition $\{u'_1, \ldots, u'_{j+1}\}$ can create v' as a pirate set. □

The previous theorem leaves open the question of whether Reed-Solomon codes with minimum distance at most $r - \frac{r}{c^2}$ might still have traceability when the IPP algorithm is used even though the TA algorithm may no longer correctly identify traitors. The following family of counterexamples illustrates that this is not generally the case. It gives examples of Reed-Solomon codes of length r and minimum distance $r - r/c^2$ which are not c-IPP.

Theorem 4. *Let s and c be positive integers with $c \geq 2$, and let p be a prime number greater than c^2. For $i = 1, \ldots, c$, let $a_i = (i-1)c$. For $i = 1, \ldots, c$, if s*

is not divisible by p*, let* $g_i(x) = x^s - i$*; otherwise let* $g_i(x) = x^s + x - i$*. Let* T *be the set of roots of all the* c^2 *polynomials* $g_i - a_j$*. Let* q *be a sufficiently high power of* p *so that* T *is a subset of the finite field* F_q*. Then* T *consists of* $c^2 s$ *distinct elements of* F_q*. Let* C *be the Reed-Solomon code in which the codewords are the evaluations at the elements of* T *of all polynomials over* F_q *of degree at most* s*. Then the dimension of the code* C *is* $s+1$*, the length* r *of the codewords is* $r = c^2 s$*, the minimum distance of* C *is* $r - r/c^2$*, and* C *is not* c*-IPP.*

Proof. We first show that T consists of $c^2 s$ distinct elements. Let $h_{ij} = g_i - a_j$. Then $h_{ij}(x) - h_{mn}(x) = -i-(j-1)c+m+(n-1)c$. If $h_{ij}(x) - h_{mn}(x) = 0$, then $m - i$ is divisible by c. Since m and i are both in the range $1, \dots, c$, they must be equal. Thus $(j-1)c = (n-1)c$, and so $j = n$. Therefore the set $\{h_{ij}\}$ consists of c^2 distinct polynomials of degree s, any two of which differ by a non-zero constant. Therefore no two can have a root in common. Further, the derivative of h_{ij} is sx^{s-1} if s is not divisible by p, and is 1 otherwise. In both cases this derivative is relatively prime to h_{ij} (in the first case, note that h_{ij} is always of the form x^s+(a non-zero constant), so it never has 0 as a root). Therefore all the roots of h_{ij} are simple. So T consists of $c^2 s$ distinct elements, and it makes sense to consider the Reed-Solomon code defined by evaluating polynomials of degree at most s at the elements of T. The code clearly has the stated parameters. The two coalitions corresponding to the polynomials in the sets $\{a_1, \dots, a_c\}$ and $\{g_1, \dots, g_c\}$ are disjoint, and each coalition can produce the pirate word defined as follows: for each β in T, the β-th entry of the pirate word is $g_i(\beta) = a_j$, for the unique i and j such that the equality holds. It follows that the code is not c-IPP. □

By evaluating the polynomials at subsets of T of size at least $s+1$ (to ensure that $k \leq r$), we can take the length r to be anything between $s+1$ and $c^2 s$. The resulting minimum distance $r - s$ is then at most $r - r/c^2$.

We remark that if s is not divisible by p, then we can always find a q that works which is a divisor of p^s.

The results in this section lead to the following questions which, while peripheral to the traitor tracing problem, are of independent interest. Is it the case that all Reed-Solomon codes of length r with minimum distance $d \leq r - r/c^2$ are not c-IPP? It is easy to see that this is false for linear codes in general. For example, one-dimensional linear codes are always both c-IPP and c-TA, but can have $d \leq r - r/c^2$ if they are not Reed-Solomon codes (for one-dimensional codes, the minimum distance d is the number of non-zero entries in the non-zero codewords; the codewords of distance less than d from the pirate lie in every coalition that can create the pirate). If the answer to the above question were yes, combining it with Theorem 1 would imply that all Reed-Solomon c-IPP codes are c-TA. We raise as an open question whether all *linear* c-IPP codes are c-TA.

5 Finding All Possible Coalitions

In this section, we describe how a coding theoretic approach can be used to amass additional piracy information: a list of all coalitions that are capable of creating a given pirate. Such information is useful in two respects. It clears all codewords not appearing in any of these coalitions of involvement in constructing the pirate word, and it constitutes useful audit information that may be helpful in the prosecution of a traitor later on. The two algorithms of this section require only that the code have minimum distance greater than $r - \frac{r}{c^2}$, and therefore are applicable to the codes in Theorem 2. The algorithms are fast when fast list decoding techniques exist. In addition, we note that for every code meeting this minimum distance requirement and having fast list decoding, the algorithms enable the IPP traitor tracing algorithm [23,2,37] to run more efficiently (as that algorithm works by intersecting all coalitions that are capable of creating a given pirate word).

At a high level, the first algorithm builds a "tree" from which all c-coalitions capable of constructing a pirate w can be extracted. At the root of the tree lie all codewords that we know must be in *every* such coalition. The children are then candidate codewords for the next member of the coalition. Branches of the tree are extended until the current coalition "covers" w (i.e., is capable of constructing w), or until it becomes clear that this is impossible (e.g., because the coalition is already of size c and still cannot create w). In the latter case that "dead-end" coalition is discarded and other branches of the tree are explored. Before describing the algorithm in more detail, we introduce some of the ideas used. If S is a subset of $\{1, \dots, r\}$ and $s = |S|$, define a map $f_S : F_q^r \to F_q^{r-s}$ by "forgetting" the entries in positions corresponding to elements of S. If C is a code, then the image code $f_S(C)$ is the *punctured code*, where we view the code C as having been punctured at the positions corresponding to the elements of S. If u is in $f_S(C)$, any codeword v such that $f_S(v) = u$ is called a *lift* of u to C.

We say that U is a *minimal* c-coalition for w if $|U| \leq c$, $w \in \mathsf{desc}(U)$, but w is not in $\mathsf{desc}(V)$ for any proper subset V of U. To obtain all coalitions of size at most c that can create w from the minimal ones, append arbitrary elements of the code.

Algorithm Sketch:
Input: Integer $c > 1$, code C of length r and minimum distance greater than $r - \frac{r}{c^2}$, pirate word $w \in \mathsf{desc}_c(C)$.
Output: A list of coalitions of size at most c that can create w, including all minimal c-coalitions for w.

The basic steps of the algorithm are as follows:

(i) Use list decoding to find all codewords $u_1, \dots, u_a \in C$ $(a \leq c)$ within distance $r - r/c$ of w. Let S be the subset of $\{1, \dots, r\}$ on which w agrees with at least one of $\{u_1, \dots, u_a\}$, and let $s = |S|$. Let $r_1 = r - s$, $c_1 = c - a$, $C_1 = f_S(C)$, and $w_1 = f_S(w)$. (Thus C_1 is the punctured code, r_1 is its length, w_1 is the word which is the image of the pirate word under the puncturing map, and c_1 is the number of coalition members still to be found.) If $r_1 = 0$, quit and output $\{u_1, \ldots, u_a\}$. Set $i = 1$.

(ii) Use list decoding to find all codewords $v_{i1}, \ldots, v_{ib_i} \in C_i$ ($b_i \leq c_i$) within distance $r_i - r_i/c_i$ of w_i. (Note that the first time this is executed, the output is non-empty.) If this outputs the empty-set, exit to Step (iii). Otherwise, let S_i be the subset of $\{1, \ldots, r_i\}$ on which w_i agrees with v_{ib_i}, and let $s_i = |S_i|$. Let $r_{i+1} = r_i - s_i$, $c_{i+1} = c_i - 1$, $C_{i+1} = f_{S_i}(C_i)$, and $w_{i+1} = f_{S_i}(w_i)$.

(iii) To create the coalitions to output, always start with $u_1, \ldots, u_a$. Then add (a lift to C of) v_{1b_1}, v_{2b_2}, and so on. Continue until the list of codewords "covers" the pirate w. When this process succeeds or dead-ends (i.e., the current list does not yet cover w, but either we cannot find any codewords within the required distance $r_i - r_i/c_i$ of w_i, or we already have c codewords in our list), then move up the "tree" of v_{ij}'s (i.e., move back through the v_{ij}'s) to find the first unexplored branch and continue from there (repeating Step (ii) with a different v_{ij} in place of v_{ib_i}). The algorithm terminates when all branches have been explored.

Analysis of the Algorithm:
The algorithm is correct because the output is clearly a list of coalitions of size at most c that can create the pirate, and includes each minimal c-coalition at least once. (In fact, it may list a coalition more than once.) Note that in Step (iii), all lifts of each v_{ij} should be considered. By Theorem 1, $u_1, \ldots, u_a$ are in every coalition that can create w. In Step (ii), if $d_i > r_i - r_i/c_i^2$ where d_i is the minimum distance of the punctured code C_i, then every coalition that can produce the original pirate w will contain some lift to the original code of some v_{ij}. Moreover, if a lift to C of v_{ij} is in some coalition that can create the original pirate w, then there exists a codeword within $r_i - r_i/c_i$ of v_{ij} (by the pigeonhole principle), and the algorithm will proceed. If Step (ii) returns the empty-set, then the current path is a dead-end. Note that list decoding a punctured code and then lifting accomplishes the same thing as erasure-and-error decoding. When C satisfies any of the sets of conditions in Theorem 2, then Step (i) can be done efficiently (time polynomial in r).

Note that the brute force method for finding all coalitions runs in time $O(crN^c)$, where N is the total number of codewords in the code (for each of the at most N^c coalitions of size at most c, compare each of the r entries of the pirate to the corresponding entry of each member of the coalition). For Reed-Solomon codes with $r = \Theta(c^2k)$, this gives a runtime of $O(c^3N^c \log N)$.

Our second algorithm is to list decode to find all codewords $u_1, \ldots, u_a$ ($1 \leq a \leq c$) within distance $r - r/c$ of the pirate (as in Step (i) above), and then use brute force to determine the remaining (at most) $c - a$ members of the coalitions. When C is a Reed-Solomon code satisfying the conditions in Theorem 2(i) with $r = \Theta(c^2k)$, the dominant term in the runtime is $O(c^3N^{c-a} \log N)$. This is clearly an improvement over brute force alone, since $a \geq 1$.

6 Future Directions: Tracing with Extra Information

In this section, we describe how other coding theoretic techniques may be applied to the traitor tracing problem when additional information about traitor behavior is available.

One possible approach to tracing traitors is to try to second-guess their strategy. For example, if you believe that one traitor has contributed more than the other members of the coalition to the pirate, you can apply bounded-distance decoding up to the error-correction bound to find such traitors very quickly. This might involve a "ringleader" or "scapegoat" scenario. If on the other hand you believe that all traitors contributed roughly equal amounts, then list decoding should be tried first. Traitors can be searched for in sequences of expanding Hamming balls around the pirate. These searches can be run in parallel or sequentially. The runtime of bounded-distance decoding up to the error-correction bound for Reed-Solomon codes is at most quadratic in the length of the codewords. Note that [32] gives a fast algorithm for list decoding Reed-Solomon codes beyond the error-correction bound (also quadratic in the codeword length), but does not go as far as the Guruswami-Sudan algorithm. It therefore will not be guaranteed to find a traitor, but would quickly find a ringleader.

In [19], list decoding is considered not just in the case of errors, but also in the case of erasures and errors (and another potentially useful case that is referred to as "decoding with uncertain receptions"). For concatenated codes, [20] also deals with the problem of decoding from errors and erasures. Building on [19], [24] presents a high-performance soft-decision list decoding algorithm. We believe that these results also have potential for use in traitor tracing problems, in cases where some additional information is known about the traitors or how they are operating.

If one has information about the traitors or their modes of operation, one can build that information into a reliability matrix, and apply soft-decision decoding algorithms to trace. For example, suppose we know that a traitor who contributed the first entry to the pirate contributed at least r/c entries to the pirate. One can use this information to construct a skewed reliability matrix. If the underlying code is a Reed-Solomon code over a finite field of size q, one can then apply the soft-decision algorithm in [24] to find such a "dominant" traitor. The channel that models this situation is a q-ary symmetric channel. The first column of the reliability matrix will have a 1 in the entry corresponding to the field element that occurs in the first position of the pirate, and 0's elsewhere. For $j > 1$, the jth column of the reliability matrix will have $1 - \epsilon$ in the entry corresponding to the field element in the jth entry of the pirate, and the other entries will all be $\frac{\epsilon}{q-1}$, where $\epsilon < \frac{q-1}{q}$ is chosen so as to optimize the soft-decision decoding algorithm in [24]. If one does not know which entry was contributed by the traitor who contributed the most, one possible search method is to choose entries at random from the pirate and apply the above strategy to search for traitors that contributed that entry.

Erasure-and-error decoding may be useful in fingerprinting or watermarking scenarios, such as those presented in [6,7,15]. In one model, a coalition creates a pirate copy of the digital content by leaving fixed all codeword entries where they all agree, and choosing the values of the remaining positions from $Q \cup \{?\}$, where Q is the alphabet. The ?'s can be viewed as erasures.

7 Conclusion

We have demonstrated that traitor tracing algorithms can be quite efficient when the construction of the traceability scheme is based on error-correcting codes and the method of tracing is based on fast list decoding algorithms. For the TA algorithm, traitors can be identified in time polynomial in r, where r is roughly $c^2 \log_q N$, rather than in time $O(N)$. In addition, list decoding on successive punctured codes gives a method for identifying all possible traitor coalitions of size at most c more efficiently than a brute force search. This is quite useful because of the additional piracy information it represents, as well as for the efficiency improvements that it enables for another traitor tracing algorithm that has garnered interest recently, the IPP algorithm. We also give evidence for a close relationship between the TA and IPP properties, for linear codes. Finally, we suggest avenues for future research, including explorations of applications of soft-decision and erasure decoding techniques to traitor tracing in scenarios where additional information has been obtained about the traitors or their mode of operation.

Acknowledgments. The authors thank Gui-Leng Feng, Tom Høholdt, Ralf Kötter, and Madhu Sudan for useful conversations.

References

1. N. Alon, J. Bruck, J. Naor, M. Naor and R. Roth. Construction of asymptotically good low-rate error-correcting codes through pseudo-random graphs. *IEEE Transactions on Information Theory* **38** (1992), 509–516.
2. A. Barg, G. Cohen, S. Encheva, G. Kabatiansky and G. Zémor. A hypergraph approach to the identifying parent property: the case of multiple parents, DIMACS Technical Report 2000-20.
3. O. Berkman, M. Parnas and J. Sgall. Efficient dynamic traitor tracing, in 11th Annual ACM-SIAM Symposium on Discrete Algorithms (SODA 2000), 586–595.
4. E. R. Berlekamp, R. J. McEliece and H. C. A. van Tilborg. On the inherent intractability of certain coding problems. *IEEE Transactions on Information Theory* **24** (1978), 384–386.
5. D. Boneh and M. Franklin. An efficient public key traitor tracing scheme, in "Advances in Cryptology – Crypto '99", *Lecture Notes in Computer Science* **1666** (1999), 338–353.
6. D. Boneh and J. Shaw. Collusion secure fingerprinting for digital data, in "Advances in Cryptology – Crypto '95", *Lecture Notes in Computer Science* **963** (1995), 452–465.
7. D. Boneh and J. Shaw. Collusion secure fingerprinting for digital data, *IEEE Transactions on Information Theory* **44** (1998), 1897–1905.

8. I. Cox, J. Kilian, T. Leighton and T. Shamoon. Secure spread spectrum watermarking for multimedia. *IEEE Transactions on Information Theory* **6** (1997), 1673–1687.
9. B. Chor, A. Fiat and M. Naor. Tracing traitors, in "Advances in Cryptology – Crypto '94", *Lecture Notes in Computer Science* **839** (1994), 480–491.
10. B. Chor, A. Fiat, M. Naor and B. Pinkas. Tracing traitors, *IEEE Transactions on Information Theory* **46** (2000), 893–910.
11. C. Dwork, J. Lotspiech and M. Naor. Digital Signets: Self-Enforcing Protection of Digital Information, in Proc. 28th ACM Symposium on Theory of Computing (STOC 1997), 489–498.
12. G.-L. Feng. Very Fast Algorithms in Sudan Decoding Procedure for Reed-Solomon Codes. Preprint.
13. G.-L. Feng. Fast Algorithms in Sudan Decoding Procedure for Hermitian Codes. Preprint.
14. A. Fiat and M. Naor. Broadcast Encryption, in "Advances in Cryptology – Crypto '93", *Lecture Notes in Computer Science* **773** (1994), 480–491.
15. A. Fiat and T. Tassa. Dynamic traitor tracing, in "Advances in Cryptology – Crypto '99", *Lecture Notes in Computer Science* **1666** (1999), 354–371.
16. E. Gafni, J. Staddon and Y. L. Yin. Efficient methods for integrating traceability and broadcast encryption, in "Advances in Cryptology – Crypto '99", *Lecture Notes in Computer Science* **1666** (1999), 372–387.
17. J. Garay, J. Staddon and A. Wool, Long-Lived Broadcast Encryption, in "Advances in Cryptology – Crypto 2000", *Lecture Notes in Computer Science* **1880** (2000), 333-352.
18. V. D. Goppa. Geometry and codes. Kluwer Academic Publishers, Dordrecht, 1988.
19. V. Guruswami and M. Sudan. Improved decoding of Reed-Solomon and algebraic-geometry codes, *IEEE Transactions on Information Theory* **45**(6) (1999), 1757–1767.
20. V. Guruswami and M. Sudan. List decoding algorithms for certain concatenated codes, in Proc. 32nd ACM Symposium on Theory of Computing (STOC 2000), 181–190.
21. T. Høholdt and R. R. Nielsen. Decoding Reed-Solomon codes beyond half the minimum distance, in Coding theory, cryptography and related areas (Guanajuato, 1998), Springer, Berlin (2000), 221–236.
22. T. Høholdt and R. R. Nielsen. Decoding Hermitian codes with Sudan's algorithm. To appear in the 13th AAECC Symposium.
23. H. D. L. Hollmann, J. H. van Lint, J-P. Linnartz and L. M. G. M. Tolhuizen. On codes with the identifiable parent property, *Journal of Combinatorial Theory A* **82** (1998), 121–133.
24. R. Koetter and A. Vardy. Algebraic soft-decision decoding of Reed-Solomon codes. Preprint. `http://www.dia.unisa.it/isit2000/lavori/455.ps`.
25. R. Kumar, S. Rajagopalan and A. Sahai. Coding constructions for blacklisting problems without computational assumptions, in "Advances in Cryptology – Crypto '99", *Lecture Notes in Computer Science* **1666** (1999), 609–623.
26. K. Kurosawa, M. Burmester and Y. Desmedt. A proven secure tracing algorithm for the optimal KD traitor tracing scheme. DIMACS Workshop on Management of Digital Intellectual Properties, April, 2000, and Eurocrypt 2000 rump session.
27. K. Kurosawa and Y. Desmedt. Optimal traitor tracing and asymmetric schemes, in "Advances in Cryptology – Eurocrypt '98", *Lecture Notes in Computer Science* **1438** (1998), 145–157.

28. J. H. van Lint. Introduction to coding theory. Third edition. Graduate Texts in Mathematics **86**, Springer-Verlag, Berlin (1999).
29. M. Naor and B. Pinkas. Efficient Trace and Revoke Schemes, to appear in Proceedings of Financial Cryptography 2000.
30. V. Olshevsky and A. Shokrollahi. A displacement structure approach to efficient decoding of algebraic geometric codes, in Proc. 31st ACM Symposium on Theory of Computing (STOC 1999), 235–244.
31. B. Pfitzmann. Trials of traced traitors, in Information Hiding, First International Workshop, *Lecture Notes in Computer Science* **1174** (1996), 49–64.
32. R. M. Roth and G. Ruckenstein. Efficient decoding of Reed-Solomon codes beyond half the minimum distance. *IEEE Transactions on Information Theory* **46** (2000), 246–257.
33. R. Safavi-Naini and Y. Wang. Sequential Traitor Tracing, in "Advances in Cryptology – CRYPTO 2000", *Lecture Notes in Computer Science* **1880** (2000), 316–332.
34. B.-Z. Shen. A Justesen construction of binary concatenated codes that asymptotically meet the Zyablov bound for low rate, *IEEE Transactions on Information Theory* **39** (1993), 239–242.
35. M. A. Shokrollahi and H. Wassermann. Decoding Algebraic-Geometric Codes Beyond the Error-Correction Bound, in Proc. 30th ACM Symposium on Theory of Computing (STOC 1998), 241–248.
36. M. A. Shokrollahi and H. Wassermann. List Decoding of Algebraic-Geometric Codes. *IEEE Transactions on Information Theory* **45** (1999), 893–910.
37. J. N. Staddon, D. R. Stinson and R. Wei. *Combinatorial properties of frameproof and traceability codes.* To appear in *IEEE Transactions on Information Theory.*
38. H. Stichtenoth. Algebraic Function Fields and Codes. Springer-Verlag, Berlin, 1993.
39. D. R. Stinson, Tran van Trung and R. Wei. Secure frameproof codes, key distribution patterns, group testing algorithms and related structures, *Journal of Statistical Planning and Inference* **86** (2000), 595-617.
40. D. R. Stinson and R. Wei. Combinatorial properties and constructions of traceability schemes and frameproof codes, *SIAM Journal on Discrete Mathematics* **11** (1998), 41–53.
41. D. R. Stinson and R. Wei. Key preassigned traceability schemes for broadcast encryption, in "Selected Areas in Cryptology – SAC '98", *Lecture Notes in Computer Science* **1556** (1999), 144–156.
42. M. Sudan. Decoding of Reed Solomon codes beyond the error-correction bound. *Journal of Complexity* **13**(1) (1997), 180–193.
43. M. Sudan. Decoding of Reed Solomon codes beyond the error-correction diameter, in Proc. 35th Annual Allerton Conference on Communication, Control and Computing (1997), 215–224.
44. M. A. Tsfasman and S. G. Vlăduţ. Algebraic-geometric codes. Kluwer Academic Publishers, Dordrecht, 1991.
45. M. A. Tsfasman, S. G. Vlăduţ and Th. Zink. Modular curves, Shimura curves, and Goppa codes, better than Varshamov-Gilbert bound. *Math. Nachr.* **109** (1982), 21–28.
46. W.-G. Tzeng and Z.-J. Tzeng. A Traitor Tracing Scheme Using Dynamic Shares, to appear in PKC2001.
47. X.-W. Wu and P. H. Siegel. Efficient List Decoding of Algebraic Geometric Codes Beyond the Error Correction Bound, submitted to *IEEE Transactions on Information Theory.*
48. F. Zane. Efficient Watermark Detection and Collusion Security, to appear in Proceedings of Financial Cryptography 2000.

Security of Reduced Version of the Block Cipher Camellia against Truncated and Impossible Differential Cryptanalysis

Makoto Sugita[1], Kazukuni Kobara[2], and Hideki Imai[2]

[1] NTT Network Innovation Laboratories, NTT Corporation
1-1 Hikari-no-oka, Yokosuka-shi, Kanagawa, 239-0847 Japan
sugita@wslab.ntt.co.jp

[2] Institute of Industrial Sciences, The University of Tokyo
4-6-1, Komaba, Meguro-ku, Tokyo, 153-8505, Japan
{kobara,imai}@imailab.iis.u-tokyo.ac.jp

Abstract. This paper describes truncated and impossible differential cryptanalysis of the 128-bit block cipher Camellia, which was proposed by NTT and Mitsubishi Electric Corporation. Our work improves on the best known truncated and impossible differential cryptanalysis. As a result, we show a nontrivial 9-round byte characteristic, which may lead to a possible attack of reduced-round version of Camellia without input/output whitening, FL or FL^{-1} in a chosen plain text scenario. Previously, only 6-round differentials were known, which may suggest a possible attack of Camellia reduced to 8-rounds. Moreover, we show a nontrivial 7-round impossible differential, whereas only a 5-round impossible differential was previously known. This cryptanalysis is effective against general Feistel structures with round functions composed of S-D (Substitution and Diffusion) transformation.

Keywords: Block Cipher Camellia, Truncated Differential Cryptanalysis, Impossible Differential Cryptanalysis

1 Introduction

Camellia is a 128-bit block cipher proposed by NTT and Mitsubishi Electric Corporation [1]. It was designed to withstand all known cryptanalytic attacks and to provide a sufficient headroom to allow its use over the next $10-20$ years. Camellia supports 128-bit block size and 128-, 192-, and 256-bit key lengths, i.e. the same interface specifications as the Advanced Encryption Standard (AES). Camellia was proposed in response to the call for contributions from ISO/IEC JTC 1/SC27 with the aim of it being adopted as an international standard. Camellia was also submitted to NESSIE (New European Schemes for Signature, Integrity, and Encryption). Furthermore, Camellia was submitted to CRYPTREC (CRYPTography Research & Evaluation Committee) in Japan and it is now being evaluated.

C. Boyd (Ed.): ASIACRYPT 2001, LNCS 2248, pp. 193–207, 2001.
© Springer-Verlag Berlin Heidelberg 2001

Like E2 [5], which was submitted to AES, Camellia uses a combination of a Feistel structure and the SPN-(Substitution and Permutation Network)-structure, but it also includes new features such as the use of improved linear transformation in SPN-structures, the change of SPN-structures from three layers into two, and the use of input/output whitening, FL and FL^{-1}. The result is improved immunity against truncated differential cryptanalysis, which was applied successfully against reduced-round version of E2 by Matsui and Tokita [12].

Truncated differential cryptanalysis was introduced by Knudsen [4], as a generalization of differential crypanalysis [3]. He defined them as differentials where only a part of the differential can be predicted. The notion of truncated differentials as introduced by him is wide, but with a byte-oriented cipher such as E2 or Camellia, it is natural to study byte-wise differentials as truncated differentials.

The initial analysis of the security of Camellia and its resistance to the truncated and impossible differential cryptanalysis is given in [1], [6]. They state that Camellia with more than 11 rounds is secure against truncated differential cryptanalysis, though they did not indicate the effective truncated differentials. Up to now, the effective cryptanalysis applicable to Camellia has been the higher order differential cryptanalysis proposed by Kawabata, et al.[7], which utilizes non-trivial 6-round higher order differentials, and the differential crypatanalysis which utilizes a 7-round differential [2].

Our analysis improves on the best known truncated and impossible cryptanalysis against Camellia. Our cryptanalysis finds a nontrivial 9-round truncated differential, which may lead to a possible attack of Camellia reduced to 11-rounds without input/output whitening, FL, or FL^{-1} by a chosen plain text scenario. Moreover, we show a nontrivial 7-round impossible differential, whereas only a 5-round impossible differentials were previously known.

The contents of this paper are as follows. In Section 2, we describes the structures of block ciphers, truncated differential probabilities, impossible differential cryptanalysis and the block cipher Camellia. In Section 3, we describe the previous work on the security of block cipher Camellia. In Section 4, we cryptanalyze Camellia by truncated differential cryptanalysis. In Section 5, we cryptanalyze Camellia by impossible differential cryptanalysis. Section 6 concludes this paper.

2 Preliminaries

In this section, we describe the general structures of block ciphers, truncated differential probabilities, impossible differential cryptanalysis and the block cipher Camellia.

2.1 Feistel Structures

Associate with a function $f : \mathrm{GF}(2)^n \to \mathrm{GF}(2)^n$, a function $D_{2n,f}(L, R) = (R \oplus f(L), L)$ for all $L, R \in \mathrm{GF}(2)^n$. $D_{2n,f}$ is called the Feistel transformation associated with f. Furthermore, for functions $f_1, f_2, \cdots, f_s : \mathrm{GF}(2)^n \to \mathrm{GF}(2)^n$, define $\psi_n(f_1, f_2, \cdots, f_s) = D_{2n,f_s} \circ \cdots \circ D_{2n,f_2} \circ D_{2n,f_1}$. We call $F(f_1, f_2, \cdots, f_s) = \psi_n(f_1, f_2, \cdots, f_s)$ the s-round Feistel structure. At this time, we call the functions $f_1, f_2, \cdots, f_s$ the round functions of the Feistel structure $F(f_1, f_2, \cdots, f_s)$.

2.2 SPN-Structures [9]

This structure consists of two kinds of layers: nonlinear layer and linear layer. Each layer has different features as follows.

Nonlinear (Substitution) layer: This layer is composed of m parallel n-bit bijective nonlinear transformations.

Linear (Diffusion) layer: This layer is composed of linear transformations over the field GF(2^n) (especially in the case of E2 and Camellia, GF(2)), where inputs are transformed linearly to outputs per word (n-bits).

Next for positive integer s, we define the s-layer SPN-structure that consists of s layers. First is a nonlinear layer, second is a linear layer, third is a nonlinear layer, $\cdots$.

2.3 Word Characteristics

We define a word characteristic function $\chi : \mathrm{GF}(2^n)^m \to \mathrm{GF}(2)^m$, $(a_1, \cdots, a_m) \longmapsto (b_1, \cdots, b_m)$ by

$$b_i = \begin{cases} 0 \text{ if } a_i = 0 \\ 1 \text{ otherwise,} \end{cases}$$

Hereafter, we call $\chi(a)$ the word characteristic of $a \in \mathrm{GF}(2^n)^m$. Especially in the case of $n = 8$, we call $\chi(a)$ the byte characteristic.

2.4 Truncated Differential Probability

Definition 1. *Let $\Delta x, \Delta y \in \mathrm{GF}(2^n)^m$ denote the input and output differences of the function f, respectively.*

$$\Delta x = (\Delta x_1, \Delta x_2, \cdots, \Delta x_m)$$
$$\Delta y = (\Delta y_1, \Delta y_2, \cdots, \Delta y_m)$$

We define the input and output truncated differential $(\delta x, \delta y) \in (\mathrm{GF}(2)^m)^2$ of the function f, where

$$\delta x = (\delta x_1, \delta x_2, \cdots, \delta x_m)$$
$$\delta y = (\delta y_1, \delta y_2, \cdots, \delta y_m)$$

by $\delta x = \chi(\Delta x), \delta y = \chi(\Delta y)$.

Let $p_f(\delta x, \delta y)$ denote the transition probability of the truncated differential induced by function f. $p_f(\delta x, \delta y)$ is defined on the truncated differential $(\delta x, \delta y)$. Truncated differential probability $p_f(\delta x, \delta y)$ is defined by

$$p_f(\delta x, \delta y) = 1/c \sum_{\chi(\Delta x)=\delta x, \chi(\Delta y)=\delta y} \Pr(x \in \mathrm{GF}(2^n)^m | f(x) \oplus f(x \oplus \Delta x) = \Delta y),$$

where c is the number of Δx that satisfy $\chi(\Delta x) = \delta x$.

2.5 Block Cipher Camellia

Fig. 1 shows the entire structure of Camellia. Fig. 2 shows its round functions, and Fig. 3. shows FL-function and FL^{-1}-function.

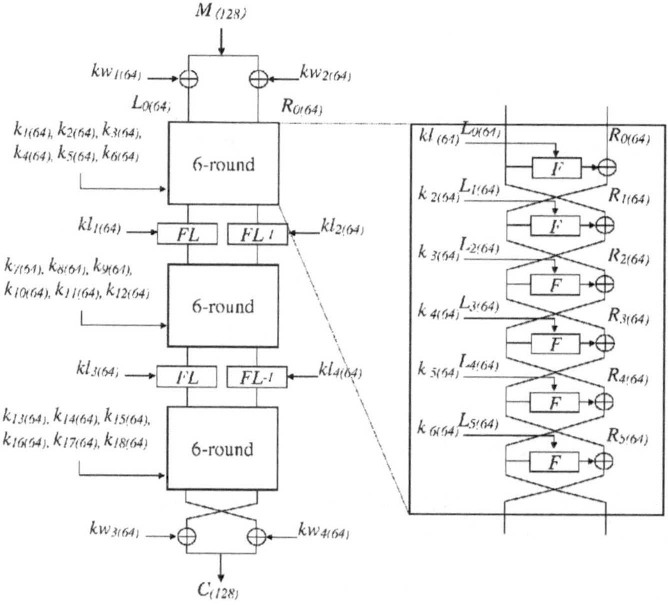

Fig. 1. Block cipher Camellia

3 Previous Security Evaluation of Camellia

3.1 Security Evaluation against Truncated Differential Cryptanalysis [10]

In [10], an algorithm to search for the effective truncated differentials of Feistel ciphers was proposed. This search algorithm consists of recursive procedures.

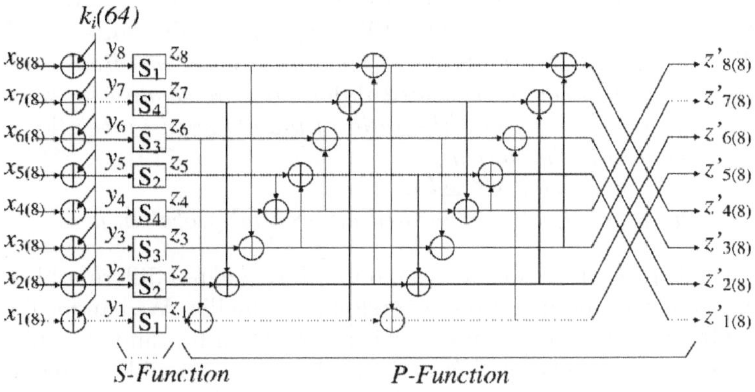

Fig. 2. F function of Camellia

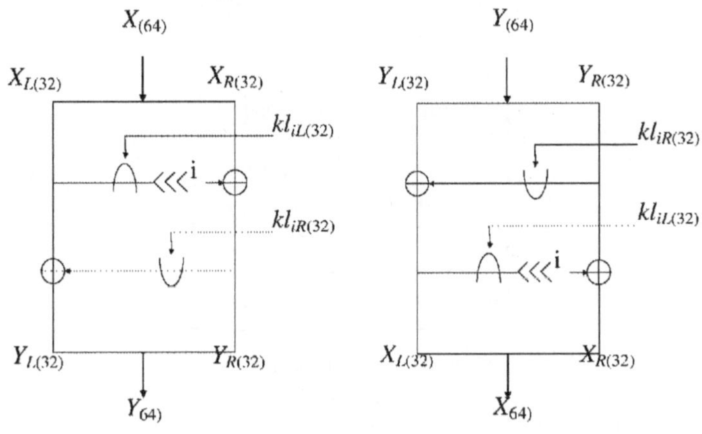

Fig. 3. FL and FL^{-1} functions of Camellia

Feistel ciphers are assumed to have S rounds and input and output block size is $2m$ bits.

Algorithm 1 *[11]*
Let $\Delta X^{(r)}, \Delta Y^{(r)} \in \mathrm{GF}(2)^m$ *be the input and output truncated difference of the r-th round functions* f_r. $(\Delta L, \Delta R)$ *is the truncated difference of the plaintext. Let* $\Pr((\Delta X^{(0)}, \Delta X^{(1)})|(\Delta L, \Delta R)))$ *be the r-round truncated differential probabilities.*

1. *Calculate all the truncated differential probabilities* $p_f(\delta x, \delta y)$ *of the round function f for all truncated differentials* $(\delta x, \delta y)$ *and save these probabilities in memory.*
2. *Select and fix* $(\Delta L, \Delta R)$. $\Pr((\Delta X^{(0)}, \Delta X^{(1)})|(\Delta L, \Delta R))$ *should be initialized as* 1 *if* $(\Delta X^{(0)}, \Delta X^{(1)}) = (\Delta L, \Delta R)$, *otherwise as* 0.

3. *Utilizing the values of* $p_f(\delta x, \delta y)$, *calculate* $\Pr((\Delta X^{(r+1)}, \Delta X^{(r+2)})|(\Delta L, \Delta R))$ *for all* $(\Delta X^{(r+1)}, \Delta X^{(r+2)})$ *from all values of* $\Pr((\Delta X^{(r)}, \Delta X^{(r+1)})|(\Delta L, \Delta R))$, *and save in memory. Repeat this from* $r = 1$ *to* S. *and save the most effective truncated differential probability in memory, where 'most effective' means that the ratio of the obtained probability to the average probability is the maximum.*
4. *Repeat 2-3 for every* $(\Delta R, \Delta L)$.
5. *return the most effective truncated differential probability.*

Using this procedure, we can search for all truncated differentials that lead to possible attacks on reduced-round version of Camellia. We cannot find any such truncated differentials for Camellia with more than 6-rounds by this algorithm. The best 6-round truncated differential that leads to possible attacks on reduced-round version of Camellia is shown in Fig. 4.

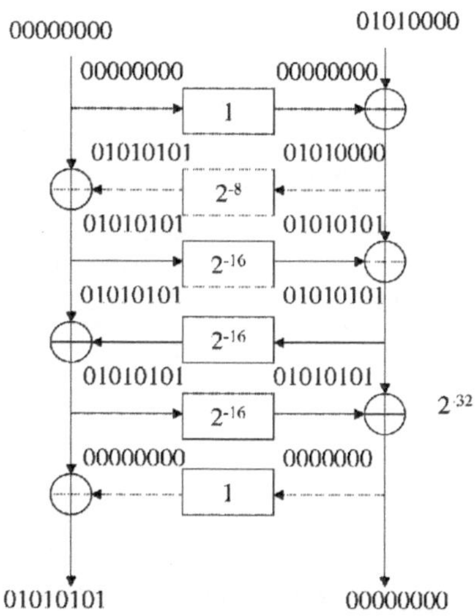

Fig. 4. 6-round truncated differential of Camellia

By this path, total probability $p \simeq 2^{-88}$, whereas the average probability, which can be obtained when entire round function is a random permutation, is 2^{-96}.

This evaluation is accurate if we take the ideal approximation model as is done in [10]. We note that this model is not always appropriate for Camellia, especially because the round function of Camellia is a 2-layer SPN, i.e. S-D (Substitution and Diffusion), not a 3-layer SPN, i.e. S-D-S. In [1] and [6], they upper-bounded the truncated differential probabilities considering this gap, and no effective truncated differentials for Camellia with more than 7-rounds (without input/output whitening, FL or FL^{-1}) are known.

4 Truncated Differential Cryptanalysis of Reduced-Round Version of Camellia without Input/Output Whitening, FL or FL^{-1}

This section indicates the truncated differentials that are effective in the cryptanalysis of reduced version of Camellia. These truncated differences cannot be found by the algorithm described in the previous section. We define notation in Fig. 5.

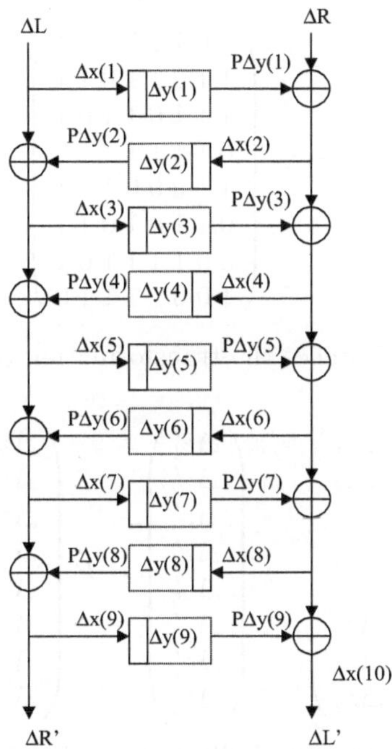

Fig. 5. Notation

First we analyze the round function of Camellia. P-function composing F-function is denoted as follows.

$$\mathrm{GF}(2^8)^8 \rightarrow \mathrm{GF}(2^8)^8$$
$$(z_1, z_2, z_3, z_4, z_5, z_6, z_7, z_8) \mapsto (z'_1, z'_2, z'_3, z'_4, z'_5, z'_6, z'_7, z'_8).$$

This transformation can be expressed by linear transformations represented by matrix P.

$$\begin{pmatrix} z_1 \\ z_2 \\ z_3 \\ z_4 \\ z_5 \\ z_6 \\ z_7 \\ z_8 \end{pmatrix} \mapsto \begin{pmatrix} z'_1 \\ z'_2 \\ z'_3 \\ z'_4 \\ z'_5 \\ z'_6 \\ z'_7 \\ z'_8 \end{pmatrix} = P \begin{pmatrix} z_1 \\ z_2 \\ z_3 \\ z_4 \\ z_5 \\ z_6 \\ z_7 \\ z_8 \end{pmatrix}$$

where

$$P = \begin{pmatrix} 1\,0\,1\,1\,0\,1\,1\,1 \\ 1\,1\,0\,1\,1\,0\,1\,1 \\ 1\,1\,1\,0\,1\,1\,0\,1 \\ 0\,1\,1\,1\,1\,1\,1\,0 \\ 1\,1\,0\,0\,0\,1\,1\,1 \\ 0\,1\,1\,0\,1\,0\,1\,1 \\ 0\,0\,1\,1\,1\,1\,0\,1 \\ 1\,0\,0\,1\,1\,1\,1\,0 \end{pmatrix}$$

This transformation induces the transformation of the difference as follows.

$$\begin{pmatrix} \Delta z_1 \\ \Delta z_2 \\ \Delta z_3 \\ \Delta z_4 \\ \Delta z_5 \\ \Delta z_6 \\ \Delta z_7 \\ \Delta z_8 \end{pmatrix} \mapsto \begin{pmatrix} \Delta z'_1 \\ \Delta z'_2 \\ \Delta z'_3 \\ \Delta z'_4 \\ \Delta z'_5 \\ \Delta z'_6 \\ \Delta z'_7 \\ \Delta z'_8 \end{pmatrix} = P \begin{pmatrix} \Delta z_1 \\ \Delta z_2 \\ \Delta z_3 \\ \Delta z_4 \\ \Delta z_5 \\ \Delta z_6 \\ \Delta z_7 \\ \Delta z_8 \end{pmatrix}$$

Next we consider the truncated differentials effective for truncated differential cryptanalysis.

When $\Delta z_1, \Delta z_2 \neq 0, \Delta z_3 = \Delta z_4 = \Delta z_5 = \Delta z_6 = \Delta z_7 = \Delta z_8 = 0$, then

$$\begin{pmatrix} \Delta z_1 \\ \Delta z_2 \\ 0 \\ 0 \\ 0 \\ 0 \\ 0 \\ 0 \end{pmatrix} \mapsto \begin{pmatrix} \Delta z_1 \\ \Delta z_1 \oplus \Delta z_2 \\ \Delta z_1 \oplus \Delta z_2 \\ \Delta z_2 \\ \Delta z_1 \oplus \Delta z_2 \\ \Delta z_2 \\ 0 \\ \Delta z_1 \oplus \Delta z_2 \end{pmatrix}$$

When $\Delta z_1 \neq \Delta z_2$, this can be expressed in terms of byte characteristics as

$$(11000000) \mapsto (11111101).$$

In this case, this transition probability (truncated differential probability) $p_1 \simeq 1$.

Utilizing the value of

$$P^{-1} = \begin{pmatrix} 0\,1\,1\,1\,0\,1\,1\,1 \\ 1\,0\,1\,1\,1\,0\,1\,1 \\ 1\,1\,0\,1\,1\,1\,0\,1 \\ 1\,1\,1\,0\,1\,1\,1\,0 \\ 1\,1\,0\,0\,1\,0\,1\,1 \\ 0\,1\,1\,0\,1\,1\,0\,1 \\ 0\,0\,1\,1\,1\,1\,1\,0 \\ 1\,0\,0\,1\,0\,1\,1\,1 \end{pmatrix},$$

when $\Delta z_1, \Delta z_2 \neq 0, \Delta z_1 \neq \Delta z_2, \Delta z_3 = \Delta z_4 = \Delta z_5 = \Delta z_1 \oplus \Delta z_2, \Delta z_6 = \Delta z_1, \Delta z_7 = 0, \Delta z_8 = \Delta z_2$, we obtain

$$\begin{pmatrix} \Delta z_1 \\ \Delta z_2 \\ \Delta z_1 \oplus \Delta z_2 \\ \Delta z_1 \oplus \Delta z_2 \\ \Delta z_1 \oplus \Delta z_2 \\ \Delta z_1 \\ 0 \\ \Delta z_2 \end{pmatrix} \mapsto \begin{pmatrix} \Delta z_2 \\ \Delta z_1 \\ 0 \\ 0 \\ 0 \\ 0 \\ 0 \\ 0 \end{pmatrix},$$

which can be expressed in terms of byte characteristics as

$$(11111101) \mapsto (11000000).$$

This transition probability (truncated differential probability) $p_2 \simeq 2^{-40}$

Utilizing these two transition probabilities, we can obtain a 9-round truncated differential that contains two different paths as in Fig. 6.

In total, the first transition probability is approximately 2^{-112} (see the evaluation in Appendix).

Similarly, we consider the other path, which is as effective as the first one. In total, this transition probability is also 2^{-112} (see also the evaluation in Appendix).

Summing the two probabilities, therefore, the truncated differential probability of

$$\Pr(\chi(\Delta L') = (11000000), \chi(\Delta R') = (00000000) | \\ \chi(\Delta L) = (00000000), \chi(\Delta R) = (11000000)) \simeq 2.0 \times 2^{-112},$$

which is approximately twice as large as the average value 2^{-112}.

Our search has not found any truncated differential more effective than this for 9-round Camellia.

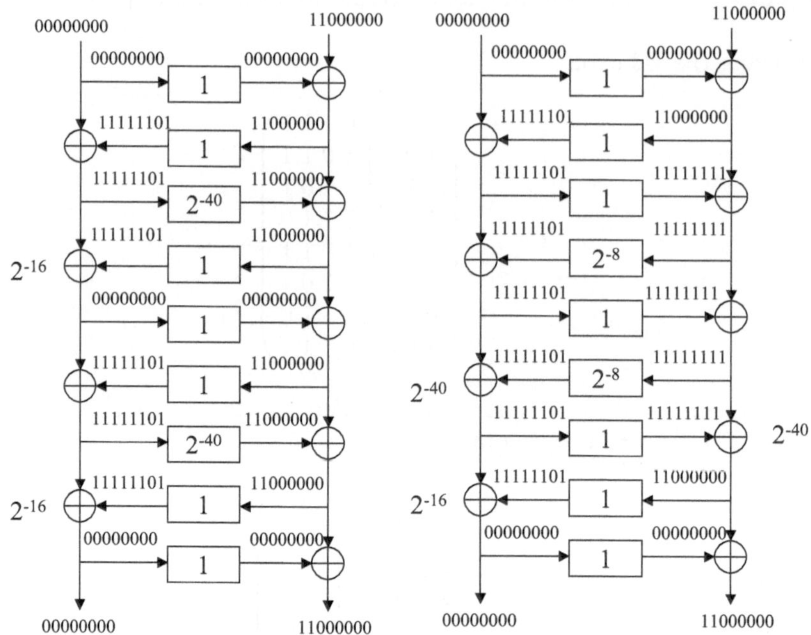

Fig. 6. 9-round byte characteristic of Camellia

5 Impossible Differential Cryptanalysis of Reduced-Round Version of Camellia without Input/Output Whitening, FL or FL^{-1}

5.1 Impossible Differential Cryptanalysis

Impossible differential means the differential that holds with probability 0, or the differential that does not exist. Using such an impossible differential, it is possible to narrow down the subkey candidates. It is known that there is at least one 5-round impossible differential in any Feistel structure with bijective round functions. Since Camellia uses the Feistel structure with FL and FL^{-1} inserted between every 6-rounds and the round function is bijective, Camellia has 5-round impossible differentials.

5.2 Impossible Differential Cryptanalysis of Reduced Camellia

In [1], they state that they have not found impossible differentials for more than 5 rounds. In this subsection, we indicate one impossible differential of a 7-round reduced-round version of Camellia without input/output whitening, FL and FL^{-1} as shown in Fig. 7.

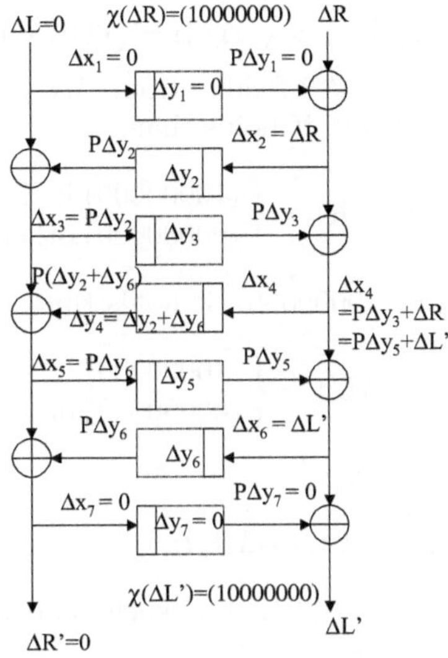

Fig. 7. 7-round impossible differential of Camellia

In this figure, we consider the byte characteristic

$$(0000000010000000) \mapsto (1000000000000000),$$

In this case, we can prove that this is an impossible differential as follows. First we assume $\chi(\Delta L) = (00000000), \chi(\Delta R) = (10000000)$, $\chi(\Delta R') = (00000000), \chi(\Delta L') = (10000000)$.

This assumption implies that

$$\Delta x_1 = \Delta y_1 = P\Delta y_1 = 0, \Delta x_2 = \Delta R, \Delta x_7 = \Delta y_7 = P\Delta y_7 = 0, \Delta x_6 = \Delta L'.$$

From

$$P\Delta y_4 = P\Delta y_2 \oplus P\Delta y_6,$$

it follows that

$$\Delta y_4 = \Delta y_2 \oplus \Delta y_6,$$

which implies

$$\chi(\Delta x_4) = \chi(\Delta y_4) = \begin{cases} (10000000) \text{ if } \Delta y_2 \neq \Delta y_6 \\ (00000000) \text{ otherwise.} \end{cases}$$

From the definition of P,

$$\chi(\Delta y_3) = \chi(\Delta x_3) = (11101001),$$

where $\Delta y_3 \neq 0$ because $\Delta x_3 \neq 0$ follows from $\Delta y_2 \neq 0$

Similarly,

$$\chi(\Delta y_5) = \chi(\Delta x_5) = (11101001),$$

where $\Delta y_5 \neq 0$ because $\Delta x_5 \neq 0$ follows from $\Delta y_6 \neq 0$.

Since $\chi(\Delta R) = (10000000)$, it holds that

$$\chi(\Delta x_4 \oplus \Delta R) = \begin{cases} (10000000) \text{ if } \Delta x_4 \neq \Delta R \\ (00000000) \text{ otherwise.} \end{cases}$$

Similarly, since $\chi(\Delta L') = (10000000)$, it holds that

$$\chi(\Delta x_4 \oplus \Delta L') = \begin{cases} (10000000) \text{ if } \Delta x_4 \neq \Delta L' \\ (00000000) \text{ otherwise.} \end{cases}$$

From Fig. 7, it holds that

$$P\Delta y_3 = \Delta x_4 \oplus \Delta R, P\Delta y_5 = \Delta x_4 \oplus \Delta L',$$

however, there is no $(t, s) \in (\mathrm{GF}(2^8)^8)^2$ such that $\chi(t) = (11101001)$, $\chi(s) = (10000000)$, $t \neq 0$ and $Pt = s$.

Thus, the truncated differential represented by

$$(0000000010000000) \mapsto (1000000000000000)$$

is impossible.

6 Conclusion

This paper evaluated the security of the block cipher Camellia against truncated and impossible differential cryptanalysis. We introduced a nontrivial 9-round truncated differential that leads to a possible attack of reduced-round version of Camellia without input/output whitening, FL or FL^{-1} in a chosen plain text scenario. Prior studies showed only a 6-round truncated differential for a possible attack against 8-round Camellia. Moreover, we showed a nontrivial 7-round impossible differential, whereas only a 5-round impossible differentials were previously known.

Acknowledgment. We would like to thank Shiho Moriai and the anonymous reviewers for their helpful comments.

References

1. K. Aoki, T. Ichikawa, M. Kanda, M. Matsui, S. Moriai, J. Nakajima and T. Tokita, "Camellia: A 128-Bit Block Cipher Suitable for Multiple Platform" `http://info.isl.ntt.co.jp/camellia/`
2. E. Biham, O. Dunkelman, V. Furman, T. Mor, "Preliminary report on the NESSIE submissions Anubis, Camellia, IDEA, Khazad, Misty1, Nimbus, Q," NESSIE public report.

3. E. Biham and A. Shamir, "Differential Cryptanalysis of DES-like Cryptosystems." Journal of Cryptology, Vol.4, No.1, pp.3-72, 1991. (The extended abstract was presented at CRYPTO'90).
4. L.R. Knudsen and T.A. Berson, "Truncated Differentials of SAFER." In Fast Software Encryption - Third International Workshop, FSE'96, Volume 1039 of Lecture Notes in Computer Science, Berlin, Heidelberg, NewYork, Springer-Verlag, 1996.
5. M. Kanda et al. "A New 128-bit Block Cipher E2," IEICE Trans. fundamentals, Vol.E83-A, No.1, Jan., 2000.
6. M. Kanda and T. Matsumoto, "Security of Camellia against Truncated Differential Cryptanalysis," In Fast Software Encryption - 8th International Workshop, FSE'00.
7. T. Kawabata, Y. Ohgaki, T. Kaneko, "A study on Strength of Camellia against Higher Order Differential Attack," Technical Report of IEICE. ISEC 2001-9, pp.55-62.
8. X. Lai, J.L. Massey and S. Murphy, "Markov Ciphers and Differential Cryptanalysis," Advances in Cryptography-EUROCRYPT '91. Lecture Notes in Computer Science, Vol. 576. Springer-Verlag, Berlin, 1992, pp.86-100.
9. A.J. Menezes, P.C. van Oorschot, S.A. Vanstone, "Handbook of Applied Cryptography", CRC Press, pp.250-250 (1997).
10. A. Moriai, M. Sugita, K. Aoki, M. Kanda, "Security of E2 against truncated Differential Cryptanalysis" Sixth Annual Workshop on Selected Areas in Cryptography (SAC'99), LNCS 1758 pp.106-117 , Springer Verlag, Berlin, 1999.
11. S. Moriai, M. Sugita and M. Kanda, "Security of E2 against truncated Differential Cryptanalysis" IEICE, Trans. fundamentals, Vol.E84-A NO.1, pp.319-325, January 2001.
12. M. Matsui, and T. Tokita, "Cryptanalysis of a Reduced Version of the Block Cipher E2" in 6-th international workshop, preproceedings FSE'99
13. K. Nyberg and L.R. Knudsen, "Provable security against a differential attack," in Advances in Cryptology - EUROCRYTO'93, LNCS 765, pp.55-64, Springer-Verlag, Berlin, 1994.
14. M. Sugita, K. Kobara, H. Imai, "Pseudorandomness and Maximum Average of Differential Probability of Block Ciphers with SPN-Structures like E2." Second AES Workshop, 1999.
15. M. Sugita, K. Kobara, H. Imai, "Relationships among Differential, Truncated Differential, Impossible Differential Cryptanalyses against Block-Oriented Block Ciphers like RIJNDAEL, E2" Third AES Workshop, 2000.
16. T. Tokita, M. Matsui, "On cryptanalysis of a byte-oriented cipher", The 1999 Symposium on Cryptography and Information Security, pp.93-98 (In Japanese), Kobe, Japan, January 1999.

Appendix: Evaluation of Truncated Differential Probability of Camellia

First we evaluate the transition probability of the first path in Fig. 6.

$$\begin{aligned}
&\Pr(\chi(P\Delta y(1)) = (00000000)|\chi(\Delta x(1)) = (00000000)) = 1 \\
&\Pr(\chi(\Delta x(2)) = (11000000)|\chi(P\Delta y(1)) = (00000000), \\
&\qquad\qquad\qquad\qquad\qquad\qquad \chi(\Delta R) = (11000000)) = 1
\end{aligned}$$

$$
\begin{aligned}
&\Pr(\chi(P\Delta y(2)) = (11111101)|\chi(\Delta x(2)) = (11000000)) \simeq 1\\
&\Pr(\chi(\Delta x(3)) = (11111101)|\chi(P\Delta y(2)) = (11111101)),\\
&\qquad \chi(\Delta L) = (00000000)) = 1\\
&\Pr(\chi(P\Delta y(3)) = (11000000)|\chi(\Delta x(3)) = (11111101)) \simeq 2^{-40}\\
&\Pr(\chi(\Delta x(4)) = (11000000)|\chi(\Delta x(2)) = \chi(P\Delta y(3)) = (11000000)) \simeq 1\\
&\Pr(\chi(P\Delta y(4)) = (11111101)|\chi(\Delta x(4)) = (11000000)) \simeq 1\\
&\Pr(\chi(\Delta x(5)) = (00000000))|\chi(\Delta x(4)) = \chi(\Delta x(2)) = (11000000),\\
&\qquad \chi(\Delta x(1)) = (00000000))\\
&= \Pr(\Delta y(4) = \Delta y(2)|\chi(\Delta x(4)) = \chi(\Delta x(2)) = (11000000)) \simeq 2^{-16}\\
&\Pr(\chi(P\Delta y(5)) = (00000000)|\chi(\Delta x(5)) = (00000000)) = 1\\
&\Pr(\chi(\Delta x(6)) = (11000000)|\chi(P\Delta y(5)) = (00000000),\\
&\qquad \chi(\Delta x(4)) = (11000000)) = 1\\
&\Pr(\chi(P\Delta y(6)) = (11111101)|\chi(\Delta x(6)) = (11000000)) \simeq 1\\
&\Pr(\chi(\Delta x(7)) = (11111101)|\chi(P\Delta y(6)) = (11111101),\\
&\qquad \chi(\Delta x(5)) = (00000000)) = 1\\
&\Pr(\chi(P\Delta y(7)) = (11000000)|\chi(\Delta x(7)) = (11111101)) \simeq 2^{-40}\\
&\Pr(\chi(\Delta x(8)) = (11000000)|\chi(P\Delta y(7)) = (11000000),\\
&\qquad \chi(\Delta x(6)) = (11000000)) \simeq 1\\
&\Pr(\chi(P\Delta y(8)) = (11111101)|\chi(\Delta x(8)) = (11000000)) \simeq 1\\
&\Pr(\chi(\Delta x(9)) = (00000000)|\chi(\Delta x(8)) = \chi(\Delta x(6)) = (11000000),\\
&\qquad \chi(\Delta x(5) = (00000000))\\
&= \Pr(\chi(P\Delta y(8) \oplus \Delta x(7)) = (00000000)|\\
&\qquad \chi(\Delta x(8)) = \chi(\Delta x(6)) = (11000000))\\
&= \Pr(\Delta y(8) = \Delta y(6)|\chi(\Delta x(8)) = \chi(\Delta x(6)) = (11000000)) \simeq 2^{-16}\\
&\Pr(\chi(P\Delta y(9)) = (00000000)|\chi(\Delta x(9)) = (00000000)) = 1\\
&\Pr(\chi(\Delta x(10)) = (11000000)|\chi(P\Delta y(9)) = (00000000),\\
&\qquad \chi(\Delta x(8)) = (11000000)) = 1
\end{aligned}
$$

In total, the transition probability is approximately 2^{-112}.

Similarly, we consider the other path in Fig. 6, which is as effective as the first one.

$$
\begin{aligned}
&\Pr(\chi(P\Delta y(1)) = (00000000)|\chi(\Delta x(1)) = (00000000)) = 1\\
&\Pr(\chi(\Delta x(2)) = (11000000)|\chi(P\Delta y(1)) = (00000000),\\
&\qquad \chi(\Delta R) = (11000000)) = 1\\
&\Pr(\chi(P\Delta y(2)) = (11111101)|\chi(\Delta x(2)) = (11000000)) \simeq 1\\
&\Pr(\chi(\Delta x(3)) = (11111101)|\chi(P\Delta y(2)) = (11111101)),\\
&\qquad \chi(\Delta R) = (00000000)) = 1\\
&\Pr(\chi(P\Delta y(3)) = (11111111)|\chi(\Delta x(3)) = (11111101)) \simeq 1
\end{aligned}
$$

$$\begin{aligned}
\Pr(\chi(\Delta x(4)) = (11111111)|\chi(\Delta x(2)) = (11000000),&\\
\chi(P\Delta y(3)) = (11111111)) &\simeq 1\\
\Pr(\chi(P\Delta y(4)) = (11111101)|\chi(\Delta x(4)) = (11111111)) &\simeq 2^{-8}\\
\Pr(\chi(\Delta x(5)) = (11111101))|\chi(P\Delta y(4)) = \chi(\Delta x(3)) = (11111101)) &\simeq 1\\
\Pr(\chi(P\Delta y(5)) = (11111101)|\chi(\Delta x(5)) = (11111111)) &\simeq 1\\
\Pr(\chi(\Delta x(6)) = (11111111)|\chi(P\Delta y(5)) = (11111111),&\\
\chi(\Delta x(4)) = (11111111)) &\simeq 1\\
\Pr(\chi(P\Delta y(6)) = (11111101)|\chi(\Delta x(6)) = (11111111)) &\simeq 2^{-8}\\
\Pr(P^{-1}\Delta x(7) \in \{x \in \mathrm{GF}(2^8)^8|\chi(x) = (11000000)\}|&\\
\chi(\Delta y(2)) = (11000000), \chi(\Delta y(4)) = \chi(\Delta y(6)) = (11111111),&\\
\chi(P\Delta y(4)) = \chi(P\Delta y(6)) = (11111101))&\\
= \Pr(\Delta y(4) \oplus \Delta y(6) \in \{x \in \mathrm{GF}(2^8)^8|\chi(x) = (11000000)\}|&\\
\chi(\Delta y(2)) = (11000000), \chi(\Delta y(4)) = \chi(\Delta y(6)) = (11111111),&\\
\chi(P\Delta y(4)) = \chi(P\Delta y(6)) = (11111101)) &\simeq 2^{-40}\\
\Pr(\chi(P\Delta y(7)) = (11111111)|\chi(\Delta x(7)) = (11111101)) &\simeq 1\\
\Pr(\chi(\Delta x(8) = (11000000)|\chi(\Delta R) = (11000000),&\\
\chi(\Delta y(3)) = \chi(\Delta y(5)) = \chi(\Delta y(7)) = (11111101))&\\
= \Pr(\Delta y(3) \oplus \Delta y(5) \oplus \Delta y(7) \in P^{-1}\{x \in \mathrm{GF}(2^8)^8|&\\
\chi(x) = (11000000)\}|&\\
\chi(\Delta y(3)) = \chi(\Delta y(5)) = \chi(\Delta y(7)) = (11111101)) &\simeq 2^{-40}\\
\Pr(\chi(P\Delta y(8)) = (11111101)|\chi(\Delta x(8)) = (11000000)) &\simeq 1\\
\Pr(\chi(\Delta x(9)) = (00000000)|\chi(\Delta y(8)) = (11000000),&\\
P^{-1}\Delta x(7) \in \{x \in \mathrm{GF}(2^8)^8|\chi(x) = (11000000)\}) &\simeq 2^{-16}\\
\Pr(\chi(P\Delta y(9)) = (00000000)|\chi(\Delta x(9)) = (00000000)) &= 1\\
\Pr(\chi(\Delta x(10)) = (11000000)|\chi(P\Delta y(9)) = (00000000),&\\
\chi(\Delta x(8) = (11000000)) &= 1
\end{aligned}$$

In total, this transition probability is also approximately 2^{-112}.

Known-IV Attacks on Triple Modes of Operation of Block Ciphers

Deukjo Hong[1], Jaechul Sung[1], Seokhie Hong[1], Wonil Lee[1], Sangjin Lee[1], Jongin Lim[1], and Okyeon Yi[2] *

[1] Center for Information Security Technologies (CIST),
Korea University, Anam Dong, Sungbuk Gu,
Seoul, Korea
{hongdj, sjames, hsh, nice, sangjin, jilim}@cist.korea.ac.kr
[2] Electronics and Telecommunications Research Institute (ETRI),
161 Gajeong-dong, Yusong-Gu, Daejon, 305-350, Korea
{oyyi@etri.re.kr}

Abstract. With chosen-IV chosen texts, David Wagner has analyzed the multiple modes of operation proposed by Eli Biham in FSE'98. However, his method is too unrealistic. We use only known-IV chosen texts to attack many triple modes of operation which are combined with cascade operations. 123 triple modes are analyzed with complexities less than E. Biham's results. Our work shows that the securities of many triple modes decrease when the initial values are exposed.

Keywords: Block cipher, mode of operation for DES, Triple DES

1 Introduction

Since the appearance of DES [7], several attacks on DES and its variants have been suggested. E. Biham and Adi Shamir introduced differential cryptanalysis of DES in 1991 and 1992 [3,4]. Mitsuru Matsui analyzed DES with linear cryptanalysis in 1993 and 1994 [5,6]. Differential cryptanalysis and linear cryptanalysis are the most powerful methods for attacking DES. These attacks have led many people in the cryptographic community to suggest stronger replacements for DES, which can be either new cryptosystems or new modes of operation for the DES. So triple DES instead of DES has been used and applied to the modes of operation for DES — ECB, CBC, OFB, and CFB. Triple DES is even more secure but slower than DES. This reason has led to consideration of multiple modes of operation combined from several consecutive applications of single modes. In hardware implementation, the multiple modes have an advantage that their speed is the same as of single modes because the single modes can be pipelined. In particular, the triple modes were expected to be as secure as triple DES although they have DES as a building block.

* Supported by a grant of Electronics and Telecommunication Research Institute, Korea.

C. Boyd (Ed.): ASIACRYPT 2001, LNCS 2248, pp. 208–221, 2001.
© Springer-Verlag Berlin Heidelberg 2001

In 1994 and 1996, E. Biham analyzed many triple modes of operation with chosen plaintexts and chosen ciphertexts, and showed that every mode considered except the the triple ECB mode is not much more secure than single modes [1,2]. Considering dictionary attacks or matching-ciphertext attacks, the commonly-used triple-DES-ECB mode when used with some outer chaining technique is not much more secure than any single modes. To solve this state of affairs, E. Biham proposed 9 new block modes and 2 new stream modes of operation for DES. The complexities of attacking these new modes are conjectured to be at least 2^{112}. The quadruple modes were conjectured to be more secure than any triple mode; furthermore, the complexity of attacking two of the quadruple modes was conjectured to be at least 2^{128}.

In 1998, D. Wagner analyzed E. Biham's proposals [8]. Using the chosen-IV chosen text queries he broke them with the complexities lower than what E. Biham has conjectured. His method utilizes an equation for an exhaustive search for a key or looks for a collision for a birthday attack. Since E. Biham's studies were premised on a more restrictive threat model that did not admit chosen-IV attacks, D. Wagner's results do not disprove E. Biham's conjectures but raise questions about the security of E. Biham's proposed modes.

D. Wagner's assumption of chosen-IV is too unrealistic, so we use known-IV chosen texts more practical than chosen-IV to re-analyze the triple modes which E. Biham analyzed. Our attacks take their place between E. Biham's and D. Wagner's in terms of controlling IVs. However, since for fixed IVs the birthday paradox is not available as D. Wagner, our attacks cannot break E. Biham's proposals. Our results show how much the security of each triple mode decreases when the initial value is exposed.

2 Preliminaries

In this section, we describe something to understand our attack. Note that the underlying block cipher of every mode throughout this paper is DES with 64 bit plaintext and 56 bit key.

We write $P_0, P_1, \cdots$ (or $C_0, C_1, \cdots$, respectively) for the blocks of the plaintext (or ciphertext, respectively). We number the keys $K_1, K_2, \cdots$ and the initial values $IV_1, IV_2, \cdots$ according to the order that the single-mode appears in the triple modes. The capital letters $A, B, \cdots$ are any fixed 64-bit values if no additional explanations for them are given.

D. Wagner chose initial values and plaintexts or ciphertexts to analyze the multiple modes which E. Biham has proposed. His method searches for some equations or collisions to apply to an exhaustive search or birthday attack. Since E. Biham's multiple modes are very secure, we think that it is very hard for anyone to find a proper method to break them. However the assumption that the attacker can choose the initial value is too unrealistic. The assumption of known-IV is more practical than that of chosen-IV because the initial values require integrity rather than secrecy.

When all initial values are known, every double mode is broken by a meet-in-the-middle attack. Furthermore, all except ECB|ECB, CBC|CBC $^{-1}$, CBC|OFB, CBC|CFB^{-1}, OFB|CBC^{-1}, OFB|OFB, OFB|CFB^{-1}, CF B|CBC^{-1}, CFB|OFB,

CFB|CFB^{-1} are broken by only two exhaustive searches for two keys. We search equations which isolate one key or two keys to analyze any triple modes. In the initial cases, such a key value is recovered with a 2^{56} exhaustive key search and then remaining two keys are with a meet-in-the middle attack. In most of the latter cases, we can apply a meet-in-the-middle attack to the equation and then find the remaining key with the exhaustive key search.

3 Known-IV Attacks

We analyze 123 out of 216 triple modes. Complete knowledge of IVs is useful in breaking the triple modes which have the feedbacks driven into certain middle parts or arranged in a particular direction. However, it hardly helps the attacker who tries to find the keys of the triple modes with the feedbacks to spread forward and backward.

A meet-in-the-middle attack for double ECB mode requires two known plaintexts. Using the one plaintext, we search some key candidates such that intermediate values are equal. Despite having the wrong key, it may make intermediate values equal with the probability of 2^{-64}. We can find the right keys with a high probability by checking them for the other plaintext.

The meet-in-the-middle attacks in our work also require two equations, two chosen plaintexts, or two chosen ciphertexts. Taking this into account, we choose the plaintexts or the ciphertexts; we classify the attacks according to the choice of the texts.

3.1 *AAB*-Attack

This method can break all triple modes in which the first two modes are ECB|ECB. We describe the attack of ECB|ECB|CBC^{-1} as an example. We choose the plaintexts (A, A, B) and obtain the ciphertexts (C_0, C_1, C_2). In Fig. 1, the intermediate values after the first ECB component and the second ECB component are (A', A', B') and (A'', A'', B''). Then the output of the third encryption box in the first block is equal to that in the second block. Therefore, we obtain the following equation.

$$E_{K_3}^{-1}(IV_3 \oplus C_0) = IV_3 \oplus C_0 \oplus C_1$$

So we may find K_3 by a 2^{56} exhaustive search, recognizing the right key value when the above equation holds. We expect no wrong key to survive the check with a high probability.

Finally, K_1 and K_2 can be recovered by the meet-in-the-middle attack with two plaintexts A and B. Consequently, we use 3 chosen plaintexts to break the ECB|ECB|CBC^{-1} mode, whereas E. Biham's method requires 2^{64} chosen plaintexts.

3.2 *AABB*-Attack

This method can break many triple modes in which the last two modes are CBC|ECB. We describe the attack of OFB|CBC|ECB as an example. We

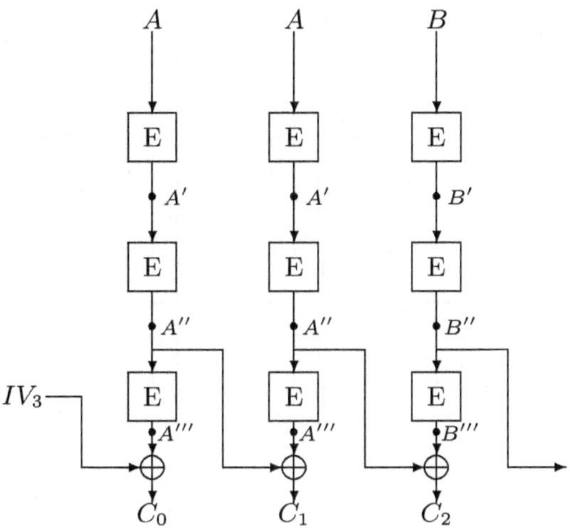

Fig. 1. Attack of ECB|ECB|CBC^{-1}

choose the ciphertexts (A, A, B, B) and obtain the corresponding plaintexts (P_0, P_1, P_2, P_3). In Fig. 2, the intermediate values entering the second encryption boxes are (A'', A'', B'', B''), where $A' = E^{-1}_{K_3}(A), B' = E^{-1}_{K_3}(B), A'' = E^{-1}_{K_2}(A')$, and $B'' = E^{-1}_{K_2}(B')$. Therefore, we obtain the following equation for the first two blocks.

$$E_{K_1}(IV_1) \oplus E_{K_1}(E_{K_1}(IV_1)) = IV_2 \oplus P_0 \oplus P_1 \oplus E^{-1}_{K_3}(A)$$

K_1 and K_3 are founded by a meet-in-the-middle attack. The right side of the above equation is computed for each of possible key values of K_3 and the result is kept in a table. Then the left side is computed under each of possible key values of K_1 and checked whether the result appears in the table. If a pair of keys (K_1, K_3) satisfies both the above and the following equations, we conclude that they are the right keys for K_1 and K_3.

$$E_{K_1}(E_{K_1}(E_{K_1}(IV_1))) \oplus E_{K_1}(E_{K_1}(E_{K_1}(E_{K_1}(IV_1))))$$

$$= P_2 \oplus P_3 \oplus E^{-1}_{K_3}(A) \oplus E^{-1}_{K_3}(B)$$

K_2 is recovered by brute force. Consequently, we use 4 chosen ciphertexts to break the OFB|CBC|ECB mode, whereas E. Biham's method requires 2^{64} chosen ciphertexts.

3.3 *AAAB*-Attack

If the last two modes are a combination of ECB, CBC, or CFB, the triple mode is vulnerable to this attack. We will describe the application of this method to the CBC|CFB|ECB. To find the keys, we choose the ciphertexts (A, A, A, B), and

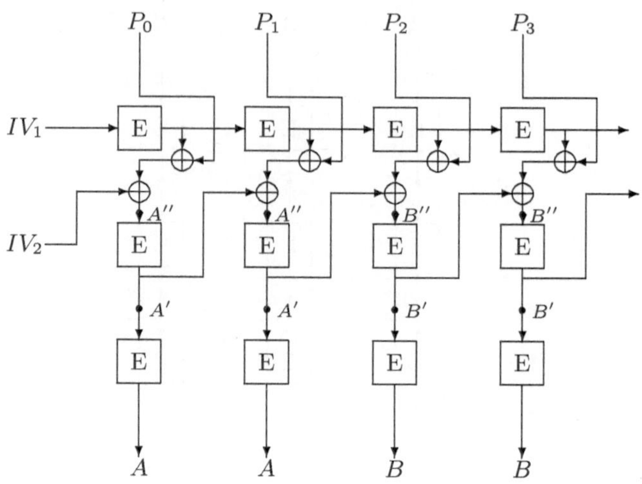

Fig. 2. Attack of OFB|CBC|ECB

obtain the corresponding plaintexts (P_0, P_1, P_2, P_3). In Fig. 3, the intermediate values after the first ECB component must be of the form $(?, F, F, ?)$. Therefore, the intermediate value entering the first encryption box in the second block is equal to that in the third block. For all the possible values of K_1, we check the following equation.

$$E_{K_1}(IV_1 \oplus P_0) \oplus E_{K_1}(E_{K_1}(IV_1 \oplus P_0) \oplus P_1) = P_1 \oplus P_2$$

Consequently, we use 4 chosen cipehrtexts to break the CBC|CFB|ECB, whereas E. Biham's method requires 2^{36} chosen ciphertexts.

3.4 *AAA*-Attack

This attack can break all triple modes in which the first two mode is ECB|OFB. We describe explain the attack of the ECB|OFB|OFB mode as an example. We choose the plaintexts (A, A, A) and obtain the corresponding ciphertexts (C_0, C_1, C_2). The following equations are obtained from the fact that all of the intermediate values after the first ECB mode are equal.

$$E_{K_2}(IV_2) \oplus E_{K_2}(E_{K_2}(IV_2)) = C_0 \oplus C_1 \oplus E_{K_3}(IV_3) \oplus E_{K_3}(E_{K_3}(IV_3))$$

$$E_{K_2}(IV_2) \oplus E_{K_2}(E_{K_2}(E_{K_2}(IV_2))) \\ = C_0 \oplus C_2 \oplus E_{K_3}(IV_3) \oplus E_{K_3}(E_{K_3}(E_{K_3}(IV_3)))$$

Then we can find K_2 and K_3 by a meet-in-the-middle attack. Consequently, we use 3 chosen plaintexts to break the ECB|OFB|OFB mode, whereas E. Biham's method requires 2^{65} chosen plaintexts.

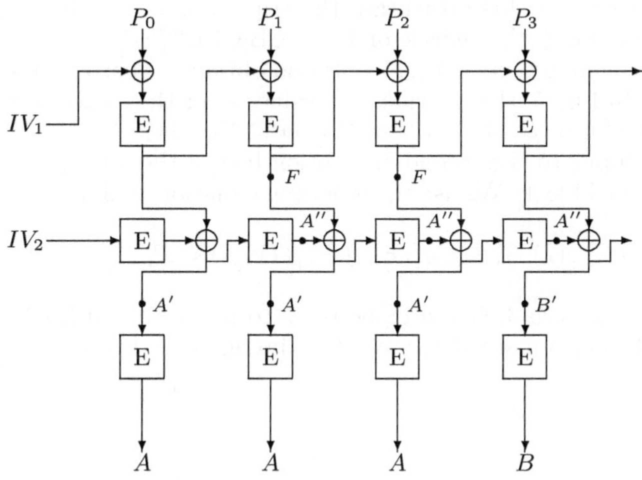

Fig. 3. Attack of CBC|CFB|ECB

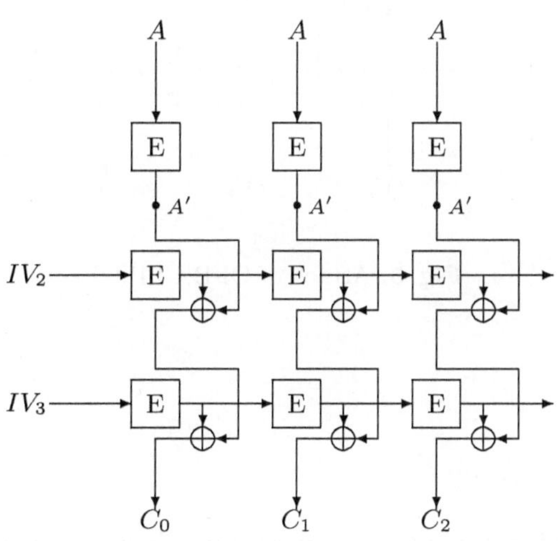

Fig. 4. Attack of ECB|OFB|OFB

3.5 *IVIVA*-Attack

The main targets of this attack are the triple modes in which the last mode is CFB. We explain the attack of the OFB|CFB|CFB mode as an example. We choose the ciphertexts (IV_3, IV_3, A) and obtain the corresponding plaintexts (P_0, P_1, P_2). In Fig. 5, the intermediate values after the second CFB component must be of the form $(B, B, ?)$, where $B = IV_3 \oplus E_{K_3}(IV_3)$. Then, the intermediate value of the input to the second encryption box in the second block is equal to that in the third block. We use the following equation to find K_3 by brute force.

$$E_{K_1}(E_{K_1}(IV_1)) \oplus E_{K_1}(E_{K_1}(E_{K_1}(IV_1))) = P_1 \oplus P_2 \oplus IV_3 \oplus A$$

Consequently, we use 3 chosen ciphertexts to break the OFB|CFB|CFB mode, whereas E. Biham's method requires 2^{66} chosen cipehrtexts.

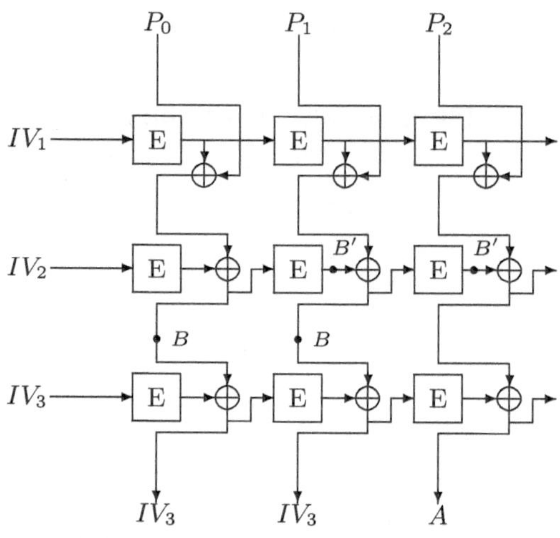

Fig. 5. Attack of OFB|CFB|CFB

3.6 *IVIVIV*-Attack

This method analyzes many triple modes in which the first mode is any single mode with a feedback and that the second mode is the OFB mode. We explain the attack of the CFB|OFB|CBC mode as a example. We choose the ciphertexts (IV_3, IV_3, IV_3) and the corresponding plaintexts (P_0, P_1, P_2). In Fig. 6, all of the intermediate values after the second OFB component are equal. We obtain the following equation.

$$E_{K_1}(IV_1) \oplus P_0 \oplus E_{K_2}(IV_2) = E_{K_1}(E_{K_1}(IV_1) \oplus P_0) \oplus P_1 \oplus E_{K_2}(E_{K_2}(IV_2))$$
$$= E_{K_1}(E_{K_1}(E_{K_1}(IV_1) \oplus P_0) \oplus P_1) \oplus P_2 \oplus E_{K_2}(E_{K_2}(E_{K_2}(IV_2)))$$

Then we can find K_1 and K_2 by a meet-in-the-middle attack. Consequently, we use 3 chosen ciphertexts to break the CFB|OFB|CBC mode, whereas E. Biham's method requires 2^{66} chosen ciphertexts. Furthermore, it takes $5 \cdot 2^{56}$ encryption times with our attack until its three keys are found, whereas it takes 2^{66} encryption times with E. Biham's attack.

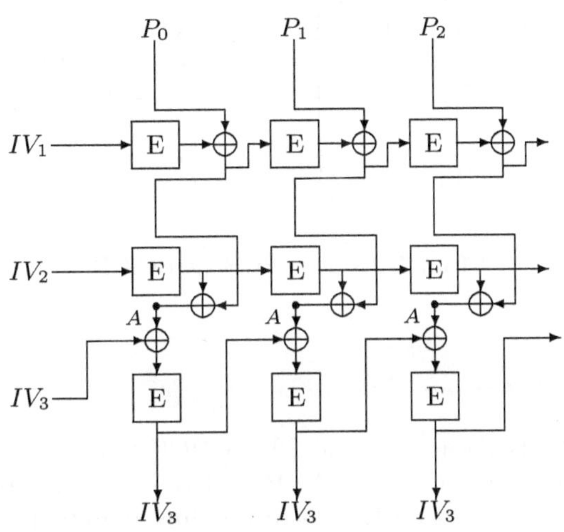

Fig. 6. Attack of CFB|OFB|CBC

3.7 *IVIVIVA*-Attack

The triple CBC mode and similar modes are broken by this method. We describe the attack of the triple CBC mode as an example. We choose the ciphertexts (IV_3, IV_3, IV_3, A) and obtain the corresponding plaintexts (P_0, P_1, P_2, P_3). In Fig. 7, the intermediate values after the first CBC component must be of the form (?, $B \oplus B'$, $B \oplus B'$, ?), where $B = IV_3 \oplus E^{-1}_{K_3}(IV_3)$ and $B' = E^{-1}_{K_2}(B)$. Therefore, the second and the third blocks in the values of the input to the first encryption boxes are equal.

$$E_{K_1}(IV_1 \oplus P_0) \oplus E_{K_1}(E_{K_1}(IV_1 \oplus P_0) \oplus P_1) = P_1 \oplus P_2$$

We may find K_1 by a 2^{56} exhaustive keysearch, recognizing the right key value when the above equation holds. Consequently, we use 4 chosen ciphertexts to break the CBC|CBC|CBC mode, whereas E. Biham's method requires 2^{34} chosen ciphertexts.

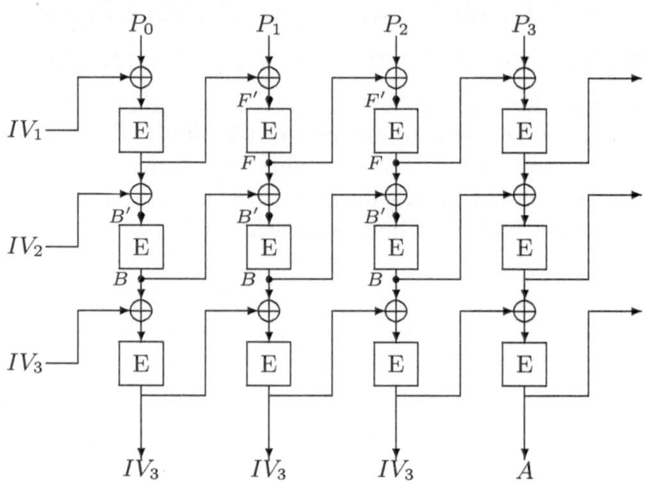

Fig. 7. Attack of CBC|CBC|CBC

3.8 *IVIVAAA*-Attack

The ECB|CBC|CFB mode and the CFB|CBC|CFB mode are broken by this method. To find the keys of the ECB|CBC|CFB mode, we choose the ciphertexts (IV_3, IV_3, A, A, A) and obtain the corresponding plaintexts $(P_0, P_1, P_2, P_3, P_4)$. In Fig. 8, the intermediate values entering the second encryption boxes must be of the form $(B, B, ?, F, F)$, where $B = E_{K_2}^{-1}(E_{K_3}(IV_3) \oplus IV_3)$ and $F = E_{K_2}^{-1}(E_{K_3}(A) \oplus A)$. Therefore, for the first and the second blocks, we obtain the following equation.

$$E_{K_1}(P_0) \oplus E_{K_1}(P_1) = IV_2 \oplus IV_3 \oplus E_{K_3}(IV_3)$$

By a meet-in-the middle attack, we find a few candidates of a pair of (K_1, K_3) from the above equation. If a candidate satisfies the following equation which we obtain for the fourth and fifth blocks, we are sure that it is the right value of (K_1, K_3).

$$E_{K_1}(P_3) \oplus E_{K_1}(P_4) = E_{K_3}(IV_3) \oplus E_{K_3}(A)$$

Then K_2 is recovered by an exhaustive search. Consequently, we use 5 chosen ciphertexts to break the ECB|CBC|CFB mode, whereas E. Biham's method requires 2^{34} chosen ciphertexts.

3.9 *IVIVAAAB*-Attack

We only apply this method to the ECB|CFB|CBC mode. To find its keys of it, we choose the ciphertexts (IV_3, IV_3, A, A, A, B) and obtain the corresponding plaintexts $(P_0, P_1, P_2, P_3, P_4, P_5)$. In Fig. 9, the intermediate values after the

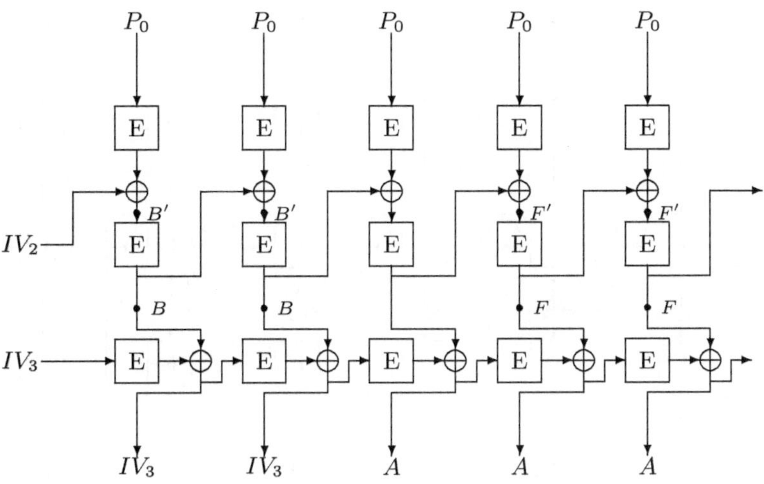

Fig. 8. Attack of ECB|CBC|CFB

second CFB component must be of the form $(F, F, ?, G, G, ?)$, where $F = IV_3 \oplus E^{-1}_{K_3}(IV_3)$, and $G = A \oplus E^{-1}_{K_3}(A)$. Therefore, for the second and third blocks, we obtain the following equation.

$$E_{K_1}(P_1) \oplus E_{K_1}(P_2) = E^{-1}_{K_3}(IV_3) \oplus E^{-1}_{K_3}(A)$$

By a meet-in-the middle attack, we find a few candidates of a pair of (K_1, K_3) from the above equation. If a candidate satisfies the following equation which we obtain for the fourth and fifth blocks, we are sure that it is the right value of (K_1, K_3).

$$E_{K_1}(P_4) \oplus E_{K_1}(P_5) = E^{-1}_{K_3}(A) \oplus E^{-1}_{K_3}(B)$$

Then K_2 is recovered by an exhaustive search. Consequently, we use 6 chosen ciphertexts to break the ECB|CFB|CBC mode, whereas E. Biham's method requires 2^{34} chosen ciphertexts.

4 Conclusion

In this paper, we have presented the attacks to break many triple modes of operation with known-IV chosen plaintexts or chosen ciphertexts. Our results require fewer texts in cryptanalysis of triple modes than E. Biham's. We have analyzed 123 among 216 triple modes of operation. If the initial values are known, the triple modes which have the feedbacks driven into certain middle parts or arranged in a direction may be much weaker than under E. Biham's assumption. They are broken with about 3-4 chosen plaintexts or ciphertexts, 2^{58} encryptions, and 2^{56} memories. However, we could not find the proper method to attack the

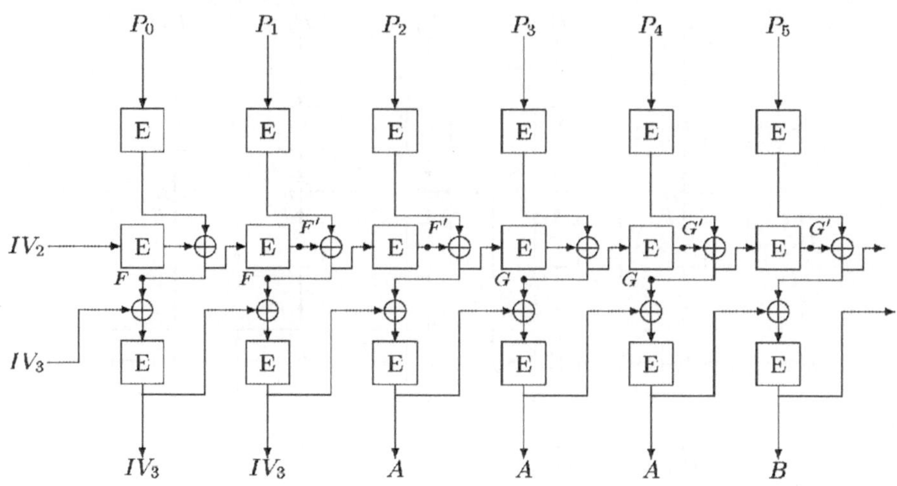

Fig. 9. Attack of ECB|CFB|CBC

others when trying to find the keys of the triple modes which have the feedbacks to spread forward and backward. We leave the problem of such triple modes open.

References

1. E. Biham. Cryptanalysis of multiple modes of operation. *Journal of Cryptology*, 011:45–58, 1998.
2. E. Biham. Cryptanalysis of triple modes of operation. *Journal of Cryptology*, 012:161–184, 1999.
3. Eli Biham and Adi Shamir. Differential cryptanalysis of DES-like cryptosystems. *J. Cryptology*, 4(1):3–72, 1991.
4. Eli Biham and Adi Shamir. Differential cryptanalysis of the full 16-round DES. In E. F. Brickell, editor, *Advances in Cryptology - CRYPTO'92*, volume 740 of *Lecture Note in Computer Science*, pages 487–496. Springer-Verlag, 1993.
5. Mitsuru Matsui. Linear cryptanalysis method for DES cipher. In T. Helleseth, editor, *Advances in Cryptology - EUROCRYPT'93*, volume 765 of *Lecture Notes in Computer Science*, pages 386–397. Springer-Verlag, 1994.
6. Mitsuru Matsui. On correlation between the order of s-boxes and the strength of DES. In *Advances in Cryptology – EUROCRYPT'94*, volume 950 of *Lecture Notes in Computer Science*, pages 366–375. Springer-Verlag, 1995.
7. National Bureau of Standards. Data Encryption Standard. FIPS Pub. 46, 1977.
8. D. Wagner. Cryptanalysis of some recently-proposed multiple modes of operation. In *FSE'98*, volume 1372 of *Lecture Notes in Computer Science*, 1998.

Appendix

In this appendix we list our result. We follow Biham's notation of the complexity, which consists of three parameters: the number of plaintexts/the number of steps of the attack(the time of encryptions)/the required memory size. 'Biham' is the result in [2] corresponding to ours. We compute some E. Biham's complexities in detail when the differences between our results and them are relatively small.

Table 1. *AAB*-attack

Mode	Complexity	Biham	Inverse
ECB|ECB|CBC	$3/4 \cdot 2^{56}/2^{56}$	$2^{33}/2^{58}/2^{56}$	CBC^{-1}|ECB|ECB
ECB|ECB|CBC^{-1}	$3/3 \cdot 2^{56}/2^{56}$	$2^{64}/2^{58}/2^{56}$	CBC|ECB|ECB
ECB|ECB|OFB	$3/4 \cdot 2^{56}/2^{56}$	$2^{64}/2^{58}/2^{56}$	OFB|ECB|ECB
ECB|ECB|CFB	$3/4 \cdot 2^{56}/2^{56}$	$2^{33}/2^{58}/2^{56}$	CFB^{-1}|ECB|ECB
ECB|ECB|CFB^{-1}	$3/4 \cdot 2^{56}/2^{56}$	$2^{64}/2^{58}/2^{56}$	CFB|ECB|ECB

Table 2. *AABB*-attack

Mode	Complexity	Biham	Inverse
ECB|CBC|ECB	$4/4 \cdot 2^{56}/2^{56}$	$5/2^{59}/-$	ECB|CBC^{-1}|ECB
CBC^{-1}|CBC|ECB	$4/4 \cdot 2^{56}/2^{56}$	$5/2^{59}/-$	ECB|CBC^{-1}|CBC
CBC|CBC|ECB	$4/4 \cdot 2^{56}/2^{56}$	$2^{34}/2^{59}/2^{33}$	ECB|CBC^{-1}|CBC^{-1}
OFB|CBC|ECB	$4/4 \cdot 2^{56}/2^{56}$	$2^{64}/5 \cdot 2^{56}/-$	ECB|CBC^{-1}|OFB
CFB^{-1}|CBC|ECB	$4/4 \cdot 2^{56}/2^{56}$	$5/2^{59}/-$	ECB|CBC^{-1}|CFB
CFB|CBC|ECB	$4/4 \cdot 2^{56}/2^{56}$	$2^{36}/2^{59}/2^{33}$	ECB|CBC^{-1}|CFB^{-1}

Table 3. *AAAB*-attack

Mode	Complexity	Biham	Inverse
ECB|CBC|CBC	$4/4 \cdot 2^{56}/2^{56}$	$2^{33}/2^{59}/2^{33}$	CBC^{-1}|CBC^{-1}|ECB
CBC^{-1}|CBC|ECB	$4/4 \cdot 2^{56}/2^{56}$	$5/2^{59}/-$	ECB|CBC^{-1}|CBC
CBC|CBC|ECB	$4/4 \cdot 2^{56}/2^{56}$	$2^{34}/2^{59}/2^{33}$	ECB|CBC^{-1}|CBC^{-1}
CFB^{-1}|CBC|ECB	$4/4 \cdot 2^{56}/2^{56}$	$5/2^{59}/-$	ECB|CBC^{-1}|CFB
CBC^{-1}|CFB|ECB	$4/4 \cdot 2^{56}/2^{56}$	$4/5 \cdot 2^{56}/-$	ECB|CFB^{-1}|CBC
CBC|CFB|ECB	$4/4 \cdot 2^{56}/2^{56}$	$2^{36}/2^{59}/2^{33}$	ECB|CFB^{-1}|CBC^{-1}
OFB|CFB|ECB	$4/5 \cdot 2^{56}/2^{56}$	$2^{64}/5 \cdot 2^{56}/-$	ECB|CFB^{-1}|OFB
CFB^{-1}|CFB|ECB	$4/4 \cdot 2^{56}/2^{56}$	$4/5 \cdot 2^{56}/-$	ECB|CFB^{-1}|CFB
CFB|CFB|ECB	$4/5 \cdot 2^{56}/2^{56}$	$2^{34}/2^{59}/2^{33}$	ECB|CFB^{-1}|CFB^{-1}
CBC^{-1}|CBC|CBC	$4/4 \cdot 2^{56}/2^{56}$	$2^{34}/2^{59}/2^{33}$	CBC^{-1}|CBC^{-1}|CBC
OFB|CBC|CBC	$4/5 \cdot 2^{56}/2^{56}$	$2^{66}/2^{59}/-$	CBC^{-1}|CBC^{-1}|OFB
CFB^{-1}|CBC|CBC	$4/4 \cdot 2^{56}/2^{56}$	$2^{34}/2^{59}/2^{33}$	CBC^{-1}|CBC^{-1}|CFB

Table 4. *AAA*-attack

Mode	Complexity	Biham	Inverse
ECB|OFB|ECB	$3/5 \cdot 2^{56}/2^{56}$	$2^{64}/5 \cdot 2^{56}/2^{56}$	itself
ECB|OFB|CBC	$3/5 \cdot 2^{56}/2^{56}$	$2^{64}/5 \cdot 2^{56}/2^{56}$	CBC^{-1}|OFB|ECB
ECB|OFB|CBC^{-1}	$3/5 \cdot 2^{56}/2^{56}$	$2^{65}/2^{65}/-$	CBC|OFB|ECB
ECB|OFB|OFB	$3/5 \cdot 2^{56}/2^{56}$	$2^{65}/2^{65}/2^{65}$	OFB|OFB|ECB
ECB|OFB|CFB	$3/5 \cdot 2^{56}/2^{56}$	$2^{64}/5 \cdot 2^{56}/2^{56}$	CFB^{-1}|OFB|ECB
ECB|OFB|CFB^{-1}	$3/5 \cdot 2^{56}/2^{56}$	$2^{65}/2^{65}/-$	CFB|OFB|ECB

Table 5. *IVIVA*-attack

Mode	Complexity	Biham	Inverse
ECB|CFB|CFB	$3/4 \cdot 2^{56}/2^{56}$	$2^{34}/2^{59}/2^{33}$	CFB^{-1}|CFB^{-1}|ECB
CBC|ECB|CBC	$3/3 \cdot 2^{56}/2^{56}$	$2^{34}/2^{59}/2^{33}$	CBC^{-1}|ECB|CBC^{-1}
CBC|ECB|CFB	$3/3 \cdot 2^{56}/2^{56}$	$2^{34}/2^{59}/2^{33}$	CFB^{-1}|ECB|CBC^{-1}
OFB|ECB|CBC	$3/4 \cdot 2^{56}/2^{56}$	$2^{64}/5 \cdot 2^{56}/-$	CBC^{-1}|ECB|OFB
CBC^{-1}|ECB|CBC	$3/4 \cdot 2^{56}/2^{56}$	$4/5 \cdot 2^{56}/-$	itself
CBC^{-1}|ECB|CFB	$3/4 \cdot 2^{56}/2^{56}$	$4/5 \cdot 2^{56}/-$	CFB^{-1}|ECB|CBC
CFB|ECB|CBC	$3/4 \cdot 2^{56}/2^{56}$	$2^{34}/2^{59}/2^{33}$	CBC^{-1}|ECB|CFB^{-1}
CBC^{-1}|CFB|CFB	$3/4 \cdot 2^{56}/2^{56}$	$2^{34}/2^{59}/2^{33}$	CFB^{-1}|CFB^{-1}|CBC
OFB|ECB|CFB	$3/4 \cdot 2^{56}/2^{56}$	$2^{64}/5 \cdot 2^{56}/-$	CFB^{-1}|ECB|OFB
OFB|CFB|CFB	$3/5 \cdot 2^{56}/2^{56}$	$2^{66}/2^{59}/-$	CFB^{-1}|CFB^{-1}|OFB
CFB|ECB|CFB	$3/4 \cdot 2^{56}/2^{56}$	$2^{34}/2^{59}/2^{33}$	CFB^{-1}|ECB|CFB^{-1}
CFB|CFB|CFB	$3/5 \cdot 2^{56}/2^{56}$	$2^{34}/2^{60}/2^{33}$	CFB^{-1}|CFB^{-1}|CFB^{-1}
CFB^{-1}|CFB|CFB	$3/4 \cdot 2^{56}/2^{56}$	$2^{34}/2^{59}/2^{33}$	CFB^{-1}|CFB^{-1}|CFB
CFB^{-1}|ECB|CFB	$3/4 \cdot 2^{56}/2^{56}$	$4/5 \cdot 2^{56}/-$	itself

Table 6. *IVIVIV*-attack

Mode	Complexity	Biham	Inverse
CBC|OFB|CBC	$3/5 \cdot 2^{56}/2^{56}$	$2^{66}/2^{66}/-$	CBC^{-1}|OFB|CBC^{-1}
CBC|OFB|CFB	$3/5 \cdot 2^{56}/2^{56}$	$2^{66}/2^{66}/-$	CFB^{-1}|OFB|CBC^{-1}
CBC^{-1}|OFB|CBC	$3/5 \cdot 2^{56}/2^{56}$	$2^{66}/5 \cdot 2^{56}/-$	itself
OFB|OFB|CBC	$3/5 \cdot 2^{56}/2^{56}$	$2^{65}/2^{65}/2^{65}$	CBC^{-1}|OFB|OFB
CBC^{-1}|OFB|CFB	$3/5 \cdot 2^{56}/2^{56}$	$2^{66}/5 \cdot 2^{56}/-$	CFB^{-1}|OFB|CBC
CFB|OFB|CBC	$3/5 \cdot 2^{56}/2^{56}$	$2^{66}/2^{66}/-$	CBC^{-1}|OFB|CFB^{-1}
OFB|OFB|CFB	$3/5 \cdot 2^{56}/2^{56}$	$2^{65}/2^{65}/2^{65}$	CFB^{-1}|OFB|OFB
CFB|OFB|CFB	$3/5 \cdot 2^{56}/2^{56}$	$2^{66}/2^{66}/-$	CFB^{-1}|OFB|CFB^{-1}
CFB^{-1}|OFB|CFB	$3/5 \cdot 2^{56}/2^{56}$	$2^{66}/5 \cdot 2^{56}/-$	itself

Table 7. *IVIVIVA*-attack

Mode	Complexity	Biham	Inverse
CBC|CBC|CBC	$4/4 \cdot 2^{56}/2^{56}$	$2^{34}/2^{60}/2^{33}$	CBC^{-1}|CBC^{-1}|CBC^{-1}
CBC|CBC|CFB	$4/4 \cdot 2^{56}/2^{56}$	$2^{34}/2^{60}/2^{33}$	CFB^{-1}|CBC^{-1}|CBC^{-1}
CBC|CFB|CBC	$4/4 \cdot 2^{56}/2^{56}$	$2^{34}/2^{60}/2^{33}$	CBC^{-1}|CFB^{-1}|CBC^{-1}
CBC|CFB|CFB	$4/4 \cdot 2^{56}/2^{56}$	$2^{34}/2^{60}/2^{33}$	CFB^{-1}|CFB^{-1}|CBC^{-1}
CBC^{-1}|CBC|CFB	$4/4 \cdot 2^{56}/2^{56}$	$5/5 \cdot 2^{56}/2^{56}$	CFB^{-1}|CBC^{-1}|CBC
CFB|CBC|CBC	$4/4 \cdot 2^{56}/2^{56}$	$2^{34}/2^{60}/2^{33}$	CBC^{-1}|CBC^{-1}|CFB^{-1}
CBC^{-1}|CFB|CBC	$4/4 \cdot 2^{56}/2^{56}$	$5/5 \cdot 2^{56}/-$	CBC^{-1}|CFB^{-1}|CBC
OFB|CFB|CBC	$4/5 \cdot 2^{56}/2^{56}$	$2^{66}/2^{59}/-$	CBC^{-1}|CFB^{-1}|OFB
CFB^{-1}|CFB|CBC	$4/4 \cdot 2^{56}/2^{56}$	$5/5 \cdot 2^{56}/-$	CBC^{-1}|CFB^{-1}|CFB
CFB|CFB|CBC	$4/4 \cdot 2^{56}/2^{56}$	$2^{34}/2^{60}/2^{33}$	CBC^{-1}|CFB^{-1}|CFB^{-1}
OFB|CBC|CFB	$4/5 \cdot 2^{56}/2^{56}$	$2^{66}/2^{59}/-$	CFB^{-1}|CBC^{-1}|OFB
CFB^{-1}|CBC|CFB	$4/4 \cdot 2^{56}/2^{56}$	$5/5 \cdot 2^{56}/-$	CFB^{-1}|CBC^{-1}|CFB

Table 8. *IVIVAAA*-attack

Mode	Complexity	Biham	Inverse
ECB|CBC|CFB	$5/4 \cdot 2^{56}/2^{56}$	$2^{34}/2^{59}/2^{33}$	CFB^{-1}|CBC^{-1}|ECB
CFB|CBC|CFB	$5/4 \cdot 2^{56}/2^{56}$	$2^{34}/2^{60}/2^{33}$	CFB^{-1}|CBC^{-1}|CFB^{-1}

Table 9. *IVIVAAAB*-attack

Mode	Complexity	Biham	Inverse
CBC|CFB|CBC	$6/5 \cdot 2^{56}/2^{56}$	$2^{34}/2^{59}/2^{33}$	CBC^{-1}|CFB^{-1}|ECB

Generic Attacks on Feistel Schemes

Jacques Patarin[1,2]

[1] CP8 Crypto Lab, SchlumbergerSema, 36-38 rue de la Princesse, BP 45, 78430 Louveciennes Cedex, France
[2] PRiSM, University of Versailles, 45 av. des États-Unis, 78035 Versailles Cedex, France

Abstract. Let A be a Feistel scheme with 5 rounds from $2n$ bits to $2n$ bits. In the present paper we show that for most such schemes A:

1. It is possible to distinguish A from a random permutation from $2n$ bits to $2n$ bits after doing at most $\mathcal{O}(2^{\frac{7n}{4}})$ computations with $\mathcal{O}(2^{\frac{7n}{4}})$ **random** plaintext/ciphertext pairs.
2. It is possible to distinguish A from a random permutation from $2n$ bits to $2n$ bits after doing at most $\mathcal{O}(2^{\frac{3n}{2}})$ computations with $\mathcal{O}(2^{\frac{3n}{2}})$ **chosen** plaintexts.

Since the complexities are smaller than the number 2^{2n} of possible inputs, they show that some generic attacks always exist on Feistel schemes with 5 rounds. Therefore we recommend in Cryptography to use Feistel schemes with at least 6 rounds in the design of pseudo-random permutations.

We will also show in this paper that it is possible to distinguish most of 6 round Feistel permutations generator from a truly random permutation generator by using a few (i.e. $\mathcal{O}(1)$) permutations of the generator and by using a total number of $\mathcal{O}(2^{2n})$ queries and a total of $\mathcal{O}(2^{2n})$ computations. This result is not really useful to attack a single 6 round Feistel permutation, but it shows that when we have to generate several pseudo-random permutations on a small number of bits we recommend to use more than 6 rounds. We also show that it is also possible to extend these results to any number of rounds, however with an even larger complexity.

Keywords: Feistel permutations, pseudo-random permutations, generic attacks on encryption schemes, Luby-Rackoff theory.

1 Introduction

Many secret key algorithms used in cryptography are Feistel schemes (a precise definition of a Feistel scheme is given in section 2), for example DES, TDES, many AES candidates, etc.. In order to be as fast as possible, it is interesting to have not too many rounds. However, for security reasons it is important to have a sufficient number of rounds. Generally, when a Feistel scheme is designed for cryptography, the designer either uses many (say ≥ 16 as in DES) very simple rounds, or uses very few (for example 8 as in DFC) more complex rounds. A natural question is: what is the minimum number of rounds required in a Feistel scheme to avoid all the "generic attacks" , i.e. all the attacks effective against

C. Boyd (Ed.): ASIACRYPT 2001, LNCS 2248, pp. 222–238, 2001.
© Springer-Verlag Berlin Heidelberg 2001

most of the schemes, and with a complexity negligible compared with a search on all the possible inputs of the permutation.

Let assume that we have a permutation from $2n$ bits to $2n$ bits. Then a generic attack will be an attack with a complexity negligible compared to $\mathcal{O}(2^{2n})$, since there are 2^{2n} possible inputs on $2n$ bits.

It is easy to see that for a Feistel scheme with only one round there is a generic attack with only 1 query of the permutation and $\mathcal{O}(1)$ computations: just check if the first half (n bits) of the output are equal to the second half of the input.

In [4] it was shown that for a Feistel scheme with two rounds there is also a generic attack with a complexity of $\mathcal{O}(1)$ chosen inputs (or $\mathcal{O}(2^{\frac{n}{2}})$ random inputs).

Also in [4], M. Luby and C. Rackoff have shown their famous result: for more than 3 rounds all generic attacks on Feistel schemes require at least $\mathcal{O}(2^{\frac{n}{2}})$ inputs, even for chosen inputs. If we call a Luby-Rackoff construction (a.k.a. L-R construction) a Feistel scheme instantiated with pseudo-random functions, this result says that the Luby-Rackoff construction with 3 rounds is a pseudorandom permutation.

Moreover for 4 rounds all the generic attacks on Feistel schemes require at least $\mathcal{O}(2^{\frac{n}{2}})$ inputs, even for a stronger attack that combines chosen inputs and chosen outputs (see [4] and a proof in [6], that shows that the Luby-Rackoff construction with 4 rounds is super-pseudorandom, a.k.a strong pseudorandom). However it was discovered in [7] (and independently in [1]) that these lower bounds on 3 and 4 rounds are tight, i.e. there exist a generic attack on all Feistel schemes with 3 or 4 rounds with $\mathcal{O}(2^{\frac{n}{2}})$ chosen inputs with $\mathcal{O}(2^{\frac{n}{2}})$ computations.

For 5 rounds or more the question remained open. In [7] it was proved that for 5 rounds (or more) the number of queries must be at least $\mathcal{O}(2^{\frac{2n}{3}})$ (even with unbounded computation complexity), and in [8] it was shown that for 6 rounds (or more) the number of queries must be at least $\mathcal{O}(2^{\frac{3n}{4}})$ (even with unbounded computations).

It can be noticed (see [7]) that if we have access to unbounded computations, then we can make an exhaustive search on all the possible round functions of the Feistel scheme, and this will give an attack with only $\mathcal{O}(2^n)$ queries (see [7]) but a gigantic complexity $\geq \mathcal{O}(2^{n2^n})$. This "exhaustive search" attack always exists, but since the complexity is far much larger than the exhaustive search on plaintexts in $\mathcal{O}(2^{2n})$, it was still an open problem to know if generic attacks, with a complexity $\ll \mathcal{O}(2^{2n})$, exist on 5 rounds (or more) of Feistel schemes.

In this paper we will indeed show that there exist generic attacks on 5 rounds of the Feistel scheme, with a complexity $\ll \mathcal{O}(2^{2n})$. We describe two attacks on 5 round Feistel schemes:

1. An attack with $\mathcal{O}(2^{\frac{7n}{4}})$ computations on $\mathcal{O}(2^{\frac{7n}{4}})$ **random** input/output pairs.
2. An attack with $\mathcal{O}(2^{\frac{3n}{2}})$ computations on $\mathcal{O}(2^{\frac{3n}{2}})$ **chosen** inputs.

For 6 rounds (or more) the problem remains open. In this paper we will describe some attacks on 6 rounds (or more) with a complexity much smaller than $\mathcal{O}(2^{n2^n})$ of exhaustive search, but still $\geq \mathcal{O}(2^{2n})$. So these attacks on 6 rounds and more are generally not interesting against a single permutation. However they may be useful when several permutations are used, i.e. they will be able to distinguish some permutation generators. These attacks show for example that when several small permutations must be generated (for example in the Graph Isomorphism scheme, or as in the Permuted Kernel scheme) then we must not use a 6 round Feistel construction.

Remark. The generic attacks presented here for 3, 4 and 5 rounds are effective against most Feistel schemes, or when the round functions are randomly chosen. However it can occur that for specific choices of the round function, the attacks, performed exactly as described, may fail. However in this case, very often there are modified attacks on these specific round functions. This point will be discussed in section 6.

2 Notations

We use the following notations that are very similar to those used in [4], [5] and [8].

- $I_n = \{0,1\}^n$ is the set of the 2^n binary strings of length n.
- For $a, b \in I_n$, $[a, b]$ will be the string of length $2n$ of I_{2n} which is the concatenation of a and b.
- For $a, b \in I_n$, $a \oplus b$ stands for bit by bit exclusive or of a and b.
- $\circ$ is the composition of functions.
- The set of all functions from I_n to I_n is F_n. Thus $|F_n| = 2^{n \cdot 2^n}$.
- The set of all permutations from I_n to I_n is B_n. Thus $B_n \subset F_n$, and $|B_n| = (2^n)!$
- Let f_1 be a function of F_n. Let L, R, S and T be elements of I_n. Then by definition

$$\Psi(f_1)[L,R] = [S,T] \overset{\text{def}}{\Leftrightarrow} \begin{cases} S = R \\ \text{and} \\ T = L \oplus f_1(R) \end{cases}$$

- Let $f_1, f_2, \ldots, f_k$ be k functions of F_n. Then by definition:

$$\Psi^k(f_1, \ldots, f_k) = \Psi(f_k) \circ \cdots \circ \Psi(f_2) \circ \Psi(f_1).$$

The permutation $\Psi^k(f_1, \ldots, f_k)$ is called "a Feistel scheme with k rounds" and also called Ψ^k.

3 Generic Attacks on 1,2,3, and 4 Rounds

Up till now, generic attacks had been discovered for Feistel schemes with 1,2,3,4 rounds. Let us shortly describe these attacks.
Let f be a permutation of B_{2n}. For a value $[L_i, R_i] \in I_{2n}$ we will denote by $[S_i, T_i] = f[L_i, R_i]$.

1 round
The attack just tests if $S_1 = R_1$. If f is a Feistel scheme with 1 round, this will happen with 100% probability, and if f is a random permutation with probability $\simeq \frac{1}{2^n}$. So with one round there is a generic attack with only 1 random query and $\mathcal{O}(1)$ computations.

2 rounds
Let choose $R_2 = R_1$ and $L_2 \neq L_1$. Then the attack just tests if $S_1 \oplus S_2 = L_1 \oplus L_2$. This will occur with 100% probability if f is a Feistel scheme with 2 rounds, and if f is a random permutation with probability $\simeq \frac{1}{2^n}$. So with two rounds there is a generic attack with only 2 chosen queries and $\mathcal{O}(1)$ computations.

Note 1: It is possible to transform this chosen plaintext attack in a known plaintext attack like the following. If we have $\mathcal{O}(2^{\frac{n}{2}})$ random inputs $[L_i, R_i]$, then with a good probability we will have a collision $R_i = R_j, i \neq j$. Then we test if $S_i \oplus S_j = L_i \oplus L_j$. Now the attack requires $\mathcal{O}(2^{\frac{n}{2}})$ random queries and $\mathcal{O}(2^{\frac{n}{2}})$ computations.

Note 2: This attack on 1 and 2 rounds was already described in [4].

3 rounds
Let ϕ be the following algorithm :

1. ϕ chooses m distinct $R_i, 1 \leq i \leq m$, and chooses $L_i = 0$ (or L_i constant) for all i, $1 \leq i \leq m$.
2. ϕ asks for the values $[S_i, T_i] = f[L_i, R_i], 1 \leq i \leq m$.
3. ϕ counts the number N of equalities of the form $R_i \oplus S_i = R_j \oplus S_j, i < j$.
4. Let N_0 be the expected value of N when f is a random permutation, and N_1 be the expected value of N when f is a $\psi^3(f_1, f_2, f_3)$, with randomly chosen f_1, f_2, f_3.
   Then $N_1 \simeq 2N_0$, because when f is a $\psi^3(f_1, f_2, f_3)$, $R_i \oplus S_i = f_2(f_1(R_i))$ so $f_2(f_1(R_i)) = f_2(f_1(R_j)), i < j$, if $f_1(R_i) \neq f_1(R_j)$ and $f_2(f_1(R_i)) = f_2(f_1(R_j))$ <u>or</u> if $f_1(R_i) = f_1(R_j)$.

So by counting N we will obtain a way to distinguish 3 round Feistel permutations from random permutations. This generic attack requires $\mathcal{O}(2^{\frac{n}{2}})$ chosen queries and $\mathcal{O}(2^{\frac{n}{2}})$ computations (just store the values $R_i \oplus S_i$ and count the collisions).

Remark. Here $N_1 \simeq 2 \cdot N_0$ when f_1, f_2, f_3 are randomly chosen. Therefore this attack is effective on most of 3 round Feistel schemes but not necessarily on all 3 round Feistel schemes. (See section 6 for more comments on this point).

4 rounds
This time, we take $R_i = 0$ (or R_i constant), and we count the number N of equalities of the form $S_i \oplus L_i = S_j \oplus L_j, i < j$. In fact, when $f = \psi^4(f_1, f_2, f_3, f_4)$, then $S_i \oplus L_i = f_3(f_2(L_i \oplus f_1(0))) \oplus f_1(0)$. So the probability of such an equality is about the double in this case (as long as f_1, f_2, f_3 are randomly chosen) than in the case where f is a random permutation (because if $f_2(L_i \oplus f_1(0)) = f_2(L_j \oplus f_1(0))$ this equality holds, and if $\beta_i = f_2(L_i \oplus f_1(0)) \neq f_2(L_j \oplus f_1(0)) = \beta_j$ but $f_3(\beta_i) = f_3(\beta_j)$, this equality also holds).

So by counting N we will obtain a way to distinguish 4 round Feistel permutations from random permutations. This generic attack requires $\mathcal{O}(2^{\frac{n}{2}})$ chosen queries and $\mathcal{O}(2^{\frac{n}{2}})$ computations (just store the values $S_i \oplus L_i$ and count the collisions).

Notes:

1. These attacks for 3 and 4 rounds have been first published in [7], and independently re-discovered in [1].
2. Here again the attack is effective against most of 4 round Feistel schemes but not necessarily on all 4 round Feistel schemes. (See section 6 for more comments on this point).

4 A Generic Attack on 5 Round Feistel Permutations with $\mathcal{O}(2^{\frac{7n}{4}})$ Random Plaintexts and $\mathcal{O}(2^{\frac{7n}{4}})$ Complexity

4.1 Notations for 5 Round Feistel Permutations

Let i be an integer. For any given i, let $[L_i, R_i]$ be a string of $2n$ bits in I_{2n}. Let

$$\Psi^5[L_i, R_i] = [S_i, T_i].$$

We introduce the intermediate variables X_i, P_i and Y_i such that:

$$\begin{cases} X_i = L_i \oplus f_1(R_i) \\ P_i = R_i \oplus f_2(X_i) \\ Y_i = X_i \oplus f_3(P_i) \end{cases}$$

So we have: $S_i = P_i \oplus f_4(Y_i)$ and $T_i = Y_i \oplus f_5(S_i)$. In other terms we have the following:

$$\begin{aligned} \Psi(f_1)[L_i, R_i] &= [R_i, X_i], \text{ as } X_i = L_i \oplus f_1(R_i) \\ \Psi(f_2)[R_i, X_i] &= [X_i, P_i], \text{ as } P_i = R_i \oplus f_2(X_i) \\ \Psi(f_3)[X_i, P_i] &= [P_i, Y_i], \text{ as } Y_i = X_i \oplus f_3(P_i) \\ \Psi(f_4)[P_i, Y_i] &= [Y_i, S_i], \text{ as } S_i = P_i \oplus f_4(Y_i) \\ \Psi(f_5)[Y_i, S_i] &= [S_i, T_i], \text{ as } T_i = Y_i \oplus f_5(S_i) \end{aligned}$$

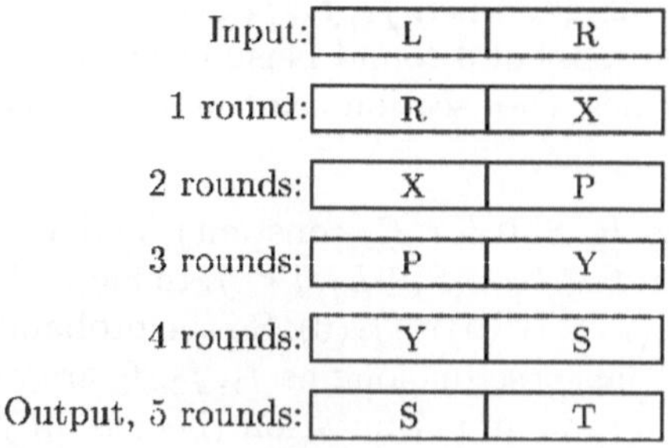

Input:	L	R
1 round:	R	X
2 rounds:	X	P
3 rounds:	P	Y
4 rounds:	Y	S
Output, 5 rounds:	S	T

Fig. 1.

We may notice that the following conditions (C) are always satisfied:

$$(\mathbf{C}) \begin{cases} R_i = R_j \Rightarrow X_i \oplus L_i = X_j \oplus L_j & \mathbf{(CR)} \\ X_i = X_j \Rightarrow R_i \oplus P_i = R_j \oplus P_j & \mathbf{(CX)} \\ P_i = P_j \Rightarrow X_i \oplus Y_i = X_j \oplus Y_j & \mathbf{(CP)} \\ Y_i = Y_j \Rightarrow S_i \oplus P_i = S_j \oplus P_j & \mathbf{(CY)} \\ S_i = S_j \Rightarrow Y_i \oplus T_i = Y_j \oplus T_j & \mathbf{(CS)} \end{cases}$$

4.2 The Attack

Let f be a permutation from B_{2n} We want to know (with a good probability) if f is a random element of B_{2n}, or if f is a Feistel scheme with 5 rounds (i.e. $f = \Phi^5(f_1, f_2, f_3, f_4, f_5)$ with f_1, f_2, f_3, f_4, f_5 being 5 functions of F_n).

The attack proceeds as follows:

Step 1: We generate m values $[S_i, T_i] = f[L_i, R_i]$, $1 \le i \le m$ such that the $[L_i, R_i]$ values are randomly chosen in I_{2n} and with $m = \mathcal{O}(2^{\frac{7n}{4}})$.

Step 2: We look if among these values, we can find 4 pairwise distinct indices denoted by $1, 2, 3, 4$ such that the following 8 equations (and 2 inequalities) are satisfied:

$$(\#) \begin{cases} R_1 = R_3 \\ R_2 = R_4 \\ L_1 \oplus L_3 = L_2 \oplus L_4 \\ S_1 = S_3 \\ S_2 = S_4 \\ S_1 \oplus S_2 = R_1 \oplus R_2 \\ T_1 \oplus T_3 = L_1 \oplus L_3 \\ T_1 \oplus T_3 = T_2 \oplus T_4 \end{cases}$$

(and with $R_1 \neq R_2$ and $L_1 \neq L_3$)

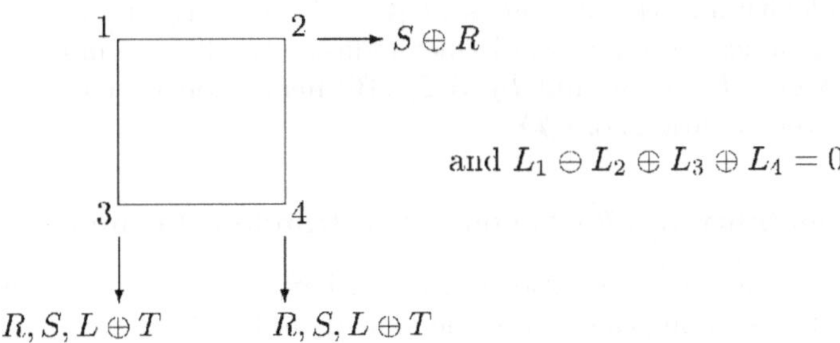

Fig. 2. A representation of the 8 equations $\#$ in L, S, R, T.

Below we explain how one can test with the complexity of $\mathcal{O}(m)$ if such indices exist.

Step 3: If such indices exist, we will guess that f is Feistel scheme with 5 rounds. If not we will say that f is not a Feistel scheme. [We will see below that the probability to find such indices is not negligible if f is a Feistel scheme with 5 rounds and $M \geq \mathcal{O}(2^{\frac{7n}{4}})$ for most of 5 round Feistel schemes].

4.3 How to Accomplish the Step 2 in $\mathcal{O}(m)$ Computations

First, we find among the $m \times m$ possibilities, all the possible indices 1 and 3 such that:

$$\begin{cases} R_1 = R_3 \\ S_1 = S_3 \\ L_1 \oplus T_1 = L_3 \oplus T_3 \end{cases}$$

It is possible to this in $\mathcal{O}(m)$ computations instead of $\mathcal{O}(m^2)$ by storing all the m values $(R_i, S_i, L_i \oplus T_i)$ in a hash table and looking for collisions. We expect to find $\frac{m^2}{2^{3n}} \ll m$ such indices (as $m \ll 2^{3n}$).

In the same way we find all the possible indices 2 and 4 such that:

$$\begin{cases} R_2 = R_4 \\ S_2 = S_4 \\ L_2 \oplus T_2 = L_4 \oplus T_4 \end{cases}$$

Each part requires $\mathcal{O}(m)$ computations and $\mathcal{O}(m)$ of memory, and, if needed, there is a tradeoff with $\mathcal{O}(m \cdot \alpha)$ computations and $\mathcal{O}(m/\alpha)$ memory.

Now we store all the values $(L_1 \oplus L_3, S_1 \oplus R_1)$ for all the indices $(1,3)$ already found. There are about $\frac{m^2}{2^{3n}} \leq m$ such values. Then we store all the values $(L_2 \oplus L_4, S_2 \oplus R_2)$ for all the indices $(2,4)$ already found. Using another birthday paradox technique, we look for the following collision:

$$\begin{cases} L_2 \oplus L_4 = L_1 \oplus L_3 \\ S_2 \oplus R_2 = S_1 \oplus R_1 \end{cases}$$

The complexity and the storage is $\mathcal{O}(\frac{m^2}{2^{3n}}) \leq \mathcal{O}(m)$ again. At the end we have at most m choices of pairwise distinct indices $(1,2,3,4)$. Among these we keep those that give $R_1 \neq R_2$ and $L_1 \neq L_3$. By inspection we check that now they satisfy all the equations of (#).

4.4 Probability of (#) When f Is a Random Permutation of B_{2n}

When f is a random permutation of B_{2n}, we have $\mathcal{O}(m^4)$ possibilities to chose the indices $1, 2, 3, 4$ among the m possible indices, and we have 8 equations to satisfy, with a probability about $\frac{1}{2^{8n}}$ to have them all true for some pairwise distinct $1, 2, 3, 4$. By inspection we check that the equations of (#) are not dependent. Thus the probability to have 4 pairwise distinct indices $1, 2, 3, 4$ that satisfy (#) is about $\frac{m^4}{2^{8n}}$ when f is a random permutation of B_{2n} (n.b. the two additional inequalities $R_1 \neq R_2$ and $L_1 \neq L_3$ change nothing). Since $m \ll 2^{2n}$ (because $m = \mathcal{O}(2^{\frac{7n}{4}})$) this probability is negligible.

4.5 Probability of (#) When f Is a Feistel Scheme with 5 Rounds

Theorem 1 *When f is a Feistel scheme with 5 rounds, the 8 equations of (#) are a logical consequence on the following 7 equations:*

$$(\Lambda)\begin{cases} R_1 = R_3 & (1)\\ R_2 = R_4 & (2)\\ L_1 \oplus L_3 = L_2 \oplus L_4 & (3)\\ S_1 = S_3 & (4)\\ X_1 = X_2 & (5)\\ P_1 = P_3 & (6)\\ Y_1 = Y_2 & (7) \end{cases}$$

Proof of Theorem 1.
We will use the facts (CR), (CX), (CP), (CY) and (CS) that have been introduced in section 4.1.

- From (1) and (CR) we get
  $X_3 = X_1 \oplus L_1 \oplus L_3$ (8)
- From (2) and (CR) we get $X_4 \oplus L_4 = X_2 \oplus L_2$, and then using (8), (5) and (3) we get
  $X_4 = X_3$ (9).
- From (5) and (CX) we get:
  $R_1 \oplus P_1 = R_2 \oplus P_2$ (10)
- From (9) and (CX) we get $R_4 \oplus P_4 = R_3 \oplus P_3$ and then from (10), (6), (1) and (2) we get:
  $P_4 = P_2$ (11)
- From (6) and (CP) we get $X_1 \oplus Y_1 = X_3 \oplus Y_3$ and then from (8) we get:
  $Y_3 = Y_1 \oplus L_1 \oplus L_3$ (12)
- From (11) and (CP) we get $X_2 \oplus Y_2 = X_4 \oplus Y_4$ and then from (12), (7), (9), (5) and (8) we get:
  $Y_4 = Y_3$ (13)
- From (7) and (CY) we get $S_1 \oplus P_1 = S_2 \oplus P_2$ and then from (10) we get:
  $S_1 \oplus S_2 = R_1 \oplus R_2$ (14)
- From (13) and (CY) we get $S_4 \oplus P_4 = S_3 \oplus P_3$ and then from (14), (4), (11), (6) and (10) we get:
  $S_4 = S_2$ (15)
- From (4) and (CS) we get $Y_1 \oplus T_1 = Y_3 \oplus T_3$ and then from (12) we get:
  $T_3 = T_1 \oplus L_1 \oplus L_3$ (16).
- From (15) and (CS) we get $Y_4 \oplus T_4 = Y_2 \oplus T_2$ and then from (13), (7), (12) and (16) we get:
  $T_4 \oplus T_2 = T_1 \oplus T_3$ (17)
- If $R_1 = R_2$ then because of (5) we have $L_1 = L_2$ and $R_1 = R_2 \Rightarrow 1 = 2$ and the indices 1 and 2 are distinct by definition. Thus
  $R_1 \neq R_2$ (18)
- Finally since $1 \neq 3$ and because of (1) we have. $L_1 \neq L_3$ (19)

So all the equations of (#) are indeed just consequences of the 7 equations (Λ) when f is a Feistel with 5 rounds. Indeed the $8+2$ conditions of (#) are now in (1), (2), (3), (4), (15), (14), (16), (17), and finally (18) and (19).

Theorem 2 *Let f be a Feistel scheme with 5 rounds, $f = \Psi^5(f_1, f_2, f_3, f_4, f_5)$. Then for most of such f, the probability to have 4 pairwise distinct indices 1,2,3,4 that satisfy # is $\geq \mathcal{O}(\frac{m^4}{2^{7n}})$, and thus is not negligible when $m \geq \mathcal{O}(2^{\frac{7n}{4}})$. Therefore the algorithm given in the section 4 is indeed a generic way to distinguish most Feistel schemes with 5 rounds from a truly random permutation of B_{2n} with a complexity of $\mathcal{O}(2^{\frac{7n}{4}})$.*

Proof.
When f_1, f_2, f_3, f_4, f_5 are randomly chosen in F_n, the probability that there exist pairwise distinct indices 1,2,3,4 chosen out of a set of m indices such that all the 7 equations (Λ) hold is $= \mathcal{O}(\frac{m^4}{2^{7n}})$. Thus from the Theorem 1 we get the Theorem 2.

Remark. Here again, the attack is effective against most of 5 round Feistel schemes, but not necessarily on all 5 round Feistel schemes. (See section 6 for more comments on that).

5 A Generic Attack on 5 Round Feistel Permutations with $\mathcal{O}(2^{\frac{3n}{2}})$ Chosen Plaintexts and $\mathcal{O}(2^{\frac{3n}{2}})$ Complexity

This attack proceeds exactly as the previous attack of the Section 4, except that now Step 1 is replaced by the following Step' 1:

Step' 1. We generate m values $f[L_i, R_i] = [S_i, T_i]$, $1 \leq i \leq m$ such that the L_i values are randomly chosen in I_n and the R_i values are randomly chosen in a subset I'_n of I_n with only $2^{\frac{n}{2}}$ elements. For example I'_n=all the strings of n bits with the first $n/2$ bits at 0.
Let $m = \mathcal{O}(2^{\frac{3n}{2}})$.

5.1 Probability of (#) When f Is a Random Permutation of B_{2n}

Now the probability that there are some indices $1, 2, 3, 4$ such that equations (#) are satisfied when f is randomly chosen in B_{2n} is about

$$\frac{m^4}{2^{\frac{n}{2}} \cdot 2^{\frac{n}{2}} 2^{6n}} = \frac{m^4}{2^{7n}}$$

(because the equations $R_1 = R_3$ and $R_2 = R_4$ have now a probability $\frac{1}{2^{\frac{n}{2}}}$ to be satisfied instead of $\frac{1}{2^n}$).

However, since here $m = \mathcal{O}(2^{\frac{3n}{2}})$, this probability $\frac{m^4}{2^{7n}}$ is still negligible.

5.2 Probability of (#) When f Is a Feistel Scheme with 5 Rounds

When f is a Feistel scheme with 5 rounds, with f_1, f_2, f_3, f_4, f_5 randomly chosen in F_n, the probability that there exist indices $1, 2, 3, 4$ chosen out of a set of m indices, such that all the 7 equations (Λ) are satisfied is about

$$\simeq \frac{m^4}{2^{\frac{n}{2}} \cdot 2^{\frac{n}{2}} 2^{5n}} = \frac{m^4}{2^{6n}}$$

(because the equations $R_1 = R_3$ and $R_2 = R_4$ have now a probability $\frac{1}{2^{\frac{n}{2}}}$ to be satisfied instead of $\frac{1}{2^n}$).

So from Theorem 1 of section 4, we see that for these functions f the probability that there exist indices 1,2,3,4 such that all the 8 equations (and 2 inequalities) # are satisfied is here generally $\geq \mathcal{O}(\frac{m^4}{2^{6n}})$.

Thus the algorithm given in this section 5 is indeed a generic way to distinguish most Feistel schemes with 5 rounds from a truly random permutation of B_{2n}, with a complexity $\mathcal{O}(2^{\frac{3n}{2}})$ and $\mathcal{O}(2^{\frac{3n}{2}})$ chosen queries.

Remark. Here again some time/memory tradeoff is possible: use $\mathcal{O}(2^{\frac{3n}{2}})$ chosen queries, $\mathcal{O}(2^{\frac{3n}{2}} \cdot \alpha)$ computations and $\mathcal{O}(2^{\frac{3n}{2}}/\alpha)$ of memory.

6 Feistel Schemes with Specific Round Functions

The problem. The generic attacks that we have presented for 3, 4 and 5 rounds are effective against most Feistel schemes, or when the round functions are randomly chosen. However it can occur that for specific choices of the round functions, these attacks, if applied exactly as described, may fail. In this cases, very often there are some other attacks, against these specific rounds functions, that are even simpler. We will illustrate this on an example pointed out by an anonymous referee of Asiacrypt'2001.

Theorem 3 (Knudsen, see [2] or [3]) *Let $[L_1, R_1]$ and $[L_2, R_2]$ be two inputs of a 5 round Feistel scheme, and let $[S_1, T_1]$ and $[S_2, T_2]$ be the outputs. Let assume that the round functions f_2 and f_3 are permutations (therefore they are* **not** *random functions of F_n). Then if $R_1 = R_2$ and $L_1 \neq L_2$ it is impossible to have simultaneously $S_1 = S_2$ and $L_1 \oplus L_2 = T_1 \oplus T_2$.*

Proof.
$R_1 = R_2 \Rightarrow X_1 \oplus X_2 = L_1 \oplus L_2$, and $S_1 = S_2 \Rightarrow Y_1 \oplus Y_2 = T_1 \oplus T_2$. Therefore if we have $L_1 \oplus L_2 = T_1 \oplus T_2$, we will have also:

$$X_1 \oplus Y_1 = X_2 \oplus Y_2.$$

Now since we have $Y_i = X_i \oplus f_3(P_i)$, we will have $f_3(P_1) = f_3(P_2)$ and since f_3 is a permutation we get $P_1 = P_2$.
Then since we have $P_i = R_i \oplus f_2[L_i \oplus f_1(R_i)]$ with $R_1 = R_2$, and since f_2 is a permutation we get

$$L_1 \oplus f_1(R_1) = L_2 \oplus f_1(R_2).$$

This is in contradiction with $R_1 = R_2$ and $L_1 \neq L_2$.

Attacks on 5 Round Feistel Schemes with f_2 and f_3 Permutations

From the above Theorem 3 we see that our attack given in section 4 and 5 against most 5 round Feistel schemes will fail when f_2 and f_3 are permutations. Indeed, the event $R_1 = R_3, L_1 \neq L_3, S_1 = S_3$ and $L_1 \oplus L_3 = T_1 \oplus T_3$ will never occur if f_2 and f_3 are permutations. However, in such a case there is an even simpler attack that comes immediately from the Theorem 3: we can randomly get m input/output values and count the number of indices $(i,j), i < j$ such that:

$$\begin{cases} R_i = R_j \\ S_i = S_j \\ L_i \oplus L_j = T_i \oplus T_j \end{cases}$$

For a random permutation this number is $\mathcal{O}(\frac{m^2}{2^{3n}})$, and for a 5 round Feistel scheme with f_2 and f_3 being permutations, it is exactly 0.
This attack requires $\mathcal{O}(2^{\frac{3n}{2}})$ random plaintext/ciphertext pairs and $\mathcal{O}(2^{\frac{3n}{2}})$ computations.

Remark: This attack can also be extended to 6 round Feistel schemes when the round functions are permutations (or "quasi-permutations"), see [2,3] for details.

Conclusion: It was known (before the present paper) that some generic attacks on 5 round Feistel schemes exist when the round functions are permutations. This particular case is interesting since two of the former AES candidates, namely DFC and DEAL, were such Feistel schemes using permutations as round functions. (More precisely they were "quasi-permutations" in DFC). The number of rounds in these functions is however ≥ 6.
In this paper we have shown a more general result that such generic attacks exist for most of 5 round Feistel schemes (even when f_2 and f_3 are **not** permutations). It can be noticed that our attack is based on specific relations on 4 points (corresponding to 4 ciphertexts), while the previous attacks were based on specific relations on only 2 points ("impossible differentials").

7 Attacking Feistel Generators

In this section we will describe what is an attack against a generator of permutations (and not only against a single permutation randomly generated by a generator of permutations), i.e. we will be able to study several permutations generated by the generator. Then we will evaluate the complexity of brute force attacks and we will notice that since all Feistel permutations have an even signature, it is possible to distinguish them from a random permutation in $\mathcal{O}(2^{2n})$.

Let G be a "k round Feistel Generator", i.e. from a binary string K, G generates a k round Feistel permutation G_K of B_{2n}.
Let G' be a truly random permutation generator, i.e. from a string K, G' generates a truly random permutation G'_K of B_{2n}.

Let G'' be a truly random even permutation generator, i.e. from a string K, G'' generates a truly random permutation G''_K of A_{2n}, with A_{2n} being the group of all the permutations of B_{2n} with even signature.

We are looking for attacks that distinguish G from G', and also for attacks that will distinguish G from G''.

Adversarial model: An attacker can choose some strings $K_1, \dots K_f$, can ask for some inputs $[L_i, R_i] \in I_{2n}$, and can ask for some $G_{K_\alpha}[L_i, R_i]$ (with K_α being one of the K_i). Here the attack is more general than in the previous sections, since the attacker can have access to many different permutations generated by the same generator.

Adversarial goal: The aim of the attacker is to distinguish G from G' (or from G'') with a good probability and with a complexity as small as possible.

Brute force attacks. A possible attack is the exhaustive search on the k round functions $f_1, \dots, f_k$ form I_n to I_n that have been used in the Feistel construction. This attack always exists, but since we have $2^{k \cdot n \cdot 2^n}$ possibilities for $f_1, \dots, f_k$, this attack requires about $2^{k \cdot n \cdot 2^n}$ computations (or $2^{\lceil \frac{k}{2} \rceil \cdot n \cdot 2^n}$ computations in a version "in the middle" of the attack) and about $k \cdot 2^{n-1}$ random queries[1] and only 1 permutation of the generator.

Attack by the signature.

Theorem 4 *If $n \geq 2$ then all the Feistel schemes from $I_{2n} \to I_{2n}$ have an even signature.*

Proof.
Let $\sigma : I_{2n} \to I_{2n}$
$[L, R] \mapsto [R, L]$.
Let f_1 be a function of F_n.
Let $\Psi'(f_1)[L, R] = [L \oplus f_1(R), R]$.
We will show that both σ and $\Psi'(f_1)$ have an even signature, so will have $\sigma \circ \Psi'(f_1) = \Psi(f_1)$, and thus by composition, all the Feistel schemes from $I_{2n} \to I_{2n}$ have an even signature.

For σ: All the cycles have 1 or 2 elements, and we have 2^n cycles with 1 element (and an even signature), and $\frac{2^{2n}-2^n}{2}$ cycles with 2 elements. When $n \geq 2$ this number is even.

For $\Psi'(f_1)$: All the cycles have 1 or 2 elements since $\Psi'(f_1) \circ \Psi'(f_1) = Id$. Moreover the number of cycles with 2 elements is $\frac{2^n \cdot k}{2}$, with k being the number of values R such that $f_1(R) \neq 0$. So when $n \geq 2$ the signature of $\Psi'(f_1)$ is even.

Theorem 5 *Let f be a permutation of B_{2n}. Then using $\mathcal{O}(2^{2n})$ computations on the 2^{2n} input/output values of f, we can compute the signature of f.*

[1] each query divides by about 2^{2n} the number of possible $f_1, \dots, f_k$

Proof.
Just compute all the cycles c_i of f, $f = \prod_{i=1}^{\alpha} c_i$ and use the formula:
signature$(f) = \prod_{i=1}^{\alpha} (-1)^{length(c_i)+1}$.

Theorem 6 *Let G be a Feistel scheme generator, then it is possible to distinguish G from a generator of truly random permutations of B_{2n} after $\mathcal{O}(2^{2n})$ computations on $\mathcal{O}(2^{2n})$ input/output values.*

Proof.
It is direct consequence of the Theorems 4 and 5 above.

Remark.
It is however probably much more difficult to distinguish G from random permutations of A_{2n}, with A_{2n} being the group of all the permutations of B_{2n} with even signature. In the next sections we will present our best attacks for this problem.

8 An Attack on 6 Round Feistel Generators in $\mathcal{O}(2^{2n})$

Attacks on 6 round Feistel. If G is a generator of 6 round Feistel permutations of B_{2n}, we have found an attack (described below) that uses a few (i.e. $\mathcal{O}(1)$) permutations from the generator G, $\mathcal{O}(2^{2n})$ computations and about $\mathcal{O}(2^{2n})$ random queries. So this attack has a complexity much smaller than the exhaustive search in $2^{63n \cdot 2^n}$. However since a permutation of B_{2n} has only 2^{2n} possible inputs, this attack has no real interest against a single specific 6 round Feistel scheme used in encryption.

It is interesting only if a few 6 round Feistel schemes are used. This can be particularly interesting for some cryptographic schemes using many permutations on a relatively small number of bits. For example in the Graph Isomorphism authentication scheme many permutations on about 2^{14} points are used (thus $n = 7$), or in the Permuted Kernel Problem PKP of Adi Shamir many permutations on about 2^6 points ($n = 3$ here). Then, we will be able to distinguish these permutations from truly random permutations with a small complexity if a 6 round Feistel scheme generator is used. And this, whatever the size of the secret key used in the generator may be. So we do not recommend to generate small pseudorandom permutations from 6 round Feistel schemes.

The Attack:
Let $[L_i, R_i]$ be an element of I_{2n}.
Let $\Psi^6[L_i, R_i] = [S_i, T_i]$. The attack proceeds as follows:

Step 1.
We choose specific permutation $f = G_K$.
We generate m values $f[L_i, R_i] = [S_i, T_i]$, $1 \le i \le m$ with the random $[L_i, R_i] \in I_{2n}$ and with $m = \mathcal{O}(2^{2n})$.

Remark: Since $m = \mathcal{O}(2^{2n})$, we cover here almost all the possible inputs $[L_i, R_i]$ for this specific permutation f.

Step 2.

We look if among these values we can find 4 pairwise distinct indices denoted by 1, 2, 3, 4 such that these 8 equations are satisfied:

$$(\#)\begin{cases} R_1 = R_3 \\ R_2 = R_4 \\ S_1 = S_2 \\ S_3 = S_4 \\ L_1 \oplus L_3 = L_2 \oplus L_4 \\ L_1 \oplus L_3 = S_1 \oplus S_3 \\ T_1 \oplus T_2 = T_3 \oplus T_4 \\ T_1 \oplus T_2 = R_1 \oplus R_2 \end{cases}$$

(and with $R_2 \neq R_1$, $S_3 \neq S_1$ and $T_1 \neq T_2$).

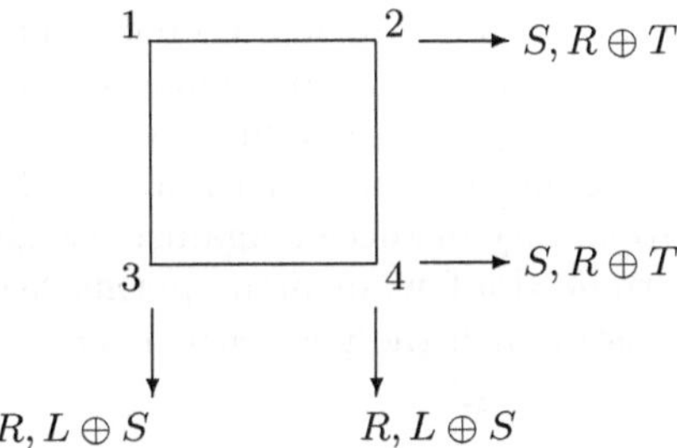

Fig. 3. A representation of the 8 equations # in L, S, R, T

It is also possible to show that all the indices that satisfy these equations can be found in $\mathcal{O}(m)$ and with $\mathcal{O}(m)$ of memory. We count the number of solutions found.

Step 3.

We try again at Step 1 with another $f = G_{K'}$ and we will do this a few times, say λ times with $\lambda = \mathcal{O}(1)$. Let α be the total number of solutions found at Step 2 for all the λ functions tested. It is possible to prove that for a generator of pseudorandom permutation of B_{2n} we have

$$\alpha \simeq \frac{\lambda m^4}{2^{8n}}.$$

Moreover it is possible to prove that for a generator of 6 round Feistel schemes the average value we get for α is

$$\alpha \geq \text{ about } \frac{2\lambda m^4}{2^{8n}}.$$

Proof.
The proof is very similar to the proof we did for Ψ^5 (due to the lack of space we do not explicit it here).

So by counting this value α we will distinguish 6 round Feistel generators from truly random permutation generators each time when $\frac{\lambda m^4}{2^{8n}}$ is not negligible, for example when $\lambda = \mathcal{O}(1)$ and $m = \mathcal{O}(2^{2n})$, as claimed.

Examples: Thus we are able, to distinguish between a few 6 round Feistel permutations taken from a generator, and a set of truly random permutations (or from a set of random permutations with an even signature) from 32 bits to 32, within approximately 2^{32} computations and 2^{32} chosen plaintexts.

9 An Attack on k Round Feistel Generators

It is also possible to extend these attacks on more than 6 rounds, to any number of rounds k. However for more than 6 rounds, as already for 6 rounds, all our attacks require a complexity and a number of queries $\geq \mathcal{O}(2^{2n})$, so they can be interesting to attack generators of permutations, but not to attack a single permutation (the probability of success against one single permutation is generally negligible, and we need a few, or many permutations from the generator, in order to be able to distinguish the generator from a truly random permutation generator).

Example of attack on a Feistel generator with k rounds. Let k be an integer. For simplicity we will assume that k is even (the proof is very similar when k is odd). Let $\lambda = \frac{k}{2} - 1$. Let G be a generator of Feistel permutations of k rounds of B_{2n}. We will consider an attack with a set of equations in (L, R, S, T) illustrated in figure 3. For simplicity we do not write all the equations explicitly.

Here we have $\mu = \lambda^2 = (\frac{k}{2} - 1)^2$ indices, and we have $4\lambda(\lambda - 1) = k^2 - 6k + 8$ equations in L, R, S, T. Here it is possible to prove that the probability that the $4\lambda(\lambda - 1)$ equations of figure 3 exist, will be about twice for a Feistel scheme with k rounds, than for a truly random permutation.

Thus, on a fixed permutation this attack succeeds with a probability in

$$\mathcal{O}\left(\frac{m^{(\frac{k}{2}-1)^2}}{2^{n \cdot 4\lambda(\lambda-1)}}\right)$$

If we take $m = \mathcal{O}(2^{2n})$ for such a permutation, it gives a probability of success in

$$\mathcal{O}\left(\frac{2^{2n(\frac{k}{2}-1)^2}}{2^{n \cdot (k^2-6k+8)}}\right)$$

So we will use $\mathcal{O}(2^{n(\frac{k^2}{2}-4k+6)})$ permutations, and the total complexity and the total number of queries on all these permutations will be $\mathcal{O}(2^{n(\frac{k^2}{2}-4k+8)})$. The total memory will be $\mathcal{O}(2^{2n})$.

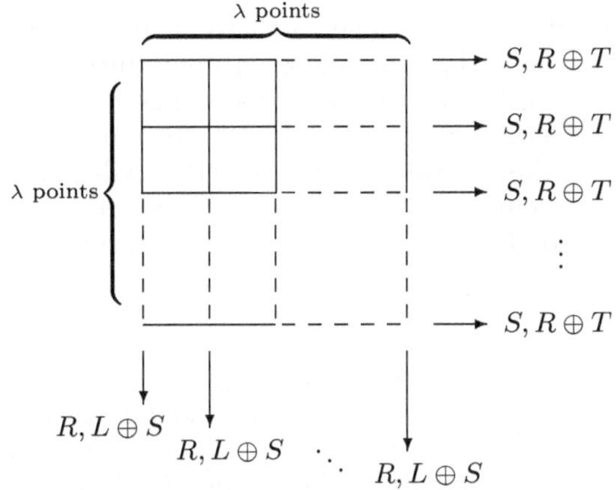

Fig. 4. Modelling the $4 \cdot \lambda(\lambda - 1)$ equations in L, R, S, T.

Examples:

- With $k = 6$ this attack uses $\mathcal{O}(1)$ permutations and $\mathcal{O}(2^{2n})$ computations (exactly as we did in section 8).
- With $k = 8$ we need $\mathcal{O}(2^{6n})$ permutations and $\mathcal{O}(2^{8n})$ computations.

10 Conclusion

Up till now, generic attacks on Feistel schemes were known only for 1,2,3 or 4 rounds. In this paper we have seen that some generic attacks also do exist on 5 round Feistel schemes. So we do not recommend to use 5 round Feistel schemes in cryptography for general purposes. Our first attack requires $\mathcal{O}(2^{\frac{7n}{4}})$ **random** plaintext/ciphertext pairs and the same amount of computation time. Our second attack requires $\mathcal{O}(2^{\frac{3n}{2}})$ **chosen** plaintext/ciphertext pairs and the same amount of computation time. For example, it is possible to distinguish most of 5 round Feistel ciphers with blocks of 64 bits, from a random permutation from 64 bits to 64 bits, within about 2^{48} chosen queries and 2^{48} computations.

We have also seen that when we have to generate several small pseudo-random permutations we do not recommend to use a Feistel scheme generator with only 6 rounds (whatever the length of the secret key may be). As an example, it is possible to distinguish most generators of 6 round Feistel permutations from truly random permutations on 32 bits, within approximately 2^{32} computations and 2^{32} chosen plaintexts (and this whatever the length of the secret key may be).

Similar attacks can be generalised for any number of rounds k, but they require to analyse much more permutations and they have a larger complexity when k increases.

Acknowledgments. I would like to thank Jean-Jacques Quisquater who allowed me to do this work, as it has been done during my invited stay at the university of Louvain-La-Neuve. I also would like to thank the anonymous referee of Asiacrypt'2001, for pointing out the references [2,3], and for observing that my attack against 5 round Feistel schemes will not in general apply as it is, against some specific round functions such as permutations. Finally I would like to thank Nicolas Courtois for his help writing this paper.

References

1. William Aiollo, Ramarathnam Venkatesan: *Foiling Birthday Attacks in Length-Doubling Transformations - Benes: A Non-Reversible Alternative to Feistel.* Eurocrypt 96, LLNCS 1070, Springer-Verlag, pp. 307-320.
2. L.R. Knudsen: *DEAL - A 128-bit Block Cipher*, Technical report #151, University of Bergen, Department of Informatics, Norway, February 1998. Submitted as a candidate for the Advanced Encryption Standard. Available at http://www.ii.uib.no/~larsr/newblock.html
3. L.R. Knudsen, V. Rijmen: *On the Decorrelated Fast Cipher (DFC) and its Theory.* Fast Software Encryption (FSE'99), Sixth International Workshop, Rome, Italy, March 1999, LNCS 1636, pp. 81-94, Springer, 1999.
4. M. Luby, C. Rackoff, *How to construct pseudorandom permutations from pseudorandom functions*, SIAM Journal on Computing, vol. 17, n. 2, pp. 373-386, April 1988.
5. Moni Naor and Omer Reingold, *On the construction of pseudo-random permutations: Luby-Rackoff revisited*, J. of Cryptology, vol 12, 1999, pp. 29-66. Extended abstract in: Proc. 29th Ann. ACM Symp. on Theory of Computing, 1997, pp. 189-199.
6. J. Patarin, *Pseudorandom Permutations based on the DES Scheme*, Eurocode'90, LNCS 514, Springer-Verlag, pp. 193-204.
7. J. Patarin, *New results on pseudorandom permutation generators based on the DES scheme*, Crypto'91, Springer-Verlag, pp. 301-312.
8. J. Patarin, *About Feistel Schemes with Six (or More) Rounds*, in Fast Software Encryption 1998, pp. 103-121.

A Compact Rijndael Hardware Architecture with S-Box Optimization

Akashi Satoh, Sumio Morioka, Kohji Takano, and Seiji Munetoh

IBM Research, Tokyo Research Laboratory, IBM Japan Ltd., 1623-14, Shimotsuruma, Yamato-shi, Kanagawa 242-8502, Japan
{akashi,e02716,chano,munetoh}@jp.ibm.com

Abstract. Compact and high-speed hardware architectures and logic optimization methods for the AES algorithm Rijndael are described. Encryption and decryption data paths are combined and all arithmetic components are reused. By introducing a new composite field, the S-Box structure is also optimized. An extremely small size of 5.4 Kgates is obtained for a 128-bit key Rijndael circuit using a 0.11-μm CMOS standard cell library. It requires only 0.052 mm^2 of area to support both encryption and decryption with 311 Mbps throughput. By making effective use of the SPN parallel feature, the throughput can be boosted up to 2.6 Gbps for a high-speed implementation whose size is 21.3 Kgates.

1 Introduction

DES (Data Encryption Standard) [14,1], which is a common-key block cipher for US federal information processing standards, has also been used as a de facto standard for more than 20 years. NIST (National Institute of Standard Technology) has selected Rijndael [2] as the new Advanced Encryption Standard (AES) [13]. Many hardware architectures for Rijndael were proposed and their performances were evaluated by using ASIC libraries [8,18,10,9] and FPGAs [3, 17,6,11,5]. However, they are simple implementations according to the Rijndael specification, and none are yet small enough for practical use. The AES has to be embeddable not only in high-end servers but also in low-end consumer products such as mobile terminals. Therefore, sharing and reusing hardware resources, and compressing the gate logic are indispensable to produce a small Rijndael circuit.

The SPN structure of Rijndael is suitable for highly parallel processing, but it usually requires more hardware resources compared with the Feistel structure used in many other ciphers developed after DES. This is because, all data is encoded in each round of Rijndael processing, while only half of data is processed at once in DES. In addition, Rijndael has two separate data paths for encryption and decryption.

In this paper, we describe a compact data path architecture for Rijndael, where the hardware resources are efficiently shared between encryption and decryption. The key arithmetic component S-Box has been implemented using

C. Boyd (Ed.): ASIACRYPT 2001, LNCS 2248, pp. 239–254, 2001.
© Springer-Verlag Berlin Heidelberg 2001

look-up table logic or ROMs in the previous approaches, which requires a lot of hardware support. Reference [16] proposed the use of composite field arithmetic to reduce the computation cost of the S-Box, but no detailed hardware implementation was provided. Therefore, we propose a methodology to optimize the S-Box by introducing a new composite field, and show its advantages in comparison to the previous work.

2 Rijndael Algorithm

Fig. 1 shows a Rijndael encryption process for 128-bit plain text data string and a 128-bit secret key, with the number of rounds set to 10. These numbers are used throughout this paper, including for our hardware implementation. Each round and the initial stage requires a 128-bit round key, and thus 11 sets of round keys are generated from the secret key. The input data is arranged as a 4×4 matrix of bytes. The primitive functions SubBytes, ShiftRows and MixColumns are based on byte-oriented arithmetic, and AddRoundKey is a simple 128-bitwise XOR operation.

SubBytes is a nonlinear transformation that uses 16 byte substitution tables (S-Boxes). An S-Box is the multiplicative inverse of a Galois field $GF(2^8)$ followed by an affine transformation. In the decryption process, the affine transformation is executed prior to the inversion. The irreducible polynomial used by a Rijndael S-Box is

$$m(x) = x^8 + x^4 + x^3 + x + 1. \tag{1}$$

ShiftRows is a cyclic shift operation of the last three rows by different offsets. MixColumns treats the 4-byte data in each column as coefficients of a 4-term polynomial, and multiplies the data modulo $x^4 + 1$ with the fixed polynomial given by

$$c(x) = \{03\}x^3 + \{01\}x^2 + \{01\}x + \{02\}. \tag{2}$$

In the decryption process, InvMixColumns multiplies each column with the polynomial

$$c^{-1}(x) = \{\texttt{0B}\}x^3 + \{\texttt{0D}\}x^2 + \{09\}x + \{\texttt{0E}\} \tag{3}$$

and InvShiftRows shifts the last three rows in the opposite direction from ShiftRows.

The key expander in Fig. 1 generates 11 sets of 128-bit round keys from one 128-bit secret key by using a 4-byte S-Box. These round keys can be prepared on the fly in parallel with the encryption process. In the decryption process, these sets of keys are used in reverse order. Therefore, all keys have to be generated and stored in registers in advance, or the final round key in the encryption process has to be pre-calculated for on-the-fly key scheduling. Because the first method requires the equivalent of a 1,408-bit register (128 bits $\times$ 11), and is not suitable

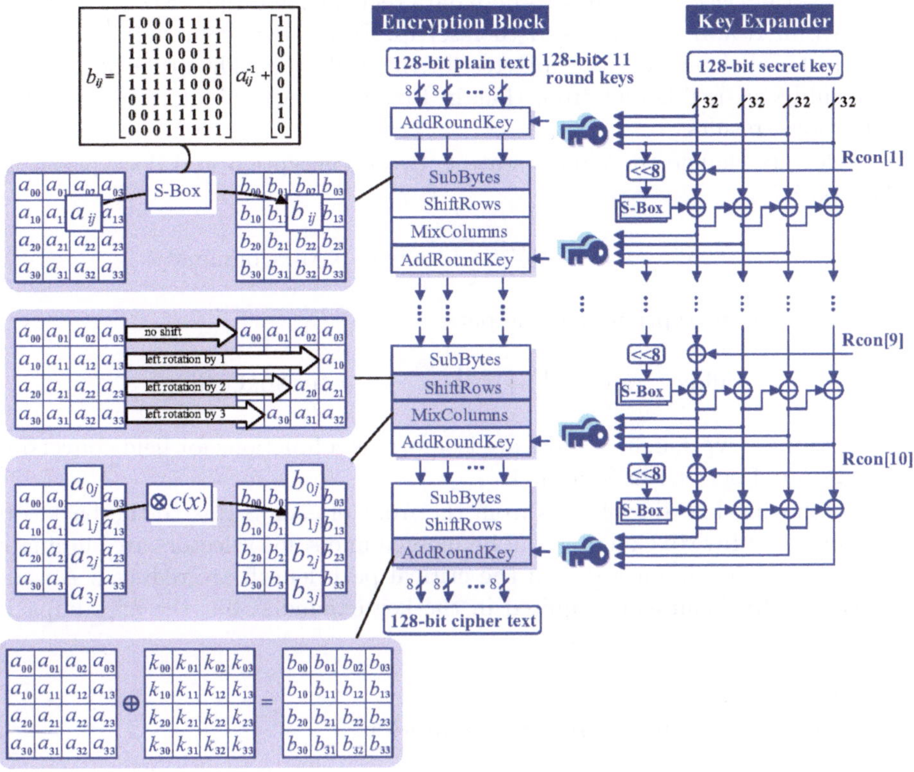

Fig. 1. Encryption process of Rijndael algorithm

for compact hardware, the second approach was chosen for the implementation described in the next section. Rcon[i] in Fig. 1 is a 4-byte value, and the lower 3 bytes are 0 for all i, and the highest byte is the bit representation of the polynomial $x^i \bmod m(x)$.

3 Data Path Architecture

3.1 Data Path Sharing between Encryption and Decryption

In order to minimize the size of our Rijndael hardware, resource sharing in the data path is fully employed as shown in Fig. 2. This circuit can execute both encryption and decryption. The 128-bit data (4 × 4 bytes) block is divided into four 32-bit columns, and is processed column by column through the 32-bit data bus. Therefore one round takes 4 clock cycles. It is not a good idea to make the bus width smaller than 32 bits, because the MixColumns operation needs 32-bits of data at one time. A smaller bus requires more registers and selectors, and resource sharing is hindered, resulting in an inefficient implementation.

The "Enc/Dec block" has 16-byte data registers, and they execute ShiftRows (or InvShiftRows) operations by themselves. Each 4-byte column is transformed by four parallel S-Boxes as SubBytes (or InvSubBytes). The order of ShiftRows and SubBytes is different from that in Fig. 1, though this does not affect the operations' results.

Selectors change the circuit state between encryption and decryption. The data path

$$\delta^{-1} \rightarrow x^{-1} \rightarrow \delta^{-1} \text{ and affine} \rightarrow \text{MixColumns}$$

is selected for encryption, and the path

$$\text{affine}^{-1} \text{ and } \delta^{-1} \rightarrow x^{-1} \rightarrow \delta^{-1} \rightarrow \text{InvMixColumns}$$

is used for decryption. δ^{-1} and δ are isomorphism functions for field conversions. Details are described in Section 4.

By moving InvMixColumns from the front of each S-Box to the back, MixColumns and InvMixColumns can be merged and some selectors are eliminated. As a result, the circuit size and the critical path length are reduced. An additional InvMixColumns is required in the key expander, but the area impact is minor.

3.2 S-Box Sharing with Key Expander

The key expander reuses the S-Boxes in the encryption/decryption block to generate a 128-bit key in each round. The S-Boxes are used once by the key expander, and four times by the encryption/decryption block, for a total of five times in every round. While the key expander uses the S-Boxes, the ShiftRows (or InvShiftRows) operation is executed simultaneously. As shown in Fig. 1, only the AddRoundKey operation is executed in the initial round, and the MixColumns (or InvMixColumns for decryption) is omitted in the final round. This operation switching is carried out by controlling the 4:1 selector at the bottom of Fig. 2. The first round key used in AddRoundKey is the initial key data stored in the key registers, and a transformation with the S-Boxes is not necessary. Therefore the first round takes four cycles, and the entire encryption process takes 54 ($= 4 + 5 \times 10$) cycles. The decryption process also takes 54 cycles. When a new secret key is provided, the key expander takes 10 cycles to generate the initial decryption key, which is the final round key in the encryption.

As described in Section 2, Rcon[i] is a 4-byte constant value, and the highest order byte is generated by modular multiplication on $GF(2^8)$. The circuit RC in Fig. 3 generates the constant values sequentially during the encryption process, starting from {01}, and RC^{-1} calculates the same values in reverse order from {36}. These circuits are also merged as shown in this figure.

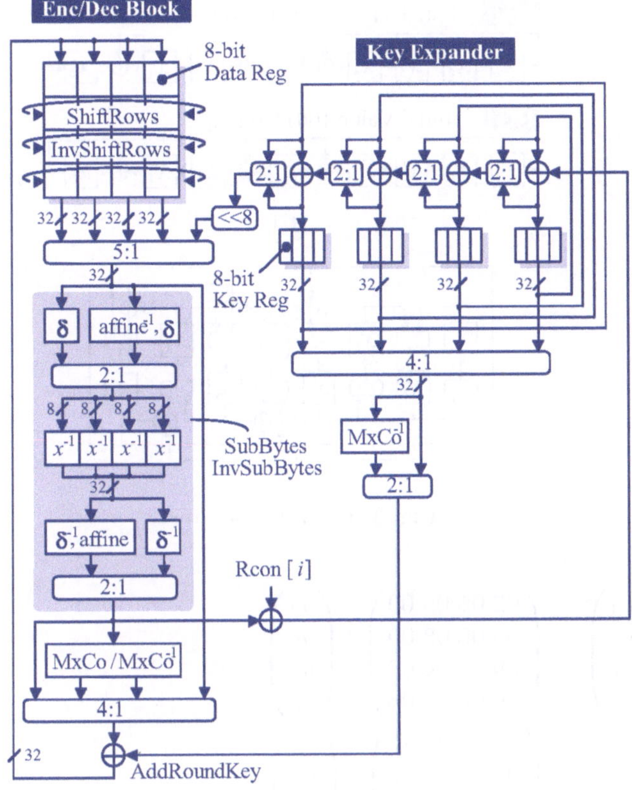

Fig. 2. Data path architecture

3.3 Factoring in MixColumns and InvMixColumns

MixColumns and InvMixColumns are modular multiplications with constant polynomials (2) and (3) that can be written as the constant matrix multiplications shown in Equations (4) and (5) respectively.

$$\begin{pmatrix} b_3 \\ b_2 \\ b_1 \\ b_0 \end{pmatrix} = \begin{pmatrix} 02\ 03\ 01\ 01 \\ 01\ 02\ 03\ 01 \\ 01\ 01\ 02\ 03 \\ 03\ 01\ 01\ 02 \end{pmatrix} \cdot \begin{pmatrix} a_3 \\ a_2 \\ a_1 \\ a_0 \end{pmatrix}$$
$$= \begin{pmatrix} 02\ 02\ 00\ 00 \\ 00\ 02\ 02\ 00 \\ 00\ 00\ 02\ 02 \\ 02\ 00\ 00\ 02 \end{pmatrix} \cdot \begin{pmatrix} a_3 \\ a_2 \\ a_1 \\ a_0 \end{pmatrix} + \begin{pmatrix} 00\ 01\ 01\ 01 \\ 01\ 00\ 01\ 01 \\ 01\ 01\ 00\ 01 \\ 01\ 01\ 01\ 00 \end{pmatrix} \cdot \begin{pmatrix} a_3 \\ a_2 \\ a_1 \\ a_0 \end{pmatrix} \tag{4}$$

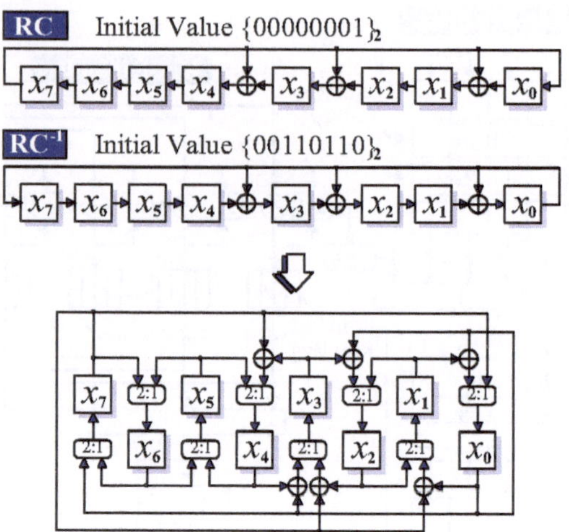

Fig. 3. Rcon[i] generator

$$\begin{pmatrix} c_3 \\ c_2 \\ c_1 \\ c_0 \end{pmatrix} = \begin{pmatrix} \texttt{0E 0B 0D 09} \\ \texttt{09 0E 0B 0D} \\ \texttt{0D 09 0E 0B} \\ \texttt{0B 0D 09 0E} \end{pmatrix} \cdot \begin{pmatrix} a_3 \\ a_2 \\ a_1 \\ a_0 \end{pmatrix}$$
$$= \begin{pmatrix} \texttt{02 03 01 01} \\ \texttt{01 02 03 01} \\ \texttt{01 01 02 03} \\ \texttt{03 01 01 02} \end{pmatrix} \cdot \begin{pmatrix} a_3 \\ a_2 \\ a_1 \\ a_0 \end{pmatrix}$$
$$+ \begin{pmatrix} \texttt{08 08 08 08} \\ \texttt{08 08 08 08} \\ \texttt{08 08 08 08} \\ \texttt{08 08 08 08} \end{pmatrix} \cdot \begin{pmatrix} a_3 \\ a_2 \\ a_1 \\ a_0 \end{pmatrix} + \begin{pmatrix} \texttt{04 00 04 00} \\ \texttt{00 04 00 04} \\ \texttt{04 00 04 00} \\ \texttt{00 04 00 04} \end{pmatrix} \cdot \begin{pmatrix} a_3 \\ a_2 \\ a_1 \\ a_0 \end{pmatrix} \tag{5}$$

$$\begin{cases} b_3 &= 02X_3 + X_1 + a_2 \\ b_2 &= 02X_2 + X_1 + a_3 \\ b_1 &= 02X_1 + X_3 + a_0 \\ b_0 &= 02X_0 + X_3 + a_1 \end{cases} \quad \begin{cases} X_3 &= a_3 + a_2 \\ X_2 &= a_2 + a_1 \\ X_1 &= a_1 + a_0 \\ X_0 &= a_0 + a_3 \end{cases} \tag{6}$$

$$\begin{cases} c_3 &= b_3 + Z_1 \\ c_2 &= b_2 + Z_0 \\ c_1 &= b_1 + Z_1 \\ c_0 &= b_0 + Z_0 \end{cases} \quad \begin{cases} Z_1 &= Y_2 + Y_1 \\ Z_0 &= Y_2 + Y_0 \end{cases} \quad \begin{cases} Y_2 &= 02(Y_1 + Y_0) \\ Y_1 &= 04(a_3 + a_1) \\ Y_0 &= 04(a_2 + a_0) \end{cases} \tag{7}$$

As seen in Equation (5), InvMixColumns contains a complete MixColumns matrix. Therefore we merged these two functions into one circuit as shown in Fig. 4. In addition, both functions can be broken into regular matrices whose non-zero elements are only one of the values $\{08, 04, 02, 01\}$. Therefore the number of common terms can be greatly reduced by factoring, finally resulting in Equations (6) and (7). The result, shown in Table 1, is that the XOR logic gates are decreased by 2/3 (from 592 XORs to 195 XORs) with only 2 XOR gates of additional delay.

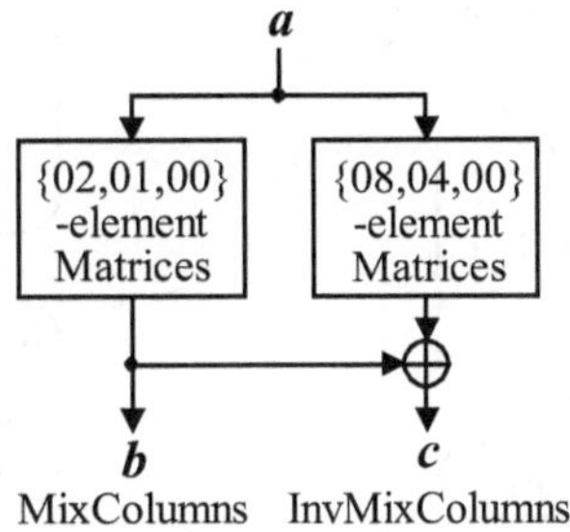

Fig. 4. MixColumns/InvMixColumns circuit

Table 1. Factoring effects of MixColumns and InvMixColumns

	Original Matrices			Our Implementation
	MixColumns	InvMixColumns	Total	
Number of XOR	152	440	592	195
Delay (gates)	3	5	5	7

4 S-Box Optimization

4.1 Structure of New S-Box

Designing a compact S-Box is one of the most critical problems for reducing the total circuit size of the Rijndael hardware. It is possible to implement the S-Box as a practical circuit based on its functional specification by using automatic logic synthesis tools, because the size of the S-Box function table is small; 256 entries × 1 byte. However, a significant reduction in the size of the S-Box was achieved in [16], by using composite field arithmetic [7]. In the following, we propose further optimization of S-Box by introducing a new composite field.

Fig. 5 shows the outline of our S-Box implementation. The most costly operation in the S-Box is the multiplicative inversion over a field A, where A is an extension field over $GF(2)$ with the irreducible polynomial $m(x)$. To reduce the cost of this operation, we adopted the following 3-stage method.

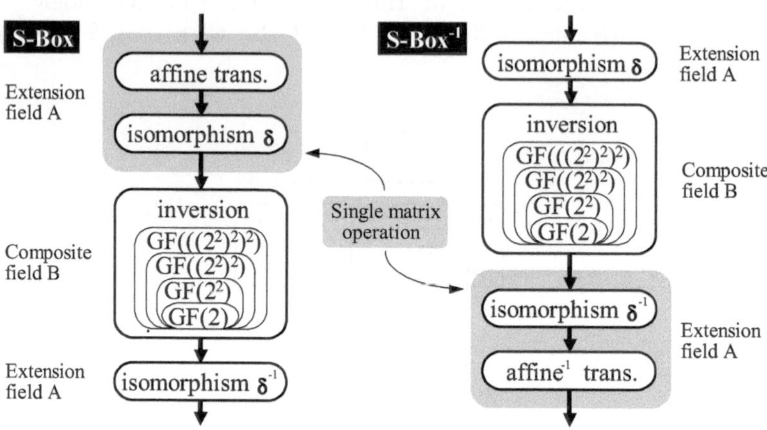

Fig. 5. The computation sequence of our S-Box implementation

(Stage 1) Map all elements of the field A to a composite field B, using an isomorphism function δ.
(Stage 2) Compute the multiplicative inverses over the field B.
(Stage 3) Re-map the computation results to A, using the function δ^{-1}.

Even though isomorphism functions are required in this method, the cost of those functions can mostly be hidden by merging them with the affine transformations.

4.2 Multiplicative Inversion over A New Composite Field

The composite field B in Stage 2 is constructed not by applying a single degree-8 extension to $GF(2)$, but by applying multiple extensions of smaller degrees. To reduce the cost of Stage 2 as much as possible, we built the composite field B by repeating degree-2 extensions under a polynomial basis using these irreducible polynomials:

$$\begin{cases} GF(2^2) & : x^2 + x + 1 \\ GF((2^2)^2) & : x^2 + x + \phi \\ GF(((2^2)^2)^2) & : x^2 + x + \lambda \end{cases} \tag{8}$$

where $\phi = \{10\}_2$, $\lambda = \{1100\}_2$. The inverter over the field above has fewer $GF(2)$ operators compared with the composite field used in [16]

$$\begin{cases} GF(2^4) & : x^2 + x + 1 \\ GF((2^4)^2) & : x^2 + x + \omega_{14} \end{cases} \tag{9}$$

where $\omega_{14} = \{1001\}_2$.

Our hardware implementation of Stage 2 is shown in Fig. 6. For any composite fields $GF((2^m)^n)$ which are constructed using a degree-n extension after a degree-m extension, computing the multiplicative inverses can be done as a combination of operations over the subfields $GF(2^n)$, using the equation described in [7,4]

$$P^{-1} = (P^r)^{-1} \cdot P^{r-1}, \text{ where } r = (2^{nm} - 1)/(2^m - 1). \tag{10}$$

In our case ($n = 2$, $m = 4$), so Equation (10) becomes

$$P^{-1} = (P^{17})^{-1} \cdot P^{16}. \tag{11}$$

The circuit in Fig. 6 is an implementation of Equation (11), with additional optimizations. In the circuit, P^{16} is computed first (note that the hardware costs for computing 2-powers over Galois fields are very small) and then P^{17} is obtained by multiplying P by P^{16} over $GF(((2^2)^2)^2)$. This operation requires only two multiplications, one addition and one constant multiplication over $GF((2^2)^2)$. Because P^{17} is always an element of $GF((2^2)^2)$ according to Fermat's Little Theorem (i.e., the upper 4 bits of P^{17} are always 0), computing the upper 4 bits of P^{17} is unnecessary [7]. $(P^{17})^{-1}$ is computed recursively over $GF((2^2)^2)$, then multiplied by P^{16} over $GF(((2^2)^2)^2)$, and finally P^{-1} is obtained. This multiplication requires fewer circuit resources than usual, because P^{17} is an element of $GF((2^2)^2)$. Note that our multipliers and inverter over subfield $GF((2^2)^2)$ are also small [15]. Further gate reduction is possible by sharing parts of the three $GF((2^2)^2)$ multipliers in Fig. 6, where common inputs are used.

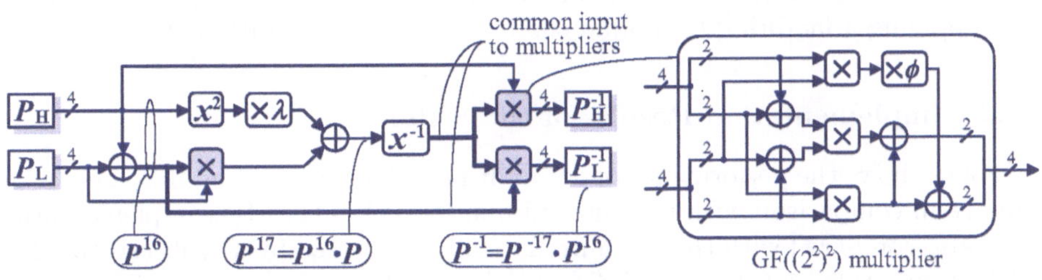

Fig. 6. Our implementation of an inverter over a composite field $GF(((2^2)^2)^2)$.

4.3 Generating Isomorphism Functions

The isomorphism functions δ and δ^{-1} are located at the both ends of the S-Boxes, and one of them is merged with an affine transformation. On the other hand, Reference [16] proposes locating these isomorphism functions at the circuit's primary input and output, and thus it cannot be merged with an affine transformation. The function is also required between the AddRoundKey and the key expander in Fig. 1. Therefore our approach is much more suitable for a reduced hardware implementation. The isomorphism functions δ and δ^{-1} in Stages 1 and 3 were constructed as follows. First, search for a generator element α in A and a generator β in B, where both α and β are roots of the same primitive irreducible polynomial. Any primitive polynomial can be applied, and here we use

$$p(x) = x^8 + x^4 + x^3 + x^2 + 1. \tag{12}$$

Once such elements are found, the definition table of the isomorphism function δ (or δ^{-1}) is immediately determined, where α^k is mapped to β^k (or β^k to α^k) for any $1 \leq k \leq 254$. The hardware implementation of these functions can be obtained by mapping only the basis elements of A (or B) into B (or A), and these mappings are described as multiplications of constant matrixes over $GF(2)$. The functions δ and δ^{-1} are as follows:

$$\delta = \begin{pmatrix} 1&1&0&0&0&0&1&0 \\ 0&1&0&0&1&0&1&0 \\ 0&1&1&1&1&0&0&1 \\ 0&1&1&0&0&0&1&1 \\ 0&1&1&1&0&1&0&1 \\ 0&0&1&1&0&1&0&1 \\ 0&1&1&1&1&0&1&1 \\ 0&0&0&0&0&1&0&1 \end{pmatrix}_2 \quad \delta^{-1} = \begin{pmatrix} 1&0&1&0&1&1&1&0 \\ 0&0&0&0&1&1&0&0 \\ 0&1&1&1&1&0&0&1 \\ 0&1&1&1&1&1&0&0 \\ 0&1&1&0&1&1&1&0 \\ 0&1&0&0&0&1&1&0 \\ 0&0&1&0&0&0&1&0 \\ 0&1&0&0&0&1&1&1 \end{pmatrix}_2 \tag{13}$$

where the least significant bits are in the upper left corners.

All of these isomorphism functions and the constant multipliers in the S-Boxes are implemented as XOR arrays, and their Boolean logic is compressed by applying a factoring technique based on a greedy algorithm [12].

4.4 Implementation Results of the S-Box

Table 2 shows the performance of our multiplicative inverter and S-Box described above in comparison with that using Equation (9). The S-Box implementations are also compared with the one automatically generated by a synthesis tool from a look-up table. A 0.11-μm CMOS standard cell library (one gate is equivalent to a 2-way NAND) is used here, and the delay time is evaluated under the worst-case conditions. The hardware size of our S-Box using the field $GF(((2^2)^2)^2)$ is 294 gates, which is about 20% smaller and slightly faster than the one using the field $GF((2^4)^2)$.

Table 2. S-Box features of proposed method. (gate = 2-way NAND)

Method	Inverter		S-Box		S-Box^{-1}	
	Area (gates)	Delay (ns)	Area (gates)	Delay (ns)	Area (gates)	Delay (ns)
Ours, Equation (8)	173	2.55	294	3.69	← merged	
Equation (9)	241	2.50	362	3.75	← merged	
Look-up Table	-	-	696	2.71	700	2.29

Our S-Box consists of affine transformations, isomorphism functions, inverters and selectors, and can be applied to both encryption and decryption. On the other hand, the look-up table method requires two different circuits, an S-Box for encryption and an S-Box^{-1} for decryption. The S-Box tables appear as random numbers to CAD tools, and therefore logic compression is very hard. As a result, a large amount of hardware, 1,396 (= 696 + 700) gates, is required for each one-byte S-Box based on the look-up table method, while our method is less than 1/4 of that size.

By applying our new composite field, merging the isomorphism functions with affine transformations, using a factoring technique, and combining the encryption and decryption paths, a very small S-Box was produced.

5 Performance Comparison in ASICs

The architecture described in Section 3 has been implemented by using 0.11-μm CMOS technology, and the extremely small size of 5.4 Kgates was obtained with a 7.62-ns cycle time (131.24 MHz) under the worst-case conditions. The gate size of each component and the critical path delay are detailed in Table 3 and Fig. 7, respectively. The function SubBytes (S-Box) occupies about 22% of the circuit area, and accounts for almost half of the delay time. The second major component is neither MixColumns nor AddRoundKey, but the selectors. The requirement to use selectors is not obvious from the Rijndael algorithm specification, where they appear as conditional branches and data selections. However, they require 1,099 (= 699 + 400) gates (20.36% of the circuit), because of the wide data width. In order to drive those selectors, drivers with high fan out are also required. Therefore, we carefully analyzed the critical data path and optimized the order of data selection, and adjusted the driver size. As a result, the delay time of the selector and driver section was reduced from 3.46 ns down to 1.95 ns, without changing the total gate count.

Using our proposed architecture, we designed and synthesized five implementations as shown in Fig. 8. Higher throughputs with higher parallelism were achieved by increasing the number of S-Boxes and the bus width. Four S-Boxes are shared between the data encryption block and the key expander in the 5- and

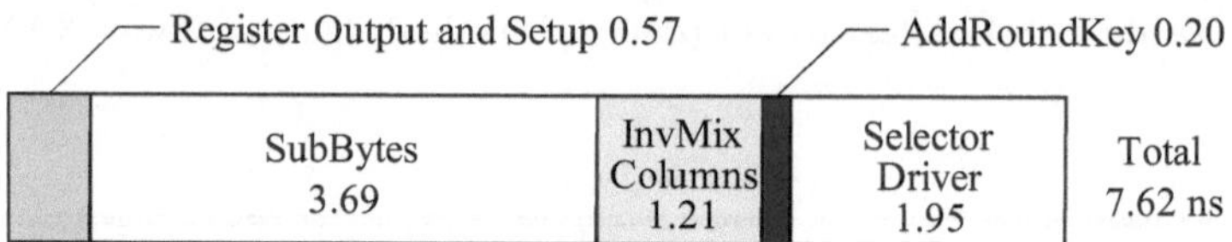

Fig. 7. Critical path delay

Table 3. Factoring effects of MixColumns and InvMixColumns

Components	Gates	%
Encryption/Decryption Block	(3,305)	(61.23)
Data Register	864	16.01
ShiftRows	160	2.96
SubBytes	1,176	21.79
MixColumns/InvMixColumns	350	6.48
AddRoundKey	56	1.04
Selector	699	12.95
Key Expander	(1,896)	(35.12)
Key Register	864	16.01
InvMixColumns	294	5.45
RC/RC^{-1}	100	1.85
XOR	238	4.41
Selector	400	7.41
Controller, Selector, Driver	197	3.65
Total	5,398	100.00

3-cycle/round versions. In the other three implementations, the key expanders have their own S-Boxes. Two circuits were synthesized from each implementation (a total of ten implementations), one optimized for size and the other for speed. The sizes and speeds are also shown in Table 4, in comparison with other ASIC implementations [8,18,10,9] under the worst-case conditions. Data and the key sizes are both 128 bits in all implementations, except that of [10], where 128-bit data and a 256-bit key (14 rounds) are used. A gate wireability of 80% is assumed to calculate the silicon area of our implementations. Reference [16] shows a throughput of 7.5 Gbps with 32 parallel cores, with a circuit size for encryption of 256 Kgates. However, this number was not evaluated by any synthesis tool, so we did not include it in the table.

It is obvious that in our implementation that more hardware resources yield higher throughput. For instance, the number of operation clock cycles can be reduced by increasing the size of the S-Box, which allows more parallel computation. Increases in fan out can also be used to increase the speed. In order to clarify the total efficiency of each implementation, we show the throughput per gate on the right side of Table 4. In general, it is not easy to compare the

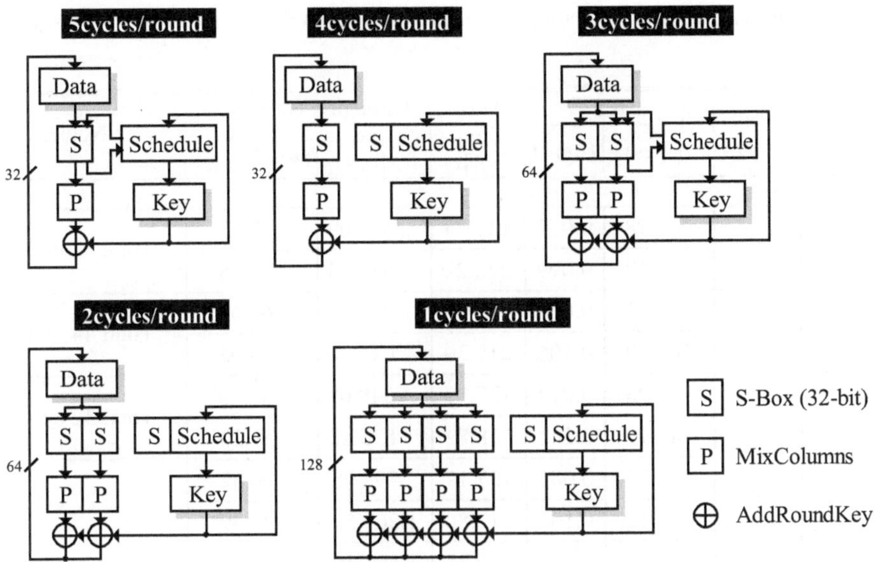

Fig. 8. Data Path Architectures of each implementation

implementations using different CAD tools and different technology libraries. However, even considering this difficulty of precise comparison, our hardware architecture is by far the best. Our smallest implementation is less than 1/6 of the 33.8 Kgates in the best previous approach [9]. Our 1-cycle/round version sets a new record for throughput at 2.6Gbps, not only in ECB mode but also in CBC cipher feedback mode. Reference [8] shows the second best throughput as 1.95 Gbps, but it uses 113.5 times as many gates, because all 11 rounds are unrolled. Throughout these comparisons of ASIC implementations, our hardware architecture has advantages in both size and speed.

6 Conclusion

In this paper, a compact yet high-speed architecture for Rijndael was proposed and evaluated through ASIC implementations. In order to minimize the hardware size, the order of the arithmetic functions was changed, and encryption and decryption data paths were efficiently combined. Logic optimization techniques such as factoring were applied to the arithmetic components, and gate counts were greatly reduced.

Our architecture provides high flexibility from a compact 32-bit bus implementation to a high-speed implementation using a 128-bit bus. The S-Box has been implemented as look-up table logic in the previous work, and has required extensive hardware resources. We introduced a new composite field $GF(((2^2)^2)^2)$ and proposed an optimization method for the S-Box. Our S-Box requires less

Table 4. Performance comparison in ASIC implementations. (worst case)

Cycles /Round	S-Box (bytes)	Area (gates)	Area (mm^2)	Max. Freq. (MHz)	Through -put (Mbps)	Throughput /Area (Kbps/gate)	Notes
Ours 0.11μm							
5	4	5,398	0.052	131.24	311.09	57.63	Total 54 cycles
		10,338	0.099	222.22	526.74	50.95	
4	8	6,292	0.060	137.55	400.15	63.60	Total 44 cycles
		10,990	0.106	219.30	637.96	58.05	
3	8	7,998	0.077	137.17	548.68	68.60	Total 32 cycles
		14,777	0.142	218.82	875.28	59.23	
2	12	8,836	0.085	137.17	798.08	90.32	Total 22 cycles
		17,016	0.163	217.86	1,267.55	74.49	
1	20	12,454	0.130	145.35	1,691.35	135.81	Total 11 cycles
		21,337	0.205	224.22	2,609.11	122.28	
[10] 0.18μm							
1	48	184,000	4.23	48	435 (256-bit key)	2.51	256-bit data and key supported Decryption not supported
[9] 0.35 μm							
1	40	33,850	-	-	509.70	15.06	
[8] 0.35 μm							
1/11	400	612,834	-	15.23	1,950.03	3.18	11 round unrolled
[18] 0.5 μm							
1	40	68,872	20.74	21.18	271.13	3.94	4 transistors/gate is assumed
1	40	160,421	33.85	47.36	605.77	3.78	

than 1/4 the size of one using a look-up table, and also showed 20% better performance in comparison with the one using a $GF((2^4)^2)$ field.

Our smallest implementation using 0.11-μm CMOS technology is 5.4 Kgates, which is less than 1/6 the size of the best hardware of previous work. Making the best use of the parallel processing allowed by Rijndael, a high-speed version obtained the best performance of 2.6 Gbps with 21.3 Kgates. Thus, our Rijndael hardware can be applied to various targets from mobile equipment to high-end security servers.

Our continuing research is to develop and evaluate even faster hardware for 10 Gbps class high-speed communication links and beyond.

Acknowledgements. We are grateful to Mr. G. Zang for supporting us generously with his CAD tool expertise, and also would like to thank Mr. A. Rudra and Dr. N. Ohba for their helpful comments on this work.

References

[1] ANSI (American National Standards Institute). *Triple Data Encryption Algorithm Modes of Operation*, 1998.

[2] J. Daemen and V. Rijmen. AES Proposal: Rijndael. NIST AES Proposal, June 1998. Available at http://csrc.nist.gov/encryption/aes /rijndael/Rijndael.pdf.

[3] A. J. Elbirt, W. Yip, B. Chetwynd, and C. Paar. An FPGA Implementation and Performance Evaluation of the AES Block Cipher Candidate Algorithm Finalists. In *The Third Advanced Encryption Standard Candidate Conference*, pages 13–27. NIST, April 2000. Available at http://csrc.nist.gov/encryption/aes/round2/conf3/papers/08-aelbirt.pdf.

[4] J.L. Fan and C. Paar. On Efficient Inversion in Tower Fields of Characteristic Two. In *International Symposium on Information Theory*, page 20. IEEE, June 1997.

[5] V. Fischer and M. Drutarovsky. Two Methods of Rijndael Implementation in Reconfigurable Hardware. In *Workshop on Cryptographic Hardware and Embedded Systems (CHES2001)*, pages 81–96, May 2001.

[6] K. Gaj and P. Chodowiec. Comparison of the Hardware Prformance of the AES Candidates using Reconfigurable Hardware. In *The Third Advanced Encryption Standard Candidate Conference*, pages 40–56. NIST, April 2000. Available at http://csrc.nist.gov/encryption/aes/round2/conf3/papers/22-kgaj.pdf.

[7] J. Guajardo and C. Paar. Efficient Algorithms for Elliptic Curve Cryptosystems. In Jr. Burton S. Kaliski, editor, *Advances in Cryptology—CRYPTO '97*, volume 1294 of *Lecture Notes in Computer Science*, pages 342–356. Springer-Verlag, August 1997.

[8] T. Ichikawa, T. Kasuya, and M. Matsui. Hardware Evaluation of the AES Finalists. In *The Third Advanced Encryption Standard Candidate Conference*, pages 279–285. NIST, April 2000. Available at http://csrc.nist.gov/encryption/aes/round2/conf3/papers/15-tichikawa.pdf.

[9] T. Ichikawa, T. Tokita, and M. Matsui. On Hardware Implementation of 128-bit Block Ciphers (III). In *2001 Symposium on Cryptography and Information Security (SCIS 2001)*, pages 669–674, January 2001. (Japanese).

[10] H. Kuo and I. Verbauwhede. Architectural Optimization for a 1.82 Gbits/sec VLSI Implementation of the AES Rijndael Algorithm. In *Workshop on Cryptographic Hardware and Embedded Systems (CHES2001)*, pages 53–67, May 2001.

[11] M. McLoone and J.V. McCanny. High performance Single-chip FPGA Rijndael Algorithm Implementations. In *Workshop on Cryptographic Hardware and Embedded Systems (CHES2001)*, pages 68–80, May 2001.

[12] S. Morioka and Y. Katayama. Design Methodology for a One-Shot Reed-Solomon Encoder and Decoder. In *International Conference on Computer Design (ICCD '99)*, pages 60–67. IEEE, October 1999.

[13] National Institute of Standards and Technology (U.S.). Advanced Encryption Standard (AES). Available at http://csrc.nist.gov/publications/drafts/dfips-AES.pdf.

[14] National Institute of Standards and Technology (U.S.). Data Encryption Standard (DES). FIPS Publication 46-3, NIST, 1999. Available at http://csrc.nist.gov/publications/fips/fips46-3/fips46-3.pdf.

[15] C. Paar. A New Architecture for a Parallel Finite Field Multiplier with Low Complexity Based on Composite Fields. *IEEE Transactions on Computers*, 45(7):856–861, July 1996.

[16] A. Rudra, P.K. Dubey, C.S. Jutla, V. Kumar, J.R. Rao, and P. Rohatgi. Efficient Rijndael Encryption Implementation with Composite Field Arithmetic. In *Workshop on Cryptographic Hardware and Embedded Systems (CHES2001)*, pages 175–188, May 2001.

[17] N. Weaver and J. Wawrzynek. A Comparison of the AES Candidates Amenability to FPGA Implementation. In *The Third Advanced Encryption Standard Candidate Conference*, pages 28–39. NIST, April 2000. Available at http://csrc.nist.gov/encryption/aes/round2/conf3/papers/13-nweaver.pdf.

[18] B. Weeks, M. Bean, T. Rozylowicz, and C. Ficke. Hardware Performance Simulation of Round 2 Advanced Encryption Standard Algorithm. Available at http://csrc.nist.gov /encryption/aes/round2/NSA-AESfinalreport.pdf.

Provable Security of KASUMI and 3GPP Encryption Mode $f8$

Ju-Sung Kang[1], Sang-Uk Shin[1], Dowon Hong[1], and Okyeon Yi[2]

[1] Section 0741, Information Security Technology Division, ETRI
161 Kajong-Dong, Yusong-Gu, Taejon, 305-350, KOREA
{jskang,shinsu,dwhong}@etri.re.kr

[2] Department of Mathematics, Kookmin University
Jeongreung3-Dong, Seongbuk-Gu, Seoul, 136-702, KOREA
oyyi@kmu.kookmin.ac.kr

Abstract. Within the security architecture of the 3GPP system there is a standardised encryption mode $f8$ based on the block cipher KASUMI. In this work we examine the pseudorandomness of the block cipher KASUMI and the provable security of $f8$. First we show that the three round KASUMI is not a pseudorandom permutation ensemble but the four round KASUMI is a pseudorandom permutation ensemble under the adaptive distinguisher model by investigating the properties of the round functions in a clear way. Second we provide the upper bound on the security of $f8$ mode under the reasonable assumption from the first result by means of the left-or-right security notion.

1 Introduction

There is a standardised encryption algorithm $f8$ within the security architecture of the 3GPP(3rd Generation Partnership Project) system and this algorithm is based on the block cipher KASUMI that produces a 64-bit output from a 64-bit input under the control of an 128-bit key[12]. To guarantee the message confidentiality over a radio access link of W-CDMA IMT-2000, $f8$ encryption mode with KASUMI has been proposed. The purpose of this work is to investigate the pseudorandomness of the block cipher KASUMI and the provable security of $f8$.

A block cipher can be regarded as a family of permutations on a message space indexed by a secret key. Luby-Rackoff[7] introduced a theoretical model for the security of block ciphers by using the notion of pseudorandom and super-pseudorandom permutations. A pseudorandom permutation can be interpreted as a block cipher that no attacker with polynomially many encryption queries can distinguish between the block cipher and the perfect random permutation. In [7], Luby and Rackoff used the DES-type transformation in order to construct a pseudorandom permutation from a pseudorandom function. They showed that the DES-type transformation with three rounds yielded $2n$-bit pseudorandom permutation under the assumption that each round function was an n-bit pseudorandom function. Sakurai-Zheng[11] showed that the three round MISTY-type transformation was not a pseudorandom permutation ensemble. MISTY-type

C. Boyd (Ed.): ASIACRYPT 2001, LNCS 2248, pp. 255–271, 2001.
© Springer-Verlag Berlin Heidelberg 2001

transformation[8,9] was another two-block structure different from DES-type. Recently, Gilbert-Minier[4] and Kang et al.[6] showed independently that the four round MISTY-type transformation was a pseudorandom permutation.

The overall structure of KASUMI is the DES-type, but its round function FO composed of three round MISTY-type transformation which is not a pseudorandom function. Thus we cannot straightforwardly apply the Luby-Rackoff's result to KASUMI. FO function within KASUMI has FI function as its component function which is composed of four round unbalanced MISTY-type transformation. We show that this is a pseudorandom permutation. And we prove that the three round KASUMI is not a pseudorandom permutation but the four round KASUMI is a pseudorandom permutation. In [6], the authors investigated the pseudorandomness of KASUMI for non-adaptive distinguishers. In this paper we consider the security model for adaptive distinguishers similar to the approach of Naor and Reingold[10] and investigate the properties of the round function of KASUMI more precisely than the previous results like [4] and [6].

On the other hand $f8$ is one of the modes of operation for block ciphers. Several modes of operation for block ciphers have been proposed to encrypt plaintext blocks more than one block and to fulfil varying application requirements. As standardized modes of operation, ECB(electronic codebook), CBC(cipher block chaining), CFB(cipher feedback) and OFB(output feedback) are known[3]. 3GPP $f8$ encryption mode can be seen as a variant of OFB mode.

Proving the security of modes of operation started by Bellare et al.[1] in 1994 who analyzed the security of CBC MAC mode. In 1997, Bellare et al.[2] introduced the security notions of the symmetric encryption scheme and proved the security of CTR mode and CBC mode. Recently, Alkassar et al.[13] analyzed the security of CFB mode and proposed the OCFB mode which improved the performance of CFB mode. In this paper we show that 3GPP $f8$ encryption mode is secure by means of the left-or-right security notion. To prove this fact we should have the assumption that the underlying block cipher KASUMI is secure. This assumption is reasonable since by the first our result we already obtain that KASUMI is a pseudorandom permutation ensemble.

2 Pseudorandomness of the Block Cipher KASUMI

2.1 Preliminaries

Let I_n denote the set of all n-bit strings and $\mathcal{P}_n$ be the set of all permutations from I_n to itself where n is a positive integer. That is, $\mathcal{P}_n = \{\pi : I_n \to I_n \mid \pi \text{ is a bijection}\}$. We define an n-bit perfect random permutation as an uniformly drawn element of $\mathcal{P}_n$.

Definition 1. *$\mathcal{P}_n$ is called the UPE(uniform permutation ensemble) if all permutations in $\mathcal{P}_n$ are uniformly distributed. That is, for any permutation $\pi \in \mathcal{P}_n$, $Pr(\pi) = \frac{1}{2^n!}$.*

We consider the following security model. Let $\mathcal{D}$ be a computationally unbounded distinguisher with an oracle $\mathcal{O}$. The oracle $\mathcal{O}$ chooses randomly a permutation π from the UPE $\mathcal{P}_n$ or from a permutation ensemble $\Lambda_n \subset \mathcal{P}_n$. For an n-bit block cipher, Λ_n is the set of permutations determined by all the secret keys. The purpose of the distinguisher $\mathcal{D}$ is to distinguish whether the oracle $\mathcal{O}$ implements the UPE $\mathcal{P}_n$ or Λ_n.

Definition 2. *Let $\mathcal{D}$ be a distinguisher, $\mathcal{P}_n$ be the UPE, and Λ_n be a permutation ensemble obtained from a block cipher. Then the advantage $ADV_{\mathcal{D}}$ of $\mathcal{D}$ is defined by*

$$ADV_{\mathcal{D}} = |Pr(\mathcal{D} \text{ outputs } 1 \mid \mathcal{O} \leftarrow \mathcal{P}_n) - Pr(\mathcal{D} \text{ outputs } 1 \mid \mathcal{O} \leftarrow \Lambda_n)| \ ,$$

where $\mathcal{O} \leftarrow \mathcal{P}_n$ and $\mathcal{O} \leftarrow \Lambda_n$ denote that $\mathcal{O}$ implements $\mathcal{P}_n$ and Λ_n, respectively.

Assume that the distinguisher $\mathcal{D}$ is restricted to make at most $poly(n)$ queries to the oracle $\mathcal{O}$, where $poly(n)$ is some polynomial in n. We call $\mathcal{D}$ is a pseudorandom distinguisher if it queries x and the oracle answers $y = \pi(x)$, where π is a randomly chosen permutation by $\mathcal{O}$. We say that $\mathcal{D}$ is a super-pseudorandom distinguisher if it is a pseudorandom distinguisher and also makes a query y and receives $x = \pi^{-1}(y)$ from the oracle $\mathcal{O}$.

Definition 3. *A function $h : \mathbb{N} \to \mathbb{R}$ is called negligible if for any constant $c > 0$ and all sufficiently large $n \in \mathbb{N}$, $h(n) < \frac{1}{n^c}$.*

Definition 4. *Let Λ_n be an efficiently computable permutation ensemble. Then Λ_n is called a PPE(pseudorandom permutation ensemble) if $ADV_{\mathcal{D}}$ is negligible for any pseudorandom distinguisher $\mathcal{D}$.*

Definition 5. *Let Λ_n be an efficiently computable permutation ensemble. Then we call Λ_n is a SPPE(super-pseudorandom permutation ensemble) if $ADV_{\mathcal{D}}$ is negligible for any super-pseudorandom distinguisher $\mathcal{D}$.*

In Definition 4 and 5, a permutation ensemble is efficiently computable if all permutations in the ensemble can be computed efficiently. See [10] for the rigorous definition of this. It is reasonable assumption that Λ_n is an efficiently computable permutation ensemble if it is obtained from an n-bit block cipher. Hence we assume that any permutation ensemble obtained from a block cipher is efficiently computable.

We define two transformations, DES-type and MISTY-type, which are obtained from two representative structures of current block ciphers. Let $\mathcal{F}_n$ denote the set of all functions from I_n to itself. We call briefly f is an n-bit function(resp. permutation) where $f \in \mathcal{F}_n$(resp. $f \in \mathcal{P}_n$).

Definition 6. *For any n-bit function $f \in \mathcal{F}_n$, $2n$-bit DES-type permutation $\mathbf{D}_f \in \mathcal{P}_{2n}$ is defined by $\mathbf{D}_f(L, R) = (R, L \oplus f(R))$, where $L, R \in I_n$.*

Definition 7. *For any n-bit permutation $f \in \mathcal{P}_n$, $2n$-bit MISTY-type permutation $\mathbf{M}_f \in \mathcal{P}_{2n}$ is defined by $\mathbf{M}_f(L, R) = (R, f(L) \oplus R)$, where $L, R \in I_n$.*

Several noticeable results about the pseudorandomness of DES-type and MISTY-type transformations are as follows. It is aware that PFE(pseudorandom function ensemble) can be similarly defined as Definition 4 by considering function space instead of permutation space.

- $\mathbf{D}_{f_2} \circ \mathbf{D}_{f_1}$ is not a $2n$-bit PPE and $\mathbf{D}_{f_3} \circ \mathbf{D}_{f_2} \circ \mathbf{D}_{f_1}$ is not a $2n$-bit SPPE, although all f_i's$(i = 1, 2, 3)$ are independently chosen from an n-bit PFE[7].
- $\mathbf{D}_{f_3} \circ \mathbf{D}_{f_2} \circ \mathbf{D}_{f_1}$ is a $2n$-bit PPE and $\mathbf{D}_{f_4} \circ \mathbf{D}_{f_3} \circ \mathbf{D}_{f_2} \circ \mathbf{D}_{f_1}$ is a $2n$-bit SPPE if all f_i's$(i = 1, 2, 3, 4)$ are independently chosen from an n-bit PFE[7].
- $\mathbf{M}_{f_3} \circ \mathbf{M}_{f_2} \circ \mathbf{M}_{f_1}$ is not a $2n$-bit PPE and $\mathbf{M}_{f_4} \circ \mathbf{M}_{f_3} \circ \mathbf{M}_{f_2} \circ \mathbf{M}_{f_1}$ is not a $2n$-bit SPPE, although each $f_i(i = 1, 2, 3, 4)$ is chosen independently from an n-bit PPE[4,11].
- $\mathbf{M}_{f_4} \circ \mathbf{M}_{f_3} \circ \mathbf{M}_{f_2} \circ \mathbf{M}_{f_1}$ is a $2n$-bit PPE and $\mathbf{M}_{f_5} \circ \mathbf{M}_{f_4} \circ \mathbf{M}_{f_3} \circ \mathbf{M}_{f_2} \circ \mathbf{M}_{f_1}$ is a $2n$-bit SPPE, where all f_i's$(i = 1, 2, 3, 4, 5)$ are independently chosen from an n-bit PPE[4,5,6].

On the other hand KASUMI is a modified version of the block cipher MISTY1[9] and we can classify the permutation of KASUMI into the following three stages:

- The overall permutation of KASUMI is a 64-bit permutation composed of the eight round DES-type permutation with the two round permutation FO and FL.
- FO function is a 32-bit permutation composed of the three round MISTY-type transformation with the round permutation FI.
- FI function is a 16-bit permutation which is composed of the four round unbalanced MISTY-type transformation obtained from 7-bit S-box $S7$ and 9-bit S-box $S9$.

First we show that FI function is a 16-bit PPE by examining the pseudorandomness of unbalanced MISTY-type transformation. Second we prove that three round KASUMI is not a 64-bit PPE but four round KASUMI is a 64-bit PPE on the base of the first result. Note that FO function is not a 32-bit PPE, so it doesn't seem that the three round DES-type permutation of KASUMI is a 64-bit PPE as the Luby-Rackoff cipher. Since the FL function is to round key mixing, we can omit FL function in order to analyze the pseudoranomness of KASUMI.

2.2 Pseudorandomness of the Unbalanced MISTY-Type Transformation

We describe simple but useful two lemmas which their proofs are given in [6].

Lemma 1. *Let π be a permutation chosen from the UPE $\mathcal{P}_n$. Then for any $x_1 \neq x_2, y \in I_n$,*

$$Pr(\pi(x_1) \oplus \pi(x_2) = y) = \begin{cases} \frac{1}{2^n - 1} & if\, y \neq 0\,, \\ 0 & otherwise. \end{cases}$$

Lemma 2. *Let π_1 and π_2 be two permutations independently chosen from the UPE $\mathcal{P}_n$. Then for any $a, b, c, d, y \in I_n$,*

$$Pr\left(\pi_1(a) \oplus \pi_1(b) \oplus \pi_2(c) \oplus \pi_2(d) = y\right) < \frac{1}{2^{n-1}}\,,\ for\ n \geq 2.$$

Now we define two unbalanced MISTY-type transformations to examine accurately the pseudorandomness of FI function.

Definition 8. *Let n and m be two positive integers such that $m \leq n$. Then for any n-bit permutation f and m-bit permutation g, two $(n+m)$-bit unbalanced MISTY-type transformations $\overline{\mathbf{M}}_f \in \mathcal{P}_{n+m}$ and $\widehat{\mathbf{M}}_g \in \mathcal{P}_{n+m}$ are defined by*

$$\overline{\mathbf{M}}_f(L,R) = (R, f(L) \oplus \overline{R}) \in I_m \times I_n\ , \quad \forall (L,R) \in I_n \times I_m$$

and

$$\widehat{\mathbf{M}}_g(L,R) = (R, g(L) \oplus \widehat{R}) \in I_n \times I_m\ , \quad \forall (L,R) \in I_m \times I_n\ ,$$

where for any n-bit vector x, $\widehat{x}$ denotes the m-bit value obtained by discarding the $n-m$ most-significant end and for any m-bit vector y, $\overline{y}$ denotes the n-bit value obtained by adding $n-m$ zero bits to the most-significant end.

Note that the FI function of KASUMI can be represented as 16-bit permutation $\widehat{\mathbf{M}}_{f_4} \circ \overline{\mathbf{M}}_{f_3} \circ \widehat{\mathbf{M}}_{f_2} \circ \overline{\mathbf{M}}_{f_1}$, where f_1, f_3 are 9-bit permutations and f_2, f_4 are 7-bit permutations. The pseudorandomness of the FI function is guaranteed by the following theorem.

Theorem 1. *Let for any positive integer n and m such that $m \leq n$, $f_1, f_3 \in \mathcal{P}_n$ and $f_2, f_4 \in \mathcal{P}_m$ be independently chosen from two n-bit and m-bit PPEs, respectively. Then the four round unbalanced MISTY-type transformation $\widehat{\mathbf{M}}_{f_4} \circ \overline{\mathbf{M}}_{f_3} \circ \widehat{\mathbf{M}}_{f_2} \circ \overline{\mathbf{M}}_{f_1}$ is an $(n+m)$-bit PPE.*

Recall that a pseudorandom distinguisher $\mathcal{D}$ can make query x and the oracle $\mathcal{O}$ answers $y = \pi(x)$, where π is a randomly chosen permutation by $\mathcal{O}$. Now we assume that $\mathcal{D}$ makes exactly q queries and refer to the sequence $\{(x^{(1)}, y^{(1)}), \cdots, (x^{(q)}, y^{(q)})\}$ of all query-answer pairs as the $\mathcal{D}$-transcript, where $q = poly(n)$. We consider an adaptive pseudorandom distinguisher as the following definition.

Definition 9. *$\mathcal{D}$ is called an adaptive pseudorandom distinguisher if it has a transcript $\{(x^{(1)}, y^{(1)}), \cdots, (x^{(q)}, y^{(q)})\}$ and a function $C_\mathcal{D}$ of $\mathcal{D}$-transcript such that for every $2 \leq i \leq q$,*

$$x^{(i)} = C_\mathcal{D}(\{(x^{(1)}, y^{(1)}), \cdots, (x^{(i-1)}, y^{(i-1)})\})$$

and

$$\textit{the ouput of } \mathcal{D} = C_\mathcal{D}(\{(x^{(1)}, y^{(1)}), \cdots, (x^{(q)}, y^{(q)})\})\ .$$

Under the adaptive distinguisher model, for any i-th query of $\mathcal{D}$ is fully determined by the first $i-1$ query-answer pairs and $\mathcal{D}$'s output is a function of its transcript. Throughout this paper we assume that all queries are distinct.

To prove the Theorem 1, we formally define a bad event and estimate its probability.

Definition 10. *For any n-bit permutation f_1 and m-bit permutation f_2, $BAD(f_1, f_2)$ is defined as the set of all $\mathcal{D}$-transcripts $\sigma = \{(x^{(1)}, y^{(1)}), \cdots, (x^{(q)}, y^{(q)})\}$ satisfying: $\exists 1 \le i < j \le q$ such that*

$$f_1(x_L^{(i)}) \oplus \overline{x_R^{(i)}} = f_1(x_L^{(j)}) \oplus \overline{x_R^{(j)}}$$

or

$$f_2(x_R^{(i)}) \oplus \widehat{f_1(x_L^{(i)})} \oplus x_R^{(i)} = f_2(x_R^{(j)}) \oplus \widehat{f_1(x_L^{(j)})} \oplus x_R^{(j)} ,$$

where $x^{(i)} = (x_L^{(i)}, x_R^{(i)}) \in I_n \times I_m$ for all $1 \le i \le q$.

Lemma 3. *Let f_1 and f_2 be chosen independently from UPE $\mathcal{P}_n$ and UPE $\mathcal{P}_m$, respectively. Then for any $\mathcal{D}$-transcript $\sigma = \{(x^{(1)}, y^{(1)}), \cdots, (x^{(q)}, y^{(q)})\}$ and $n \ge m \ge 2$,*

$$Pr(\sigma \in BAD(f_1, f_2)) < (q^2 - q)\left(\frac{1}{2^n} + \frac{1}{2^m}\right) .$$

Proof. By definition, $\sigma \in BAD(f_1, f_2)$ if there exist $1 \le i < j \le q$ such that either

$$f_1(x_L^{(i)}) \oplus \overline{x_R^{(i)}} = f_1(x_L^{(j)}) \oplus \overline{x_R^{(j)}}$$

or

$$f_2(x_R^{(i)}) \oplus \widehat{f_1(x_L^{(i)})} \oplus x_R^{(i)} = f_2(x_R^{(j)}) \oplus \widehat{f_1(x_L^{(j)})} \oplus x_R^{(j)} .$$

For any given i and j, we estimate probabilities of these two events. We have the following three cases.
Case 1: $x_L^{(i)} \ne x_L^{(j)}$ and $x_R^{(i)} = x_R^{(j)}$. Since f_1 is a permutation,

$$Pr\left(f_1(x_L^{(i)}) \oplus \overline{x_R^{(i)}} = f_1(x_L^{(j)}) \oplus \overline{x_R^{(j)}}\right) = Pr\left(f_1(x_L^{(i)}) = f_1(x_L^{(j)})\right) = 0 .$$

Observe that, by the similar result to Lemma 1

$$Pr\left(f_2(x_R^{(i)}) \oplus \widehat{f_1(x_L^{(i)})} \oplus x_R^{(i)} = f_2(x_R^{(j)}) \oplus \widehat{f_1(x_L^{(j)})} \oplus x_R^{(j)}\right)$$
$$= Pr\left(\widehat{f_1(x_L^{(i)})} = \widehat{f_1(x_L^{(j)})}\right) = 2^n \cdot \frac{2^{n-m} \cdot (2^n - 2)!}{2^n!} = \frac{2^{n-m}}{2^n - 1} .$$

Case 2: $x_L^{(i)} = x_L^{(j)}$ and $x_R^{(i)} \ne x_R^{(j)}$. In this case the probability of the first event is equal to $Pr(\overline{x_R^{(i)}} = \overline{x_R^{(j)}}) = 0$. By Lemma 1, the probability of the second event is estimated as

$$Pr\left(f_2(x_R^{(i)}) \oplus f_2(x_R^{(j)}) = x_R^{(i)} \oplus x_R^{(j)}\right) = \frac{1}{2^m - 1} .$$

Case 3: $x_L^{(i)} \ne x_L^{(j)}$ and $x_R^{(i)} \ne x_R^{(j)}$. By Lemma 1, the probability of the first event is estimated as

$$Pr\left(f_1(x_L^{(i)}) \oplus f_1(x_L^{(j)}) = \overline{x_L^{(i)}} \oplus \overline{x_L^{(j)}}\right) = \frac{1}{2^n - 1} .$$

Similarly, by Lemma 2, the probability of the second event is also estimated as

$$Pr\left(\widehat{f_1(x_L^{(i)})} \oplus \widehat{f_1(x_L^{(j)})} \oplus f_2(x_R^{(i)}) \oplus f_2(x_R^{(j)}) = x_R^{(i)} \oplus x_R^{(j)}\right) < \frac{1}{2^m - 1} ,$$

since $n \geq m \geq 2$.

Hence, for any case, we obtain that

$$Pr\left(f_1(x_L^{(i)}) \oplus \overline{x_R^{(i)}} = f_1(x_L^{(j)}) \oplus \overline{x_R^{(j)}}\right) < \frac{1}{2^{n-1}}$$

and

$$Pr\left(f_2(x_R^{(i)}) \oplus \widehat{f_1(x_L^{(i)})} \oplus x_R^{(i)} = f_2(x_R^{(j)}) \oplus \widehat{f_1(x_L^{(j)})} \oplus x_R^{(j)}\right) < \frac{1}{2^{m-1}} .$$

Therefore

$$Pr\left(\sigma \in BAD(f_1, f_2)\right) < \binom{q}{2}\left(\frac{1}{2^{n-1}} + \frac{1}{2^{m-1}}\right) < (q^2 - q)\left(\frac{1}{2^n} + \frac{1}{2^m}\right) . \quad \square$$

Definition 11. *Let Λ_{n+m} be the $(n+m)$-bit permutation ensemble obtained from $\Lambda_{n+m}(f_1, f_2, f_3, f_4) = \widehat{\mathbf{M}}_{f_4} \circ \overline{\mathbf{M}}_{f_3} \circ \widehat{\mathbf{M}}_{f_2} \circ \overline{\mathbf{M}}_{f_1}$. Then $T_{\mathcal{P}_{n+m}}$ and $T_{\Lambda_{n+m}}$ are defined by the random variables such that $T_{\mathcal{P}_{n+m}}$ is the $\mathcal{D}$-transcript when the oracle $\mathcal{O}$ implements the UPE $\mathcal{P}_{n+m}$ and $T_{\Lambda_{n+m}}$ is the $\mathcal{D}$-transcript when the oracle $\mathcal{O}$ implements the permutation ensemble Λ_{n+m}.*

Lemma 4. *Let Λ_{n+m} be the $(n+m)$-bit permutation ensemble of all $\Lambda_{n+m}(f_1, f_2, f_3, f_4)$ such that $f_1, f_3 \in \mathcal{P}_n$ and $f_2, f_4 \in \mathcal{P}_m$ are independently chosen from the n-bit and m-bit UPEs, respectively. Then for any $\mathcal{D}$-transcript $\sigma = \{(x^{(1)}, y^{(1)}), \cdots, (x^{(q)}, y^{(q)})\}$,*

$$\left|Pr\left(T_{\Lambda_{n+m}} = \sigma \mid \sigma \notin BAD(f_1, f_2)\right) - Pr\left(T_{\mathcal{P}_{n+m}} = \sigma\right)\right| < \varepsilon_{n,m,q} ,$$

where

$$\varepsilon_{n,m,q} = \frac{1}{2^{n+m}(2^n - 1)(2^m - 1) \cdots (2^n - q + 1)(2^m - q + 1)} .$$

Proof. For any possible $\mathcal{D}$-transcript we have that

$$Pr\left(T_{\mathcal{P}_{n+m}} = \sigma\right) = \frac{(2^{n+m} - q)!}{2^{n+m}!} .$$

Consider any specific n-bit permutation f_1 and m-bit permutation f_2 such that $\sigma \notin BAD(f_1, f_2)$. Note that $T_{\Lambda_{n+m}} = \sigma$ if and only if for all $1 \leq i \leq q$, $y^{(i)} = \Lambda_{n+m}(x^{(i)})$. Since $\Lambda_{n+m} = \widehat{\mathbf{M}}_{f_4} \circ \overline{\mathbf{M}}_{f_3} \circ \widehat{\mathbf{M}}_{f_2} \circ \overline{\mathbf{M}}_{f_1}$,

$$y^{(i)} = \Lambda_{n+m}(x^{(i)}) \Leftrightarrow f_3(L_2^{(i)}) = y_L^{(i)} \oplus \overline{R_2^{(i)}} \in I_n \text{ and } f_4(R_2^{(i)}) = \widehat{y_L^{(i)}} \oplus y_R^{(i)} \in I_m ,$$

where $(L_2^{(i)}, R_2^{(i)}) = \widehat{\mathbf{M}}_{f_2} \circ \overline{\mathbf{M}}_{f_1}(x_L^{(i)}, x_R^{(i)})$. By definition of $BAD(f_1, f_2)$, if $\sigma \notin BAD(f_1, f_2)$, then $L_2^{(i)} \neq L_2^{(j)}$ and $R_2^{(i)} \neq R_2^{(j)}$ for all $1 \leq i \neq j \leq q$. Therefore, since f_3 and f_4 are independently chosen from the UPEs $\mathcal{P}_n$ and $\mathcal{P}_m$, respectively, we obtain that

$$Pr\left(T_{\Lambda_{n+m}} = \sigma \mid \sigma \notin BAD(f_1, f_2)\right) = \frac{(2^n - q)!}{2^n!} \cdot \frac{(2^m - q)!}{2^m!} ,$$

which complete the assertion. □

<u>*Proof of Theorem 1*</u>: It suffices to show the assertion under the assumption that $f_1, f_3 \in \mathcal{P}_n$ and $f_2, f_4 \in \mathcal{P}_m$ be independently chosen from two n-bit and m-bit UPEs, respectively. Let Λ_{n+m} be the $(n+m)$-bit permutation ensemble of all $\Lambda_{n+m}(f_1, f_2, f_3, f_4) = \overline{\mathbf{M}}_{f_4} \circ \widehat{\mathbf{M}}_{f_3} \circ \overline{\mathbf{M}}_{f_2} \circ \widehat{\mathbf{M}}_{f_1}$ and Θ be the set of all $\mathcal{D}$-transcripts σ such that the output of $\mathcal{D}$ is $C_{\mathcal{D}}(\sigma) = 1$. Then

$$\begin{aligned}
& ADV_{\mathcal{D}} \\
&= \left|Pr\left(C_{\mathcal{D}}(T_{\Lambda_{n+m}}) = 1\right) - Pr\left(C_{\mathcal{D}}(T_{\mathcal{P}_{n+m}}) = 1\right)\right| \\
&\leq \sum_{\sigma\in\Theta} Pr(\sigma \notin BAD(f_1, f_2)) \\
&\qquad \cdot \left|Pr\left(T_{\Lambda_{n+m}} = \sigma \mid \sigma \notin BAD(f_1, f_2)\right) - Pr\left(T_{\mathcal{P}_{n+m}} = \sigma\right)\right| && (1) \\
&\quad + \sum_{\sigma\in\Theta} Pr\left(T_{\Lambda_{n+m}} = \sigma, \sigma \in BAD(f_1, f_2)\right) && (2) \\
&\quad + \sum_{\sigma\in\Theta} Pr\left(\sigma \in BAD(f_1, f_2)\right) \cdot Pr(T_{\mathcal{P}_{n+m}} = \sigma). && (3)
\end{aligned}$$

By Lemma 4, the term (1) is bounded above by $\varepsilon_{n,m,q}$ and by Lemma 3, the value of (3) is bounded by

$$\max_{\sigma\in\Theta} Pr(\sigma \in BAD(f_1, f_2)) \cdot Pr\left(\cup_{\sigma\in\Theta}\{T_{\mathcal{P}_{n+m}} = \sigma\}\right) < (q^2 - q)\left(\frac{1}{2^n} + \frac{1}{2^m}\right) .$$

On the other hand, by Lemma 3, the value of (2) is estimated as

$$\begin{aligned}
& \sum_{\sigma\in\Theta} Pr\left(T_{\Lambda_{n+m}} = \sigma, \sigma \in BAD(f_1, f_2)\right) \\
&= \sum_{\sigma\in\Theta} Pr(T_{\Lambda_{n+m}} = \sigma) \cdot Pr\left(\sigma \in BAD(f_1, f_2) \mid T_{\Lambda_{n+m}} = \sigma\right) \\
&< (q^2 - q)\left(\frac{1}{2^n} + \frac{1}{2^m}\right) .
\end{aligned}$$

Therefore we can conclude that

$$ADV_{\mathcal{D}} < 2(q^2 - q)\left(\frac{1}{2^n} + \frac{1}{2^m}\right) + \varepsilon_{n,m,q} ,$$

which is negligible. □

2.3 Pseudorandomness of KASUMI

From Theorem 1, it becomes a reasonable assumption that FI function of KASUMI is a PPE. In order to investigate the pseudorandomness of KASUMI, we use a simplified figure of KASUMI. The four round simplified KASUMI is illustrated in Figure 1, where $x = (x_1, x_2, x_3, x_4)$ denotes a $4n$-bit input value, $w = (w_1, w_2, w_3, w_4)$, $y = (y_1, y_2, y_3, y_4)$, and $z = (z_1, z_2, z_3, z_4)$ denote corresponding outputs of the two, three, and four round KASUMI, respectively. Each of x_i, w_i, y_i, and z_i is an n-bit value. By the following theorem, we obtain the fact that three round of KASUMI is insufficient to be a PPE.

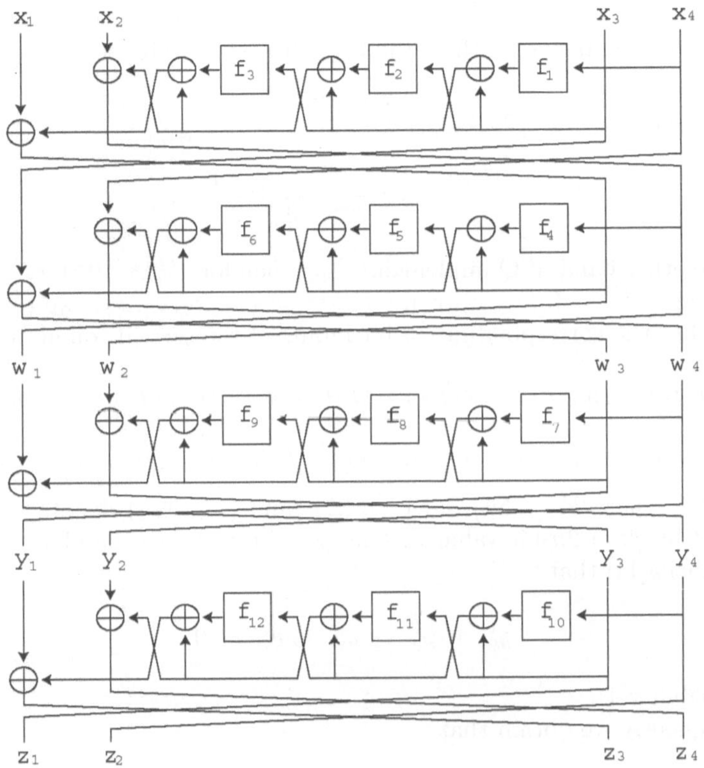

Fig. 1. Simplified four round KASUMI

Theorem 2. *The three round simplified KASUMI is not a $4n$-bit PPE though f_i's$(i = 1, \cdots, 9)$ of Figure 1 are independently chosen from an n-bit PPE.*

Proof. Let Λ_{4n} be the set of all permutations over I_{4n} obtained from the three round simplified KASUMI. Consider a distinguisher $\mathcal{D}$ such as follows:

1. $\mathcal{D}$ chooses four $4n$-bit queries $x^{(1)}$, $x^{(2)}$, $x^{(3)}$, and $x^{(4)}$ such that

$$x^{(1)} = (0, 0, x_3, x_4)\ ,\ \ x^{(2)} = (x_1, 0, x_3, x_4)\ ,$$

$$x^{(3)} = (0, x_2, x_3, x_4)\ ,\ \ x^{(4)} = (x_1, x_2, x_3, x_4)\ ,$$

   where $x_1 \neq 0 \neq x_2$ and x_3, x_4 are fixed n-bit values.
2. $\mathcal{D}$ sends these four queries to the oracle $\mathcal{O}$ and receives the corresponding answers $(y_1^{(i)}, y_2^{(i)}, y_3^{(i)}, y_4^{(i)})(i = 1, 2, 3, 4)$ from the oracle.
3. $\mathcal{D}$ outputs 1 if and only if

$$y_2^{(1)} \oplus y_2^{(2)} \oplus y_2^{(3)} \oplus y_2^{(4)} = 0\ .$$

If the oracle implements the UPE $\mathcal{P}_{4n}$, then we obtain that

$$\begin{aligned} Pr(\mathcal{D} \text{ outputs } 1 \mid \mathcal{O} \leftarrow \mathcal{P}_{4n}) &\leq \frac{2^{4n}(2^{4n}-1)(2^{4n}-2)2^{3n}(2^{4n}-4)!}{2^{4n}!} \\ &= \frac{2^{3n}}{2^{4n}-3} \leq \frac{1}{2^{n-1}}\ . \end{aligned}$$

On the other hand, if $\mathcal{O}$ implements Λ_{4n}, then for $x^{(1)} = (0, 0, x_3, x_4)$, $x^{(2)} = (x_1, 0, x_3, x_4)$, $x^{(3)} = (0, x_2, x_3, x_4)$, and $x^{(4)} = (x_1, x_2, x_3, x_4)$, we can see from Figure 1 that the corresponding $2n$-bit inputs of the second round are

$$(F_1(x_3, x_4)|_L, F_1(x_3, x_4)|_R)\ ,\ \ (F_1(x_3, x_4)|_L, x_1 \oplus F_1(x_3, x_4)|_R)\ ,$$

$$(x_2 \oplus F_1(x_3, x_4)|_L, F_1(x_3, x_4)|_R)\ ,\ \ (x_1 \oplus F_1(x_3, x_4)|_L, x_2 \oplus F_1(x_3, x_4)|_R)$$

respectively, where $F_1 = \mathbf{M}_{f_3} \circ \mathbf{M}_{f_2} \circ \mathbf{M}_{f_1}$ and $(x|_L, x|_R)$ denote the left and right n-bit block of $2n$-bit value x. Thus we obtain by the similar argument of Sakurai-Zheng[11] that

$$y_2^{(1)} \oplus y_2^{(2)} \oplus y_2^{(3)} \oplus y_2^{(4)} = 0$$

with probability 1.

Consequently we obtain that

$$\begin{aligned} ADV_{\mathcal{D}} &= |Pr(\mathcal{D} \text{ outputs } 1 \mid \mathcal{O} \leftarrow \mathcal{P}_{4n}) - Pr(\mathcal{D} \text{ outputs } 1 \mid \mathcal{O} \leftarrow \Lambda_{4n})| \\ &\geq 1 - \frac{1}{2^{n-1}}\ , \end{aligned}$$

which is non-negligible. □

The following theorem guarantees that the four or more round KASUMI is a pseudorandom permutation ensemble.

Theorem 3. *If f_i's$(i = 1, 2, \cdots, 12)$ in Figure 1 are independently chosen from an n-bit PPE, then the four round KASUMI is a $4n$-bit PPE.*

From Figure 1, we can see that the second round output w_3 and w_4 are depend on $f_1, \cdots, f_6$ and $f_1, \cdots, f_5$, respectively. So we set

$$w_3 = w_3^{f_1,\cdots,f_6}(\mathbf{x}) \text{ and } w_4 = w_4^{f_1,\cdots,f_5}(\mathbf{x}) ,$$

where $\mathbf{x} = (x_1, x_2, x_3, x_4) \in I_{4n}$ is an input value of KASUMI. As the similar work to previous section, we define bad event needed to prove Theorem 3.

Definition 12. *For every n-bit permutations $f_1, \cdots, f_6$, $BAD(f_1, \cdots, f_6)$ is defined as the set of all D-transcripts $\sigma = \{(\mathbf{x}^{(1)}, \mathbf{y}^{(1)}), \cdots, (\mathbf{x}^{(q)}, \mathbf{y}^{(q)})\}$ satisfying: $\exists 1 \le i < j \le q$ such that*

$$w_3^{f_1,\cdots,f_6}(\mathbf{x}^{(i)}) = w_3^{f_1,\cdots,f_6}(\mathbf{x}^{(j)}) \text{ or } w_4^{f_1,\cdots,f_5}(\mathbf{x}^{(i)}) = w_4^{f_1,\cdots,f_5}(\mathbf{x}^{(j)}) .$$

Lemma 5. *Let $f_1, \cdots, f_6$ be chosen independently from UPE $\mathcal{P}_n$. Then for any $\mathcal{D}$-transcript $\sigma = \{(\mathbf{x}^{(1)}, \mathbf{y}^{(1)}), \cdots, (\mathbf{x}^{(q)}, \mathbf{y}^{(q)})\}$,*

$$Pr(\sigma \in BAD(f_1, \cdots, f_6)) \le \frac{q^2 - q}{2^n - 1} .$$

Proof. Let $\alpha_k^{(i)}$ be the n-bit input value of $f_k (k = 1, \cdots, 6)$ when the query of $\mathcal{D}$ is $\mathbf{x}^{(i)} = (x_1^{(i)}, x_2^{(i)}, x_3^{(i)}, x_4^{(i)}) (i = 1, \cdots, q)$. For example,

$$\begin{aligned} \alpha_3^{(i)} &= x_3^{(i)} \oplus f_1(x_4^{(i)}) , \\ \alpha_5^{(i)} &= x_1^{(i)} \oplus x_3^{(i)} \oplus f_1(x_4^{(i)}) \oplus f_2(x_3^{(i)}) \oplus f_3(x_3^{(i)} \oplus f_1(x_4^{(i)})) . \end{aligned}$$

Then it is easy to show that if $\alpha_k^{(i)} \ne \alpha_k^{(j)}$ for some $k = 1, \cdots, 6$, by Lemma 1

$$Pr\left(w_3^{f_1,\cdots,f_6}(\mathbf{x}^{(i)}) = w_3^{f_1,\cdots,f_6}(\mathbf{x}^{(j)})\right) \le \frac{1}{2^n - 1} ,$$

otherwise ($\alpha_k^{(i)} = \alpha_k^{(j)}$, for all $k = 1, \cdots, 6$) we obtain that this probability is zero, since in this case $w_3^{f_1,\cdots,f_6}(\mathbf{x}^{(i)}) = w_3^{f_1,\cdots,f_6}(\mathbf{x}^{(j)})$ implies to $\mathbf{x}^{(i)} = \mathbf{x}^{(j)}$ which contradicts to the assumption that all queries are distinct. By the similar argument we can also show that $Pr(w_4^{f_1,\cdots,f_5}(\mathbf{x}^{(i)}) = w_4^{f_1,\cdots,f_5}(\mathbf{x}^{(j)})$ has the same upper bound. □

Lemma 6. *Let Λ_{4n} be the 4n-bit permutation ensemble obtained from the four round KASUMI of Figure 1 where all f_i's($i = 1, \cdots, 12$) are independently chosen from the n-bit UPE. Then for any $\mathcal{D}$-transcript $\sigma = \{(\mathbf{x}^{(1)}, \mathbf{y}^{(1)}), \cdots, (\mathbf{x}^{(q)}, \mathbf{y}^{(q)})\}$,*

$$|Pr\,(T_{\Lambda_{4n}} = \sigma \mid \sigma \notin BAD(f_1, \cdots, f_6)) - Pr\,(T_{\mathcal{P}_{4n}} = \sigma)| < \varepsilon'_{n,q} ,$$

where

$$\varepsilon'_{n,q} = \frac{1}{2^{4n}(2^n - 1)^4 \cdots (2^n - q + 1)^4} .$$

Proof. For any possible $\mathcal{D}$-transcript we have that

$$Pr\left(T_{\mathcal{P}_{4n}} = \sigma\right) = \frac{(2^{4n} - q)!}{2^{4n}!} .$$

In Figure 1, by considering four paths $w_3 \to f_8 \to z_2$, $w_3 \to f_8 \to f_{10} \to f_{12} \to z_3$, $w_4 \to f_7 \to f_9 \to z_1$, and $w_4 \to f_7 \to f_9 \to f_{11} \to z_4$, we can obtain that

$$Pr\left(T_{\Lambda_{4n}} = \sigma \mid \sigma \notin BAD(f_1, \cdots, f_6)\right) = \left(\frac{(2^n - q)!}{2^n!}\right)^4 ,$$

which complete the proof of this lemma. □

Proof of Theorem 3: From Lemma 5 and 6, Theorem 3 is proved straightforwardly by the similar process in the proof of Theorem 1. □

3 Provable Security of the Encryption Mode $f8$

To guarantee the message confidentiality over the wireless link of W-CDMA for 3GPP, $f8$ encryption mode has been proposed, which is based on the block cipher KASUMI[12]. In this section we examine the provable security of the 3GPP encryption mode $f8$ under the assumption that the underlying block cipher is a pseudorandom permutation. Note that this assumption is reasonable from the result of previous section.

3.1 Notions of Security for a Symmetric Encryption Mode

Symmetric encryption scheme is defined as a triple of algorithms, $(\mathcal{K}, \mathcal{E}, \mathcal{D})$, where $\mathcal{K}$ is the probabilistic algorithm for key generation, $\mathcal{E}$ is the probabilistic algorithm which encrypts the plaintext M with the key K and outputs the ciphertext C, and $\mathcal{D}$ is the the deterministic algorithm which decrypts the ciphertext C with the key K and outputs the corresponding plaintext M. Here M is selected in a set of messages. Bellare et al.[2] considered four notions for security of symmetric encryption modes. "Real-or-random indistinguishability" and its variant "left-or-right indistinguishability" were first introduced. "Find-then-guess security" and "semantic security" which are the notions for the asymmetric encryption scheme, were adapted to the symmetric setting. They also investigated the relation among these notions of security[2]. Real-or-random and left-or-right indistinguishability were equivalent up to a small constant factor in the reduction. Also these notions had a security-preserving reduction to find-then-guess security. However the reduction from find-then-guess security to left-or-right indistinguishability was not security-preserving. It had security-preserving reductions between find-then-guess and semantic security.

Here we analyze the security of 3GPP $f8$ mode by applying the notion of left-or-right indistinguishability, since the left-or-right security implies good reductions to the other three definitions as described above. Left-or-right indistinguishability is a strong form of chosen-plaintext security. It considers two different games. In either game a query is a pair (x_1, x_2) of equal-length strings from the given message space. In either game a random key $a \in K$ is selected at random and fixed for duration of the game. In Game 1, an oracle receiving (x_1, x_2) responds with $\mathcal{E}_a(x_1)$. In Game 2, it responds with $\mathcal{E}_a(x_2)$. Thus Game 1 provides a "left" oracle and Game 2 provides a "right" oracle. An encryption scheme is secure if a reasonable adversary cannot obtain significant advantage in distinguishing Game 1 and 2.

Definition 13 (Left-or-right indistinguishability[2]). *Encryption scheme $(\mathcal{K}, \mathcal{E}, \mathcal{D})$ is said to be $(t, q, \mu; \epsilon)$-secure, in left-or-right sense, if for any adversary $\mathcal{A}$ who runs in time at most t and makes at most q oracle queries, totaling at most μ bits,*

$$ADV_{\mathcal{A}}^{lr} \stackrel{\text{def}}{=} \left| Pr_{a \leftarrow K}\left(\mathcal{A}^{\mathcal{E}_a(\mathcal{O}(1,(\cdot,\cdot)))} = 1\right) - Pr_{a \leftarrow K}\left(\mathcal{A}^{\mathcal{E}_a(\mathcal{O}(2,(\cdot,\cdot)))} = 1\right)\right| \leq \epsilon\,.$$

Encryption scheme $(\mathcal{K}, \mathcal{E}, \mathcal{D})$ is $(t, q, \mu; \epsilon)$-break, in left-or-right sense, if for an adversary $\mathcal{A}$ who runs in time at most t and makes at most q oracle queries, totaling at most μ bits, $ADV_{\mathcal{A}}^{lr} > \epsilon$.

In the above definition $\mathcal{A}^{\mathcal{E}_a(\mathcal{O}(1,(\cdot,\cdot)))}$ and $\mathcal{A}^{\mathcal{E}_a(\mathcal{O}(2,(\cdot,\cdot)))}$ indicate $\mathcal{A}$ with an oracle $\mathcal{O}$ which returns $y = \mathcal{E}_a(x_1)$ and $y = \mathcal{E}_a(x_2)$, respectively, in response to query (x_1, x_2). And $Pr_{a \leftarrow K}(\mathcal{A}^{\mathcal{E}_a(\mathcal{O}(i,(\cdot,\cdot)))} = 1)$ $(i = 1, 2)$ denotes the probability that the adversary $\mathcal{A}$ with an oracle $\mathcal{O}(i, (\cdot, \cdot))$ $(i = 1, 2)$ outputs 1 when a key a is chosen randomly from the key space K.

The encryption mode $f8$ is based on the block cipher KASUMI and this is a pseudorandom permutation ensemble by referring to last section. Let $\mathcal{B}_l$ be the function family obtained from a block cipher with l-bit input/output values. To analyze the provable security of $f8$ mode, we need more rigorous definition about PPE than Definition 4.

Definition 14. *A permutation family $\mathcal{B}_l$ is said to be a $(t, q; \epsilon)$-secure PPE if for any distinguisher $\mathcal{D}$ who makes at most q oracle queries and runs in time at most t, $ADV_{\mathcal{D}} \leq \epsilon$.*

3.2 Security of $f8$ Encryption Mode

In this subsection, we prove the security of 3GPP $f8$ encryption mode by using the notion of left-or-right security. The underlying function of the encryption mode is fixed to a PPE $\mathcal{B}_l$ with l-bit input/output length. Let $a \in K$ be the key shared between the two parties who run the encryption scheme. It will be used to specify the function $g = \mathcal{B}_l[a]$ and $g' = \mathcal{B}_l[a \oplus KM]$ determined by the key a and $a \oplus KM$, respectively, where KM is an 128-bit fixed constant. We describe rigorously the encryption mode $f8$ as the following scheme. This scheme is also illustrated in Figure 2.

The scheme $f8^g(x)$ works as follows:

Function $f8^g(x)$
$IV \leftarrow g'(Count||Direction||Bearer||0\ldots0)$

$Reg_1 = IV$
for $i = 1, \ldots, n$ do
  $o_i = g(Reg_i)$
  $y_i = o_i \oplus x_i$
  $Reg_{i+1} = IV \oplus i \oplus o_i$
return $(y_1 \ldots y_n)$

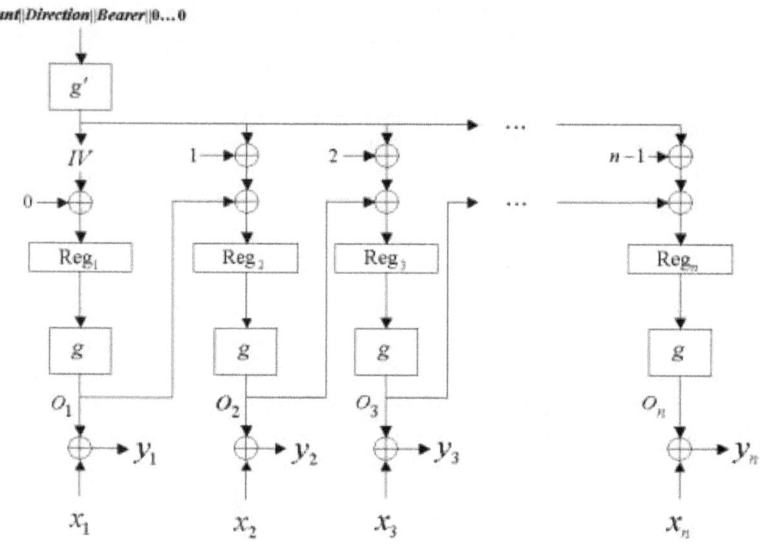

Fig. 2. 3GPP $f8$ encryption mode

In the above scheme *Count* is an encryption sequence number of 32-bit length depending on the time, *Bearer* is a 5-bit bearer identifier, *Direction* is an 1-bit direction identifier, and $0\ldots0$ denotes the padding so that the length of the input is an l-bit. The difference between OFB and $f8$ mode is that an initial nonce $ctr = (Count||Direction||Bearer||0\ldots0)$ is not sent to the receiver and $g'(ctr)$ is applied to the underlying function g, instead of ctr in a cleartext.

We consider two function family. $f8^{\mathcal{P}_l}$ is the set of all functions $f8^g$, where g is chosen from the UPE $\mathcal{P}_l$, and $f8^{\mathcal{B}_l}$ is the set of all functions $f8^g$, where g is chosen from the PPE $\mathcal{B}_l$. We first derive an upper bound on the success of any adversary trying to break the $f8^{\mathcal{P}_l}$ in the left-or-right sense. Next we examine the security of $f8^{\mathcal{B}_l}$. The basic idea for proving the security of $f8$ is that left-or-right security breaks down at the first repetition of the value of *Reg*. If $Reg_i = Reg_j$ for $i \neq j$, then also $o_i = o_j$. Hence $y_i \oplus y_j = x_i^b \oplus x_j^b$ $(b = 1, 2)$. Thus b is revealed if $x_i^1 \oplus x_j^1 \neq x_i^2 \oplus x_j^2$.

Lemma 7. *Let $\mathcal{A}$ be any adversary attacking $f8^{\mathcal{P}_l}$ in the left-or-right sense, making at most q queries, totaling at most μ bits. Then*

$$ADV_{\mathcal{A}}^{lr} \leq \delta_{f8^{\mathcal{P}_l}} \stackrel{\text{def}}{=} \frac{\mu/l \cdot (\mu/l - 1)}{2^{l+1}} .$$

Proof. Let $(x_1^1, x_1^2), \ldots, (x_q^1, x_q^2)$ be the oracle queries of the adversary $\mathcal{A}$, each consisting of a pair of equal length messages. These queries are random variables that depend on the coin tosses of $\mathcal{A}$ and responses of the oracle to previous queries. Let $ctr_i = (Count_i||Direction_i||Bearer_i||0\ldots0)$ and $IV_i = g'(ctr_i) \in \{0,1\}^l$, associated to (x_i^1, x_i^2) as computed by the oracle, for $i = 1, \ldots, q$. Let n_i be the number of blocks in the i-th query, $x_i^b = x_i^b[1] \cdots x_i^b[n_i]$ $(b \in \{1,2\})$ be the i-th query message, and $y_i = y_i[1] \cdots y_i[n_i]$ be the response of the oracle to the i-th query message. $Reg_i = Reg_i[1] \cdots Reg_i[n_i]$ is the contents of the register Reg in the i-th query, where $Reg_i[j]$ $(j \in \{1, \ldots, n_i\})$ denotes the content of the register corresponding to the j-th block of the i-th query message. We set $o_i[j]$ is a value computed by applying $Reg_i[j]$ to the function g. Let $Pr_1(\cdot)$ denote the probability in Game 1 providing the adversary $\mathcal{A}$ with the left oracle, and $Pr_2(\cdot)$ denote the probability in Game 2 providing the adversary $\mathcal{A}$ with the right oracle.

Let C be the collision event, i.e., $Reg_i[k] = Reg_j[k']$ whenever $(i,k) \neq (j,k')$, for all $i,j = 1, \ldots, q$ and $k = 1, \ldots, n_i$ and $k' = 1, \ldots, n_j$. The event C^c, complement of C, depends on IV_i, $o_i[k]$ and k for each query. Since g and g' are chosen from the UPE $\mathcal{P}_l$, IV_i and $o_i[k]$ are random and independent of the message given to the oracle. Thus the collision probability does not depend on b, and the following equation holds:

$$Pr_1(C^c) = Pr_2(C^c) . \tag{4}$$

For the same reason, if no collision occurs, the adversary outputs 1 with the same probability for Game 1 and Game 2 because each ciphertext block given to the adversary is independent of any previous ciphertext blocks and of message blocks. Namely, the following holds:

$$Pr_1(\mathcal{A} = 1 \mid C^c) = Pr_2(\mathcal{A} = 1 \mid C^c) . \tag{5}$$

Therefore, by using the equation (4) and (5), we can write the adversary's advantage as follows:

$$\begin{aligned} ADV_{\mathcal{A}}^{lr} &= \big|Pr_1(\mathcal{A} = 1) - Pr_2(\mathcal{A} = 1)\big| \\ &= \big|Pr_1(\mathcal{A} = 1 \mid C) \cdot Pr_1(C) + Pr_1(\mathcal{A} = 1 \mid C^c) \cdot Pr_1(C^c) \\ &\quad - Pr_2(\mathcal{A} = 1 \mid C) \cdot Pr_2(C) - Pr_2(\mathcal{A} = 1 \mid C^c) \cdot Pr_2(C^c)\big| \\ &= \big|(Pr_1(\mathcal{A} = 1 \mid C) - Pr_2(\mathcal{A} = 1 \mid C))Pr_1(C)\big| \\ &\leq Pr_1(C) . \end{aligned}$$

Given the equation (4) we drop the subscript in talking about the probability of C and write the above just as $Pr(C)$. Now we want to compute the upper bound of $Pr(C)$. The adversary does not know the contents of the register Reg because she does not know $IV_i = g(ctr_i)$. Hence the adversary does not identify the collision $Reg_i[k] = Reg_j[k']$ ($(i,k) \neq (j,k')$ for all $i,j = 1,\dots,q$ and $k = 1,\dots,n_i$ and $k' = 1,\dots,n_j$). However the adversary knows the values $o_i[k]$ since she knows the queried message block $x_i[k]$ and the answered ciphertext block $y_i[k]$. Then she can identity $o_i[k] = o_j[k']$. Since g is a permutation, the output collision, $o_i[k] = o_j[k']$, implies the following:

$$o_i[k] = o_j[k'] \;\Leftrightarrow\; g(Reg_i[k]) = g(Reg_j[k']) \;\Leftrightarrow\; Reg_i[k] = Reg_j[k'] \,.$$

Thus, to compute the upper bound of $Pr(C)$ we compute the probability of the output collision event, T, i.e., $o_i[k] = o_j[k']$ whenever $(i,k) \neq (j,k')$, for all $i,j = 1,\dots,q$ and $k = 1,\dots,n_i$ and $k' = 1,\dots,n_j$. We define the stream B as

$$B = o_1[1]\dots o_1[n_1]o_2[1]\dots o_2[n_2]\dots o_q[1]\dots o_q[n_q].$$

That is, B is the output values of g until the n_q-th encryption of the last q-th query. The length of B is $Q = l \cdot \sum_{i=1}^{q} n_i \leq \mu$ bits. We first compute the number of streams with a collision $o_i[k] = o_j[k']$ for every possible pair (i,k) and (j,k') ($(i,k) \neq (j,k'), 1 \leq i,j \leq q, k = 1,\dots,n_i, k' = 1,\dots,n_j$). As $o_i[k] = o_j[k']$, there are 2^l possible values for the both values. The remaining $Q - 2l$ bits have 2^{Q-2l} possibilities. Thus the number of streams with a collision is 2^{Q-l}. There are $(\mu/l)(\mu/l-1)/2$ possible pairs (i,k) and (j,k'). Hence the number η of streams B with at least one collision is less than $(\mu/l)(\mu/l-1)2^{Q-l-1}$. The stream B has 2^Q possibilities. Thus

$$Pr(T^c) = \frac{(2^Q - \eta)}{2^Q} \geq 1 - \frac{(\mu/l)(\mu/l-1)}{2^{l+1}} \,.$$

This implies the following because of $Pr(C) = Pr(T)$:

$$Pr(C) \leq \frac{(\mu/l)(\mu/l-1)}{2^{l+1}} \,. \quad \Box$$

In the practical situation, because the underlying block cipher g is modeled as a pseudorandom permutation, we prove the security of 3GPP $f8$ mode using a pseudorandom permutation. This is derived from the Lemma 7.

Theorem 4. *Let $\mathcal{B}_l$ be a $(t',q';\epsilon')$-secure PPE with l-bit input/output length. Then $f8^{\mathcal{B}_l}$ scheme is $(t,q,\mu;\epsilon)$-secure in the left-or-right sense. Here $q = q'$, $\mu = q'l$, $t = t' - c\frac{\mu}{l}(l+l)$ and $\epsilon = 2\epsilon' + \delta_{f8^{\mathcal{P}_l}}$, where $c > 0$ is a small constant and $\delta_{f8^{\mathcal{P}_l}} = \frac{(\mu/l)(\mu/l-1)}{2^{l+1}}$.*

Proof. The details of this proof are omitted since it is similar to the proof of Theorem 12 in [2] by replacing pseudorandom function with pseudorandom permutation. □

4 Conclusion

In this work we examined the pseudorandomness of the block cipher KASUMI and the provable security of $f8$. We proved that FI function within KASUMI composed of four round unbalanced MISTY-type structure was a pseudorandom permutation. And we showed that the three round KASUMI was not a permutation ensemble but the four round KASUMI was a pseudorandom permutation ensemble under the adaptive distinguisher model. Moreover we provided the upper bound on the security of $f8$ encryption mode under the reasonable assumption from the first result by means of the left-or-right security notion.

References

1. M. Bellare, J. Kilian, and P. R. Rogaway, *The security of cipher block chaining message authentication codes*, Advances in Cryptology-Crypto '94, LNCS 839, Springer-Verlag, 1994, pp. 341-358.
2. M. Bellare, A. Desai, E. Jokipii, and P. Rogaway, *A Concrete Security Treatment of Symmetric Encryption: Analysis of the DES Modes of Operation*, 38th Symposium on Foundations of Computer Science(FOCS), IEEE Computer Society, 1997, pp. 394-403.
3. FIPS PUB 81, DES Modes of Operation, Federal Information Processing Standards Publication 81, December 2, 1980.
4. H. Gilbert and M. Minier, *New results on the pseudorandomness of some block cipher constructions*, Preproceedings of Fast Software Encryption workshop 2001, (2001, Yokohama), pp. 260-277.
5. T. Iwata, T. Yoshino, T. Yuasa, and K. Kurosawa, *Round security and super-pseudorandomness of MISTY type structure*, Preproceedings of Fast Software Encryption workshop 2001, (2001, Yokohama), pp. 245-259.
6. J. S. Kang, O. Y. Yi, D. W. Hong, and H. S. Cho, *Pseudorandomness of MISTY-type transformations and the block cipher KASUMI*, ACISP2001, LNCS 2119, Springer-Verlag, 2001, pp. 60-73.
7. M. Luby and C. Rackoff, *How to construct pseudorandom permutations and pseudorandom functions*, SIAM J. Comput., Vol. 17, 1988, pp. 189-203.
8. M. Matsui, *New permutation of Block Ciphers with Provable Security against Differential and Linear Cryptalaysis*, Fast Software Encryption, LNCS 1039, Springer-Verlag, 1996, pp. 205-218.
9. M. Matsui, *New Block Encryption Algorithm MISTY*, Fast Software Encryption'97, LNCS 1267, Springer-Verlag, 1997, pp. 54-68.
10. M. Naor and O. Reingold, *On the construction of pseurandom permutations: Luby-Rackoff revisited*, J. Cryptology, Vol. 12, 1999, pp. 29-66.
11. K. Sakurai and Y. Zheng, *On non-pseudorandomness from block ciphers with provable immunity against linear cryptanaysis*, IEICE Trans. Fundamentals, Vol. E80-A, No. 1, 1997, pp. 19-24.
12. 3G TS 35.201, *Specification of the 3GPP confidentiality and integrity algorithm; Document 1: f8 and f9 specifications*, available at http://www.3gpp.org
13. A. Alkassar, A. Geralay, B. Pfitzmann, and A. R. Sadeghi, *Optimized Self-Synchronizing Mode of Operation*, Preproceedings of 8th Fast Software Encryption Workshop, April 2, 2001, pp. 82-96.

Efficient and Mutually Authenticated Key Exchange for Low Power Computing Devices*

Duncan S. Wong and Agnes H. Chan

College of Computer Science, Northeastern University,
Boston, MA 02115, U.S.A.
{swong,ahchan}@ccs.neu.edu

Abstract. In this paper, we consider the problem of mutually authenticated key exchanges between a low-power client and a powerful server. We show how the Jakobsson-Pointcheval scheme proposed recently [15] can be compromised using a variant of interleaving attacks. We also propose a new scheme for achieving mutually authenticated key exchanges. The protocol is proven correct within a variant of Bellare-Rogaway model [3,4]. This protocol gives the same scalability as other public-key based authenticated key exchange protocols but with much higher efficiency and fewer messages. It only takes 20 msec total computation time on a PalmPilot and has only three short messages exchanged during the protocol.

1 Introduction

The goal of a mutually authenticated key exchange protocol (MAKEP) between two communicating parties is to provide them with some assurance that they know each other's true identity and at the same time to have the two parties end up sharing a common key known only to them. This common key, also known as session key, can then be used to provide privacy and data integrity during the session. In this paper, we focus our attention on the design and analysis of MAKEPs for the two parties in which one of them is strictly limited in both computational power and memory capacity while the other is as powerful as a conventional desktop computer. We call the low-power party as the client and the powerful one as the server. Such a low-power client could be a Personal Digital Assistant (PDA), a cellular phone or a smart card in real applications. A powerful server could be a base station or the security center of a wireless network.

Although there is a long history of designing MAKEPs and many protocols have been proposed for various kinds of distributed systems, they seldom designed for such an unbalanced system setup. For symmetric-key based MAKEPs, [21,22,3,4,16], two communicating parties share a long-lived key or have a third party involved during runtime. In the first case, each party has to maintain a set of distinct keys for communicating with different parties. In the later case, a centralized trusted party is required to be present whenever the protocol is

* This work was sponsored by the U. S. Air Force under contract F30602-00-2-0518.

C. Boyd (Ed.): ASIACRYPT 2001, LNCS 2248, pp. 272–289, 2001.
© Springer-Verlag Berlin Heidelberg 2001

executed. Hence key management and scalability are two major issues when deploying the schemes in practice. For public-key based MAKEPs, [9,7,1,11], on the other hand, high computational complexity is required on both communicating parties. For example, a 512-bit modular exponentiation on a 16MHz Palm V requires over one minute of pure computation as shown in [27].

Recently several schemes [26,15] have been proposed for systems with unbalanced compututional power. In [26], we proposed two MAKEPs which attain efficiency on the client side and provide scalability for most systems. However, the schemes do not give as much scalability as a pure public-key based MAKEP does. In [15], Jakobsson and Pointcheval proposed a MAKEP which improves on efficiency by using precomputation. However, the protocol does not scale well and is susceptible to a variant of interleaving attacks as shown in Sect. 3.

In this paper, we propose a new scheme which not only gives the same scalability as other public-key based schemes do, but also requires only three short messages in a single protocol run. It takes only 20 msec computation time during a run of the protocol in the case where the client is a 16MHz PalmPilot. We also show that it is secure within a variant of Bellare-Rogaway model [3,4] and provides a reasonable amount of forward secrecy.

The remainder of the paper is organized as follows. In Sect. 2, we describe some notations used throughout this paper. Then we present a variant of interleaving attacks and show how it can be used to compromise the MAKEP proposed in [15]. In Sect. 4, we introduce a new protocol referred to as the Client-Server MAKEP. A formal security analysis of the protocol is given in Sect. 5 and the performance is examined in Sect. 6. We conclude with some discussions on other properties of the protocol in Sect. 7.

2 Preliminaries

Let $\mathcal{E}_K$ and $\mathcal{D}_K$ denote the encryption and decryption transformations under the symmetric key K respectively. For public key systems, each entity A has a public key PK_A and a private key SK_A. For simplicity of notation, we use $\mathcal{E}_{PK_A}$ to denote the public key encryption. Sig_{TA} is a *secret* signing algorithm and Ver_{TA} is the *public* verification algorithm of a trusted authority TA. A certificate $Cert_A$ of an entity A is denoted by

$$Cert_A = <\mathrm{ID}_A,\ m,\ Sig_{\mathrm{TA}}(\mathrm{ID}_A, m)>,$$

where ID_A uniquely identifies A, and m is some message being certified. Usually m is A's public key with some other information such as a serial number and an expiration date. By $< X >$, we mean an appropriate encoding of X. A *nonce*, denoted by $r \leftarrow \{0,1\}^l$, is a l-bit random number. We use $x \in_R S$ to denote that x is chosen uniformly at random from the set S.

Throughout the paper, we assume that G is a cyclic group of prime order q and g is a generator of G. For simplicity, we only consider the case where G is a subgroup of $\mathbb{Z}_p^*$, the multiplicative group of the integers modulo a prime p. However the discussion applies equally well to any group of prime order in which the discrete logarithm problem is computationally intractable. We also assume that the domain parameters (g, p, q) are publicly known. This assumption can be dropped when the information is exchanged via certificates. An example of a certificate which contains the domain parameters will be given in Sect. 4.

3 A Variant of Interleaving Attacks — Hijacking Attack

As described in [26], a challenge-response based MAKEP should have two primitive elements. The first one is two challenge-response pairs which provide the mutual authentication for both participating parties. The second one is the 'binding' of each party's encrypted secret session key contribution to the corresponding challenge number sent by its partner. The bindings have to be stringent enough to guarantee the *freshness* of the session key and to counteract various attacks. These notions are later formalized in Definition 2. To illustrate their importance, we consider a recently proposed protocol [15] for mutual authentication and key exchange between a low-power client and a powerful server. The protocol is shown in Fig. 1, where H_0, H_1 and H_2 denote some cryptographic hash functions.

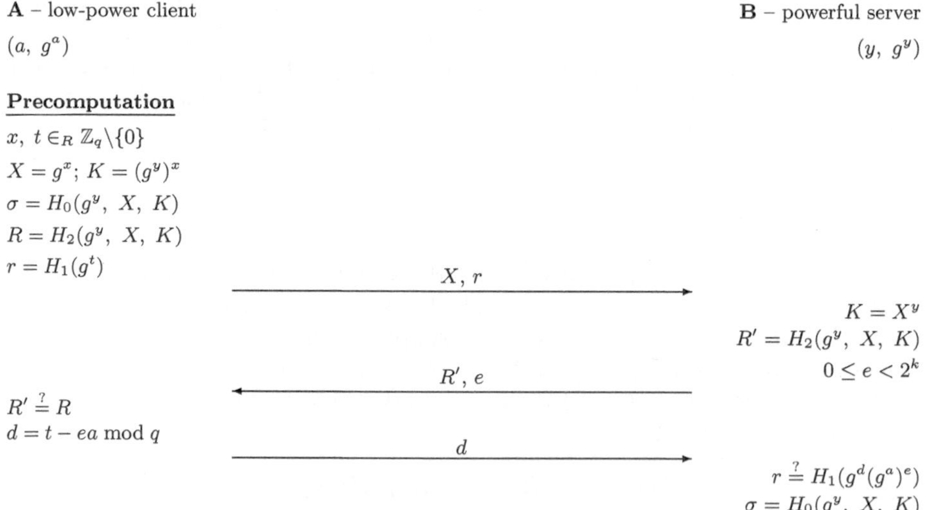

A – low-power client
$(a,\ g^a)$
B – powerful server
$(y,\ g^y)$
Precomputation
$x,\ t \in_R \mathbb{Z}_q \backslash \{0\}$
$X = g^x;\ K = (g^y)^x$
$\sigma = H_0(g^y,\ X,\ K)$
$R = H_2(g^y,\ X,\ K)$
$r = H_1(g^t)$
$X,\ r$
$K = X^y$
$R' = H_2(g^y,\ X,\ K)$
$0 \leq e < 2^k$
$R',\ e$
$R' \stackrel{?}{=} R$
$d = t - ea \bmod q$
d
$r \stackrel{?}{=} H_1(g^d(g^a)^e)$
$\sigma = H_0(g^y,\ X,\ K)$

Fig. 1. Jakobsson-Pointcheval MAKEP

In the protocol, A is a low-power client and B is a powerful server. Each of them has a public key pair. A's private key is a which is chosen randomly in $\mathbb{Z}_q^*$ and g^a is her public key. It is assumed that A knows B's authentic public key, and vice versa. Before the protocol begins, A precomputes a session key σ from K where K is a secret to be shared with B using the Diffie-Hellman key exchange technique [9]. (This is done through X in the first message from A to B in the protocol.) She also precomputes the expected response of B which is denoted as R and a value r which is intended to be used for client authentication. Now when A runs the protocol with B, A authenticates B (server authentication) by checking if the incoming message (second message of the protocol) contains a value denoted by R' which is equal to R. We note that B's response R' 'binds' the challenge number X of A to his computed secret K. For client authentication, B chooses an independent random number e and sends it to A. A then computes a value d and sends it back to B. After B receives it, he then computes $H_1(g^d(g^a)^e)$ and compares it with the value of r received from A. We note that B's challenge number e has never been bounded to any secret value or any previous messages of the current session. This oversight creates a vulnerability for the protocol.

We now show that this protocol is vulnerable to an attack pictured in Fig. 2 where E denotes an adversary, S1 and S2 denote two parallel sessions to be established with B.

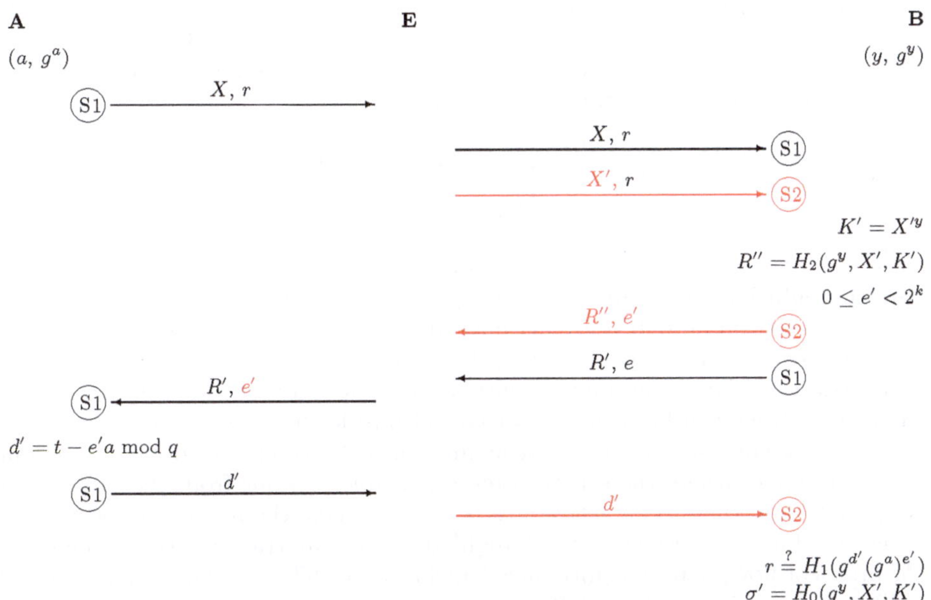

Fig. 2. An attack on Jakobsson-Pointcheval MAKEP

In this attack, we assume that E is an active adversary who is 'sitting' in between A and B and can intercept and inject messages. We also assume that B accepts multiple session connections from different instances of a single party simultaneously. In the figure, each message is associated with a session denoted by a circle containing the session number. For example, the message $< X,\ r >$ from A to E indicates that the message is for session S1. Session S1 begins when A sends the first message $< X,\ r >$ to B but it is also eavesdropped by E. E immediately creates a new message which contains an integer $X' \in G$ and r. E then sends this message to B alleging that it comes from another instance of A. B thinks that A wants to establish two sessions, denoted by S1 and S2, with him at the same time. He then sends back two messages, one for session S1 and the other for session S2. After receiving the two messages from B, E constructs a message which contains the correct response R' for session S1 and the challenge number e' of B for session S2. This message is sent to A by E alleging that it comes from B. Since A thinks that she is establishing only one session with B, she believes that the message must be the response of B for session S1. A then verifies the response and computes an outgoing message d'. However this message is relayed to B as the response of A for session S2. Meanwhile session S1 remains incomplete and will finally be terminated by B after a timeout.

Through this attack, the client authentication is compromised. A believes that she has established a secure session S1 with B sharing a secret key $K = X^y$, while B believes that he has established a different secure session S2 with A sharing a different secret key $K' = X'^y$. Furthermore E can choose randomly a value $x' \in \mathbb{Z}_q \setminus \{0\}$ and computes the pair (X', K') as $g^{x'}$ and $(g^y)^{x'}$ respectively. Hence she can decrypt all the encrypted messages sent from B to A.

One possible patch to the authentication flaw is to modify the computation of d and the verification process to the following:

$$d = t - h(e, K)a \bmod q$$
$$r \stackrel{?}{=} H_1(g^d (g^a)^{h(e,K)})$$

where h is a cryptographic hash function. Another possible solution is to change the computations of R and R' to $H_2(g^y, X, K, e)$. The tradeoff is that A can no longer precompute R and this may affect the performance of the protocol. We call this attack a 'hijacking attack' because an adversary E can hijack A's conversations with B and impersonate her in a *new* run of the protocol. In fact, hijacking attack can be seen as a variant of interleaving attacks [5,10,19].

To ensure our protocol is secure against such an attack, we adopt the communication model where the server accepts multiple sessions from the same client in parallel and there is an active adversary. Details of the model will be given in Sect. 5.1. Furthermore, this attack highlights the importance of the authenticity of protocol flow, a notion formalized in [3] as *Matching Conversations*, which will be summarized in Sect. 5.2.

4 The Protocol

In this section, we propose a new scheme which is designed for mutually authenticated key exchange between a low-power client and a powerful server and we call it the Client-Server MAKEP. In this scheme, each party has a long-lived public key pair. For the powerful server B, we use SK_B and PK_B to denote the corresponding private and public keys respectively. Our description of the protocol can apply to any public key cryptographic algorithm but in practice, one that requires less memory and can do efficient encryption is preferred since the protocol requires the client to do one public-key encryption. We assume that the public key of the server is publicly known. For the low-power client A, $a \in_R \mathbb{Z}_q \backslash \{0\}$ is the private key and g^a is the public key. A also has a certificate obtained from the Trusted Authority (TA). The certificate is given by

$$Cert_A = < \mathrm{ID}_A,\ g,\ p,\ q,\ g^a,\ Sig_{\mathrm{TA}}(\mathrm{ID}_A, g, p, q, g^a) >$$

As described in Sect. 2, the domain parameters (g, p, q) can be removed from the certificate if they are publicly known. The protocol is illustrated in Fig. 3 and is described as follows.

Client-Server MAKEP

1. A selects $r_A \leftarrow \{0,1\}^k$, $b \in_R \mathbb{Z}_q \backslash \{0\}$ and computes $x = \mathcal{E}_{PK_B}(r_A)$ and $\beta = g^b$.
2. A sends $Cert_A$, β and x to B.
3. B checks $Cert_A$ by running Ver_{TA} and verifies that $1 < \beta < p$ and $\beta^q \equiv 1 \pmod p$. If any check fails, B terminates the protocol run with failure. Otherwise, B decrypts x and obtains r_A.
4. B selects $r_B \leftarrow \{0,1\}^k$, computes $\mathcal{E}_{r_A}(r_B,\ \mathrm{ID}_B)$ and sends it to A. B also computes $\sigma = r_A \oplus r_B$ and destroys r_A, r_B from his memory.
5. A decrypts the incoming message under r_A and checks if the decrypted message contains a proper coding of ID_B with some number. If the check fails, A terminates the protocol run with failure. Otherwise, A denotes the number as r_B.
6. A computes $\sigma = r_A \oplus r_B$ and $y \equiv ah(\sigma) + b \pmod q$ where $h : \{0,1\}^* \rightarrow \mathbb{Z}_q \backslash \{0\}$ is a cryptographic hash function. Then A sends y to B. A also computes $K = H(\sigma)$ as the session key and accepts the connection. She destroys r_A, r_B and σ from her memory.
7. B verifies if $g^y \equiv (g^a)^{h(\sigma)}\beta \pmod p$. If it is false, B terminates the protocol run with failure. Otherwise, B computes $K = H(\sigma)$ as the session key and accepts the connection. He also destroys σ from his memory.

In step 3, B verifies that $1 < \beta < p$ and $\beta^q \equiv 1 \pmod p$. This process is called public key validation [7]. It is a very important security measure in practice for protecting the system from several subtle attacks such as small subgroup attacks [18,17] and identity element attack [7]. $H : \{0,1\}^* \rightarrow \{0,1\}^k$ is a hash function instantiating a public random oracle [2]. It is also called a key derivation function [14] here because it is used to derive the session key K from the shared secret σ. One reason for doing this is to destroy the algebraic relationships between the session key K and the nonces $(r_A,\ r_B)$. Another reason is to mix together *strong* bits and potential *weak* bits of σ where weak bits are certain bits of information about σ that can be correctly predicted with non-negligible advantage.

5 Security Analysis

To prove that the protocol described above is secure, we use a variant of the Bellare and Rogaway's model [3,4]. The approach we take closely follows the approach of [6].

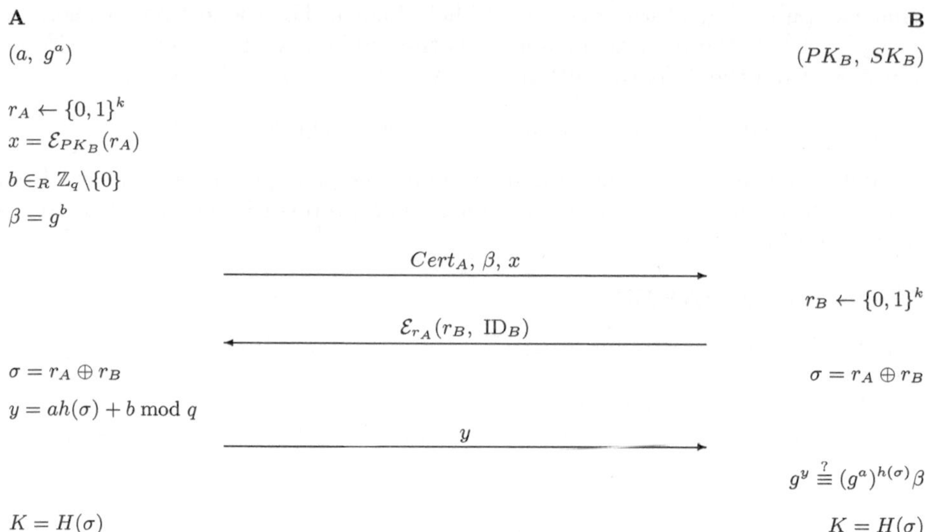

Fig. 3. Client-Server MAKEP

A distributed system in our model has a set I_C of clients and a set I_S of servers. I_C (or I_S) is a set of identities which defines the clients (or servers) who can participate in the protocol. The cardinalities of these two sets may be any two polynomial functions of a security parameter, k. The adversary E is *not* in I_C or I_S.

Definition 1. *A C-S MAKEP is a triple $P = (\Pi, \Psi, LL)$ of probabilistic polynomial-time computable functions (each with respect to its first argument). Π specifies how a honest client behaves; Ψ specifies how a honest server behaves; and LL specifies the initial distribution of clients' and servers' long-lived keys. The domain and range of these functions are briefly described as follows.*

$(m, \delta, \sigma) = \Pi(1^k, A, B, SK_A, PK_A, PK_B, conv, r)$ *where*

- $1^k \in \mathbb{N}$ — *the security parameter;*
- $A \in I_C$ — *client's identity;*
- $B \in I_S$ — *(intended) server's identity;*
- $SK_A \in \{0,1\}^*$ — *long-lived secret key of A.*
- $PK_A \in \{0,1\}^* \cup \{*\}$ — *long-lived public key of A; (Note that the value "*" refers to that the client does not have a long-lived public key. This is the case when the client is using a symmetric encryption algorithm in the protocol.)*
- $PK_B \in \{0,1\}^* \cup \{*\}$ — *long-lived public key of B;*
- $conv \in \{0,1\}^*$ — *conversation so far;*
- $r \in \{0,1\}^\omega$ — *random coin flips;*
- $m \in \{0,1\}^* \cup \{*\}$ — *the next message sent to B;*

$\delta \in \{\mathbf{A}, \mathbf{R}, *\}$ — *the decision where* **A** *represents accept,* **R** *represents reject and* $*$ *refers to that no decision has been made so far;*
$\sigma \in \{0,1\}^* \cup \{*\}$ — *the* private *output; it is the session key when it is string-valued.*

$(m, \delta, \sigma) = \Psi(1^k, B, A, SK_B, PK_B, PK_A, conv, r)$ *where the notations are similarly defined as those for* Π.

$(SK, PK) = LL(1^k, t, r)$ *where* $t \in \{\mathbf{client}, \mathbf{server}\}$ *is the type of the entity.*

In our protocol, if the value t of LL is **client**, SK is denoted by a four-tuple (g, p, q, a) and PK is denoted by another four-tuple (g, p, q, b) where p and q are prime such that $q \mid p-1$. The length of p is polynomial in k. g is an element in $\mathbb{Z}_p^*$ of order q. a is chosen randomly from $\mathbb{Z}_q \backslash \{0\}$ and $b = g^a \bmod p$. If t is **server**, the value returned by LL will be the public key pair of some asymmetric key encryption algorithm chosen by the protocol.

5.1 Communication Model (Adversarial Model)

ADVERSARY E. All communication among interacting parties is under the control of an adversary. In particular, the adversary can read, inject, modify, delete, delay and replay messages in the model. The adversary can also start up entirely new "instances" of any of the parties at any time. Hence there could be multiple *sessions* engaged in the system at the same time. This model gives the adversary the capability to launch attacks such as reflection and interleaving attacks suggested in [20,5,10,19] and the hijacking attack described in Sect. 3.

Formally, an adversary E is a probabilistic machine which equips with an oracle denoted by LL_{PK} and an *infinite* collection of oracles Π_{ij}^s and Ψ_{ji}^t, for $i \in I_C$, $j \in I_S$ and $s, t \in \mathbb{N}$. Oracle LL_{PK}, which will be described below, models the event that public keys of all entities (both clients and servers) are publicly known. Oracle Π_{ij}^s models the instance s of client i attempting to agree on a shared session key with server j. Oracle Ψ_{ji}^t models the instance t of server j attempting to agree on a shared session key with client i. The model uses oracle queries to capture E's attacks such that (1) E writes queries on a special tape; (2) the corresponding oracle reads the tape automatically; and (3) E gets back a response in unit time (i.e. oracle query is treated as one step in an algorithm).

RUNNING THE PROTOCOL. A generic execution of a protocol between a client instance and a server instance is called a *run* of the protocol. We model one run of the protocol as conducting one experiment in the presence of an adversary E using a security parameter k. It is described as follows.

1. **Initialization**
   Toss coins for LL, E and all oracles Π_{ij}^s and Ψ_{ji}^t.
2. **Run** E
   E may make oracle queries and the queries are answered as described in Table 1.

Table 1. The queries which E can ask of its oracles

	On query of	Return
1	$(\texttt{SendClient}, i, j, s, x)$	$\Pi^{m\delta}(1^k, i, j, SK_i, PK_i, PK_j, conv\Pi^s_{ij}, coins\Pi^s_{ij})$ And then set $conv\Pi^s_{ij} \leftarrow conv\Pi^s_{ij} \cdot x \cdot m$
2	$(\texttt{SendServer}, j, i, t, x)$	$\Psi^{m\delta}(1^k, j, i, SK_j, PK_j, PK_i, conv\Psi^t_{ji}, coins\Psi^t_{ji})$ And then set $conv\Psi^t_{ji} \leftarrow conv\Psi^t_{ji} \cdot x \cdot m$
3	$(\texttt{RevealClient}, i, j, s)$	$\Pi^{\sigma}(1^k, i, j, SK_i, PK_i, PK_j, conv\Pi^s_{ij}, coins\Pi^s_{ij})$
4	$(\texttt{RevealServer}, j, i, t)$	$\Psi^{\sigma}(1^k, j, i, SK_j, PK_j, PK_i, conv\Psi^t_{ji}, coins\Psi^t_{ji})$
5	$(\texttt{CorruptClient}, i, SK, PK)$	$< SK_i, PK_i, conv\Pi^s_{ij}, coins\Pi^s_{ij} >_{j,s}$ And then set $SK_i \leftarrow SK$; $PK_i \leftarrow PK$
6	$(\texttt{CorruptServer}, j, SK, PK)$	$< SK_j, PK_j, conv\Psi^t_{ji}, coins\Psi^t_{ji} >_{i,t}$ And then set $SK_j \leftarrow SK$; $PK_j \leftarrow PK$
7	$(\texttt{RequestPublic})$	$< PK_i, PK_j >_{i,j}$
8	$(\texttt{Test}, i, j, s)$	Choose at random a bit θ. If $\theta = 0$ return $r \leftarrow \{0,1\}^k$; if $\theta = 1$ return $\Pi^{\sigma}(1^k, i, j, SK_i, PK_i, PK_j, conv\Pi^s_{ij}, coins\Pi^s_{ij})$.
9	Other queries (meaningless)	λ

When the adversary makes `SendClient` or `SendServer` query, oracle Π^s_{ij} or Ψ^t_{ji} calculates the answer using the description of function Π or Ψ respectively. Hence E gets the message m, and the decision **A**/**R**/* which means that she can "see" when an oracle accepts. When the adversary makes `RevealClient` or `RevealServer` query, she gets back the private output of the corresponding oracle. The most severe type of loss for a player is when the player's complete internal state becomes known to the adversary. To model this possibility, we allow `CorruptClient` and `CorruptServer` queries, from which E learns the internal state of a player, including the private key SK_i, and substitutes some new value SK for the player's long-lived key. From that point on, we assume that all other players will use the revised long-lived keys of the corrupted players. When E writes (`RequestPublic`) on its query tape, we think of that query as being answered by the oracle LL_{PK}. The return is simply all the public keys of both clients and servers.

We define that Π^s_{ij} is

1. *accepted* if $\Pi^{\delta}(1^k, i, j, SK_i, PK_i, PK_j, conv\Pi^s_{ij}, coins\Pi^s_{ij}) = \mathbf{A}$;
2. *opened* if there has been a `RevealClient` query;
3. *unopened* if it is not opened;
4. *corrupted* if there has been a `CorruptClient` query; and
5. *uncorrupted* if it is not corrupted.

Similar notations apply to the server instances.

BENIGN ADVERSARIES. We use the term, *benign adversary*, to model a reliable channel. This is used to show that a protocol is 'well-defined' which means that the protocol provides the two communicating parties (two oracles) the same session key at the end of a protocol run (when the oracles have accepted). For every $(i, j, s, t) \in I_C \times I_S \times \mathbb{N} \times \mathbb{N}$, there exists an (i, j, s, t)-benign adversary which is deterministic and always performs a single run of the protocol between Π^s_{ij} and Ψ^t_{ji} by faithfully relaying flows between these two oracles.

5.2 Definition of Security

With such a powerful adversary described above, a C-S MAKEP is considered secure if:

1. The protocol provides an instance of a client Π^s_{ij} with some assurance that she is involved in a *real-time* communication with an instance of a server Ψ^t_{ji}, and vice versa.
2. No adversary can learn anything about a session key which is held by an uncorrupted, unopened but accepted client instance Π^s_{ij} (or respective server instance Ψ^t_{ji}) and the corresponding server instance Ψ^t_{ji} (or respective client instance Π^s_{ij}) is uncorrupted and unopened but not necessarily accepted.

Before we give a formal definition of security, we need two more tools described as follows.

MATCHING CONVERSATIONS. Matching conversations [3] provide the necessary formalism to define the assurance provided to one player that she has been involved in a *real-time communication* with another player. The formal definition of matching conversations is given in Appendix A. Here, we describe the notion of $\texttt{No-Matching}^E(k)$. This notation is used to specify the event that when protocol P is run against adversary E, there exists an uncorrupted oracle which has accepted but there is no other oracle which has engaged in a matching conversation with this oracle.

PROTECTING FRESH SESSION KEYS. The notion that no adversary can learn information about fresh session keys [3,4] is formalized by using the polynomial indistinguishability approach. Specifically at the end of an experiment, the adversary should not be able to gain more than a negligible advantage on distinguishing the actual fresh session key from a random number sampled from $\{0,1\}^k$. This idea is formalized by using the type 8 query shown in Table 1. We make the following modifications to the experiment.

- The `Test` query must be the adversary's last meaningful query, and it must be asked of a fresh oracle.
- To answer the query, the oracle flips a fair coin $\theta \leftarrow \{0,1\}$. If $\theta = 0$, the oracle returns a key sampled at random from $\{0,1\}^k$. If $\theta = 1$, it returns the session key.
- The adversary's job is to guess θ and outputs a bit `Guess`.

Let $\texttt{Good-Guess}^E(k)$ be the event that $\texttt{Guess} = \theta$. Then we define

$$advantage^E(k) = 2 \cdot \Pr[\texttt{Good-Guess}^E(k)] - 1.$$

Now we give a formal definition of security which is modified from [6]:

Definition 2. *A C-S MAKEP $P = (\Pi, \Psi, LL)$ is secure if:*

1. *(Well-Defined Protocol)* *In the presence of the benign adversary on Π_{ij}^s and Ψ_{ji}^t, both oracles always accept holding the same session key, which is uniformly distributed over $\{0,1\}^k$;*

and for any adversary E:

2. *(Real-time Partner)* *If two uncorrupted oracles have matching conversations, then both oracles accept and hold the same session key;*
3. *(Authenticity of Flow)* *The probability of $\texttt{No-Matching}^E(k)$ is negligible;*
4. *(Protecting Fresh Session Key)* *$advantage^E(k)$ is negligible.*

Here we use the conventional definition of the negligible function, that is, a real-valued function $\epsilon(k)$ is *negligible* if for every $c > 0$ there exists a $k_c > 0$ such that $\epsilon(k) < k^{-c}$ for all $k > k_c$.

Before we can show that our scheme meets the conditions of Definition 2, we need to specify the cryptographic primitives on which the security of our scheme relies. The primitives used in Client-Server MAKEP are public key encryption scheme, symmetric key encryption scheme and discrete logarithm problem. Their security definitions are derived from [12,13] and are given in Appendix A.

Theorem 1. *The Client-Server MAKEP (described in Sect. 4) is a secure C-S MAKEP provided that the discrete logarithm problem is intractable, a secure symmetric key encryption scheme exists and a secure public key encryption scheme exists.*

The proof of this theorem appears in Appendix B.

6 Performance

In Sect. 4, we present the protocol without specifying any particular public key cryptographic algorithm for both TA and the server B. The choice of which is up to the target applications by taking into consideration of their systems' capabilities and constraints. The performance evaluation given here is therefore based on the number of times the cryptographic operations have to be performed, the sizes of the messages, the total number of messages sent in each protocol run and the memory requirement. We also restrict our attention to the efficiency of the client side only. Throughout this section, we also use the measurement results given in [27] to estimate the speed of the protocol running on a 16MHz Palm V with Palm OS version 3.3.

Speed. The client is required to compute one public key encryption, one symmetric key decryption, one modular exponentiation, one modular multiplication, one modular addition, two hashes and two random number generations. On the average, SHA-1 only takes 0.9 msec to digest a 128-bit binary string on the Palm V. Therefore we can ignore the time taken for hashing in our evaluation. In practice, hash functions are also used to generate pseudo-random numbers.

Thus their generation speed is comparable to that of hashing and can be ignored. For symmetric key decryption, both SSC2 [28] and ARC4 (Alleged RC4) only take about one millisecond each to decrypt a 256-bit ciphertext. Even for a block cipher, like Rijndael [8], it takes less than three milliseconds. The public key encryption is also doable if we choose a public-key cryptographic algorithm with very efficient encryption process. For example, it takes 710 msec to do a 512-bit RSA encryption when the value of the public exponent is three. If the Rabin cryptosystem is used, the encryption process will comprise one modular addition and one modular multiplication. By ignoring the overhead of doing any appropriate encoding of the plaintext, it takes only 110 msec to perform a 512-bit encryption.

We notice that in step 1 of the protocol described in Sect. 4, all the parameters can be prepared offline as precomputation. Hence only the following operations have to be done by the client during the runtime of the protocol:

1. one symmetric key decryption
2. one modular multiplication
3. one modular addition

We find that a 160-bit modular addition and multiplication can be done in 0.29 msec and 15 msec respectively on a 16MHz Palm V. Thus if k is 128, the length of ID_B is 128 bits and the length of q is 160 bits, the time taken to do the computation is less than 20 msec.

Network and Storage Efficiency. There are only three messages exchanged in a single run of the protocol. If 1024-bit RSA is chosen to be the public key cryptographic algorithm for the TA, 512-bit RSA for the server, the length of p is 512 bits and the domain parameters are publicly known, the length of $Cert_A$ would be 208 bytes. The sizes of the three messages would be 336 bytes, 32 bytes and 20 bytes respectively. Therefore, this scheme is also very suitable for wireless communications. For storage, the client needs to store a, $Cert_A$, PK_{TA}, PK_B, ID_B, r_A, x, b, β and (g, p, q) if precomputation is applied. The total memory requirement for storing these parameters is 940 bytes. The actual memory requirement depends on the specific cryptographic algorithms and their parameters set in each target application. We notice that much less memory is required if G is the group of points of an elliptic curve over a finite field and an elliptic curve cryptosystem is used by both the TA and the server.

7 Concluding Remarks

Forward Secrecy. It is clear that if the server's private key is compromised, then all the session keys from the earlier runs can be recovered from the transcripts. However, the corruption of the client may not help to reveal the session keys. Hence our scheme provides half forward secrecy [7]. Since the client may be a weak device while the server can be a strong and secure entity which support much stronger security measures than the client, we believe that forward secrecy on the client side is a much more important feature than that on the server side.

On the other side, consider that all previous sessions have been compromised, that is, the adversary E knows $\{\sigma_i\}_{0<i<n}$ where σ_i is the "algebraic" session key of the i-th session, E may not be able to reveal the client's private key, a, because of the unknown random values of b in these sessions. In brief, since r_A and r_B are nonces and no single party can control the values of the session keys, we assume that $\sigma_i \neq \sigma_j$ for $i \neq j$. Hence if h is a cryptographic hash function, the probability that $h(\sigma_i) = h(\sigma_j)$ is negligible. Similarly, since b is randomly chosen from $\mathbb{Z}_q \backslash \{0\}$, we assume that $b_i \neq b_j$ for $i \neq j$. If E wants to obtain a, E needs to solve the following equations:

$$y_i = ah(\sigma_i) + b_i \pmod q \qquad \text{for } 0 < i < n$$

with unknowns a and $\{b_i\}_{0<i<n}$. It can be seen that there are $(q-1)$ sets of possible solutions. Hence it is no easier than solving the discrete logarithm problem.

Similarly, to know the random values of b for a couple of sessions alone may not be enough to reveal the client's private key either. It is required that both b and $h(\sigma)$ of a particular session are known in order to reveal a. Alternatively, two sessions with the same value of b and having the values of $h(\sigma)$ known are also being able to compute the client's private key, namely $a \equiv (y_1 - y_2)/(h(\sigma_1) - h(\sigma_2)) \pmod q$.

Precomputation. As we mentioned earlier, the scheme benefits from the precomputation technique to significantly reduce the computational requirement during the runtime of the protocol. In most of the applications, the client can conduct the precomputation during idle time. This technique also helps to reduce the peak power consumption by averaging out most of the computations over time.

Scalability. The protocols proposed in [15,26] are either server-specific or limited by the memory capacity and the precomputation overhead. Server-specific means that the client needs to pre-determine the server she wants to communicate and has to do some precomputations offline. In this scheme, the precomputation is optional and also applications have the flexibility to choose the extent of precomputation. For example, the client can pre-select the values of r_A and b and precompute β. During runtime, she only needs to compute x using a public key encryption before sending out the first message. As mentioned in Sect. 6, it can still be done efficiently if we choose a public-key cryptographic algorithm with very efficient encryption process. In this way, the values precomputed by the client do not depend on any specific server. Therefore it gives full scalability that other public-key MAKEPs provide but with higher efficiency and fewer messages.

Hash function h. The hash function h specified in the Client-Server MAKEP is to output an integer in $\mathbb{Z}_q \backslash \{0\}$. As the client authentication is essentially the same as Schnorr's identification scheme [23,24], the requirement of h can be loosed to output a binary string in $\{0, \cdots, 2^t - 1\}$ where 2^{-t} governs the success rate of an adversary to launch a *crooked* proof attack, which is described in the proof of Claim 3.

References

1. Ashar Aziz and Whitfield Diffie. A secure communcations protocol to prevent unauthorized access – privacy and authentication for wireless local area networks. *IEEE Personal Communications*, First Quarter 1994.
2. Mihir Bellare and Phillip Rogaway. Random oracles are practical: A paradigm for designing efficient protocols. In *First ACM Conference on Computer and Communications Security*, pages 62–73, Fairfax, 1993. ACM.
3. Mihir Bellare and Phillip Rogaway. Entity authentication and key distribution. In Douglas R. Stinson, editor, *Proc. CRYPTO 93*, pages 232–249. Springer, 1994. Lecture Notes in Computer Science No. 773.
4. Mihir Bellare and Phillip Rogaway. Provably secure session key distribution– the three party case. In *Proc. 27th ACM Symp. on Theory of Computing*, pages 57–66, Las Vegas, 1995. ACM.
5. R. Bird, I. Gopal, A. Herzberg, P. Janson, S. Kutten, R. Molva, and M. Yung. Systematic design of two-party authentication protocols. In J. Feigenbaum, editor, *Proc. CRYPTO 91*, pages 44–61. Springer, 1992. Lecture Notes in Computer Science No. 576.
6. Simon Blake-Wilson, Don Johnson, and Alfred Menezes. Key agreement protocols and their security analysis. In *Sixth IMA International Conference on Cryptography and Coding*, pages 30–45, 1997. Lecture Notes in Computer Science No. 1355.
7. Simon Blake-Wilson and Alfred Menezes. Authenticated Diffie-Hellman key agreement protocols. In *5th annual international workshop, SAC'98*, pages 339–361. Springer-Verlag, 1998. Lecture Notes in Computer Science No. 1556.
8. J. Daemen and V. Rijmen. AES Proposal: Rijndael. *AES Algorithm Submission*, Sep 1999. http://www.nist.gov/aes.
9. W. Diffie and M. E. Hellman. New directions in cryptography. *IEEE Trans. Inform. Theory*, IT-22:644–654, November 1976.
10. Whitfield Diffie, Paul C. Van Oorschot, and Michael J. Wiener. Authentication and authenticated key exchanges. *Designs, Codes, and Cryptography*, 2(2):107–125, June 1992.
11. Alan O. Freier, Philip Karlton, and Paul C. Kocher. *The SSL Protocol Version 3.0.* INTERNET-DRAFT, Nov 1996. www.netscape.com/eng/ssl3/draft302.txt.
12. S. Goldwasser and S. Micali. Probabilistic encryption. *JCSS*, 28(2):270–299, April 1984.
13. Shafi Goldwasser and Mihir Bellare. *Lecture Notes on Cryptography.* www-cse.ucsd.edu/users/mihir/papers/gb.html, 1996.
14. IEEE. *P1363: Standard Specifications For Public Key Cryptography*, Nov 1999. Draft P1363 / D13.
15. Markus Jakobsson and David Pointcheval. Mutual authentication for low-power mobile devices. In P. Syverson, editor, *Proceedings of Financial Cryptography 2001.* Springer-Verlag, February 2001.
16. J. Kohl and C. Neuman. *The Kerberos Network Authentication Service (V5).* IETF RFC1510, Sep 1993.
17. Laurie Law, Alfred Menezes, Minghua Qu, Jerry Solinas, and Scott Vanstone. An efficient protocol for authenticated key agreement. Technical Report CORR 98-05, University of Waterloo, 1998.
18. C. Lim and P. Lee. A key recovery attack on discrete log-based schemes using a prime order subgroup. In *Proc. CRYPTO 97*, pages 249–263. Springer, 1997. Lecture Notes in Computer Science No. 1294.

19. Alfred J. Menezes, Paul C. van Oorschot, and Scott A. Vanstone. *Handbook of Applied Cryptography.* CRC Press LLC, 1997.
20. C. Mitchell. Limitations of challenge-response entity authentication. *Electronics Letters*, 25(17), Aug 1989.
21. R. M. Needham and M. D. Schroeder. Using encryption for authentication in large networks of computers. *Communications of the ACM*, 21(12):993–999, December 1978.
22. Dave Otway and Owen Rees. Efficient and timely mutual authentication. *Operating Systems Review*, 21, Jan 1987.
23. C. P. Schnorr. Efficient identification and signatures for smart cards. In G. Brassard, editor, *Proc. CRYPTO 89*, pages 239–252. Springer, 1990. Lecture Notes in Computer Science No. 435.
24. C. P. Schnorr. Efficient signature generation by smart cards. *Journal of Cryptology*, 4(3), 1991.
25. Victor Shoup. Lower bounds for discrete logarithms and related problems. In Walter Fumy, editor, *Proc. EUROCRYPT 97*, pages 256–266. Springer, 1997. Lecture Notes in Computer Science No. 1233.
26. Duncan S. Wong and Agnes H. Chan. Mutual authentication and key exchange for low power wireless communications. to appear in IEEE MILCOM 2001 Conference Proceedings, Oct 2001.
27. Duncan S. Wong, Hector Ho Fuentes, and Agnes H. Chan. The performance measurement of cryptographic primitives on palm devices. to appear in the Proceedings of the 17th Annual Computer Security Applications Conference, Dec 2001.
28. Muxiang Zhang, Christopher Carroll, and Agnes H. Chan. The software-oriented stream cipher SSC2. *Fast Software Encryption Workshop 2000*, 2000.

A Definitions

For matching conversations, we use the same definition as given in [3]. Without loss of generality, we may assume the number of flows R in the protocol to be ood. Let E be an adversary. For any oracle Π^s_{ij} or Ψ^t_{ji}, its *conversation* can be captured by a sequence:

$$C = (\tau_1, \alpha_1, \beta_1),\ (\tau_2, \alpha_2, \beta_2), \cdots, (\tau_m, \alpha_m, \beta_m).$$

This sequence encodes that at time τ_1 the oracle was asked α_1 and responded with β_1; at time $\tau_2 > \tau_1$, the oracle was asked α_2 and answered β_2; finally, at time τ_m, it was asked α_m and answered β_m. At time τ_m, adversary E terminates without asking any more queries. If oracle Π^s_{ij} (or Ψ^t_{ji}) has $\alpha_1 = \lambda$, it is called an *initiator oracle*; otherwise it is called a *responder oracle.*

Definition 3 ([3]). *Let P be a R-flow protocol, where $R = 2\rho - 1$ is the number of flows. Run P in the presence of an adversary E and consider two oracles, an initiator oracle, and a responder oracle, that engage in conversations C and C' respectively.*

1. *C' is said to be a* matching conversation *to C if there exist $\tau_0 < \tau_1 < \cdots < \tau_{R-1}$ and $\alpha_1, \beta_1, \cdots, \beta_{\rho-1}, \alpha_\rho$ such that C is prefixed by:*

$$(\tau_0, \lambda, \alpha_1),\ (\tau_2, \beta_1, \alpha_2), \cdots, (\tau_{2\rho-2}, \beta_{\rho-1}, \alpha_\rho)$$

and C' is prefixed by:

$$(\tau_1, \alpha_1, \beta_1),\ (\tau_3, \alpha_2, \beta_2), \cdots, (\tau_{2\rho-3}, \alpha_{\rho-1}, \beta_{\rho-1}).$$

2. *C is said to be a* matching conversation *to C' if there exist $\tau_0 < \tau_1 < \cdots < \tau_R$ and $\alpha_1, \beta_1, \cdots, \beta_{\rho-1}, \alpha_\rho$ such that C' is prefixed by:*

$$(\tau_1, \alpha_1, \beta_1),\ (\tau_3, \alpha_2, \beta_2), \cdots, (\tau_{2\rho-3}, \alpha_{\rho-1}, \beta_{\rho-1}),\ (\tau_{2\rho-1}, \alpha_\rho, *)$$

and C is prefixed by:

$$(\tau_0, \lambda, \alpha_1),\ (\tau_2, \beta_1, \alpha_2), \cdots, (\tau_{2\rho-2}, \beta_{\rho-1}, \alpha_\rho).$$

If C is a matching conversation to C' and C' is a matching conversation to C, then the two oracles are said to have had matching conversations.

The following definitions are derived from [12,13]. They are just briefly introduced here.

Definition 4. *A public key encryption scheme is a triple, $(\mathcal{G}, \mathcal{E}', \mathcal{D}')$, of probabilistic polynomial-time algorithms satisfying the following conditions:*

1. *key generation algorithm : $(e, d) \leftarrow \mathcal{G}(1^k)$ where k is the security parameter, e is the public key, and d is the corresponding private key.*
2. *encryption algorithm : $c \leftarrow \mathcal{E}'_e(1^k, m)$ where $m \leftarrow \{0,1\}^l$ is the message, $c \in \{0,1\}^*$ is the ciphertext and l is polynomial in k.*
3. *decryption algorithm : $m \leftarrow \mathcal{D}'_d(1^k, c)$.*

Definition 5. *A public key encryption scheme $(\mathcal{G}, \mathcal{E}', \mathcal{D}')$ is secure (polynomial time indistinguishable) if for every PPT algorithm E and for every polynomial Q, for all sufficiently large k,*

$$\begin{aligned}\Pr[E(1^k, e, m_0, m_1, c) = m \mid\ & (e,d) \leftarrow \mathcal{G}(1^k);\ m_0 \leftarrow \{0,1\}^k;\ m_1 \leftarrow \{0,1\}^k; \\ & m \leftarrow \{m_0, m_1\};\ c \leftarrow \mathcal{E}'_e(m)] \\ < \frac{1}{2} + \frac{1}{Q(k)} &\end{aligned}$$

B Proof of Theorem 1

Proof. (sketch) We prove the security of the protocol by establishing each condition of Definition 2.

<u>**Condition 1 and 2:**</u> The first two conditions follow immediately from the description of the Client-Server MAKEP and the assumption that H is a hash function instantiating a random oracle [2].

<u>**Condition 3:**</u> We prove this by contradiction. Assume that E is an arbitrary adversary and that $\Pr[\texttt{No-Matching}^E(k)]$ is non-negligible, then we show that certain cryptographic primitives which have assumed to be secure would be broken. We divide the proof into several subsections.

SERVER AUTHENTICATION

Consider the first two messages of the protocol shown in Fig. 3.

Claim 1 *If there exists a secure pair of symmetric key encryption scheme and public key encryption scheme, then upon receiving* $x = \mathcal{E}_{PK_B}(r_A)$*, only B can compute the second message,* $\mathcal{E}_{r_A}(r_B, \mathrm{ID}_B)$*; that is, for every PPT algorithm E, there is a negligible function* $\epsilon(k)$ *such that for sufficiently large k,*

$$\Pr[E(1^k, x, \mathrm{ID}_B, PK_B) = \mathcal{E}_{r_A}(r, \mathrm{ID}_B) \mid r_A \leftarrow \{0,1\}^k;\ x \leftarrow \mathcal{E}_{PK_B}(r_A)] \leq \epsilon(k)$$

where r is some random number of length k.

This can be shown by contradiction. Suppose that on inputs 1^k, $\mathcal{E}_{PK_B}(r_A)$, ID_B and PK_B, E computes $\mathcal{E}_{r_A}(r, \mathrm{ID}_B)$ without asking an oracle of B (i.e. without knowing SK_B) where r is some random number of length k. We can construct a machine C to break the public key encryption scheme. The following is only a sketch of the complete proof. Formally we should simulate an adversary's point of view completely like the proofs given in [3,4,6].

$C =$ "On inputs 1^k, PK_B, m_0, m_1 and $c = \mathcal{E}_{PK_B}(m)$ where $m \leftarrow \{m_0,\ m_1\}$:

1. We simulate E's view and answer all the oracle queries involved.
2. For an initiator oracle, set $x = c$ (i.e. $r_A = m$) and denote the identity of the corresponding responder as ID_B.
3. For the second query of the initiator oracle (here we think the query contains $\mathcal{E}_{r_A}(r, \mathrm{ID}_B)$), decrypt the query under m_0 and check if the decrypted message contains a proper coding of ID_B with some number.
4. If it is true, output m_0; otherwise decrypt the query under m_1 and check the validity of the decrypted message again.
5. If it passes, output m_1; otherwise give up.

Claim 2 *The first two messages of Client-Server MAKEP are sent in the correct order.*

Let the adversary E makes $Q(k)$ oracle calls. If we assume that B produces $\mathcal{E}_{r_A}(r_B, \mathrm{ID}_B)$ *before* A sends out $\mathcal{E}_{PK_B}(r_A)$, then the probability that B is queried with the correct value of $\mathcal{E}_{PK_B}(r_A)$ is at most $Q(k) \cdot 2^{-k}$ which is negligible.

CLIENT AUTHENTICATION

Claim 3 *Assuming the discrete logarithm problem is intractable, then only A can compute the correct pair* $(\beta,\ y)$ *such that* $g^y \equiv (g^a)^{h(\sigma)}\beta \pmod{p}$.

The authentication mechanism is similar to Schnorr's identification scheme [23, 24]. Its security is based on the intractability of the discrete logarithm problem where the problem instance is $\log_g g^a$. To see the forgery probability of the client authentication, we consider an adversary E who impersonates A by choosing some b, guessing the correct value of $h(\sigma)$ (which may be obtained from the guessed value of r_B or σ instead) and sending $\beta = g^b(g^a)^{-h(\sigma)}$ and then $y = b$ to B. This is called *crooked proof attack*. The probability of success for this attack is $1/\Psi$ where $\Psi = \min(|\mathcal{H}|,\ 2^k)$ and $\mathcal{H}$ denotes the range of h. In the original papers, Schnorr showed that this success rate cannot be increased unless computing the discrete logarithm is easy. Detailed security analysis can be referred to the original papers as well as [25].

Claim 4 *A cannot send y out before B sends* $\mathcal{E}_{r_A}(r_B, \mathrm{ID}_B)$.

The proof is similar to that for Claim 2.

Condition 4: As all the messages are generated by the intended parties in the right order in a single run of the protocol, it is obvious that an adversary cannot obtain any information of r_A or r_B from the messages provided that there exists a secure public key encryption scheme and a secure symmetric key encryption scheme. □

Provably Authenticated Group Diffie-Hellman Key Exchange – The Dynamic Case

Emmanuel Bresson[1], Olivier Chevassut[2,3*], and David Pointcheval[1]

[1] École Normale Supérieure, 75230 Paris Cedex 05, France
http://www.di.ens.fr/~{bresson,pointche},
{Emmanuel.Bresson,David.Pointcheval}@ens.fr.
[2] Lawrence Berkeley National Laboratory, Berkeley, CA 94720, USA,
http://www.itg.lbl.gov/~chevassu, OChevassut@lbl.gov.
[3] Université Catholique de Louvain, 31348 Louvain-la-Neuve, Belgium.

Abstract. Dynamic group Diffie-Hellman protocols for Authenticated Key Exchange (AKE) are designed to work in a scenario in which the group membership is not known in advance but where parties may join and may also leave the multicast group at any given time. While several schemes have been proposed to deal with this scenario no formal treatment for this cryptographic problem has ever been suggested. In this paper, we define a security model for this problem and use it to precisely define Authenticated Key Exchange (AKE) with "implicit" authentication as the fundamental goal, and the entity-authentication goal as well. We then define in this model the execution of a protocol modified from a dynamic group Diffie-Hellman scheme offered in the litterature and prove its security.

1 Introduction

1.1 The Group Diffie-Hellman Key Exchange

Group Diffie-Hellman schemes for Authenticated Key Exchange are designed to provide a pool of players communicating over a public network and holding long-lived secrets with a session key to be used to achieve multicast message confidentiality or multicast data integrity. In this paper, we consider the scenario in which the group membership is not known in advance – *dynamic* rather than *static* – where parties may join and leave the multicast group at any given time.

After the initialization phase, and throughout the lifetime of the multicast group, the parties need to be able to engage in a conversation after each change in the membership at the end of which the session key is updated to be sk'. The secret value sk' is only known to the party in the multicast group during the period when sk' is the session key. The adversary may generate repeated

* The second author was supported by the Director, Office of Science, Office of Advanced Scientific Computing Research, Mathematical Information and Computing Sciences Division, of the U.S. Department of Energy under Contract No. DE-AC03-76SF00098. This document is report LBNL-48202.

C. Boyd (Ed.): ASIACRYPT 2001, LNCS 2248, pp. 290–309, 2001.
© Springer-Verlag Berlin Heidelberg 2001

and arbitrarily ordered changes in the membership for subsets of parties of his choice.

The above scenario is a distributed application in which up to one hundred parties work together in order to get a task done where many of the parties may be sending data to the multicast group [13]. Examples of such applications include replicated server [22], audio-video conferencing [21] and collaborative tools [2].

Several papers [3,19,20,29] have addressed this scenario and one of its incarnations is the system offered in [1]. However these protocols, and this existing system, are based on or use an informal approach and do not rely on proofs of security. These approaches are several years later often found to be flawed and, indeed, weaknesses have already been discovered for some protocols [24]. One way to improve the security of the protocols is to complete formal proofs and thus avoid many of the weaknesses.

1.2 The Security Notions

In the paradigm of provable security [25] one identifies a concrete cryptographic problem to solve (like the group Diffie-Hellman key exchange) and defines a formal model for this problem. The model captures the capabilities of the adversary and the capabilities of the players. Within this model one defines security goals to capture what it means for a group Diffie-Hellman scheme to be secure. And, for a particular scheme one exhibits a proof of its security. The security proof aims to show that the scheme actually achieves the claimed security goals under computational assumptions.

The fundamental security goal for a group Diffie-Hellman scheme to achieve is Authenticated Key Exchange (with "implicit" authentication) identified as AKE. In AKE, each player is assured that no other player aside from the arbitrary pool of players can learn the session key. Another stronger highly desirable goal for a group Diffie-Hellman scheme to provide is Mutual Authentication (MA). In MA, each player is assured that only its partners actually have possession of the distributed session key.

With these security goals in hand the security of a group Diffie-Hellman scheme can be analyzed in the standard model or in an idealized model of computation (ideal-hash model [7,14], ideal-cipher model [5], generic model [27]). Previous security analyses in the ideal-hash model, the so-called random-oracle model [7,14] wherein the cryptographic hash functions (like SHA or MD5) are viewed as random functions, provide satisfactorily convincing guarantees of security for numerous cryptographic schemes [8,15,26] although not at the same level as those in the standard model.

1.3 Contributions

This paper provides major contributions to the solution of the group Diffie-Hellman key exchange problem. We present the first formal model to help manage the complexity of definitions and proofs for the authenticated group Diffie-

Hellman key exchange when the group membership is *dynamic*. This model is equipped with some notions of dynamicity in the group membership where the various types of attacks are modeled by queries to the players. This model does not yet encompass attacks involving multiple player's instances activated concurrently and simultaneously by the adversary. Also, in order to be correctly formalized, the intuition behind mutual authentication requires cumbersome definitions of session IDS and partner IDS which may be skipped at the first reading.

We start with the model and definitions introduced in [11] and extend them to deal with the authenticated *dynamic* group Diffie-Hellman key exchange. We define the partnering, freshness of session key and measures of security for AKE. In this model we define the execution of a protocol, we refer to it as AKE1, modified from [3] and show that it can be proven secure under reasonable and well-defined intractability assumptions.

Our paper is organized as follows. In the remainder of this section we summarize the related work. In Section 2 we define our security model. We use it in Section 3 to define the security definitions that should be satisfied by a group Diffie-Hellman scheme. We present the AKE1 protocol in Section 4 and justify its security in the random oracle model. Finally in Section 5 we briefly deal with MA in the random oracle model.

1.4 Related Work

Many group Diffie-Hellman protocols [3,4,12,16,18,28,30,31] aim to distribute a session key among the multicast group members for a scenario in which the membership is *static* and known in advance. However these protocols are not well-suited for a scenario in which members join and leave the multicast group at a relatively high rate. Fortunately, these protocols can be extended to address this latter scenario and several papers [3,19,20,29] have shown how to do so. The protocol presented in [3] has been found to be flawed in [24] and the other papers assume authenticated links, or more specifially do not consider the AKE and MA goals as part of the protocols. These goals need to be addressed separately.

A first step has already been taken toward a formal treatment of the authenticated Diffie-Hellman key exchange problem in the multi-party setting. Indeed, we presented in [11] the first formal model for this problem for a scenario in which the membership is *static*. The model was derived from Bellare et al.'s model of distributed computing [5,17]. Addressed in detail were the AKE and MA goals. For each we presented a definition, a protocol and a proof that the protocol achieves these goals.

2 The Model

In this section we formalize the group Diffie-Hellman key exchange and the adversary's capabilities. In our formalization, the players do not deviate from the protocol, the adversary is not a player and the adversary's capabilities are modeled by various queries. These queries provide the adversary a capability to initialize a multicast group via Setup-queries, add players to the multicast group via Join-queries, and remove players from the multicast group via Remove-queries.

2.1 Protocol Participants

We fix a nonempty set $\mathcal{U}$ of players that can participate in a group Diffie-Hellman key exchange protocol P. The number n of players is polynomial in the security parameter k. Also, when we mean a specific player of $\mathcal{U}$ we use U_i while when we mean a not fixed member of $\mathcal{U}$ we use U without any index.

We also consider a nonempty subset of $\mathcal{U}$ which we call the *multicast group* $\mathcal{I}$. And in $\mathcal{I}$ a player U_{GC}, the so-called "group controller", initiates the addition of players to $\mathcal{I}$ or the removal of players from $\mathcal{I}$. U_{GC} is trusted to do only this.

2.2 Long-Lived Keys

Each player $U \in \mathcal{U}$ holds a long-lived key LL_U which is either a pair of matching public/private keys or a symmetric key. Associated to protocol P is a LL-key generator $\mathcal{G}_{LL}$ which at initialization generates LL_U and assigns it to U.

2.3 Generic Group Diffie-Hellman Schemes

A group Diffie-Hellman scheme P for $\mathcal{U}$ is defined by four algorithms: (the session key SK is known by any player in $\mathcal{I}$ but unknown to any player not in $\mathcal{I}$.)

- the *key generation algorithm* $\mathcal{G}_{LL}$ which has an input of 1^k, where k is the security parameter, provides each player in $\mathcal{U}$ with a long-lived key LL_U. $\mathcal{G}_{LL}$ is a probabilistic algorithm.
- the *setup algorithm* which has an input of a set of players $\mathcal{J}$, sets variable $\mathcal{I}$ to be $\mathcal{J}$ and provides each player U in $\mathcal{I}$ with a session key SK_U. The setup algorithm is an interactive multi-party protocol between some players of $\mathcal{U}$.
- the *remove algorithm* which has an input of a set of players $\mathcal{J}$, updates variable $\mathcal{I}$ to be $\mathcal{I}\backslash\mathcal{J}$ (the set of all players in $\mathcal{I}$ that are not in $\mathcal{J}$) and provides each player U in this updated set with an updated session key SK_U. The remove algorithm is an interactive multi-party protocol between some players of $\mathcal{U}$.
- the *join algorithm* which has an input of a set of players $\mathcal{J}$, updates variable $\mathcal{I}$ to be $\mathcal{I}\cup\mathcal{J}$ and provides each player U in this updated set with an updated session key SK_U. The join algorithm is an interactive multi-party protocol between some players of $\mathcal{U}$.

An execution of P consists of running the *key generation* algorithm once, and then many times the *setup, remove* and *join* algorithms. We will also use the term *operation* to mean one of the algorithms: *setup, remove* or *join.*

Session IDS. We define the session IDS (SIDS) for player U_i in an execution of protocol P as $\text{SIDS}(U_i) = \{\text{SID}_{ij} : j \in ID\}$ where SID_{ij} is the concatenation of all flows that U_i exchanges with player U_j in executing an operation. Therefore, U_i setsSK_{U_i} to 0 and $\text{SIDS}(U_i)$ and $\emptyset$ before executing an operation. (SIDS is publicly available.)

Accepting and Terminating. A player U accepts when it has enough information to compute a session key SK_U. At any time a player U who is in "expecting state" can accept and it accepts at most once in executing an operation. As soon as U accepts in executing an operation, SK and SIDS are defined. Now once having accepted U has not yet terminated this execution. Player U may want to get confirmation that its partners in this execution have actually computed SK or that they are really the ones it wants to share a session key with. As soon as U gets this confirmation message, it terminates the execution of this operation - it will not send out any more messages and remains in a "stand by" state until the next operation.

2.4 Security Model

Queries. The adversary $\mathcal{A}$ interacts with the players U by making various queries. There are seven types of queries. The Setup, Join and Remove queries may at first seem useless since, using Send queries, the adversary already has the ability to initiate a *setup*, a *remove* or a *join* operation. Yet these queries are essential for properly dealing with the dynamic case. To deal with sequential membership changes, these three queries are only available if all the players in $\mathcal{U}$ have terminated. We now explain the capability that each kind of query captures.

- Setup($\mathcal{J}$): This query models adversary $\mathcal{A}$ initiating the *setup* operation. The query is only available to adversary $\mathcal{A}$ if all the players in $\mathcal{U}$ have terminated and are thus in a "stand by" state.. $\mathcal{A}$ gets back from the first player U in $\mathcal{J}$ the flow initiating the *setup* execution. Other players are aware of the *setup* and move to an "expecting state" but do not reply any message.
- Remove($\mathcal{J}$): This query models adversary $\mathcal{A}$ initiating the *remove* operation. The query is only available to adversary $\mathcal{A}$ if all the players in $\mathcal{U}$ have terminated. $\mathcal{A}$ gets back from the group controller U_{GC} the flow initiating the *remove* execution. Other players are aware of the *remove* operation but do not reply. They move from a "stand by" state to an "expecting state".
- Join($\mathcal{J}$): This query models adversary $\mathcal{A}$ initiating the *join* operation. The query is only available to adversary $\mathcal{A}$ if all the players in $\mathcal{U}$ have terminated. $\mathcal{A}$ gets back from the group controller U_{GC} the flow initiating the *join* execution. Other players are aware of the *join* operation but do not reply. They move from a "stand by" state to an "expecting state".
- Send(U, m): This query models adversary $\mathcal{A}$ sending a message to a player. The adversary $\mathcal{A}$ gets back from his query the response which player U would have generated in processing message m (this could be the empty string if the

message is uncorrect or unexpected). If player U has not yet terminated and the execution of protocol P leads to accepting, variable SIDS(U) is updated as explained above.

– Reveal(U): This query models the attacks resulting in the misuse of the session key, which may then be revealed. The query is only available to adversary $\mathcal{A}$ if player U has accepted. The Reveal-query unconditionally forces player U to release SK_U which is otherwise hidden to the adversary.
– Corrupt(U): This query models the attacks resulting in the player U's LL-key been revealed. $\mathcal{A}$ gets back LL_U but does not get any internal data of U executing P.
– Test(U): This query models the semantic security of the session key SK, namely the following game $\mathbf{Game}^{ake}(\mathcal{A}, P)$ between adversary $\mathcal{A}$ and the players U involved in an execution of the protocol P. The Test-query is only available if U is **Fresh** (see Section 3). In the game $\mathcal{A}$ asks any of the above queries however it can only ask a Test-query once. Then, one flips a coin b and returns sk_U if $b = 1$ or a random string if $b = 0$. At the end of the game, adversary $\mathcal{A}$ outputs a bit b' and *wins* the game if $b = b'$.

Executing the Game. Choose a protocol P with a session-key space **SK**, and an adversary $\mathcal{A}$. The security definitions take place in the context of making $\mathcal{A}$ play $\mathbf{Game}^{ake}(\mathcal{A}, P)$. P determines how players behave in response to messages from the environment. $\mathcal{A}$ sends these messages: she controls all communications between players; she can repeatedly initiate in a non-concurrent way but in arbitrary order sequential changes in the membership for subsets of players of her choice; she can at any time force a player U to divulge SK or more seriously LL_U. This game is initialized by providing coin tosses to $\mathcal{G}_{LL}$, $\mathcal{A}$, all U, and running $\mathcal{G}_{LL}(1^k)$ to set LL_U. Then

1. Initialize any U with $\mathrm{SIDS} \leftarrow \textsc{null}, \mathrm{PIDS} \leftarrow \textsc{null}, \mathrm{SK} \leftarrow \textsc{null}$,
2. Initialize adversary $\mathcal{A}$ with 1^k and access to all U,
3. Run adversary $\mathcal{A}$ and answer queries made by $\mathcal{A}$ as defined above.

3 The Definitions

In this section we present the definitions that should be satisfied by a group Diffie-Hellman scheme. We define the partnering from the session IDS and use it to define security measurements that an adversary will defeat the security goals. We also recall that a function $\varepsilon(k)$ is *negligible* if for every $c > 0$ there exists a $k_c > 0$ such that for all $k > k_c$, $\varepsilon(k) < k^{-c}$.

3.1 Partnering Using SIDS

The partnering captures the intuitive notion that the players with which U_i has exchanged messages in executing an operation, are the players with which U_i believes it has established a session key. Another simple way to understand the notion of partnering is that U_j is a partner of U_i in the execution of an operation, if U_j and U_i have directly exchanged messages or there exists some sequence of players that have directly exchanged messages from U_j to U_i.

In an execution of P, or in $\mathbf{Game}^{ake}(\mathcal{A}, P)$, we say that players U_i and U_j are **directly partnered** if both players accept and $\text{SIDS}(U_i) \cap \text{SIDS}(U_j) \neq \emptyset$ holds. We denote the direct partnering as $U_i \leftrightarrow U_j$.

We also say that players U_i and U_j are **partnered** if both players accept and if, in the graph $G_{SIDS} = (V, E)$ where $V = \{U_i \ : \ i = 1, \ldots, |\mathcal{I}|\}$ and $E = \{(U_i, U_j) \ : \ U_i \leftrightarrow U_j\}$ the following holds:

$$\exists k > 1, \prec U_1, U_2, \ldots, U_k \succ \text{ with } U_1 = U_i,\ U_k = U_j,\ U_{i-1} \leftrightarrow U_i.$$

We denote this partnering as $U_i \leftrightsquigarrow U_j$.

We complete in polynomial time (in $|V|$) the graph G_{SIDS} to obtain the graph of partnering: $G_{PIDS} = (V', E')$, where $V' = V$ and $E' = \{(U_i, U_j) \ : \ U_i \leftrightsquigarrow U_j\}$, and then define the partner IDS for oracle U_i as:

$$\text{PIDS}(U_i) = \{U_j \ : \ U_i \leftrightsquigarrow U_j\}$$

3.2 Freshness

A player U is **Fresh**, in the current operation execution, (or holds a **Fresh** SK) if the following two conditions are satisfied. First, nobody in $\mathcal{U}$ has ever been asked for a Corrupt-query from the beginning of the game. Second, in the current operation execution, U has accepted and neither U nor its partners $\text{PIDS}(U)$ have been asked for a Reveal-query.

Let's also recall that forward-secrecy entails that loss of a LL-key does not compromise the semantic security of previously-distributed session keys.

3.3 Security Notions

AKE Security. In an execution of P, we say an adversary $\mathcal{A}$ *wins* if she asks a single Test-query to a **Fresh** player U and correctly guesses the bit b used in the game $\mathbf{Game}^{ake}(\mathcal{A}, P)$. We denote the **ake advantage** as $\mathsf{Adv}_P^{ake}(\mathcal{A})$; the advantage is taken over all bit tosses. (The advantage is twice the probability that $\mathcal{A}$ will defeat the AKE security goal of the protocol minus one[1].) Protocol P is an $\mathcal{A}$**-secure AKE** if $\mathsf{Adv}_P^{ake}(\mathcal{A})$ is negligible.

MA Security. In an execution of P, we say adversary $\mathcal{A}$ violates mutual authentication (MA) if there exists an operation execution wherein a player U terminates holding $\text{SIDS}(U)$, $\text{PIDS}(U)$ and $|\text{PIDS}(U)| \neq |\mathcal{I}| - 1$. We denote the **ma success** as $\mathsf{Succ}_P^{ma}(\mathcal{A})$ and say protocol P is an $\mathcal{A}$**-secure MA** if $\mathsf{Succ}_P^{ma}(\mathcal{A})$ is negligible.

[1] $\mathcal{A}$ can trivially defeat AKE with probability $1/2$, multiplying by two and substracting one rescales the probability.

Therefore to deal with mutual authentication, we consider a new game, we denote $\mathbf{Game}^{ma}(\mathcal{A}, P)$, wherein the adversary exactly plays the same way as in the game $\mathbf{Game}^{ake}(\mathcal{A}, P)$ with the same player accesses but with a different goal: to violate the mutual authentication.

Secure Signature Schemes. A signature scheme is defined by the following [26]:

- Key generation algorithm $\mathcal{G}$. On input 1^k with security parameter k, the algorithm $\mathcal{G}$ produces a pair (K_p, K_s) of matching public and secret keys. Algorithm $\mathcal{G}$ is probabilistic.
- Signing algorithm Σ. Given a message m and (K_p, K_s), Σ produces a signature σ. Algorithm Σ might be probabilistic.
- Verification algorithm V. Given a signature σ, a message m and K_p, V tests whether σ is a valid signature of m with respect to K_s. In general, algorithm V is not probabilistic.

The signature scheme is (t, ε)-**CMA-secure** if there is no adversary $\mathcal{A}$ which can get a probability greater than ε in mounting an existential forgery under an adaptively Chosen-Message Attack (CMA) within time t. We denote this probability ε as $\mathsf{Succ}_{\Sigma}^{cma}(\mathcal{A})$.

3.4 Diffie-Hellman Problems

Computational Diffie-Hellman Assumption (CDH). Let $\mathbb{G}$ be a cyclic group $< g >$ of prime order q and x_1, x_2 chosen at random in $\mathbb{Z}_q$. A (T, ε)-CDH-attacker in $\mathbb{G}$ is a probabilistic Turing machine Δ running in time T that given (g^{x_1}, g^{x_2}), outputs $g^{x_1 x_2}$ with probability at least ε. We denote this probability by $\mathsf{Succ}_{\mathbb{G}}^{cdh}(\Delta)$. The CDH problem is (T, ε)-**intractable** if there is no (T, ε)-attacker in $\mathbb{G}$.

Group Computational Diffie-Hellman Assumption (G-CDH). Let $\mathbb{G}$ be a cyclic group $< g >$ of prime order q and a polynomial-bounded integer n. Let I_n be $\{1, \ldots, n\}$, $\mathcal{P}(I_n)$ be the set of all subsets of I_n and Γ be a subset of $\mathcal{P}(I_n)$ such that $I_n \notin \Gamma$.

We define the *Group Diffie-Hellman distribution* relative to Γ as:

$$G\text{-}CDH_{\Gamma} = \Big\{ \bigcup_{J \in \Gamma} (J, g^{\prod_{j \in J} x_j}) \mid x = (x_1, \ldots, x_n) \in_R \mathbb{Z}_p^n \Big\}.$$

If $\Gamma = \mathcal{P}(I) \backslash \{I_n\}$, we say that $G\text{-}CDH_{\Gamma}$ is the **Full** *Generalized Diffie-Hellman distribution* [9,23,30].

Given Γ, a (T, ε)-G-CDH$_{\Gamma}$-attacker in $\mathbb{G}$ is a probabilistic Turing machine Δ running in time T that given $G\text{-}CDH_{\Gamma}$ outputs $g^{x_1 \cdots x_n}$ with probability at least ε. We denote this probability by $\mathsf{Succ}_{\mathbb{G}}^{gcdh}(\Delta)$. The G-CDH$_{\Gamma}$ problem is (T, ε)-**intractable** if there is no (T, ε)-G-CDH$_{\Gamma}$-attacker in $\mathbb{G}$.

Random Self-Reducibility of CDH and G-CDH. In a prime-order group $\mathbb{G}$, the CDH and G-CDH are random self-reducible problems [23]. Informally, this property means that solving the problem on any original instance $\mathcal{D}$ can be reduced to solving the problem on a random instance $\mathcal{D}'$. This requires an efficient way to generate the random instances $\mathcal{D}'$ from the original instance $\mathcal{D}$ and an efficient way to compute the solution to the problem on $\mathcal{D}'$ from the solution to the problem on $\mathcal{D}$.

Certainly the most common is the additive random self-reducibility of the CDH and G-CDH problems. We examplify this property for the G-CDH problem. Given, for example, an instance $\mathcal{D} = (g^a, g^b, g^c, g^{ab}, g^{bc}, g^{ac})$ for any a, b, c it is possible to generate a random instance

$$\mathcal{D}' = (g^{(a+\alpha)}, g^{(b+\beta)}, g^{(c+\gamma)}, g^{(a+\alpha).(b+\beta)}, g^{(b+\beta).(c+\gamma)}, g^{(a+\alpha).(c+\gamma)})$$

where α, β and γ are random numbers in $\mathbb{Z}_q$; however the cost of such a computation may be high. And given the solution $z = g^{(a+\alpha).(b+\beta).(c+\gamma)}$ to the instance $\mathcal{D}'$ it is possible to recover the solution g^{abc} to the random instance $\mathcal{D}$ (i.e. $g^{abc} = z(g^{ab})^{-\gamma}(g^{ac})^{-\beta}(g^{bc})^{-\alpha}(g^a)^{-\beta\gamma}(g^b)^{-\alpha\gamma}(g^c)^{-\alpha\beta}g^{-\alpha\beta\gamma}$). It is, in effect, easy to see that such a reduction works only if $\mathcal{D}$ is the **Full** Generalized DH distribution and that its cost increases exponentially with the size of $\mathcal{D}$.

The other one is the multiplicative random self-reducibility of the CDH and G-CDH problems. The property holds if $\mathbb{G}$ is a prime-order cyclic group. We examplify this property for the G-CDH problem. Given, for example, an instance $\mathcal{D} = (g^a, g^b, g^{ab}, g^{ac})$ for any a, b, c it is easy to generate a random instance $\mathcal{D}' = (g^{a\alpha}, g^{b\beta}, g^{ab\alpha\beta}, g^{ac\alpha\gamma})$ where α, β and γ are random numbers in $\mathbb{Z}_q^*$. And given the solution $g^{a\alpha b\beta c\gamma}$ to the instance $\mathcal{D}'$ it is easy to see that the solution g^{abc} to the random instance $\mathcal{D}$ can be efficiently computed (i.e. $g^{abc} = \left(g^{a\alpha b\beta c\gamma}\right)^{(\alpha\beta\gamma)^{-1}}$). Such a reduction is efficient and only requires a linear number of modular exponentiations.

Adversary's Resources. The security is formulated as a function of the amount of resources the adversary $\mathcal{A}$ expends. The resources are:

- T-time of computing;
- $q_s, q_r, q_c, Q_S, Q_R, Q_J$ numbers of Send, Reveal, Corrupt, Setup, Remove and Join queries the adversary $\mathcal{A}$ respectively makes.

By notation $\mathsf{Adv}(T, \ldots)$ or $\mathsf{Succ}(T, \ldots)$, we mean the maximum values of $\mathsf{Adv}(\mathcal{A})$ or $\mathsf{Succ}(\mathcal{A})$ respectively, over all adversaries $\mathcal{A}$ that expend at most the specified amount of resources.

4 A Secure Authenticated Group Diffie-Hellman Scheme

In the following theorem and proof we assume the random oracle model [6] and denote $\mathcal{H}$ a hash function from $\{0,1\}^*$ to $\{0,1\}^\ell$, where ℓ is a security parameter.

U_1 U_2 U_3 U_4

$x_1 \stackrel{R}{\leftarrow} [1, q-1]$ $x_2 \stackrel{R}{\leftarrow} [1, q-1]$ $x_3 \stackrel{R}{\leftarrow} [1, q-1]$ $x_4 \stackrel{R}{\leftarrow} [1, q-1]$

$X_1 := \{g, g^{x_1}\}$ $Fl_1 := \{\mathcal{I} \| X_1\}$ $\xrightarrow{[Fl_1]_{U_1}}$

$V(Fl_1) \stackrel{?}{=} \text{True}$ $X_2 := \{g^{x_2}, g^{x_1}, g^{x_1 x_2}\}$ $Fl_2 := \{\mathcal{I} \| X_2\}$ $\xrightarrow{[Fl_2]_{U_2}}$

$V(Fl_2) \stackrel{?}{=} \text{True}$ $X_3 := \{g^{x_2 x_3}, g^{x_1 x_3}, g^{x_1 x_2}, g^{x_1 x_2 x_3}\}$ $Fl_3 := \{\mathcal{I} \| X_3\}$ $\xrightarrow{[Fl_3]_{U_3}}$

$V(Fl_3) \stackrel{?}{=} \text{True}$ $X_4 := \{g^{x_2 x_3 x_4}, g^{x_1 x_3 x_4}, g^{x_1 x_2 x_4}, g^{x_1 x_2 x_3}\}$ $Fl_4 := \{X \| \mathcal{I} \| X_4\}$ $\xleftarrow{[Fl_4]_{U_4}}$

$\xleftarrow{[Fl_4]_{U_4}}$

$\xleftarrow{[Fl_4]_{U_4}}$

$V(Fl_4) \stackrel{?}{=} \text{True}$ $K := (g^{x_2 x_3 x_4})^{x_1}$ $sk_{U_1} := \mathcal{H}(\mathcal{I} \| Fl_4 \| K)$

$V(Fl_4) \stackrel{?}{=} \text{True}$ $K := (g^{x_1 x_3 x_4})^{x_2}$ $sk_{U_2} := \mathcal{H}(\mathcal{I} \| Fl_4 \| K)$

$V(Fl_4) \stackrel{?}{=} \text{True}$ $K := (g^{x_1 x_2 x_4})^{x_3}$ $sk_{U_3} := \mathcal{H}(\mathcal{I} \| Fl_4 \| K)$

$K := (g^{x_1 x_2 x_3})^{x_4}$ $sk_{U_4} := \mathcal{H}(\mathcal{I} \| Fl_4 \| K)$

Fig. 1. Algorithm SETUP1. An example of an honest execution with 4 players: $\mathcal{J} = \{U_1, U_2, U_3, U_4\}$. The multicast group is $\mathcal{I} = \{U_1, U_2, U_3, U_4\}$ and the shared session key is $sk = \mathcal{H}(\mathcal{I} \| Fl_4 \| g^{x_1 x_2 x_3 x_4})$. The partner IDS for U_1 is $pids_{U_1} = \{U_2, U_3, U_4\}$, for U_2 is $pids_{U_2} = \{U_1, U_3, U_4\}$, for U_3 is $pids_{U_3} = \{U_1, U_2, U_4\}$ and for U_4 is $pids_{U_4} = \{U_1, U_3, U_4\}$.

The session-key space **SK** associated to this protocol is $\{0,1\}^\ell$ equipped with a uniform distribution. The arithmetic is in a finite cyclic group $\mathbb{G} =< g >$ of order a k-bit prime number q and the operation is denoted multiplicatively. This group could be a prime subgroup of $\mathbb{Z}_p^*$, or it could be an (hyper)-elliptic curve based group.

4.1 Description

The AKE1 protocol consists of the SETUP1, REMOVE1 and JOIN1 algorithms. As illustrated by an AKE1 execution in Figures 1, 2 and 3 (an execution with more steps can be found in the full version [10]), this is a protocol wherein the players are arranged in a ring, and wherein each player saves the set of values it receives in the down-flow of SETUP1, REMOVE1, JOIN1. In effect, in the subsequent removal of players from $\mathcal{I}$ any player U could be selected as U_{GC} and so will need these values to execute REMOVE1.

Unlike [3], this is a protocol wherein the player with the highest-index in $\mathcal{I}$ is the group controller, the flows are signed using the long-lived key LL_U, the names of the players are in the protocol flows, and the session key SK is $sk = \mathcal{H}(\mathcal{I} \| Fl_{max(\mathcal{I})} \| g^{x_1 \cdots x_{\max(\mathcal{I})}})$; $Fl_{max(\mathcal{I})}$ is the down-flow, SIDS and PIDS are appropriately defined. The notion of index models "pre-existing" relationships among players: for example, it may capture different levels of reliability (i.e. the higher the index is, the more reliable the player). This is also a protocol,

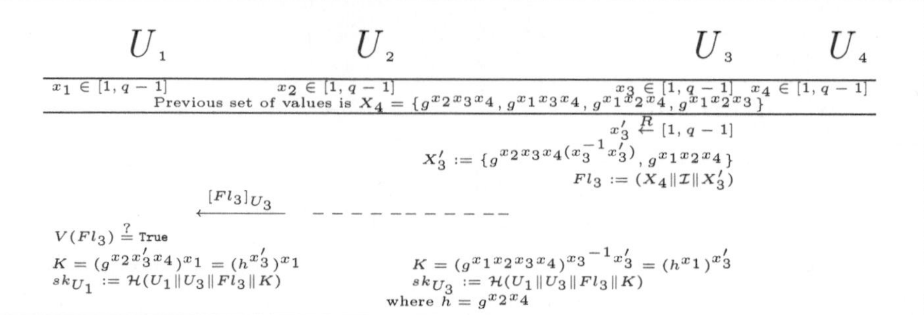

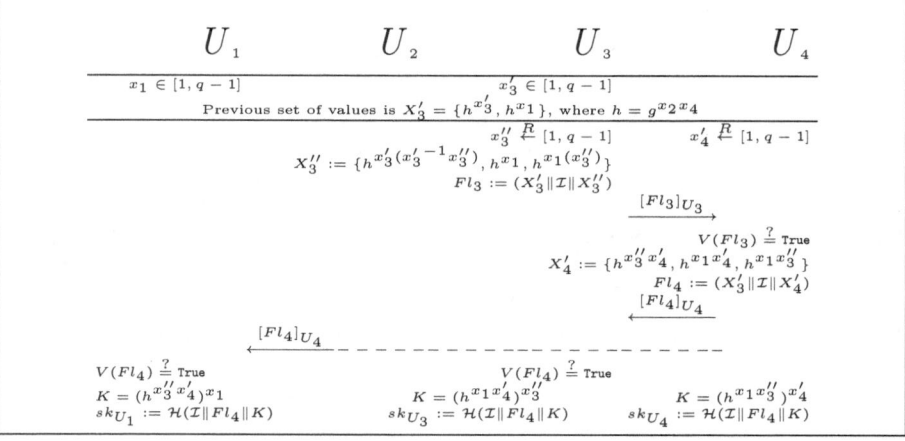

Fig. 2. Algorithm REMOVE1. An example of an honest execution with 4 players: $\mathcal{I} = \{U_1, U_2, U_3, U_4\}$, $\mathcal{J} = \{U_2, U_4\}$. The new multicast group is $\mathcal{I} = \{U_1, U_3\}$, $U_{GC} = U_3$ and the shared session key is $sk = \mathcal{H}(\mathcal{I} \| Fl_3 \| g^{x_1 x_2 x'_3 x_4})$, the partner IDS for U_1 is $pids_{U_1} = \{U_3\}$, for U_3 is $pids_{U_3} = \{U_1\}$.

U_1 U_2 U_3 U_4

$x_1 \in [1, q-1]$ $x'_3 \in [1, q-1]$

Previous set of values is $X'_3 = \{h^{x'_3}, h^{x_1}\}$, where $h = g^{x_2 x_4}$

U_3: $x''_3 \stackrel{R}{\leftarrow} [1, q-1]$
$X''_3 := \{h^{x'_3 (x'^{-1}_3 x''_3)}, h^{x_1}, h^{x_1 (x''_3)}\}$
$Fl_3 := (X'_3 \| \mathcal{I} \| X''_3)$
$[Fl_3]_{U_3} \rightarrow U_4$

U_4: $x'_4 \stackrel{R}{\leftarrow} [1, q-1]$
$V(Fl_3) \stackrel{?}{=}$ True
$X'_4 := \{h^{x''_3 x'_4}, h^{x_1 x'_4}, h^{x_1 x''_3}\}$
$Fl_4 := (X'_3 \| \mathcal{I} \| X'_4)$
$[Fl_4]_{U_4}$ (broadcast)

U_1: $V(Fl_4) \stackrel{?}{=}$ True, $K = (h^{x''_3 x'_4})^{x_1}$, $sk_{U_1} := \mathcal{H}(\mathcal{I} \| Fl_4 \| K)$

U_3: $V(Fl_4) \stackrel{?}{=}$ True, $K = (h^{x_1 x'_4})^{x''_3}$, $sk_{U_3} := \mathcal{H}(\mathcal{I} \| Fl_4 \| K)$

U_4: $K = (h^{x_1 x''_3})^{x'_4}$, $sk_{U_4} := \mathcal{H}(\mathcal{I} \| Fl_4 \| K)$

Fig. 3. Algorithm JOIN1. An example of an honest execution with 4 players: $\mathcal{I} = \{U_1, U_3\}$, $\mathcal{J} = \{U_4\}$ and $U_{GC} = U_3$. The new multicast group is $\mathcal{I} = \{U_1, U_3, U_4\}$ and the shared session key is $sk = \mathcal{H}(\mathcal{I} \| Fl_4 \| g^{x_1 x_2 x''_3 (x_4 x'_4)})$ The partner IDS for U_1 is $pids_{U_1} = \{U_3, U_4\}$, for U_3 is $pids_{U_3} = \{U_1, U_4\}$ and for U_4 is $pids_{U_4} = \{U_1, U_3\}$.

unlike [3], where the set of values from the down-flow is included in the flows of REMOVE1 and JOIN1, which avoids replay attacks.

Algorithm SETUP1. The algorithm consists of two stages: up-flow and down-flow. The multicast group $\mathcal{I}$ is set to $\mathcal{J}$. As illustrated by the example in Figure 1, in the up-flow the player U_i receives a set (Y, Z) of intermediate values, with

$$Y = \bigcup_{0<m<i} \{Z^{1/x_m}\} \text{ and } Z, \text{ where } Z = g^{\prod_{0<t<i} x_t}.$$

Player U_i chooses at random a private value x_i, raises the values in Y to the power of x_i and then concatenates with Z to obtain his intermediate values

$$Y' = \bigcup_{0<m\leq i} \{Z'^{1/x_m}\}, \text{ where } Z' = Z^{x_i} = g^{\prod_{0<t\leq i} x_t}.$$

Player U_i then forwards the values (Y', Z') to the next player in the ring. The down-flow takes place when $U_{max(\mathcal{I})}$ receives the last up-flow. At that point $U_{max(\mathcal{I})}$ performs the same steps as a player in the up-flow but broadcasts the set of intermediate values Y' only. In effect, the value Z' computed by $U_{max(\mathcal{I})}$ will lead to the session key sk, since $Z' = g^{\prod_{0<t\leq n} x_t}$. Players in $\mathcal{I}$ compute sk and accept.

Algorithm REMOVE1. This algorithm consists of a down-flow only. The multicast group $\mathcal{I}$ is first set to $\mathcal{I}\backslash\mathcal{J}$. As illustrated in Figure 2, the group controller U_{GC} (i.e. player with the highest-index in $\mathcal{I}\backslash\mathcal{J}$) generates a random value x'_{GC} and removes from the saved previous broadcast the values destinated to the players in $\mathcal{J}$. U_{GC} then raises all the remaining values in which x_{GC} appeared to the power of $(x_{GC}^{-1}.x'_{GC})$ and broadcasts the result. (x_{GC} is U_{GC}'s previous secret value.) Players in $\mathcal{I}$ compute sk and accept. Players in $\mathcal{J}$ erase any internal data. U_{GC} erases x_{GC} and x_{GC}^{-1} while internally saving x'_{GC}.

Algorithm JOIN1. This algorithm consists of two stages: up-flow and down-flow. As illustrated in Figure 3, the group controller U_{GC} (i.e. player with the highest-index in $\mathcal{I}$) generates a random value x'_{GC}, raises the values from the saved previous broadcast in which x_{GC} appears to the power of $(x_{GC}^{-1}.x'_{GC})$ and obtains a set of values Y'. (x_{GC} is U_{GC}'s previous secret exponent.) U_{GC} also computes the value Z' by raising the last value in Y' to x'_{GC}. U_i then forwards the values (Y', Z') to the first joining player in $\mathcal{J}$. From that point JOIN1 will work as the SETUP1 algorithm. Upon receiving the brodcast flow players in $\mathcal{I} \cup \mathcal{J}$ erase previous session keys, compute sk and accept. The multicast group $\mathcal{I}$ is then set to $\mathcal{I} \cup \mathcal{J}$.

4.2 Security Result

Theorem 1. *Let P be the AKE1 protocol,* ***SK*** *be the session-key space and $\mathcal{G}$ be the associated LL-key generator. Let $\mathcal{A}$ be an adversary against the AKE security of P within a time bound T, on a multicast group of size s among the n players in $\mathcal{U}$, after $Q = Q_S + Q_J + Q_R$ interactions with the parties, q_s send-queries and q_h hash-queries. Then we have:*

$$\mathsf{Adv}_P^{ake}(T, Q, q_s, q_h) \leq 2Q \cdot \binom{n}{s} \cdot s \cdot q_h \cdot \mathsf{Succ}_{\mathbb{G}}^{gcdh_{\Gamma_s}}(T') + 2n \cdot \mathsf{Succ}_{\Sigma}^{cma}(T', Q + q_s)$$

where $T' \leq T + (Q + q_s)nT_{exp}(k)$; $T_{exp}(k)$ is the time of computation required for an exponentiation modulo a k-bit number and Γ_s corresponds to the elements the adversary $\mathcal{A}$ can possibly view:

$$\Gamma_s = \bigcup_{2 \leq j \leq s-2} \{\{i \mid 1 \leq i \leq j, i \neq l\} \mid 1 \leq l \leq j\} \bigcup \{\{i \mid 1 \leq i \leq s, i \neq k, l\} \mid 1 \leq k, l \leq s\}.$$

Let us just highlight the main ideas. We consider an adversary $\mathcal{A}$ attacking the protocol P and then "breaking" the AKE security. $\mathcal{A}$ would have carried out her attack in different ways: (1) she may have gotten her advantage by forging a signature with respect to some player's long-lived public key. We will then use $\mathcal{A}$ to build a forger by "guessing" for which player $\mathcal{A}$ will produce her forgery, (2) she may have broken the scheme without altering the content of the flows. We will use it to solve an instance of the G-CDH problem, by "guessing" the moment at which $\mathcal{A}$ will make the Test-query and by injecting into the game the elements from the instance of G-CDH received as input.

To work (2) requires two things. We first "guess" the moment of the Test-query which means that we have to "guess": the number of operations that will occur before the adversary makes the Test-query and the membership of the multicast group when the adversary makes the Test-query. Second, based on this guess we "embed" the instance of G-CDH into the protocol. We generate many random instances from the original instance of G-CDH using the (multiplicative) random self-reducibility property of the G-CDH problem[2]. Indeed, the group Diffie-Hellman secret key relative to these random instances can efficiently be computed from the group Diffie-Hellman secret relative to the original instance.

The specific structure of Γ_s (see figure 4 for Γ_4) makes the simulation perfectly indistinguishable from the adversary point of view if our guesses are all correct.

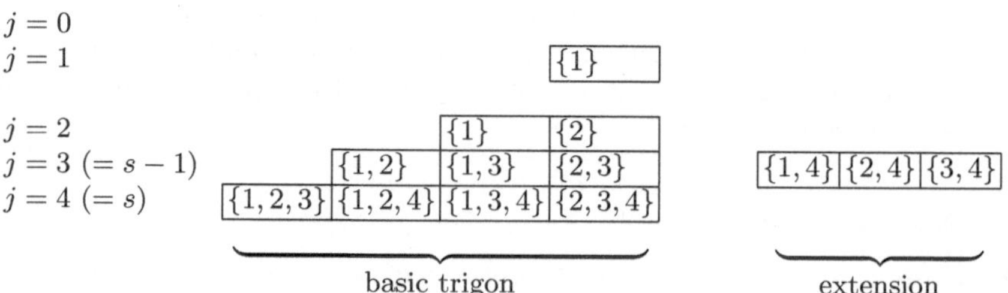

Fig. 4. Extended Trigon for Γ_4

But then, because of the random oracle $\mathcal{H}$, to have any information about the session key the adversary wants to test, she has to have asked for $\mathcal{H}(\mathcal{I} \| Fl_{last} \| K)$,

[2] The multiplicative random self-reducibility will lead to a far more efficient reduction than the additive one would do.

where K is the value we are looking for. Therefore, if the adversary has some advantage in breaking the AKE security, this value K can be found in the list of the queries asked to $\mathcal{H}$. The details of the simulation can be found in appendix A.

4.3 AKE1 in Practice

We want our results to be practical. This means that when system designers choose a scheme they will take into account its security but also its efficiency in terms of computation, communication, ease of integration and so on. However, if provable security is achieved at the cost of a loss of efficiency, system designers will often prefer the heuristic schemes.

AKE1 is to date the first group Diffie-Hellman scheme to exhibit a proof that it achieves a strong notion of security. It is secure in the random oracle model under the G-CDH assumption. It thus provides stronger security guarantees than other schemes [3,12,18] while being more efficient than [3]. However security proofs for existing schemes or slight variants may show up.

On the integration front, the question that may be raised is what happens when several groups merge to form a larger group. A scenario that occurs in practice when a network failure partitions the multicast group in several disjoints sub-groups which will later need to merge when the network is be repaired [1]. The most efficient way in terms of computation and communication is to add players from the smaller sub-groups into the largest of the merging sub-groups. That is, U_{GC} is chosen as the player with the highest-index in the largest merging sub-group and the players from the smaller sub-groups are added via the JOIN1 algorithm.

5 Mutual Authentication

The well-known approach [5] for turning an AKE protocol into a protocol that provides mutual authentication (MA) is to use the shared session key to construct a simple "authenticator" for the other parties. We have described in [11] the transformation for turning an AKE group Diffie-Hellman scheme into a protocol providing MA and justified its security in the random-oracle model. We turn an AKE *dynamic* group Diffie-Hellman scheme into a protocol providing MA by simply applying the transformation MA described in [11] to the setup, join and remove algorithms respectively.

6 Conclusion and Further Research

This paper provides the first formal treatment of the authenticated group Diffie-Hellman key exchange problem in a scenario in which the membership is *dynamic* rather than static. Addressed in this paper were two security goals of the group Diffie-Hellman key exchange: the authenticated key exchange and the mutual authentication. For each we presented a definition, a protocol and a security proof in the random oracle model that the protocol meets its goals.

The model introduced in this paper captures attacks that are realistic threats in practice. However the model does not yet capture "more serious" attacks: indeed, it does not recognize multiple player's instances the adversary may activate in concurrent and simultaneous sessions. A typical research topic is to enhance our model to capture these attacks and to investigate in this more stringent setting the security of the protocols presented in this paper. We are currently extending our model to encompass these attacks.

The security reduction presented for AKE1 in this paper does not inject much of the security of the group computational Diffie-Hellman problem and signature scheme: actually, the reduction is exponential in s. This leads one to use a larger security parameter or to limit the maximum size of the group. Another research direction is to find a security proof that would achieve a better security bound. We believe it is possible and are currently working on it.

Acknowledgements. The authors thank Deborah Agarwal and Jean-Jacques Quisquater for many insightful discussions and comments on an early draft of this paper. The authors also thank the anonymous referees for their useful comments.

References

1. D. A. Agarwal, O. Chevassut, M.R. Thompson, and G. Tsudik. An Integrated Solution for Secure Group Communication in Wide-Area Networks. In *Proc. of 6th IEEE Symposium on Computers and Communications*, 2001.
2. D. A. Agarwal, S. R. Sachs, and W. E. Johnston. The Reality of Collaboratories. *Computer Physics Communications*, 10(issue 1-3):pages 270–299, coverdate May 1998.
3. G. Ateniese, M. Steiner, and G. Tsudik. New Multiparty Authentication Services and Key Agreement Protocols. *IEEE Journal of Selected Areas in Communications*, April 2000.
4. K. Becker and U. Wille. Communication Complexity of Group Key Distribution. In *5th ACM Conference on Computer and Communications Security*, pages 1–6, November 1998.
5. M. Bellare, D. Pointcheval, and P. Rogaway. Authenticated Key Exchange Secure Against Dictionary Attacks. In B. Preneel, editor, *Proc. of Eurocrypt '00*, volume 1807 of *Lecture Notes in Computer Science*, pages 139–155. Springer-Verlag, 2000.
6. M. Bellare and P. Rogaway. Entity Authentification and Key Distribution. In D.R. Stinson, editor, *Proc. of Crypto '93*, Lecture Notes in Computer Science. Springer-Verlag, 1993.
7. M. Bellare and P. Rogaway. Random Oracles are Practical: a Paradigm for Designing Efficient Protocols. In *Proc of ACM CCS '93.* ACM Press, 1993.
8. M. Bellare and P. Rogaway. The Exact Security of Digital Signatures: How to sign with RSA and Rabin. In U. Maurer, editor, *Proc of Eurocrypt'96*, Lecture Notes in Computer Science. Springer-Verlag, 1996.
9. D. Boneh. The Decision Diffie-Hellman Problem. In *Third Algorithmic Number Theory Symposium*, volume 1423 of *Lecture Notes in Computer Science*, pages 48–63. Springer-Verlag, 1998.

10. E. Bresson, O. Chevassut, and D. Pointcheval. Provably Group Diffie-Hellman Key Exchange – The Dynamic Case. Technical report, December 2001. Full version of this paper, available at `http://www.di.ens.fr/~pointche`.
11. E. Bresson, O. Chevassut, D. Pointcheval, and J. J. Quisquater. Provably Group Diffie-Hellman Key Exchange. In *Proc. of 8th ACM Conference on Computer and Communications Security*, Nov 2001.
12. M. Burmester and Y. Desmedt. A Secure and Efficient Conference Key Distribution System. In A. De Santis, editor, *Proc of Eurocrypt' 94*, volume 950 of *Lecture Notes in Computer Science*, pages 275–286. Springer-Verlag, 1995.
13. R. Canetti, J. Garay, G. Itkis, D. Micciancio, M. Naor, and B. Pinkas. Issues in Multicast Security: A Taxonomy and Efficient Constructions. In *Proc. of INFOCOM '99*, March 1999.
14. R. Canetti, O. Goldreich, and S. Halevi. The Random Oracle Methodology, Revisited. In *Proc of. Symposium on the Theory of Computing (SOC)*. ACM, March 1998.
15. E. Fujisaki, T. Okamoto, D. Pointcheval, and J. Stern. RSA-OAEP is Secure under the RSA Assumption. In *Proc of. Crypto'01*, August 2001.
16. I. Ingemarsson, D. Tang, and C. Wong. A Conference Key Distribution System. In *IEEE Transactions on Information Theory*, volume 28(5), pages 714–720, September 1982.
17. M. Jakobsson and D. Pointcheval. Mutual Authentication for Low-Power Mobile Devices. In *Proc. of Financial Cryptography '2001*, 2001.
18. M. Just and S. Vaudenay. Authenticated Multi-Party Key Agreement. In *Proc. of ASIACRYPT'96*, volume 1163 of *Lecture Notes in Computer Science*, pages 36–49. Springer-Verlag, 1996.
19. Y. Kim, A. Perrig, and G. Tsudik. Simple and Fault-Tolerant Key Agreement for Dynamic Collaborative Group. In *Proc. of ACM Conference on Computer and Communications Security (CCS-7)*, November 2000.
20. Y. Kim, A. Perrig, and G. Tsudik. Communication-Efficient Group Key Agreement. In *Proc. of International Federation for Information Processing (IFIP SEC 2001)*, June 2001.
21. S. McCanne and V. Jacobson. vic: A Flexible Framework for Packet Video. In *ACM Multimedia '95*, pages 511–522, November 1995.
22. L.E. Moser, P.M. Melliar-Smith, and P. Narasimhan. Consistent Object Replication in the Eternal System. *Theory and Practice of Object Systems*, 4(2):pages 81–92, 1998.
23. M. Naor and O. Reingold. Number-Theoretic Constructions of Efficient Pseudo-Random Functions. In *Proc. of 38th IEEE FOCS Symposium*, pages 458–467, 1997.
24. O. Pereira and J. J. Quisquater. A Security Analysis of the Cliques Protocols Suites. In *14-th IEEE Computer Security Foundations Workshop*. IEEE Computer Society Press, June 2001.
25. D. Pointcheval. Secure Designs for Public-Key Cryptography based on the Discrete Logarithm. *To appear in Discrete Applied Mathematics*, Elsevier Science, 2001.
26. D. Pointcheval and J. Stern. Security Arguments for Digital Signatures and Blind Signatures. *J. of Cryptology*, 13(3):361–396, 2000.
27. V. Shoup. Lower Bounds for Discrete Logarithms and Related Problems. In W. Fumy, editor, *Proc. of Eurocrypt '97*, volume 1233 of *Lecture Notes in Computer Science*, pages 256–266. Springer-Verlag, 1997.

28. D. Steer, L. Strawczynski, W. Diffie, and M. Wiener. A Secure Audio Teleconference System. In S. Goldwasser, editor, *Proc. of Crypto' 88*, volume 403 of *Lecture Notes in Computer Science*, pages 520–528. Springer-Verlag, 1988.
29. M. Steiner, G. Tsudik, and M. Waidner. Key Agreement in Dynamic Peer Groups. In *IEEE Transactions on Parallel and Distributed Systems*, August 2000.
30. M. Steiner, G. Tsudik, and M. Waidner. Diffie-Hellman Key Distribution Extended to Groups. In *ACM CCS'96*, March 1996.
31. Wen-Guey Tzeng. A Practical and Secure Fault-Tolerant Conference-Key Agreement Protocol. In *Proc. of PKC2000*, Lecture Notes in Computer Science. Springer-Verlag, 2000.

A Proof of Theorem 1

Let $\mathcal{A}$ be an adversary that can get an advantage ε in breaking the AKE security of protocol P within time T. We construct from it a (T'', ε'')-forger $\mathcal{F}$ and a (T', ε')-G-CDH$_{\Gamma_s}$-attacker Δ.

Forger $\mathcal{F}$. Let's assume that $\mathcal{A}$ breaks the protocol P by forging, with probability greater than ν, a signature with respect to some player's (public) LL-key (Of course before $\mathcal{A}$ corrupts U). We construct from it a (T'', ε'')-forger $\mathcal{F}$ which outputs a forgery (σ, m) with respect to a given (public) LL-key K_p, produced by $\mathcal{G}_{LL}(1^k)$. This forger works exactly as in [11]. A detailed description can be found in the full version [10].

G-CDH$_{\Gamma_s}$-attacker Δ. Let's assume that $\mathcal{A}$ breaks the protocol P without producing a forgery. Here, with probability smaller than ν, the (valid) flows signed using LL$_U$ come from player U and not from $\mathcal{A}$ (Of course before $\mathcal{A}$ corrupts U). The replay attacks involving the flows of JOIN1 and REMOVE1 do not also need to be considered since the values from the previous broadcast are included in these flows. One may then worry about replay attacks against SETUP1, however SETUP1 has already been proved to be secure for concurrent executions by Bresson et al. [11].

We now construct from $\mathcal{A}$ a (T', ε')-G-CDH$_{\Gamma_s}$-attacker Δ that receives as input an instance $\mathcal{D}$ of G-CDH$_{\Gamma_s}$ with random size s and outputs the Diffie-Hellman secret value (i.e $g^{x_1 \cdots x_s}$) relative to this instance. More precisely, a G-CDH$_{\Gamma_s}$ with size $s \in [1, n]$ and Γ_s of the form

$$\Gamma_s = \bigcup_{2 \leq j \leq s-2} \{\{i \,|\, 1 \leq i \leq j, i \neq l\} \,|\, 1 \leq l \leq j\} \bigcup \{\{i \,|\, 1 \leq i \leq s, i \neq k, l\} \,|\, 1 \leq k, l \leq s\}.$$

This in turn leads to an instance $\mathcal{D} = (S_1, S_2, \ldots, S_{s-2}, S_{s-1}, S_s)$ wherein: S_j, for $2 \leq j \leq s-2$ and $j = s$, is the set of all the $j-1$-tuples one can build from $\{1, \ldots, j\}$; but S_{s-1} is the set of all $s-2$ tuples one can build from $\{1, \ldots, s\}$.

The aim of the simulation is to have all the elements of S_s, embedded into the protocol when the adversary $\mathcal{A}$ asks the Test-query. In this case, $\mathcal{A}$ will not be able to get any information about the value sk of the session key without having previously queried the random hash oracle $\mathcal{H}$ on the Diffie-Hellman secret value $g^{x_1 \dots x_s}$. Thus, to break the security of P the adversary $\mathcal{A}$ would have to have asked a query of the form $\mathcal{H}(\mathcal{I}, Fl_{last}, g^{x_1 \dots x_s})$ which as a consequence will be in the list of queries asked to $\mathcal{H}$.

To reach this aim Δ has to guess several values: $c_0, \mathcal{I}_0$ and i_0. We now describe what these values are used for and we will return to the formal simulation later on.

Δ first picks at random in $[1, Q]$ the number of operations c_0 that will occur before $\mathcal{A}$ asks the Test-query and embeds the elements of S_s into the operation that will occur at c_0. However Δ can not embed all the elements of S_s at c_0 since, contrary to SETUP1, in JOIN1 and REMOVE1 the players are not all added to the group at c_0. Δ rather embeds the elements from S_1 to S_s in the order the players are added to the group[3] but only for the players that will belong to the group at c_0. Thus, Δ also chooses at random s index-values u_1 through u_s in $[1, n]$ that it hopes will make up the group membership at c_0.

Δ also needs to cope with protocol executions wherein the players u_i, $1 \leq i \leq s$, are repeatedely added and removed from the group in order to have several times before reaching c_0 the group membership be $\mathcal{I}_0$. If, in effect, Δ embeds all the elements of S_s into the protocol execution the first time the group membership is $\mathcal{I}_0$, Δ is neither able to compute the Diffie-Hellman secret value involved nor the session key value sk needed to answer to the Reveal-query.

To be able to answer, Δ does not in fact embed S_s into the broadcast flow of the operation which updates the group membership to be $\mathcal{I}_0$ but embeds truly random values. Δ guesses the player u_{i_0} from $\mathcal{I}_0$ who will embed S_s into the broadcast flow of the operation that occurs at c_0[4] but generates a truly random exponent and uses it to embed truly random values for the operations that occur before c_0 and after c_0. The index i_0 is set as follows. If the c_0-th operation is JOIN1 then i_0 is the last joining player's index, otherwise i_0 is the group controller's index $\max(\mathcal{I}_0)$.

We now show that the above simulation and the random self-reducibility of G-CDH allows Δ to answer all the queries until $\mathcal{A}$ asks the Test-query at c_0. Since Δ embeds elements of S_i when a player u_i from $\mathcal{I}_0$ (except u_{i_0}) is added to the group and Δ does not remove it when u_i leaves, each protocol flow consists of a random self-reduction on one line (line 0, i.e. S_0 down to line $s-1$, i.e. $_{s-1}$) of the basic trigon. The trigon is illustrated on Figure 4. Thus, Δ can derivate the value sk of the session key from one of the values in the line below (line 1, i.e. S_1 up to line s, i.e. S_s).

However, Δ also needs to be able to answer to all queries after c_0 and more specifically the Reveal-queries. To this aim, Δ has to un-embed the element S_s from the protocol and do it in the operation that occurs at $c_0 + 1$. However

[3] More precisely, Δ keeps in some variable $\mathcal{T}$ the order of arrival for the players in $\mathcal{I}_0$, in order to know which elements of the trigon have to be used for each player. The variable $\mathcal{T}$ is reset whenever a Setup occurs.

[4] Δ may also embed a self-reduced element generated from S_s into the broadcast flow.

depending on the operation that occurs at $c_0 + 1$, Δ may not be able to do it for player u_{i_0}. This is the reason why the line S_{s-1} has to contain all the possible $(s-2)$-tuples: extension of the basic trigon illustrated on Figure 4. For the operations that will occur after $c_0 + 1$, Δ uses truly random exponents for all the players including those in $\mathcal{I}_0$. Thus, after $c_0 + 1$ all the protocol flows involve elements in S_{s-1} and S_s only.

This brief description completes the proof. The full behavior of the simulator is on Figure 5, with example on Figure 6. The probability analyses can be found in the full paper [10].

$\mathsf{Setup}(\mathcal{J})$	Reset $\mathcal{T}$ to 0 Increment c Update $\mathcal{I} \leftarrow \mathcal{J}$ $u \leftarrow \min(\mathcal{J})$ • $c < c_0 : u \in \mathcal{I}_0, u \neq i_0 \Rightarrow$ simulate using RSR according to $\mathcal{T}$ • $c = c_0 : \mathcal{J} \neq \mathcal{I}_0 \Rightarrow$ output "`Fail`" $\mathcal{J} = \mathcal{I}_0, u = i_0 \Rightarrow$ simulate using RSR according to $\mathcal{T}$ `Else` proceed as in P using $r_u \stackrel{R}{\leftarrow} \mathbb{Z}_q^*$
$\mathsf{Join}(\mathcal{J})$	Increment c $u \leftarrow \max(\mathcal{I})$ Update $\mathcal{I} \leftarrow \mathcal{I} \cup \mathcal{J}$ • $c < c_0 : u \in \mathcal{I}_0, u \neq i_0$ simulate using RSR according to $\mathcal{T}$ • $c = c_0 : \mathcal{I} \neq \mathcal{I}_0 \vee \max(\mathcal{J}) \neq i_0 \Rightarrow$ output "`Fail`" $\mathcal{I} = \mathcal{I}_0 \Rightarrow$ simulate using RSR according to $\mathcal{T}$ `Else` proceed as in P using $r_u \stackrel{R}{\leftarrow} \mathbb{Z}_q^*$
$\mathsf{Remove}(\mathcal{J})$	Increment c Update $\mathcal{I} \leftarrow \mathcal{I} \backslash \mathcal{J}$ $u \leftarrow \max(\mathcal{I})$ • $c < c_0 : u \in \mathcal{I}_0, u \neq i_0$ simulate using RSR according to $\mathcal{T}$ • $c = c_0 : \mathcal{I} \neq \mathcal{I}_0 \Rightarrow$ output "`Fail`" $\mathcal{I} = \mathcal{I}_0 \Rightarrow$ simulate using RSR according to $\mathcal{T}$ `Else` proceed as in P using $r_u \stackrel{R}{\leftarrow} \mathbb{Z}_q^*$
$\mathsf{Send}(U_i, m)$	• $c < c_0 : i \in \mathcal{I}_0, i \neq i_0 \Rightarrow$ simulate using RSR according to $\mathcal{T}$ • $c = c_0 : i \in \mathcal{I}_0 \Rightarrow$ simulate using RSR according to $\mathcal{T}$ `Else` proceed as in P using $r_i \stackrel{R}{\leftarrow} \mathbb{Z}_q^*$

Fig. 5. Game$^{ake}(\mathcal{A}, P)$. The multicast group is $\mathcal{I}$. The Test-query is "guessed" to be made: after c_0 operations, the multicast group is $\mathcal{I}_0$, and the last joining player is U_{i_0}. In the variable $\mathcal{T}$, Δ store which exponents of instance $\mathcal{D}$ have been injected in the game so far. RSR holds for *random self-reducibility.*

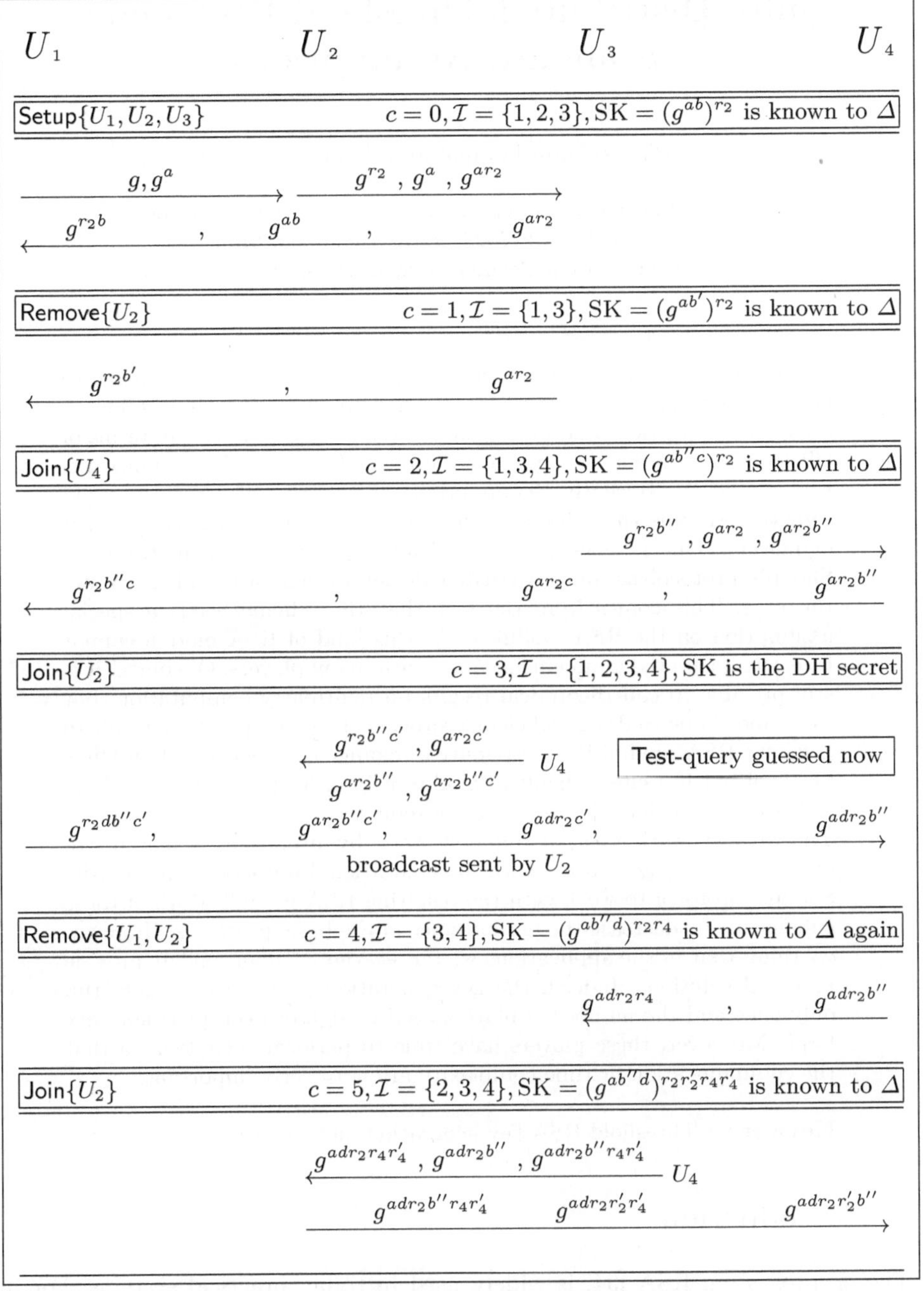

Fig. 6. An example of an execution of the protocol P=AKE1 with the adversary. We represent the simulation by Δ according to the following "guesses": $c_0 = 3, s = 4, \mathcal{I}_0 = \{1,2,3,4\}, i_0 = 2$. We denote by b, b', b'' etc. some blinding exponents used in the self-reduction of G-CDH (think b'' as being $b\beta''$, e.g.). Also note that when rejoining the group at steps $c = 3$ and $c = 5$, U_2 does not "remove" its random exponent.

Fully Distributed Threshold RSA under Standard Assumptions

Pierre-Alain Fouque and Jacques Stern

École Normale Supérieure, Département d'Informatique
45, rue d'Ulm, F-75230 Paris Cedex 05, France
{Pierre-Alain.Fouque,Jacques.Stern}@ens.fr

Abstract. The aim of this article is to propose a *fully distributed* environment for the RSA scheme. What we have in mind is highly sensitive applications and even if we are ready to pay a price in terms of efficiency, we do not want any compromise of the security assumptions that we make. Recently Shoup proposed a practical RSA threshold signature scheme that allows to share the ability to sign between a set of players. This scheme can be used for decryption as well. However, Shoup's protocol assumes a trusted dealer to generate and distribute the keys. This comes from the fact that the scheme needs a special assumption on the RSA modulus and this kind of RSA moduli cannot be easily generated in an efficient way with many players. Of course, it is still possible to call theoretical results on multiparty computation, but we cannot hope to design efficient protocols. The only practical result to generate RSA moduli in a distributive manner is Boneh and Franklin's protocol but it seems difficult to modify it in order to generate the kind of RSA moduli that Shoup's protocol requires.

The present work takes a different path by proposing a method to enhance the key generation with some additional properties and revisits Shoup's protocol to work with the resulting RSA moduli. Both of these enhancements decrease the performance of the basic protocols. However, we think that in the applications we target, these enhancements provide practical solutions. Indeed, the key generation protocol is usually run only once and the number of players used to sign or decrypt is not very large. Moreover, these players have time to perform their task so that the communication or time complexity are not overly important.

Keywords: Threshold RSA key generation and signature

1 Introduction

The cryptosystem RSA [34] is widely used in today practical systems. For instance, a lot of PKI products are based on it. In such systems, the protection of the root key needs strong security requirements. Therefore, threshold protocols can be used to share the signature capabilities among a subset of people rather than to give the power of signing to only one person. Moreover, this kind of protocols can withstand stronger adversaries than "centralized" cryptosystems.

C. Boyd (Ed.): ASIACRYPT 2001, LNCS 2248, pp. 310–330, 2001.
© Springer-Verlag Berlin Heidelberg 2001

Indeed, threshold cryptography can cope with *break-ins* adversaries that have the ability to corrupt people and read the memory of servers [6]. These adversaries are stronger than "normal" adversaries that can only read exchanged messages. In a "centralized" cryptosystem, if one *break-ins* adversary attacks the memory, he then knows all the key and the system is down. As this kind of attacks done by intruders (hackers, Trojan horses) or by corrupted insiders are very common and frequently easy to perform, systems must be protected against them. In threshold cryptography, the secret key is split into shares and each share is given to one of a group of servers. However, in order to be sure that at no moment the key is entirely in one machine, one can also distribute the key generation phase. Consequently, we say that a cryptosystem is *fully distributed* if it is distributed from the key generation to the signature or decryption phase.

In the case of discrete-log based cryptosystems, known solutions exist to distribute DSS, El Gamal, Cramer-Shoup [20,38,7]. Moreover, a protocol to distribute a discrete-log key has been first proposed by Pedersen in [27]. This protocol has been further revisited to solve a security flaw [21,15]. Therefore, discrete-log cryptosystems are fully distributed. However, a *fully distributed* version of RSA is a more challenging and important task.

In this paper we propose new techniques to fully distribute RSA. This solves an open problem where one needs to cope with requirements that do not match. On one hand, at Eurocrypt'00, Shoup describes a practical threshold signature scheme in [37] where the primes of the RSA modulus should be safe. On the other hand, Boneh and Franklin at Crypto '97 [4] describe a protocol to share the key generation of an RSA modulus. However, the generation of *safe* modulus seems to be hard with this protocol. The present work takes a different path by proposing a method to enhance the key generation with some additional properties and revisits Shoup's protocol to work with the resulting RSA moduli.

1.1 Why It Is Important to Share Shoup's Threshold RSA ?

Shoup threshold RSA signature scheme [37] presents interesting features. First of all, it is secure and robust in the random oracle model assuming the RSA problem is hard. Next, the signature share generation and verification are completely non-interactive and finally, the size of an individual signature share is bounded by a constant times the size of the RSA modulus. However, this scheme requires a trusted dealer to generate the keys and distribute the shares of the secret key among ℓ servers.

When a message m has to be signed by a quorum of at least $t+1$ servers, where $2t+1 \leq \ell$, a special server, called the *combiner*, forwards the message m or $x = H(m)$ to all servers. Then, each server computes its signature share along with a proof of correctness. Finally, the combiner selects a subgroup of $t+1$ servers by checking the proofs and combines the $t+1$ related signature shares to generate the signature s.

Efficient communication model against active adversary. The main characteristic of Shoup's protocol in relation to previous proposals [17,16,32] is the following. In the discrete-log case, it is easy to compute inverses mod q, if we note q the order of the group G generated by g, because q is public. With RSA, we cannot disclose inverses of a known value mod $\varphi(N)$ without revealing the factorization of N unless we use a special algebraic structure, called a module, as in [35,19]. We can note that computations in such structure can be done efficiently if we consider [25]. If we do not want to use a module, we face the problem of computing inverses when we use polynomial sharing in order to compute the Lagrange coefficients. Consequently some authors in [17,32] have proposed additive sharings to avoid this calculation. Therefore, as they need all shares to generate the signature, they devise strategies to cope with corrupted or crashed servers. Different strategies can be used to reconstruct the lost shares by non-corrupted servers : either using two different sharings (additive and polynomial) as in [32] or using probabilistic assignments as in [17]. With additive sharing, $d = \sum_{i=1}^{\ell} d_i \bmod \varphi(N)$, the combiner easily computes the signature s from the ℓ correct signature shares $s_i = x^{d_i} \bmod N$, where $x = H(m)$ is the message to be signed, by using the formula :

$$s = \prod_{i=1}^{\ell} s_i \left(= \prod_{i=1}^{\ell} x^{d_i} = x^{\sum_{i=1}^{\ell} d_i} = x^d \bmod N \right) \tag{1}$$

The main drawbacks of these techniques are the size of the key shares because of the need of different sharings and the use of protocols to reconstruct the bad signature shares in the presence of active (malicious) players.

In [16], the authors proposed the first proven scheme based on polynomial sharing, which is based on Desmedt and Frankel's scheme [13]. However in the case of active adversaries, which are allowed to send bad shares, the protocol has to be rewind at most t times, to remove the bad servers as the signature shares depend on the subgroup of $t+1$ servers enabling the reconstruction of the signature. Let $\Delta = \ell!$. The shares of d are such that $\Delta | d_i$ and $d_i = f(i)$ where f is a polynomial of degree t and of constant term equals to d. If we denote by S the subgroup of $t+1$ servers, let Lagrange coefficients to be ${\lambda'}_{i,j}^{S} = \prod_{j' \in S \setminus j} \frac{i-j'}{j-j'}$. Therefore, $d = \sum_{i \in S} {\lambda'}_{0,i}^{S} d_i \bmod \varphi(N)$ from the Lagrange formula. There are two problems. First of all, ${\lambda'}_{i,j}$ cannot be computed in $\mathbb{Z}_{\varphi(N)}$ since $(j-j')$ could be even and not invertible mod$\varphi(N)$. Next, the combiner has to compute

$$s = \prod_{i \in S} s_i^{{\lambda'}_{0,i}^{S}} \left(= \prod_{i \in S} s^{{\lambda'}_{0,i}^{S} d_i} = s^{\sum_{i \in S} {\lambda'}_{i,0}^{S} d_i} = s^d \bmod N \right) \tag{2}$$

But, as ${\lambda'}_{i,j}^{S}$'s are not integers and the combiner cannot compute roots modulo a composite number, otherwise it can solve the RSA Problem, he cannot compute equation (2). The key idea is to note that $\Delta \times {\lambda'}_{i,j}^{S}$ are integers. Therefore, if we write $d_i = \Delta \times d'_i$, then, for a group S of $t+1$ servers, each one can compute $l_{0,i}^{S} = \Delta \times {\lambda'}_{0,i}^{S} \in \mathbb{Z}$ and $s'_i = x^{{\lambda'}_{i,0}^{S} d_i} = x^{l_{0,i}^{S} \times {d'}_i} \bmod N$. Finally, the combiner computes

$$s = \prod_{i \in S} s'_i \left(= \prod_{i \in S} x^{l^S_{0,j} \times d'_i} = \prod_{i \in S} x^{\Delta \times \lambda'^S_{0,i} \times d'_i} = \prod_{i \in S} x^{\lambda'^S_{0,i} d_i} = x^d \bmod N \right) \quad (3)$$

If the signature is not valid, the combiner removes the bad servers thanks to a proof of robustness, defines another group S and rewind the protocol. It is then obvious that after t trials, all bad servers are removed from the set S and the signature will be correct. However, this redefinition of the subgroup does not seem very nice and Shoup and others have proposed a new trick to avoid this problem.

Shoup in [37] and Miyazaki, Sakurai and Yung in [26] solve this problem by using a well-known lemma to extract an e-root of w modulo a composite number from a e-root of a known power of w [24] without any secret. The solution is to multiply Lagrange coefficients by Δ such that they are integers : $\lambda^S_{i,j} = \Delta \times \lambda'^S_{i,j} \in \mathbb{Z}$ and $\Delta d = \sum_{i \in S} \lambda^S_{0,i} d_i$, if we denote by S a subset of $t+1$ elements. Therefore, if we let $s_i = x^{d_i}$, the combiner can compute signatures and change of group (compute new Lagrange coefficients according to the group of $t+1$ servers) without asking new signature shares to the servers. He computes equation (2) using $\lambda^S_{i,j}$ and gets $s^\Delta = \prod_{i \in S} s_i^{\lambda^S_{0,i}} (= x^{\Delta d}) \bmod N$. Finally, the combiner can compute a e-root of x^Δ with the previous formula and can recover a e-root of x using the well-known lemma. Consequently, if we use Shoup's scheme, there is no need to generate d_i such that $\Delta | d_i$ as it is done in [18,12].

Even if the protocol of Frankel *et al.* in [16] proposed a fully distributed version of RSA, it is less elegant than Shoup's one and it will be nice to share this protocol. Moreover, this scheme proposes others improvements that are valuable such as the proof of robustness. This proof uses safe primes, like Gennaro *et al.*'s one [20] and avoids the drawbacks of a special relation between prover and verifier. Furthermore, in [16], the authors describe an interactive protocol which leads to a less efficient protocol than Shoup's one which uses non-interactive zero-knowledge proofs. Therefore, we face the problem of distributing this non-interactive proof.

Proof of robustness and use of safe primes. As we said before, a second key point in Shoup's signature scheme is the *proof of correctness*, which guarantees the *robustness of the scheme*. Robustness means that corrupted servers should not be able to prevent uncorrupted servers from signing. This property is attractive for threshold protocols in presence of active adversaries that can modify the behavior of servers. In Shoup's scheme, the proof of correctness requires an RSA modulus built with safe primes. In the proof of correctness, servers must prove that they raise x to the correct power, namely d_i, their share of the secret. To this end, each server i has a verification key $v_i = v^{d_i} \bmod N$ and makes a proof that $\log_v v_i = \log_x s_i (= d_i \bmod \varphi(N))$. The problems are : $\mathbb{Z}_N{}^*$ is not a cyclic group, its order is unknown, such generator v do not exist and elements of maximal order cannot be easily found. However, Shoup noted that if we use

RSA moduli with safe primes, then the group of squares in $\mathbb{Z}_N{}^*$ is cyclic and it is easy to find generators. Consequently, the proof of correctness can be made non-interactively and correctly proved without further assumptions. Finally, safe prime moduli are also used in the key generation protocol in order to guarantee the secrecy of Shamir's secret sharing.

Shared Generation of RSA Keys. This raises the question of generating RSA moduli for Shoup's threshold scheme without a trusted dealer. There exist protocols that generate RSA keys in a distributive manner [4,18,9,10,3,30,23]. Boneh and Franklin in [4] designed such protocol for the generation of an RSA modulus in the honest-but-curious model. Later, Frankel, MacKenzie and Yung in [18] made this algorithm robust against malicious servers. In [30], Poupard and Stern also provided a protocol to compute a shared modulus for two players only. Finally, Gilboa in [23] has extended Poupard and Stern method. As we can note, the Boneh and Franklin protocol is most efficient but no protocol is known to efficiently create shared safe RSA moduli.

1.2 Outline of the Paper

We begin by presenting the problem in section 2, *i.e.* where the properties of safe primes are used in Shoup's protocol, and in section 3 the security model. Next, in section 4, we describe how to enhance the Boneh-Franklin scheme to generate RSA moduli having special requirements. In section 5 we show that Shoup's protocol is still secure against passive adversary and in section 6, we show a new proof of correctness making Shoup scheme robust against active adversary. Finally, in section 7 we present practical parameters for our scheme.

1.3 Notations and Definitions

Throughout this paper, we use the following notation: for any integer $N = pq$, where $n = \log(N)$ is a security parameter, as well as k, ℓ, t, k', k_1 and k_2,

- we use Q_N to denote the group of squares in $\mathbb{Z}_N{}^*$,
- we use $\varphi(N)$ to denote the Euler totient function, i.e. the cardinality of $\mathbb{Z}_N{}^*$,
- we use $\lambda(N)$ to denote Carmichael's lambda function defined as the largest order of the elements of $\mathbb{Z}_N{}^*$.

Let $p = 2p' + 1$ and $q = 2q' + 1$ where in general $p' = \prod_{p_i} p_i{}^{e_i}$ and $q' = \prod_{q_j} q_j{}^{e_j}$. Set $M = p'q'$. Finally, a prime number p is a *safe prime* if p and p' are both prime. A RSA modulus $N = pq$ is called a *safe prime modulus* if p and q are both safe primes.

2 The Problem

As we will see in the following, safe primes are used in the key generation in order to prove that Shamir secret sharing scheme [36] is secure in the ring $\mathbb{Z}_M$, and not in a finite field, and in the proof of correctness. Let us explain the second problem as it is less obvious.

2.1 What Is the Problem?

Robustness guarantees that even if t malicious players send false signature shares, the signature scheme still correctly generates a signature s. This property is needed since otherwise combination faces the problem of selecting the correct shares.

For example, the combiner receives signature shares from the servers and has to generate the correct signature. One way for him is to pick at random $t+1$ signature shares, generate the possible signatures s' and test whether s' is a valid signature of m. If s' passes the verification protocol, the correct signature has been found, otherwise, the combiner has to test another group of $t+1$ signature shares. Since the combiner cannot guess where the bad shares are, it might face an exponential number of trials. Therefore, it is necessary to devise an efficient test in order to check whether a player has correctly answered a request. Shoup has proposed an efficient proof to achieve such check non-interactively and the same kind of proof appears in [32,19] but still requires safe prime modulus.

In order to avoid the generation of shared safe moduli, which appears currently out of reach, this paper proposes a tradeoff between the requirements of the RSA modulus for the signature and decryption protocols and the requirements at key generation. Independly of our work, Damgard and Koprowski have recently considered the same problem in [12]. They revisited Shoup's paper and used non-standard assumptions to show that the proof of correctness works without other requirements on the RSA modulus. They use [18] to generate standard RSA moduli.

In our work, we consider environments where high security is required such as electronic voting schemes. Therefore, we prefer to use protocols based on standard assumptions. We believe that standard assumptions and security proofs are needed to build secure protocols. Several electronic schemes [14,11,1] have been based on Paillier cryptosystem which is related to RSA. The techniques developed in the paper can be used to *fully share* this cryptosystem.

2.2 Our Results

We prove that Shoup's protocol can be modified to work with RSA moduli having special properties under standard assumptions and that these moduli can be jointly generated.

Safe prime moduli are needed in the proof of robustness and in the key generation. Moreover, different characteristics of these numbers are used in the proof of robustness. Indeed, Shoup's protocol uses two important properties of the subgroup Q_N of squares of $\mathbb{Z}_N{}^*$ when N is a safe modulus. On one hand, this subgroup is cyclic and on the other hand, its order M does not have small prime factors. The cyclic group is used to show the **existence of the discrete log** in the proof of correctness. The use of safe primes allows to guarantee that, with overwhelming probability, a random element in Q_N is a **generator**.

Our first observation relates the structure of Q_N with $\gcd(p-1, q-1)$ and the search for generators in this group to the prime factor decomposition of $\frac{p-1}{2}$ and

$\frac{q-1}{2}$. In particular, if $\frac{p-1}{2}$ and $\frac{q-1}{2}$ have no small prime factors, then with high probability few randomly chosen elements generate the entire group Q_N. If we choose enough random elements $(g_1, \ldots, g_k)$, we can guarantee that the group generated by $\langle g_1, \ldots, g_k \rangle$ is all of Q_N with high probability. Such techniques have already been used by Frankel *et al.* in [18] and a precise treatment has been given by Poupard and Stern in [31]. Moreover, using a nice trick of Gennaro *et al.* which first appeared in [22] and the protocol recently proposed by Catalano *et al.* in [8], the calculation of $\gcd(p-1, q-1)$ can be performed in a distributed way. These methods allow to keep key generation and signature efficient.

In this paper, we show how to jointly construct RSA moduli such that the subgroup Q_N is cyclic, which guarantees the existence of discrete logs and of generators of Q_N. Moreover, the order M of this group does not have small prime factors less than some sieving bound B. Checking such primes does not exceedingly increase the running time of the key generation algorithm.

3 Security Model

3.1 The Network

We assume a group of ℓ servers connected to a broadcast medium, and that messages sent on the communication channel instantly reach every party connected to it.

3.2 The Adversary

The adversary is computationally bounded and it can corrupt servers at any moment by viewing the memories of corrupted servers (passive adversary), and/or modifying their behavior (active adversary). The adversary decides on whom to corrupt at the start of the protocol (static adversary). We also assume that the adversary corrupts no more than t out of ℓ servers throughout the protocol, where $\ell \geq 2t + 1$.

3.3 Formal Definition

A RSA threshold signature scheme consists of the four following components :

- A *key generation algorithm* takes as input security parameters n,the number k of elements to generate Q_N, the number ℓ of signing servers, the threshold parameter t and a random string ω; it outputs a public key (N, e) where n is the size in bits of N, the private keys $d_1, \ldots d_\ell$ only known by the correct server and for each $u \in [1, k]$ a list $v_u, v_{u,1} = {v_u}^{d_1}, \ldots v_{u,\ell} = {v_u}^{d_\ell} \bmod N$ of verification keys.
- A *share signature algorithm* takes as input the public key (N, e), an index $1 \leq i \leq \ell$, the private key d_i and a message m; it outputs a signature share $s_i = x^{d_i} \bmod N$, where $x = H(m)$ and $H(.)$ is a hash-and-pad function, and a proof of its validity $proof_i$ (for all $u \in [1, k]$, $\log_{v_u} v_{u,i} = \log_x s_i$).

- A *combining algorithm* takes as input the public key (N, e), a message m, a list $s_1, \dots s_\ell$ of signature shares, for each $u \in [1, k]$ the list $v_u, v_{u,1}, \dots v_{u,\ell}$ of verification keys and a list $proof_1, \dots proof_\ell$ of validity proofs; it outputs a signature s or fails.
- A *verification algorithm* takes as input the public key (N, e), a message m, a signature s; it outputs a bit b indicating whether the signature is correct or not.

3.4 The Players and the Scenario

Our game includes the following players : a combiner, a set of ℓ servers P_i, an adversary and users asking signature. All are considered as probabilistic polynomial time Turing machines. We consider the following scenario :

- At the initialization phase, the servers use the distributed key generation algorithm to create the public, private and verification keys. The public key (N, e) and all the verification keys v_u's and $v_{u,i}$'s are published and each server obtains its share d_i of the secret key d.
- To sign a message m, the combiner first forwards m to the servers. Using its secret key d_i and its verification keys $v_u, v_{u,i}$ for $u \in [1, k]$, each server runs the share signature algorithm and outputs a signature share s_i together with a proof of validity of the share signature $proof_i$. Finally, the combiner uses the combining algorithm to generate the signature, provided enough signature shares are available and valid.

3.5 Properties of Threshold Signature Schemes

The two properties of a t out of ℓ threshold signature scheme of interest to us are robustness and unforgeability. As we already mentioned, robustness guarantees that even if up to t malicious players send false signature shares, the scheme still returns a correct signature. This property is useful only in the presence of active adversaries.

Unforgeability guarantees that any subset of $t+1$ players can generate a signature s, but disallows the generation by fewer than t players. This unforgeability property should hold even if some subset of less than t players are corrupted and collude. This property expresses the security of the signature scheme and is useful in the presence of passive or active adversary.

3.6 The Games

In this section, we describe the security notions for threshold key generation and threshold signing protocols. We have to show that the information revealed during the key generation and the signing protocols does not release secret information to the adversary.

Game for the threshold key generation. The correctness of the key generation requires that the probability of the secret keys d, p, q, and the public key (N, e) seem be uniformly distributed to the adversary.
The secrecy of the key generation means that if there exists an adversary $\mathcal{A}$ which corrupts at most t servers at the beginning of the game, then he cannot obtain more information on the secret key held by uncorrupted players.

Game for the threshold signing protocol. The secrecy of the signing protocol means that if there exists an adversary $\mathcal{A}$ which corrupt at most t servers at the beginning of the game and even if he can obtain signatures on messages adaptively chosen, then he cannot forge a signature on a new message.

4 Enhancing the Boneh-Franklin Scheme to Generate RSA Moduli with Special Requirements

The aim of this section is to generate RSA moduli such that the group of squares is a cyclic group whose order has no small prime factors, namely $N = pq$, $p = 2p' + 1$ and $q = 2q' + 1$, $\gcd(p', q') = 1$ and no primes $p < B$ divises neither p' nor q'. In section 6, we will prove that this group can be generated with few random elements. Moreover, we use here a sieving method to simultaneously improve the key generation protocol, the probability of finding a set of generators of Q_N, and to make secure Shamir Secret Sharing Scheme. We also present a protocol to compute the GCD of a known value and a shared value and prove the robustness and the secrecy of this new distributed key generation protocol.

4.1 A New Distributed RSA Modulus Generation

In [4] Boneh and Franklin present a protocol for generating a shared RSA modulus. We describe this protocol here with our adaptation.

1. In the first step, each server picks at random two values p_i and q_i in the interval $[\lfloor 2^{(n-1)/2} \rfloor, \lfloor \frac{2^{n/2}-1}{\ell} \rfloor[$ according to [39], where n is the size in bits of the modulus N. Then, we use a sieving algorithm in order to discard $p_1 + \ldots + p_\ell$ and $q_1 + \ldots + q_\ell$ that have small prime factors and if $p_1 + \ldots + p_\ell - 1$ or $q_1 + \ldots + q_\ell - 1$ have small prime factors. The servers check whether $\gcd(p-1, 4P) \stackrel{?}{=} 2$ and whether $\gcd(4P, q-1) \stackrel{?}{=} 2$, where $P = \prod_{2 < p_i < B} p_i$ and B is the sieving bound using the GCD algorithm that we describe below.
2. Then the BGW protocol [2,4] is run to compute the product N of $p_1 + \ldots + p_\ell$ and $q_1 + \ldots + q_\ell$. They also compute the product $\varphi(N) = (p-1)(q-1)$ and check whether $\gcd(\varphi(N), N-1) \stackrel{?}{=} 1$ using the GCD algorithm.
3. Next, the parties perform a primality test similar to the Fermat test modulo N. The practicality of this test is based on the empirical results of [33] where Rivest showed that if a sieving algorithm is first performed, the Miller-Rabin primality test is not needed as pseudoprimes are rare according to

Pomerance's conjectures [28,29]. Moreover, carmichael numbers are avoided due to a trick similar to Soloway-Strassen primality test. We set $p = p_1 + \ldots + p_\ell$ and $q = q_1 + \ldots + q_\ell$.

4.2 Computing the gcd of a Public Value and a Shared Secret Value

We briefly recall the protocol presented by Catalano *et al.* [8] for inverting a public value e modulo a shared value φ. The basic trick stems from the observation that $\gcd(e, \varphi) = \gcd(e, \varphi + Re)$ where R is a large integer used to mask the shared and secret value $\varphi = \varphi_1 + \ldots + \varphi_\ell$. Server i chooses a random integer $r_i \in_R [0..2^{n+k'}]$, where k' is a security parameter, computes $c_i = \varphi_i + er_i$ and forwards c_i to all other servers. Each server can compute $c = \sum_i c_i = \varphi + eR$ if we set $R = \sum_i r_i$. This value can be publicly known and then, all servers can compute $\gcd(e, c)$ which is equal to $\gcd(e, \varphi)$ and u and v such that $eu + cv = 1$ when $\gcd(e, \varphi) = 1$. Then, it is easy to show that if we replace c by $\varphi + eR$, we obtain $e(u + Rv) + \varphi v = 1$. Hence, $u + Rv$ is the inverse of e modulo φ. In this case, if we note d the inverse of e mod φ, each server assigns its share of the inverse d to $d_i = vr_i$ and the first server to $d_1 = u + vr_1$.

We have presented here the protocol in the honest-but-curious model. But this protocol can be made robust following [8]. We can also note that this algorithm allows to compute the gcd of a known value and a shared one. We call this protocol the GCD algorithm.

4.3 Efficient Sieving Algorithm Improving the Generation of Random Number without Small Factors

In this subsection, we present the sieving algorithm used in phase 1 of protocol 4.1 and we show how to generate N such that neither $p' = \frac{p-1}{2}$ nor $q' = \frac{q-1}{2}$ have no small prime factors. Our method uses a new distributed sieving protocol designed by Boneh, Malkin and Wu in [5] that we patch in order to create p such as neither p, nor p' has a small prime factor less than B. Moreover, we show how to withstand malicious adversaries. We denote by P the product of all odd small primes up to B.

1. Each server picks a random integer a_i in the range $[1, P]$ such that a_i is relatively prime to P. Then, since each a_i is a random integer relatively prime to P, their product $a = a_1 \times \ldots \times a_\ell \bmod P$ is also relatively prime to P.
2. The servers perform a protocol to convert the multiplicative sharing of a to an additive sharing of $a = b_1 + \ldots + b_\ell$ using the BGW protocol.
3. Each server picks a random $r_i \in [\lfloor \frac{\sqrt{2}.2^{n/2-1}}{P} \rfloor, \lfloor \frac{2^{n/2}-1}{P\ell} \rfloor[$ and sets $p_i = r_i P + b_i$.

Clearly, $p = \sum p_i \equiv a \bmod P$ and hence p is not divisible by any prime smaller than B. We can note that $p = RP + a$ where $R = \sum_i r_i$. This sieve works only for B s.t. $P < p$. We can increase the bound B to B_1 by checking whether $\gcd(P', p) = 1$ where $P' = \prod_{B \leq p \leq B_1} p$ thanks to the GCD algorithm.

In order to also prove that $p' = \frac{p-1}{2}$ has no prime factors less than B, one has to check whether $\gcd(p-1, P) = \gcd(2p', P) = 1$ and $\gcd(p-1, 4P) = 2$ to test the power of 2. If we denote P' by $4P$, we can perform a single test $\gcd(p-1, P') = 2$.

To distribute this test in the honest-but-curious model, the first server sets its parts p_1 to $p_1 - 1$ and we use the distributed GCD protocol described in section 4.2. It is also possible to make this test robust in presence of malicious players as it is explained in appendix 9.1.

Finally, the protocol that transforms the multiplicative sharing of a into an additive sharing can also be made robust as it uses the BGW protocol. This transformation calls ℓ times the BGW protocol. At the beginning, $b_{i,0} = 0$ for all $i \in \{0, \ldots, \ell\}$. Then, for $i = 1$ to ℓ, $u_i = a_i$ and $u_j = 0$ for $\forall j \neq i$, and the BGW protocol performs $(b_{1,i-1} + \ldots + b_{\ell,i-1}) \times a_i = (b_{1,i-1} + \ldots + b_{\ell,i-1})(u_1 + \ldots + u_\ell) = b_{1,i} + \ldots + b_{\ell,i}$.

Theorem 1. *The key generation protocol of Boneh-Franklin and the sieving protocol allow to generate RSA moduli such that the order M of the group Q_N does not contain small prime factors less than B.*

It is obvious to see that the use of the sieving method to guess p_i's and q_i's allows to improve the first step of Boneh-Franklin's protocol and speeds up the running time of this algorithm since this avoids many rewindings in phase 3 of Boneh-Franklin. Moreover, this sieving protocol is adapted to take into account the small factors in the factorization of p' and q'.

4.4 The Key Generation of N Such That Q_N Is Cyclic

Here, we show how to generate N such that the group Q_N is cyclic. To guarantee this property, we use the fact that the product of two cyclic groups which orders are coprime is a cyclic group. The following lemma and the GCD protocol enables to check that p' and q' are coprime in a distributed way. First we prove a lemma which has been used in another form in [22].

Lemma 1. *Let $N = pq$ an RSA modulus,* $\gcd(p-1, q-1) | \gcd(N-1, \varphi(N))$ *and the square free part of* $\gcd(N-1, \varphi(N))$ *divides* $\gcd(p-1, q-1)$.

See appendix section 9.2 for a proof of this lemma.

Corollary 1. *If* $\gcd(N-1, \varphi(N)) = 2$, *then* $\gcd(p-1, q-1) = 2$.

Proof. If $\gcd(N-1, \varphi(N)) = 2$, then as $\gcd(p-1, q-1) | \gcd(N-1, \varphi(N))$, $\gcd(p-1, q-1) = 2$ since $\gcd(p-1, q-1)$ cannot be equal to 1. These last verification can be jointly done using the GCD algorithm described in section 4.2 □

Theorem 2. *The key generation protocol of Boneh-Franklin and the GCD protocol allow to generate RSA moduli such that the group Q_N is cyclic of order $M = p'q'$, where $N = pq$, $p = 2p' + 1$, $q = 2q' + 1$ and neither p' nor q' have prime factors smaller than B. The iteration number of this protocol with respect to the Boneh-Franklin protocol is on average $4 \times e^\gamma \ln(B)$.*

Proof. Following section 4.3 and corollary1, we can assume that we get an RSA modulus such that $p-1$ and $q-1$ have all theirs prime divisors greater than B, they do not have common divisors, *i.e.*, $\gcd(p-1, q-1) = 2$ and $\frac{p-1}{2}$ and $\frac{q-1}{2}$ do not have small prime factors. As the product of cyclic groups whose order are coprime is a cyclic group, the groups of squares in $\mathbb{Z}_p{}^*$ and in $\mathbb{Z}_q{}^*$ are cyclic, and so the group Q_N is also cyclic. This allows to guarantee that there exists a cyclic subgroup in $\mathbb{Z}_N{}^*$ of order $M = p'q'$.

We can estimate the iteration number of this algorithm with respect to phase 1 of the Boneh-Franklin protocol. First, it is a well-known fact that $\lim_{n\to\infty} \Pr_{(p',q')\in[1,n[^2}[\gcd(p',q') = 1] = \frac{6}{\pi^2} > 1/2$ assuming prime numbers in $[2^{\frac{n-1}{2}}, 2^{\frac{n}{2}}[$ are uniformly distributed. Moreover, the only slowing factor at the key generation is the check that $\gcd(P', p-1) = 2$, where $P' = 4P$. We can note that $\Pr_{p'}[\gcd(2p', P') = 2] = \Pr_{p'}[2 \nmid p' \wedge 3 \nmid p' \wedge \ldots \wedge B \nmid p'] = (1-\frac{1}{2})(1-\frac{1}{3})\ldots(1-\frac{1}{B}) = \prod_{p_i \leq B}(1-\frac{1}{p_i}) \approx \frac{1}{e^\gamma \ln(B)}$ according to the second theorem of Mertens, where γ is the Euler constant. Therefore, we have to run this algorithm $2 \times (e^\gamma \ln(B) + e^\gamma \ln(B))$ on average in order to get such RSA moduli. □

4.5 Proofs of Security and Robustness

For security and robustness proofs see [4,18,8] or the extended version.

4.6 Distributed Generation of the Keys in Shoup's Protocol

Once N is generated, let prime e be the first prime greater than $4\Delta^2$ so that each server can compute it. Then, Catalano *et al.* protocol's [8] is run to generate a shared secret key d in a distributed manner. At the end of the protocol, each server can compute its verification keys as $v_{u,i} = v_u^{\Delta d_i}$ for random v_u computed as $y_u^2 \bmod N$ where y_u is the concatenation of $H(N\|i)$ for sufficiently many i's to get the correct security parameter in the random oracle model.

5 Security of Shoup Protocol against Passive Adversary

In this part we show that Shoup's protocol is secure with the RSA moduli generated in previous section against static and passive adversary.

5.1 Key Generation

At the end of the key generation protocol, we want to know if the information an adversary can collect, helps him to get useful information on the secret key d. Let $d_{i_1}, \ldots, d_{i_t}$, t shares of the secret key obtained by the adversary from the t correupted servers.

Information revealed by the shares d_{i_j}. For each d_{i_j}, we can note that

$$d_{i_j} = f(i_j) = a_0 + a_1 i_j + \ldots + a_t i_j^t \bmod M$$

If we add a $(t+1)^{\text{th}}$ equation, $a_0 = d$, we obtain the following linear system :

$$\begin{cases} a_0 + a_1 i_1 + \ldots + a_t {i_1}^t = d_{i_1} \bmod M \\ a_0 + a_1 i_2 + \ldots + a_t {i_2}^t = d_{i_2} \bmod M \\ \ldots \\ a_0 + a_1 i_t + \ldots + a_t {i_t}^t = d_{i_t} \bmod M \\ a_0 + 0 + \ldots + 0 = d \end{cases}$$

or with matrix $I.A = D \bmod M$

$$\begin{pmatrix} 1 & i_1 & i_1^2 & \ldots & i_1^t \\ 1 & i_2 & i_2^2 & \ldots & i_2^t \\ \ldots & & \ddots & & \ldots \\ 1 & i_t & i_t^2 & \ldots & i_t^t \\ 1 & 0 & 0 & \ldots & 0 \end{pmatrix} \cdot \begin{pmatrix} a_0 \\ a_1 \\ \vdots \\ a_t \end{pmatrix} = \begin{pmatrix} d_{i_1} \\ d_{i_2} \\ \vdots \\ d_{i_t} \\ d \end{pmatrix} \bmod M$$

The matrix I is a Vandermonde matrix. The determinant of such matrix is $\det(I) = \prod_{1 \leq j < k \leq t+1} (i_k - i_j) \bmod M$, where $i_{t+1} = 0$. As the i_j's are distinct in $\mathbb{Z}_M$, $i_k - i_j \neq 0 \bmod M$ since $\ell < B$. Hence, each factor $(i_k - i_j)$ is invertible modulo M, so $\det(I)$ is invertible modulo M. Therefore, all values of d are possible. Hence a group of t players cannot get information on d from shares of d. We see here that the sieving algorithm performed is important to avoid leakage of information on d. Consequently, $\ell < B$.

Information revealed from the verification keys. For all $u \in [1, k]$, the verification keys $v_{u,i}$'s of non-corrupted servers do not reveal any information as they can easily be simulated from the parts of the corrupted servers. The simulator chooses at random $y_u \in {\mathbb{Z}_N}^*$, computes $v_u = y_u^{2\Delta e} \bmod N$. Hence, $v_u^d = y_u^{2\Delta}$. We can note that $\Delta d_i = \sum_{k=1}^{t+1} \lambda_{i,k}^S d_j \bmod \varphi(N)$ if we denote by S a group of $t+1$ values and if we define the Lagrange coefficients as $\lambda_{i,j}^S = \Delta \times \prod_{j' \in S \backslash j} \frac{i-j'}{j-j'} \in \mathbb{Z}$. The simulator can then compute for all $u \in [1, k]$, $v_u^{\Delta d_i} = y_u^{2\Delta\lambda_{i,0}^S} \times \prod_{j=1}^t v_u^{\lambda_{i,i_j}^S d_{i_j}} \bmod N$, where $S = \{0, i_1, \ldots, i_t\}$. Hence a group of t players cannot get information from the validation keys of non-corrupted servers.

5.2 Signing Protocol

To generate a signature on message m, each server i computes $x = H(m)$, $s_i = x^{2\Delta d_i} \bmod N$ and sends s_i to the combiner without any proof because we are in the honest-but-curious model. The combiner selects a set S of $t+1$ values and computes $w = \prod_{j=1}^{t+1} s_{i_j}^{2\lambda_{0,i_j}^S} \bmod N$. It then follows that $w^e = x^{4\Delta^2}$ as

$s_{i_j}^2 = x^{4\Delta d_{i_j}}$. Applying a well-known lemma, we can extract e-root of x from w which is a e-root of $x^{4\Delta^2}$ as $4\Delta^2$ is a known value using the extended Euclidean algorithm on e and $4\Delta^2$. As we are in the honest-but-curious model, the signature s is always correct.

We can prove the following theorem :

Theorem 3. *In the random oracle model, the signing protocol described above is a secure threshold signature scheme (non-forgeable) assuming the standard RSA signature scheme is secure.*

Proof. Similar to [37]. □

6 Enhancing the Shoup Scheme against Active Adversary

The aim of this section is to revisit the proof of correctness originally designed by Shoup to cover the case of RSA moduli generated as in the section 4. This uses a method by which we generate the entire group of squares with few random elements with high probability.

6.1 Proof of Correctness

Let N be a modulus such that $N = pq$ and $p = 2p' + 1$ and $q = 2q' + 1$ where p' and q' have no small prime factors and $\gcd(p-1, q-1) = 2$. Accordingly Q_N is cyclic and there exists a generator g in Q_N. Thus, the discrete log of any element ${s_i}^2$ in basis g exists, where $s_i = x^{2\Delta d_i}$ and $\Delta = \ell!$. As we will see in section 6.3, we can denote by $v_1, \ldots, v_k$ a k-tuple of random elements in $(Q_N)^k$ such that with high probability, this tuple generates the whole group Q_N of order $M = p'q'$, *i.e.* for each $x \in Q_N$, there exists $(a_1, \ldots, a_k) \in [0, M[^k$ such that $x = \prod_{i=1}^k {v_i}^{a_i} \bmod N$.

Each server i has a k-tuple of verification keys $v_{1,i} = {v_1}^{d_i} \bmod N, \ldots, v_{k,i} = {v_k}^{d_i} \bmod N$. He computes a signature share, $s_i = x^{2d_i\Delta} \bmod N$, where d_i is the ith signature share of d and proves that

$$\log_{v_1}(v_{1,i}) = \ldots = \log_{v_k}(v_{k,i}) = \log_{x^{4\Delta}}({s_i}^2)(= d_i \bmod M)$$

The value ${s_i}^2$ is a square and is an element of Q_N.

Now, we describe the proof of "correctness" and still let $d_i \in [0, M[$ be the secret share of a server, and A and B' two integers such that $\log(A) \geq \log(B'Mh) + k_2$ where B' and k_2 are security parameters and h is the number of rounds. Finally, k_1 is a parameter such that the cheating probability $1/{B'}^h$ is $< 1/2^{k_1}$. Whereas security parameter k_1 controls the completeness and statistical zero-knowledge results, security parameter k_2 controls the soundness result. We present the protocol for one round ($h = 1$).

The prover chooses a random r in $[0, A[$. Then, he computes $t = (v'_1, \ldots, v'_k, x') = (v_1^r, \ldots, v_k^r, x^{4\Delta r})$. Let e be the first $b' = \log(B') - 1$ bits of the hash value

$$e = [H(v_1, \ldots, v_k, x^{4\Delta}, v_{1,i}, \ldots, v_{k,i}, {s_i}^2, v'_1, \ldots, v'_k, x')]_{b'}$$

if we denote by $[x]_{b'}$ the first b' bits of x. Next, the prover calculates z where $z = r + ed_i$. The proof is the pair $(e, z) \in [0, B'[\times[0, A[$. To check it, the verifier has to compute whether

$$e = [H(v_1, \dots, v_k, x^{4\Delta}, v_{1,i}, \dots, v_{k,i}, s_i{}^2, v_1{}^z v_{1,i}{}^{-e}, \dots, v_k{}^z v_{k,i}{}^{-e}, x^{4\Delta z} s_i{}^{-2e})]_{b'}$$

and verify whether $0 \leq z < A$.

6.2 Security Analysis of the Proof of Correctness

Proof of Completeness

Theorem 4. *The execution of the protocol between a prover who knows the secret d_i and a verifier is successful with overwhelming probability if $B'Mh/A$ is negligible where h is the number of rounds.*

Proof. If the prover knows a secret $d_i \in [0, M[$ and follows the protocol, he fails only if some $z \geq A$. For any value $x \in [0, M[$ the probability of failure of such event taken over all possible choices of r is smaller than $B'M/A$. Consequently the execution of the proof is successful with probability $\geq (1 - \frac{B'M}{A})^h \geq 1 - \frac{B'Mh}{A}$. □

Proof of Soundness. Let us focus on soundness of the interactive proof system.

Lemma 2. *If the verifier accepts the proof, with probability $\geq 1/B' + \varepsilon$ where ε is a non-negligible quantity, then using the prover as a "black-box" it is possible to compute σ and τ such that $|\sigma| < A$ and $|\tau| < B'$ such that $v_1{}^\sigma = v_{1,i}{}^\tau, \dots, v_k{}^\sigma = v_{k,i}{}^\tau, x^{4\Delta\sigma} = s_i{}^{2\tau}$.*

Proof. If we rewind the adversary and get two valid proofs for the same commitment t, (e, z) and (e', z'), we have for $u = 1, \dots, k$ *i.e.* for all verification keys, $v_u{}^r = v_u{}^z v_{u,i}^{-e} = v_u{}^{z'} v_{u,i}^{-e'}$. So, we obtain $v_u{}^\sigma = v_{u,i}{}^\tau \bmod N$ if we set $\sigma = z' - z$ and $\tau = e - e'$. Therefore we can write $v_1{}^\sigma = v_{1,i}{}^\tau, \dots, v_k{}^\sigma = v_{k,i}{}^\tau, x^{4\Delta\sigma} = s_i{}^{2\tau}$. □

Theorem 5. ***(Soundness)*** *Assume that some probabilistic polynomial Turing machine $\tilde{P}$ is accepted with non-negligible probability. If $B' < B$, $h \times \log(B') = \theta(k_1)$, $k = \theta(k_1/\log(B))$ and $\log(A)$ is a polynomial in k_1 and $\log(N)$, we can prove that $x^{4\Delta d_i} = s_i{}^2$ and so s_i is a correct signature share.*

Proof. By the previous lemma we can assume that we have τ and σ such that $v_u{}^\sigma = v_u{}^{d_i \tau}$ for $u = 1, \dots, k$ and $x^{4\Delta\sigma} = s_i{}^{2\tau}$.

Then, we can write $x^{4\Delta}$ with the set of generators of Q_N since it is a square : $x^{4\Delta} = v_1{}^{\beta_1} \times \dots \times v_k{}^{\beta_k}$.

Consequently if we raise this equation to the power σ, we obtain $x^{4\Delta\sigma} = v_1{}^{\sigma\beta_1} \times \dots \times v_k{}^{\sigma\beta_k}$. But, $x^{4\Delta\sigma}$ is equal to $s_i{}^{2\tau}$ and $v_1{}^{\sigma\beta_1} \times \dots \times v_k{}^{\sigma\beta_k}$ is equal to $(v_1{}^{\beta_1} \times \dots \times v_k{}^{\beta_k})^{\tau d_i}$ as $v_u{}^\sigma = v_u{}^{d_i \tau}$ for $u = 1, \dots, k$.

Therefore, $s_i{}^{2\tau} = (x^{4\Delta})^{\tau d_i}$ with $|\tau| < B'$. We can simplify this equation by τ if τ is coprime with $p'q'$. So we obtain $x^{4\Delta d_i} = s_i{}^2$ if $B' < B$.

Let $\tilde{\pi}(k_1)$ the probability of success of $\tilde{P}$. If $\tilde{\pi}(k_1)$ is non-negligible, there exists an integer c such that $\tilde{\pi}(k_1) \geq 1/{k_1}^c$ for infinitely many values k_1. The probability for $\tilde{P}$ to generate a correct signature share while the v_is generate the group Q_N is larger than $\tilde{\pi}(k_1) - 2 \times \frac{2}{k-1} \times \frac{1}{B^{k-1}}$ according to the result of the section 6.3. So, if $k = \theta(k_1/\log(B))$, for infinitely many values k_1, $2 \times \frac{2}{k-1} \times \frac{1}{B^{k-1}} \leq 1/3{k_1}^c$.

Furthermore, for k_1 large enough, $1/{B'}^h < 1/3{k_1}^c$ if $h \times \log(B') = \theta(k_1)$. So by taking $\varepsilon = \tilde{\pi}(k_1)/3$ in lemma 2 we conclude that it is possible to obtain (σ, τ) in polynomial time $O(1/\varepsilon) = O({k_1}^c)$. □

Proof of Statistical Zero-Knowledge

Proof. Furthermore, we can prove that if A is much larger than $B' \times N$, the protocol statistically gives no information about the secret. In the random oracle model where the attacker has a full control of the values returned by the hash function H, we define the first b' bits of the value of H at

$$(v_1, \ldots, v_k, x^{4\Delta}, v_{1,i}, \ldots, v_{k,i}, {s_i}^2, {v_1}^z {v_{1,i}}^{-e}, \ldots, {v_k}^z {v_{k,i}}^{-e}, x^{4\Delta z} {s_i}^{-2e})$$

to be e. With overwhelming probability, the attacker has not yet defined the random oracle at this point so the adversary $\mathcal{A}$ cannot detect the fraud. □

6.3 Choice of Parameters

In this section we prove that with high probability we generate the entire square group Q_N with only few random elements.

Theorem 6. *With probability greater than* $1 - 2 \times \frac{2}{k-1} \times \frac{1}{B^{k-1}}$, *a random k-tuple* $(v_1, \cdots, v_k)$ *generates* Q_N.

See appendix section 9.3 for a proof of this theorem.

7 Practical Parameters for the Scheme

In the key generation we can test whether p, q, p' and q' are divisible by small primes $\leq B$ and $\gcd(p', q') = 1$. We can assume that B is the first prime greater than 2^{16}. The loss in the key generation phase is a factor 80 on average. Indeed, we have seen in the proof of theorem 2 section 4.4 that $\Pr_{p'}[p'$ has no small prime factors $\leq B] \approx \frac{1}{e^\gamma \ln(B)}$. If we fix B to 2^{16}, we can assume that $\Pr_{p'}[p'$ has no small prime factors $\leq B] > \frac{1}{20}$. Therefore, to generate p and q such that neither p' nor q' have small prime factors and such that $\gcd(p-1, q-1) = 2$, we have to run on average $2 \times (20+20) = 80$ times the first phase of Boneh-Franklin's protocol. This factor is not critical as this algorithm is run only once.

In the proof of correctness, if we want to have a security parameter of 2^{80}, we choose $B' = 2^{16} < B$. Hence, we have to choose $h = 5$ rounds. To generate the group of squares with probability greater than $1 - 2^{80}$, we need $u = 6$ verification keys. Therefore, we need 30 proofs of correctness but is is acceptable in the applications that we have in mind.

8 Conclusion

In this paper, we have showed how to avoid safe prime RSA modulus in Shoup's proof of robustness such that the proof remains correct. We consider environments where high security is required such as electronic voting schemes, and therefore, we need protocols using standard assumptions and we are ready to pay the price for it.

Basically, we use three different techniques allowing to prove that :

- the group of square is cyclic,
- we generate p and q such that p' and q' do not contain small prime factors, which allows us to generate the group Q_N and make Shamir Secret Sharing Scheme secure
- we generate a set of generators of Q_N by picking at random different generators in Q_N.

Finally, we show how to adapt Shoup proof in order to work with different elements that generate Q_N instead of a single one.

References

1. O. Baudron, P.A. Fouque, D. Pointcheval, G. Poupard, and J. Stern. Practical Multi-Candidate Election System. In *PODC '01*. ACM, 2001.
2. M. Ben-Or, S. Goldwasser, and A. Widgerson. Completeness theorems for non-cryptographic fault-tolerant distributed computing. In *Proceedings of the 20th STOC*, ACM, pages 1–10, 1988.
3. S. Blackburn, S. Blake-Wilson, S. Galbraith, and M. Burmester. Shared Generation of Shared RSA Keys. Technical report, University of Waterloo, Canada, February 1998. CORR-98-19.
4. D. Boneh and M. Franklin. Efficient Generation of Shared RSA keys. In *Crypto '97*, LNCS 1233, pages 425–439. Springer-Verlag, 1997.
5. D. Boneh, M. Malkin, and T. Wu. Experimenting with Shared Generation of RSA keys. In *Internet Society's 1999 Symposium on Network and Distributed System Security (SNDSS)*, pages 43–56, 1999.
6. R. Canetti, R. Gennaro, A. Herzberg, and D. Naor. Proactive Security : Long-term Protection Against Break-ins. *CryptoBytes*, 3(1), Spring 1997.
7. R. Canetti and S. Goldwasser. An Efficient Threshold Public Key Cryptosystem Secure Against Adaptive Chosen Ciphertext Attack. In *Eurorypt '99*, LNCS 1592, pages 90–106. Springer-Verlag, 1999.
8. D. Catalano, R. Gennaro, and S. Halevi. Computing Inverses over a Shared Secret Modulus. In *Eurocrypt '00*, LNCS 1807, pages 190–207. Springer-Verlag, 2000.
9. C. Cocks. Split Knowledge Generation of RSA Parameters. In *Cryptography and Coding : 6th IMA Conference*, LNCS 1355, pages 89–95. Springer-Verlag, 1997.
10. C. Cocks. Split Generation of RSA Parameters with Multiple Participants. Technical report, CESG, 1998. Available at `http://www.cesg.gov.uk`.
11. I. Damgård and M. Jurik. A Generalisation, a Simplification and Some Applications of Paillier's Probabilistic Public-Key System. In *PKC '01*, LNCS 1992, pages 119–136. Springer-Verlag, 2001.

12. I. Damgård and M. Koprowski. Practical Threshold RSA Signatures Without a Trusted Dealer. In *Eurocrypt '01*, LNCS 2045, pages 152–165. Springer-Verlag, 2001.
13. Y. Desmedt and Y. Frankel. Shared Generation of Authenticators and Signature. In *Crypto '91*, LNCS 576, pages 457–469. Springer-Verlag, 1991.
14. P. A. Fouque, G. Poupard, and J. Stern. Sharing Decryption in the Context of Voting or Lotteries. In *Financial Crypto '00*, LNCS. Springer-Verlag, 2000.
15. P. A. Fouque and J. Stern. One Round Threshold Discrete-Log Key Generation without Private Channels. In *PKC '01*, LNCS 1992. Springer-Verlag, 2001.
16. Y. Frankel, P. Gemmell, P. MacKenzie, and M. Yung. Optimal Resilience Proactive Public-Key Cryptosystems. In *FOCS '97*, pages 384–393, 1997.
17. Y. Frankel, P. Gemmell, P. MacKenzie, and M. Yung. Proactive RSA. In *Crypto '97*, pages 440–454, 1997.
18. Y. Frankel, P. MacKenzie, and M. Yung. Robust Efficient Distributed RSA Key Generation. In *STOC '98*, pages 663–672, 1995.
19. R. Gennaro, S. Jarecki, H. Krawczyk, and T. Rabin. Robust and Efficient Sharing of RSA Functions. In *Crypto '96*, LNCS 1109, pages 157–172. Springer-Verlag, 1996.
20. R. Gennaro, S. Jarecki, H. Krawczyk, and T. Rabin. Robust Threshold DSS Signatures. In *Eurocrypt '96*, LNCS 1070, pages 425–438. Springer-Verlag, 1996.
21. R. Gennaro, S. Jarecki, H. Krawczyk, and T. Rabin. Secure Distributed Key Generation for Discrete-Log Based Cryptosystems. In *Eurocrypt '99*, LNCS 1592, pages 295–310. Springer-Verlag, 1999.
22. R. Gennaro, D. Micciancio, and T. Rabin. An Efficient Non-Interactive Statistical Zero-Knowledge Proof System for Quasi-Safe Prime Products. In *Proc. of the Fifth ACM Conference on Computer and Communications Security '98*. ACM, 1998.
23. N. Gilboa. Two Party RSA Key Generation. In *Crypto '99*, LNCS 1666. Springer-Verlag, 1999.
24. L. C. Guillou and J.-J. Quisquater. A Practical Zero-Knowledge Protocol Fitted to Security Microprocessor Minimizing Both Transmission and Memory. In *Eurocrypt '88*, LNCS 330, pages 123–128. Springer-Verlag, 1988.
25. B. King. Improved Methods to Perform Threshold RSA. In *Asiacrypt '00*, LNCS 1976, pages 359–372. Springer-Verlag, 2000.
26. S. Miyazaki, K. Sakurai, and M. Yung. On Threshold RSA-signing with no dealer. In *ICICS '99*, LNCS 1787. Springer-Verlag, 1999.
27. T.P. Pedersen. A Threshold Cryptosystem without a Trusted Party. In *Eurocrypt'91*, LNCS 547, pages 522–526. Springer-Verlag, 1991.
28. C. Pomerance. On the distribution of pseudoprimes. In *Mathematics of Computation*, 37(156), pages 587–593, 1981.
29. C. Pomerance. Two methods in elementary analytic number theory. pages 135–161. Kluwer Academic Publishers, 1989.
30. G. Poupard and J. Stern. Generation of Shared RSA Keys by Two Parties. In *Asiacrypt '98*, LNCS 1514, pages 11–24. Springer-Verlag, 1998.
31. G. Poupard and J. Stern. Short Proofs of Knowledge for Factoring. In *PKC '00*, LNCS 1751, pages 147–166. Springer-Verlag, 2000.
32. T. Rabin. A Simplified Approach to Threshold and Proactive RSA. In *Crypto '98*, LNCS 1462, pages 89–104. Springer-Verlag, 1998.
33. R. Rivest. Finding Four Million Large Random Primes. In *Crypto '90*, LNCS 537, pages 625–626. Springer-Verlag, 1991.

34. R. Rivest, A. Shamir, and L. Adleman. A Method for Obtaining Digital Signatures and Public Key Cryptosystems. *Communications of the ACM*, 21(2):120–126, February 1978.
35. A. De Santis, Y. Desmedt, Y. Frankel, and M. Yung. How to share a function securely. In *STOC '94*, pages 522–533. ACM, 1994.
36. A. Shamir. How to Share a Secret. *Communications of the ACM*, 22:612–613, November 1979.
37. V. Shoup. Practical Threshold Signatures. In *Eurocrypt '00*, LNCS 1807, pages 207–220. Springer-Verlag, 2000.
38. V. Shoup and R. Gennaro. Securing Threshold Cryptosystems against Chosen Ciphertext Attack. In *Eurocrypt '98*, LNCS 1403, pages 1–16. Springer-Verlag, 1998. cf. the extended version for the Journal of Cryptology, available at `http://www.shoup.net/papers/`.
39. R.D. Silverman. Fast Generation of Random, Strong RSA Primes. RSA Laboratories, May 1997.

9 Appendix

9.1 Robustness of the Distributed Sieving Protocol

To resist against such players, we first run a "sum-to-poly" algorithm as described in [16,18]. When the polynomial sharing of p is obtained, one can note that $\sum_{j\in S\setminus\{0\}} \lambda_{0,j}^S = 1$, where $\lambda_{0,j}^S$ denotes the Lagrange coefficient of the jth server. Therefore, server i can set its new polynomial share to $f(i)-1$ if $f(i)$ denotes its polynomial share. Indeed, if $p = f(0) = \sum_{j\in S\setminus\{0\}} \lambda_{0,j}^S f(j)$, then

$$p-1 = f(0)-1 = \sum_{j\in S\setminus\{0\}} \lambda_{0,j}^S f(j) - \sum_{j\in S\setminus\{0\}} \lambda_{0,j}^S = \sum_{j\in S\setminus\{0\}} \lambda_{0,j}^S (f(j)-1)$$

Next, the GCD protocol can also be applied with a polynomial sharing of the secret value $\varphi = p-1$.

9.2 Proof of Lemma 1

Proof. We can note that $\varphi(N) = N-p-q+1 = (N-1)-(p-1)-(q-1)$. So,

$$(N-1)-\varphi(N) = (p-1)+(q-1)$$

Consequently, $\gcd(N-1,\varphi(N)) = \gcd((N-1)-\varphi(N),\varphi(N)) = \gcd((p-1)+(q-1),\varphi(N))$. If we note $a = p-1$ and $b = q-1$, we have to compare $\gcd(a+b,ab)$ and $\gcd(a,b)$. It is easy to see that $\gcd(a,b) | \gcd(a+b,ab)$, because if $f|\gcd(a,b)$, $f|a$ and $f|b$, so $f|a+b$ and $f|ab$.
But let $f|\gcd(a+b,ab)$. As,

$$\gcd(a+b,ab) = \gcd(a+b,ab-a(a+b)) = \gcd(a+b,-a^2) = \gcd(a+b,a^2)$$

We can assume that $f|(a+b)$ and $f|a^2$. If f is a prime number, $f|a$ and as $f|(a+b)$, $f|\gcd(a,b)$. If f is not a prime number but a power of some prime number, say $f = f'^{\alpha}$, we have $f'^{\alpha}|a^2$ and $\alpha = 2\beta$. Hence, $f'^{\beta}|a+b$ and $f'^{\beta}|a$, so $f'^{\beta}|\gcd(a,b)$. □

9.3 Proof of Theorem 6

Theorem 5. *With probability greater than* $1-2\times\frac{2}{k-1}\times\frac{1}{B^{k-1}}$, *a random* k*-tuple* $(v_1,\ldots,v_k)$ *generates* Q_N.

Let us first define additional notations. If $(v_1,\ldots,v_k)$ is a k-tuple of $(Q_N)^k$, we use $\langle v_1,\ldots,v_k\rangle$ to denote the subgroup of Q_N that is generated by the v_i's, *i.e.*, $\langle v_1,\ldots,v_k\rangle=\{x\in Q_N|\exists(\lambda_1,\ldots,\lambda_k)\ x=\prod_{i=1}^k v_i^{\lambda_i} \bmod N\}$.

We also denote by $\zeta(k)$ the Riemann Zeta function defined by $\zeta(k)=\sum_{d=1}^{+\infty}\frac{1}{d^k}$ for any integer $k\geq 2$. If $n=q_1{}^{e_1}\times q_2{}^{e_2}\times\ldots\times q_j{}^{e_j}$, we denote by $\varphi_k(n)$ $n^k\times(1-\frac{1}{q_1{}^k})(1-\frac{1}{q_2{}^k})\ldots(1-\frac{1}{q_j{}^k})$, the generalization of the Euler function in the case of k generators. Finally, if n has no prime factors less than B, we define $\zeta_B(k)$ has $\sum_{d=B}^{+\infty}\frac{1}{d^k}$.

To find a generator v of Q_N, we have to find a v such that $v \bmod p$ generates Q_p and $v \bmod q$ generates Q_q.

We estimate the probability that $x\in Q_N$ is a generator of Q_N. The probability to catch such number depends on the factorization of the order p' of Q_p and q'. Yet, even if $M=p'q'$ has no small factors, the probability is to obtain such generator is not overwhelming. Indeed, if we pick a random element v in Q_p, the probability that v is a generator of Q_p is

$$\mathsf{Pr}=\Pr_{v\in Q_p}(\langle v\rangle=Q_p)=\frac{\varphi(p')}{p'}=\prod_{p_i\geq B,p_i|p-1}(1-\frac{1}{p_i})\leq 1-\frac{1}{p_1}$$

and if $p_1\leq 2B$, we can bound the probability by $\leq 1-\frac{1}{2B}$. The probability that $B\leq p_1\leq 2B$ is equal to the probability that p' is divisible by at least one prime that belongs in $[B,2B]$. So, $\Pr_{p_1}[B\leq p_1\leq 2B]=\sum_{B\leq q_i\leq 2B,q_i prime}\frac{1}{q_i}\geq\frac{1}{2B}\times(\pi(2B)-\pi(B))$ if we denote by $\pi(x)$ the number of primes between 2 and x. If $B=2^{16}$, with probability $\geq 1/26$, $\mathsf{Pr}\leq\frac{1}{2^{17}}$. Consequently, we cannot say that this probability is overwhelming.

However, if we allow to choose several random elements in Q_N, then the subgroup $\langle v_1,\ldots,v_k\rangle$ is a equal to Q_N with high probability. A k-tuple $(v_1,\ldots,v_k)$ is a set of generators of Q_N if $(v_1 \bmod p,\ldots,v_k \bmod p)$ is a set of generators of Q_p and if $(v_1 \bmod q,\ldots,v_k \bmod q)$ is a set of generators of Q_q. Hence, the number of k-tuples of $(Q_N)^k$ that generate Q_N is the number of these k-tuples viewed as elements of $(Q_p)^k$ that generate Q_p and viewed as elements of $(Q_q)^k$ that generate Q_q.

There are $p'=\frac{p-1}{2}$ elements in Q_p. To generate this cyclic subgroup of $\mathbb{Z}_p{}^*$ (since it is a subgroup of a cyclic group), there are $\varphi(p')$ such generators.

The analysis made by Poupard and Stern in [31] can be extended in our context as it is true in general cyclic groups and not only in $\mathbb{Z}_{p^e}{}^*$. Let us now present a preliminary lemma.

Lemma 3. *The number of* k*-tuples of* $(Q_p)^k$ *that generate* Q_p *is* $\varphi_k(p')$.

Proof. Let $(v_1,\ldots,v_k)$ be k-tuple of $(Q_p)^k$ and v be a generator of Q_p; for $i=1,\ldots,k$, we define $\alpha_i\in\mathbb{Z}_{p'}$ by the relation $v^{\alpha_i}=v_i \bmod p$. We first notice

that $(v_1, \ldots, v_k)$ generates Q_p if and only if the ideal generated by $\alpha_1, \ldots, \alpha_k$ in the ring $\mathbb{Z}_{p'}$ is the entire ring. Bezout equality shows that this event occurs iff $\gcd(\alpha_1, \ldots, \alpha_k, p') = 1$.
Now, we count the k-tuples $(\alpha_1, \ldots, \alpha_k) \in (Q_p)^k$ such that $\gcd(\alpha_1, \ldots, \alpha_k, p') = 1$. Let $\prod_{i=1}^{t'} {q_i}^{f_i}$ the prime factorization of p'. Then, $\gcd(x, \prod_{i=1}^{t'} {q_i}^{f_i}) = 1 \iff \forall i \leq t'$, $\gcd(x, {q_i}^{f_i}) = 1 \iff \forall i \leq t'$, $\gcd(x \bmod {q_i}^{f_i}, {q_i}^{f_i}) = 1$.

Using the Chinese remainder theorem, the problem reduces to counting the number of k-tuples $(\beta_1, \ldots, \beta_k)$ of $(\mathbb{Z}_{{q_i}^{f_i}})^k$ such that $\gcd(\beta_1 \bmod {q_i}^{f_i}, \ldots, \beta_k \bmod {q_i}^{f_i}, {q_i}^{f_i}) = 1$ for $i = 1, \ldots, t'$. The k-tuples that do not verify this relation for a fixed index i are of the form $(q_i\gamma_1, \ldots, q_i\gamma_k)$ where $(\gamma_1, \ldots, \gamma_k) \in {\mathbb{Z}_{{q_i}^{f_i - 1}}}^k$ and there are exactly ${q_i}^{k(f_i-1)}$ such k-tuples.
Finally, the number of k-tuples of $(\mathbb{Z}_{p'})^k$ such that $\gcd(\alpha_1, \ldots, \alpha_k, p') = 1$ is $\prod_{i=1}^{t'}({q_i}^{kf_i} - {q_i}^{k(f_i-1)}) = \prod_{i=1}^{t'} \varphi_k({q_i}^{f_i}) = \varphi_k(p')$ since φ_k is multiplicative. □

Now, we return back to the proof of the theorem 6. Let us first introduce a notation : for any integer x, let S_x be the set of the indices i such that p_i is a factor of x. From the previous lemma, we know that the probability for a k-tuple of $(Q_p)^k$ to generate Q_p is $\frac{\varphi_k(p')}{{p'}^k}$. Lemma 3 shows that pr is equal to the product $\prod_{i \in S_{p'}} 1 - \frac{1}{{p_i}^k}$. The inverse of each term $1 - \frac{1}{{p_i}^k}$ can be expanded in power series : $(1 - \frac{1}{{p_i}^k})^{-1} = \sum_{j=0}^{\infty} (1/{p_i}^k)^j$. The probability pr is a product of series with positive terms, $\mathsf{pr} = (\prod_{i \in S_{p'}} \sum_{\alpha_i=0}^{\infty} \frac{1}{{p_i}^{\alpha_i k}})^{-1}$ so we can distribute terms and obtain that pr^{-1} is the sum of $1/d^k$ where d ranges over integers whose prime factors are among p_is, $i \in S_{p'}$. This sum is smaller than the unrestricted sum $\sum_{d=1}^{\infty} 1/d^k = \zeta(k)$. Finally, we obtain $\mathsf{pr} > 1/\zeta(k)$.

In our case, neither p' nor q' have prime factors less than B, therefore : $1 + \zeta_B(k) = 1 + \sum_{d=B}^{\infty} 1/d^k < 1 + 1/B^k + \int_B^{\infty} dx/x^k = 1 + \frac{k-1+B}{k-1} \times \frac{1}{B^k}$. Since for all $x > -1$, $1/(1+x) \geq 1 - x$, $1/(1 + \zeta_B(k)) > 1 - \frac{k-1+B}{k-1} \times \frac{1}{B^k}$.

Therefore, the number of k-tuples of $(Q_p)^k$ that generate Q_p is $\varphi_k(p')$ and

$$\Pr_{(v_1, \ldots, v_k) \in (Q_p)^k}\{\langle v_1, \ldots, v_k \rangle = Q_p\} = \frac{\varphi_k(p')}{{p'}^k} > \frac{1}{1 + \zeta_B(k)} > 1 - \frac{k+B-1}{k-1} \times \frac{1}{B^k}$$

Consequently, with probability greater than $1 - 2 \times \frac{2}{k-1} \times \frac{1}{B^{k-1}}$, the k-tuple $(v_1, \ldots, v_k)$ generates Q_p and Q_q and therefore Q_N. For example, with $k = 6$ and $B = 2^{16}$, this probability is larger than $1 - 1/2^{80}$.

Adaptive Security in the Threshold Setting: From Cryptosystems to Signature Schemes

Anna Lysyanskaya and Chris Peikert

Laboratory for Computer Science
Massachusetts Institute of Technology
Cambridge, MA 02139, USA
{anna,cpeikert}@theory.lcs.mit.edu

Abstract. Threshold cryptosystems and signature schemes give ways to distribute trust throughout a group and increase the availability of cryptographic systems. A standard approach in designing these protocols is to base them upon existing single-server systems having the desired properties.

Two recent (single-server) signature schemes, one due to Gennaro et al., the other to Cramer and Shoup, have been developed which are provably secure using only standard number-theoretic hardness assumptions. Catalano et al. proposed a statically secure threshold implementation of these schemes. We improve their protocol to make it secure against an adaptive adversary, thus providing a threshold signature scheme with stronger security properties than any previously known.

As a tool, we also develop an adaptively secure, erasure-free threshold version of the Paillier cryptosystem.

1 Introduction

The goal of threshold cryptography [14,15] is to enable a cluster of cooperating servers to securely and efficiently implement such cryptographic tasks as signing and decrypting. A threshold cryptographic system should remain functional and secure even when a fraction (say, almost one half) of the servers become malicious. This problem is well-motivated in practice.

One of the strongest adversarial models in this setting is the so-called *adaptive erasure-free* model. In this setting (1) the adversary corrupts servers over time depending on its entire view of the computation; and (2) upon becoming corrupted, the players have to hand over to the adversary their entire computation history; i.e., nothing can be erased.

Although results in general multi-party computation guarantee feasibility [6, 5,10], they cannot be directly applied without incurring a considerable computation penalty. In contrast, threshold protocols are tailor-made for a specific task at hand and are therefore much more practical.

Securing threshold cryptographic systems against adaptive attacks has been the subject of extensive recent research [7,17,21]. Erasure-free solutions have also been considered [21]. However, none of these papers considered the question of

C. Boyd (Ed.): ASIACRYPT 2001, LNCS 2248, pp. 331–350, 2001.
© Springer-Verlag Berlin Heidelberg 2001

constructing adaptively secure threshold versions of signature schemes provably secure against adaptive chosen message attacks [18,12].

On the other hand, statically secure threshold versions of the Gennaro et al. [18] and Cramer-Shoup [12] signature schemes has been proposed by Catalano et al. [8]. Unlike the adaptive adversary, the static adversary's corruption strategy is independent of the computation history and can be assumed to be fixed in advance. It is known that statically secure protocols are not necessarily adaptively secure [6,5,10]. While Catalano et al. suggest that it is possible to turn their statically-secure solution into an adaptively secure one, they do not give an explicit construction.

In this paper, we extend the protocol of Catalano et al. and obtain the first construction of erasure-free adaptively secure threshold version of the Cramer-Shoup signature scheme. Our results apply as well to the Gennaro et al. [18] signature scheme. Practical threshold signature schemes with this level of security have not been exhibited before.

The general structure of our results. The protocol of Catalano et al. is constructed as follows: first, they give a secure protocol for the honest-but-curious adversary, i.e. the adversary whose corruption strategy may be adaptive, but who cannot force the players he controls to deviate from their respective protocols. Then, they show how to secure it against an adversary that forces the players to deviate from their protocols arbitrarily. In this second step, they are only able to exhibit static security.

Our starting point is the honest-but-curious protocol of Catalano et al. In order to make it adaptively secure in the erasure-free model, we utilize the techniques due to Cramer et al. [11].

Cramer et al. show how, given a threshold cryptosystem with certain desirable properties, to securely (in the static model) compute any arithmetic circuit. The general structure of their construction is as follows: the inputs to the circuit are given in encrypted form. The circuit is evaluated gate-by-gate. To evaluate each gate, the players run a corresponding protocol. For example, for the multiplication gate, the players run a special protocol that, on input two ciphertexts $E(a)$ and $E(b)$, produces a ciphertext $E(ab)$. Once all the gates are evaluated, the players jointly decrypt the ciphertext that corresponds to the output of the last gate. Cramer et al. [11] provide a statically secure multiplication protocol and a statically secure threshold cryptosystem with the required properties.

We extend the results of Cramer et al. in two ways: (1) we observe that their multiplication protocol is adaptively secure under a weaker definition that allows probability of failure, provided that the underlying threshold cryptosystem is adaptively secure; and (2) we exhibit an adaptively secure threshold cryptosystem with the required properties, namely, we show how to secure the Fouque et al. [16] version of the Paillier cryptosystem [24] against the adaptive adversary.

We then plug the resulting adaptively secure multiplication protocol into a slight modification of the honest-but-curious protocol of Catalano et al.

This approach has more general applications. It is intuitively simpler to construct protocols secure against the honest-but-curious adversary than ones secure against the active adversary. In fact the first results in secure multi-party computation were of this flavor [19]. Guided by our example, one can convert a protocol for the honest-but-curious case into one that is secure against an active and adaptive adversary in the erasure-free model, at only a small cost in efficiency.

2 The Adaptive Adversary Model

In this paper, we use a standard model [7] to describe the execution of protocols and the capabilities of the adversary. We assume the existence of l parties communicating over a synchronous broadcast channel with rushing, where up to a threshold $t < l/2$ of them may be corrupted. The value k will represent the security parameter.

A t-limited *adaptive adversary* may choose to corrupt any party at any point over the lifetime of the system, as long as it does not corrupt more than t parties in total. The choices may be based on everything the adversary has seen up to that point (all broadcast messages and the computation histories of all other corrupted parties). When an adaptive adversary corrupts a party, it is given the entire computation history of that party and takes control of its actions for the life of the system. Note that this prohibits the honest parties from making *erasures* of their internal states at any time.

Summary of definitions and techniques. As expected, security of a protocol is defined in the adaptive model using the simulation paradigm. For any adaptive adversary $\mathcal{A}$, there must exist a simulator $\mathcal{S}$ which interacts with $\mathcal{A}$ to provide a view which is computationally indistinguishable from the adversary's view of the real protocol. The main difficulty in designing secure protocols in the adaptive model is in being able to fake the messages of the honest parties such that there are consistent internal states that can be supplied to the adversary when it chooses to corrupt new parties. In fact, we will design the protocols such that the simulator can supply consistent states on behalf of all honest parties except one, which we call the "single inconsistent party," [7] and denote P_S. We stress that the inconsistent party is chosen at random by the global simulator, and remains the same throughout all simulator subroutines.

We will design simulators that supply a suitable view to the adversary *provided P_S is not corrupted*, or said another way, the adversary's view will be indistinguishable from a real invocation *up to the point at which P_S is corrupted* (if ever). Of course, if P_S does become corrupted, then we may assume that the adversary detects the simulation perfectly; we call the probability of this event the *error* of the protocol. In order to make a reduction from a single-server signature scheme to a threshold version, we will require that the error be non-negligibly smaller than one. In particular, because $2t < l$, the error of our protocols will be less than $1/2$.

In all of our protocols, any deviation that is detectable by all honest parties will cause the misbehaving party to be excluded from all protocols for the life of the system. Upon detecting a dishonest party, the others restart *only the current protocol* from the beginning. Intuitively, this strategy prevents an adversary from gaining some advantage by failing to open its commitments after witnessing the honest parties' behavior. This rule will apply in each round of every protocol, even when not stated explicitly. In our simulators, we will be explicit about when misbehaving parties cause the protocol to be restarted, and when they cause the adversary to be rewound.

A note about the round-efficiency of this rule: the number of rounds of a single protocol execution is bounded only by a constant multiple of the threshold t (since one corrupt party may force a restart every time). However, the adversary can force a total of only $O(t)$ extra rounds to be executed over all invocations, which is a negligible amortized cost over the life of the system. (This assumes that all protocols are constant-round when no malicious parties are present, which will be the case.)

3 Tools

In this section we address a special kind of zero-knowledge proof called a *Σ-protocol* [11]. First we summarize Σ-protocols in the two-party setting, then we demonstrate how to implement them in a multiparty setting using *trapdoor commitments*. A reader familiar with the work of Cramer et al. [11] can skip to section 4.

Two-party Σ-protocols. The two-party Σ-protocols we use here are, in summary, honest-verifier perfect zero-knowledge proofs of knowledge with perfect completeness in which the knowledge extractor needs only two different conversation in order to extract a witness. We refer the reader to Cramer et al. [11] for formal definitions.

Trapdoor commitments. A *trapdoor commitment scheme* is much like a regular commitment scheme: a party P commits to a value by running some probabilistic algorithm on the value. The commitment gives no information about the committed value. At some later stage, P *opens* the commitment by revealing the committed value and the random coins used by the commitment algorithm. P must not be able to find a different value (and corresponding random string) that would yield the same commitment.

Trapdoor commitment schemes have one additional property: there exists a *trapdoor value* which allows P to construct commitments that he can open arbitrarily, such that this cheating is not detectable. Cramer et al. [11] provide a formal definition.

Multiparty Σ-protocols. The goal of a multiparty Σ-protocol is for many parties to make claims of knowledge such that all parties will be convinced. If all players are honest-but-curious, a naive way of achieving this goal is to make each prover participate in a separate (two-party) Σ-protocol with each of the other players. However, this approach incurs significant communication overhead, and it is not secure against an active adversary, since Σ-protocols are only honest-verifier zero-knowledge.

Part of an efficient multiparty Σ-protocol involves choosing a shared k-bit challenge string, though no particular distribution is required[1]. We simply need two different invocations to generate different challenge strings, except with negligible probability. The challenges are generated as follows: in a preprocessing phase, generate a key for a collision-resistant hash function from $\{0,1\}^l$ to $\{0,1\}^k$. To generate a challenge, each party contributes one random bit, then the hash function is applied to the concatentation of these bits. If two identical challenges are created, then either the inputs to the hash function were identical (which happens with probability at most $1/2^{l/2}$ since at least half the parties are honest), or a collision in the hash function is found.

The complete description of a multiparty Σ-protocol is as follows: in a preprocessing phase, a public key k_i for a trapdoor commitment scheme is generated for each P_i, and is distributed to all the parties by a key-distribution protocol which hides the trapdoor values. In a single proof phase, some subset P' of parties contains the parties who are to prove knowledge.

1. Each $P_i \in P'$ computes a_i, the first message of the two-party Σ-protocol. It then broadcasts a commitment $c_i = C(a_i, r_i, k_i)$, where r_i is chosen randomly by P_i.
2. A challenge r is generated by the parties, as described above. This single challenge will be used by all the provers.
3. Each $P_i \in P'$ computes the answer z_i to the challenge r, and broadcasts a_i, r_i, z_i.
4. Every party can check every proof by verifying that $c_i = C(a_i, r_i, k_i)$ and that (a_i, r, z_i) is an accepting conversation in the two-party Σ-protocol.

Cramer et al. [11] prove the security of this protocol against a static adversary. We have shows that it is also secure in the adaptive setting, using the single inconsistent party technique. We refer the reader to the full version [22] of this paper for the proof.

4 Threshold Signatures Using a Threshold Cryptosystem

Suppose an adaptive-chosen-message-secure signature scheme (such as Cramer-Shoup [12]) and a semantically secure cryptosystem (such as Paillier [24]) are given. Our signature scheme will be constructed as follows: besides the key pair (PK, SK) for a secure signature algorithm, the key generation algorithm also generates the key pair (E, D) for some semantically secure cryptosystem. The public key for the resulting signature scheme will be $(PK, E, E(SK))$, while the secret key is simply the secret key of the underlying signature scheme, i.e., SK. The signature and verification algorithms are the same as in the signature scheme given. It is easy to see, by a hybrid argument, that this resulting signature scheme is secure against the adaptive chosen message attack.

In the following sections, we will describe secure protocols for key generation and signing, and give proofs of security for these protocols.

[1] Thanks to an anonymous referee for suggesting this improvement, due to Nielsen [23].

4.1 Key Generation

Recall the Cramer-Shoup signature scheme. The public key of the signer is a tuple (N, h, x, e', H), where: $N = pq$ is an RSA modulus such that $p = 2p' + 1$, $q = 2q' + 1$, and p',q' are both primes (p and q are called *safe* primes); the values $h, x \in \mathbb{Z}_n^*$ are both quadratic residues modulo N; e' is a random prime number; H is a collision-resistant hash function. The signature on a message m is a tuple (e, y', y) such that: $e \neq e'$ is a random prime number; y' is a random quadratic residue modulo N; $y^e = xh^{H(x')}$ where $x' = \frac{y'^{e'}}{h^{H(m)}} \bmod N$.

The key generation algorithm for the Cramer-Shoup cryptosystem generates a public key $PK = (N, h, x, e', H)$ and a secret key $SK = \phi(N)$. Our public key will also include a Paillier public key (g, n) (where n is a product of two safe primes, and g has order n modulo n^2), and a ciphertext $E(\phi(N)) = g^{\phi(N)} r^n \bmod n^2$ for a random $1 \leq r \leq n^2$.

Our key generation protocol will not be efficient, but since it is only carried out once, this efficiency penalty can be ignored. It will proceed in two steps. In Step 1, the parties will run a general multi-party computation to generate the public key $(PK, E, E(SK))$, as follows: each party P_i will contribute a random string r_i. The resulting key will be computed using the circuit for single-server key generation with randomness obtained by the exclusive-or of the r_is: $R = \oplus_{i=1}^{l} r_i$. The inputs to Step 2 are the values $\{r_i\}$ (i.e., the coins from Step 1) and fresh random bits $\{r'_i\}$. Then, using general MPC, the parties will compute the auxiliary information, emulating the one-server algorithm provided in section 5.2. For the underlying general MPC we can use the protocol due to Cramer et al. [10], which is adaptively secure and tolerates any number of corruptions below one half of the servers. In order to implement secure channels required by Cramer et al. [10], we use the non-committing encryption technique due to Canetti et al. [6] and Damgård and Nielsen [13].

The protocol described above will be secure: suppose we are given a target public key $(PK, E, S(SK))$. Our goal is to construct a simulator S which, on input the identity of an inconsistent party P_S, simulates the adversary's view of the computation provided the adversary does not corrupt P_S. We can use the simulator S_{MPC} as a subroutine. For Step 1, S will give S_{MPC} the value $(PK, E, S(SK))$ as the target output. We will also supply it with some random coins for parties that are corrupted at the beginning. As more parties get corrupted over time, S_{MPC} will request that we provide it with their inputs. S will just provide some more random coins each time that happens. If the adversary ever tries to corrupt the inconsistent party P_S, S aborts. Since the actual randomness of the algorithm is the exclusive-or of the coins of all parties, the resulting view will be correct. For Step 2, we will first run the one-server algorithm for generating the simulated auxiliary information and the simulated secret information for all but one party P_S. This one-server algorithm is described in section 5.2. We will then run the simulator S_{MPC} in the same way as for Step 1.

4.2 Computing a Signature

Signature generation is done in three steps: (1) generation of (y', e); (2) generation of a verifiable additive sharing of $e^{-1} \bmod \phi(N)$; and (3) computation of y such that $y^e = xh^{H(x')}$ where $x' = \frac{y'^{e'}}{h^{H(m)}} \bmod N$, i.e. computation of $(xh^{H(x')})^{1/e} \bmod N$.

Adaptively secure erasure-free threshold protocols for selecting a random number already exist (see, for example, the one due to Jarecki and Lysyanskaya [21]). These can be employed for performing Step 1.

Suppose a secure protocol for computing an additive sharing of $e^{-1} \bmod \phi(N)$ has been performed. Let d_i denote the share held by player P_i. Suppose it is backed up with a public ciphertext $E(d_i)$. Each player P_i computes $x' = \frac{y'^{e'}}{h^{H(m)}} \bmod N$ and reveals $(xh^{H(x')})^{d_i}$, and proves that this was done correctly by invoking the Σ-protocol for proving equality of discrete logarithms. (If a player fails, his d_i is decrypted so that the other players can compute whatever is needed without him.) This takes care of Step 3.

Therefore, the only challenging piece is the computation of a verifiable additive sharing of e^{-1}, i.e., Step 2. Following the example of Catalano et al., we cast this as the *modular inversion problem.* The problem is as follows: suppose $E(\phi)$ is public. On input e, the task is to compute an additive sharing of the value $d \equiv e^{-1} \bmod \phi$, with public backup encryptions of each share. For the problem at hand, the value ϕ is, of course, $\phi(N)$.

Simulating signature generation given a signature. Suppose the simulators for each specific steps are given (simulators for steps 1 and 3 are known; the one for step 2 is given below). Here is how we simulate the signature generation. Our input is a signature on message m: (e, y', y). First, we run the simulator for Step 1 and simulate the distributed generation of (e, y'). Then we run the simulator for Step 2 and arrive at a verifiable additive sharing of e^{-1}. Finally, we run the simulator for Step 3 to simulate raising the value $xh^{H(x')}$ to the power e^{-1} to obtain y.

Background for the modular inversion protocol. Catalano et al. [8] present two versions of a modular inversion protocol which are secure against a static adversary. The first is private but not robust, while the second adds robustness at the cost of more complexity. Here we give an adaptively secure version, based on their simpler protocol. Our protocol requires $O(lk)$ bits to be broadcast, which is the same cost as the protocol of Catalano et al.

We assume the existence of a homomorphic threshold cryptosystem, defined in Appendix B. We denote an encryption of a message x as $\overline{x}$ when the public key is clear from the context. We also assume a trapdoor commitment scheme as described in section 3.

Using an adaptively secure multiparty Σ-protocol, the MULT protocol from Cramer et al. [11] is secure against an adaptive adversary as well, because its simulator only uses a single inconsistent party.

A preliminary subprotocol. First we assume existence of a secure protocol MAD (meaning "multiply, add, decrypt") which has the following specification:

(1) public inputs $w, \overline{x}, \overline{y}, \overline{z}$ to all parties, (2) public output $F = wx + yz$ for all parties.

Given a suitable homomorphic threshold cryptosystem, MAD can be implemented using the secure MULT and DECRYPT protocols. We give the protocol and a proof of security in Appendix A.

Two preliminary Σ-protocols. In the inversion protocol, each party provides a ciphertext and must prove that it is an encryption of zero. In addition, each party must publish a ciphertext and prove that the corresponding plaintext lies within a specified range. We describe both of these proofs for the Paillier cryptosystem in section 5.1.

The inversion protocol. The INVERT protocol has the following specification:

- common public input $(pk, e, N, \overline{\phi}, \{k_i\})$. Here pk is the public key of the homomorphic cryptosystem, e is a prime to be inverted modulo the secret ϕ, N is an upper bound on the value of ϕ, and $\{k_i\}$ is the set of all public trapdoor commitment keys.
- secret input sk_i, the ith secret key share, to party P_i.
- common public output $\overline{d_i}$ where d_i is described below.
- secret output d_i from each party. The $\{d_i\}$ constitute an additive sharing of the inverse, i.e. $\sum_{P_i \in P} d_i = e^{-1} \bmod \phi$.

The protocol proceeds as follows:

1. Each P_i publishes a random encryption $\overline{0_i}$ of zero, and proves that it is valid (see section 4.2). All parties internally compute $\overline{\phi_B} = (\boxplus \overline{0_i}) \boxplus \overline{\phi}$.
2. Each P_i chooses random λ_i from the range $[0 \ldots N^2]$, and random r_i from the range $[0 \ldots N^3]$, and encrypts them to get $\overline{\lambda_i}$ and $\overline{r_i}$, respectively.
3. Each P_i broadcasts a commitment to his ciphertexts $\overline{\lambda_i}$ and $\overline{r_i}$.
4. **Step R.** Each P_i decommits by broadcasting $\overline{\lambda_i}$ and $\overline{r_i}$, and the random strings used to generate the commitments.
5. Each party proves that its λ_i and r_i values are within the proper respective intervals: each party first publishes commitments to both values, then proves that the committed values are the same as their respective plaintexts, and finally proves that the committed values are within range.
6. Each party proves knowledge of its plaintexts λ_i and r_i using a multiparty Σ-protocol. Let $\lambda = \sum_{i \in P} \lambda_i$, $R = \sum_{i \in P} r_i$, and $F = Re + \lambda\phi$.
7. The parties run the MAD protocol on e, $\overline{R}$, $\overline{\lambda}$, and $\overline{\phi_B}$, where $\overline{R} = \boxplus_{i \in P} \overline{r_i}$, $\overline{\lambda} = \boxplus_{i \in P} \overline{\lambda_i}$ by addition of ciphertexts. This protocol securely computes the value $F = Re + \lambda\phi$ as the common output.
8. Each party determines whether $(e, F) = 1$. Because e is prime, $(e, F) \neq 1$ only if e divides λ, which happens with probability about $1/e$ because at least one λ_i is chosen at random. If $(e, F) \neq 1$, the parties repeat the protocol. Otherwise, all parties compute a, b such that

$$\begin{aligned} aF + be = 1 &\iff aRe + a\lambda\phi + be = 1 \\ &\iff (aR + b) \equiv e^{-1} \bmod \phi. \end{aligned}$$

P_i's share is $d_i = ar_i$ for $i > 1$, and $d_1 = ar_1 + b$ for $i = 1$. Note that any party can use the homomorphic properties of the cryptosystem to compute an encryption $\overline{d_i}$ for any $i \in P$, because the values of a and b are known to all parties, as well as encryptions $\overline{r_i}$ for all $i \in P$.

Theorem 1 (Security of inversion protocol). *For $t < l/2$,* INVERT *is an adaptively t-secure protocol for computing an additive sharing of* $e^{-1} \bmod \phi$.

Proof: We will assume a secure key-generation protocol for the homomorphic cryptosystem. We describe the construction of such a protocol for a threshold Paillier cryptosystem in section 5. We will also assume a secure key-generation protocol for the trapdoor commitment scheme.

Let k_P be the public commitment keys for all the parties. Let P_S be the inconsistent party and t_S be its trapdoor value determined by the simulator for the key-generation protocol. Given $\mathcal{A}$, we will construct a simulator subroutine $\mathcal{S}_{Invert}$ which takes input $(\mathcal{A}, pk, e, N, \overline{\phi}, k_P, P_S, t_S)$.

$\mathcal{S}_{Invert}$ operates as follows:

1. For each honest P_i except P_S, honestly publish and prove validity of a random encryption of zero. For P_S, publish a blinding of $\overline{N} \boxminus \overline{\phi}$ and run $\mathcal{E}_\Sigma$ (see section 3) with the trapdoor value t_S to give a false proof of validity (do not extract witnesses from the corrupt parties, however). If any parties gave invalid proofs, restart the protocol and exclude them. At this stage, $\phi_B = N$, and all of the parties hold an encryption of N instead of an encryption of ϕ.
2. Through the decommitment phase, behave honestly. That is, choose random λ_i and r_i for each honest party, commit to their ciphertexts, and decommit honestly. If any parties fail to decommit, restart the protocol and exclude them.
3. During the round in which the parties prove plaintext knowledge, use the subroutine $\mathcal{E}_{POPK}$ to determine the values λ_i and r_i for all corrupted parties P_i who supplied valid proofs. If any parties fail to give valid proofs, restart the protocol and exclude them.
4. Set $R' = \sum_{i \in P} r_i$ and $\lambda' = \sum_{i \in P} \lambda_i$. Run $\mathcal{S}_{Mad}$ on $e, \boxplus_{i \in P} \overline{r_i}, \boxplus_{i \in P} \overline{\lambda_i}, \overline{\phi_B} = \overline{N}$, and $F' = R'e + \lambda' N$.
5. Proceed exactly according to the protocol, repeating if $(e, F') \neq 1$.

It is clear that the simulator runs in expected polynomial time. It remains to be shown that the output of the simulator is computationally indistinguishable from the output of a real run of the protocol. Let us assume for now that this is not the case, and that there is an adversary $\mathcal{A}$ which can distinguish between a real-life execution of INVERT and an interaction with $\mathcal{S}_{Invert}$ with non-negligible advantage. We will provide a reduction that uses $\mathcal{A}$ to break the semantic security of the cryptosystem, thus establishing a contradiction. The reduction will employ a *hybrid simulator* interacting with $\mathcal{A}$.

Consider the simulator $\mathcal{S}_{Hybrid}$ which receives the public key pk of the homomorphic cryptosystem, the public commitment keys k_P, the identity of the inconsistent party P_S, and its commitment trapdoor value t_S. In addition, it is supplied with $N, e, \phi, \overline{\phi}$, a ciphertext $\overline{b}$ where b is either 0 or 1, and an auxiliary input representing the state of the adversary $\mathcal{A}$. The adversary's interaction with $\mathcal{S}_{Hybrid}$, and its resulting decision (whether the interaction was real or simulated), will determine with non-negligible probability whether b was 0 or 1. The hybrid simulator works as follows:

1. For each $P_i \neq P_S$, publish a random encryption of zero and proves its validity. For P_S, publish $(N - \phi) \boxdot \overline{b}$ and give a false proof of validity using $\mathcal{E}_\Sigma$ and the trapdoor t_S. Let $\overline{\phi_B}$ be as in the INVERT protocol.
2. For all uncorrupted $P_i \neq P_S$, honestly choose λ_i and r_i and commit to their ciphertexts. For P_S, choose λ_S and r_S from the proper ranges, but use the commitment trapdoor t_S to create cheating commitments. Receive all commitments from the corrupt parties. Let the set of corrupt parties at this point be called C. Let $\lambda_H = \sum_{i \notin C} \lambda_i$ and $R_H = \sum_{i \notin C} r_i$.
3. For all honest $P_i \neq P_S$, decommit the ciphertexts honestly. For P_S, open the cheating commitments as $\overline{\lambda_S}$ and $\overline{r_S}$. If any parties fail to open their commitments, restart the protocol (excluding those parties forever).
4. Honestly prove plaintext knowledge for all honest parties, and use $\mathcal{E}_{POPK}$ to extract λ_i and r_i for all $P_i \in C$ who provided valid proofs. If any parties fail to give valid proofs, restart the protocol (excluding those parties forever). Let $\lambda_C = \sum_{i \in C} \lambda_i$, and $R_C = \sum_{i \in C} r_i$. Solve for λ'_H and R'_H such that $F = (\lambda_C + \lambda_H)\phi + (R_C + R_H)e = (\lambda_C + \lambda'_H)N + (R_C + R'_H)e$. We shall prove that such λ'_H and R'_H are easy to compute, and are statistically indistinguishable from λ_H and R_H (respectively). Then execute the following loop:
   a) Rewind the adversary to **Step R** in the protocol. For all honest parties $P_i \neq P_S$, again honestly decommit to their ciphertexts $\overline{\lambda_i}$, $\overline{r_i}$. For P_S, open the cheating commitments as blinded ciphertexts $\overline{\lambda'_S}$ and $\overline{r'_S}$, where $\overline{\lambda'_S} = \overline{\lambda_S} \boxplus ((\lambda'_H - \lambda_H) \boxdot \overline{b})$, and $\overline{r'_S} = \overline{r_S} \boxplus ((R'_H - R_H) \boxdot \overline{b})$. Receive decommitments from each corrupt party (which, if valid, must be the same as the earlier valid decommitment). If any parties refuse to open their commitments, go to the beginning of the loop.
   b) Honestly prove plaintext knowledge on behalf of all honest parties $P_i \neq P_S$, and use $\mathcal{E}_{POPK}$ and the commitment trapdoor t_S to provide fake proofs of plaintext knowledge for $\overline{\lambda'_S}$ and $\overline{r'_S}$. Also receive proofs of plaintext knowledge from the corrupt parties. If any parties give invalid proofs, go to the beginning of the loop.
   c) Exit the loop.
5. Run $\mathcal{S}_{Mad}$ on e, $\overline{R}$ and $\overline{\lambda}$ (as in the real protocol), $\overline{\phi_B}$, and the value F. Finish according to the protocol.

First we must analyze the running time of the hybrid simulator. It suffices to show that the loop is executed a polynomial number of times in expectation. Note that the loop is only reached if all parties open their commitments and prove plaintext knowledge correctly. Let ϵ_0 be the probability that the the adversary behaves in this way, given that P_S publishes random encryptions $\overline{\lambda_S}$, $\overline{r_S}$, and let

ϵ_1 be defined similarly, given that P_S publishes random encryptions $\overline{\lambda'_S}$, $\overline{r'_S}$. By semantic security, ϵ_0 is negligibly close to ϵ_1. The contribution of the loop to the expected running time, therefore, is negligibly more than ϵ_0 times the expected number of times the loop is executed (which is $1/\epsilon_0$), so the contribution of the loop is $O(1)$.

We now prove the correctness of the reduction. Certainly if the adversary corrupts any party besides P_S, the hybrid can supply a valid computation history because it is acting honestly on behalf of that party. We now show that if $b = 0$, the output of $\mathcal{S}_{Hybrid}$ is indistinguishable from a real run of the INVERT protocol. Similarly, if $b = 1$, the output is indistinguishable from the output of $\mathcal{S}_{Invert}$. Therefore an adversary that can detect a simulation of INVERT can be used to break the semantic security of the underlying cryptosystem.

First, assume that $b = 0$. Then it is easy to verify that the hybrid acts honestly on behalf of all the uncorrupted parties, and in the first round P_S indeed publishes a random encryption of zero, so $\phi_B = \phi$. The only deviation from the real protocol occurs in the creation of cheating commitments for P_S and in the proofs of plaintext knowledge, but these commitments are computationally indistinguishable from honest commitments. Because $\lambda'_S = \lambda_S$ and $r'_S = r_S$, the behavior of P_S is indistinguishable from an honest party's behavior in the real protocol.

Now assume that $b = 1$. Then P_S publishes a random encryption of $N - \phi$ as in the simulation, and $\phi_B = N$. Note that all λ_i, r_i belonging to honest parties are chosen uniformly except for λ'_S and r'_S. But as we will show, the distributions of those variables are statistically indistinguishable from the respective uniform distributions. So in fact the behavior of P_S in the hybrid is indistinguishable from its behavior under $\mathcal{S}_{Invert}$.

It only remains to be proven that λ_S, r_S are similarly-distributed with λ'_S, r'_S (respectively), which we do here. We assume for simplicity that $N - \phi = O(\sqrt{N})$, as is the case when $\phi = \phi(N)$ and N is the product of two large primes of approximately equal size. First we state the following lemma:

Lemma 1 ([8]). *Let x, y be two integers such that $(x, y) = 1$ and A, B two integers such that $A < B$, $x, y < A$, and $B > Ax$. Then every integer z in the closed interval $[xy - x - y + 1, Ax + By - xy + x + y - 1]$ can be written as $ax + by$ where $a \in [0, A]$ and $b \in [0, B]$. Furthermore, there exists a polynomial-time algorithm that on input x, y, and z, outputs such a and b.*

Let us denote λ as $\lambda_C + \lambda_H$, λ' as $\lambda_C + \lambda'_H$, R as $R_C + R_H$, and R' as $R_C + R'_H$. We apply this lemma twice, first with $x = \phi$, $y = e$, and again with $x = N$, $y = e$ to conclude that any integer F in the interval $[\delta, \Delta]$ can be written both as $\lambda\phi + Re$, and as $\lambda' N + R'e$, where $\lambda, \lambda' \in [0, nN^2]$ and $R, R' \in [0, nN^3]$. Here $\delta = Ne - e + 1$, and $\Delta = n(N^2\phi + N^3 e) - \phi e + \phi + e - 1$.

Now, given any fixed λ_C, R_C (the sums of the adversaries' chosen values in the protocol) and any λ_H (respectively, R_H) distributed as the sum of at least $n/2$ honestly-chosen uniform values from $[0, N^2]$ (respectively, $[0, N^3]$), it is easy to see by Chernoff bounds that the probability that F falls outside the range $[\delta, \Delta]$ is negligible since both bounds fall far away from the mean of F.

Now suppose $F \in [\delta, \Delta]$ and λ_C, R_C are fixed as in the protocol. Given a pair λ_H, R_H such that $F = (\lambda_C + \lambda_H)\phi + (R_C + R_H)e$, we present an efficient mapping that produces λ'_H, R'_H such that $F = (\lambda_C + \lambda'_H)N + (R_C + R'_H)e$. That is, $\lambda\phi - \lambda' N = (R' - R)e$. Since $(N, e) = 1$, for any given λ there exists a unique and efficiently-computable $\lambda' \in [\lambda, \lambda + e - 1]$ such that $\lambda\phi - \lambda' N$ is a multiple of e. This determines the value $\lambda'_H - \lambda_H + \lambda_S = \lambda'_S$ (one of the values published by the first honest party in the hybrid simulator), and from that we can solve for $R' - R + r_S = r'_S$ (the other published value).

We need only show that λ_S, λ'_S and r_S, r'_S are close enough in a statistical sense, i.e. that their differences are small relative to the sizes of the intervals from which they are drawn. Indeed, $\frac{|\lambda' - \lambda|}{N^2} \leq \frac{e}{N^2} \leq \frac{1}{N}$ and

$$|r_1 - r'_1| = \frac{|\lambda\phi - \lambda' N|}{e} = \left|\frac{(\lambda - \lambda')\phi}{e} + \frac{\lambda'(\phi - N)}{e}\right| \leq \left|\phi - \frac{nN^2\sqrt{N}}{e}\right| \leq nN^2\sqrt{N}$$

Thus $\frac{|r_1 - r'_1|}{N^3} \leq \frac{n}{\sqrt{N}}$ which again is negligible. This completes the proof.

5 An Adaptively Secure Threshold Paillier Cryptosystem

We introduce the following notation: for any $n \in \mathbb{N}$, $\lambda(n)$ denotes Carmichael's lambda function, defined as the largest order of the elements of $\mathbb{Z}_n^*$. It is known that if the prime factorization of an odd integer n is $\prod_{i=1}^{k} q_i^{f_i}$, then $\lambda(n) = \text{lcm}_{i=1\ldots k}(q_i^{f_i - 1}(q_i - 1))$.

Our protocols will make use of two tools: Shamir secret sharing over the integers [26], and proofs of discrete log equality in groups of unknown order [9, 4].

5.1 The Paillier Cryptosystem

The Paillier cryptosystem [24] is based on *composite-degree residuosity classes*, and has the desired homomorphic properties. It is based upon the Carmichael lambda function in $\mathbb{Z}_{n^2}^*$ and two useful facts regarding it: for all $w \in \mathbb{Z}_{n^2}^*$, $w^{\lambda(n)} = 1 \bmod n$, and $w^{n\lambda(n)} = 1 \bmod n^2$. Here we recall the cryptosystem.

Key generation. Let $n = pq$ where p, q are primes. Let $g = (1 + n)^a b^n \bmod n^2$ for random $a, b \in \mathbb{Z}_n^*$. The public key is (n, g) and the secret key is $\lambda(n)$.

Encryption. To encrypt a message $M \in \mathbb{Z}_n$, randomly choose $x \in \mathbb{Z}_n^*$ and compute the ciphertext $c = g^M x^n \bmod n^2$.

Decryption. To decrypt c, compute $M = \frac{L(c^{\lambda(n)} \bmod n^2)}{L(g^{\lambda(n)} \bmod n^2)} \bmod n$ where the domain of L is the set $S_n = \{u < n^2 \;:\; u = 1 \bmod n\}$ and $L(u) = \frac{u-1}{n}$.

Other useful properties. The Paillier cryptosystem is *homomorphic*, in the sense of the definition in Appendix B. Cramer et al. [11] provide Σ-protocols for proof of plaintext knowledge and proof of correct multiplication. We also require a proof that a ciphertext is an encryption of zero; is merely a proof of nth residuosity modulo n^2. Such a proof and is virtually identical to a zero-knowledge proof of quadratic residuosity mod n as given by, for example, Goldwasser et al. [20] Finally, we require a proof that, given a ciphertext, the corresponding plaintext lies within a specified range. Boudot [2] describes such a proof for committed values, and a proof of equality between a committed value and a ciphertext in the Paillier cryptosystem can be constructed using standard techniques (see, for example, Camenisch and Lysyanskaya [3]).

The security of the scheme is based upon the composite residuosity class problem, which is exactly the problem of decrypting a ciphertext. Semantic security can be proven based on the hardness of detecting nth residues mod n^2.

Fouque et al. [16] present a threshold version of the Paillier cryptosystem, using techniques developed by Shoup [27] for threshold RSA signatures. The version presented there is known to be secure only in the static adversary model, assuming the semantic security of the non-threshold version.

5.2 An Adaptively Secure Threshold Version

Here we present the novel result of a threshold Paillier cryptosystem which is secure in the adaptive adversary model, based upon the security of Paillier's cryptosystem and the existence of trapdoor commitment schemes. This cryptosystem is inspired by the statically-secure threshold version presented in Fouque et al. [16]

Description of the protocols. Recall $\Delta = l!$, where l is the number of parties.
Key generation. We first describe key generation in terms of an l-party function on input k, the security parameter. This function is evaluated by a trusted party, who distributes the proper values to the parties.

Choose an integer n, the product of two strong primes p, q of length k such that $p = 2p' + 1$ and $q = 2q' + 1$, and $\gcd(n, \phi(n)) = 1$. Set $\lambda = 2p'q' = \lambda(n)$. Choose random $(a, b) \leftarrow \mathbb{Z}_n^* \times \mathbb{Z}_n^*$, and let $g = (1+n)^a b^n \bmod n^2$. The secret key is the value $\beta\lambda$ for a random $\beta \leftarrow \mathbb{Z}_n^*$, which is shared additively as follows: for all parties P_i but one, choose random $s_i \leftarrow \mathbb{Z}_{n\lambda}$, and choose the last s_i such that $\sum_{i \in P} s_i = \beta\lambda \bmod n\lambda$. The public key is the triple (g, n, θ), where $\theta = a\beta\lambda$ mod n. To compute public verification keys, choose a random public square v from $\mathbb{Z}_{n^2}^*$, and let $v_i = v^{s_i} \bmod n^2$. In addition, compute polynomial backups for each s_i as follows: let $a_{i,0} = \Delta s_i$, and choose random $a_{i,j} \leftarrow [-\Delta^2 n^3/2, \ldots, \Delta^2 n^3/2]$, then define a polynomial *over the integers* $f_i(X) = \sum_{j=0}^{t} a_{i,j} X^j$ (so that $f_i(0) = \Delta s_i$). To each party P_j, give the values $f_i(j)$ for all i. Finally, compute public commitments for these backup shares using any perfectly-hiding commitment scheme, such as Pedersen's [25]. Let the public value $w_{i,j}$ be a commitment to $f_i(j)$ under public key k_j and random string $r_{i,j}$, and give $r_{i,j}$ to party P_j.

It is well known [19,1,6,10] that for any l-party function, there is an adaptively secure protocol which evaluates it. Therefore there is a simulator which, given all the outputs of the function (excluding any values belonging only to P_S), interacts with the adversary and gives it a suitable view of the key generation protocol. In section 5.2 we describe how to provide suitably-distributed inputs

to this simulator. This key generation protocol may be very inefficient, but it is only executed once to initialize the threshold cryptosystem.
Encryption. To encrypt a message $M \in \mathbb{Z}_n$, pick random $x \leftarrow \mathbb{Z}_n^*$ and compute the ciphertext $c = g^M x^n \bmod n^2$.
Computing decryption shares. Player P_i computes his decryption share $c_i = c^{s_i} \bmod n^2$, and proves via a Σ-protocol that c_i^2 (in base c^2) and v_i (in base v) have the same discrete log s_i in $\mathbb{Z}_{n^2}^*$.
Combining shares. If any party P_i refuses to publish his c_i, or gives an invalid proof, then the other parties reconstruct his secret share s_i as follows. Each party P_j publishes its backup share $f_i(j)$ and random string $r_{i,j}$, and all parties verify that $w_{i,j}$ matches those values. Because there are at least $t+1$ honest parties, each party may pick some $t+1$ honestly-published values $f_i(j)$, and by interpolation, discover $s_i = f_i(0)/\Delta$ and compute $c_i = c^{s_i} \bmod n^2$.

Now each party has a correct value $c_i = c^{s_i} \bmod n^2$, for all i. The message can be computed by each party as follows:

$$\frac{L\left(\prod_{i \in P} c_i\right)}{\theta} = \frac{L\left(c^{\sum s_i \bmod n\lambda}\right)}{\theta} = \frac{L(c^{\beta\lambda})}{\theta} = \frac{L(g^{\beta\lambda M})}{\theta} = \frac{a\beta\lambda M}{\theta} = M \bmod n$$

since the value $\theta = a\beta\lambda \bmod n$ is part of the public key.
Simulating decryption. The input to the decryption simulator is a tuple $(\{s_i\}, \{v_P\}, \{a_{i,j}\}, \{w_{i,j}\}, \{k_i\}, g, n, \theta, v, c, M, P_S, t_S, \mathcal{A})$. The sets and (g, n, θ) are simulated values corresponding to those in the real protocol; c is the ciphertext to be decrypted and M is its decryption; P_S is the identity of the single inconsistent party and t_S is its commitment trapdoor; $\mathcal{A}$ is an arbitrary input corresponding to the state of the adversary before the protocol execution. In the next section, we describe how these simulated values can be generated from only a public key from the single-server Paillier cryptosystem.

The simulator acts honestly on behalf of all uncorrupted parties P_i (excluding P_S) by publishing $c_i = c^{s_i} \bmod n^2$ and proving correctness of the decryption shares. On behalf of P_S, the simulator publishes $c_S = (1 + M\theta n) \prod_{i \neq S} c^{-s_i} \bmod n^2$ and provides a false proof of correctness using t_S. If any corrupted party fails to provide a correct decryption share, the simulator honestly interpolates that party's secret share as in the decryption protocol, and proceeds normally. The simulator then honestly computes the plaintext by multiplying the published shares, yielding $(1+M\theta n) \bmod n^2$, applying L, and dividing by θ to get common output M.

The view of the adversary under the simulation is statistically indistinguishable from a real run of the protocol, provided that all public inputs are suitably simulated. If the adversary corrupts any party P_j (other than P_S), that party's behavior over every invocation of the protocol is consistent with the secret s_j revealed to the adversary. In addition, the adversary is entitled to see $f_i(j)$ and $r_{i,j}$, for all i. When $j \neq S$, the values are consistent with anything else the adversary has seen. For $i = S$, we prove below that with high probability, any set of at most t values $f_S(j)$ is distributed similarly regardless of the value being shared, and therefore the simulated values $f_i(j)$ are statistically indistinguishable from those in a real run.

Simulating key generation. We now show that the outputs of the key generation function can be simulated (up to statistical closeness), given a public key (g', n) and the identity of the single inconsistent party P_S. (It is sufficient to simulate every value produced by the key generation function, except the secret share s_S belonging to P_S. This is because the entire simulation is aborted if the adversary ever attempts to corrupt P_S, so we need not simulate its private data.) When these values are given to the simulator for the key-generation protocol, it generates a suitable view for the adversary.

Choose random $(x, y, \theta) \leftarrow \mathbb{Z}_n^* \times \mathbb{Z}_n^*$ and let $g = (g')^x y^n \bmod n^2$. Choose random $\alpha \leftarrow \mathbb{Z}_n^*$, and let $v = g^{2\alpha}$. Then for each player, choose random $s_i \leftarrow [0, \ldots, \lfloor n/2 \rfloor - 1]$, and create verification shares $v_i = v^{s_i} \bmod n^2$ for all parties but P_S. For P_S, set $v_S = (1 + 2\alpha\theta n) v^{-\sum_{i \neq S} s_i} \bmod n^2$. Finally, create commitments $w_{i,j}$ honestly (from polynomials with free terms Δs_i and random coefficients) for all i and j, and random $r_{i,j}$.

First, note that the statistical difference between the uniform distributions on $[0, \ldots, \lfloor n/2 \rfloor - 1]$ and $\mathbb{Z}_{n\lambda}$ is $O(n^{-1/2})$, so any set of at most $l-1$ secret keys s_i is statistically indistinguishable between a real and simulated run. Both g and θ are uniformly chosen from their respective domains, and are identically-distributed with their respective values in the real protocol. In addition, v is a random element of Q_{n^2}, the cyclic group of squares mod n^2. Because $|Q_{n^2}| = pqp'q'$, and $\phi(pqp'q') = (p-1)(q-1)(p'-1)(q'-1)$, v is a generator of Q_{n^2} with high probability, and is identically-distributed with its value in the real protocol.

Note that any set of at most $l-1$ simulated verification keys v_i is statistically close to a real set. However, in the real protocol with a fixed v, the values of $l-1$ verification shares induce a distribution upon the last (because the values of $l-1$ secret shares s_i induce a distribution upon the last). That is, it is necessary and sufficient that $\prod_{i \in P} v_i = v^{\beta\lambda} \bmod n^2$ for some uniformly-chosen β from $\mathbb{Z}_n^*$. In the simulation, we choose $\prod_{i \in P} v_i = v^{\beta\lambda} = (1 + 2\alpha\theta n) \bmod n^2$ without knowing λ but just by randomly choosing θ, which induces a uniform distribution upon β as desired.

Finally, we note that the simulated set $\{w_{i,j}\}$ is identically-distributed to its counterpart in the real protocol, by the perfect-hiding of the commitment scheme.

It remains to be shown that the simulated values $f_i(j)$ for all i and for the adversary's chosen j are indistinguishable from those in a real run. It is clear that the $f_i(j)$ are identically distributed for $i \neq S$, because the simulator behaves honestly. It is also obvious that the points of different polynomials are independent. We therefore show that with high probability, the values $f_S(j)$ seen by the adversary are consistent with a polynomial having free term $\Delta \hat{s}_S$ and coefficients from the proper range, for any value of $\hat{s}_S$.

Let $f_S(X)$ be the polynomial used in the simulation, that is, $f_S(X) = \Delta s_S + \sum_{i=1}^t a_{S,j} X^j$ where the $a_{S,j}$ are randomly chosen. Say that the adversary has corrupted a set of parties C, with $|C| \leq t$. We wish to find a polynomial $\hat{f}_S(X)$ such that $\hat{f}_S(0) = \Delta \hat{s}_S$ for an arbitrary $\hat{s}_S$, and $\hat{f}_S(i) = f_S(i)$ for $i \in C$. Consider a polynomial $h(X)$ such that $h(0) = \Delta(\hat{s}_S - s_S)$, and $h(i) = 0$ for $i \in C$. Then we have $\hat{f}(X) \equiv f(X) + h(X)$. By interpolation,

$$h(X) = \sum_{i \in C} h(i) \cdot \prod_{j \neq i, j \in \{0\} \cup C} \frac{z-j}{i-j} = \Delta(\hat{s}_S - s_S) \prod_{j \in C} \frac{z-j}{-j}$$

so the coefficient of X^i in $h(X)$ is: $\Delta(\hat{s}_S - s_S) \sum_{B \subseteq C, |B|=i} \frac{\prod_{j \in B}(-j)}{\prod_{j \in C}(-j)} \in \mathbb{Z}$ which is bounded in absolute value by

$$\sum_{B \subseteq C, |B|=i} \Delta(\hat{s}_S - s_S) \leq \Delta(\hat{s}_S - s_S) \binom{t}{i} \leq \frac{\Delta(\hat{s}_S - s_S)t!}{i!(t-i)!} \leq \Delta(\hat{s}_S - s_S)t! \leq \Delta^2 n^2/2$$

since $\hat{s}_S, s_S \in \{0, \ldots n^2/2\}$.

Now the coefficients of $\hat{f}(X)$ are outside of the desired range only if any of the coefficients of $f(X)$ are outside of $[-\Delta^2(n^3 - n^2)/2, \ldots, \Delta^2(n^3 - n^2)/2]$. By the union bound, this happens with probability at most t/n, which is negligible. In addition, there is a bijection between the coefficients of f and the coefficients of $\hat{f}$ when s_S, $\hat{s}_S$, and C are fixed. Therefore the distribution of the coefficients of $\hat{f}$ is statistically close to uniform, as desired.

A reduction from the original cryptosystem. With these simulations in hand, the reduction from one-server semantic security to threshold semantic security is straightforward. Assume there is an adversary that can break the security of the threshold cryptosystem. Given a public key (g', n) for the single-server Paillier cryptosystem, we first simulate the key generation protocol and any decryptions as described above. (Recall that the public key of the threshold cryptosystem is $(g = (g')^x y^n \bmod n^2, n, \theta)$ for some uniformly-chosen x, y, θ.) The adversary then outputs two messages m_0, m_1, which we send to an oracle, who responds with a random encryption c of m_b for some random bit b. We compute $\chi = c^x \bmod n^2$ (where x is the value chosen by the key generation simulator) and give it to the adversary. By assumption, the adversary can distinguish with non-negligible advantage whether χ is an encryption of m_0 or m_1 under (g, n, θ). This is equivalent to whether c is an encryption of m_0 or m_1 under (g', n), hence we have broken the semantic security of the original cryptosystem. This completes the reduction.

Acknowledgements. We are indebted to Ron Rivest for valuable discussions. We would also like to thank the anonymous referees for their detailed and thoughtful comments. Anna Lysyanskaya acknowledges the support of an NSF graduate fellowship, the Lucent Technologies GRPW program, and the Merrill-Lynch grant given to R. L. Rivest. Chris Peikert is supported by an MIT Presidential Fellowship, sponsored by Akamai Technologies.

References

1. Michael Ben-Or, Shafi Goldwasser, and Avi Wigderson. Completeness theorems for non-cryptographic fault-tolerant distributed computation. In *Proc. 20th Annual ACM Symposium on Theory of Computing (STOC)*, pages 1–10, 1988.

2. Fabrice Boudot. Efficient proofs that a committed number lies in an interval. In Bart Preneel, editor, *Advances in Cryptology — EUROCRYPT 2000*, volume 1807 of *Lecture Notes in Computer Science*, pages 431–444. Springer Verlag, 2000.
3. Jan Camenisch and Anna Lysyanskaya. An identity escrow scheme with appointed verifiers. In Joe Kilian, editor, *Advances in Cryptology — CRYPTO 2001*, volume 2139 of *Lecture Notes in Computer Science*, pages 388–407. Springer Verlag, 2001.
4. Jan Camenisch and Markus Michels. A group signature scheme based on an RSA-variant. Technical Report RS-98-27, BRICS, Departement of Computer Science, University of Aarhus, November 1998. Preliminary version in:*Advances in Cryptology — ASIACRYPT '98*, vol. 1514 of *LNCS*.
5. Ran Canetti. Security and composition of multi-party cryptographic protocols. *Journal of Cryptology*, 13(1):143–202, 2000.
6. Ran Canetti, Uri Feige, Oded Goldreich, and Moni Naor. Adaptively secure multi-party computation. In *Proceedings of the 28th Annual ACM Symposium on Theory of Computing*, pages 639–648, 1996.
7. Ran Canetti, Rosario Gennaro, Stanislaw Jarecki, Hugo Krawczyk, and Tal Rabin. Adaptive security for threshold cryptosystems. In *Advances in Cryptology—CRYPTO 99*. Springer-Verlag, 1999.
8. Dario Catalano, Rosario Gennaro, and Shai Halevi. Computing inverses over a shared secret modulus. In Bart Preneel, editor, *Advances in Cryptology — EUROCRYPT 2000*, volume 1807 of *Lecture Notes in Computer Science*, pages 190–206. Springer Verlag, 2000.
9. David Chaum and Torben Pryds Pedersen. Wallet databases with observers. In Ernest F. Brickell, editor, *Advances in Cryptology — CRYPTO '92*, volume 740 of *Lecture Notes in Computer Science*, pages 89–105. Springer-Verlag, 1993.
10. Ronald Cramer, Ivan Damgård, Stefan Dziembowski, Martin Hirt, and Tal Rabin. Efficient multiparty computations secure against an adaptive adversary. In *Advances in Cryptology—EUROCRYPT 99*, pages 311–326. Springer-Verlag, 1999.
11. Ronald Cramer, Ivan Damgård, and Jesper Buus Nielsen. Multiparty computation from threshold homomorphic encryption. In Birgit Pfitzmann, editor, *Advances in Cryptology — EUROCRYPT 2001*, Lecture Notes in Computer Science. Springer Verlag, 2001.
12. Ronald Cramer and Victor Shoup. Signature schemes based on the strong RSA assumption. In *Proc. 6th ACM Conference on Computer and Communications Security*, pages 46–52. ACM press, nov 1999.
13. Ivan Damgård and Jesper Buus Nielsen. Improved non-committing encryption schemes based on a general complexity assumption. In Mihir Bellare, editor, *Advances in Cryptology — CRYPTO '00*, volume 1880 of *Lecture Notes in Computer Science*, pages 432–450. Springer Verlag, 2000.
14. Yvo Desmedt. Society and group oriented cryptography. In *Advances in Cryptology—CRYPTO 87*. Springer-Verlag, 1987.
15. Yvo Desmedt and Yair Frankel. Threshold cryptography. In *Advances in Cryptology — CRYPTO '89*, volume 435 of *Lecture Notes in Computer Science*, pages 307–315. Springer-Verlag, 1990.
16. P. Fouque, G. Poupard, and J. Stern. Sharing decryption in the context of voting or lotteries. In *Financial Cryptography 2000*, Lecture Notes in Computer Science. Springer Verlag, 2000.
17. Yair Frankel, Philip MacKenzie, and Moti Yung. Adaptively-secure optimal-resilience proactive RSA. In *Advances in Cryptology—ASIACRYPT 99*. Springer-Verlag, 1999.

18. Rosario Gennaro, Shai Halevi, and Tal Rabin. Secure hash-and-sign signatures without the random oracle. In Jacques Stern, editor, *Advances in Cryptology — EUROCRYPT '99*, volume 1592 of *Lecture Notes in Computer Science*, pages 123–139. Springer Verlag, 1999.
19. Oded Goldreich, Silvio Micali, and Avi Wigderson. How to play any mental game or a completeness theorem for protocols with honest majority. In *Proc. 19th Annual ACM Symposium on Theory of Computing (STOC)*, pages 218–229, 1987.
20. Shafi Goldwasser, Silvio Micali, and Charles Rackoff. The knowledge complexity of interactive proof systems. In *Proc. 27th Annual Symposium on Foundations of Computer Science*, pages 291–304, 1985.
21. Stanisław Jarecki and Anna Lysyanskaya. Adaptively secure threshold cryptography: introducing cocurrency, removing erasures. In Bart Preneel, editor, *Advances in Cryptology — EUROCRYPT 2000*, volume 1807 of *Lecture Notes in Computer Science*, pages 190–206. Springer Verlag, 2000.
22. Anna Lysyanskaya and Chris Peikert. Adaptive security in the threshold setting: From cryptosystems to signature schemes. Manuscript. Available from `http://eprint.iacr.org`.
23. Jesper Buus Nielsen. Personal communication.
24. Pascal Paillier. Public-key cryptosystems based on composite residuosity classes. In Jacques Stern, editor, *Advances in Cryptology — EUROCRYPT '99*, volume 1592 of *Lecture Notes in Computer Science*, pages 223–239. Springer Verlag, 1999.
25. Torben Pryds Pedersen. Non-interactive and information-theoretic secure verifiable secret sharing. In Joan Feigenbaum, editor, *Advances in Cryptology - CRYPTO '91*, volume 576 of *Lecture Notes in Computer Science*, pages 129–140. Springer Verlag, 1992.
26. Adi Shamir. How to share a secret. *Communications of the ACM*, 22(11):612–613, November 1979.
27. Victor Shoup. Practical threshold signatures. In Bart Preneel, editor, *Advances in Cryptology — EUROCRYPT '00*, volume 1807 of *Lecture Notes in Computer Science*, pages 207–220. Springer Verlag, 2000.

A The Mad Protocol

The Mad protocol takes common inputs $w, \overline{x}, \overline{y}, \overline{z}$ and returns common output $F = wx + yz$. It is implemented as follows:

1. Each party publishes a trapdoor-commitment to a random string r_i for use in the multiplication-by-ring-element algorithm.
2. The parties open their commitments, and compute r as the exclusive-or of all properly-decommitted strings.
3. Each party runs the multiplication-by-ring-element algorithm on inputs w and $\overline{x}$ with random string r, yielding a common random ciphertext $\overline{wx}$.
4. The parties enter the Mult protocol on $\overline{y}, \overline{z}$, yielding common random ciphertext $\overline{yz}$.
5. Each party uses the deterministic addition-of-ciphertexts algorithm to compute a common input $\overline{wx + yz}$ to the Decrypt protocol, yielding common output $F = wx + yx$, as desired.

For the proof of security, we refer the reader to the full version [22] of the paper.

B Homomorphic Threshold Encryption

Here, for self-containment, we provide a modification of the definitions given by Cramer et al. [11]. The modification is that we require security in the adaptive setting.

B.1 Threshold Cryptosystems

Here we define threshold encryption schemes and their security properties.

Definition 1. *An* adaptively-secure threshold cryptosystem *for parties* $P = \{P_1, \ldots, P_l\}$ *with* threshold $t < l$ *and security parameter* k *is a 5-tuple* $(K, \mathrm{KG}, M, E, \textsc{Decrypt})$ *having the following properties:*

1. *(Key space) The* key space $K = \{K_{k,l}\}_{k,l \in \mathbb{N}}$ *is a family of finite sets of the form* $(pk, sk_1, \ldots, sk_l)$. *We call* pk *the* public key *and* sk_i *the* private key share *of party* P_i. *For* $C \subseteq P$ *we denote the family* $\{sk_i\}_{i \in C}$ *by* sk_C.
2. *(Key generation) There exists an adaptively* t*-secure* key generation l*-party protocol* KG *which, on input* 1^k, *computes, in probabilistic polynomial time, public output* pk *and secret output* sk_i *for party* P_i, *where* $(pk, sk_1, \ldots, sk_l) \in K_k$. *We write* $(pk, sk_1, \ldots, sk_l) \leftarrow \mathrm{KG}(1^k)$ *to represent this process.*
3. *(Message sampling) There exists some probabilistic polynomial-time algorithm which, on input* pk, *outputs a uniformly random element from a* message space M_{pk}. *We write* $m \leftarrow M_{pk}$ *to describe this process.*
4. *(Encryption) There exists a probabilistic polynomial-time algorithm* E *which, on input* pk *and* $m \in M_{pk}$, *outputs an encryption* $\overline{m} = E_{pk}(m)[r]$ *of* m. *Here* r *is a uniformly random string used as the random input, and* $E_{pk}(m)[r]$ *denotes the encryption algorithm run on inputs* pk *and* m, *with random tape containing* r.
5. *(Decryption) There exists an adaptively* t*-secure protocol* Decrypt *which, on common public input* (c, pk) *and secret input* sk_i *for each uncorrupted party* P_i, *where* sk_i *is the secret key share of the public key* pk *(as generated by* KG*) and* $c = E_{pk}(m)[r]$ *is an encrypted message for some* r, *returns* m *as a common public output.*
6. *(Threshold semantic security) For all probabilistic circuit families* $\{S_k\}$ *(the* message sampler*) and* $\{D_k\}$ *(called the* distinguisher*), all constants* $c > 0$, *all sufficiently large* k, *and all* $C \subseteq P$ *such that* $|C| \leq t$,

$$\begin{aligned}\Pr[\ &(pk, sk_1, \ldots, sk_l) \leftarrow \mathrm{KG}(1^k); (m_0, m_1, s) \leftarrow S_k(pk, sk_C); i \stackrel{R}{\leftarrow} \{0,1\};\\ &e \leftarrow E(pk, m_i); b \leftarrow D_k(s, e) : b = i] < 1/2 + 1/k^c\end{aligned}$$

B.2 Homomorphic Properties

We also need the cryptosystem to have the following homomorphic properties:

1. (Message ring) For all public keys pk, the message space M_{pk} is a ring in which we can compute efficiently using the public key only. We denote the ring $(M_{pk}, \cdot_{pk}, +_{pk}, 0_{pk}, 1_{pk})$. We require that the identity elements 0_{pk} and 1_{pk} be efficiently computable from the public key.

2. ($+_{pk}$-homomorphic) There exists a polynomial-time algorithm which, given public key pk and encryptions $\overline{m}_1 \in E_{pk}(m_1)$ and $\overline{m}_2 \in E_{pk}(m_2)$, outputs a uniquely-determined encryption $\overline{m} \in E_{pk}(m_1 +_{pk} m_2)$. We write $\overline{m} = \overline{m}_1 \boxplus \overline{m}_2$. Likewise, there exists a polynomial-time algorithm for performing subtraction: $\overline{m} = \overline{m}_1 \boxminus \overline{m}_2$.
3. (Multiplication of a ciphertext by a ring element) There exists a probabilistic polynomial-time algorithm which, on input pk, $m_1 \in M_{pk}$ and $\overline{m}_2 \in E_{pk}(m_2)$, outputs a random encryption $\overline{m} \leftarrow E_{pk}(m_1 \cdot_{pk} m_2)$. We assume that we can multiply a ring element from both the left and right. We write $\overline{m} \leftarrow m_1 \boxdot \overline{m}_2 \in E_{pk}(m_1 \cdot_{pk} m_2)$ and $\overline{m} \leftarrow \overline{m}_1 \boxdot m_2 \in E_{pk}(m_1 \cdot_{pk} m_2)$. Let $(m_1 \boxdot \overline{m}_2)[r]$ denote the unique encryption produced by using r as the random coins in the multiplication-by-ring-element algorithm.
4. (Addition of a ciphertext and a ring element) There exists a probabilistic polynomial-time algorithm which, on input pk, $m_1 \in M_{pk}$ and $\overline{m}_2 \in E_{pk}(m_2)$, outputs a uniquely-determined encryption $\overline{m} \in E_{pk}(m_1 +_{pk} m_2)$. We write $\overline{m} = m_1 \boxplus \overline{m}_2$.
5. (Blindable) There exists a probabilistic polynomial-time algorithm B which, on input pk and $\overline{m} \in E_{pk}(m)$, outputs an encryption $\overline{m}' \in E_{pk}(m)$ such that $\overline{m}' = E_{pk}(m)[r]$, where r is chosen uniformly at random.
6. (Check of ciphertextness) By C_{pk} we denote the set of possible encryptions of any message, under the public key pk. Given $y \in \{0,1\}^*$ and pk, it is easy to check whether $y \in C_{pk}$.
7. (Proof of plaintext knowledge) Let $L_1 = \{(pk, y) : pk \text{ is a public key} \wedge y \in C_{pk}\}$. There exists a Σ-protocol for proving the relation R_{POPK} over $L_1 \times (\{0,1\}^*)^2$ given by $R_{POPK} = \{((pk, y), (x, r)) : x \in M_{pk} \wedge y = E_{pk}(x)[r]\}$. Let $\mathcal{E}_{POPK}$ be the simulator for this Σ-protocol, which is just a special case of $\mathcal{E}_\Sigma$ described in section 3.
8. (Proof of correct multiplication) Let $L_2 = \{(pk, x, y, z) : pk \text{ is a public key} \wedge x, y, z \in C_{pk}\}$. There exists a Σ-protocol for proving the relation R_{POCM} over $L_2 \times (\{0,1\}^*)^3$ given by $R_{POCM} = \{((pk, x, y, z), (d, r_1, r_2)) : y = E_{pk}(d)[r_1] \wedge z = (d \boxdot x)[r_2]\}$.

We call any such scheme meeting these additional requirements a *homomorphic threshold cryptosystem.*

From these properties, it is clear how to perform addition of two ciphertexts: use the $+_{pk}$ algorithm, following by an optional blinding step. The remaining operation to be supported is secure multiplication of ciphertexts. That is, given $\overline{a}$ and $\overline{b}$, determine a ciphertext $\overline{c}$ such that $c = a \cdot_{pk} b$, without leaking any information about a, b, or c. Cramer et al. [11] give the MULT protocol for multiplication, and prove its security against a static adversary.

Threshold Cryptosystems Secure against Chosen-Ciphertext Attacks

Pierre-Alain Fouque and David Pointcheval

École Normale Supérieure, Département d'Informatique
45, rue d'Ulm, F-75230 Paris Cedex 05, France
{Pierre-Alain.Fouque,David.Pointcheval}@ens.fr

Abstract. Semantic security against chosen-ciphertext attacks (IND-CCA) is widely believed as the correct security level for public-key encryption scheme. On the other hand, it is often dangerous to give to only one people the power of decryption. Therefore, threshold cryptosystems aimed at distributing the decryption ability. However, only two efficient such schemes have been proposed so far for achieving IND-CCA. Both are El Gamal-like schemes and thus are based on the same intractability assumption, namely the Decisional Diffie-Hellman problem.
In this article we rehabilitate the twin-encryption paradigm proposed by Naor and Yung to present generic conversions from a large family of (threshold) IND-CPA scheme into a (threshold) IND-CCA one in the random oracle model. An efficient instantiation is also proposed, which is based on the Paillier cryptosystem. This new construction provides the first example of threshold cryptosystem secure against chosen-ciphertext attacks based on the factorization problem. Moreover, this construction provides a scheme where the "homomorphic properties" of the original scheme still hold. This is rather cumbersome because homomorphic cryptosystems are known to be malleable and therefore not to be CCA secure. However, we do not build a "homomorphic cryptosystem", but just keep the homomorphic properties.

Keywords: Threshold Cryptosystems, Chosen-Ciphertext Attacks

1 Introduction

1.1 Chosen-Ciphertext Security

Semantic security against chosen-ciphertext attacks represents the correct security definition for a cryptosystem [31,41,4]. Therefore a lot of works [26,25, 38,34] have recently proposed schemes to convert any one-way function into a cryptosystem secure according to this security notion.

Before this notion, Naor and Yung in [33] proposed a weaker security notion that they called lunch-time attack (a.k.a. indifferent, or non-adaptive, chosen-ciphertext attack). The adversary can only ask decryption of ciphertexts before he receives the target ciphertext. Naor and Yung [33] presented a conversion to secure schemes against chosen-ciphertext attack in a lunch-time scenario. They

C. Boyd (Ed.): ASIACRYPT 2001, LNCS 2248, pp. 351–368, 2001.
© Springer-Verlag Berlin Heidelberg 2001

used non-interactive zero-knowledge proof systems (proofs of membership [9,8]) to show the consistency of the ciphertext, but not to prove that the people who built the ciphertext necessarily “knew its decryption”.

Later Rackoff and Simon [41] refined this construction replacing the non-interactive zero-knowledge proofs of membership by non-interactive zero-knowledge proofs of knowledge. Therefore, when encrypting a message, one furthermore appends a non-interactive proof of knowledge of the plaintext, which leads to (adaptive) chosen-ciphertext secure cryptosystems. Indeed, the sender proves that he knows the plaintext and thus CCA is reduced to CPA.

A similar notion has thereafter been defined, the so-called “plaintext-awareness” [7,4], which means that when someone builds a valid ciphertext, he necessarily “knows” the corresponding plaintext. Therefore, a decryption oracle is unuseful for an adversary. But this latter notion is meaningful only in the random oracle model [6].

For few years, several efficient schemes have been proposed which achieve this high security level. Most of them have only been proven in the random oracle model [7,27,48,36,25,26,38,34] using the plaintext-awareness property, but only one in the standard model [14].

1.2 Threshold Cryptosystems

On the one hand, in public-key cryptography in general, the ability of decrypting or signing is restricted to the owner of the secret key. This means that only one people has all the power. Whereas in some situations, such an ability should not be given to only one people, but shared among a group of users, such that a minimal number of them, the threshold, is needed to sign or decrypt.

On the other hand, the goal of cryptography is to withstand attackers. In the case of break-ins, *i.e.* adversary that can enter into a computer and steal the secret key, public-key systems in general are not protected against exposure of the secret key. As this kind of attacks done by intruders (hackers, Trojan horses) or by corrupted insiders are very common and frequently easy to perform, systems must be protected against them. Threshold cryptography can solve this problem by distributing trust among several components or servers. The secret key is then split into shares and each share is given to one of a group of servers.

First, the key generation process has to be distributed, in order to generate the shares of each server, without trusted party. This has been done in both the discrete logarithm [37,30,21], and the RSA [10,24,20] settings. For signature schemes, the signing process has been distributed in both environments [43,29, 28,22,40,47] as well.

For distributing the decryption process, similar techniques can be used, until one just wants to prevent chosen-plaintext attacks from passive adversaries (see below for precise definitions). However, when we want to prevent chosen-ciphertext attacks, in general, servers cannot start decryption before knowing whether the ciphertext is valid or not because an attacker can be one of these servers and in case of invalid ciphertexts, he had learned some information.

Consequently, when we try to share a cryptosystem, we should not wait until the end of the decryption to know whether the servers can really decrypt or not. Therefore, we have to integrate some proof of validity of the ciphertext that should be publicly verifiable. Unfortunately, most of all the known cryptosystems secure against chosen-ciphertext attacks are not suitable. Indeed, in the decryption processes, the alleged plaintext is decrypted, and the redundancy is checked just before returning the plaintext. Since the redundancy involves a hash function, the final check cannot be done efficiently in a distributed way.

1.3 Related Work

There are two methods to distribute the decryption process of a cryptosystem. Whereas the first one uses randomness, the second follows the model described by Lee and Lim in [32] where the usual decryption process for attaining cryptosystems immune against CCA is reversed: the receiver starts checking whether the ciphertext is valid before decrypting.

The first method has been proposed by Canetti and Goldwasser in [12]. In the Cramer-Shoup cryptosystem [14], the receiver can check the validity of a ciphertext by using one part of the secret key, before decrypting the valid ciphertext using the second part of the secret key. Therefore, one can think that it is easy to share this cryptosystem. Canetti and Goldwasser [12] succeeded in distributing this cryptosystem. But instead of checking the validity of the ciphertext in a first round and decrypting it according to the validity, they proposed a new strategy with only one round. The servers decrypt any ciphertext submitted and the decryption process is randomized. The servers compute $m \cdot (v'/v)^s$ where s is a random shared between the servers (part of the secret key), v the proof inside the ciphertext, and v' the proof calculated by the servers. In the centralized version, the decryption process verifies whether $v = v'$ or not. In the distributed version, if the proof is correct, $(v/v')^s = 1$ and the decryption gives m, otherwise it returns a random value. Nobody knows if the decrypted message is correct or not if there is no redundancy in the plaintext m. A solution is to decrypt twice the same ciphertext. If the results are the same, the message was well-formed. The main drawback is that the servers must keep in the secret key a sharing of a random s and hence, the length of the key is linear in the size of the number of decrypted messages. Consequently, even if the basic method with two rounds appears to be slower, it has nice features in term of storage and avoids the need of a protocol to compute a shared random.

This method is unfortunately specific to the Cramer-Shoup cryptosystem. The second method used by Shoup and Gennaro [48] follows Lee and Lim paper [32], with the El Gamal [17] cryptosystem, but in the random oracle model [6]. First, they tried to add a non-interactive zero-knowledge proof of knowledge of discrete logarithm, using the Schnorr signature [44]. But they remarked that the decryption simulation without the secret key would require an exponential time, because of a combinatorial explosion of the forking lemma [39]. This explosion can be avoided under stronger assumption [45]. They finally used non-interactive zero-knowledge proofs of membership (as in [33]) to avoid the rewinding, and thus the combinatorial explosion in the decryption simulation. In fact, the simulation of the decryption process cannot rewind the machine. The problem is the same as in the resettable zero-knowledge setting. Therefore, the same techniques of proof of membership in a hard language can be used [5]. We can note here that the proof of knowledge of Rackoff and Simon is actually a proof of membership. In this cryptosystem, there are two keys as in [33] : one which belongs to the receiver but the other one belongs to the sender. Since the prover has one of the two keys, he can decrypt and obtain the plaintext. Therefore, the proof turns to be a proof of knowledge for a specific sender. The sender can then decrypt messages and since it is a proof of membership we can simulate the proof without using rewinding technique.

1.4 The Basic Tool: Non-interactive Zero-Knowledge Proof Systems

The model proposed by Naor and Yung strongly uses non-interactive zero-knowledge proofs of language membership in the common random string setting. Because of that, they had to restrict the power of the security model to lunch-time attacks since the adversary could use the target ciphertext and generate a new proof of membership. If the proof was correct, the decryption oracle decrypts it. But Naor and Yung cannot prove that the proof of membership cannot be changed by someone who does not know a witness. Indeed, they did not use any non-malleable property for the non-interactive zero-knowledge proof. Recently, this property has been considered [42], but only for theoretical proof systems.

In this paper, we use the idealized assumption of the random oracle model [6], which assumes that some functions behave like truly random functions. This allows to build efficient non-interactive zero-knowledge proofs, without the common random string setting, which achieve a weaker notion than non-malleability, but strong enough for our purpose, the *simulation soundness* [42].

Simulation Soundness. Let us consider any language $\mathcal{L}$, and a non-interactive zero-knowledge proof system for $\mathcal{L}$. For any adversary $\mathcal{A}$, with access to a proof $p^\star$, for a word $x^\star$, in or out of $\mathcal{L}$, we consider her ability to forge a new proof p, for a word out of $\mathcal{L}$. Therefore, for any adversary $\mathcal{A}$, we consider

$$\mathsf{Succ}^{\mathsf{sim-nizk}}(\mathcal{A}) = \Pr[(x,p) \leftarrow \mathcal{A}(Q) \mid x \in \bar{\mathcal{L}} \wedge (x,p) \notin Q],$$

having access to a bounded list Q of proven words $(x^\star, p^\star)$, where the word $w^\star$ is any word (in or out of the language $\mathcal{L}$) and $p^\star$ an accepted proof for $w^\star$. We denote by $\bar{\mathcal{L}}$ the complement of $\mathcal{L}$, and thus all the words out of the language $\mathcal{L}$.

More generally, we denote by $\mathsf{Succ}^{\mathsf{sim-nizk}}(t)$ the maximal success probability over any adversary, with running time bounded by t, in forging a new accepted proof for an invalid word, even after having seen a bounded number of accepted proofs on (in)valid words. In our situation, this bounded number will just be one.

This is a stronger notion than the classical soundness for non-interactive zero-knowledge proofs, but a weaker than non-malleability. Indeed, Sahai [42] showed that non-malleability of non-interactive zero-knowledge proofs implies this notion, that he calls *simulation soundness.*

As we see in the sequel, in the random oracle model, we can provide efficient proofs which achieve this security level.

1.5 Our Solution

Fujisaki and Okamoto [26] proposed a generic conversion from any IND-CPA cryptosystem into an IND-CCA one, in the random oracle model [6]. In this paper, we revisit the twin-encryption technique of Naor and Yung [33], by providing a generic conversion from any IND-CPA cryptosystem into an IND-CCA one with publicly verifiable validity of the ciphertext (in front of the same kind of adversary, see below). Namely, this conversion provides threshold cryptosystems strongly secure. We furthermore present practical instantiations in the random oracle model, which achieve IND-CCA against active and adaptive adversaries.

2 Security Model

2.1 The Network

We assume a group of ℓ (probabilistic) servers, all connected to a common broadcast medium, called the communication channel. It can be an asynchronous channel like the Internet.

2.2 The Adversary

The adversary is computationally bounded and it can corrupt servers at any time by viewing the memories of corrupted servers (passive adversary), and/or modifying their behavior (active adversary). The adversary decides on whom to corrupt at the start of the protocol (static adversary). We also assume that the adversary corrupts no more than t out of ℓ servers throughout the protocol, where $\ell \geq 2t + 1$.

2.3 Threshold Cryptosystems

A t out of ℓ threshold cryptosystem consists of the following components:

- A *key generation algorithm* $\mathcal{K}$ that takes as input a security parameter in unary notation 1^k, the number ℓ of decryption servers, and the threshold parameter t; it outputs a *public key* pk, a list $\mathsf{sk}_1, \ldots, \mathsf{sk}_\ell$ of private keys (which represents a sharing of the private key sk) and a list $\mathsf{vk}_1, \ldots, \mathsf{vk}_\ell$ of verification keys.
- An *encryption algorithm* $\mathcal{E}$ that takes as input the public key pk and a cleartext m, and outputs a ciphertext c.

- Several *decryption algorithms* $\mathcal{D}_i$ (for $1 \leq i \leq \ell$) that take as input the public key pk, the private key sk_i, a ciphertext c, and output a *decryption share* σ_i (which may include a verification part to achieve robustness).
- A *recovery algorithm* that takes as input the public key pk, a ciphertext c, and a list $\sigma_1, \ldots, \sigma_\ell$ of decryption shares (or at least $t+1$ of them), together with the verification keys $\mathsf{vk}_1, \ldots, \mathsf{vk}_\ell$, and outputs a cleartext m or rejects if less than $t+1$ decryption shares are correct in the case of active adversaries. All users can run this algorithm.

2.4 Security Notions

In this section, we define the game an adversary plays and tries to win in order to achieve the goal of the attack. Adversary against threshold cryptosystems tries to attack the two following properties :

- Security of the underlying primitive. In the case of cryptosystem, it means one-wayness, semantic security [31], or non-malleability [16].
- Robustness. This means that corrupted players should not be able to prevent uncorrupted servers from decrypting ciphertexts. This notion is useful only in the presence of active adversaries. In other terms, it means that the decryption service is available even if the adversary can send bad decryption shares.

A user who wants to decrypt a ciphertext c sends it to a special server, called the *combiner*, who forwards it to all servers. The servers start checking the validity of the ciphertext, then compute a decryption share σ_i and eventually return it to the combiner. This latter combines the decryption shares to obtain the plaintext m and returns it to the user. If we want to withstand active adversaries, the combiner must decide when he receives decryption shares σ_i whether they are valid or not. A nice way is to use checking protocols [23], and verification keys are consequently needed. The goal of checking protocols is to allow each server to prove to others that it has achieved its task correctly.

Semantic Security. In the following, we focus on the semantic security [31] goal, denoted IND, and forget any other security notions (one-wayness and non-malleability.) Therefore, the game to consider is the following :

1. The key generation algorithm $\mathcal{K}$ is run. The adversary therefore receives the public key pk. With this public key, the adversary has the ability to encrypt any plaintext of his choice (hence the basic "chosen-plaintext attack").
2. The adversary chooses two cleartexts m_0 and m_1. These are given to an "encryption oracle" that chooses $b \in \{0, 1\}$ at random, encrypts m_b and gives the ciphertext c to the adversary.
3. At the end of the game, the adversary outputs $b' \in \{0, 1\}$. We say that the adversary wins the game if $b' = b$.

Semantic security against chosen-plaintext attack means that for any polynomial time bounded adversary, $b' = b$ with probability only negligibly greater than $1/2$.

Chosen Ciphertext Attacks. A stronger attack is usually considered, the so-called chosen-ciphertext attack [41], in which the adversary is given a full access to the decryption oracle $\mathcal{D}_{\mathsf{sk}}$, feeding it with any ciphertext. It therefore obtains the corresponding plaintext, or the "reject" answer. There is the trivial restriction not to ask the challenge ciphertext.

Threshold Security. The above attacks are the classical attacks in the standard (non-threshold) setting of the cryptosystem. Even if it is a threshold one, the view of the adversary is the same as if there would be only one secret key. However, in the threshold setting, we have to consider the leakage of decryption shares. To this aim, we give a new oracle access to the adversary: the adversary is given a full access to the decryption oracles $\mathcal{D}_{\mathsf{sk}_i}$, but feeding them with a valid pair of plaintext-ciphertext. It therefore obtains the decryption share σ_i. If the pair is not valid (the ciphertext does not encrypt the given plaintext) the oracle may output anything [19]. This is therefore the basic security notion (for both IND-CPA and IND-CCA) in the threshold setting: IND-TCPA and IND-TCCA respectively.

As explained in the motivation of threshold cryptosystems, such a scheme should resist to the corruption of some servers. Therefore, we have to consider this situation, which means that the adversary has control of some servers:

- still playing honestly — the adversary is thus a **passive** adversary. He has access to any internal data of some servers, but cannot modify their behavior.
- or modifying their behavior — the adversary is then an **active** adversary.

To sum up, we have several possible mixes of attacks and adversaries: the chosen-plaintext (CPA) or chosen-ciphertext (CCA) attacks, performed by passive (-Passive) or active (-Active) adversaries. According to the choice of corrupted servers, we consider adaptive or non-adaptive adversaries. Non-adaptive adversaries make their choice first (before anything else), whereas adaptive ones make their choice along the attack, adaptively. It has been proven that passive and adaptive adversaries are equivalent to passive and non-adaptive adversaries, when the number of servers is logarithmic [11].

One may remark that in the particular case where $\ell = 1$ and $t = 0$, we are back to the classical situation, where passive/active and (non)-adaptive adversaries are meaningless.

3 Generic Conversions into IND-CCA Cryptosystems

In this section, we revisit the twin-encryption paradigm proposed by Naor and Yung [33], while assuming that $(\mathcal{K}, \mathcal{E}, \mathcal{D})$ is a (possibly threshold) cryptosystem which already achieves semantic security against chosen-plaintext attacks (IND-CPA or IND-TCPA, in the threshold setting). Then, we provide a new scheme which prevents CCA (or TCCA, resp.) whatever the kind of adversary.

3.1 Generic Conversion GC

The Key Generation: $\mathbf{K}(1^k)$ runs twice $\mathcal{K}(1^k)$ to get two public keys $(\mathsf{pk}, \mathsf{pk}')$, which represent the new public key $\mathbf{PK}$. The same way, one defines the new set of secret keys as $\mathbf{SK} = \{\mathbf{SK}_i\}_{1 \le i \le \ell} = \{\mathsf{sk}, \mathsf{sk}'\} = \{\mathsf{sk}_i, \mathsf{sk}'_i\}_{1 \le i \le \ell}$ and the new set of verification keys $\mathbf{VK} = \{\mathbf{VK}_i\}_{1 \le i \le \ell} = \{\mathsf{vk}, \mathsf{vk}'\} = \{\mathsf{vk}_i, \mathsf{vk}'_i\}_{1 \le i \le \ell}$.

Encryption of m

- one first encrypts twice m under pk and pk', $a_0 = \mathcal{E}_{\mathsf{pk}}(m)$ and $a_1 = \mathcal{E}_{\mathsf{pk}'}(m)$;
- one then builds a proof that both ciphertexts encrypt the same plaintext under the keys pk and pk' respectively, $c = \mathsf{Proof}[\mathsf{pk}, \mathsf{pk}', \mathcal{D}_{\mathsf{sk}}(a_0) = \mathcal{D}_{\mathsf{sk}'}(a_1)]$.

Partial Decryption of (a_0, a_1, c)

- the server checks the validity of the proof c;
- it computes both decryption shares of the ciphertexts a_0 and a_1 (only one could be enough, but the same random choice should be done by all the servers).

It is then possible to reconstruct the plaintext, using the recovery algorithm.

With this generic construction, it is not clear that the proof c does not leak any information (as remarked in [33]), furthermore such a proof can seldom be done efficiently in the standard model. However, the random oracle model allows to make efficient non-interactive zero-knowledge proofs [39].

3.2 Non-interactive Zero-Knowledge Proofs

In order to make the following proof to work, we need a strong security notion about the proof c on the language

$$\mathcal{L} = \{(\mathsf{pk}, \mathsf{pk}', \mathcal{E}_{\mathsf{pk}}(m), \mathcal{E}_{\mathsf{pk}'}(m)) \mid \forall m\},$$

called *simulation soundness* [42].

Indeed, we want that any adversary $\mathcal{A}$, having seen a pair $(x^\star, c^\star)$, where $x^\star = (\mathsf{pk}, \mathsf{pk}', \mathcal{E}_{\mathsf{pk}}(m), \mathcal{E}_{\mathsf{pk}'}(m'))$ (with $m = m'$ but also possibly $m \neq m'$) and $c^\star$ an accepted proof for $x^\star$, has a negligible success probability in forging a new proof c for a word $x \notin \mathcal{L}$:

$$\mathsf{Succ}^{\mathsf{sim-nizk}}(\mathcal{A}) = \Pr[(x, c) \leftarrow \mathcal{A}(x^\star, c^\star) \mid x \in \bar{\mathcal{L}} \wedge (x, c) \neq (x^\star, c^\star)].$$

The idea behind this success probability is that the adversary should not be able to build a new proof from previous ones, excepted for valid words (which means in $\mathcal{L}$). Indeed, one cannot avoid the adversary to build an accepted proof for a correct word chosen by herself, and in such a case the ciphertext is valid.

Furthermore, the adversary has access to a proof for a word in $\mathcal{L}$, or maybe out of $\mathcal{L}$, because the simulator will sometimes create an accepted proof for a word that is not in $\mathcal{L}$. Such a proof should not give any further information to the adversary either.

The proof c convinces everybody that the ciphertext is valid before starting the decryption. In the security proof, the decryption simulator knows one secret key. But the challenge ciphertext will not necessarily be a valid one (possibly with two distinct encrypted messages). Thanks to the random oracle model, it is still possible to simulate, in an indistinguishable way, an accepted proof even for such a wrong string, under the assumption of the intractability of the problem of deciding membership (a weaker assumption than the semantic security of the underlying cryptosystem).

Finally, we present some practical non-interactive zero-knowledge proofs, which are easily proven to be simulation-sound using the forking lemma technique [39].

3.3 Security Proof

We show that from any adversary $\mathcal{A}$ against IND-CCA of twin scheme, we can build an adversary $\mathcal{B}$ against IND-CPA of the original scheme, first only considering passive adversaries.

3.4 Passive Adversaries

Theorem 1. *Given an IND-CPA (or IND-TCPA) cryptosystem $\mathcal{S}$, the twin conversion provides an IND-CCA (or IND-TCCA, resp.) cryptosystem $\mathcal{S}_{tw}$, in the random oracle model.*

Proof. Our proof proceeds by reduction. Given a (t, ε)-adversary $\mathcal{A}$ against our scheme $\mathcal{S}_{tw}$ in the sense of IND-CCA, we build a (t', ε')-attacker $\mathcal{B}$ against scheme $\mathcal{S}$ where $t' = t$ and $\varepsilon' = (\varepsilon - 9 \cdot \mathsf{Succ}^{\mathsf{sim-nizk}}(t))/4$.

First of all, one can note that if a (classical) cryptosystem is IND-CPA, then if we encrypt the same message under two different public keys, the resulting twin-cryptosystem is still IND-CPA. This result can be shown by applying hybrid techniques [31] and it has already been formally proven in [3,2], with a advantage loss (divided by 2).

Now, we show how to make the reduction. The attacker $\mathcal{B}$ receives a given public key pk and we show how this attacker can use the adversary $\mathcal{A}$ that breaks IND-CCA to win the game (IND-CPA). The simulator $\mathcal{B}$ runs $\mathcal{K}(1^k)$ and gets $(\mathsf{pk}', \mathsf{sk}' = \{\mathsf{sk}'_i\})$. He tosses a coin b, and sets $\mathsf{pk}_b = \mathsf{pk}$, while $\mathsf{pk}_{1-b} = \mathsf{pk}'$. Then, he sends $(\mathsf{pk}_0, \mathsf{pk}_1)$ to $\mathcal{A}$.

At the step 2 in the game, the adversary $\mathcal{A}$ outputs two messages m_0, m_1. The simulator $\mathcal{B}$ sends them to the challenger: the challenger chooses at random a bit b' and encrypts $m_{b'}$ under $\mathcal{E}_{\mathsf{pk}_b}$, yielding to $a^\star_b = \mathcal{E}_{\mathsf{pk}_b}(m_{b'})$.

Then, $\mathcal{B}$ tosses a new coin b'' at random and computes $a^\star_{1-b} = \mathcal{E}_{\mathsf{pk}_{1-b}}(m_{b''})$ and sends to the adversary the target ciphertext $y^\star = (a^\star_0, a^\star_1, c^\star)$, where $c^\star$ is a simulated proof of correctness of $a^\star_0$ and $a^\star_1$, which can be done in an indistinguishable way in the random oracle model, under the intractability of the decision problem: do $a^\star_0$ and $a^\star_1$ encrypt the same message?

Now, we show how to simulate the decryption oracle. Adversary $\mathcal{A}$ can perform queries $y = (a_0, a_1, d)$ to the decryption oracle, at any time, where $a_i = \mathcal{E}_{\mathsf{pk}_i}(m)$ and c is a proof of correctness of the ciphertext. The simulator $\mathcal{B}$ easily decrypts a_{1-b}, as he knows the secret keys related to $\mathsf{pk}_{1-b} = \mathsf{pk}'$. If the proof is correct we know that a_0 and a_1 encrypt the same value m. This simulation is perfect. If the proof is not correct, but accepted, the adversary had broken the simulation soundness, after having seen only one proof.

Finally, $\mathcal{A}$ answers with a bit $b^\star$, which is output by $\mathcal{B}$. Since the simulation may not be perfect, the adversary may never stop. In this latter case, after a time-out, $\mathcal{B}$ flips a coin $b^\star$. This latter has won if $b^\star = b'$, and thus with probability

$$\frac{\varepsilon'+1}{2} = \Pr[b^\star = b' \wedge \mathsf{NIZK}] + \Pr[b^\star = b' \wedge \neg\mathsf{NIZK}] \geq \Pr[b^\star = b' \mid \mathsf{NIZK}] \cdot \Pr[\mathsf{NIZK}].$$

In the above formula, NIZK denotes the event that none of the proofs sent by the adversary to the decryption oracle breaks the simulation soundness, after having possibly seen one proof.

Indeed, if the adversary can forge proofs of membership, for wrong words, the simulator will always answer with the message encrypted under pk'. Therefore, the adversary can decide which key has the simulator.

However, under the assumption NIZK, saying that the adversary did not forge a wrong proof, our simulation of the decryption oracle is perfect. Then, using the notation pr for probabilities under this assumption:

- in the case $b'' = b'$, the simulation is perfect. Indeed, the challenge ciphertext is a valid ciphertext, and all the decryption queries are valid ciphertexts (under the NIZK assumption). And thus, the advantage is greater than $\varepsilon/2$, thanks to results about multicast encryption [2,3] (excepted a possible advantage in the real game thanks to an attack on the soundness). Thus
$$\mathsf{pr}[b^\star = b' \mid b'' = b'] \geq \frac{\varepsilon/2+1}{2} - \Pr[\neg\mathsf{NIZK}] = \frac{\varepsilon}{4} + \frac{1}{2} - \Pr[\neg\mathsf{NIZK}].$$
- in the case $b'' \neq b'$, even a powerful adversary that can decrypt a_0 and a_1, will obtain m_0 and m_1. Therefore, he cannot get any advantage. However, the adversary who detects it may choose to never stop, or to cheat. If she decides to never stop, the time-out makes $\mathcal{B}$ to flip a coin. If she tries to cheat, she has no information about b'. Then, $\mathsf{pr}[b^\star = b' \mid b'' \neq b'] = 1/2$.

Therefore,

$$\begin{aligned}\frac{\varepsilon'+1}{2} &\geq \left(\frac{\mathsf{pr}[b^\star = b' \mid b'' = b'] + \mathsf{pr}[b^\star = b' \mid b'' \neq b']}{2}\right) \cdot \Pr[\mathsf{NIZK}] \\ &\geq \frac{1}{2} \cdot \left(\frac{\varepsilon}{4} + 1 - \Pr[\neg\mathsf{NIZK}]\right) \cdot \Pr[\mathsf{NIZK}] \geq \frac{1}{2} \cdot \left(\frac{\varepsilon}{4} + 1 - \frac{9}{4} \cdot \Pr[\neg\mathsf{NIZK}]\right).\end{aligned}$$

And thus,

$$\varepsilon' = 2\Pr[b^\star = b'] - 1 \geq \frac{\varepsilon - 9 \cdot \Pr[\neg\mathsf{NIZK}]}{4}.$$

In order to upper bound $\Pr[\neg\mathsf{NIZK}]$, we play the same game but knowing the two secret keys. Then, as soon as the adversary produces an accepted proof for an invalid word, we detect it, and thus output it. This breaks the simulation soundness with time t: $\Pr[\neg\mathsf{NIZK}] \leq \mathsf{Succ}^{\mathsf{sim-nizk}}(t)$. □

3.5 Active Adversaries

It is clear that the proof still holds whatever the adversary is, even in the threshold setting. We provided a rigorous proof without any corruption. But if the underlying scheme already prevents IND-TCPA against passive or active adversaries, the new one even prevents IND-TCCA against the same kind of adversaries.

4 Examples

The first example of semantically secure cryptosystem with easy proofs of equality of plaintexts is certainly the El Gamal cryptosystem [17]. Even if more efficient threshold versions have already been proposed [48] (even in the standard model [12]), we apply the first conversion on it.

The second example will provide the first RSA-based threshold cryptosystem secure under chosen-ciphertext attacks, even against active and adaptive adversaries. It is based on the Paillier's cryptosystem [35,19]. Another version to share Paillier cryptosystem appears in [15].

In this part, we describe the cryptosystems and we insist on the proofs of membership which are specific.

4.1 The El Gamal Cryptosystem

Description of the El Gamal Cryptosystem. Let p be a strong prime, such that $q|p-1$ is also a large prime, and g be an element of $\mathbb{Z}_p^*$ of order q. We thus denote by G the subgroup of $\mathbb{Z}_p^*$ of the elements of order q. It is spanned by g. Let $y = g^x$ be the public key corresponding to the secret key x. To encrypt a message $M \in G$, randomly choose $r \in \mathbb{Z}_q$ and compute the ciphertext $(M.y^r, g^r)$. To decrypt a ciphertext $a = (\alpha, \beta)$, the receiver computes α/β^x. It is well-known that the semantic security of El Gamal is based on the Decisional Diffie-Hellman (DDH) problem [49].

IND-CPA Threshold Version of El Gamal Cryptosystem. The secret key x is split with Shamir secret sharing scheme. Each server has a share sk_i of the secret key sk and a verification key $\mathsf{vk}_i = g^{\mathsf{sk}_i}$. To decrypt a ciphertext $a = (\alpha, \beta)$, each server computes a decryption share $\beta_i = \beta^{\mathsf{sk}_i}$, and proves that $\log_g \mathsf{vk}_i = \log_\beta \beta_i$. The combiner selects a set S of $t+1$ correct shares and computes

$$\beta^x = \prod_{i \in S} \beta_i^{\lambda_{0,i}^S} \bmod p$$

where $\lambda_{i,0}^S$ denote the symbol of Lagrange. Finally, the combiner computes $\alpha/\beta^x \bmod p$ to recover the plaintext. One can easily show that if an adversary can break the semantic security of this cryptosystem, one can build an attacker that can break the semantic security of El Gamal, and thus the DDH assumption.

IND-CCA Threshold Version of El Gamal Cryptosystem. We can therefore apply previous twin conversion. One still gets one group G, with a generator g of prime order. Then the key generation algorithm is run twice and the public keys are $y_0 = g^{x_0}$ and $y_1 = g^{x_1}$. To encrypt a message M, the sender computes $a_0 = (M \cdot y_0^r, g^r) = (\alpha_0, \beta_0)$ and $a_1 = (M \cdot y_1^s, g^s) = (\alpha_1, \beta_1)$.

The proof of equality of plaintexts consists in proving the existence of r and s such that $\beta_0 = g^r$, $\beta_1 = g^s$ and $\alpha_0/\alpha_1 = y_0^r y_1^{-s}$.

To this aim, one chooses random $a, b \in \mathbb{Z}_q$, and computes $A = g^a$, $B = g^b$ and $C = y_0^a y_1^b$. Then, one gets the random challenge $e \in \mathbb{Z}_q$ from a hash function which is assumed to behave like a random oracle: $e = H(g, y_0, y_1, a_0, a_1, A, B, C)$. Eventually, one computes $\rho = a - re \bmod q$ and $\sigma = b + se \bmod q$. This proof can be easily verified by $A = g^\rho \beta_0^e$, $B = g^\sigma \beta_1^{-e}$, and $C = y_0^\rho y_1^\sigma (\alpha_0/\alpha_1)^e$, or equivalently by

$$e = H(g, y_0, y_1, a_0, a_1, g^\rho \beta_0^e, g^\sigma \beta_1^{-e}, y_0^\rho y_1^\sigma (\alpha_0/\alpha_1)^e),$$

where the proof consists of the tuple (e, ρ, σ).

The decryption process is straightforward, using the same technique as presented above, but twice, after having checked the validity of the ciphertext.

Security Analysis. The basic threshold El Gamal cryptosystem is clearly IND-CPA. The generic conversion makes then the new proposal to be IND-TCCA, but under the condition that the above proof of equality of plaintexts is simulation-sound. We thus have to prove it.

First, we have to be able to build a list Q of accepted proofs for words in and out of the language. This can easily be done, thanks to the random oracle property of H: one chooses ρ, σ and e in $\mathbb{Z}_q$, and defines

$$H(g, y_0, y_1, a_0, a_1, g^\rho \beta_0^e, g^\sigma \beta_1^{-e}, y_0^\rho y_1^\sigma (\alpha_0/\alpha_1)^e) \leftarrow e.$$

Now, let us assume that with access to this list of proofs, an adversary is able to forge a new proof for a wrong word $(\mathsf{pk}_0, \mathsf{pk}_1, a_0, a_1)$, with probability ν, within time t. Since everything is included in the query to the random oracle H, we can apply the forking lemma [39], which claims that

Lemma 1. *Let $\mathcal{A}$ be a probabilistic polynomial time Turing machine which can ask q_h queries to the random oracle, with $q_h > 0$. We assume that, within the time bound t, $\mathcal{A}$ produces, with probability $\nu \geq 7q_h/q$, a new accepted proof for a wrong word $(\mathsf{pk}_0, \mathsf{pk}_1, a_0, a_1)$, $(g, y_0, y_1, a_0, a_1; A, B; e; \rho, \sigma)$. Then, within time*

$t' \leq 16q_h t/\nu$, and with probability $\nu' \geq 1/9$, a replay of this machine outputs two accepted proofs of a wrong word $(\mathsf{pk}_0, \mathsf{pk}_1, a_0, a_1)$:

$$(g, y_0, y_1, a_0, a_1; A, B; e_0; \rho_0, \sigma_0) \text{ and } (g, y_0, y_1, a_0, a_1; A, B; e_1; \rho_1, \sigma_1),$$

with $e_0 \neq e_1 \bmod q$.

Let us assume that the adversary has not broken the collision intractability of H, then

$$g^{\rho_0}\beta_0^{e_0} = g^{\rho_1}\beta_0^{e_1}, \quad g^{\sigma_0}\beta_1^{-e_0} = g^{\sigma_1}\beta_1^{-e_1}$$
$$y_0^{\rho_0}y_1^{\sigma_0}(\alpha_0/\alpha_1)^{e_0} = y_0^{\rho_1}y_1^{\sigma_1}(\alpha_0/\alpha_1)^{e_1}$$

and thus,

$$\beta_0 = g^{\rho}, \ \beta_1 = g^{\sigma}, \text{ and } \alpha_0/\alpha_1 = y_0^{\rho}y_1^{-\sigma},$$

where

$$\rho = \frac{\rho_1 - \rho_0}{e_0 - e_1} \bmod q, \text{ and } \sigma = \frac{\sigma_0 - \sigma_1}{e_0 - e_1} \bmod q.$$

Since $\alpha_0 = M_0 y_0^{\rho}$, and $\alpha_1 = M_1 y_1^{\sigma}$, we eventually get $M_0 = M_1$, which means that the word is in the language, unless one has broken the collision intractability for H. But under the random oracle assumption, to get a probability greater than 1/9 to find a collision, one has to have asked more than $\sqrt{q}/3$ queries to H, using the birthday paradox, and thus

$$\frac{16q_h t}{\nu} \geq t' \geq \frac{\sqrt{q}}{3}\tau,$$

where τ is the time required for an evaluation of H. This leads to

$$\mathsf{Succ}^{\mathsf{sim-nizk}}(t) \leq \nu \leq 48\frac{q_h}{\sqrt{q}}\frac{t}{\tau}.$$

This proves the soundness of the proof system. But since this lemma still holds, even for an adversary with auxiliary information (the list Q), it furthermore proves the simulation soundness.

4.2 The Paillier Cryptosystem

Review of the Basic Cryptosystem. The Paillier cryptosystem is based on the properties of the Carmichael lambda function in $\mathbb{Z}_{n^2}^*$. We recall here the main two properties: for any $w \in \mathbb{Z}_{n^2}^*$,

$$w^{\lambda(n)} = 1 \bmod n, \quad \text{and} \quad w^{n\lambda(n)} = 1 \bmod n^2$$

Let n be an RSA modulus $n = pq$, where p and q are prime integers. Let g be an integer of order $n\alpha$ modulo n^2. The public key is $\mathsf{pk} = (n, g)$ and the secret key is $\mathsf{sk} = \lambda(n)$. To encrypt a message $M \in \mathbb{Z}_n$, randomly choose $x \in \mathbb{Z}_n^*$ and compute the ciphertext $c = g^M x^n \bmod n^2$. To decrypt c, compute

$$M = \frac{L(c^{\lambda(n)} \bmod n^2)}{L(g^{\lambda(n)} \bmod n^2)} \bmod n,$$

where the L function takes elements from the set $\mathcal{U}_n = \{u < n^2 \mid u = 1 \bmod n\}$ and computes $L(u) = (u-1)/n$. The semantic security is based on the difficulty to distinguish n^{th} residues modulo n^2. We refer to [35] for details.

IND-CPA Threshold Version of Paillier Cryptosystem. We recall that $\Delta = \ell!$ where ℓ is the number of servers.

Key Generation Algorithm. Choose an integer n, product of two safe primes p and q, such that $p = 2p' + 1$ and $q = 2q' + 1$ and $\gcd(n, \varphi(n)) = 1$. One can note that the safe prime requirement can be avoided [20] using Shoup protocol [47] without using safe primes. This allows to fully share Paillier cryptosystem from the key generation protocol to the decryption process as it appears difficult to generate RSA moduli with safe prime modulus using [10]. However, for the clarity of the description we use RSA moduli with safe primes. Set $m = p'q'$. Let β be an element randomly chosen in $\mathbb{Z}_n^*$.

The secret key $\mathsf{sk} = \beta \times m$ is shared with the Shamir scheme [46] modulo mn. Let v be a square that generates with overwhelming probability the cyclic group of squares in $\mathbb{Z}_{n^2}^*$. The verification keys vk_i are obtained with the formula $v^{\Delta \mathsf{sk}_i} \bmod n^2$.

Encryption Algorithm. To encrypt a message M, randomly pick $x \in \mathbb{Z}_n^*$ and compute $c = g^M x^n \bmod n^2$.

Partial Decryption Algorithm. The ith player P_i computes the decryption share $c_i = c^{2\Delta \mathsf{sk}_i} \bmod n^2$ using his secret share sk_i. He makes a proof of correct decryption which assures that $c^{4\Delta} \bmod n^2$ and $v^{\Delta} \bmod n^2$ have been raised to the same power sk_i in order to obtain c_i^2 and vk_i.

Recovery Algorithm. If less than $t + 1$ decryption shares have valid proofs of correctness the algorithm fails. Otherwise, let S be a set of $t+1$ valid shares and compute the plaintext using the Lagrange interpolation on the exponents (which is possible since exponents are multiplied by $\Delta = \ell!$, and thus no modular root extraction is required.)

In [19], they proved the following theorem.

Theorem 2. *Under the decisional composite residuosity assumption and in the random oracle model, the threshold version of Paillier cryptosystem is IND-TCPA against active but non-adaptive adversaries.*

Even if their definition of threshold security (the partial decryption oracles behavior) is not the same, the security result still holds within our model.

IND-CCA Threshold Version of Paillier Cryptosystem. We can therefore apply previous twin conversion.

Key Generation Algorithm. Choose, for $j = 0, 1$, an integer n_j, product of two safe primes p_j and q_j. Set $m_j = (p_j - 1)(q_j - 1)/4$. Let β_j be an element randomly chosen in $\mathbb{Z}_{n_j}^*$.

The secret keys $\mathsf{sk}_j = \beta_j \times m_j$ are shared with the Shamir scheme [46] modulo $m_j n_j$. Let v_j be a square that generates all the cyclic group of squares in $\mathbb{Z}_{n_j^2}^*$. The verification keys $\mathsf{vk}_{i,j}$ are obtained with the formula $v_j^{\Delta \mathsf{sk}_{i,j}} \bmod n_j^2$.

Encryption Algorithm. To encrypt a message M, randomly pick $x_j \in \mathbb{Z}_{n_j}^*$ and compute $a_j = g_j^M x_j^{n_j} \bmod n_j^2$. Furthermore compute a proof that a_0 and a_1 encrypt the same value: Let r be a randomly chosen element in $[0, A[$, and random elements $\alpha_j \in \mathbb{Z}_{n_j}^*$. Compute $y_j = g_j^r \alpha_j^{n_j} \bmod n_j^2$. Let e be the hash value $H(g_0, g_1, a_0, a_1, y_0, y_1)$ where H is a hash function which outputs values in the range $[0, B[$. Then, compute $z = r + e \times M$, $u_j = \alpha_j x_j^e \bmod n_j$ A proof of equality is the tuple

$$(e, z, u_0, u_1) \in [0, B[\times[0, A[\times \mathbb{Z}_{n_1}^* \times \mathbb{Z}_{n_2}^*$$

It is checked by the equation

$$e = H(g_0, g_1, a_0, a_1, g_0^z u_0^{n_0} / a_0^e \bmod n_0^2, g_1^z u_1^{n_1} / a_1^e \bmod n_1^2)$$

The decryption process is the same as in [19]. Furthermore, the above proof can be shown to be simulation-sound, using the same technique as for the El Gamal scheme, thanks to the forking lemma [39].

It is amazing to note that the Generic Conversion of Paillier cryptosystem keeps the homomorphic properties, namely that $\mathcal{E}(M_1 + M_2) \equiv \mathcal{E}(M_1) \times \mathcal{E}(M_2)$ and $\mathcal{E}(M)^k \equiv \mathcal{E}(kM)$. For example, in voting scheme, such as [15,1], the authority can check the universally checkable proofs of validity of ciphertext and compute the tally. However, the result will no longer be a ciphertext that withstands CCA.

5 Conclusion

In this paper we have constructed generic conversions to threshold cryptosystems secure against chosen-ciphertext attacks from any cryptosystems secure against CPA. We have proposed the first version of threshold cryptosystems CCA-secure which rely on the factorization problem. A new version of Paillier cryptosystem based on a new assumption related to RSA appears in [13]. By applying our techniques, one can also share this cryptosystem under their new assumption. This provides the second threshold cryptosystem secure under CCA based on RSA.

However, as it is noted in [48], it appears to be difficult to share RSA. It seems even difficult to share OAEP-RSA without redundancy, which is a cryptosystem which achieves IND-CPA, but in the random oracle model. Indeed, the proof of membership appears to be odd and not practical.

Acknowledgement. We would like to thank Masayuki Abe for fruitful discussions.

References

1. O. Baudron, P.A. Fouque, D. Pointcheval, G. Poupard, and J. Stern. Practical Multi-Candidate Election System. In *PODC '01*. ACM, 2001.

2. O. Baudron, D. Pointcheval, and J. Stern. Extended Notions of Security for Multicast Public Key Cryptosystems. In *Proc. of the 27th ICALP*, LNCS 1853, pages 499–511. Springer-Verlag, Berlin, 2000.
3. M. Bellare, A. Boldyreva, and S. Micali. Public-key Encryption in a Multi-User Setting: Security Proofs and Improvements. In *Eurocrypt '2000*, LNCS 1807, pages 259–274. Springer-Verlag, Berlin, 2000.
4. M. Bellare, A. Desai, D. Pointcheval, and P. Rogaway. Relations among Notions of Security for Public-Key Encryption Schemes. In *Crypto '98*, LNCS 1462, pages 26–45. Springer-Verlag, Berlin, 1998.
5. M. Bellare, M. Fischlin, S. Goldwasser, and S. Micali. Identification Protocols Secure against Reset Attacks. In *Eurocrypt '2001*, LNCS 2045, pages 495–511. Springer-Verlag, Berlin, 2001.
6. M. Bellare and P. Rogaway. Random Oracles Are Practical: a Paradigm for Designing Efficient Protocols. In *Proc. of the 1st CCS*, pages 62–73. ACM Press, New York, 1993.
7. M. Bellare and P. Rogaway. Optimal Asymmetric Encryption – How to Encrypt with RSA. In *Eurocrypt '94*, LNCS 950, pages 92–111. Springer-Verlag, Berlin, 1995.
8. M. Blum, P. Feldman, and S. Micali. Non-Interactive Zero-Knowledge and its Applications. In *Proc. of the 20th STOC*, pages 103–112. ACM Press, New York, 1988.
9. M. Blum, P. Feldman, and S. Micali. Proving Security against Chosen-Ciphertext Attacks. In *Crypto '88*, LNCS 403, pages 256–268. Springer-Verlag, Berlin, 1989.
10. D. Boneh and M. Franklin. Efficient Generation of Shared RSA Keys. In *Crypto '97*, LNCS 1294, pages 425–439. Springer-Verlag, Berlin, 1997.
11. R. Canetti, I. Damgård, S. Dziembowski, Y. Ishai, and T. Malkin. On Adaptive vs. Non-adaptive Security of Multiparty Protocols. In *Eurocrypt '2001*, LNCS 2045, pages 262–279. Springer-Verlag, Berlin, 2001.
12. R. Canetti and S. Goldwasser. An Efficient Threshold PKC Secure Against Adaptive CCA. In *Eurocrypt '99*, LNCS 1592, pages 90–106. Springer-Verlag, Berlin, 1999.
13. D. Catalano, R. Gennaro, N. Howgrave-Graham, and P. Q. Nguyen. Paillier's Cryptosystem Revisited. In *ACM CCS '2001*, ACM Press, 2001.
14. R. Cramer and V. Shoup. A Practical Public Key Cryptosystem Provably Secure against Adaptive Chosen Ciphertext Attack. In *Crypto '98*, LNCS 1462, pages 13–25. Springer-Verlag, Berlin, 1998.
15. I. Damgård and M. Jurik. A Generalisation, a Simplification and Some Applications of Paillier's Probabilistic Public-Key System. In *PKC '2001*, LNCS 1992, pages 119–137. Springer-Verlag, Berlin, 2001.
16. D. Dolev, C. Dwork, and M. Naor. Non-Malleable Cryptography. *SIAM Journal on Computing*, 30(2):391–437, 2000.
17. T. El Gamal. A Public Key Cryptosystem and a Signature Scheme Based on Discrete Logarithms. *IEEE Transactions on Information Theory*, IT–31(4):469–472, July 1985.
18. A. Fiat and A. Shamir. How to Prove Yourself: Practical Solutions of Identification and Signature Problems. In *Crypto '86*, LNCS 263, pages 186–194. Springer-Verlag, Berlin, 1987.
19. P. A. Fouque, G. Poupard, and J. Stern. Sharing Decryption in the Context of Voting or Lotteries. In *Financial Cryptography '2000*, LNCS. Springer-Verlag, Berlin, 2000.

20. P. A. Fouque and J. Stern. Fully Distributed Threshold RSA under Standard Assumptions. In *Asiacrypt '2001*, LNCS, Springer-Verlag, Berlin, 2001.
21. P. A. Fouque and J. Stern. One Round Threshold Discrete-Log Key Generation without Private Channels. In *PKC '2001*, LNCS 1992, pages 300–316. Springer-Verlag, Berlin, 2001.
22. Y. Frankel, P. Gemmel, Ph. MacKenzie, and M. Yung. Optimal-Resilience Proactive Public-Key Cryptosystems. In *Proc. of the 38th FOCS*, pages 384–393. IEEE, New York, 1997.
23. Y. Frankel, P. Gemmell, and M. Yung. Witness Based Cryptographic Program Checking and Robust Function Sharing. In *Proc. of the 28th STOC*, pages 499–508. ACM Press, New York, 1996.
24. Y. Frankel, P. MacKenzie, and M. Yung. Robust Efficient Distributed RSA Key Generation. In *Proc. of the 30th STOC*, pages 663–672. ACM Press, New York, 1998.
25. E. Fujisaki and T. Okamoto. Secure Integration of Asymmetric and Symmetric Encryption Schemes. In *Crypto '99*, LNCS 1666, pages 537–554. Springer-Verlag, Berlin, 1999.
26. E. Fujisaki and T. Okamoto. How to Enhance the Security of Public-Key Encryption at Minimum Cost. *IEICE Transaction of Fundamentals of Electronic Communications and Computer Science*, E83-A(1):24–32, January 2000.
27. E. Fujisaki, T. Okamoto, D. Pointcheval, and J. Stern. RSA–OAEP is Secure under the RSA Assumption. In *Crypto '2001*, LNCS. Springer-Verlag, Berlin, 2001.
28. R. Gennaro, S. Jarecki, H. Krawczyk, and T. Rabin. Robust and Efficient Sharing of RSA Functions. In *Crypto '96*, LNCS 1109, pages 157–172. Springer-Verlag, Berlin, 1996.
29. R. Gennaro, S. Jarecki, H. Krawczyk, and T. Rabin. Robust Threshold DSS Signatures. In *Eurocrypt '96*, LNCS 1070, pages 425–438. Springer-Verlag, Berlin, 1996.
30. R. Gennaro, S. Jarecki, H. Krawczyk, and T. Rabin. Secure Distributed Key Generation for Discrete-Log Based Cryptosystems. In *Eurocrypt '99*, LNCS 1592, pages 295–310. Springer-Verlag, Berlin, 1999.
31. S. Goldwasser and S. Micali. Probabilistic Encryption. *Journal of Computer and System Sciences*, 28:270–299, 1984.
32. C.H. Lim and P.J. Lee. Another Method for Attaining Security Against Adaptively Chosen Ciphertext Attacks. In *Crypto '93*, LNCS 773, pages 287–296. Springer-Verlag, Berlin, 1994.
33. M. Naor and M. Yung. Public-Key Cryptosystems Provably Secure against Chosen Ciphertext Attacks. In *Proc. of the 22nd STOC*, pages 427–437. ACM Press, New York, 1990.
34. T. Okamoto and D. Pointcheval. REACT: Rapid Enhanced-security Asymmetric Cryptosystem Transform. In *CT - RSA '2001*, LNCS 2020, pages 159–175. Springer-Verlag, Berlin, 2001.
35. P. Paillier. Public-Key Cryptosystems Based on Discrete Logarithms Residues. In *Eurocrypt '99*, LNCS 1592, pages 223–238. Springer-Verlag, Berlin, 1999.
36. P. Paillier and D. Pointcheval. Efficient Public-Key Cryptosystems Provably Secure against Active Adversaries. In *Asiacrypt '99*, LNCS 1716, pages 165–179. Springer-Verlag, Berlin, 1999.
37. T. Pedersen. A Threshold Cryptosystem without a Trusted Party. In *Eurocrypt '91*, LNCS 547, pages 522–526. Springer-Verlag, Berlin, 1992.
38. D. Pointcheval. Chosen-Ciphertext Security for any One-Way Cryptosystem. In *PKC '2000*, LNCS 1751, pages 129–146. Springer-Verlag, Berlin, 2000.

39. D. Pointcheval and J. Stern. Security Arguments for Digital Signatures and Blind Signatures. *Journal of Cryptology*, 13(3):361–396, 2000.
40. T. Rabin. A Simplified Approach to Threshold and Proactive RSA. In *Crypto '98*, LNCS 1462, pages 89–104. Springer-Verlag, Berlin, 1998.
41. C. Rackoff and D. R. Simon. Non-Interactive Zero-Knowledge Proof of Knowledge and Chosen Ciphertext Attack. In *Crypto '91*, LNCS 576, pages 433–444. Springer-Verlag, Berlin, 1992.
42. A. Sahai. Non-Malleable Non-Interactive Zero-Knowledge and Chosen-Ciphertext Security. In *FOCS '99*, LNCS 2139. IEEE, 1999.
43. A. De Santis, Y. Desmedt, Y. Frankel, and M. Yung. How to Share a Function Securely. In *Proc. of the 26th STOC*, pages 522–523. ACM Press, New York, 1994.
44. C. P. Schnorr. Efficient Identification and Signatures for Smart Cards. In *Crypto '89*, LNCS 435, pages 235–251. Springer-Verlag, Berlin, 1990.
45. C. P. Schnorr and M. Jakobsson. Security of Signed ElGamal Encryption. In *Asiacrypt '2000*, LNCS 1976, pages 458–469. Springer-Verlag, Berlin, 2000.
46. A. Shamir. How to Share a Secret. *Communications of the ACM*, 22:612–613, November 1979.
47. V. Shoup. Practical Threshold Signatures. In *Eurocrypt '2000*, LNCS 1807, pages 207–220. Springer-Verlag, Berlin, 2000.
48. V. Shoup and R. Gennaro. Securing Threshold Cryptosystems against Chosen Ciphertext Attack. In *Eurocrypt '98*, LNCS 1403, pages 1–16. Springer-Verlag, Berlin, 1998.
49. Y. Tsiounis and M. Yung. On the Security of El Gamal based Encryption. In *PKC '98*, LNCS. Springer-Verlag, Berlin, 1998.

Oblivious Polynomial Evaluation and Oblivious Neural Learning

Yan-Cheng Chang[1] and Chi-Jen Lu[2]

[1] ROC Airforce, Taiwan r88023@csie.ntu.edu.tw
[2] Institute of Information Science, Academia Sinica, Taipei, Taiwan
cjlu@iis.sinica.edu.tw

Abstract. We study the problem of Oblivious Polynomial Evaluation (OPE). There are two parties, Alice who has a polynomial P, and Bob who has an input x. The goal is for Bob to compute $P(x)$ in such way that Alice learns nothing about x and Bob learns only what can be inferred from $P(x)$. Previously existing protocols are based on some intractability assumptions that have not been well studied [15,14], and these protocols are only applicable for polynomials over finite fields. In this paper, we propose efficient OPE protocols which are based on Oblivious Transfer only. Unlike that of [15], slight modifications to our protocols immediately give protocols to handle multi-variate polynomials and polynomials over floating-point numbers. Many important real-world applications deal with floating-point numbers, instead of integers or arbitrary finite fields, and our protocols have the advantage of operating directly on floating-point numbers, instead of going through finite field simulation as that of [14]. As an example, we give a protocol for the problem of Oblivious Neural Learning, where one party has a neural network and the other, with some training set, wants to train the neural network in an oblivious way.

1 Introduction

Assume that there are two parties, Alice who has a function f and Bob who has an input x. They want to collaborate in a way for Bob to compute $f(x)$ such that Alice learns nothing about x and Bob learns only what can be inferred from $f(x)$. A protocol achieving this task for any function f and any input x is called an Oblivious Function Evaluation protocol. The remarkable results of Yao [17] and Goldreich, Micali, and Wigderson [9] showed that such protocols exist, under some standard cryptographic assumptions. Their protocols use a Boolean circuit to represent the function f and then simulate the computation of this circuit in some oblivious way. The computational or communicational overhead of their protocols depends only linearly on the circuit size of the function f, which is the best one can expect from a complexity-theoretical point of view. However, their protocols are far from being practical in general, and this problem still needs a lot of work to be done. One line of research is to study cases when different representations of functions can lead to more efficient simulation.

C. Boyd (Ed.): ASIACRYPT 2001, LNCS 2248, pp. 369–384, 2001.
© Springer-Verlag Berlin Heidelberg 2001

Noar and Pinkas [15] considered polynomials over finite fields. Note that any function from m bits to m bits can be represented by a polynomial over a finite field $GF(2^m)$, but its degree could go as high as $2^m - 1$. So one would like to focus on those functions that can be represented by low degree polynomials. This turns out to have several interesting applications [15,8,14,12]. The scheme proposed in [15] is much more efficient than the conventional way of going through oblivious circuit evaluation, but its security is based on two assumptions. One assumption is the existence of a secure Oblivious Transfer protocol while the other, proposed by themselves, is the intractability of a Noisy Polynomial Interpolation Problem. Bleichenbacher and Nguyen [3] later showed that this new assumption may be much weaker than expected and suggested the use of a possibly stronger intractability assumption on a Polynomial Reconstruction Problem. Still, no one can say how hard this problem is as it is not that well-studied. Recently, Lindell and Pinkas [14] mentioned a not-yet-published OPE protocol, which is also based on some newly proposed assumption. The assumption is that the Decisional Diffie-Hellman Assumption, denoted as DDH, also holds over the group $\mathbb{Z}^*_{n^2}$, where n is the product of two large primes. Contrary to the well studied DDH over $\mathbb{Z}^*_n$ [2], more research may need to be done before one can have some confidence on this new assumption. As there may be doubt on the security of both existing OPE protocols, a more satisfactory solution is certainly welcome.

As in [15,14], we will focus on the case with semi-honest parties, who may be curious but still follow the protocol. The malicious case can be handled in some standard way using commitments and zero-knowledge proofs, which will only be briefly mentioned. We will propose three OPE protocols of different flavors. Compared to previous ones, the security of our first two protocols is only based on a well-accepted cryptographic assumption, namely, the existence of a secure 1-out-of-2 oblivious transfer protocol, denoted as OT^2_1. For polynomials of degree d over a finite field $\mathbb{F}$, our first protocol uses $d \log |\mathbb{F}|$ invocations of OT^2_1 while [15] needs $(2kd+1) \log m$ invocations of OT^2_1 for some unspecified integers k and $m \gg d$ depending on their proposed assumption.[1] Note that for the problem in their assumption to be intractable, at least m must be very large just to prevent a brute-force algorithm that tries every possibility. So, even with their additional security concern, their protocol is better than ours only when $|\mathbb{F}| > m^{2k}$, i.e. when $|\mathbb{F}|$ is very large. Moreover, other than carrying out OT's, our protocol involves only extremely simple computation. Our second protocol is less efficient than our first one, but we include it here as the technique for achieving security seems interesting and may have other applications. Our third protocol involves a third party who does not collude with others but may be curious, and our protocol is perfectly secure, without any cryptographic assumption. Unlike that of [15], all our protocols can immediately handle multi-variate polynomials.

One attractive feature of our protocols is that they can be modified very easily to handle floating-point numbers. This is not the case for existing OPE

[1] Actually they use $2kd + 1$ invocations of 1-out-of-m oblivious transfer, denoted as OT^m_1. It is known that one OT^m_1 can be simulated by $\log m$ calls to OT^2_1, together with several evaluations of a pseudo-random function [15].

protocols which rely on some specific properties of finite fields. Many important applications in real life involve numerical computation over floating-point numbers, instead of over integers or arbitrary finite fields. There is no efficient mapping known that embeds floating-point numbers into finite fields where arithmetics can be carried out easily. The approach of [14] is to scale floating-point numbers up to integers with some book-keeping, apply some OPE protocol over integers, and then do a normalization to get back floating-point numbers. This extra work could complicate their algorithm design and slow down the performance a little. We show how our OPE protocols over finite fields can be easily modified to operate directly on floating-point numbers, and we believe that such protocols are more likely to have practical applications.

In addition to computing functions obliviously, some computational tasks may also involve security issues and people may want to perform them in some oblivious way. We use machine learning as an example, and demonstrate the applicability of our OPE protocol over floating-point numbers. Lindell and Pinkas [14] considered the scenario where two parties, each holding a private database, want to jointly construct a decision tree that classifies entries in both databases, using a so-called ID3 algorithm. Such kind of learning is not robust to changes in the sense that changes to a database may cause the whole process to be run again. We use neural network as our learning model and consider the following scenario. Alice has a neural network which is trained to some degree and she uses it to serve the classification requests from other parties. Alice wants to keep her neural network secret, while others want to keep their requests secret. This is the task of oblivious neural computing. At some point, another party Bob with a set of training examples wants to help Alice's neural network get better, maybe for his own good later. Alice wants to have a secure learning process so that Bob learns nothing from her, while Bob also wants to keep his training set secret. Later, other parties having their own training set can help Alice too, and Alice's neural network can adapt in an incremental way. This is the task of oblivious neural learning. We will apply our OPE protocol over floating-point numbers, and derive protocols for oblivious neural computing and oblivious neural learning.

The rest of the paper is organized as follows. In Section 2, we give definitions and tools that will be used later. Three OPE protocols are proposed in Section 3. We derive OPE protocols for floating-point numbers in Section 4. In Section 5, we show oblivious protocols for neural computing and learning.

2 Preliminaries

For a positive integer n, let $[n]$ denote the set $\{1, \ldots, n\}$. For an n-dimensional vector v, let v_i, for $i \in [n]$, denote the component in the i'th dimension, and we write $v = (v_1, \ldots, v_n) = (v_i)_{i \in [n]}$. Fix a security parameter τ, so that numbers about $2^{-\tau}$ are considered *negligible* and circuits of sizes about 2^{τ} are considered *infeasible*. For a distribution D over a set S, let $D(i)$, for $i \in S$, denote the probability of i according to D, and define $D(A)$, for $A \subseteq S$, to be $\sum_{i \in A} D(i)$.

Definition 1. *Let D and D' be two distributions over a set S. Their distance is defined as $d(D, D') = \max_{A \subseteq S} d_A(D, D')$, with $d_A(D, D') = |D(A) - D'(A)|$.*

Note that $d(D, D') = \frac{1}{2} \sum_{i \in S} |D(i) - D'(i)|$, which is a useful way for calculating $d(D, D')$.

Definition 2. *Let D and D' be two distributions. They are statistically indistinguishable, denoted as $D \stackrel{s}{\equiv} D'$, if $d(D, D')$ is negligible. They are computationally indistinguishable, denoted as $D \stackrel{c}{\equiv} D'$, if $d_A(D, D')$ is negligible for any subset A decided by a circuit of feasible size.*[2]

We will assume that parties in our protocols have only circuits of feasible sizes for computation unless mentioned otherwise. So we will focus on computational security, and the default distinguishability will be the computational one.

An important cryptographic primitive is the 1-out-of-2 oblivious transfer, denoted as OT_1^2. There are several variants which are all equivalent, and the one most suited for us is the following *string* version of OT_1^2. Let $\mathbb{F}$ be a set.

Definition 3. *An OT_1^2 protocol has two parties,* Sender *who has input $(x_0, x_1) \in \mathbb{F}^2$ and* Chooser *who has a choice $c \in \{0, 1\}$. The protocol is* correct *if the Chooser learns x_c for any (x_0, x_1) and c. The protocol is* secure *if both conditions below are satisfied for any (x_0, x_1) and c:*

- *Chooser cannot distinguish the distribution of Sender's messages from that induced by Sender having a different value of x_{1-c}.*
- *Sender cannot distinguish the distributions of Chooser's messages induced by c and $1 - c$.*

Similarly one can define OT_1^k for any $k \geq 3$, with Sender having k elements and Chooser wanting to learn one. We will use OT_1^k, for $k \geq 2$, to denote an assumed correct and secure OT_1^k protocol. It is known that the existence of OT_1^2 implies the existence of OT_1^k for any $k \geq 3$ [5,15].

Definition 4. *A protocol for* oblivious polynomial evaluation *has two parties, Alice who has a polynomial P over some finite field $\mathbb{F}$ and Bob who has an input $x_* \in \mathbb{F}$. An OPE protocol is* correct *if Bob learns $P(x_*)$ for any x_* and P. It is* secure *if both conditions below are satisfied for any x_* and P:*

- *Alice cannot distinguish the distribution of Bob's messages from that induced by Bob having a different x'_*.*
- *Bob cannot distinguish the distribution of Alice's messages from that induced by Alice having a different P' with $P'(x_*) = P(x_*)$.*

We say that a party in a protocol is *semi-honest* if the party follows the protocol but may try to learn more information than he or she should. We only focus on semi-honest parties in this paper. The case of *malicious* parties can

[2] Note that for A decided by a circuit C, $d_A(D, D') = |\mathbf{P}_{x \in D}[C(x) = 1] - \mathbf{P}_{x \in D'}[C(x) = 1]|$.

be handled in a standard way, using commitments and zero-knowledge proofs, which will only be briefly sketched for our first protocol.

Suppose D and D' are two distributions depending on distributions E and E' respectively. For any possible outcome t of E and E', let $(D|E = t)$ and $(D'|E' = t)$ denote the distributions of D and D' conditioned on $E = t$ and $E' = t$ respectively. Here is a useful lemma for showing $D \stackrel{c}{\equiv} D'$, which will be used several times in our security proofs later.

Lemma 1. *$D \stackrel{c}{\equiv} D'$ provided $E \stackrel{s}{\equiv} E'$ and $(D|E = t) \stackrel{c}{\equiv} (D'|E' = t)$ for any t.*

Proof. Let C be a circuit which outputs 1 with probabilities p and p' with respect to D and D'. Let p_t and p'_t denote the corresponding probabilities with respect to $(D|E = t)$ and $(D'|E' = t)$. Let $q_t = E(t)$ and $q'_t = E'(t)$. Then

$$\begin{aligned} |p - p'| &= |\sum_t q_t p_t - \sum_t q'_t p'_t| \\ &\leq \sum_t |q_t p_t - q_t p'_t| + \sum_t |q_t p'_t - q'_t p'_t| \\ &\leq \sum_t q_t |p_t - p'_t| + \sum_t |q_t - q'_t| \end{aligned}$$

So if $\sum_t |q_t - q'_t|$ is negligible and each $|p_t - p'_t|$ is negligible, then $|p - p'|$ is negligible.

Some cases later have identical E and E', and we only need to check each $|p_t - p'_t|$.

A family H of functions from S_1 to S_2 is said to satisfy a *pair-wise independent property* if for any distinct $\alpha, \alpha' \in S_1$,

$$\mathbf{P}_{h \in H}[h(\alpha) = h(\alpha')] = \frac{1}{|S_2|}.$$

Let $(H, H(S_1))$ denote the distribution of $(h, h(v))$ with random $h \in H$ and random $v \in S_1$, and let (H, S_2) denote the uniform distribution over $H \times S_2$. We will use the following lemma, which is a special case of the so-called Leftover Hash Lemma [10,11].

Lemma 2. *Let H be any family of functions from S_1 to S_2 satisfying the pair-wise independent property. Then $d((H, H(S_1)), (H, S_2)) \leq \sqrt{|S_2|/|S_1|}$.*

A proof of this lemma is given in the appendix for completeness.

3 Oblivious Polynomial Evaluation Protocols

We will present three OPE protocols of different flavors in this section. Assume that both parties have agreed that polynomials are over a finite field $\mathbb{F}$ and have degrees at most d. The set of such polynomials can be identified with the set $\mathbb{T} = \mathbb{F}^{d+1}$ in a natural way. Suppose now Alice has a polynomial $P(x) = \sum_{i=0}^{d} a_i x^i \in \mathbb{T}$ and Bob has $x_* \in \mathbb{F}$.

3.1 The First Protocol for OPE

To make the picture clear, we only discuss the case $\mathbb{F} = GF(p)$ for some prime p. The generalization to $GF(p^k)$ with $k > 1$ is straightforward. Let $m = \lceil \log_2 |\mathbb{F}| \rceil$. Each coefficient a_i in the polynomial can be represented as $a_i = \sum_{j \in [m]} a_{ij} 2^{j-1}$ with $a_{ij} \in \{0, 1\}$. For $i \in [d]$ and $j \in [m]$, let $v_{ij} = 2^{j-1} x_*^i$. Note that for each $i \in [d]$, $\sum_{j \in [m]} a_{ij} v_{ij} = a_i x_*^i$. The idea is to have Bob prepare $(v_{ij})_{j \in [m]}$ and have Alice get those v_{ij} with $a_{ij} = 1$, in some secret way. This is achieved by having Bob prepare the pair $(r_{ij}, v_{ij} + r_{ij})$ for a random noise r_{ij}, and having Alice get what she wants via OT_1^2. Note that what Alice obtains is $a_{ij} v_{ij} + r_{ij}$. Here is our first protocol, basing only on the existence of secure OT_1^2.

Protocol 1

1. Bob prepares dm pairs $(r_{ij}, v_{ij} + r_{ij})_{i \in [d], j \in [m]}$, with each r_{ij} chosen randomly from $\mathbb{F}$.
2. For each pair $(r_{ij}, v_{ij} + r_{ij})$, Alice runs an independent OT_1^2 with Bob to get r_{ij} if $a_{ij} = 0$ and $v_{ij} + r_{ij}$ otherwise.
3. Alice sends to Bob the sum of a_0 and those dm values she got. Bob subtracts $\sum_{i,j} r_{ij}$ from it to obtain $P(x_*)$.

Lemma 3. *Protocol 1 is correct when parties are semi-honest.*

Proof. The sum Bob obtains in Step 3 is $a_0 + \sum_i \sum_j (a_{ij} v_{ij} + r_{ij}) = P(x_*) + \sum_{i,j} r_{ij}$.

Lemma 4. *Protocol 1 is secure when parties are semi-honest.*

Proof. First, we prove Alice's security. Suppose P and P' are two distinct polynomials with $P(x_*) = P'(x_*) = y_*$. According to Lemma 1, it suffices to show that for any fixed $(r_{ij})_{i \in [d], j \in [m]}$, Alice's respective message distributions D and D' induced by P and P' are indistinguishable. Note that the last message from Alice is $y_* + \sum_{i,j} r_{ij}$ for both P and P' and can be ignored. So we focus on Alice's dm messages from the dm independent executions of OT's. For $0 \le k \le dm$, let D_k denote the distribution with the first k messages from D and the remaining messages from D'. Assume that there exists a distinguisher C for D and D'. A standard argument shows that C can also distinguish $D_{k_0 - 1}$ and D_{k_0} for some k_0. Note that Alice must select different elements from that pair in the k_0'th OT, as otherwise the two distributions are identical. Then one can break Chooser's security in OT_1^2 when Sender has this input, because with Chooser's messages for different choices replacing the k_0'th message of $D_{k_0 - 1}$, we get exactly $D_{k_0 - 1}$ and D_{k_0}, which can be distinguished by C. As OT_1^2 is assumed to be secure, D and D' are indistinguishable, and Alice is secure.

Next, we prove Bob's security. Note that Bob sends dm messages to Alice for the dm independent executions of OT's. Let $x_* \neq x'_*$, let E and E' be

Bob's respective message distributions, and let E_k denote the distribution with the first k messages from E and the remaining messages from E'. Suppose a distinguisher for E and E' exists. Then it can also distinguish E_{k_0-1} and E_{k_0} for some k_0. The pairs in that k_0'th OT have the forms $(r, v+r)$ and $(r', v'+r')$, for some fixed v and v' and for random r and r'. Alice's polynomial is fixed, so which element to choose in that k_0'th OT is also fixed. Suppose Alice chooses the first one in that pair. Then according to Lemma 1, there is a fixed r_0 such that E_{k_0-1} conditioned on Bob having $(r_0, v+r_0)$ and E_{k_0} conditioned on Bob having $(r_0, v'+r_0)$ are distinguishable. Similarly as before, one can distinguish Sender's messages when Sender has $(r_0, v+r_0)$ and $(r_0, v'+r_0)$ respectively and Chooser selects the first element, which violates Sender's security in OT_1^2. The case when Alice chooses the second one in that pair can be argued similarly, by noticing that the distribution $(r, v+r)$ and the distribution $(-v+r, r)$ are identical. As OT_1^2 is assumed to be secure, so is Bob.

Theorem 1. *Protocol 1 is correct and secure when parties are semi-honest.*

Note that only dm invocations of OT_1^2 are required and they can be done concurrently. Also observe that if OT_1^2 can achieve perfect security for Chooser (e.g. [1]) in the information-theoretical sense, then so is Protocol 1 for Alice.

A slight modification to Protocol 1 can handle the case of malicious parties. The only complication is to enforce a malicious Bob to prepare dm pairs that are *consistent* in the sense that there is some x_* such that $v_{ij} = 2^{j-1}x_*^i$ for every i and j, which can be achieved as follows. Bob sends his commitments of dm pairs to Alice, Alice uses OT_1^2 to have her dm choices decommitted, and Bob uses a zero-knowledge proof to convince Alice that those dm pairs are consistent. All these can be done using, for example, the methods in [13].

3.2 The Second Protocol for OPE

The idea of our second protocol is to have Alice hide the random shares of her polynomial P among other random polynomials, have Bob evaluate all of them on his input x_*, and then have Alice select those values corresponding to the shares, which sum to $P(x_*)$. Recall that $\mathbb{T} = \mathbb{F}^{d+1}$. Let $n = \log|\mathbb{T}| + 2\tau$. For $P \in \mathbb{T}$ and $R = (R_1, \ldots, R_n) \in \mathbb{T}^n$, define the function $h_{R,P} : \{0,1\}^n \to \mathbb{T}$ as

$$h_{R,P}(\alpha) = P - \sum_{i \in [n]} \alpha_i R_i.$$

It's easy to check that for any $P \in \mathbb{T}$, the class $H_P = \{h_{R,P} : R \in \mathbb{T}^n\}$ satisfies the pair-wise independent property. Here is our second OPE protocol, which is also based on OT_1^2 only.

Protocol 2

1. Alice generates random $R \in \mathbb{T}^n$ and $\alpha \in \{0,1\}^n$ and sends $(R_1, \ldots, R_n, h_{R,P}(\alpha))$ to Bob. Let $R_{n+1} = h_{R,P}(\alpha)$ and $\alpha_{n+1} = 1$.
2. Bob generates random $r \in \mathbb{F}^{n+1}$ and prepares $n+1$ pairs $(r_i, R_i(x_*) + r_i)_{i \in [n+1]}$.
3. For pair i, Alice runs an OT_1^2 with Bob to get r_i if $\alpha_i = 0$ and $R_i(x_*) + r_i$ otherwise.
4. Alice sends the sum of the $n+1$ values to Bob. Bob subtracts $\sum_{i=1}^{n+1} r_i$ from it to get $P(x_*)$.

Theorem 2. *Protocol 2 is correct and secure when parties are semi-honest.*

Proof. The correctness is obvious because the sum what Bob obtains in Step 4 is $\sum_{i=1}^{n+1} \alpha_i R_i(x_*) + \sum_{i=1}^{n+1} r_i = P(x_*) + \sum_{i=1}^{n+1} r_i$. Bob's security proof is almost identical to that of Protocol 1, so we only prove Alice's security here.

Fix any two polynomials $P, P' \in \mathbb{T}$, let D and D' denote Alice's respective message distributions, and let E and E' be Alice's respective message distributions in Step 1. According to Lemma 1, it suffices to show $E \overset{s}{\equiv} E'$ and $(D|E=t) \overset{c}{\equiv} (D'|E'=t)$ for each $t \in \mathbb{T}$. Using an argument similar to that in Protocol 1, one can show $(D|E=t) \overset{c}{\equiv} (D'|E'=t)$ for each $t \in \mathbb{T}$ as otherwise one can break Chooser's security in OT_1^2. Note that the family H_P satisfies the pair-wise independent property and E is the distribution $(H_P, H_P(\{0,1\}^n))$. With $n = \log|\mathbb{T}| + 2\tau = (d+1)m + 2\tau$, Leftover Hash Lemma [10,11] guarantees that the distance between E and the uniform distribution is at most $\sqrt{|\mathbb{T}|2^{-n}} = 2^{-\tau}$, which is negligible. Similarly E' also has a negligible distance to the uniform one. So $d(E, E')$ is negligible and $E \overset{s}{\equiv} E'$. According to Lemma 1, Alice is secure.

Note that there are $(n+1)\log|\mathbb{T}| = O(dm(dm+\tau))$ bits sent in Step 1, $O(dm+\tau)$ executions of OT_1^2 in Step 3, and m bits sent in Step 4.

3.3 A Protocol for 3-Party OPE

Here we show how to remove the use of OT_1^2 with the help a third party Clark. As a result, our protocol does not rely on any cryptographic assumption and is information-theoretically secure when no collusion exists. Again, we assume that Alice has a polynomial $P \in \mathbb{T}$, Bob has $x_* \in \mathbb{F}$ and only Bob learns $P(x_*)$. Now the security must also hold against Clark so that the messages he receives altogether look completely random to him; i.e.,

- *Clark cannot distinguish the uniform distribution from the joint distribution of messages he receives from Alice and Bob.*

Note that our model is slightly different from that of Feige, Kilian, and Naor [7], who have Clark as the party to receive the result. Here is the protocol.

Protocol 3

1. Bob sends random $(r_i)_{i\in[k]} \in \mathbb{F}^k$ to Alice. He also sends $(x'_i = x^i_* + r_i)_{i\in[k]}$ to Clark.
2. Alice sends random $(s_i)_{0\le i\le k} \in \mathbb{F}^{k+1}$ to Bob. She also sends $a'_0 = a_0 + s_0 - \sum_{i\in[k]} a_i r_i$ and $(a'_i = a_i + s_i)_{i\in[k]}$ to Clark.
3. Clark sends $y = a'_0 + \sum_{i\in[k]} a'_i x'_i$ to Bob, and Bob gets $P(x_*) = y - (s_0 + \sum_{i\in[k]} x'_i s_i)$.

Theorem 3. *Protocol 3 is correct and perfectly secure provided no collusion exists,*

Proof. The correctness is easy to verify. What Alice or Clark receives is completely random. Bob receives random $(s_i)_{0\le i\le k}$ in Step 2, and receives $P(x_*) + s_0 + \sum_{i\in[k]}(x^i_* + r_i)s_i$ in Step 4, so he sees the same distribution for any polynomial P' with $P'(x_*) = P(x_*)$. So each party is perfectly secure as long as no collusion exists.

3.4 Generalizations

It is not hard to see that all the protocols in this section can be easily extended to deal with multi-variate polynomials. In particular, we can solve an interesting special case: Alice has $a = (a_i)_{i\in[n]} \in \mathbb{F}^n$ while Bob has $x = (x_i)_{i\in[n]} \in \mathbb{F}^n$ and wants to learn the inner product $a \cdot x = \sum_{i\in[n]} a_i x_i$.

We have only considered the setting where Alice and Bob have their own inputs and Bob gets the final result. Later we will see a variation with each input and output shared by the two parties. We call this *computing with random shares.* Let's use the inner product function as an example. Suppose that Alice has $u, v \in \mathbb{F}^n$ and Bob has $u', v' \in \mathbb{F}^n$. They want to compute the inner product of $u + u'$ and $v + v'$, and produce random shares, one for each party, that sum to the inner product. This generalization can be reduced to the original problem in the following way. Note that $(u + u') \cdot (v + v')$ is equal to

$$(u \cdot v) + (u \cdot v' + v \cdot u') + (u' \cdot v').$$

Now Alice generates a random $r \in \mathbb{F}$ and prepares the $2(n + 1)$-dimensional vector

$$a = (-r + u \cdot v, u_1, \ldots, u_n, v_1, \ldots, v_n, 1),$$

while Bob prepares the $2(n + 1)$-dimensional vector

$$x = (1, v'_1, \ldots, v'_n, u'_1, \ldots, u'_n, u' \cdot v').$$

Bob can obtain $a \cdot x = -r + (u + u') \cdot (v + v')$ using a protocol for the original problem, and each party now holds a random share of the inner product $(u+u') \cdot (v + v')$. The variation for multi-variate polynomials can be handled similarly.

4 Oblivious Polynomial Evaluation for Floating-Point Numbers

4.1 Floating-Point Number System

We first give the definition of a floating-point number system.

Definition 5. *A floating-point number is a rational number* $b = \pm \sum_{j=1}^{2m} b_j 2^{m-j}$ *for some* m, *with* $b_j \in \{0, 1\}$. *Let* $\hat{m}$ *denote the floating-point number system containing all such numbers together with standard arithmetic operations.*

Such a floating-point number can be represented by $2m+1$ bits: m bits for the fractional part, m bits for the integral part, and 1 bit for the sign. Unlike finite fields, operations in a floating-point number system are not closed and errors may occur because of the limitation of finite precision. An *underflow* occurs when the produced number needs more bits for the fractional part, and a *rounding* takes place to convert it into the nearest number in the floating-point number system. An *overflow* occurs when the produced number needs more bits for the integral part, and the result is left undefined.

When we want to hide an element v of a finite field $\mathbb{F}$ in our previous protocols, we generate a pair $(r, r+v)$ with a random $r \in \mathbb{F}$, so that any element of the pair itself looks completely random. There is a slight complication for floating-point numbers, but it can be easily fixed.

Lemma 5. *Suppose* $v, v' \in \hat{\ell}$ *for some* ℓ *and suppose* $k \geq \ell + \tau + 1$. *The distributions of* $v + r$ *and* $v' + r'$ *with random* $r, r' \in \hat{k}$ *have a negligible distance.*

Proof. The distance is at most $\frac{|v-v'|}{2(2^k-2^{-k})+1} \leq \frac{2^{\ell+1}}{2^k} \leq 2^{-\tau}$.

4.2 An OPE Protocol for Floating-Point Numbers

Assume Alice holds $P(x) = \sum_{i=0}^{d} a_i x^i$, where $a_i \in \hat{m}$, and Bob holds $x_* \in \hat{m}$. For each i, let $|a_i| = \sum_{j=1}^{2m} a_{ij} 2^{m-j}$, with $a_{ij} \in \{0, 1\}$. All our previous protocols can be easily modified for floating-point numbers, and here we only demonstrate one, which comes from Protocol 1. We will use OT_1^3, which can be implemented by 2 executions of OT_1^2 [15]. Let $k = (d+1)m + \tau + 1$ and $n = k + \log(2dm)$. Parties agree on the floating-point system $\hat{k}$ for random numbers, and the floating-point system $\hat{n}$ for all arithmetics so that no underflow or overflow will ever occur. Let $v_{ij} = 2^{m-j} x_*^i$.

Protocol 4

1. Bob prepares $2dm$ 3-tuples $(r_{ij}, v_{ij} + r_{ij}, -v_{ij} + r_{ij})_{i \in [d], j \in [2m]}$, with each r_{ij} chosen randomly from $\hat{k}$.
2. For each 3-tuple $(r_{ij}, v_{ij} + r_{ij}, -v_{ij} + r_{ij})$, Alice runs an OT_1^3 with Bob to get r_{ij} if $a_{ij} = 0$, $v_{ij} + r_{ij}$ if $a_{ij} = 1 \wedge a_i > 0$, and $-v_{ij} + r_{ij}$ otherwise.
3. Alice sends to Bob the sum of a_0 and those $2dm$ values she got. Bob subtracts $\sum_{i,j} r_{ij}$ from it to obtain $P(x_*)$.

Note that all the arithmetic are carried out in the system $\hat{n}$, which is large enough to guarantee that no error ever occurs. Then it's not hard to verify the correctness of this protocol, while its security is guaranteed by the following.

Lemma 6. *Protocol 4 is secure when parties are semi-honest.*

Proof. Alice's security proof is almost identical to that of Protocol 1, so we only discuss Bob's security here. Let $x_*, x'_* \in \hat{m}$, let E and E' be Bob's respective message distributions, and let E_k denote the distribution with the first k messages from E and the remaining messages from E'. Suppose E_{k_0-1} and E_{k_0} can be distinguished, for some k_0, and the 3-tuples in that k_0'th OT have the forms $(r, v+r, -v+r)$ and $(r', v'+r', -v'+r')$, for random r and r' and for some fixed v and v'. Let $\ell = (d+1)m$ and note that $v, v' \in \hat{\ell}$ because $2^{m-j}x^i \in \hat{\ell}$ for any $x \in \hat{m}$, $i \in [d]$ and $j \in [2m]$. Then according to Lemma 5, no matter which element Alice chooses, the two distributions of that element have a negligible distance. Using Lemma 1 and adapting Bob's security proof for Protocol 1, one can show that E and E' are indistinguishable.

Note that the generalizations discussed in Section 3.4 also hold for floating-point numbers, and we have the following theorem.

Theorem 4. *Oblivious protocols exist for the problem of multi-variate polynomial evaluation (with random shares) over floating-point numbers.*

5 Oblivious Neural Learning

5.1 Neural Computing and Learning

There are several variants of the neural network model. We only demonstrate our result via 2-layer feedforward neural networks with back-propagation learning. Other variants can be handled similarly.

A 2-layer feedforward neural network has an internal layer of J nodes, with the j'th node having a weight vector $u_j = (u_{j1}, \ldots, u_{jI})$, and an output layer of K nodes, with the k'th node having a weight vector $w_k = (w_{k1}, \ldots, w_{kJ})$. Each node is associated with an activation function $f(z) = a\tanh(bz)$ (the hyperbolic tangent function). The network takes an input vector $x = (x_1, \ldots, x_I)$ and produces an output vector $o = (o_1, \ldots, o_K)$ in the following way.

Neural Computing

1. Compute $y_j = f(u_j \cdot x)$, for $j \in [J]$. Let $y = (y_1, \ldots, y_J)$.
2. Compute $o_k = f(w_k \cdot y)$, for $k \in [K]$.

The output vector o may not be correct, and a learning algorithm adjusts the weights according to how the vector o differs from the correct output vector d. The pair (x, d) constitutes a training example. The back-propagation learning (BP-Learning) algorithm adjusts the weights in the following way, with γ being some learning constant.

BP-Learning

1. Compute $\delta_{ok} = \frac{b}{a}(d_k - o_k)(a^2 - o_k^2)$, for $k \in [K]$.
2. Compute $\delta_{yj} = \frac{b}{a}(a^2 - y_j^2)\sum_{k=1}^{K} \delta_{ok} w_{kj}$, for $j \in [J]$.
3. Update $w_{kj} = w_{kj} + \gamma \delta_{ok} y_j$, for $k \in [K], j \in [J]$.
4. Update $u_{ji} = u_{ji} + \gamma \delta_{yj} x_i$, for $i \in [I], j \in [J]$.

The process above can be repeated for a set of training examples.

5.2 Oblivious Neural Computing and Learning

Now we want to carry out neural computing and neural learning in an oblivious way between two parties, Alice and Bob. Oblivious neural computing can be defined in a way similar to oblivious polynomial evaluation, except with Alice's polynomial replaced by a neural network. For oblivious neural learning, Bob has a set of training examples and wants to train Alice's neural network so that Bob knows nothing about Alice's neural network while Alice knows only what is implied by the weight changes. We need to be careful about Bob's security, as Alice's neural network has $IJ + JK$ weights and that many weight changes may reveal a lot to Alice. So we do not let Alice know the weight changes induced by each training example, and only let her get the overall weight changes after the training of all examples. Now a learning protocol is secure for Bob if Alice cannot distinguish two training sets that give the same overall weight changes. Note in practice, neural learning typically involves large training sets.

Another scenario is for Bob to keep random shares of those final weights, as long as he is willing to help Alice serve requests from other parties for oblivious neural computing. Later when another party wants to continue the training of Alice's neural network, Bob only needs to help with his shares for the first training example, and his duty is off after that. Contrary to the previous scenario, Alice cannot learn anything about Bob's training set in this way.

5.3 Oblivious Activation Function Evaluation

Here we discuss options for evaluating the activation function $f(z) = a\tanh(bz) = a(1 - \frac{2}{1+e^{2bz}})$ in an oblivious way. We will rely on an protocol for oblivious circuit evaluation [17,9,16], denoted as OCE, which is efficient for small circuits. Assume that Alice has x while Bob has y, and they want to generate random shares of $f(x+y)$ for Alice and Bob. One way is to use an OCE directly, if one can accept that the circuit for f is reasonably small. For cases allowing a large b, $f(z)$ is close to the threshold function, which has a very simple circuit, and again we can use OCE directly. Otherwise, we will approximate f in a piece-wise way by low degree polynomials and then apply our OPE protocol for it, which is described in the following. As f is smooth, there are intervals $I_0 = (-\infty, \ell_0], I_1 = (\ell_0, \ell_1], \ldots, I_n = (\ell_{n-1}, \infty)$, and degree-$d$ polynomials $P_0, P_1, \ldots, P_n$ such that

$$f(z) \approx P_i(z) \text{ for } z \in I_i,$$

for some small n and d, which seem good enough for practical purposes.[3] Let I be the function such that $I(z) = i$ for $z \in I_i$, which has a rather simple circuit and thus an efficient OCE protocol. Let $P_{i,x}(y) = P_i(x+y)$. Here is the oblivious protocol for evaluating the activation function.

Protocol 5

1. Alice generate random r_1. Bob runs OCE with Alice to get $r_2 = I(x + y) - r_1$.
2. Alice generate random s_1 and prepares the polynomial

$$Q_x(a, y) = -s_1 + \sum_{i=0}^{n} \frac{\prod_{j \neq i}(a + r_1 - j)}{\prod_{j \neq i}(i - j)} P_{i,x}(y).$$

Bob runs OPE with Alice for $s_2 = Q_x(r_2, y)$.

Note that Alice has s_1 and Bob has s_2 with $s_1 + s_2 = P_i(x + y)$ for $x + y \in I_i$, so the protocol is correct. The security proof is again similar to previous ones.

5.4 Oblivious Neural Algorithms

First we need to determine the possible range of floating-point numbers that can ever occur during computation. Then we can determine an appropriate floating-point number system $\hat{k}$ for random numbers and a system $\hat{n}$ for error-free arithmetics. Here is the protocol for oblivious neural computing which uses the OCE and OPE protocols with random shares.

Protocol 6

1. For $j \in [J]$, Alice and Bob compute random shares s_{j1}, s_{j2} of the inner product $u_j \cdot x$, and then compute random shares y_{j1}, y_{j2} of $y_j = f(s_{j1} + s_{j2})$. Let $y = (y_1, \ldots, y_J)$.
2. For $k \in [K]$, Alice and Bob compute random shares t_{k1}, t_{k2} of $w_k \cdot y$, and then compute random shares o_{k1}, o_{k2} of $o_k = f(t_{k1} + t_{k2})$.

At the end, Alice can send her shares o_{k1} to Bob for him to obtain the output vector o. This is not needed for oblivious learning. Note that the protocol still works when the each weight vector is shared by two parties instead of owned by Alice, which is the case in oblivious learning.

[3] For example, the error can be bounded by 2×10^{-6} with $n = 9$, $d = 9$, $\ell_0 = -7$, $\ell_8 = 7$, $P_0 = -1$, and $P_9 = 1$.

Theorem 5. *Oblivious neural computing can be achieved by Protocol 5.*

Proof. The correctness is easy to verify. The security relies on the security of the protocol for oblivious polynomial evaluation with random shares and the protocol for oblivious evaluation of the activation function. Any breaking of Protocol 5's security gives a way for breaking one of the protocols which has been shown to be secure.

An oblivious neural learning protocol can be derived similarly. Now only the protocol for OPE with random shares is needed.

Protocol 7

1. Alice and Bob compute random shares of each $\delta_{ok} = \frac{b}{a}(d_k - o_k)(a^2 - o_k^2)$.
2. Alice and Bob compute random shares of each $\delta_{yj} = \frac{b}{a}(a^2 - y_j^2)\sum_{k=1}^{K} \delta_{ok} w_{kj}$.
3. Alice and Bob compute random shares of each $w_{kj} = w_{kj} + \gamma\delta_{ok} y_j$.
4. Alice and Bob compute random shares of each $u_{ji} = u_{ji} + \gamma\delta_{yj} x_i$.

The learning process can be repeated for a set of training examples. At the end of the whole process, Bob reveals his shares of those weights obtained in the last iteration, and Alice derives the resulting neural network. The correctness is easy to verify. The security can be proved similarly as before. Now Alice cannot distinguish among training sets that give the same overall weight changes. So we have the following theorem.

Theorem 6. *Oblivious neural learning can be achieved by the combination of Protocol 6 and Protocol 7.*

As discussed before, an alternative scenario is not to have Bob give away his final shares to Alice, but for him to help Alice for her future task. In this way, Alice only obtains random shares of her new weights after each training example, including the final one. So each training example is secure and now Alice learns nothing about Bob's training set.

Acknowledgements. We would like to thank Prof. Yuh-Dauh Lyuu for his help.

References

1. M. Bellare and S. Micali, Non-interactive oblivious transfer and applications, in: Proc. CRYPTO '89, Lecture Notes in Computer Science, Vol. 435 (Springer, 1990), pp. 547–557.
2. D. Boneh, Decision Diffie-Hellman problem, in: Proc. Algorithmic Number Theory 1998, Lecture Notes in Computer Science, Vol. 1423 (Springer, 1998), pp. 48–63.

3. D. Bleichenbacher and P. Nguyen, Noisy polynomial interpolation and noisy chinese remaindering, in: Proc. EUROCRYPT 2000, Lecture Notes in Computer Science, Vol. 1807 (Springer, 2000), pp. 53–69.
4. G. Brassard, D. Chaum, and C. Crepeau, Minimum disclosure proofs of knowledge, Journal of Computer and System Sciences 37(2), 1988, pp. 156–189.
5. G. Brassard, C. Crepeau, and J. M. Robert, Information theoretical reductions among disclosure problems, in: Proc. 27th Ann. IEEE Symp. Foundations of Computer Science, 1986, pp. 168–173.
6. D. Chaum, C. Crepeau, and I. Damgard, Multiparty unconditionally secure protocols (extended abstract), in: Proc. 20th Ann. ACM Symp. Theory of Computing, 1988, pp. 11–19.
7. U. Feige, J. Kilian, and M. Naor, A minimal model for secure computation, in: Proc. 26th Ann. ACM Symp. Theory of Computing, 1994, pp. 554–563.
8. Niv Gilboa, Two party RSA key generation, in: Proc. CRYPTO '99, Lecture Notes in Computer Science, Vol. 1666 (Springer, 1999), pp. 116–129.
9. O. Goldreich, S. Micali, and A. Wigderson, How to play any mental game or a completeness theorem for protocols with honest majority, in: Proc. 19th Ann. ACM Symp. Theory of Computing, 1987, pp. 218–229.
10. J. Håstad, R. Impagliazzo, L. Levin, and M. Luby, Construction of a pseudorandom generator from any one-way function, SIAM Journal on Computing 28(4), 1999, pp. 1364–1396.
11. R. Impagliazzo and D. Zuckerman, How to recycle random bits, in: Proc. 30th Ann. IEEE Symp. Foundations of Computer Science, 1989, pp. 248–253.
12. Y. Ishai and E. Kushilevitz, Randomizing polynomials: a new representation with applications to round-efficient secure computaion, in: Proc. 41st Ann. IEEE Symp. Foundations of Computer Science, 2000, pp. 294–304.
13. J. Kilian, Founding cryptography on oblivious transfer, in: Proc. 20th Ann. ACM Symp. Theory of Computing, 1988, pp. 20–31.
14. Y. Lindell and B. Pinkas, Privacy preserving data mining, in: Proc. CRYPTO 2000, Lecture Notes in Computer Science, Vol. 1880 (Springer, 2000), pp. 36–54.
15. M. Naor and B. Pinkas, Oblivious transfer and polynomial evaluation, in: Proc. 31st Ann. ACM Symp. Theory of Computing, 1999, pp. 245–254.
16. T. Sander, A. Young, and M. Yung, Non-interactive cryptocomputing for NC^1, in: Proc. 40th Ann. IEEE Symp. Foundations of Computer Science, 1999, pp. 554–567.
17. A. C. Yao, How to generate and exchange secrets, in: Proc. 27th Ann. IEEE Symp. Foundations of Computer Science, 1986, pp. 162–167.
18. J. M. Zurada, Introduction to Artificial Neural Systems, PWS Publishing, 1994.

A Proof of Lemma 2

Let $\ell = |H||S_2|$. From Cauchy-Schwartz, $\sum_{h,v} |\mathbf{P}_{g,u}[(g, g(u)) = (h, v)] - 1/\ell|$ is at most

$$\begin{aligned}
&\sqrt{\ell}\sqrt{\sum_{h,v}(\mathbf{P}_{g,u}[(g, g(u)) = (h, v)] - 1/\ell)^2} \\
&= \sqrt{\ell \sum_{h,v} \mathbf{P}_{g,u}[(g, g(u)) = (h, v)]^2 - 2 + 1}
\end{aligned}$$

$$
\begin{aligned}
&= \sqrt{\ell \mathbf{P}_{h,h',u,u'}[(h,h(u)) = (h',h'(u'))] - 1} \\
&= \sqrt{\ell \mathbf{P}_{h,h'}[h = h']\mathbf{P}_{h,u,u'}[h(u) = h(u')] - 1} \\
&\leq \sqrt{\ell |H|^{-1}(\mathbf{P}_{u,u'}[u = u'] + \mathbf{P}_{h,u,u'}[h(u) = h(u')|u \neq u']) - 1} \\
&= \sqrt{|S_2|\,(1/|S_1| + 1/|S_2|) - 1} \\
&= \sqrt{|S_2|/|S_1|}.
\end{aligned}
$$

Mutually Independent Commitments

Moses Liskov[1], Anna Lysyanskaya[1], Silvio Micali[1], Leonid Reyzin[2], and Adam Smith[1]

[1] Laboratory for Computer Science
Massachusetts Institute of Technology
Cambridge, MA 02139, USA
{mliskov,anna,asmith}@theory.lcs.mit.edu
[2] Boston University
Department of Computer Science
Boston, MA 02215, USA
reyzin@bu.edu

Abstract. We study the two-party commitment problem, where two players have secret values they wish to commit to each other. Traditional commitment schemes cannot be used here because they do not guarantee independence of the committed values. We present three increasingly strong definitions of independence in this setting and give practical protocols for each. Our work is related to work in non-malleable cryptography. However, the two-party commitment problem can be solved much more efficiently than by using non-malleability techniques.

1 Introduction

We consider the scenario in which two players have some private values in mind, and want to commit these values to one another. In these circumstances, simply using commitment schemes on each side does not provide sufficient security. While this approach guarantees that the two commitments will each be hiding and binding, it does not guarantee their independence.

For example, if Alice is selling something to Bob, and commits to a (her lowest price) by publishing $c(a)$, then Bob can commit to a as his highest bid (without knowing the value), by copying $c(a)$. Thus, Bob will force Alice to always sell at her lowest price. Though this is an obvious and easily preventable attack, more sophisticated ones exist. For example, if the commitment scheme being used is that of Pedersen [Ped91], Bob could, without risking detection, copy (or indeed add an arbitrary constant to) Alice's value.

Independence of committed values is quite fundamental to secure two-party protocols. Indeed, in any protocol to which both parties have inputs which they are unwilling to reveal at the outset, the inputs must be committed (so that the parties cannot change their minds later) and independent (so that each party's influence of the outcome is limited to the choice of its own input).

THE TWO-PARTY COMMITMENT PROBLEM. In our setting, Alice and Bob have secret values a and b, respectively. They want to commit their values to each other. Informally, we want the following security properties to hold:

C. Boyd (Ed.): ASIACRYPT 2001, LNCS 2248, pp. 385–401, 2001.
© Springer-Verlag Berlin Heidelberg 2001

- *Hiding:* A dishonest party cannot discover the honest party's value.
- *Binding:* A dishonest party cannot open his or her commitment in more than one way.
- *Non-correlation:* A dishonest party cannot commit to a value that is in some significant way correlated to the honest party's value.

We formalize the last property in three increasingly stronger definitions.

- *Mutually independent announcement*: Non-correlation is guaranteed given that the parties open their commitments.
- *Mutually independent commitment*: Non-correlation is guaranteed once the commitments are exchanged and accepted.
- *Mutually independent and aware commitment*: Each party is guaranteed to *know* his or her own value once the commitments are exchanged and accepted. This property, combined with the hiding property of the commitment, actually guarantees non-correlation.

We also give practical protocols that satisfy these definitions. Specifically, we give a two-round[1] mutually independent announcement protocol based on the existence of one-way permutations. We give two mutually independent commitment protocols: a two-round protocol based on the assumption that subexponentially hard one-way permutations exist, and a three-round protocol based on the assumption that dense cryptosystems exist. Finally, we give a seven-round mutually independent and aware commitment protocol based on the discrete logarithm assumption. With the exception of an eleven-round mutually independent and aware protocol we present to elucidate the definitions, all the protocols we present are efficient enough to be useful in practice.

1.1 Mutual Independence versus Other Notions

Mutually independent commitments provide a new approach to an important cryptographic problem: how ensure that *secret* and *committed* values are *independent.* This problem has been addressed before in other settings.

Independence in the Multi-party Setting. In protocols involving more than two parties, it has long been recognized that independence of committed values is fundamental to the very notion of security: a player who can correlate his input to those of other players (without necessarilly knowing them) may be able to change the outcome of the protocol in his favor. In that setting, the problem of independence was introduced by Chor, Goldwasser, Micali and Awerbuch [CGMA85], who solved it using verifiable secret-sharing protocols.

[1] The round complexity we refer to is the number of rounds required for the commit stage of the protocols. All the protocols we present have one-round reveal stages, except the mutually independent announcement protocol, which has a two-round reveal stage.

Subsequent impovements to their solution were made by Chor and Rabin [CR87] and by Gennaro [Gen95].

Our two-party setting, while similar to the multi-party setting at first glance, is actually quite different: the multi-party protocols assume that a majority of players are honest. This allows "committed information" to actually be distributed among multiple players. Because we have only two players, we cannot assume an honest majority without trivializing the problem. Thus, each player in our setting will have *all* the committed information from the other player.

Non-malleability of Commitment Schemes. It has also long been recognized that the hiding property of a commitment scheme (carried out between two parties, a sender and a receiver) does not prevent an adversary from committing to a value related to someone else's commitment. In the two-party commitment setting, the notion of non-malleability was introduced to address this problem.

Defined by Dolev, Dwork and Naor [DDN00], non-malleability for commitment schemes captures the following intuitive notion: if Alice commits a value to Bob, and Bob commits a value to Charlie (using the same commitment scheme), then Bob's committed value should be independent of Alice's. Thus, this is a setting with two honest parties (Alice and Charlie) who are unaware of each other, and one adversary (Bob). Because Alice and Charlie are unaware of each other, Bob can arbitrarily vary the timing of the two interactions in which he is involved. This, in particular, implies that Bob can always just copy Alice's committed value by simply being a "transparent intermediary." Copying committed values is, in fact, explicitly permitted in the definiton of [DDN00].

In our setting there are three crucial differences. First, Alice and Charlie are, in a sense, the same person. (This, in particular, prevents Bob from arbitrarily scheduling the exections of the two commitment protocols and thus copying the committed value.) Second, we are not restricted to using the same commitment scheme for the two commitments. Finally, either party in our setting can be the adversary, and independence needs to be ensured both ways.

While, as we describe in the next section, non-malleable commitment schemes may be used to provide mutually independent commitments, the mutually independent commitment problem can be solved more efficiently in other ways.

1.2 Relevance of Prior Solutions

As we pointed out above, solutions in the multiparty setting seem inapplicable to our setting, because they assume an honest majority of players. Non-malleable commitments, on the other hand, can address our problem.

In fact, any commitment protocol non-malleable *with respect to commitment* (i.e., in which non-malleability is assured even if the adversary never sees Alice's decommitted value) can be used to provide mutually independent commitments: simply run two copies of the protocol in parallel, one from Alice to Bob and the other from Bob to Alice. Either party can detect if the other is copying the transcript, and thus prevent it from copying the commitment[2]. However, only one non-malleable commitment protocol is known that does not require extra set-up assumptions: the one of [DDN00]. It is quite impractical, requiring a non-constant number of rounds (it will, however, achieve mutually independent and *aware* commitments in our setting). In contrast, we present simple constant-round protocols that solve the problem.

A number of much simpler non-malleable commitment schemes (constructed using either non-malleable encryption [DDN00,CS98,Sah99] or directly [DIO98, FF00,DKOS01,CF01]) are known, all requiring trusted public file to be set up ahead of time. Because we are interested in a two-party scenario, we are unwilling to assume the existence of trusted public parameters.

Moreover, some of the above commitment schemes [DIO98,FF00,DKOS01] achieve only a weaker security notion called non-malleability *with respect to opening*. That is, it may be possible for Bob to commit to a value related to Alice's, but he won't know how to open it. Using such a protocol in our setting will achieve mutually independent *announcement*, but not necessarily *commitment* (in particular, some of the schemes are perfectly hiding, and then it is unclear how the committed value can be defined prior to opening).

1.3 Applicability of Mutually Independent Commitments

The protocols we present are for the two-party model. We intend our notions to be useful as essential building blocks in secure two-party computation protocols.

As we have discussed, the question of mutual independence also naturally arises in the multi-party setting, and has been previously studied. By focusing on the two-party case exlusively, we obtain protocols that are more efficient and conceptually simpler. Applicability of our techniques in other settings is a subject of further study.

2 Definitions

2.1 Notation

NEGLIGIBLE FUNCTIONS. The expression negl(k) is used to denote any function f that is negligible in k; that is, for any positive polynomial q, $f(k) = o(1/q(k))$.

PROBABILITY .[3] If S is a probability space, then "$x \leftarrow S$" denotes assigning to x an element randomly selected according to S. If F is a finite set, then the notation "$x \leftarrow F$" denotes the algorithm that chooses x uniformly from F.

[2] The original definition of [DDN00] did not prevent Bob from copying the committed value while *not* copying the transcript. However, any non-malleable commitment scheme can be modified to preclude this possibilty [KOS], thus leaving Bob only one way to copy the commitment: by copying the transcript exactly.

[3] This notation closely follows that of [BDMP91] and [GMR88].

If p is a predicate, the notation $\Pr[x \leftarrow S; y \leftarrow T; \cdots : p(x, y, \cdots)]$ denotes the probability that $p(x, y, \cdots)$ will be true after the ordered execution of the algorithms $x \leftarrow S$; $y \leftarrow T; \cdots$. The notation $[x \leftarrow S; y \leftarrow T; \cdots : (x, y, \cdots)]$ denotes the probability space over $\{(x, y, \cdots)\}$ generated by the ordered execution of the algorithms $x \leftarrow S$, $y \leftarrow T, \cdots$.

PROTOCOLS. The schemes discussed in this paper are protocols $P = (A, B)$ run between two parties, A and B. Both A and B are probabilistic polynomial-time interactive Turing machines (ppITMs). Given (1) a *security parameter* 1^k, which is available to both parties; (2) *inputs* (a, b), where a is private to A and b is private to B; and (3) *random tapes* (r_A, r_B), where r_A is private to A and r_B is private to B, protocol P computes in a sequence of rounds, alternating between A-rounds and B-rounds. In an A-round (respectively, B-round) only A (only B) is active and sends a string that will become an available input to B (to A) in the next B-round (A-round). We will divide P into two stages: the *commit* stage $P_C = (A_C, B_C)$, and the *reveal* stage $P_R = (A_R, B_R)$ (state information for A and B is saved between the stages). At the end of the commit stage, A_C and B_C will each output "accept" or "reject." At the end of the reveal stage, A_R will output the value β that B revealed to it, which is a string or a special symbol "reject", and B_R will similarly output the value α. For notational convenience, we will assume that if the output of the commit stage is "reject," then so is the output of the reveal stage—i.e., if a party did not accept the commitment, it will not accept its revealing, either. The terms "output of A" and "output of B" shall mean "output of A_R" and "output of B_R."

We will also consider the situation in which one of the two parties is dishonest. The dishonest party, denoted by A' or B', is also a ppITM. A dishonest party can, of course, simply stop the protocol before the other party produces an output. In such a case, for notational convenience, we will consider the honest party's output to be "reject."

TRANSCRIPTS, VIEWS, AND OUTPUTS.[4] Letting E be an execution of a protocol (A, B) on inputs $(1^k, a, b, r_A, r_B)$, we make the following definitions:

- The *transcript* of E consists of the sequence of messages exchanged by A and B, and is denoted by $\text{TRANS}^{A,B}(1^k, a, b, r_A, r_B)$ (for notational convenience, we will include the outputs of A and B into the transcript);
- The *view of A* consists of the triplet $(1^k, a, r_A, t)$, where t is E's transcript, and is denoted by $\text{VIEW}_A^{A,B}(1^k, a, b, r_A, r_B)$;
- The *output of A* is denoted by $\text{OUT}_A^{P_A, P_B}(a, b, r_A, r_B)$;
- The output and view of B are defined similarly and denoted by $\text{VIEW}_B^{A,B}(a, b, r_A, r_B)$ and $\text{OUT}_B^{P_A, P_B}(a, b, r_A, r_B)$;
- We use the symbol $\cdot$ in place of r_A and r_B in the above notation to denote the distribution induced on the transcripts, views and outputs when r_A or r_B is selected at random. Thus, for example, $\text{TRANS}^{A,B}(1^k, a, b, \cdot, \cdot)$, is a

[4] We borrow much of our protocol notation from [BMM99] and [GMR85].

probability space of transcripts, with probabilities induced by selecting r_A and r_B at random and executing (A, B) on $(1^k, a, b, r_A, r_B)$.

The output of each party is, of course, computed based solely on that party's view. Therefore, we denote by $\mathrm{OUT}_A(1^k, a, r_A, t)$ the output of A as computed on the particular view (note that A is not assumed to be interacting with anyone in this case). We use similar notation for the output of B.

Finally, we will denote the transcript for the commit stage by t_C, the transcript for the reveal stage by t_R, and the combined transcript by $t = t_C \circ t_R$.

2.2 Mutually Independent Announcement

A protocol (A, B) is a *mutually independent announcement* if the following properties hold:

- *A-completeness.* If A and B are honest, then A can can commit and reveal her value successfully:

  $$\begin{aligned}&\forall a, b\\ &\Pr[\alpha \leftarrow \mathrm{OUT}_B^{A,B}(1^k, a, b, \cdot, \cdot) :\\ &\quad \alpha = a] = 1 - \mathrm{negl}(k)\end{aligned}$$

- *A-soundness.* This property prevents a dishonest B' from influencing which value A commits to. That is, if the honest A is interacting with a dishonest B'_C during the commit stage and outputs "accept," then A would only reveal a during the reveal stage, at least with an honest B_R.

  $$\begin{aligned}&\forall t_C, t_R, a, b, r_A, r_B,\\ &\mathrm{OUT}_{A_C}(1^k, a, r_A, t_C) = \text{"accept"} \wedge\\ &\mathrm{OUT}_{B_R}(1^k, b, r_B, t_C \circ t_R) = \alpha \Rightarrow\\ &\alpha = a\end{aligned}$$

- *Computational A-hiding.* No adversary B', interacting only with A_C, the commit stage of A, can break the GM-security [GM84] of A's commitments:

  $$\begin{aligned}&\forall (a_0, a_1)\ \forall B'\\ &\Pr[v \leftarrow \{0,1\};\\ &\quad z \leftarrow \mathrm{OUT}_{B'}^{A_C,B'}(a_v, (a_0, a_1), \cdot, \cdot) :\\ &\quad z = v] < 1/2 + \mathrm{negl}(k)\end{aligned}$$

- *Perfect A-binding.* If the commit stage B_C of B outputs "accept," then the reveal stage B_R will accept only one revealed value; moreover, this value depends only on the transcript of the reveal stage, not on the private input of B:

  $$\begin{aligned}&\forall t_C, b, b', r_B, r'_B, t_R, t'_R, \alpha, \alpha',\\ &\mathrm{OUT}_{B_C}(1^k, b, r_B, t_C) = \text{"accept"} \wedge\\ &\mathrm{OUT}_{B_C}(1^k, b', r'_B, t_C) = \text{"accept"} \wedge\\ &\mathrm{OUT}_{B_R}(1^k, b, r_B, t_C \circ t_R) = \alpha \wedge\\ &\mathrm{OUT}_{B_R}(1^k, b', r'_B, t_C \circ t'_R) = \alpha' \Rightarrow\\ &\alpha = \alpha'.\end{aligned}$$

– *A-non-correlation at opening.* To define non-correlation at opening, we use techniques similar to those used in defining commitment schemes that are non-malleable with respect to opening ([DDN00,DIO98,FF00]). The definition essentially states that for any polynomial-time relation R, any adversary B' that engages in a protocol with $A(1^k, a, r_A)$ and then opens his committed value as β, has no more chance of achieving $R(a, \beta)$ than a simulator who does not engage in any interaction with A at all. Note, of course, that because we are defining non-correlation at opening (rather than at commitment), B' may already know a before revealing β, and thus may simply refuse to reveal depending on a. If B' refuses to reveal, A will output "reject." There is no getting around the fact that B' can correlate "reject" to a. Thus, as for non-malleability, we explicitly require that $R(a, \text{"reject"}) = 0$, so that forcing A to reject is not considered correlating to a better than a simulator. We call such polynomial-time relations *allowable*. Note that, unlike the definitions of non-malleability, we do not require that the relation be non-reflexive—that is, our definitions do not even allow B' to copy the commitment of A.

$$\begin{aligned}
&\forall B' \, \exists S \, \forall \text{ allowable } R \, \forall \text{ efficiently sampleable } \mathcal{D} \\
&\Pr[a \leftarrow \mathcal{D}; \\
&\quad \beta \leftarrow \mathrm{OUT}_A^{A,B'}(1^k, a, -, \cdot, \cdot): \\
&\quad R(a, \beta) = 1] < \\
&\Pr[a \leftarrow \mathcal{D}; \\
&\quad \beta \leftarrow S(1^k, \mathcal{D}): \\
&\quad R(a, \beta) = 1] + \mathrm{negl}(k)
\end{aligned}$$

– *B-completeness, B-soundness, computational B-binding, perfect B-binding, B-non-correlation at opening.* Defined the same way as for the above, with A and B reversed.

2.3 Mutually Independent Commitment

Mutually independent commitments are defined the same way as mutually independent announcements, except for the non-correlation property, which is defined as follows.

– *A-non-correlation at commitment.* Because our commitments are perfectly B-binding, at the end of the commit stage, there is at most one value β that a (dishonest) B' can reveal. Moreover, this value is determined uniquely by the transcript t_C of the commit stage, provided that A outputs "accept." Let $U_B(t_C)$ be this value, or $\perp$ if no such value exists or if A outputs "reject" (recall that we included A's output into the transcript, by definition). Note that, by B-hiding, $U_B(t_C)$ is not efficiently computable, at least when B is honest. A-non-correlation at commitment requires that $U_B(t_C)$ be not correlated to a, for any polynomial-time relation R. Unlike the case for non-correlation at opening, we do not require that $R(a, \perp) = 0$ —that is, B' should not even be able to correlate to a the fact that no valid decommitment exists. This stronger requirement is justified in this case, because B' does not get to see a before deciding whether a decommitment should exist.

$$
\begin{aligned}
&\forall B'\ \exists S\ \forall \text{ poly-time } R\ \forall \text{ efficiently samplable } \mathcal{D}\\
&\Pr[a \leftarrow \mathcal{D};\\
&\quad t_C \leftarrow \mathrm{TRANS}^{A_C,B'_C}(1^k, a, \mathcal{D}, \cdot, \cdot) :\\
&\quad R(a, U_B(t_C)) = 1] <\\
&\Pr[a \leftarrow \mathcal{D};\\
&\quad \beta \leftarrow S(1^k, \mathcal{D}) :\\
&\quad R(a, \beta) = 1] + \mathrm{negl}(k)
\end{aligned}
$$

- *B-non-correlation at commitment.* Defined similarly, with A and B reversed.

2.4 Mutually Independent and Aware Commitment

In addition to the properties of mutually independent commitments defined above, we want to capture the strong notion that B, if he accepts the commitment stage, is assured that A "knows" the value she committed to. We mean "knowledge" in the sense of the existence of a knowledge extractor E, in the tradition of the definitions of a proof of knowledge [TW87,FFS88]. Note, however, that unlike proofs of knowledge, where an NP-witness y is being extracted for a predetermined statement x, in our case, no predetermined statement exists. What is being extracted—the commited-to value—is determined only by the transcript t_C of the commitment stage. Thus, our definition gives E the view of the dishonest party (which includes the transcript of the conversation), and E has to extract, given oracle access to the dishonest party, the committed value.

- *A-awareness.* Similarly to the definition of A-non-correlation at commitment, given a transcript t_C of the commitment stage, let $U_A(t_C)$ be the unique value α that A' can reveal, or $\perp$ if no such value exists or if B output "reject" at the end of the commitment stage. Because the view $V_{A'_C}$ of A includes t_C, we will use $U_A(V_{A'_C})$ to mean $U_A(t_C)$.

$$
\begin{aligned}
&\exists E\ \forall b\ \forall A'\\
&\Pr[V_{A'_C} = \mathrm{VIEW}_{A'_C}^{A'_C,B_C}(1^k, -, a, \cdot, \cdot);\\
&\quad a \leftarrow E^{A'}(V_{A'_C});\\
&\quad \alpha = U_A(V_{A'_C}) :\\
&\quad a = \alpha] > 1 - \mathrm{negl}(k)
\end{aligned}
$$

- *B-awareness.* Defined similarly, with A and B reversed.

Note that A-awareness, combined with B-hiding, implies B-non-correlation at commitment, because we can simply use E in place of the simulator S. Therefore, aware commitments are automatically mutually independent.

3 Protocols

3.1 Mutually Independent Announcement

Theorem 1. *If one-way permutations exist, then there exists a protocol for mutually independent announcements with a two-round commit stage and a two-round reveal stage*[5].

Proof sketch. The protocol is simplicity itself. Let c be a perfectly binding non-interactive commitment scheme, which can be constructed based on any one-way permutation [GL89]. The commit stage consists of two rounds:

1. Alice sends $c(a)$ to Bob
2. Bob sends $c(b)$ to Alice

The reveal stage likewise consists of two rounds:

1. Bob opens his commitment
2. Alice opens her commitment

The completeness, soundness, binding, and hiding properties are easy to see. It is clear that A non-correlation at opening holds, since after step 1 of the reveal stage, Bob still cannot understand Alice's commitment, so if he could correlate then he would break the hiding property of c. On the other hand, B non-correlation at opening holds, since Alice commits first. Thus, her only option is to refuse to open her commitment, which, be definition, can only hurt her chances of being correllated. □

It is worth noting that while non-malleable commitments "with respect to opening" can be used for mutually independent announcement, the solution they offer is far more complex than this. This elegantly illustrates the point that the problem we solve requires less security than the problem that non-malleable commitments solve.

3.2 Mutually Independent Commitment

All the remaining protocols we present have a one-round reveal stage under the imperfect synchronization assumption. That is, each player sends only one message in the reveal stage, and the order does not matter: we allow the honest players to not wait to receive a message before sending one. Note that we do not assume that the messages are actually sent simultaneously: dishonest players can always wait for receipt of a message before sending theirs.

[5] This protocol can be modified to be based only on one-way functions, but this requires a 3-round commit stage: we simply use the construction of Naor [Nao91] to make a commitment scheme based on one-way functions, which requires a round to set up.

From this point on, when we refer to the number of rounds a protocol requires, we mean only the rounds in the commitment stage.

We present two protocols for mutually independent commitment: a two-round protocol based on the assumption that subexponentially hard one-way permutations exist, and a three-round protocol based on the assumption that 'dense' cryptosystems exist.

Two Round Protocol. A subexponentially hard one-way permutation is one for which there exists an $\epsilon > 0$ such that the permutation remains one-way even against adversaries that run in time 2^{n^ϵ}, where n is the security parameter. We note that, based on the current state of the art in factoring and discrete logarithm techniques, it is reasonable to assume that both RSA and exponentiation in a large prime-order subgroup of Z_p^* are subexponentially hard one-way permutation with some $\epsilon \leq 1/3$ (because the best known attacks against them take time $2^{O(n^{1/3}(\log n)^{2/3})}$).

Theorem 2. *If subexponentially hard one-way permutations exist, then there exists a two-round mutually independent commitment protocol.*

Proof sketch. Let c be a subexponentially secure non-interactive commitment scheme: i.e., one that is semantically secure against adversaries that run in time 2^{n^ϵ} for some $\epsilon > 0$, where n is the security parameter (such a commitment scheme can be constructed based on subexponentially hard one-way permutations). Assume that, for security parameter n, a commitment can be forced open in time 2^{n^δ}, for some $\delta > 0$ (this must be true for some δ, because one should be able to simply enumerate all the possible decommitment strings).

Let k be the security parameter for our scheme, and $K = k^{2\delta/\epsilon}$. The protocol is, again, very simple:

1. Alice commits to a using c with security parameter K.
2. Bob commits to b using c with security parameter k,

In the reveal stage, Alice and Bob reveal their values. It is clear that this scheme is complete, sound, hiding and binding. It is also clear that Alice cannot correlate her value to Bob's, since Alice is bound to her value before Bob commits to his. On the other hand, if Bob could correlate his value to Alice's, we could force open his commitment in time $2^{k^\delta} = 2^{K^\epsilon/2}$, and then use b to break the subexponentially strong semantic security of Alice's commitment in time $2^{K^\epsilon/2}$, which is a contradiction, because Alice's security parameter is K. □

Three Round Protocol. This protocol assumes the existence of dense, perfectly faithful cryptosystems. Following [DP92], a δ-*dense* cryptosystem is defined by modifying the definition of a secure cryptosystem [GM84] as follows: first, we add the requirement that a public key generated by the key generation algorithm is distributed uniformly over $\{0,1\}^{p(k)}$, for some polynomial p in the security parameter k; and second we require security for only a δ-fraction of public keys. It was observed by [DDP00] that the assumption of the existence of δ-dense cryptosystems, for a non-negligible δ, is equivalent to the existence of $(1-\epsilon)$-dense cryptosystems, for any negligible ϵ. We will actually need the latter. They can be constructed based on the ElGamal cryptosystem, for example.

Theorem 3. *If dense cryptosystems exist, then there exists a three-round mutually independent commitment scheme.*

Proof sketch. Let ϵ be a negligible function, and let (G, E, D) be a $(1-\epsilon)$-dense cryptosystem. Let $p(k)$ be the length of the public key for a security parameter k. Let c be a perfectly binding non-interactive commitment scheme. The commit stage is as follows:

1. Alice generates a random $p(k)$-bit string, R_A, and sends $c(R_A)$ to Bob.
2. Bob sends $c(b)$ to Alice, and sends a random $p(k)$ bit string R_B to Alice.
3. Alice computes $\mathrm{PK} = R_A \oplus R_B$, $C = E(\mathrm{PK}, a)$, and sends PK, C to Bob. Note that Alice does *not* open her commitment $c(R_A)$ at this step.

In the reveal stage, Bob opens his commitment to b, and Alice opens her commitment to R_A and reveals her value a and the random bits used to come up with C. Bob checks if R_A was revealed correctly, if PK indeed equals $R_A \oplus R_B$, and if the random bits were correct. If any of these checks fail, Bob rejects.

Completeness, soundness, binding, and B-hiding are easy to prove. A-hiding is proved as follows. Suppose B' is able to break the semantic security of the commitment of A. Then we will build a machine to break the semantic security of the dense cryptosystem. The machine will be given, as input, a public key PK and a ciphertext C. The machine will simulate A to B': it will commit to a random string R_A in the first round, and receive $c(b)$ and R_B in the second round. In the third round, it will ignore the first two rounds, and simply send the PK and C that were input to it. Note that B' should not be able to tell that $\mathrm{PK} \neq R_A \oplus R_B$—otherwise, it would be violating the hiding property of $c(R_A)$. Therefore, B' will "behave the same way" as with the true A, and thus would break the semantic security of the ciphertext C.

A-non-correlation at commitment is simple to prove: B' has no information about a at the time it has to commit to b.

B-non-correlation at commitment is proved as follows. Suppose A' can correlate to b. Then we will build a machine M that breaks the hiding property (semantic security) of $c(R_B)$, as follows. M receives a commitment c to some unknown value b. It will generate a key pair $(\mathrm{PK}', \mathrm{SK}')$ for the encryption scheme, and run A', simulating B to it by sending it c and a random string R_B in the second round. In the third round, A' will send PK to M. M will compute $R = \mathrm{PK} \oplus R_B$ (note that, if A' computed PK faithfully, then $R = R_A$), and $R'_B = \mathrm{PK}' \oplus R$. M will then rewind A' to the end of the first round, run it again, this time sending c and R'_B to A'. If A' again computes the public key faithfully, then it will encrypt a with PK', for which M knows the corresponding secret key SK′. This will allow M to recover a, which is correlated to the unknown committed value b, and thus will allow M to break the semantic security of c.

Of course, when M runs A' in this manner and A' does not compute the public key faithfully, then M will fail. However, if A' computes the public key faithfully with only a negligible probability, then the commitment of A' is invalid with all but a negligible probability, so A' is not correlating to b any better than a simulator who just outputs $\perp$ all the time. If, on the other hand, A' computes the public key faithfully with probability better than negligible, then M will break the semantic security of c with probability better than negligible, as well. □

3.3 Mutually Independent and Aware Commitment

We present two protocols for mutually independent and aware commitment. The first protocol, previously known in the folklore, uses non-interactive perfectly binding commitments and general zero-knowledge arguments of knowledge (arguments, as opposed to proofs, are sound only if the prover is computationally bounded, which suffices for our case). Specifically, to minimize the number of rounds, we use the 5-round protocol of [FS89], which is based on one-way permutations.

This protocol is not practical, because it uses general zero-knowledge proofs of NP statements. We present it here for didactic purposes: it clearly illustrates the notion of mutually independent and aware commitments.

Theorem 4. *If one-way permutations exist, then there exists an 11-round mutually independent and aware commitment protocol.*

Proof sketch. Let c be a commitment scheme.

The commit stage proceeds as follows:

1. Alice publishes a commitment $c(a)$ to her value.
2. Bob publishes a commitment $c(b)$ to his value.
3. Bob uses the [FS89] ZK argument of knowledge to prove to Alice that he knows how to open his commitment.
4. Alice uses the [FS89] ZK argument of knowledge to prove to Bob that she knows how to open her commitment.

This takes eleven rounds, since Bob can send $c(b)$ and the first round of his proof in the same message.

It is easy to show that this protocol is complete, binding, and sound. If it is not (say) A-hiding, then whatever B' breaks A-hiding can break the commitment scheme, because the zero-knowledge argument of knowledge can be simulated. Awareness follows simply by using the extractor for the proof of knowledge. □

The second protocol is much more efficient than the first. It requires just seven rounds, each of which takes only a few modular exponentiations. It relies on the hardness of discrete logarithms. In its simplest version, it assumes that there exists an easily indexable sequence of "safe" primes and generators $(p_1, g_1), (p_2, g_2), \ldots, (p_k, g_k), \ldots$, one pair for every value of the security parameter k, such that $p_i = 2q_i + 1$ (where q_i is a prime), g_i is a generator of the subgroup of order q_i in $Z_{p_i}^*$, and discrete logarithm is hard in that subgroup.[6] To simplify notation, we will assume the security parmater k is fixed, and will simply use p, q, g in place of p_k, q_k, g_k when describing our protocol.

With a loss of efficiency, our protocol can be modified to be based on general assumptions rather than the hardness of discrete logarithms.

Theorem 5. *Assuming the hardness of discrete logarithms, there exists a seven-round mutually independent and aware commitment scheme.*

Proof sketch. For clarity, we will present our protocol for commitments to single-bit messages first, and then explain how it can modified for longer messages. Let H be a hardcore predicate for discrete log (in particular, [BM84] prove that the sign of the exponent minus $(p-1)/2$ is hardcore).

Let C denote a perfectly hiding trapdoor commitment scheme based on the discrete logarithm assumption. To be specific, we use the scheme of Pedersen [Ped91], in which one has two bases (generators), g and $h = g^\alpha$, and commits to a value v by publishing $g^v h^r$, for a random r. The scheme is binding because decommitting in two different ways allows one to find α. On the other hand, the scheme is trapdoor because knowing α allows one to decommit to any v'.

The commit stage of our protocol proceeds as follows.

1. Alice randomly generates an element g_a of order q and $\alpha \in Z_q$, and computes $h_a = g_a^\alpha$. She sends (g_a, h_a) to Bob, to be used by him as bases for the Pedersen commitment scheme.
2. a) Bob likewise generates g_b, β and h_b, which he sends to Alice to be used by her as bases for the Pedersen commitment scheme.
   b) Bob generates k_b, k_b' such that $H(k_b) \oplus H(k_b') = b$, and sends g^{k_b} and $g^{k_b'}$ to Alice.
   c) Bob generates a random $r_b \in Z_q$, computes g^{r_b}, and then commits to g^{r_b} using Pedersen commitments with bases g_a and h_a. He sends the resulting commitment $C_a(g^{r_b})$ to Alice.
3. a) Alice generates k_a and k_a' such that $H(k_a) \oplus H(k_a') = a$, and sends $g^{k_a}, g^{k_a'}$ to Bob.
   b) Alice generates a random r_a, and, just like Bob, commits to g^{r_a} using Pedersen commitments with bases g_b and h_b. She sends the resulting commitment $C_b(g^{r_a})$ to Bob.
   c) Alice generates and sends to Bob a random $c_a \in Z_q$.
4. a) Bob generates and sends to Alice a random c_b in Z_q.
   b) Bob decommits g^{r_b} from $C_a(g^{r_b})$ and sends the decommitment to Alice.

[6] This assumption can be relaxed by having the parties provide the parameters to each other; moreover, we do not need primes of the form $2q_i + 1$; primes of the form $k_i q_i + 1$, for sufficiently long prime q_i, would suffice. For the sake of clarity, however, we do not present our protocol that way.

c) Bob computes $d_b = c_a k_b + r_b$, and sends d_b to Alice.

5. a) Alice checks the decommitment of g^{r_b} and verifies that $g^{d_b} = (g^{k_b})^{c_a} g^{r_b}$. If the checks fail, she outputs "reject" and stops.
   b) Alice decommits g^{r_a} from $C_b(g^{r_a})$ and sends the decommitment to Bob.
   c) Alice computes $d_a = c_b k_a + r_a$, and sends d_a to Bob.
   d) Alice sends α to Bob.
6. a) Bob checks the decommitment of g^{r_a} and verifies that $g_a^d = (g^{k_a})^{c_b} g^{r_a}$. If the checks fails, he outputs "reject" and stops.
   b) Bob also checks that $g_a^\alpha = h_a$. If not, he outputs "reject" and stops.
   c) Bob sends k'_b and β to Alice.
7. a) Alice checks that $g_b^\beta = h_b$. If not, she outputs "reject" and stops.
   b) Alice checks k'_b received from Bob against $g^{k'_b}$ that was sent to her in step 2. If they do not agree, she outputs "reject" and stops. Otherwise, she outputs "accept."
   c) Alice sends k'_a to Bob.
8. Bob checks k'_a against $g^{k'_a}$ that was sent to him in Step 3. If the agree, he outputs "accept." Otherwise, he outputs "reject."

In the reveal stage, Alice reveals k_a and Bob reveals k_b. The value a is then calculated as $H(k_a) \oplus H(k'_a)$. b is calculated similarly.

Intuition. The following may help explain what is happening in this protocol with respect to Alice's commitment (Bob's commitment is, of course, similar). In step 3(a), Alice commits to her bit a by splitting it into two parts, k_a and k'_a. In steps 3(b), 4(a) and 5(c), she proves knowledge of k_a using a Schnorr [Sch89] three-round proof of knowledge for discrete logarithms. The only difference from the Schnorr proof is that the initial message of Alice's proof of knowledge is committed using the trapdoor commitment that Bob set up for Alice in step 2(a), and Alice reveals that message in step 5(b), at the end of the proof of knowledge. Bob reveals the trapdoor for the commitment scheme in step 6(c). Only after Alice gets the trapdoor does she reveal k'_a in step 7(c).

The reason for not using Schnorr's proof of knowledge directly is that it is not known to be simulatable. However, if the simulator knows Bob's trapdoor, then we can simulate the proof. As we explain in detail below, Bob can refuse to reveal the trapdoor, but then the simulator does not have to reveal k'_a. (We note that the idea of revealing the trapdoor to allow the simulation to go through has been used before in a number of protocols, and seems to have first appeared in [CDM00].)

Security. It should be clear that the protocol is complete, sound, and binding. Awareness is fairly easy to show: the extractor (say, for Bob) need only run the protocol through where c_a is sent by Alice, and repeatedly try substituting different challenges. If Bob ever answers two different challenges, k_b can be recovered. Then, since Bob reveals k'_b in the protocol, b can be determined.

The hiding property is a little more difficult to demonstrate. Suppose there is a B' which can find A's secret values. There are two cases.

case i: In this case, with a non-negligible probability, when Bob does not give the correct β, he still can still distinguish whether A was committing to a 0 or a 1. In this case, we can break the hard-core predicate. Suppose we are given $z = g^x$ and we are asked to find $H(x)$ with good probability. Then we first of all randomly choose k_a and run the protocol as Alice faithfully, except that we give z in place of $g^{k'_a}$. If Bob returns the correct β we output a coin flip. Otherwise, we get Bob's guess a and output $a \oplus H(k'_a)$. Note that since we either output a coin flip, or Bob doesn't return the correct β, we never actually reach step 7, so it doesn't matter that we don't know x.

case ii: In this case, when Bob does not give the correct β, he can only distinguish with negligible probability. Thus, if Bob does give β correctly, he must be able to distinguish. So, since he is able to distinguish with non-negligible probability, he must give β with non-negligible probability. Thus, we run honestly until Bob gives β. If it is incorrect, we output a coin flip. Otherwise, we rewind and give z in place of g^{k_a} and fake the proof of knowledge (which we can do now that we have the trapdoor.) If Bob then completes the protocol, we take his guess a and output $a \oplus H(k'_a)$. Otherwise, we output a coin flip. This will give us an edge in guessing the most significant bit of the discrete logarithm of z.

LONGER MESSAGES. In order to extend this protocol to longer messages, there are two techniques. The first is the obvious one: run the protocol many times in parallel (though we can collapse some rounds together: for instance, only one pair g_a, h_a is needed). We can do better than this by relaxing our assumptions and assuming that the discrete logarithm problem has more hardcore bits. That is, if A's secret is n bits long and $H_n(x)$ returns n hardcore bits x, then we simply modify the protocol so that Alice generates k_a and k'_a such that $H_n(k_a) \oplus H_n(k'_a) = a$. □

Acknowledgements. We would like to thank the anonymous referees for their detailed comments.

Moses Liskov was supported by a grant from the Merrill Lynch corporation. Anna Lysyanskaya was supported by an NSF graduate fellowship, the Lucent Technologies GRPW program, and the Merrill-Lynch grant given to R.L.Rivest. The work of Leonid Reyzin was performed while at MIT, and was supported by a grant from the NTT corporation. Adam Smith was supported by U.S. Army Research Office Grant DAAD19-00-1-0177.

References

[BDMP91] Manuel Blum, Alfredo De Santis, Silvio Micali, and Giuseppe Persiano. Noninteractive zero-knowledge. *SIAM Journal on Computing*, 20(6):1084–1118, December 1991.

[BM84] M. Blum and S. Micali. How to generate cryptographically strong sequences of pseudo-random bits. *SIAM Journal on Computing*, 13(4):850–863, November 1984.

[BMM99] A. Beimel, T. Malkin, and S. Micali. The all-or-nothing nature of two-party secure computation. In Michael Wiener, editor, *Advances in Cryptology—CRYPTO '99*, volume 1666 of *Lecture Notes in Computer Science*, pages 80–97. Springer-Verlag, 15–19 August 1999.

[CDM00] Ronald Cramer, Ivan Damgård, and Philip MacKenzie. Efficient zero-knowledge proofs of knowledge without intractability assumptions. In *Public Key Cryptography (PKC 2000)*, pages 354–372. Springer-Verlag, 2000.

[CF01] R. Canetti and M. Fischlin. Universally composable commitments. In Joe Kilian, editor, *Advances in Cryptology—CRYPTO 2001*, Lecture Notes in Computer Science. Springer-Verlag, 19–23 August 2001.

[CGMA85] B. Chor, S. Goldwasser, S. Micali, and B. Awerbuch. Verifiable secret sharing and achieving simultaneity in the presence of faults. In *26th IEEE Symposium on Foundations of Computer Science*, pages 383–395, 1985.

[CR87] Benny Chor and Michael Rabin. Achieving independence in logarithmic number of rounds. In *Principles of Distributed Computing (PODC 87)*, pages 260–268. ACM, 1987.

[CS98] R. Cramer and V. Shoup. A practical public key cryptosystem provably secure against chosen ciphertext attack. In Hugo Krawczyk, editor, *Advances in Cryptology—CRYPTO '98*, volume 1462 of *Lecture Notes in Computer Science*. Springer-Verlag, 23–27 August 1998.

[DDN00] D. Dolev, C. Dwork, and M. Naor. Nonmalleable cryptography. *SIAM*, 30:391–437, 2000.

[DDP00] Alfredo De Santis, Giovanni Di Crescenzo, and Giuseppe Persiano. Necessary and sufficient assumptions for non-interactive zero-knowledge proofs of knowledge for all np relations. In U. Montanari, J. D. P. Rolim, and E. Welzl, editors, *Automata Languages and Programming: 27th International Colloquim (ICALP 2000)*, volume 1853 of *Lecture Notes in Computer Science*, pages 451–462. Springer-Verlag, July 9–15 2000.

[DIO98] G. Di Crescenzo, Y. Ishai, and R. Ostrovsky. Non-interactive and non-malleable commitment. In *Proceedings of the Thirtieth Annual ACM Symposium on Theory of Computing*, Dallas, Texas, 23–26 May 1998.

[DKOS01] G. Di Crescenzo, J. Katz, R. Ostrovsky, and A. Smith. Efficient and non-interactive non-malleable commitment. In Birgit Pfitzmann, editor, *Advances in Cryptology—EUROCRYPT 2001*, volume 2045 of *Lecture Notes in Computer Science*, pages 40–59. Springer-Verlag, 6–10 May 2001.

[DP92] Alfredo De Santis and Giuseppe Persiano. Zero-knowledge proofs of knowledge without interaction. In *33rd Annual Symposium on Foundations of Computer Science*, pages 427–436, Pittsburgh, Pennsylvania, 24–27 October 1992. IEEE.

[FF00] M. Fischlin and R. Fischlin. Efficient non-malleable commitment schemes. In Mihir Bellare, editor, *Advances in Cryptology—CRYPTO 2000*, volume 1880 of *Lecture Notes in Computer Science*. Springer-Verlag, 20–24 August 2000.

[FFS88] Uriel Feige, Amos Fiat, and Adi Shamir. Zero-knowledge proofs of identity. *Journal of Cryptology*, 1(2):77–94, 1988.

[FS89] Uriel Feige and Adi Shamir. Zero knowledge proofs of knowledge in two rounds. In G. Brassard, editor, *Advances in Cryptology—CRYPTO '89*, volume 435 of *Lecture Notes in Computer Science*, pages 526–545. Springer-Verlag, 1990, 20–24 August 1989.

[Gen95] Rosario Gennaro. Achieving independence efficiently and securely. In *Principles of Distributed Computing (PODC 95)*, pages 130–136. ACM, 1995.

[GL89] O. Goldreich and L. Levin. A hard-core predicate for all one-way functions. In *Proceedings of the Twenty First Annual ACM Symposium on Theory of Computing*, pages 25–32, Seattle, Washington, 15–17 May 1989.

[GM84] S. Goldwasser and S. Micali. Probabilistic encryption. *Journal of Computer and System Sciences*, 28(2):270–299, April 1984.

[GMR85] Shafi Goldwasser, Silvio Micali, and Charles Rackoff. Knowledge complexity of interactive proofs. In *Proceedings of the Seventeenth Annual ACM Symposium on Theory of Computing*, pages 291–304, Providence, Rhode Island, 6–8 May 1985.

[GMR88] Shafi Goldwasser, Silvio Micali, and Ronald L. Rivest. A digital signature scheme secure against adaptive chosen-message attacks. *SIAM Journal on Computing*, 17(2):281–308, April 1988.

[KOS] J. Katz, R. Ostrovsky, and A. Smith. Personal Communication.

[Nao91] Moni Naor. Bit commitment using pseudorandomness. *Journal of Cryptology*, 4(2):151–158, 1991.

[Ped91] Torben Pryds Pedersen. Non-interactive and information-theoretic secure verifiable secret sharing. In J. Feigenbaum, editor, *Advances in Cryptology—CRYPTO '91*, volume 576 of *Lecture Notes in Computer Science*, pages 129–140. Springer-Verlag, 1992, 11–15 August 1991.

[Sah99] Amit Sahai. Non-malleable non-interactive zero-knowledge and adaptive chosen-ciphertext security. In *40th Annual Symposium on Foundations of Computer Science*, New York, October 1999. IEEE.

[Sch89] C. P. Schnorr. Efficient identification and signatures for smart cards. In J.-J. Quisquater and J. Vandewalle, editors, *Advances in Cryptology—EUROCRYPT 89*, volume 434 of *Lecture Notes in Computer Science*, pages 688–689. Springer-Verlag, 1990, 10–13 April 1989.

[TW87] Martin Tompa and Heather Woll. Random self-reducibility and zero knowledge interactive proofs of possession of information. In *28th Annual Symposium on Foundations of Computer Science*, pages 472–482, Los Angeles, California, 12–14 October 1987. IEEE.

Efficient Zero-Knowledge Authentication Based on a Linear Algebra Problem MinRank

Nicolas T. Courtois[1,2,3]*

[1] CP8 Crypto Team, SchlumbergerSema, BP 45
36-38 rue de la Princesse, 78430 Louveciennes Cedex, France
[2] SIS, Toulon University, BP 132, F-83957 La Garde Cedex, France
[3] Projet Codes, INRIA Rocquencourt, BP 105, 78153 Le Chesnay - Cedex, France
`courtois@minrank.org`
`http://www.minrank.org/minrank/`

Abstract. A Zero-knowledge protocol provides provably secure entity authentication based on a hard computational problem. Among many schemes proposed since 1984, the most practical rely on factoring and discrete log, but still they are practical schemes based on NP-hard problems. Among them, the problem SD of decoding linear codes is in spite of some 30 years of research effort, still exponential. We study a more general problem called MinRank that generalizes SD and contains also other well known hard problems. MinRank is also used in cryptanalysis of several public key cryptosystems such as birational schemes (Crypto'93), HFE (Crypto'99), GPT cryptosystem (Eurocrypt'91), TTM (Asiacrypt'2000) and Chen's authentication scheme (1996).
We propose a new Zero-knowledge scheme based on MinRank. We prove it to be Zero-knowledge by black-box simulation. An adversary able to fraud for a given MinRank instance is either able to solve it, or is able to compute a collision on a given hash function.
MinRank is one of the most efficient schemes based on NP-complete problems. It can be used to prove in Zero-knowledge a solution to any problem described by multivariate equations. We also present a version with a public key shared by a few users, that allows anonymous group signatures (a.k.a. ring signatures).

Keywords: Zero-knowledge, identification, entity authentication, MinRank problem, NP-complete problems, multivariate cryptography, rank-distance codes, syndrome decoding (SD), group signatures, ring signatures.

1 Introduction

The general problem we address is the classical problem of interactive entity authentication. It is known since Fiat-Shamir [5] that solving this problem combined with a cryptographic hash function also allows non-interactive authentication, for example digital signatures.

* The work described in this paper has been supported by the French Ministry of Research under RNRT Project "Turbo-signatures".

C. Boyd (Ed.): ASIACRYPT 2001, LNCS 2248, pp. 402–421, 2001.
© Springer-Verlag Berlin Heidelberg 2001

The notion of Zero-knowledge identification has been formalized by Goldwasser, Micali and Rackoff in [18]. In such a scheme a Prover proves his identity to a Verifier. Provided the underlying problem is difficult, we prove that there is no interactive strategy for the Verifier communicating with the Prover, to extract any information whatsoever on the prover's secret. Several such schemes have been proposed since the original Fisher-Micali-Rackoff scheme (1984), and the most practical ones are Fiat-Shamir, Guillou-Quisquater and Schnorr schemes. Unfortunately they rely on problems that are (believed) not NP-hard such as factoring or discrete log. Still there are schemes using an NP-hard problem and still practical, for example PKP by Shamir [31], CLE by Stern [35] or PPP by Pointcheval [27]. However the most interesting schemes are in our opinion the schemes related to coding, as the decoding problem(s) are believed intractable even since the 1970s [2]. There were many proposals [34,40,20,16,4] and the best of them is the scheme SD by Stern [34,40]. The simplest decoding problem is the problem of Syndrome Decoding (SD) and consists of finding a small weight vector in an affine subspace of a linear space. Similarly the MinRank problem is a problem of finding a linear (or affine) combination of given matrices that has a small rank. Both problems are NP-hard. Moreover SD have withstood more than 20 years of extensive research on the cryptanalysis of the McEliece cryptosystem [22] and all the known attacks for SD are still exponential, [1,3,21, 36,40]. MinRank in fact contains SD and thus is also probably exponential. It also contains the decoding problem for rank-distance codes of Gabidulin, used in public-key authentication scheme of Chen [4] cryptanalysed in [37,11], and also used in the public-key encryption scheme GPT [14]. The MinRank problem, not always named so, has many applications in cryptanalysis of various schemes such as Shamir's birational schemes [30,6,7] cryptanalysed by Coppersmith, Stern and Vaudenay solving a MinRank with a small rank. Similarly Goubin and Courtois broke the TTM cryptosystem in [19]. In [32] Shamir and Kipnis reduced the cryptanalysis of Hidden Field Equations (HFE) scheme [24] to MinRank.

In the present paper we present a new Zero-knowledge protocol, for MinRank. More precisely we show have to prove in Zero-knowledge an ability to compute (or have) MinRank solutions. We may build instances that have only one solution, and for those it will also be a proof of knowledge. We show that the scheme can also be applied to prove in Zero-knowledge a solution to **any** other problem expressed as a system of multivariate equations over a finite field.

The paper is organized as follows: First we recall the basic requirements of a Zero-knowledge protocol. Then in §3 defines MinRank and studies related hard problems. The §4 shows how to build secure instances for practical use, evaluated with all the 5 attacks currently known for MinRank. In the §5 we describe key generation and setup of the MinRank identification which is described in §6. The following §7 gives proofs of completeness, soundness and Zero-knowledge. Then in §8 we analyse the performance of the scheme and in §8.2 we compare it to other schemes based on NP-complete problems. In Appendix C we compute useful probability distributions for ranks of matrices. The Appendix B contains various practical improvements to the scheme, notably reducing the fraud probability form 2/3 to 1/2. Finally, the Appendix C shows that MinRank allows to achieve authentication and signature, for any small subgroup, of a given group of users sharing the same public key.

2 Zero-Knowledge Protocols

An interactive protocol involves two entities/strategies: the Prover (P) and the Verifier (V) that will be two probabilistic Turing machines. The Verifier and prover interact and at the end the Verifier gives an answer: Accept or Refuse.

In known Zero-knowledge protocols, there is a possibility of fraud: a cheater is usually able to answer to some types of questions (for which he was prepared in advance) but not for all of them. The protocols are designed in such a way that an answer to one question gives no information (Zero-knowledge), while answering all the questions is proved to reveal Prover's secret (Soundness). The security is in fact based on the impossibility by the Prover to predict Verifier's questions. If we iterate the protocol, the global fraud probability becomes then as small as we want.

A Zero-knowledge identification scheme should be: **complete**, **sound** and **Zero-knowledge**:

Completeness. The legitimate Prover gets always accepted.

(Computational) Soundness. An illegitimate Prover will be rejected with some fixed probability. We usually show the Prover that always succeeds can be used to extract the Prover's secret (a knowledge extractor).

Zero-knowledge. It is much stronger that saying the Verifier learns merely nothing about the secret. We demand that no Verifier strategy, can extract any information from the Prover, even in several interactions. It gives provable security against active attacks. Proofs are made by simulation using the Verifier as an oracle, or black-box, and therefore this definition has been called black box (computational) Zero-knowledge, as formalized by Goldreich, and Oren [17]:

Definition 1 (Black box Zero-knowledge, [17]). *A strategy P is told to be black box Zero-knowledge on inputs from S (common input) if there exists an efficient simulating algorithm U so that for every feasible Verifier strategy V, the two following probability ensembles are computationally indistinguishable:*

- $\{(P,V)(x)\}_{x\in S} \stackrel{def}{=}$ *all the outputs of V when interacting with P on a common input $x \in S$.*
- $\{U(V)(x)\}_{x\in S} \stackrel{def}{=}$ *the output of U using V as a black box, on $x \in S$.*

The definition above is strong and still realistic: all well-known Zero-knowledge protocols are proven in this model.

3 The MinRank Problem

Let M_0; $M_1, \ldots, M_m$ be some $\eta \times n$ matrices over a ring R. The problem MinRank(η, n, m, r, R) is to find a solution $\alpha \in R^m$ such that:

$$Rank(\sum_i \alpha_i M_i - M_0) \leq r.$$

3.1 Related Problems

This version of the MinRank, is a generalized version of one among many NP-complete rank problems studied in [23] and [10]. In our scheme R will be a finite field $GF(q)$.

MinRank over a field can be defined in terms of codes: it is a decoding problem for a kind of subfield subcode of Gabidulin's linear rank-distance code over $GF(q^\eta)$ [13,11,37]. Currently one of the two best known attacks to decode rank distance codes is based on MinRank [11,37]. Therefore MinRank is essential to the security of Chen and GPT public key schemes [14,4,11]. MinRank also appears in attacks known on the HFE [32,8,10], TTM cryptosystem [19] and Shamir's birational signature scheme [30,6,7]. Finally, as we show in §3.3, MinRank contains the SD problem for ordinary codes that underlies the security of McEliece [22] and various identification schemes [34,40,16,20].

MinRank over rings should also be mentioned. MinRank over $\mathbb{Z}$ might be broken by the widely-used LLL algorithm. Indeed, when all the M_i are diagonal of size up to 300×300, the problem is to find a vector in a lattice with a small number of non-zero elements, and this problem is closely related to the well known lattice reduction problem that has numerous applications in cryptography. Still MinRank over $\mathbb{Z}$ is undecidable in general, because it can encode any set of diophantine equations (Tenth Hilbert's problem) [23].

3.2 Encoding NP Problems as MinRank

The problem of proving in Zero-knowledge that a system of equations over a finite field has a solution has already been solved in [12] under RSA **or** DL intractability. Our solution is based on an NP-complete problem.

Theorem 1 (Determinant Universality, Valiant 1979). *Any set of multivariate equations over a ring can be encoded as a determinant of a matrix with entries being constants or variables.*

It was first shown by Valiant [38]. For a simpler, and still effective proof see [23]. Both give an **effective** algorithm to encode any set of multivariate polynomial equations as a MinRank. However the size of matrices it gives seems hard to improve, for m equations of degree d with n variables we need matrices of width about mn^d.

From now we always suppose that $R = GF(q)$. Solving multivariate quadratic equations over a field is NP-hard [26], thus:

3.3 MinRank Is NP-Hard

The proof of [23], however, gives instances of MinRank in which the size of the matrices will be polynomial in the number of matrices. It might seem that MinRank is less secure with m matrices $n \times n$ and m and n being of the same order of magnitude. We are going to show a reduction from an NP-complete problem that gives instances that are known to be hard both in theory in practice, with m, n and r being of the same order of magnitude. We reduce from the Syndrome Decoding problem of a linear error correcting code that is NP-complete. The proof for the case $q = 2$ is to be found in [2], and an extension to the arbitrary field is sketched in [39], page 1764. Let (n, k, d) be an error correcting code. The encoding is trivial: each of the lines of the generating matrix will be put on the diagonal of a $n \times n$ matrix M_i that will have all 0's elsewhere. Similarly M_0 contains the fixed codeword to decode. Solving MinRank with rank r is then equivalent to correcting r errors.

4 MinRank Instances and Attacks

4.1 Preliminary Requirement

The instance of MinRank should be chosen in such a way that the probability it has many solutions (apart from those we might put by construction) should be small. One possible way of achieve this is an explicit reduction from an instance of another problem that has only one solution, as for example in §3.2.

Another way is to choose parameters such that the probability it has a solution, given in Appendix B, is small, and thus we will be able to build instances with one (constructed) solution that are unlikely to have (m)any more. In this case, as we show in section A, we need to have

$$m \leq m_{max} \quad \text{with} \quad m_{max} \stackrel{def}{=} \eta n + r^2 - (\eta + n)r + 1$$

4.2 Known Attacks

We assume $\eta \geq n$. [1] There are five attacks known for the problem MinRank. Let ω be the exponent of the Gaussian reduction $2 \leq \omega < 3$, in practice $\omega \simeq 3$.

Exhaustive search. It is $q^m r^\omega$, see [10] for details.

Attacking square MinRank with $r \approx n$. In some cases the exhaustive search may break MinRank [2]. For example we consider a MinRank with m matrices $n \times n$ and with $r = n - s$. Then we have $m_{max} = r^2 + n^2 - 2nr = s^2$. A randomly generated MinRank with such parameters can be solved in about q^{s^2}, which can be quite small. However if the MinRank with $m >> m_{max}$ is generated from a reduction from another problem (see §3.2) having not too many solutions, it is still secure.

[1] The problem is symmetric with respect to transposition of matrices with swapping η and n and by inspection we verify that all the complexities given in the present paper are already given for the better of the two cases.

[2] This attack was suggested to me by prof. Claus P. Schnorr.

Attack Using Sub-matrices. This simple attack works **only if** $r << n$, not the case in this paper, and was first used by Coppersmith, Stern and Vaudenay in [6,7]. It was then described in details and used in [8,9].

MQ-solving Attacks. Another attack that works only for $r << n$ is due to Shamir and Kipnis [32]. It reduces MinRank to the MQ problem, i.e. to a system of **M**ultivariate **Q**uadratic equations. If $r << n$ the system is overdefined, and surprisingly such a system will be solved in expected polynomial time [32]. Improved algorithms will give roughly about $n^{\mathcal{O}(r)}$, see [10,11,8,33].

Since we will never have $r << n$, both these attacks fail.

The Kernel Attack is the best attack for the parameter sets we propose. It is due to Louis Goubin and described in [19] with a complexity of $q^{\lceil \frac{m}{n} \rceil r} m^{\omega}$ for $n = \eta$. A more general version described in [10] and [11] gives

$$Min\left(\; q^{\lceil \frac{m}{n} \rceil r} \;\; , \;\; q^{\lfloor \frac{m}{n} \rfloor r + (m \bmod n)} \; \right) \cdot m^{\omega}.$$

For small r there are further improvements described in [11].

The "Big m" Attack. This attack designed for $m >> n$ and is described in [11] and [10]. It is trivial and consists of constraining as many entries of the matrix M, as possible to 0. It runs in

$$q^{Max(0,\eta(n-r)-m)} \left(\eta(n-r)\right)^{\omega}.$$

The Syndrome Attack. Another attack for $m >> n$ and is described in [11] and [10]. It is not very practical and gives about

$$q^{Max\left(\frac{\eta n - m - 1}{2},(\eta+n)r/2-m-r^2/4\right)} \cdot \mathcal{O}(r\eta n)$$

Hard Instances: All the attacks known for MinRank described above are exponential in general. In a work in progress, [11] it is conjectured that for fixed $\eta = n$ the best security of $q^{\frac{4}{27}n^2}$ is achieved with $r = n/3$ and $m \approx \frac{4}{9}n^2$. If m is fixed, one may also build instances as close as we want to the exhaustive search if we put $n > 3\sqrt{m}$ and as big as possible, and with $r = n - \sqrt{m}$.

4.3 Practical Parameter Choices

We propose six sets of parameters A-F that use square matrices ($\eta = n$) and work either over $GF(2)$ or over $GF(65521)$, the biggest prime that fits in 16 bits. In the following table we compare the complexity of all known attacks described

above for A-F, give the communication complexity computed following §B.2, as well as the probability that it has a solution computed computed in §A.
For comparison we also include two MinRank instances that appear in the Shamir-Kipnis attack on HFE cryptosystem [3] [32], given for the HFE Challenge 1 [24,9] and for a subsystem of Quartz [4] [25].

Cryptosystem	MinRank identification						HFE	
Parameter set	A	B	C	D	E	F	Chall. 1	Quartz
m	10	10	10	81	121	190	80	103
n	6	7	11	19	21	29	80	103
η	6	7	11	19	21	29	80	103
r	3	4	8	10	10	15	7	8
q	65521	65521	65521	2	2	2	2^{80}	2^{103}
$Pr_\alpha[Rank \leq r]$	0.6	0.6	0.6	0.6	0.6	2^{-6}	$< 2^{-10^5}$	
20×Comm. [Kb]	1.94	2.99	4.86	2.17	2.36	3.13		

Attack								
Brute force	2^{168}	2^{168}	2^{170}	2^{81}	2^{134}	2^{205}	2^{80}	2^{103}
Kernel	$\mathbf{2^{106}}$	$\mathbf{2^{122}}$	$\mathbf{2^{138}}$	$\mathbf{2^{64}}$	$\mathbf{2^{81}}$	$\mathbf{2^{128}}$	2^{577}	2^{844}
Big m	2^{108}	2^{205}	2^{399}	2^{113}	2^{135}	2^{243}	2^{461k}	2^{997k}
Syndrome	2^{118}	2^{312}	2^{1002}	2^{151}	2^{172}	2^{339}	2^{252k}	2^{530k}
Sub-matrices	∞	∞	∞	∞	∞	∞	$\mathbf{2^{97}}$	$\mathbf{2^{114}}$
MQ	∞	∞	∞	∞	∞	∞	2^{152}	2^{188}

5 Setup of MinRank Identification

5.1 Key Setup

The public key are $1+m$ matrices $\eta \times n$ over a finite field $GF(q)$, M_0; $M_1, \ldots, M_m$. Let $r < n$. To generate a random hard [5] instance we pick $1+m-1$ (pseudo-)random matrices M_0; $M_1, \ldots, M_{m-1}$. We chose a random M of rank r and we "adapt" M_m. For this we pick a random $\alpha \in GF(q)^m$ such that $\alpha_m \neq 0$ and M_m is computed as:

$$M_m = (M + M_0 - \sum \alpha_i M_i)/\alpha_m$$

[3] The brute force workfactors given in the table for HFE correspond to the direct brute force attack on HFE itself, not on MinRank that would give much more.

[4] Since it is only a subsystem, an attack on MinRank does not really break Quartz.

[5] The instances of MinRank generated here are such that the matrices, and a linear combination that yields a small rank, are all random and uniformly distributed. It is believed to give hard instances most of the time with respect to all the attacks from section 4.2. It might change if a better way to produce hard instances is known. The same problem is an issue for **any** cryptosystem based on an NP-complete problem: there is a difference between an NP-complete problem in general, and the actual instances in the samplable distribution generated by a finite-length algorithm.

In practice, we generate M and $M_1, \ldots, M_{m-1}$ out of a pseudo-random generator with a seed of 160 bits. It is better to pick all M_i invertible, but it's not necessary. We may use the well-known LU method [6] to generate a deterministic pseudo-random invertible matrix. In order to generate M, first we generate a matrix L which is random invertible matrix $r \times r$, completed with 0's to an $\eta \times n$ matrix. Then a random couple of invertible matrices S and T is applied $M = SLT$, see Lemma 1.

The secret key. It is the solution $\alpha \in GF(q)^n$ such that

$$Rank(\sum \alpha_i \cdot M_i - M_0) = r.$$

Key sizes. All the public key is generated out of a pseudo-random generator with a seed of 160 bits, except M_m that is transmitted. The size of the public key is thus only $160 + n\eta \log_2 q$ bits. The secret key requires only additional $m \log_2 q$ bits to store α.

6 MinRank Identification Scheme

We use a collision-intractable one-way hash function H for commitments that is supposed to be behave as a random oracle. The Prover is going to convince the Verifier of his knowledge of α (and M).

The Prover chooses two random invertible matrices S, T that are $\eta \times \eta$ and $n \times n$, and a totally random $\eta \times n$ matrix X. We call $\overline{STX}$ the triple (S, T, X). Then, he picks a random combination β_1 of the M_i:

$$N_1 = \sum \beta_{1i} \cdot M_i$$

He puts and $N_2 = M + M_0 + N_1$ and uses his secret expression of M to get:

$$N_2 = \sum \beta_{2i} \cdot M_i$$

We have $\beta_2 - \beta_1 = \alpha$, but each of β_i (taken separately) is random and uniformly distributed. Each of the N_i is just a random combination of the M_i.

One Round of Affine MinRank Identification:

1. The Prover sends to the Verifier:

$$\overrightarrow{H(\overline{STX}),\ H(TN_1S + X),\ H(TN_2S + X - TM_0S)}$$

[6] This method is known to give a slight bias, but it seems easy tor repair for example by multiplying a few such matrices and permuting columns.

2. The Verifier chooses a query $\mathcal{Q} \in \{0,1,2\}$ and sends $\mathcal{Q}$ to the Prover.

$$\xleftarrow[\mathcal{Q} \in \{0,1,2\}]{\qquad\qquad\qquad\qquad}$$

3. If $\mathcal{Q} = 0$ the Prover gives the following values:

$$\xrightarrow[(TN_1S + X),\ (TN_2S + X - TM_0S)]{\qquad\qquad\qquad\qquad}$$

Verification $\mathcal{Q} = 0$: The Verifier accepts if $H(TN_1S + X) and H(TN_2S + X - TM_0S)$ are correct and if

$$(TN_2S + X - TM_0S) - (TN_1S + X) = TMS$$

is indeed a matrix of rank r.

3' If $\mathcal{Q} = 1, 2$ the Prover reveals:

$$\xrightarrow[\overline{STX},\ \beta_{\mathcal{Q}}]{\qquad\qquad\qquad\qquad}$$

Verification $\mathcal{Q} = 1, 2$: The Verifier checks if S and T are invertible and $H(\overline{STX})$ is correct. Then he computes

$$TN_{\mathcal{Q}}S = \sum \beta_{\mathcal{Q}i}\ TM_iS$$

and verifies $H(TN_1S + X)$ or $H(TN_2S + X - TM_0S)$.

6.1 Completeness

It is clear that a legitimate Prover that knows α always succeeds.

6.2 Soundness

We will show that a false Prover is rejected with probability $\frac{1}{3}$. Let C (Charlie or the Cheater), be an expected polynomial time Turing machine. We suppose that there is such a false Prover C that can answer all the questions $\mathcal{Q}$. In fact the proof below shows that such a Prover **will either be able to compute a collision for H, or be able to solve the given instance of the NP-complete problem MinRank** [7].

Proof: C commits (with H) to the values of TN_1S+X and TN_2S+X. For $\mathcal{Q} = 1$ and 2 he proves that he has indeed generated them in the form $X+T(\sum \beta_{1i}M_i)S$ and $X+T(\sum \beta_{2i}M_i)S$. In both cases we verify $H(\overline{STX})$ and we are certain that he used the same X, S and T. Finally when $\mathcal{Q} = 0$ we will verify the rank of the following matrix is indeed r:

[7] Here it can be just any instance of MinRank, however in the practical authentication the public key is generated in a specific way, see note 1 on the bottom of page 408.

$$\left(T(\sum \beta_{2i} M_i)S - TM_0S + X\right) - \left(T(\sum \beta_{1i} M_i)S + X\right) =$$

$$= \sum_{i=1}^{m} (\beta_{2i} - \beta_{1i}) \cdot TM_iS - TM_0S$$

When $\mathcal{Q} = 1$ or 2 we check that S and T are invertible, thus

$$\sum_{i=1}^{m} (\beta_{2i} - \beta_{1i}) \cdot M_i - M_0$$

is also of rank r. Thus the Prover knows a solution to MinRank $\alpha = (\beta_2 - \beta_1)$, i.e. either the secret key α or an equivalent one. □

One can see that the fraud probability for several rounds is:

$$Pr_{fraud} = \left(\frac{2}{3}\right)^{\#\text{rounds}}.$$

For details and an improvement to $\left(\frac{1}{2}\right)^{\#\text{rounds}}$ see B.1 and B.1.

7 Black-Box Zero-Knowledge of MinRank

Let the Prover strategy P be a probabilistic average polynomial time Turing machine. We suppose that H is a random function (oracle).The simplicity of MinRank makes very easy to show it is Zero-knowledge.

- In cases $\mathcal{Q} = 1, 2$ we only disclose random unrelated variables S, T, $\beta_{\mathcal{Q}}, X$.
- The case $\mathcal{Q} = 0$: disclosing (TN_1S+X) and (TN_2S-TM_0S+X) is equivalent to disclosing $(TN_1S + X)$ and their difference $TN_2S - TM_0S - TN_1S = TMS$.

  Since X is completely random, (TN_1S+X) is a random matrix independent from TMS. As for TMS, we show that it is a uniformly distributed matrix of rank r:

Lemma 1. *Let M be a $\eta \times n$ matrix of rank r. Let S and T be two uniformly distributed random invertible matrices $\eta \times \eta$ and $n \times n$. Then TMS is uniformly distributed among all $\eta \times n$ matrices of rank r.*

Proof sketch: All the $\eta \times n$ matrices M of rank r are equivalent modulo invertible variable changes and can be written as:

$$M = S' \cdot \begin{pmatrix} Id_{r\times r} & 0_{r\times(n-r)} \\ 0_{(\eta-r)\times r} & 0_{(\eta-r)\times(n-r)} \end{pmatrix} \cdot T'$$

7.1 The Exact Proof of Zero-Knowledge by Simulation

We construct a simulator U with oracle access to V, see Def. 1:

1. $U(V)$ chooses a random query $\mathcal{Q} = 1, 2$. He will prepare to answer to questions 0 **and** $\mathcal{Q}$.
2. He chooses $N = \sum \delta_i M_i$ with a random δ.
3. He picks up $\overline{STX} = (S, T, X)$ with invertible S and T.
4. He picks up a random matrix R of rank r.
5. Let $N_{\mathcal{Q}} = N$ and $N_{3-\mathcal{Q}} = N + (-1)^{\mathcal{Q}+1}(R + M_0)$. Now $N_2 - N_1 = R + M_0$.
6. He asks for Verifier's query on his commitment:

$$\mathcal{Q}' = V\left(H(\overline{STX}),\ H(TN_1S + X),\ H(TN_2S - TM_0S + X)\right) \in \{0, 1, 2\}.$$

7. He repeats steps 1-6 about 2 times (**rewinding**), until he does get one of the two queries he has prepared to answer:

$$\mathcal{Q}' \in \{0, \mathcal{Q}\}$$

8. If $\mathcal{Q}' = 0$ the simulator $U(V)$ reveals $(TN_2S + X - TM_0S)$ and $(TN_1S + X)$ with indeed a difference TRS of rank r.

8' If $\mathcal{Q}' = \mathcal{Q}$ the simulator $U(V)$ reveals $\overline{STX}$ and δ, that were indeed used to construct the committed $TN_{\mathcal{Q}}S + X[-TM_0S]$.

8 Performance of the Scheme

8.1 Communication Complexity

We assume that hash values are computed with SHA-1. Thus we need $3 \cdot 160 + 2$ bits for the first two passes.
We note that the values of $\overline{STX} = (S, T, X)$ does not need to be transmitted, they are in practice generated using a pseudorandom generator out of a seed of 160 bits, using the method we described in §5 to generate pseudorandom invertible matrices S and T. [8]
The last pass requires $2n\eta \log_2 q$ bits in the case $\mathcal{Q} = 0$. In the two other cases it requires $160 + m \log_2 q$ bits. The weighted average bit complexity for the whole scheme is $3 \cdot 160 + 2 + \frac{2}{3} \cdot 160 + \frac{2}{3}(n\eta + m) \log_2 q$.

This is to be multiplied by the number of rounds which is ≥ 35 for the round fraud probability of 2/3. In the Appendix C we show how achieve 1/2 instead (which will require only 20 rounds) and present several other improvements. Our best scheme (cf. B.2 and B.3) gives a communication complexity as low as :

$$\text{Comm. [in bits]} = 2 \cdot 160 + \left(4 \cdot 160 + 8 + \frac{n\eta + m}{2} \log_2 q\right) \cdot \#\text{rounds}$$

[8] Such modifications make the security depend on an additional assumption. It seems to be a quite weak and plausible assumption. For example here (S, T, X) should be indistinguishable from random.

8.2 Comparison with Other Schemes

The following table compares different Zero-knowledge protocols based on NP-complete problems based on previous work of Pointcheval [28].

	PKP Shamir	SD Stern	Chen [4] Chen	CLE Stern	PPP Pointcheval	MinRank (A) Author
matrix	16 x 34	256 x 512	32 x 16	24 x 48	101 x 117	6 x 6
field	$\mathbb{F}_{251}$	$\mathbb{F}_2$	$\mathbb{F}_{65535}$	$\mathbb{F}_{257}$	$\mathbb{F}_2$	$\mathbb{F}_{65521}$
passes	5	3	5	3/5	3/5	3
impersonation probability	$\frac{1}{2}$	$\frac{2}{3}$	$\frac{1}{2}$	$\frac{2}{3}/\frac{1}{2}$	$\frac{3}{4}/\frac{2}{3}$	$\frac{2}{3}/\frac{1}{2}$
rounds	20	35	20	35/20	48/35	35/20
impersonation global	10^{-6}	10^{-6}	10^{-6}	10^{-6}	10^{-6}	10^{-6}
public key [bits]	272	256	256	80	149	735
secret key [bits]	128	512	512	80	117	160
best attack	2^{60}	2^{70}	2^{53}	2^{73}	2^{61}	$\mathbf{2^{106}}$
bits send/round	665	954	1553	940/824	896/1040	1075/694
global [Kbytes]	**1.62**	4.08	3.79	4.01/**2.01**	5.25/4.44	4.6/**1.94**

9 Conclusion and Perspectives

We described a new MinRank authentication scheme. It is proven Zero-knowledge and relies on a linear algebra problem MinRank. This NP-hard problem contains in a very natural way some famous problems such as Syndrome Decoding. Both these problems are believed hard on average and all the known algorithms are exponential.

It is possible to use MinRank to prove in Zero-knowledge a knowledge of a solution for any problem expressed as a set of multivariate equations over a finite field (see 3.2). However, the encoding will not always be practical.

Among known schemes based on NP-complete problems MinRank is one of the most efficient, though several schemes are not much worse.

MinRank also allows to share the public key among several users in such a way that any small subgroup can identify itself or produce signatures.

Acknowledgments. I would like to thank prof. Ernst M. Gabidulin, prof. Jacques Patarin, prof. Claus P. Schnorr and dr. Louis Goubin for helpful remarks.

References

1. Alexander Barg: *Handbook of coding theory, Chapter 7: Complexity Issues in Coding Theory;* North Holland, 1999.
2. E.R. Berlekamp, R.J. McEliece, H.C.A. van Tilborg: *On the inherent intractability of certain coding problems*; IEE Trans. Inf. Th., IT-24(3), pp. 384-386, May 1978.

3. Anne Canteaut, Florent Chabaud: *A new algorithm for finding minimum-weight words in a linear code: application to McEliece's cryptosystem and to BCH Codes of length 511;*
4. Kefei Chen: *A new identification algorithm.* Cryptography Policy and algorithms conference, vol. 1029, LNCS, Springer-Verlag, 1996.
5. Amos Fiat, Adi. Shamir: *How to prove yourself: Practical solutions to identification and signature problems.* In Advances in Cryptology, Crypto '86, pp. 186-194, Springer-Verlag, 1987.
6. Don Coppersmith, Jacques Stern, Serge Vaudenay: *Attacks on the birational permutation signature schemes*; Crypto 93, Springer-Verlag, pp. 435-443.
7. Don Coppersmith, Jacques Stern, Serge Vaudenay, *The Security of the Birational Permutation Signature Schemes*, in Journal of Cryptology, 10(3), pp. 207-221, 1997.
8. Nicolas Courtois: *The security of Hidden Field Equations (HFE)*; Cryptographers' Track Rsa Conference 2001, San Francisco 8-12 April 2001, LNCS2020, Springer-Verlag.
9. The HFE cryptosystem home page: `http://hfe.minrank.org`.
10. Nicolas Courtois: *The security of cryptographic primitives based on multivariate algebraic problems: MQ, MinRank, IP, HFE*; PhD thesis, September 25th 2001, Paris 6 University, France. Mostly in French. Available at `http://www.minrank.org/phd.pdf`
11. Nicolas Courtois and Ernst M. Gabidulin.: *Security of cryptographic schemes based on rank problems*; work in progress.
12. Ronald Cramer, Ivan Damgård: *Zero-Knowledge Proofs for Finite Field Arithmetic or: Can Zero-Knowledge be for Free?* Crypto'98, LNCS 1642, pp. 424-441, Springer Verlag. See `http://www.brics.dk/RS/97/27/`
13. Ernst M. Gabidulin. *Theory of codes with maximum rank distance.* Problems of Information Transmission, 21:1-12, 1985.
14. Ernst M. Gabidulin, A. V. Paramonov, O. V. Tretjakov: *Ideals over a Non-Commutative Ring and their Applications in Cryptology. Eurocrypt 1991, pp. 482-489.*
15. *Ernst M. Gabidulin, Alexei V. Ourivski: Modified GPT PKC with Right Scrambler.* WCC 2001, Paris, France, Daniel Augot and Claude Carlet Editor.
16. Marc Girault: *A (non-practical) three pass identification protocol using coding theory;* Advances in cryptology, AusCrypt'90, LNCS 453, pp. 265-272.
17. Oded Goldreich, Y. Oren. Definitions and properties of Zero-knowledge proof systems. Journal of Cryptology 1994, vol.7, no.1, pp.1-32.
18. S. Goldwasser, S. Micali and C. Rackoff, *The knowledge Complexity of interactive proof systems*; SIAM Journal of computing, 1997, Vol. 6, No.1, pp.84.
19. Louis Goubin, Nicolas Courtois *Cryptanalysis of the TTM Cryptosystem*; Advances of Cryptology, Asiacrypt'2000, 3-9 December 2000, Kyoto, Japan, Springer-Verlag.
20. Sami Harari. *A new authentication algorithm.* In Coding Theory and Applications, volume 388, pp.204-211, LNCS, 1989.
21. P. J. Lee and E. F. Brickell. *An observation on the security of McEliece's public-key cryptosystem;* In Advances in Cryptology , Eurocrypt'88, LNCS 330, pp. 275–280. Springer-Verlag, 1988.
22. R.J. McEliece: *A public key cryptosystem based on algebraic coding theory*; DSN Progress Report42-44, Jet Propulsion Laboratory, 1978, pp. 114-116.

23. Jeffrey O. Shallit, Gudmund S. Frandsen, Jonathan F. Buss: *The Computational Complexity of Some Problems of Linear Algebra problems*, BRICS series report, Aaarhus, Denmark, RS-96-33, available on the net http://www.brics.dk/RS/96/33/.
24. Jacques Patarin: *Hidden Fields Equations (HFE) and Isomorphisms of Polynomials (IP): two new families of Asymmetric Algorithms*; Eurocrypt'96, Springer Verlag, pp. 33-48.
25. Jacques Patarin, Louis Goubin, Nicolas Courtois: Quartz, *128-bit long digital signatures*; Cryptographers' Track Rsa Conference 2001, San Francisco 8-12 April 2001, LNCS2020, Springer-Verlag.
26. Jacques Patarin, Louis Goubin, Nicolas Courtois, + papers of Eli Biham, Aviad Kipnis, T. T. Moh, et al.: *Asymmetric Cryptography with Multivariate Polynomials over a Small Finite Field;* known as 'orange script', compilation of papers with added material. Available from JPatarin@slb.com.
27. David Pointcheval: *A new Identification Scheme Based on the Perceptrons Problem*; In Advances in Cryptology, Proceedings of Eurocrypt'95, LNCS 921, pp.319-328, Springer-Verlag.
28. David Pointcheval: *Les preuves de connaissance et leurs preuves de sécurité*, PhD thesis, December 1996, Caen University, France.
29. Ronald R. Rivest, Adi Shamir and Yael Tauman: *How to leak a secret* ; Asiacrypt 2001, LNCS, Springer-Verlag.
30. Adi Shamir: *Efficient signature schemes based on birational permutations*; Crypto'93, Springer-Verlag, pp. 1-12.
31. Adi Shamir: *An efficient Identification Scheme Based on Permuted Kernels*, In Advances in Cryptology, Crypto'89, LNCS 435, pp.606-609, Springer-Verlag.
32. Adi Shamir, Aviad Kipnis: *Cryptanalysis of the HFE Public Key Cryptosystem*; In Advances in Cryptology, Proceedings of Crypto'99, Springer-Verlag, LNCS.
33. Nicolas Courtois, Adi Shamir, Jacques Patarin, Alexander Klimov, *Efficient Algorithms for solving Overdefined Systems of Multivariate Polynomial Equations*, Eurocrypt'2000, LNCS 1807, Springer-Verlag, pp. 392-407.
34. Jacques Stern: *A new identification scheme based on syndrome decoding*; Crypto'93, LNCS 773, pp.13-21, Springer-Verlag.
35. Jacques Stern: *Designing identification schemes with keys of short size*; In Advances in Cryptology, Proceedings of Crypto'94, LNCS 839, pp.164-73, Springer-Verlag.
36. Jacques Stern: *A method for finding codewords of small weight;* Coding Theory and Applications, LNCS 434, pp.173-180, Springer-Verlag.
37. Jacques Stern, Florent Chabaud: *The cryptographic security of the syndrome decoding problem for rank distance codes.* In Advances in Cryptology, Asiacrypt'96, LNCS 1163, pp. 368-381, Springer-Verlag.
38. L.G. Valiant: *Completeness classes in algebra.* In Proc. Eleventh Ann. ACM Symp. Theor. Comp., pp. 249-261, 1979.
39. Alexander Vardy: *The intractability of computing the minimum distance of a code; IEEE Transactions on Information Theory, Nov 1997, Vol.43, No. 6; pp. 1757-1766.*
40. Pascal Véron, *Problème SD, Opérateur Trace, Schémas d'Identification et Codes de Goppa*; PhD thesis in french, Toulon University, France, july 1995.

A Probability Distribution of Ranks

Following [13] the probability that a random matrix $\eta \times n$ is of rank r is

$$P(\eta, n, r) = \frac{(q^n-1)\cdot \ldots \cdot (q^n - q^{r-1})}{(q^r-1)\cdot \ldots \cdot (q^r - q^{r-1})} \cdot \frac{(q^\eta-1)\cdot \ldots \cdot (q^\eta - q^{r-1})}{q^{\eta n}}.$$

If $r \le min(n, \eta)$ it is non-zero, and when all the $n, \eta, r \to \infty$ we get the following approximation:

$$p(\eta, n, r) \simeq \mathcal{O}(q^{(\eta+n)r - r^2 - \eta n})$$

The probability that a random matrix $\eta \times n$ is of rank $> r$ is about:

$$(1 - \sum_{s=0}^{r} q^{(\eta+n)s - s^2 - \eta n}) \approx (1 - q^{(\eta+n)r - r^2 - \eta n})$$

There are $\frac{q^m-1}{q-1}$ non-collinear combinations α of the M_i. The probability that all of them give $Rank(\sum_i \alpha_i M_i - M_0) > r$ with $r \le min(n, \eta)$ is about:

$$Pr_\alpha[Rank \le r](\eta, n, r) = 1 - (1 - q^{(\eta+n)r - r^2 - \eta n})^{\frac{q^m-1}{q-1}}$$

We want to evaluate the value m_{max} such that for $m \le m_{max}$ we expect to have solutions for a random MinRank, and such that for $m \approx m_{max}$, we expect to have one solution on average. Therefore:

$$(q^{m_{max}} - 1)/(q-1) \cdot \mathcal{O}(q^{(\eta+n)r - r^2 - \eta n}) \approx 1$$
$$m_{max} \stackrel{def}{=} \eta n + r^2 - (\eta + n) r + 1$$

B Achieving Fraud Probability 1/2

We present a technique to achieve the fraud probability 1/2 instead of 2/3. It has the following interesting features:

- It requires additional assumption (of type one-wayness of a function).
- Should this assumption fail, the scheme is still at least as secure as before, only with a worse impersonation probability.

The principle of the "trick" is to replace some random choices by a deterministic procedure so that they are still random but cannot be chosen. We add an additional "verifiable" requirement on generation of some values, and thus we eliminate some fraud scenarios (but not others). Then we modify the probabilities of different questions in order to balance the probabilities for the remaining fraud scenarios.

We consider any Zero-knowledge protocol in which a Prover picks up 2 values β_1 and β_2 such that $\beta_2 - \beta_1 = \alpha$ is a given (usually secret) value. Usually we will generate β_1 at random and compute β_2, which enables fraud scenarios in which the adversary may chose a value for one out of β_1, β_2. We want to avoid this. Let F be a function with a following properties:

(1) It is very hard to compute an inverse $F^{-1}(y)$ for a given random y.
(2) It is very easy to compute two solutions x and x' such that $F(x') - F(x)$ is a given value Δy and $x' = x + \Delta x$ with a given constant Δx.

Example 1: $F : x \mapsto x^2 \bmod N$, N being an RSA modulus. The inversion problem (1) is as hard as factoring.

Example 2: $F : GF(q)^n \to GF(q)^n$ is a set of random quadratic equations over a finite field. The inversion problem (1) is called MQ, is NP-hard very difficult in practice [26,33].

In both examples, (2) is a linear problem easily solved.

We note that each of the above examples is applied with an operation '+' that belongs to a different group. Only the first example can be used for MinRank, as our '+' will be the component-by-component addition in the finite field.

B.1 Application to MinRank Scheme

Let $F : GF(q)^{n\eta} \to GF(q)^{n\eta}$ be a public fixed random set of quadratic equations. In the modified MinRank scheme, the Prover picks up two 160-bit seeds Z and $\overline{STX}$. Let $\Delta y = \text{Expand}(Z)$ and $(S,T,X) = \text{Expand}(\overline{STX})$ be the output of a pseudo-random generator. He solves

$$(S) \begin{cases} F(T(\sum \beta_{2i} M_i)S - TM_0S + X) - F(T(\sum \beta_{1i} M_i)S + X) = \text{Expand}(Z) \\ \qquad\qquad\qquad\qquad \beta_2 - \beta_1 \qquad\qquad\qquad\qquad\qquad = \alpha \end{cases}$$

The first equation becomes linear in β_1 after substitution of $\beta_2 = \beta_1 + \alpha$. He gets m linear equations with m variables β_{1i}. If there is no solution (β_1, β_2) found, he tries again with a new Z.

Verification that the Prover Follows the Scenario: If $\mathcal{Q} = 0$, the Prover will send an additional value Z. The Verifier will check that $F(TN_2S + X) - TM_0S - F(TN_1S + X) = \text{Expand}(Z)$. In the previous version of MinRank scheme possible fraud scenarios were:

01 Try to be able to answer $\mathcal{Q} = 0$ and 1.
It is easy to produce two matrices, seemingly $T(\sum \beta_{1i} M_i)S + X$ and $(T(\sum \beta_{2i} M_i)S - TM_0S + X)$, such that only one of them is really constructed in such a form, and the other is adjusted to get a difference of rank r.
02 Try to be able to answer $\mathcal{Q} = 0$ and 2 in the same way.
12 Try to be able to answer $\mathcal{Q} = 1$ and 2: We pick up any $\overline{STX}, \beta_1, \beta_2$ and produce a genuine $T(\sum \beta_{\mathcal{Q}i} M_i)S + X[-TM_0S]$.
0 Try to be able to answer $\mathcal{Q} = 0$ only. For this we just give any matrices that have a difference with rank r.
1 Try to be able to answer $\mathcal{Q} = 1$ only. For this we produce $T(\sum \beta_{1i} M_i)S + X$ in the required form.
2 Try to be able to answer $\mathcal{Q} = 2$ only. As above.

The new version excludes the scenarios (01) and (02). Let us see why on the example of scenario (01). We assume that a false Prover wants to answer $\mathcal{Q} = 0$ and 1. He may try the following possibilities:

a. Since S, T and X are always obtained as Expand($\overline{STX}$), if we cheat and have not selected them in this way, we are only able to answer $\mathcal{Q} = 0$.
b. He may try to pick up β_1. Since F is one way (the NP-chard problem MQ), he will be unable to produce a matrix R such that $F(Q) - F(T(\sum \beta_{1i} M_i)S + X) = \text{Expand}(Z)$.
c. Another way is to try find R of rank r and write the $n\eta$ equations with m variables $(\sum \beta_{2i} M_i - M_0) - (\sum \beta_{1i} M_i) = R$. However to find a solution is hard because $\alpha = \beta_2 - \beta_1$ would allow him to solve an instance of MinRank.

Still an adversary has the capacity to answer all possible questions separately: fraud scenarios (0), (1) and (2).

Resulting Changes in the Protocol. Now we may modify the probabilities. The question $\mathcal{Q} = 0$ is asked with probability 1/2 and $\mathcal{Q} = 1, 2$ with probability 1/4 each. The following table shows the probabilities of success for all fraud scenarios.

Fraud scenario	0	1	2	01	02	12	012
Pr[Success] before	$\frac{1}{3}$	$\frac{1}{3}$	$\frac{1}{3}$	$\frac{2}{3}$	$\frac{2}{3}$	$\frac{2}{3}$	0
now	$\frac{1}{2}$	$\frac{1}{4}$	$\frac{1}{4}$	0	0	$\frac{1}{2}$	0

A false Prover is detected with probability 1/2. Now only 20 instead of 35 rounds are needed to achieve the security of 10^{-6}.

Note: We obtained a more efficient authentication scheme with an added computational assumption based on the NP-hard problem MQ. This problem is believed very hard [33], but if it wasn't then the scenarios (01) and (02) will be possible again and the fraud probability will be 3/4. The MinRank scheme will remain secure, but with worse fraud probability, or equivalently, it will require more iterations.

Further Improvements. First we remark that if $\mathcal{Q} = 0$, it is not necessary at all to transmit the two values $TN_2S - TM_0S + X$ and $TN_1S + X$. In fact it is enough to transmit their difference TMS and Z that is already among the values that are transmitted. The values of $TN_2S - TM_0S + X$ and $TN_1S + X$ can be then recovered by the Verifier that has to solve a system similar to (B.1.(S)). We saved a transfer of one matrix $\eta \times n$.

Another improvement is to use only one seed $\overline{STXZ}$ with:

$$(S, T, X, Z) = \text{Expand}(\overline{STXZ})$$

B.2 The Modified Version MinRank-v2

Now we integrate all improvements in order to have a general view. The prover chooses a random seed of 160-bits $\overline{STXZ}$. Let

$$(S, T, X, Z) = \text{Expand}(\overline{STXZ})$$
$$\Delta y = \text{Expand}(Z)$$

Now the Prover solves:

$$(S) \begin{cases} F(T(\sum \beta_{2i} M_i)S - TM_0S + X) - F(T(\sum \beta_{1i} M_i)S + X) = \text{Expand}(Z) \\ \beta_2 - \beta_1 \qquad\qquad\qquad\qquad = \alpha \end{cases}$$

If there is no solution, (β_1, β_2), we try again a small number of times. with a different seed $\overline{STXZ}$. Then in each round of authentication:

1. The Prover sends to the Verifier:

$$\xrightarrow{\hspace{6cm}}$$
$$H(\overline{STXZ}),\ H(TN_1S + X),\ H(TN_2S + X - TM_0S)$$

2. The Verifier chooses a query $\mathcal{Q}$, such that $\mathcal{Q} = 0$ with probability 1/2, and $\mathcal{Q} \in \{1, 2\}$ with probability 1/4 each. He sends $\mathcal{Q}$ to the Prover.

$$\xleftarrow{\hspace{6cm}}$$
$$\mathcal{Q} \in \{0, 1, 2\}$$

3. If $\mathcal{Q} = 0$, the Prover gives the following values:

$$\xrightarrow{\hspace{6cm}}$$
$$TMS,\ Z$$

Verification $\mathcal{Q} = 0$: The Verifier will compute the $(TN_1S + X)$ and $(TN_2S + X - TM_0S)$, see B.1 Then he will accept if $H(TN_1S + X)$ and $H(TN_2S + X - TM_0S)$ are correct, and if $Rank(TMS) = r$.

3' In the case $\mathcal{Q} = 1, 2$, the Prover reveals:

$$\xrightarrow{\hspace{6cm}}$$
$$\overline{STXZ},\ \beta_{\mathcal{Q}}$$

Verification $\mathcal{Q} = 1, 2$: The Verifier checks if S and T are invertible and if $H(\overline{STXZ})$ is correct. Then he computes

$$TN_{\mathcal{Q}}S = \sum \beta_{\mathcal{Q}i}\ TM_iS$$

and verifies the correctness of $H(TN_1S + X)$ or $H(TN_2S + X - TM_0S)$.

B.3 Improvements in the Communications

As in 8.1 we compute the communication complexity of the new version. By inspection we see that it becomes:

$$\left(3 \cdot 160 + 2 + \frac{n\eta + m}{2} \log_2 q\right) \cdot \#\text{rounds}$$

Remark: The value of 160 bits for a length of seeds and commitments is appropriate for the security level of 2^{80} and should be increased otherwise. For example for a security level 2^{SF} we should use $2SF$ bits. So we get

$$\left(6SF + 2 + \frac{n\eta + m}{2}\log_2 q\right) \cdot \#\text{rounds}$$

Chaining random seeds. It is also possible to save on the size of random seeds used in the scheme and use one single seed A_0 of $2SF$ bits for the whole scheme. Each time we compute a seed A_i as the following:

$$A_i = H(A_0||i||b_1, \ldots, b_7)$$

with an appropriate length hash function and with 7 random bits b_i, as the seed $\overline{STXZ} = A_i$ will only work in sec. B.2 with a probability different than 1. Thus we may try again for b_i in order to have a working seed. With $2^7 = 128$ tries we have a negligible probability to never find an appropriate seed. The main seed A_0 is only given **at the end**, after all rounds of authentication, and only then all the verifications are carried. Now, with the exception of A_0, each round requires only $4SF + 7 + 2 + \frac{n\eta+m}{2}\log_2 q$ bits. Thus we get a communication complexity of

$$2SF + \left(4SF + 9 + \frac{n\eta + m}{2}\log_2 q\right) \cdot \#\text{rounds}.$$

C Group Authentication/Signatures with MinRank

It is easy to produce almost totally random instances of MinRank with several users, each of which has one solution to MinRank and no information about other solutions. We pick $1+m$ [pseudo-]random matrices $M_0; M_1, \ldots, M_m$. Each user i has the right to pick up a matrix U_i such that $U_i - M_0$, plus some randomly chosen linear combination of the $M_1 \ldots M_m$, has a small rank. It can be done for an unlimited (in practice) number of users. Then the set of matrices: $M_0; M_1, \ldots, M_m$; with the $\{U_i | i \in G\}$ is the public key for any small [9] subgroup G. Now any member of the group G, can use the MinRank authentication scheme to anonymously prove his membership.

C.1 Ring Signatures with MinRank

A well known method (see [5]) that transforms a Zero-knowledge protocol into a signature scheme will also apply to MinRank. This in turn can be combined with the above multi-user setting. We obtain an anonymous group signature scheme known as **a ring signature scheme** [29], with the following properties:

[9] Here the total number of matrices m can be very big: attacks such as the "big m attack" described in §2 or in [11,10] will only apply to a smaller m', the maximum cardinal of a subgroup used.

- Each group member signs with his own private key (no shared secrets).
- He may sign on behalf on **any** subgroup of users that contains himself.
- There is no central authority.
- The user within the group that signs is anonymous (inside the group).
- Security is based on the NP-hard problem MinRank.
- At any moment we may introduce a new user and remove a user.
- Selective repudiation of signatures: introducing a new user U' and invalidating his public key can be used as a mean to repudiate all signatures made with this user included in the subgroup. The repudiation is controlled by the person who knows the secret key of U' and publishes it.

Responsive Round Complexity and Concurrent Zero-Knowledge

Tzafrir Cohen[1], Joe Kilian[2], and Erez Petrank[1]

[1] Dept. of Computer Science, Technion - Israel Institute of Technology, Haifa 32000, Israel, {tzafrir|erez}@cs.technion.ac.il
[2] Yianilos Labs, joe@pnylab.com

Abstract. The number of communication rounds is a classic complexity measure for protocols; reducing round complexity is a major goal in protocol design. However, when the communication time is inconstant, and in particular, when one of the parties intentionally delays its messages, the round complexity measure may become meaningless. For example, if one of the rounds takes longer than the rest of the protocol, then it does not matter if the round complexity is bounded by a constant or by a polynomial. In this paper, we propose a complexity measure called *responsive round complexity*. Loosely speaking, a protocol has responsive round complexity m with respect to Party A, if it makes the following guarantee. If A's longest delay in responding to a message in a run of the protocol is t, then, in that run, the overall communication time is at most $m \cdot t$. The logic behind this definition is that if a party responds quickly to a message, whether it has a good connection or it just chooses not to delay its messages, then this party deserves to get an overall quicker running time. Responsive round complexity is particularly interesting in a setting where a party may gain something by delaying its messages. In this case, the delaying party does not deserve the same response time as another party that behaves nicely.

We demonstrate the significance of responsive round complexity by presenting a new protocol for concurrent zero-knowledge. The new protocol is a black-box concurrent zero knowledge proof for all languages in NP with round complexity $\tilde{O}(\log^2 n)$ but responsive round complexity $\tilde{O}(\log n)$. While the round complexity of the new protocol is similar to what is known from previous works, its responsive round complexity is a significant improvement: all known concurrent zero-knowledge protocols require $\tilde{O}(\log^2 n)$ rounds. Furthermore, in light of the known lower bounds, the responsive round complexity of this protocol is basically optimal.

Keywords: Zero-knowledge, concurrent zero-knowledge, cryptographic protocols.

1 Introduction

In this work, we study a new measure related to the round complexity of protocols. We propose a notion of *responsive round complexity* that properly relates

C. Boyd (Ed.): ASIACRYPT 2001, LNCS 2248, pp. 422–441, 2001.
© Springer-Verlag Berlin Heidelberg 2001

the running time of the protocol with the response time of each of the parties. Finally, we show how to improve state-of-the-art concurrent zero-knowledge protocols with respect to their responsive round complexity, and obtain an (almost) optimal protocol with respect to its responsive round complexity.

1.1 Round Complexity

The number of rounds in a run of a protocol can be a major time-consuming component. Therefore, round complexity is one of the important complexity measures of protocols. However, a protocol's round complexity is not always directly proportional to its time complexity. The reason is that communication rounds do not always have the same length. Thus, for example, if the length of one of the rounds exceeds the accumulative length of all the other rounds, the round complexity does not tell us anything about the time complexity.

The difference in the length of communication rounds may be a result of two different reasons. One is that the network is unstable and communication times vary during the run of the protocol. The second possible reason is that one of the parties may delay its messages. It is sometimes useful for a party to delay its answer in the protocol until something happens. For example, it may delay its answer until it obtains information from another source, or it may try to foil timing assumptions made by other parties in the protocol.

We propose a new complexity measure called *responsive round complexity*. Our intention is to relate the overall running time of a party to its response time. By this measure, each party gets a guarantee on the overall communication time, which relates to the longest delay it imposes on the run of the protocol. A party that always responds quickly gets a good guarantee on the overall communication time, and a party that sometimes responds slowly gets a poor guarantee on the overall communication time.

In this extended abstract we concentrate on the two-party case. An extension of the definition to multi-party protocols is straightforward.

Definition 1.1. Response time: *We say that the response time of a party A in round i of a specific run σ of a protocol Π is t, if Π in run σ tells A to send a message in round i, and t is the length of the time interval starting from the time B sent its message in round $i-1$ and ending at the time B received A's response of round i. (If A is not supposed to send a message in round i, then its response time is 0 for round i.) The response time of A in run σ of protocol Π is the maximum over all rounds i of A's response time in round i.*

Definition 1.2. Responsive time complexity: *We say that a protocol Π has responsive round complexity m with respect to Party A, if for any possible run σ of Protocol Π, the overall communication time does not exceed $t \cdot m$ where t is A's response time in Run σ of Protocol Π.*

Note that if all rounds are equally long in all runs of the protocol, then the responsive round complexity measure with respect to each of the two parties equals the (standard) round complexity measure[1].

Our primary interest in this notion is for cases when a party actually delays its messages to gain something; our goal is to develop protocols in which purposeful delays merely punishes the delayer. However, we note that the guarantee of our protocol also holds for networks with unstable or heterogeneous communication links. Parties will be (unfairly) punished for network delays beyond their control, but this punishment will be roughly proportional to the inherent delays. Thus, someone with a slow connection will obtain slow service, but will not be starved, as would be the case if a protocol simply timed-out on slow participants.

We demonstrate the usefulness of the new notion by using it for analyzing concurrent zero-knowledge protocols and constructing a new protocol with an (almost) optimal responsive round complexity.

1.2 Concurrent Zero Knowledge

Zero-knowledge interactive proofs as presented by Goldwasser, Micali, and Rackoff [16] are proofs that yield no knowledge but the validity of the proven assertion. These proof systems have proven important tools for a variety of cryptographic applications. However, the original definition of zero-knowledge considers security only in a restricted scenario in which the prover and the verifier execute the proof disconnected from the rest of the computing environment.

In recent years, several papers have studied the affect of a modern computing environment on the security of zero-knowledge. In particular, many computers today are connected through networks in which connections are maintained in parallel asynchronous sessions. It would be common to find several connections (such as FTP, Telnet, an internet browser, etc.) running together on a single workstation. Can zero-knowledge protocols be trusted in such an environment?

Zero-knowledge in a concurrent environment was first explored by Feige [12], and by Dwork, Naor, and Sahai [10]. Dwork, Naor and Sahai denoted zero-knowledge protocols that are robust to asynchronous composition *concurrent zero-knowledge* protocols. They observed that several known zero-knowledge proofs, with a straightforward adaptation of their original simulation to the asynchronous environment, may cause the simulator to work exponential time. Thus, it seems that the zero-knowledge property does not necessarily carry over to the asynchronous setting.

Kilian, Petrank, and Rackoff [19] gave the first lower bound for concurrent zero-knowledge, showing that any language that has a 4-round black-box concurrent zero-knowledge interactive proof or argument is in BPP. Thus, a large class of known zero-knowledge interactive proofs and arguments for non-trivial languages do not remain zero-knowledge in an asynchronous environment. Rosen

[1] Here, we adopt the measure by which a round consists of two messages: one from Party A to Party B and the other is the response of B to A.

[25] has improved this lower bound from from 4 rounds to 7. Canetti, Kilian, Petrank and Rosen [5] have substantially improved the lower bound to $\tilde{\Omega}(\log k)$.[2] The parameter k is the security parameter. A polynomial in k bounds the length of the inputs, the number of proofs that may start concurrently, and the time complexity that the parties spend in the protocol.

On the other hand, Richardson and Kilian [24] exhibited a concurrent zero-knowledge proof for any language in NP. Their protocol requires polynomially many rounds in k. Kilian and Petrank [18] substantially narrow the gap between the upper bound and the lower bounds. Using a different simulator, they provide a tighter security analysis for the Richardson-Kilian protocol, and show that it remains concurrent zero knowledge when run with only $\omega(\log^2 k)$ rounds.

How do these results translate to responsive round complexity? Zero-knowledge is about providing security to the prover. Thus, we expect the prover to follow the protocol and not delay its answers. The verifier is the bad guy, who may choose to deviate from what the protocol dictates in order to get knowledge from the prover. Thus, the verifier may delay its answers, and we would like to investigate how protocols behave in this case. It seems fair to provide quicker service (overall communication time) to verifiers that respond quickly and do not delay their answers. Verifiers that do delay their answers may get an overall slower run of the protocol. Responsive round complexity guarantees that the overall communication time is proportional to the longest delay of the verifier.

Looking at the best known upper bound protocol in [18], it is easy to see that the responsive round complexity with respect to the verifier is equivalent to its round complexity in stable networks. The verifier may simply keep its response time steady, and then the two measures equate. This is the best known protocol with respect to responsive time complexity, and it has responsive time complexity of any function m satisfying $m = \omega(\log^2 k)$.

If we look at the best known lower bound in [5], it provides a specific schedule such that if the protocol does not have enough rounds, no black box simulator can simulate it in this schedule. In the demonstrated schedule each verifier has its own response time, but each of the verifiers does not change its response time during the proof. Thus, the lower bound holds also for responsive round complexity, and we cannot do better than $\tilde{\Omega}(\log k)$.

In this paper we present a new concurrent zero-knowledge proof for all languages in NP that has responsive round complexity m for any function $m = \omega(\log k)$. Namely, the responsive round complexity of this protocol can be set to any function asymptotically larger than $\log k$. Thus, we get an algorithm whose responsive round complexity is almost optimal (up to a factor of at most $O(\log^2 \log k)$). Thus, any verifier that does not delay its messages (or even just does not change the delay from round to round) is guaranteed a round complexity of $\tilde{O}(\log k)$. Verifiers that do delay their messages get a protocol whose running time is at most $\tilde{O}(\log k)$ times the longest delay they choose to use.

---

[2] The "twiddle" notation neglects multiplicative factors that are polylogarithmic in the main term.

In a recent breakthrough work, Barak [1] gives a concurrent zero-knowledge for NP which is not black-box and requires only a constant number of rounds. A slight drawback of this protocol is that the maximum number k of concurrent sessions tolerated must be predetermined in advance, and the communication required by this protocol is proportional to the chosen polynomial. Our protocol, like previous black-box protocols, is robust against any polynomial number of concurrent sessions, and its overall communication is independent of the number of sessions.

1.3 Techniques

One set of previous protocols [24,18] ignore the timing of the messages and consider only their order; black-box simulatable protocols exist in the general model, though with high round complexity. Another approach is taken by the protocols of [10,11]. In this approach, strong restrictions are enforced (or assumed) on the ratio between the slowest and longest response times, simplifying the task of producing a simulation, and allowing for constant-round protocols. One interpretation (and implementation) of this restriction is that verifiers with slow response times are treated as malicious; their responses are rejected.

Our approach is intermediate between these two approaches. As with the latter approach, we do take response times into consideration, but as with the first approach we place no restriction on these delays. Instead, we monitor the delays and "punish" verifiers with long delays, though in a proportionate fashion. Each verifier has an associated response time that is doubled when the verifier does not respond in time. Thus, there are $O(\log k)$ sets of verifiers, each set containing verifiers responding at around the same time. The prover may delay its answers to each verifier to match its delays with those of the verifier. For each set, we use techniques similar to those in [10,11] to simulate the conversations with verifiers in this set. We then show that simulating the $O(\log k)$ sets together is still doable in polynomial time.

1.4 Contributions

The first contribution of this work is in proposing the notion of responsive round complexity. We feel that this notion may be useful in settings when one of the parties may gain something from delaying its responses. A guarantee on the responsive round complexity provides a guarantee on the time complexity such that each party "gets what it deserves".

Our second contribution is in providing a concurrent black-box zero-knowledge protocol with almost optimal responsive round complexity. Our design uses the protocol of [24,18] as a subroutine; its main technical contribution is a method for restarting this subroutine so as to obtain a better protocol in practice.

1.5 Related Work

Our notion of responsive-round complexity is of course related to the vast literature on distributed algorithms, and continues the program of studying zero-knowledge in distributed settings.[3] We do point out the difference between our notion and the most commonly used distributed model. In the standard distributed model, an adversary can speed up responses in worst case fashion; we require that all parties give a "correct" output by the end of the protocol. In our model, we impose additional requirement on when individual parties finish (give a final output); parties whose responses have been sped up may have to finish long before the end of the protocol as a whole. (Here, the "protocol" is the collective set of interactive proofs)

Several recent works have overcome the difficulty of the asynchronous setting by putting limits on the asynchronisity of the system (timing assumptions) [10, 11,6,9] or by making some set-up assumptions on the environment (such as a public key infrastructure) [7,4].

1.6 Terminology

Some words on the terminology we are using. By zero-knowledge we mean *computational* zero-knowledge, i.e., the distribution output by the simulation is polynomial-time indistinguishable from the distribution of the views of the verifier in the original interaction. Our proof is black-box zero-knowledge. The proof will be perfectly sound, i.e., we will construct an interactive proof, yet it will be possible to run the prover in polynomial time given a witness to the NP assertion that the prover is making.

1.7 Guide to the Paper

In Section 2 we go over the preliminaries. We state our main result in Sect. 3. We provide an overview on the protocol and proof in Sect. 4. The protocol itself is presented in Sect. 5, the simulator to the protocol is presented in Sect. 6, and the analysis of the simulator is given in Sect. 7.

2 Preliminaries

2.1 Zero-Knowledge Proofs

Let us recall the concept of interactive proofs, as presented by [16]. For formal definitions and motivating discussions the reader is referred to [16].

Definition 2.1. *A protocol between a (computationally unbounded)* prover P *and a (probabilistic polynomial-time)* verifier V *constitutes an* interactive proof *for a language* L *if there exists a negligible function* ε *such that*

[3] Indeed, we would not be surprised if quite similar definitions have been proposed in this literature.

- **Completeness:** *If* $x \in L$ *then* $\Pr[(P, V)(x) \text{ accepts}] \geq 1 - \varepsilon(|x|)$.
- **Soundness:** *If* $x \notin L$ *then for any prover* P^*

$$\Pr[(P^*, V)(x) \text{ accepts}] \leq \varepsilon(|x|) .$$

Brassard, Chaum, and Crépeau [2] introduced a modification of interactive proofs, called *arguments*, in which the prover is also polynomial time bounded. Thus, the soundness property is modified to be guaranteed only for probabilistic polynomial time provers P^*.

Let $(P, V)(x)$ denote the random variable that represents V's view of the interaction with P on common input x. The view contains the verifier's random tape as well as the sequence of messages exchanged between the parties.

We briefly recall the definition of black-box zero-knowledge [16,23,15,17]. The reader is referred to [17] for more details and motivation.

Definition 2.2. *A protocol* (P, V) *is* computational zero-knowledge *(resp.,* statistical zero-knowledge*) over a language* L*, if there exists an oracle polynomial time machine* S *(simulator) such that for any polynomial time verifier* V^* *and for every* $x \in L$*, the distribution of the random variable* $S^{V^*}(x)$ *is polynomially indistinguishable from the distribution of the random variable* $(P, V^*)(x)$ *(resp., the statistical difference between* $M(x)$ *and* $(P, V)(x)$ *is a negligible function in* $|x|$*).*

In this paper, we concentrate on black-box computational zero-knowledge, and use *zero-knowledge* as shorthand for *black-box computational zero-knowledge.*

2.2 Bit Commitments

We include a short and informal presentation of commitment schemes. For more details and motivation, see [22]. A commitment scheme involves two parties: The *sender* and the *receiver.* These two parties are involved in a protocol which contains two phases. In the first phase the sender commits to a bit, and in the second phase it reveals it. A useful intuition to keep in mind is the "envelope implementation" of bit commitment. In this implementation, the sender writes a bit on a piece of paper, puts it in an envelope and gives the envelope to the receiver. In a second (later) phase, the *reveal* phase, the receiver opens the envelope to discover the bit that was committed on. In the actual digital protocol, we cannot use envelopes, but the goal of the cryptographic machinery used, is to simulate this process.

More formally, a commitment scheme consists of two phases. First comes the *commit* phase and then we have the *reveal* phase. We make two security requirements which (loosely speaking) are:

Secrecy: At the end of the *commit phase*, the receiver has no knowledge about the value committed upon.

Binding property: It is infeasible for the sender to pass the commit phase successfully and still have two different values which it may reveal successfully in the reveal phase.

Various implementations of commitment schemes are known, each has its advantages in terms of security (i.e., binding for the receiver and secrecy for the receiver), the assumed power of the two parties etc.

Two-round commitment schemes with perfect secrecy can be constructed from any collection of claw-free permutations; see [22]. It is shown in [2] how to commit to bits with statistical security, based on the intractability of certain number-theoretic problems. Dåmgard, Pedersen and Pfitzmann [8] give a protocol for efficiently committing to and revealing strings of bits with statistical security, relying only on the existence of collision-intractable hash functions. This scheme is quite practical and we adopt it for the verifiers in our protocol. For the prover, we use a commitment scheme whose binding is information theoretic and security is computational. Such schemes can be constructed from any one-way function, see [20]. For simplicity, we simply speak of committing to and revealing bits when referring to the protocols of [8] for the verifier and [20] for the prover. We will need to use the properties of the commitment schemes in the concurrent setting.

Theorem 2.3. *The security of the bit commitments in [20] and [21] holds also in the concurrent setting.*

Proof. By definition, the binding property must be robust to asynchronous composition. Otherwise, the committer may play a mental game in which his real stand-alone commitment is part of an asynchronous game which he simulates, and then defeat the binding property in the normal stand-alone world.

As for the secrecy, a similar argument may be more complicated, since the receiver cannot simulate the behavior of the committer. Specifically, the committer has some information that the receiver does not have: the value of the committed string, which may be used in the other commitments. However, in our proof, the committer commits on uniformly chosen random strings. (And on nothing else.) Thus, if the committer follows the protocol, then the receiver is able to simulate the rest of the environment and the above argument holds for secrecy as well. □

2.3 Witness Indistinguishability

Witness indistinguishable proofs were presented in [13]. The motivation was to provide a cryptographic mechanism whose notion of security is similar though weaker than zero-knowledge, it is meaningful and useful for cryptographic protocols, and the security is preserved in an asynchronous composition. A witness indistinguishable proof is a proof for a language in NP such that the prover is using some witness to convince the verifier that the input is in the language, yet, the view of the verifier in case the prover uses witness w_1 or witness w_2 is polynomial time indistinguishable. Thus, the verifier gets no knowledge on which witness was used in the proof. The formal definition follows. For further discussion and motivation the reader is referred to [13].

2.4 Black-Box Simulation

The initial definition of zero-knowledge [17] requires that for any probabilistic polynomial time verifier $\hat{V}$, a simulator $S_{\hat{V}}$ exists that simulates $\hat{V}$'s view. Oren [23] proposes a seemingly stronger, "better behaved" notion of zero-knowledge, known as *black-box* zero-knowledge. The basic idea behind black box zero-knowledge is that instead of having a new simulator $S_{\hat{V}}$ for each possible verifier, we have a single probabilistic polynomial time simulator S that interacts with each possible $\hat{V}$. Furthermore, S is not allowed to examine the internals of $\hat{V}$, but must simply look at $\hat{V}$'s input/output behavior. That is, it can have conversations with $\hat{V}$ and use these conversations to generate a simulation of $\hat{V}$'s view that is computationally indistinguishable from $\hat{V}$'s view of its interaction with P.

For further definitions and motivations the reader is referred to [23]

2.5 Concurrent Zero-Knowledge

Following [10], we consider a setting in which a polynomial time adversary controls many verifiers simultaneously. The adversary $\mathcal{A}$ takes as input a partial conversation transcript of a prover interacting with several verifiers concurrently, where the transcript includes the local times on the prover's clock when each message was sent or received by the prover. The output of $\mathcal{A}$ will be a tuples of the form (V, α, t), indicating that P receives message α from a verifier V at time t on $P's$ local clock. The adversary may either output a new tuple as above, or wait for P to output its next message to one of the verifiers. The time that is written by the adversary in the tuple, must be greater than all times previously used in the system (by messages sent to P or by P). The view of the adversary on input x in such an interaction (including all messages and times, and the verifiers random tapes) is denoted $(P, \mathcal{A})(x)$.

Definition 2.4. *We say that a proof or argument system (P, V) for a language L is (computational)* concurrent zero-knowledge *if there exists a probabilistic polynomial time oracle machine S (the simulator) such that for any probabilistic polynomial time adversary $\mathcal{A}$, the distributions $(P, A)(x)$ and $S^{\mathcal{A}}(x)$ are computational indistinguishable over the strings that belong to the language L.*

In what follows, we will usually refer to the adversary $\mathcal{A}$ as the *adversarial verifier* V^* or just the *verifier* V^*. All these terms mean the same.

In our setting, the simulator will simulate a predetermined time interval which is polynomial in k. We assume that while rewinding the verifier, the simulator may also set its clock to the required rewound time.

2.6 The Complexity Parameters

In this paper, we simplify the discussion by using a single security parameter k. Our proof has (in worst case) $\omega(\log^2 k)$ rounds and it has responsive round complexity $\omega(\log k)$. The zero-knowledge simulation is guaranteed for a polynomial (in k) number of concurrent proofs. Also, the running time of the protocol is polynomial in k. We will measure time by the smallest time units that are relevant in this setting. For example, one may think of the time unit as the minimal time a round in the protocol may take. But we may also use a much smaller time unit: the time of a computer cycle. In any of these time units, it holds that the running time of the protocol is polynomial (in k).

3 Main Result

Our main result is the existence of black-box concurrent zero-knowledge interactive proof for all languages in NP with responsive round complexity m for any m satisfying $m = \omega(\log k)$. We state this explicitly in the following theorem.

Theorem 3.1. *Assume there exist secure two-round commitment schemes with statistical secrecy and secure two-round commitment schemes with statistical binding (such schemes follow from the existence of a family claw-free permutation pairs). Let k be a complexity parameter bounding the size of the input. The verifier is polynomial time in k, and the concurrent proof may contain a polynomial (in k) number of proofs concurrently. Then there exists a black-box concurrent zero-knowledge interactive proof for all languages in NP, with:*

- *responsive round complexity $m(k)$, for any function $m(k)$ satisfying $m(k) = \omega(\log k)$, and*
- *a worst case round complexity of $m(k) \cdot \log k$.*

4 Overview of Protocol and Proof

We start with the protocol in [24,18]. We choose the following parameters for this protocol: the preamble consists of m rounds for $m = \omega(\log k)$ (recall that a round consists of a message sent from the prover to the verifier followed by a response of the verifier). The body of the proof consists of a low error, constant round, auxiliary-input witness-indistinguishable interactive proof for NP in which the prover can be efficient given the witness to the proven assertion. The zero-knowledge protocol of [14] will do.

When a new copy of the protocol is initiated by the verifier, the verifier in the new protocol is associated with a response time which is initially the minimal possible response time, say the time of a computer cycle. When the verifier fails to respond within this time, the time associated with this verifier is doubled and the verifier is notified that it must start again with the doubled time. In this case we say that the verifier has been reset and has gone one level up. This may happen at most $O(\log k)$ times since at some such level the response time becomes greater than the running time of the adversary, or bigger than the time interval that has to be simulated. The verifiers may be viewed as working in levels of responsiveness. Level i contains all verifiers with response time at most $\beta_i = 2^i$ and greater than β_{i-1}. The prover treats each verifier independently in light of its associated response time or level. For each verifier, the prover delays its answer according to its associated delay β in a manner yet to be discussed.

The completeness and soundness of the interactive proof hold as in [24,18]. The worst case number of rounds for this protocol happens when the verifier goes through m steps in each level and then delays its last message and is reset while going up to the next level. This yields $m \cdot O(\log k)$ rounds in the worst case. But since each level takes double the time of the previous level, the overall interaction time is dominated by the time of the highest level interaction, and is $O(m)$.

It remains to prove that the interactive proof is zero-knowledge. The delays imposed by the prover are similar to those suggested in [10]. Thus, simulating all protocols at the same level becomes possible in a way similar to that in [10]. In fact, these verifiers may be viewed as adhering to timing constraints. The delay imposed by the prover are not more than twice β_i for a verifier in level i and thus, do not increase the protocol time too much. It remains to show that rewinding protocols at higher levels do not force too many rewinding at lower levels. This is obtained with some care in the setting of the prover delays and by the fact that there are at most a logarithmic number of levels.

The reason we need a logarithmic number of rounds and cannot do with a constant number of rounds for each level as in [10] is the relation between the various levels. We allow ourselves one rewind only to any interval we wish to rewind. Any other constant will do, but rewinding a super-constant number of times (or polynomial as in [10]) will make the overall simulation time super-polynomial. Note also that this is an inherent problem since the lower bound in [5] uses verifiers that in each specific copy of the proof do not modify their response time. Thus the lower bound holds also for responsive round complexity and we cannot do with asymptotically less than $\log k/\log\log(k)$ responsive round complexity.

5 The Zero-Knowledge Protocol

We start by presenting the protocol. It consists of a preamble of m rounds where m is any function satisfying $m = \omega(\log k)$ and a body consisting of a (not concurrent) constant round zero-knowledge proof. If this were the full picture, we would get that the overall number of rounds is dominated by m and is thus almost logarithmic. However, we sometimes let the prover say "RESET". This happens only during the preamble, and is caused by a long delay in the verifier response. When such a delay occurs, the protocol starts from the beginning with a delay parameter doubled. At this point we say the the proof has gone up one *level*. Generally a proof is at level i if it has gone through i resets.

To see that the overall round complexity is $m \cdot O(\log k)$ it is enough to note that the maximum number of resets is logarithmic. This is true since the delay can only be doubled a logarithmic number of times. The logarithm is in the length of the simulated period. We denote this length by Δ and measure it in units of β_0, i.e., the time of a computer cycle. In Figure 1 we describe the protocol. This is the protocol presented in [24,18] enhanced with time monitoring and possible

Step V-0:	V →P: V Selects m strings, $v_1, \ldots, v_m \in \{0,1\}^n$ uniformly and independently at random, and send Commit $(v_1), \ldots,$ Commit (v_m) to the prover.
Step P-1:	P →V: Send Commit (p_1) exactly T after the Step V-0 message was sent.
Step V-1:	V →P: Reveal v_1.
⋮	
Step P-j:	P →V: If V's message from Step V-$(j-1)$ was received more than T time units after P's message from Step P-$(j-1)$ was sent then goto RESET. Else, send Commit (p_j) exactly $2T$ time units after P's round $(j-1)$ message was sent.
Step V-j:	V →P: Reveal v_j.
⋮	
Step P-m:	V →P: If V's message from Step V-$(m-1)$ was received more than T time units after P's message from Step P-$(m-1)$ was sent then goto RESET. Else, send Commit (p_j) exactly $2T$ time units after P's round $(j-1)$.
Step V-m:	V →P: Reveal v_m.
Proof body:	P waits T time units and then proves to V in zero-knowledge that $x \in L$ or that $\exists i$, $1 \leq i \leq m$, such that $p_i = v_i$. (No delays or time monitoring is used during the course of this proof.)
End of proof	
RESET:	
P →V:	A reset message with parameter $2T$. Both P and V continue by setting $T = 2T$ and starting the protocol from Step V-0 again.

Fig. 1. The protocol

resets. All commitments from the verifier to the prover are statistically secret and all commitments from the prover to the verifier are statistically binding.

Theorem 5.1. *If the zero-knowledge proof used in the body of the protocol has completeness error ϵ_c and soundness error ϵ_s then our interactive proof as in Fig. 1 has completeness error ϵ_c and soundness error at most $\varepsilon_s + \varepsilon$ for some negligible fraction ε.*

Proof. Clearly, the completeness error cannot increase. As for the soundness, the prover may gain extra strength by managing to set $p_i = v_i$ for some $1 \leq i \leq m$. However, since the verifier is using statistically hiding commitment scheme this may happen with negligible probability only, and we are done. □

Lemma 5.2. *The protocol has responsive round complexity* $5m$.

Proof. We show that it holds for the preamble. The additional constant number of rounds in the body of the proof cannot increase the responsive round complexity since the prover answers with no delays at that stage of the protocol.

Consider a proof that ended the preamble at level ℓ, i.e., had ℓ resets. (We will discuss later the case that the preamble has not ended at all within the time

Δ.) If $\ell = 0$ then the proof had m rounds, the length of each equals the minimum possible response time, that was actually matched by the verifier. Otherwise, we have $\ell \neq 0$. The proof was last reset at level $\ell - 1$ which means that the verifier did not respond within time $\beta_{\ell-1} = 2^{\ell-1} = \frac{1}{2}\beta_\ell$. Thus the response-time of the verifier is at least $\beta_{\ell-1}$. We now compute the overall communication time and show that it is smaller than $5m \cdot \beta_{\ell-1}$.

At each of the levels $i = 1, 2, \ldots, \ell - 1$ the protocol ran for at most m rounds. At level ℓ we assume it finished the preamble and thus had m rounds. Summing over all the communication times during the preamble we get that the communication time is bounded by

$$\sum_{i=1}^{\ell} m \cdot \beta_i = m \cdot \sum_{i=1}^{\ell} 2^i \leq m \cdot 2^{i+1} = 4 \cdot m \cdot \beta_{\ell-1} .$$

We bound the additional communication time of the proof body by $m\beta_{\ell-1}$. This is correct for the constant round body if the verifier does not pose a delay longer than $\beta_{\ell-1}$; if it does, the responsive round complexity may only decrease.

Last, we deal with the case that the verifier does not finish. We assume that the simulation time Δ is much larger than the running time of the adversarial verifier. Thus, a particular verifier that has not yet responded will never respond and its responsive round complexity is much better than $5m$. □

We next show that the protocol is concurrent zero-knowledge, by presenting a simulator for the concurrent interaction.

6 The Simulator

We present a black box simulation of the above protocol. We assume the worst, i.e., that there is one adversary that controls all verifiers (whose number is polynomial in k). This adversary deviates from the protocol as it wishes and is limited only by being a polynomial time machine. The simulator interacts with this adversary (or with these verifiers) and its goal is to produce a transcript distribution which is indistinguishable from the real interaction between the adversary and the original prover P. Note that each message in the transcript is associated with a time telling when it is produced after the beginning of the interaction.

The simulator simulates the body of the proof simply by playing the real prover. The reason it may do that is that it rewinds each of the verifiers so that it manages to get a round i in which $p_i = v_i$. After that we say that this particular copy of the proof has been "solved", or that this particular verifier has been *neutralized.* Our goal is to ensure that there will be enough rewinding so that all proofs will be solved, while taking care that the rewinding does not exceed polynomial time.

The difficulty in the construction and in describing the simulator lies in the rewinding schedule. Other than that the operation of the simulator is quite simple. The simulator runs the adversary on a randomly chosen random string while performing all rewinds in the rewind schedule. The simulator breaks the entire sequence of time steps into sections. Each section is simulated twice by the simulator. The first run is used to obtain information, and the second run is used to produce the actual output transcript. During simulation of each such section, the simulator recursively divides it into smaller subsections.

During the first time a section is simulated, the simulator records the strings revealed by the verifiers during this run. Then, while running the second run of the rewind, the simulator solves all proofs that may be solved by setting p_i to equal v_i for known values of v_i's. The second run of the section is used to produce the transcript obtained thus far. When a body of a proof arrives, if the proof has been solved, then the simulator acts as the prover while proving the existence of i such that $p_i = v_i$ (the simulator has a witness to this fact). If the proof has not been solved, the simulation aborts and declares failure.

We will show that the probability that any of the proofs remains unsolved is negligible. Thus, the simulator rarely fails. When it does not fail, its output will be indistinguishable from the real interaction. One difference between the simulated transcripts and the real ones is in the preambles: in the simulation there is an i with $p_i = v_i$. But by the secrecy of the commitment schemes this difference cannot be detected by a polynomial-time bounded machine. Note that these strings are never revealed, avoiding difficulties arising when partial subsets are revealed. The second difference is in the witness used in the bodies of the proofs. However, a zero-knowledge proof is witness indistinguishable. This property is preserved in a concurrent setting and is thus indistinguishable by a polynomial time distinguisher.

It remains to show that there exists a rewinding schedule by which the simulator is efficient and still all proofs are solved with overwhelming probability.

6.1 The Rewinding Schedule

The schedule of the rewinds is given as a pseudo-code in Fig. 2 and is illustrated in Fig. 3.

The X axis represents the time (as viewed by the verifiers or listed in the output transcript produced by the simulator), and the numbers in the graph represent the X coordinate (=the time) of an event. The Y axis represents the order of events of the simulator itself. The advances of the simulation are shown as thick arrows, whereas the rewinds are shown as thin backward arrow.

In the example of Fig. 3 the top-level run has exactly two recursive sections. At the top level this is not always the case, but in any other level the recursion is invoked exactly twice. The top level of this run is $\log \Delta$ (all logarithms are base 2), where Δ is the length of the interval we simulate. In the example $\Delta = 4$ and the top level is 2. The first section starts at the beginning and ends when the simulator advances from 1 to 2 after the after the seventh rewind (the second $(1 \leftarrow 3)$). The second section begins in the advancement from 2 to 4 and ends at the end of the simulation.

During the run of the top-level sections there are also rewinds of lower levels. In this example there is only one lower level: level 1. For each rewind of level

```
// Recall that Δ is the overall simulation time interval.
top_level = log(Δ)

// This is a recursive algorithm. Top-level call follows:
simul(0, Δ, top_level)
output transcript

// Definition of recursive function:
simul(location, length, level)
  β = 2^level
  if (level < 1)
    // here comes the simulation of interaction with V* for time interval length
    return
  else
    // recursively run lower-level simulations
    for i = 0 to length/β - 1
      simul(location + i·β, β,level-1)
      simul(location + (i+1)·β, β,level-1)
      rewind_to_location(location + i·β)
      simul(location + i·β, β, level-1)
    end for
  end if
```

Fig. 2. Description of the Rewinding Schedule

ℓ there are 6 rewinds of level $\ell - 1$ (4 before the actual level-ℓ rewind, and two after it). Thus before the first level-2 rewind ($(0 \leftarrow 4)$) there are 4 level-1 rewinds ($(0 \leftarrow 2)$, $(1 \leftarrow 3)$, $(2 \leftarrow 4)$ and $(3 \leftarrow 5)$) and more two after it ($(0 \leftarrow 2)$ and $(1 \leftarrow 3)$).

Note that the rewinding schedule does not depend on the schedule of proofs as determined by the adversarial verifier. It may be the case that no proof ran and the simulator would still behave the same. The rewinding schedule depends on the time only.[4]

6.2 The Effects of the Rewinds

Each proof may start its i-th level run in an arbitrary point in time. However, by the delay of β_i imposed by the prover, they all have the same look during the preamble: The prover sends a message, then the verifier responds within time β_i and then the prover sends its next message exactly $2\beta_i$ time units after its

[4] We remark that one may obtain better efficiency by checking if the rewind is helpful to the simulation and avoid rewinding it it's not. Even if one does not try to check the messages, a scrutiny of the schedule may lead to other improvements. However all we care about is that the simulation is polynomial time and this is guaranteed by our simple non-optimized simulation procedure.

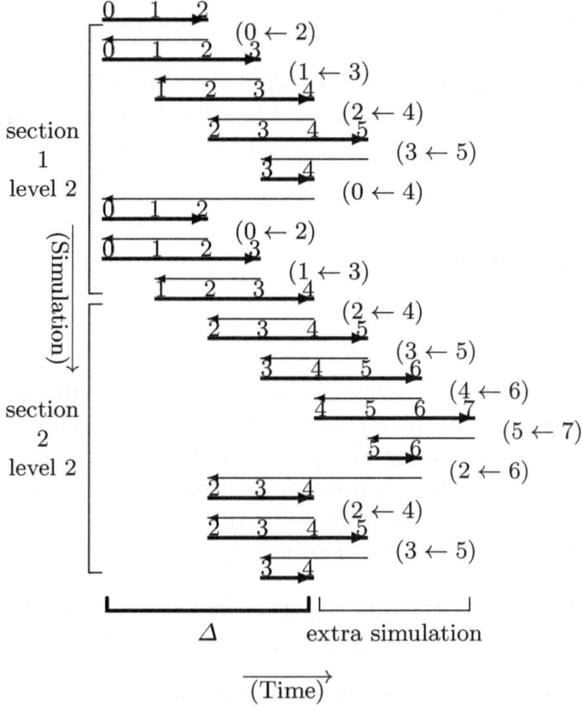

Fig. 3. The rewind schedule of two rounds **of** level 2

previous message. Thus, the time between two prover messages is always $2\beta_i$ and the response time of the verifier is less than β_i.

A *rewind* operation at level i that makes the simulator run the interval $(T, T + 2\beta_i)$ twice is meant to solve all proofs of level i in which the verifier sent a preamble message "`Reveal` v_ℓ" (for some $1 < \ell \leq m$) in response to a prover message that was sent in-between the times T and $T + \beta_i$. Note that in this case the verifier must respond before $T + 2\beta_i$ and thus simulator has learned the value of v_ℓ. Since p_ℓ is still within the rewind. The simulator may modify the commitment on p_ℓ in the second run of the rewind and commit on $p_\ell = v_\ell$. For example, the rewind $(0 \leftarrow 4)$ may solve proofs that run in the top level and whose prover has sent a message between the times 0 and 2.

Had this always worked, we wouldn't need so many preamble rounds. A couple of them would have been enough. However, here is what may go wrong: the verifier may delay its answer in the first run of the rewind interval, thus getting a reset message from the prover (simulator), yet, in the second run, provide an answer in time. In this case, the simulator would not know the value of v_ℓ in the second run since it was not exposed in the first run. The second run is the one that prevails and written to the final transcript. Thus, solving the proof in this round of the preamble fails in this case. In Sect. 7 below, we argue that this happens with constant probability in each round and with negligible probability in all the m rounds of the preamble. Note that setting $p_\ell = v_\ell$ in any one of the rounds suffices to solve the proof.

It remains to analyze the probability that the simulator succeeds in solving each of the proofs (before getting to the proof body) and to verify that the rewinding schedule results in a polynomial time simulator. This analysis is provided in the following section.

7 Analysis of the Simulator

7.1 Efficiency

We start by showing that the rewinding schedule of Section 6.1 results in a polynomial time simulator. Note that each step of the simulation, i.e., committing on strings, revealing them, and playing the prover in the proof-body are all polynomial time. Thus, if the rewinding is polynomial time, we get that the whole algorithm is efficient. We will actually show that the number of rewinds is polynomial. Since each rewind time is polynomial this is enough.

Lemma 7.1. *The overall number of rewinds during the simulation run on time interval Δ is at most Δ^3.*

Proof. We use the recursive description of the rewinding schedule as in Figure 2. Consider a run of a time interval t at level ℓ. Using the notation of Figure 2 this is a run of $simul(location, t, \ell)$. Note that the number of rewinds is independent of the location. Thus, we denote the number of rewinds in this run by $X(t, \ell)$. In a run of t time units at level ℓ there are $\frac{t}{\beta_\ell}$ iterations of the main loop of the *simul* procedure. In each iteration there is a level-ℓ rewind and 3 calls to $simul(\cdot, \beta_\ell, \ell - 1)$ are performed (recall that $\beta_\ell = 2^\ell$). Thus,

$$X(t, \ell) \;\leq\; \frac{t}{\beta_\ell} \cdot (1 + 3 \cdot X(2\beta_{\ell-1}, \ell - 1)) \;\; .$$

This recursion inequality gives the bound: $X(t, \ell) \;\leq\; \frac{t}{\beta_\ell} \cdot 7^{\ell-1}$. At the top level, $t = \Delta, \ell = \log \Delta$, and we obtain

$$\frac{t}{\beta_\ell} \cdot 7^{\ell-1} = \frac{\Delta}{2^{\log \Delta}} \cdot 7^{\log \Delta} = 7^{\log \Delta} \leq \Delta^3 \;\; ,$$

as required. □

7.2 Indistinguishability

We show that no polynomial-time algorithm can distinguish the output of the simulator from V^*'s view of its interaction with the original prover, P.

Assume, first, that the simulator always manages to solve all proofs before getting to the bodies of the proofs. We show later that this assumption holds with overwhelming probability.

We separate the discussion of the preambles and the proof bodies. The difference in the *preambles* is that in the simulation one or more of the rounds has $p_i = v_i$. In the real interaction, this seldom happens. The difference in the bodies is that the simulator always proves that "$\exists i$ such that $p_i = v_i$" whereas the original prover (almost) always proves that $x \in L$.

The prover never opens its commitments on the p_i's. By the secrecy of the commitment scheme, a polynomial time distinguisher cannot tell between preambles generated by the simulators and real preambles. Since this is the case, the adversarial verifier itself cannot distinguish between the first and the second runs of a rewind. We will use this fact to show that the simulator solves all proofs with high probability.

Finally, the *proof bodies* are witness indistinguishable. By [13] this property holds also in the concurrent setting. Thus, an efficient distinguisher cannot tell between using a witness to "$\exists i$ such that $p_i = v_i$" and using a witness to "$x \in L$" and we are done.

It remains to show that the simulator may fail to solve one of the proofs only with negligible probability. We first argue (in Claim 7.1) that for each proof, each round of its preamble that appears in its final level is rewound. We then argue (in Claim 7.2) that a rewind of such a round does not solve the proof with probability at most 1/3. Since all rewinds are rewound independently, and since solving the proof in one of them is enough, and since there are $m = \omega(\log k)$ such rewinds before the body of the proof, we get that the proof remains unsolved with probability $(1/3)^m$, which is negligible. Any of the proofs may be run a polynomial number of times by the simulator (since intervals are rewound) and there are a polynomial number of proofs. By the summation bound, the probability that any of these proofs is not solved by the end of the preamble remains negligible.

Claim 7.1. *If a proof preamble terminates at level ℓ, then each of its m rounds at level ℓ is rewound.*

Proof Sketch: By the delays posed by the prover, each round takes exactly $2\beta_\ell$ time units. For a rewind $(T \leftarrow T + 2\beta_\ell)$ to properly rewind a proof round, both the prover's message and the verifier's message have to be within the interval $(T, T + 2\beta_\ell)$.

By the requirement of the ℓ-th level, the answer of the verifier must arrive within β_ℓ. Thus if the prover has sent $Commit(p_i)$ at the interval $(T, T + \beta_\ell)$, it is guaranteed that the verifier's reply (Reveal v_i) will arrive within the rewind interval, and that the next prover message will be sent after the rewind interval ends. Thus the round will be properly rewound by a rewind $(T \leftarrow T + 2\beta_\ell)$.

It remains to show that any message that the prover sends on level ℓ has an associated level-ℓ rewind. Details are omitted. □

Claim 7.2. *If in a proof Π at level ℓ the prover's message for round i is sent at time T, then a rewind $(T' \leftarrow T' + 2\beta_\ell)$, where $0 \leq T' - T < \beta_\ell$, solves Π in this rewind with probability at least $2/3$.*

Proof. If the verifier answers in time during both runs of the rewind then the proof is solved: the first run reveals the value v_j (for round j of the proof) and in the second run of the rewind the prover may commit on a modified $p_j = v_j$. Note that the verifier cannot modify v_j, except with negligible probability, since it is committed to this value as of the first round of the proof Π. If the verifier does not answer in the second run of the rewind, then it is actually reset into level $\ell + 1$ and this proof does not need to be solved at level ℓ. The only bad case is when the verifier delays its answer in the first run, but does not delay it in the second run of the rewind. In this case, the simulator does not learn the value of v_j in the first run and thus, cannot set $p_j = v_j$ in the second run.

What is the probability of this bad incident? Since the prover commits to p_j, the verifier cannot tell if $p_j = v_j$, it cannot tell between the first and second run, except with negligible probability in which the secrecy of the commitment scheme fails. Suppose the verifier delays it's message beyond β_ℓ with probability p at the first run. It then delays its message with probability at least $p - \varepsilon$ for some negligible fraction ε in the second run. The probability that the simulator does not solve the proof is thus at most $p \cdot (1 - p + \varepsilon) \leq 1/4 + \varepsilon \leq 1/3$, and we are done. □

References

1. Boaz Barak: How to Go Beyond The Black-Box Simulation Barrier. To appear in IEEE, *Proceedings of the 41st Annual Symposium on Foundations of Computer Science*, October, 2001.
2. Brassard, G., Chaum, D., Crépeau, C.: Minimum disclosure proofs of knowledge. JCSS **37** (1988) 156–189
3. Canetti, R., Goldreich, O., Goldwasser, S., Micali, S.: Resettable zero-knowledge. Record 99-22, Theory of Cryptography Library (1999) received October 25th, 1999. Supercedes Theory of Cryptography Library Record 99-15.
4. Canetti, R., Goldreich, O., Goldwasser, S., Micali, S.: Resettable zero-knowledge (extended abstract). In ACM, ed.: Proceedings of the thirty second annual ACM Symposium on Theory of Computing: Portland, Oregon, May 21–23, [2000], New York, NY, USA, ACM Press (2000) 235–244 see also [3].
5. Canetti, R., Kilian, J., Petrank, E., Rosen, A.: Concurrent zero-knowledge requires $\tilde{\Omega}(\log n)$ rounds. In: Proceedings of the thirty third annual ACM Symposium on Theory of Computing, ACM Press (2001)
6. Crescenzo, G.D., Ostrovsky, R.: On concurrent zero-knowledge with pre-processing. In Wiener, M., ed.: Advances in Cryptology – CRYPTO ' 99. Lecture Notes in Computer Science, International Association for Cryptologic Research, Springer-Verlag, Berlin Germany (1999) 485–502
7. Damgård, I.B.: Efficient concurrent zero-knowledge in the auxiliary string model. In Preneel, B., ed.: Advances in Cryptology – EUROCRYPT ' 2000. Lecture Notes in Computer Science, Brugge, Belgium, Springer-Verlag, Berlin Germany (2000) 418–430
8. Damgård, Pedersen, T.P., Pfitzmann, B.: On the existence of statistically hiding bit commitment schemes and fail-stop signatures. In Stinson, D.R., ed.: Proc. CRYPTO 93, Springer (1994) 250–265 Lecture Notes in Computer Science No. 773.

9. Dwork, C., Naor, M.: Zaps and their applications. In IEEE, ed.: Proceedings of the 41st Annual Symposium on Foundations of Computer Science: proceedings: 12–14 November, 2000, Redondo Beach, California, IEEE Computer Society Press (2000) 283–293
10. Dwork, C., Naor, M., Sahai, A.: Concurrent zero-knowledge. In ACM, ed.: Proceedings of the thirtieth annual ACM Symposium on Theory of Computing: Dallas, Texas, May 23–26, 1998, New York, NY, USA, ACM Press (1998) 409–418
11. Dwork, C., Sahai, A.: Concurrent zero-knowledge: Reducing the need for timing constraints. Lecture Notes in Computer Science **1462** (1998) 442–457
12. Feige, U.: Alternative models for zero knowledge interactive proofs. PhD thesis, Weizmann Institute of science (1990)
13. Feige, U., Shamir, A.: Witness indistinguishable and witness hiding protocols. In ACM, ed.: Proceedings of the twenty-second annual ACM Symposium on Theory of Computing, Baltimore, Maryland, May 14–16, 1990, New York, NY, USA, ACM Press (1990) 416–426
14. Goldreich, O., Kahan, A.: How to construct constant-round zero-knowledge proof systems for NP. Journal of Cryptology: the journal of the International Association for Cryptologic Research **9** (1996) 167–189
15. Goldreich, O., Krawczyk, H.: On the composition of Zero-Knowledge Proof systems. SICOMP **25** (1996) 169–192
16. Goldwasser, S., Micali, S., Rackoff, C.: The knowledge complexity of interactive systems. SIAM Journal of Computing **18** (1989) 186–208
17. Goldwasser, S., Micali, S., Rackoff, C.: The knowledge complexity of interactive proof-systems. In: ACM Symposium on Theory of Computing (STOC '85), Baltimore, USA, ACM Press (1985) 291–304
18. Kilian, J., Petrank, E.: Concurrent zero-knowledge in poly-logarithmic rounds. In: Proceedings of the thirty third annual ACM Symposium on Theory of Computing, ACM Press (2001)
19. Kilian, J., Petrank, E., Rackoff, C.: Lower bounds for zero knowledge on the Internet. In IEEE, ed.: 39th Annual Symposium on Foundations of Computer Science: proceedings: November 8–11, 1998, Palo Alto, California, 1109 Spring Street, Suite 300, Silver Spring, MD 20910, USA, IEEE Computer Society Press (1998) 484–492
20. Naor, M.: Bit commitment using pseudorandomness. Journal of Cryptology **4** (1991) 151–158
21. Naor, M., Yung, M.: Universal one-way hash functions and their cryptographic applications. In: 21th Annual Symposium on Theory of Computing (STOC), ACM Press (1988) 33–43
22. Goldreich, O.: Foundation of cryptography — fragments of a book. Available from the *Electronic Colloquium on Computational Complexity (ECCC)* `http://www.eccc.uni-trier.de/eccc/`, February 1995. (1995)
23. Oren, Y.: On the cunning powers of cheating verifiers: Some observations about zero knowledge proofs. In Chandra, A.K., ed.: Proceedings of the 28th Annual Symposium on Foundations of Computer Science, Los Angeles, CA, IEEE Computer Society Press (1987) 462–471
24. Richardson, R., Kilian, J.: On the concurrent composition of zero-knowledge proofs. Lecture Notes in Computer Science **1592** (1999) 415–431
25. Rosen, A.: A note on the round-complexity of concurrent zero-knowledge. In: CRYPTO: Proceedings of Crypto. (2000)

Practical Construction and Analysis of Pseudo-Randomness Primitives

Johan Håstad*[1] and Mats Näslund[2]

[1] NADA
Royal Institute of Technology
SE-10044 Stockholm, Sweden
johanh@nada.kth.se
[2] Communications Security Lab
Ericsson Research
SE-16480 Stockholm, Sweden
mats.naslund@era.ericsson.se

Abstract. We give a careful, fixed-size parameter analysis of a standard [1,4] way to form a pseudorandom generator by iterating a one-way function and then pseudo-random functions from said generator, [3]. We improve known bounds also asymptotically when many bits are output each iteration and we find all auxiliary parameters efficiently. The analysis is effective even for security parameters of sizes supported by typical block ciphers and hash functions. This enables us to construct very practical pseudorandom generators with strong properties based on plausible assumptions.

1 Introduction

One of the most fundamental cryptographic primitives is the *pseudo random generator*, a deterministic algorithm that expands a few truly random bits to long "random looking" strings. Having such implies (among other things) *semantically secure* crypto systems, [5], secure key-generation for asymmetric cryptography etc.

A sound theory of pseudo randomness did not emerge until the seminal works of Blum and Micali, [1], and Yao, [15]. Therefore, constructions in the early 80's were still "ad-hoc", and many of them later turned out to be completely insecure. In a theoretical sense the area was closed when, in [6], it was shown that necessary and sufficient conditions for the existence of a pseudo-random generator is the existence of another fundamental primitive: the *one-way function;* a function easy to compute, but hard to invert. We do not know if such functions exist, but many strong candidates exist, such as a good block cipher (mapping keys to cipher-texts, keeping the plaintext fixed), hash functions, etc. Still, the construction in [6] is complex, requiring key-sizes of millions of bits to give reasonable security guarantees, and an "ad-hoc" approach is still therefore

* Work partially supported by the Göran Gustafsson foundation and NSF grant CCR-9987077.

C. Boyd (Ed.): ASIACRYPT 2001, LNCS 2248, pp. 442–459, 2001.
© Springer-Verlag Berlin Heidelberg 2001

often used in practice. Thus, a construction with provable properties, useful in practice is highly desirable.

The reason for the ineffectiveness of the theoretical constructions is that one-wayness is in itself not a strong property. A function may be hard to invert but still have very undesirable properties. For instance, even if f is one-way, most of x may still be easily deduced from $f(x)$. Paradigms for generator construction typically iterate f, and one-wayness may be lost in this process, etc. Thus, basing pseudo-randomness on one-wayness alone appears to require elaborate constructions. However, if one assumes only a little more than one-wayness, e.g. that the function f is also a permutation, the situation becomes much more favorable and reasonably practical constructions can be found from the work of Blum and Micali mentioned above, and later work by Goldreich and Levin [4]. In [1] it is shown that if f is a permutation and has at least a single bit of information, $b(x)$, that does not leak via $f(x)$, then a pseudo-random generator can be built. In [4], then, it is shown that every one-way function, in particular ones being permutations, have such a hard bit $b(x)$. In this paper we make a careful analysis of this transformation from a one-way function to a pseudorandom generator, see Sect. 3. We add new elements of the analysis when we output $m > 1$ bits for each iteration of f, significantly improving the dependence on m. First, we (non-uniformly) reduce inversion of f to distinguishing the generator from randomness, given some auxiliary parameters. We then give efficient sampling procedures to determine the values of these parameters, giving a uniform inversion algorithm, see Sect. 3.1. Values of the parameters that give almost as strong results as the existential bounds can, for most parameter values, be found in time less than the time needed for successive inversions.

A related primitive are the *pseudo-random functions*; functions that can not be distinguished from random functions on the same domain/range. Goldreich, Goldwasser, and Micali, [3], showed how such could be built from a pseudo random generator. In Sect. 3.2, we apply the same kind of fixed parameter analysis to their construction and use it to further enhance our generator.

Our explicit theorems allow us to construct a generator that is efficient in practice based on the assumption that e.g. Rijndael (mapping keys to ciphertexts, fixing a plaintext) remains hard to invert even when iterated, see Sect. 4.

2 Preliminaries

2.1 Notation

The length of binary string x is denoted $|x|$, and by $\{0,1\}^n$ we denote the set of x such that $|x| = n$. We write $\mathcal{U}_n$ for the uniform distribution on $\{0,1\}^n$. Except otherwise noted, log refers to logarithm in base 2.

Let $G : \{0,1\}^n \to \{0,1\}^{L(n)}$ and let A be an algorithm with binary output. We say that A is a $(L(n), T(n), \delta(n))$-*distinguisher* for G, if A runs in time $T(n)$ and $|\Pr_{x \in \mathcal{U}_n}[A(G(x)) = 1] - \Pr_{y \in \mathcal{U}_{L(n)}}[A(y) = 1]| \geq \delta(n)$. (We call $\delta(n)$ the *advantage* of A.) If no such A exists, G is called $(L(n), T(n), \delta(n))$-*secure*. Finally, recall that a function $\nu(n)$ is *negligible* if for all c, $\nu(n) \in o(n^{-c})$.

Our model of computation is slightly generous but realistic. We assume that simple operations like arithmetical operations and exclusive-ors on small[1] size integers can be done in unit time.

2.2 Pseudo-Random Generators from One-Way Permutations

Suppose we have a one-way function, that in addition is a permutation. Furthermore, suppose that we have a family of 0/1-functions, $B = \{b_i\}$, $b_i(x) \in \{0,1\}$, which are efficiently computable such that given $f(x)$, $b_i(x)$ is computationally indistinguishable from a random 0/1 coin toss. Note that one-wayness of f is necessary since otherwise $b_i(x)$ can be computed by first inverting f. We then say that B is a (family of) *hard-core functions* for f. The following construction, due to Blum and Micali [1], now shows how to construct a pseudo-random generator (PRG): choose x_0 (the seed), let $x_{i+1} = f(x_i)$, then output $g(x_0) = b_1(x_1), b_2(x_2), \ldots$ as the generator output.

Theorem (Blum-Micali, '84). *Suppose there is an efficient algorithm D that distinguishes (with non-negligible advantage) $g(x)$ from a completely random string. Then, there is an efficient algorithm P and an i such that given $f(x)$, P predicts $b_i(x)$ with non-negligible advantage.*

Due to the iterative construction, f must not loose one-wayness under iteration. This can be guaranteed if f is a permutation, or, heuristically if f is randomly chosen, see Theorem 1. Assumptions along these lines have been proposed by Levin in [8] and were in fact the first conditions to be proved to be both necessary and sufficient for the existence of pseudorandom generators.

This leaves us with one question: which one-way functions (if any) have hard-cores, and if so, what do these hard-cores look like?

2.3 A Hard-Core for Any One-Way Function

A fixed 0/1-function, b, can never be a general hard-core that works for *every* one-way function: given a one-way function f, the one-way function $f'(x) = f(x), b(x)$ provides a counter example. In 1989, Goldreich and Levin [4] proved, by introducing extra randomness, that *any* one-way function can be modified to have hard-cores.[2] Perhaps surprisingly, the hard-cores they found are also extremely simple to describe. If r, x are binary strings of length n, let r_i (and x_i) denote the ith bit of r (and x), fixing an order left-to-right, or right-to-left. Let $B \triangleq \{b_r(x) \mid r \in \{0,1\}^n\}$ where

$$b_r(x) \triangleq \langle r, x\rangle_2 = r_1 \cdot x_1 + r_2 \cdot x_2 + \cdots + r_n \cdot x_n \bmod 2,$$

that is, the inner product mod 2.

[1] We need words of size n where n is size of the input on which we apply our one-way function, e.g. $n = 128$ or 256 for a typical block cipher.

[2] We again stress that this does not automatically imply that a PRG can be built from any one-way function, as the construction by Blum and Micali only works for one-way permutations.

Theorem (Goldreich-Levin, '89). *Suppose there is an efficient algorithm A, that given $f'(x) = f(x), r$ for randomly chosen r, x, distinguishes (with non-negligible advantage) $b_r(x)$ from a completely random bit. Then there exists an efficient algorithm B, that inverts $f(x)$ on random x with non-negligible probability.*

If f is a one-way function, existence of such A would be contradictory.

As established already in [4], a way to improve efficiency in a PRG construction would be to extract more than one bit per iteration of f. It is possible to output as many as $m \in O(\log n)$ (where $n = |x|$) bits, by multiplying the binary vector x by a random $m \times n$ binary matrix, R. Denote the set of all such matrices $\mathcal{M}_m$, and our functions are $\{B_R^m(x) \mid R \in \mathcal{M}_m\}$. That is, $B_R^m(x) \triangleq R \cdot x \bmod 2$. The above thus leads to a general construction, given any one-way function.

3 The Construction and Its Security

3.1 The Basic PRG

Definition 1. *Let n, and m, L, λ be integers such that $L = \lambda m$ and let $f : \{0,1\}^n \to \{0,1\}^n$. The generator $BMGL^f_{n,m,L}(x, R)$ stretches $n + nm$ bits to L bits as follows. The input is interpreted as $x_0 = x$ and $R \in \mathcal{M}_m$. Let $x_i = f(x_{i-1})$, $i = 1, 2, \ldots, \lambda$ and let the output be $\{B_R^m(x_i)\}_{i=1}^{\lambda}$.*

A proof of the practical security for a concrete f and fixed n, m, requires a very exact analysis, and that analysis is the bulk of this paper. To begin with, we would like to relate the difficulty of inverting an iterated function f to that of distinguishing outputs of $BMGL^f_{n,m,L}$ from random bits. This is is made difficult by the fact that we no longer require f to be a permutation. However, under one additional and natural assumption on the "behavior" of f, we can bring the analysis one step further, relating the security of $BMGL^f_{n,m,L}$ more directly to the difficulty of inverting f itself. Our measure of success is as follows.

Definition 2. *For a function $f : \{0,1\}^n \to \{0,1\}^n$, let $f^{(i)}(x)$ denote f iterated i times, $f^{(i)}(x) \triangleq f(f^{(i-1)}(x))$, $f^{(0)}(x) \triangleq x$.*

Let A be a probabilistic algorithm which takes an input from $\{0,1\}^n$ and has output in the same range. We then say that A is a (T, δ, i)-inverter for f if when given $y = f^{(i)}(x)$ for an x chosen uniformly at random, in time T with probability δ it produces z such that $f(z) = y$.

Note that the number z might be on the form $f^{(i-1)}(x')$ but this is not required. It is interesting to investigate what happens for a random function.

Theorem 1. *Let A be an algorithm that tries to invert a black box function $f : \{0,1\}^n \to \{0,1\}^n$, and makes T calls to the oracle for f. If A is given $y = f^{(i)}(x)$ for a random x, then the probability (over the choice of f and x) that A finds a z such that $f(z) = y$ is bounded by $T(i+1)2^{-n}$. On the other hand, there is an algorithm that using at most T oracle calls outputs a correct z except with probability at most $(1 - (i+1)2^{-n})^{T-i} + i^2 2^{-n}$.*

Proof (sketch). For the lower bound on the required number of oracle calls, consider the process of computing $f^{(i)}(x)$ and let W be the values occuring in this process. If an inverter does not obtain any $w \in W$, there is no correlation between the inverter and the evaluation process. If the inverter makes T calls to the oracle, the probability of obtaining a $w \in W$ is at most $(i+1)T2^{-n}$ and this can be formalized.

To construct an inverter, first assume that the $i+1$ values seen under the evaluation of $f^{(i)}(x)$ are distinct. This happens except with probability (over random f) $\binom{i+1}{2}2^{-n} \leq i^2 2^{-n}$ and if it does not happen we simply give up. Now consider the following inverter. It is given $y = f^{(i)}(x)$. Start by setting $x_0 = 0^n$ and $x_j = f(x_{j-1})$ for $j = 1, 2, \ldots$. Continue this process until either $x_j = y$ (and it is done) or x_j is a value it has seen previously. In the latter case it changes x_j to a random value it has not seen previously and continues. Each value it sees is a random value and if it ever gets one of the $i+1$ values in W, it finds the y within at most i additional evaluations of f. The probability of not finding such a good value in the $T-i$ first steps is at most $(1-(i+1)2^{-n})^{T-i}$. □

Consider for instance the block cipher Rijndael [13] as a one-way function (fixing a message, mapping keys to cipher-texts). It is reasonable to expect that Rijndael is almost as hard to invert as a random function, so that the best achievable time over success ratio to invert it after being iterated i times would be, by the above, not too much smaller than $2^n/i$. The security is now defined as follows.

Definition 3. *A* σ-secure one-way function *is an efficiently computable function $f : \{0,1\}^n \to \{0,1\}^n$, such that the average time over success ratio for inverting the ith iterate is at most $\sigma 2^n/i$. That is, f cannot be (T, δ, i)-inverted for any $T/\delta < \sigma 2^n/i$.*

A block cipher, $f(k,p)$, $|p| = |k| = n$, is called σ-secure if the function $f_p(k)$, for fixed, known plaintext p, is a σ-secure one-way function of the key k.

Hence, for our "practical" choice, f = Rijndael, we expect it to be about 1-secure in the above terminology. Note also that if f is a permutation, only the case $i = 1$ is of interest and we have a standard notion of security.

Security of the Generator. Our objective is to show that if $BMGL^f_{n,m,L}$ is not (L, T, δ)-secure for "practical" values of L, T, δ, then there is also a practical attack on the underlying one-way function f. In particular, we show the following theorem:

Theorem 2. *Suppose that $G = BMGL^f_{n,m,L}$ is based on an n-bit function f, computable by E operations, and that G produces L bits in time S. Suppose that this generator can be (L, T, δ)-distinguished. Then, setting $\delta' = \frac{\delta m}{L}$, there exists integers $i \leq L/m \triangleq \lambda$, $0 \leq j \leq 2\log\delta'^{-1}$, such that for $k = \max(m, 1 + \log((2n+1)\delta'^{-2}) - j)$, f can be $(T', d_j/2, i)$-inverted, where d_j is given by (7) and (8), and T' equals*

$$(1+o(1))2^{m+k}(2m+k+1+T+S+E)(n+1).$$

Values of i and j such that f can be $((8+o(1))T', d_j/16, i)$-inverted can, with probability at least 1/4, be found in time $O(\delta'^{-2}(T+S))$.

The time-success ratio for most ranges of δ and T is worst when the value of j is small. For $j \in O(1)$ and $m, k, E \leq S \leq O(T)$ the ratio is $O(n^2 L^2 \delta^{-2} 2^m T)$. The preprocessing time (to find i, j) is small compared to the running time except in the cases when j is large. In those cases the time to find j is still smaller than the running time of the inverter while the running time to find i might be larger for some choices of the parameters.

A similar result could be obtained from the original works by Blum-Micali and Goldreich-Levin, but we are interested in a tight result and hence we have to be more careful than in [4] were, basically, any polynomial time reduction from inverting f to distinguishing the generator would be enough. Optimizations of the original proof also appeared in [9], but are not stated explicitly.

The proof of Theorem 2 has two main components. We first show (Lemma 1 below) that a distinguisher for BMGL can be turned into a distinguisher for $B_R^m(f^{(i-1)}(x))$, given $R, f^{(i)}(x)$, for some i. Then we show (Theorem 3) how this latter distinguisher is converted to an inverter for $f^{(i)}$.

We thus start with the following lemma.

Lemma 1. *Let $L = \lambda m$. Suppose that $BMGL^f_{n,m,L}$ runs in time $S(L)$. If this generator is not $(L, T(L), \delta)$-secure, then there is an algorithm $P^{(i)}$, $1 \leq i \leq L/m$ that, using $T(L) + S(L)$ operations, given $f^{(i)}(x), R$, for random $x \in \mathcal{U}_n$, $R \in \mathcal{M}_m$, distinguishes $B_R^m(f^{(i-1)}(x))$ from $\mathcal{U}_m$ with advantage $\delta' \triangleq \frac{\delta m}{L}$.*

$P^{(i)}$ depends on an integer i, and using $c_1 \delta'^{-2}(T(L)+S(L))$ operations, where c_1 is the constant given by (5), a value of i achieving advantage $\delta_i \geq \delta'/2$ can be found with probability at least $1/2$.

We conjecture that the time needed to find i is optimal up to the value of the constant c_1. Even if a good value i was found at no cost, the straightforward way by sampling to verify that it actually is as good as claimed would take time $\Omega(\delta'^{-2}(T(L) + S(L)))$. It is not difficult to see that the below proof can be modified to find an i with δ_i arbitrarily close to δ'. The cost is simply an increase in the constant c_1.

Assuming for the moment the following Lemma (a proof is found in the Appendix), we can use it to show Lemma 1.

Lemma 2. *Let F be a function $F : \{0,1\}^n \times \mathcal{M}_m \to (\{0,1\}^m)^\lambda$, computable in time $\leq S$. Let H^i be the distribution on $(\{0,1\}^m)^\lambda$ induced by replacing the first im bits of $F(x,R)$ by random bits.*

Suppose that H^0 $(= F(x,R))$ and H^λ $(= (\mathcal{U}_m)^\lambda)$ are distinguishable with advantage δ, by an algorithm D running in time T. Then, a value of $i < \lambda$ for which H^i, H^{i+1} can be distinguished with advantage $\delta/(2\lambda)$, can with probability at least $\frac{1}{2}$, be found in time $c_1 \delta'^{-2}(T + S)$ where c_1 is an absolute constant.

For the moment, just note that the existence of such an i (and even slightly better advantage) follows directly from the triangle inequality.

Proof. The proof uses the so called *universality of the next-bit-test*, by Yao [15], see also [1].

We assume we know the good value of i as in Lemma 2. Let $F(x, R) = BMGL^f_{n,m,\lambda m}(x, R)$. On input $f^{(i)}(x), R, \gamma$, where γ is either random, or, equal to $B_R^m(f^{(i-1)}(x))$ we do as follows. We easily generate an element according to distribution H^{i+1} as in Lemma 2, with the exception that the $i+1$st m-bit block is assigned the value γ. We feed this value to D and answers as it does. We see that precisely depending on whether γ is random or not, we run D on an input from H^i, or, from H^{i+1} and the lemma follows. □

We now give the theorem of Goldreich and Levin [4] trying to be careful with our estimates and construction. Apart from the value of the constants we have an improvement over previous results in the dependence on the parameter m. While previous constructions would yield a factor proportional to 2^{2m} we decrease this to 2^m. The improvement is due to the fact that we treat the case of general m directly rather than reducing it to the case $m = 1$ (see later discussion).

The second main step towards Theorem 2 is:

Theorem 3. *Fix x. Suppose there is an algorithm, P, using T operations, when given random R distinguishes $B_R^m(x)$ from random strings of length m with advantage at least ϵ where ϵ is given. Then, for $k \triangleq \max(m, \log(\epsilon^{-2}(2n+1)))$, we can in time*

$$(1 + o(1))2^{m+k}(2m + k + 1 + T)(n + 1)$$

produce a list of $2^{k+m}(n + 1)$ values such that the probability that x appears in this list is at least $1/2$.

As we understand, a statement similar (upto a constant), for the special case of $m = 1$, can be derived from [9]. In most application one has $m \leq \log(\epsilon^{-2}(2n+1))$ and thus the latter value of k should be considered standard.

We now collect the last pieces for the proof of Theorem 2 by proving the above Theorem 3 which, in turn, relies on the following prelimnaries.

Lemma 3. *Fix any $x \in \{0,1\}^n$. For $m < k$, from $m + k$ randomly chosen $a_0, \ldots, a_{m-1}$ and $b_0, \ldots, b_{k-1} \in \{0,1\}^n$, it is possible in time $2m2^k + k^2 + m + 4k$ to generate a set of 2^k uniformly distributed, pairwise independent matrices $R^1, \ldots, R^{2^k} \in \mathcal{M}_m$. Furthermore, there is a collection of $m \times (m + k)$ matrices $\{M_j\}_{j=1}^{2^k}$ and a vector $z \in \{0,1\}^{m+k}$ such $B_{R^j}^m(x) = M_j z$ for all j.*

The proof is given in the Appendix. The construction generalizes that of Rackoff for the case $m = 1$, see [2]. If $k < m$, we use $k' = m$ above and then simply only take the first 2^k matrices.

Lemma 4. *Let P be an algorithm, mapping pairs $\mathcal{M}_m \times \{0,1\}^m \to \{0,1\}$, whose running time is T, let R^j, M_j be the matrices generated as described in Lemma 3 and let $S = \{S_j\}_{j=1}^{2^k}$ be an arbitrary matrix set in $\mathcal{M}_m$.*

In time $2^{m+k}(2m+k+T)$ it is possible to compute 2^{m+k} values, $c_1, \ldots, c_{2^{m+k}}$ such that for at least one l we have $c_l = E_j[P(R^j + S_j, B_{R^j}^m(x))]$. The value of l is independent of S.

The role of the set S is explained shortly.

Proof. First run P on all the 2^{m+k} possible inputs of form $(R^j + S_j, r)$ and record the answers: $\{P(R^j + S_j, r)\}$. A fixed value of l above corresponds to a value of the $m + k$ bits z_l in Lemma 3. Let us assume that z_l is the correct choice, i.e. $B_{R^j}^m(x) = M_j z_l$. We define

$$c_l \triangleq 2^{-k} \sum_{j=0}^{2^k-1} P(R^j + S_j, M_j z_l) = 2^{-k} \sum_{j=0}^{2^k-1} \sum_{r=0}^{2^m-1} P(R^j + S_j, r)\Delta(r, M_j z_l), \quad (1)$$

where $\Delta(r, r') = 1$ if $r = r'$ and 0 otherwise. The naive way to calculate this number would require time 2^{2k+m} but we can do better using the Fast Fourier transform. First note that $\Delta(r, r') = 2^{-m} \sum_{\alpha \subseteq [0..m-1]} (-1)^{\langle r \oplus r', \alpha \rangle_2}$. This implies that the sum (1) equals

$$c_l = 2^{-(m+k)} \sum_{j,r,\alpha} P(R^j + S_j, r)(-1)^{\langle r \oplus M_j z_l, \alpha \rangle_2}$$
$$= 2^{-(m+k)} \sum_{j,\alpha} (-1)^{\langle M_j z_l, \alpha \rangle_2} \sum_{r} P(R^j + S_j, r)(-1)^{\langle r, \alpha \rangle_2}.$$

Let $Q(j, \alpha)$ be the inner sum and fix a value of j. Notice that each α-value then correspond to a Fourier transform and hence the 2^m different numbers $Q(j, \alpha)$ can be calculated in time $m2^m$ for this fixed j and hence all the numbers $Q(j, \alpha)$ can be computed in time $m2^{k+m}$. Finally we have

$$c_l = 2^{-(m+k)} \sum_{j,\alpha} (-1)^{\langle M_j z_l, \alpha \rangle_2} Q(j, \alpha) = 2^{-(m+k)} \sum_{j,\alpha} (-1)^{\langle z_l, M_j^T \alpha \rangle_2} Q(j, \alpha),$$

where M_j^T is the transpose. But this is just a rearrangement (induced by M_j^T) of the standard Fourier-transform of size 2^{k+m} and can be computed with $(k + m)2^{k+m}$ operations. The lemma follows. □

We prove now that we can compute useful information about x.

Lemma 5. *Let P, T, x and ϵ be as in Theorem 3. Then for any set of N vectors $\{v_i\}_{i=1}^N \subset \{0,1\}^n$ and any $k \geq m$ we can in time $(1 + o(1))2^{m+k}(2m + k + T + 1)(N + 1)$ produce a set of lists $\{b_i^{(j)}\}_{i=1}^N$, $j = 1, 2, \ldots, 2^{k+m}(N + 1)$ such that with probability $1/2$ we have for at least one j, $\langle x, v_i \rangle_2 = b_i^{(j)}$, except for at most $\frac{N}{\epsilon^2 2^{k-1}}$ of the N possible values of i.*

Proof. Start by randomly generating the 2^k matrices $\{R^j\}$ as shown in Lemma 3. Now repeat the process below for each $i = 1, \ldots, N$. Select 2^k (pairwise) independent random strings $s_j^i \in \{0,1\}^m$, and let S_j^i be the $m \times n$ matrix defined by $S_j^i \triangleq s_j^i \otimes v_i$ (the outer product, i.e. $(S_j^i)_{k,l} = (s_j^i)_k \cdot (v_i)_l$). Notice that by linearity

$$(R^j + S_j^i)x = R^j x + s_j^i \langle v_i, x \rangle_2, \tag{2}$$

which is $B_{R^j}^m(x)$ if $\langle v_i, x \rangle_2 = 0$, and a random string otherwise.

As described in Lemma 4, we now compute the values $\{c_l^i\}$.

$$c_l^i = 2^{-k} \sum_{j=0}^{2^k-1} P(R^j + S_j^i, M_j z_l).$$

Focus on the correct choice for l. If $\langle v_i, x \rangle_2 = 0$, then c_l^i is the average of a uniformly random, pairwise independent sample of the distinguisher P on inputs of the form $\{P(R, B_R^m(x))\}$. On the other hand, if $\langle v_i, x \rangle_2 = 1$, it is a sample of $\{P(R, u)\}$ over random u.

Suppose p_R is the probability that P outputs 1 when the m bits are picked as $B_R^m(x)$ and let p_U be the same probability when the m bits are picked randomly.

Let $p \triangleq (p_R + p_U)/2$. Note that we do not know the value of p. We deal with this problem later, so for the moment suppose we do.

We guess that $\langle v_i, x\rangle_2 = 0$ if $c_l^i \geq p$ and $\langle v_i, x\rangle_2 = 1$ otherwise. The choice is correct unless the average of 2^k pairwise independent Boolean variables is at least $\epsilon/2$ away from its mean. By Chebychev's inequality the probability that this happens is bounded by $2^{-k}\epsilon^{-2}$.

This implies that for the correct values of l and p, the expected number of errors is $2^{-k}\epsilon^{-2}N$, and by Markov's inequality, with probability at least at $1/2$ it is below $2^{1-k}\epsilon^{-2}N$. There are 2^{k+m} possible values of l and once l is fixed the only information on p needed is for which $i \in [1..N]$ we have $c_l^i \geq p$ (if any). Thus, there are only $N+1$ such choices.

The time needed to construct the matrices is negligible, computing the values c_l^i can be done it time $2^{k+m}(2m+k+T)N$, and at most time $2^{k+m}(N+1)$ is needed to output the final lists. □

We finally establish Theorem 3.

Proof (of Theorem 3). Set $k = \max(m, \log(\epsilon^{-2}(2n+1)))$. We apply Lemma 5 with $N = n$, and let $\{v_i\}_{i=1}^n$ be the unit vectors so that $\langle v_i, x\rangle_2$ gives the ith bit of x. With probability $1/2$ one list gives all inner-products correctly and hence determine x. □

We can now use Theorem 3 and Lemma 1 to establish Theorem 2, see the Appendix.

Instead of applying Lemma 5 with the unit vectors we can, as suggested in [2], use it with $\{v_i\}$ describing the words of an error correcting code, e.g. a suitable Goppa-code, [10]. (Similar ideas appears in [8].) If we have code words of length N, containing n information bits, and we are able to efficiently correct e errors we get the following variant of Theorem 3:

Theorem 4. *Fix x. Suppose there is an algorithm, P, that using T operations given R distinguishes $B_R^m(x)$ from random strings of length m with advantage ϵ where ϵ is given. Suppose further we have a linear error correcting code, with n information bits, N message bits that is able to correct e errors in time T_C. Then setting $k = \max(m, \log(\epsilon^{-2}(2N+1)/e))$ we can in time*

$$(1 + o(1))2^{m+k}(2m + k + 1 + T + T_C)(N + 1)$$

produce a list of $2^{k+m}(N+1)$ numbers such that the probability that x appears in this list is at least $1/2$.

Proof. We apply Lemma 5 with the given value of k and $\{v_i\}_{i=1}^N$ given by the row vectors of the generator matrix of the error correcting code. Running the decoding algorithm on each obtained "codeword" gives a list as claimed. □

Similar to Theorem 2, this translates to the quality of the inverter. We only state the resulting algorithm in existential form using O-notation.

Theorem 5. *Suppose we have a linear error correcting code with n information bits, $O(n)$ message bits that is able to correct $\Omega(n)$ errors in time T_C and that $G = BMGL_{n,m,L}^f$ is based on an n-bit function f, computable by E operations, and that G produces L bits in time S. If G can be (L, T, δ)-distinguished then,* *with* $\delta' = \frac{\delta m}{L}$, *there is an* $i \leq L/m \triangleq \lambda$ *and* $0 \leq j \leq 2\log\delta'^{-1}$ *such that for* $k = \max(m, O(1) + 2\log\delta'^{-1} - j)$ *such that* f *can be* $(T', \Omega(2^{-j/2}(j+1)^{-2}), i)$*-inverted where* T' *equals*

$$O(2^{k+m}(k + m + S + T + E + T_C)n).$$

In particular, this implies that the asymptotic time-success ratio decreases by a factor n for the parameters discussed after Theorem 2.

3.2 Applying the GGM Construction

As shown, the BMGL generator can produce any number of output bits. We here investigate an alternative way, inspired by a construction of *pseudo random functions* due to Goldreich, Goldwasser, and Micali, [3]. It has the advantage that we iterate f fewer times and hence the assumption needed for security is weaker.

The construction can be based on any PRG, $G : \{0,1\}^n \to \{0,1\}^{2n}$, though we for concreteness think of $G = G(x, R) = BMGL^f_{n,m,2n}(x, R)$ for some f. For simplicity of notation, we shall exclude R from it, keeping in mind that probabilities should be taken also over the choice of R. First, let us assume that we know in advance how may output bits that are desired. We apply [3] to obtain $2^d n$ output bits (where d is given) from $n(m+1)$-bits.

Definition 4. *Fix* $n, d \in \mathbb{N}$. *Let* $G(x)$ *be a generator, stretching* n *bits to* $2n$ *bits, and let* $G_0(x)$ $(G_1(x))$ *be the first (last)* n *bits of* $G(x)$. *For* $x \in \{0,1\}^n$, $s \in \{0,1\}^d$ *put* $g_x(s) \triangleq G_{s_d}(G_{s_{d-1}}(\cdots G_{s_2}(G_{s_1}(x))\cdots))$, *and define* $GGM^G_{d,n} : \{0,1\}^n \to \{0,1\}^{2^d n}$ *by*

$$GGM^G_{d,n}(x) \triangleq g_x(00\ldots0), g_x(00\ldots1), \cdots, g_x(11\ldots1)$$

(the concatenation of g_x *applied to all d-bit inputs).*

The construction can be pictured as a full binary tree $T = (V, E)$ of depth d. Associate $v \in V$ with its breadth-first order number; the root is 1 and the children of v are $2v, 2v+1$. Given x, the root is first labeled by $\mathcal{L}(1) = x$. For a non-leaf v labeled $\mathcal{L}(v) = y \in \{0,1\}^n$, label its children by $\mathcal{L}(2v) = G_0(y)$, $\mathcal{L}(2v+1) = G_1(y)$, respectively. The output of $GGM^G_{d,n}$ is simply the concatenation of all the "leaves" of the tree.

Notice an advantage of the above method in the case that $G = BMGL^f_{n,m,2n}$. To produce $L = 2^d n$ bits, each application of G iterates f $2n/m$ times instead of $2^d n/m$, which, in light of Theorem 1, retains more of the one-wayness of f.

Lemma 6. *Suppose that* D_1 *is a* $(2^d n, T, \delta)$*-distinguisher for* $GGM^G_{d,n}(x)$ *where* G *can be computed in time* S. *Then, there is an integer* $i \leq 2^d$ *and algorithm* D^i *that is an* $(2n, T + 2^d S, 2^{-d}\delta)$*-distinguisher for* G.

D^i *depends on* i, *and a value of* i *achieving advantage* $\delta_i \geq 2^{-(d+1)}\delta$ *can be found with probability at least* $1/2$ *in time* $c_1 2^{2d}\delta^{-2}(T + 2^d S)$ *where* c_1 *is the constant given by (5).*

Proof (sketch). Consider the binary tree T, describing a computation of $GGM^G_{d,n}$ as above. The tree has depth d, 2^d-1 internal vertices and 2^d leaves. We construct hybrid distributions $H^0, \ldots, H^{2^d-1}$ on the vertex-labels of such trees. Again, associate each $v \in V$ by its breadth-first order number. Then, H^i is defined by a simulation algorithm, $GGM^i(x)$, which on input x, assigns labels as follows. Assign the root, $v = 1$, the label x. For $v \in V$, $v = 1, 2, \ldots, i$, label v's children by letting $\mathcal{L}(2v), \mathcal{L}(2v+1)$ be independent, random n-bit strings. Then, for $v = i+1, \ldots, 2^d - 1$: $\mathcal{L}(2v) = G_0(\mathcal{L}(v))$, $\mathcal{L}(2v+1) = G_1(\mathcal{L}(v))$. Finally return the labels of the leaves in T.

Observe that H^{2^d-1} gives the uniform distribution over the node labels (in particular, over the leaves) and H^0 labels the vertices exactly as $GGM^G_{n,d}$ does on a random seed x. Since D_1 distinguishes $GGM^G_{d,n}(x)$ from random $2^d n$-bit strings with advantage δ, for some $i \leq 2^d$, it must be the case that D_1 distinguishes H^i, H^{i+1} with advantage at least $2^{-d}\delta$.

Finding i is now done in complete analogy with Lemma 2, letting the function F there correspond to the node labeling.

We now construct D^i: when D^i gets input $\gamma \in \{0,1\}^{2n}$, it selects random x and feeds D_1 a value y, computed as $GGM^{i+1}(x)$ with the following exception: $i+1$ is not assigned any label[3], and the children of $i+1$ are assigned the left/right n-bit half of γ respectively. It is not too hard to see that if γ is random, we give D_1 a value according to exactly the same distribution as H^{i+1}, whereas if $\gamma = G(x')$, D_1 is given a value from the same distribution as $GGM^i(x)$, i.e. H^i. Thus, by returning D_1's answer to y, D^i's advantage equals that of D_1. □

Unknown Output Length. If the length of the "stream" is unknown beforehand, we let the basic generator G expand n bits to $3n$ bits. Apply the tree-construction as above, labeling left/right children by the first, respectively second n-bit substring of G's output. The remaining n bits are used to produce an output at each vertex as we traverse the tree breadth-first. The analysis is analogous. To save memory, the traversal can be implemented in iterative depth-first fashion.

3.3 Concrete Examples

What does all this say? Suppose that we base the construction on Rijndael$(x) \triangleq$ Rijndael$_x(p)$ (for a fixed plaintext p) and that we want to generate $L = 2^{30}$ bits, applying our construction with $m = 32$ (32 bits per iteration). One choice of parameters gives the following corollary.

Corollary 1. *Consider $G = BMGL^{Rijndael}_{256,32,2^{30}}$ (using key/block length 256) and where Rijndael is computable by E operations, and assume that G runs in time S. If G can be $(2^{30}, T, 2^{-32})$-distinguished, then there is $i < 2^{25}$, and $0 \leq j \leq 114$ such that setting $k = \max(32, 123 - j)$, Rijndael can be (T', d_j, i)-inverted (d_j given by (7) and (8)) for $T' = 2^{41+k}(65 + k + T + S + E)$.*

[3] As the labels of non-leaves are never exposed, one can conceptually think of the process as labeling $i+1$ afterwards.

Similarly, setting $G' = BMGL^{Rijndael}_{256,32,512}$ *and then using* $GGM^{G'}_{22,256}$ *(to generate the same length outputs), the result holds for some* $i < 16$.

This is simply substituting the parameters and noting that the $o(1)$ in Theorem 2 comes from disregarding the time to construct the matrices described in Lemma 3 and for the current choice of parameters using $(1 + o(1))(n + 1) \leq 2^9$ is an overestimate.

Assuming we have a simple statistical test such as Diehard tests, [11], or those by Knuth, [7], it is reasonable to assume[4] that $65 + k + T + E \leq S$. From the first part of the corollary, then, the essential part of computing the generator comes from the 2^{25} computations of Rijndael and we end up with a time for the inverter equivalent to at most 2^{67+k} Rijndael computations. The maximum of $2^k(d_j/2)^{-1}$ is obtained for $j = 5$ in which case it equals $2^{124} \cdot 7.5 \leq 2^{127}$. We conclude that in this case we get a time-success ratio that is equivalent to at most 2^{194} computations of Rijndael and since $i \leq 2^{25}$, Rijndael would not be 2^{-37}-secure.

Alternatively, bootstrapping the BMGL construction by the GGM method, we conclude from the second part of the corollary that such a test would mean that Rijndael cannot be even 2^{-57}-secure. Thus, though somewhat more cumbersome to implement, the GGM method is more security preserving.

If we want to find the values of i and j efficiently the ratio increases by a factor 2^6. Note that for the case with small j the time needed to find i and j is much smaller than the running time of the inverter.

4 Discussion

4.1 Choice of f

To implement the generator in practice, we suggest to base the one-way function on Rijndael. First of all it is widely believed to be secure and has shown to be very efficient. (A trial implementation of BMGL gives speeds in the range 2 – 10Mb/s on a standard PC, depending on choice of m.) Secondly, as our construction requires that the block size of the cipher is equal to the key size, the fact that Rijndael supports both 128 and 256-bit block size is advantageous, as it makes it possible to vary the security parameter (key size).

Again note that the one-way function we suggest to use is to fix a message, p, let the input be the encryption key, x, and the output the cipher-text. To obtain a permutation and at the same time increased speed, it might appear to be better to have the mapping from clear-text to crypto-text and iterate $f_x(p)$ rather than $f_p(x)$. The problem is that this is by definition *not* a one-way function: anybody that can compute it can invert it. A possibility is also to use an efficient cryptographic hash function as f.

4.2 Decreasing Seed Size

The impact on security of varying m is clearly visible in the above theorems. Though increasing speed, a practical problem with a large m is the seed size; nm

[4] Common "practical" tests are almost always much faster than the generator tested.

bits specifies a matrix R. First note though, that the security does not depend on the fact that R is secret; only that it is random.

It is possible to decrease the number of bits to only n by instead of binary matrix multiplication, performing a multiplication by a random element in the finite field $\mathbb{F}_{2^n}$, and selecting any fixed set of m bits of this, see [12]. A drawback of this construction is that instead of the direct reduction from a distinguisher for $B_R^m(x)$ to a predictor for $\langle v_i, x\rangle_2$ (Lemma 5), the restricted sample-space of elements makes us need to use the so called *Computational XOR-Lemma*, [14]. Unfortunately, this reduces the initial δ-advantage of the distinguisher to a $2^{-m}\delta$-advantage for the predictor for $\langle v_i, x\rangle_2$, and when the smoke clear we lose a factor 2^m in the running time of the inverter.

An alternative, suffering the same security drawback, is to pick R as a random Toeplitz matrix, specified by $n + m - 1$ bits, [4].

5 Summary and Conclusions

We have given a careful security analysis of a very natural pseudorandom generator. Apart from optimizing known constructions and analysis we have introduced a new analysis method when several bits are output for each iteration of the one-way function.

Another common method to derive PRGs from a block cipher is to run it in *counter mode*. Though addmitedly simpler, the proof of such constructions relies on the assumption that the core, f, is a pseudo-random function. The strictly weaker type of security assumption we have proposed (a function being one-way on its iterates), although it has been proposed before by Levin, is for the first time made in a quantitative sense and we believe that this concept will be useful for future study of one-way functions.

Acknowledgment. We thank Bernd Meyer, Gustav Hast, and anonymous reviewers of different versions of this paper for helpful comments.

References

1. M. Blum and S. Micali: *How to Generate Cryptographically Strong Sequences of Pseudo-random Bits.* SIAM Journal on Computing, **13**(4), 850–864, 1984.
2. O. Goldreich: *Modern Cryptography, Probabilistic Proofs and Pseudo-randomness.* Springer-Verlag, 1999.
3. O. Goldreich, S. Goldwasser and S. Micali: *How to Construct Random Functions.* J. ACM, **33**(4), 792–807, 1986.
4. O. Goldreich and L. A. Levin: *A Hard Core Predicate for any One Way Function.* Proceedings, 21st ACM STOC, 1989, pp. 25–32.
5. S. Goldwasser and S. Micali: *Probabilistic encryption.* J. Comput. Syst. Sci., **28**(2), 270–299, 1984.
6. J. Håstad, R. Impagliazzo, L. A. Levin, and M. Luby: *Pseudo Random Number Generators from any One-way Function.* SIAM Journal on Computing, **28**, 1364–1396, 1999.

7. D. Knuth: *Seminumerical algorithms,* (2 ed.), Volume 2 of *The art of computer programming,* Addison-Wesley, 1982.
8. L. Levin: *One-way Functions and Pseudorandom Generators.* Combinatorica **7**, 357–363, 1987.
9. L. Levin: *Randomness and Non-determinism.* J. Symb. Logic, **58**(3), 1102–1103, 1993.
10. F. J. MacWilliams and N. J. A. Sloane: *The Theory of Error Correcting Codes.* North-Holland, 1977.
11. G. Marsaglia: *The Diehard statistical Tests.* http://stat.fsu.edu/~geo/diehard.html
12. M. Näslund: *Universal Hash Functions & Hard-Core Bits.* Proceedings, Eurocrypt '95, LNCS 921, pp. 356–366, Springer Verlag.
13. J. Daemen and V. Rijmen: *AES Proposal: Rijndael.* www.nist.gov/aes/
14. U. V. Vazirani and V. V. Vazirani: *Efficient and Secure Pseudo-Random Number Generation.* Proceedings, 25th IEEE FOCS, 1984, pp. 458–463.
15. A. C. Yao: *Theory and Applications of Trapdoor Functions.* Proceedings, 23rd IEEE FOCS, 1982, pp. 80–91.

A Additional Proofs

Proof (of Lemma 2). Let δ_i be D's advantage on H^i, H^{i+1}. The problem is that even though $\mathrm{E}_i[\delta_i] = \delta/\lambda \triangleq \delta'$, there is a large number of possibilities for the individual δ_i. Basically, these possibilities all lie between the two extreme cases: (1) There are a few large δ_i, while most are close to 0. (2) All δ_i are about the same, but none is very large. Suppose we try random i's. In the first case, we may need to try many i, but it can be done with a rather low sampling accuracy. In the second case, we expect to find a fairly good i rather quickly, but we need a higher precision in the sampling. The idea is therefore to divide the sampling into a number stages, $\{S(j)\}_{j\geq 0}$, each with different sampling accuracy. Stage $S(j)$ chooses some random i-values and samples D on inputs generated from H^i, H^{i+1}. As soon as a sufficiently "good" i is detected, the procedure terminates. Below we quantify the needed accuracy and the criterion for selecting the good i.

For $j \in \{0, 1, \ldots, -2\log\delta'\}$ let a_j be the fraction of i such that $\delta_i \geq 2^{(j-1)/2}\delta'$. By the assumption of the lemma we have

$$a_0 + \sum_{j=1}^{\infty} a_j(2^{(j-1)/2} - 2^{(j-2)/2}) \geq 1 - 2^{-1/2}. \tag{3}$$

Define b_0 to be $\lceil 4(1-2^{-1/2})^{-1}\rceil$ and

$$b_j = \lceil 4(1-2^{-1/2})^{-1}(2^{(j-1)/2} - 2^{(j-2)/2})\rceil = \lceil 2^{(j+3)/2}\rceil,$$

for $j > 0$. The b_j-values, together with a parameter T_j now define the sampling accuracy. Given these values, we determine i as follows.

In stage $S(j)$, $j = -2\log\delta', -2\log\delta' - 1, \ldots, 0$ choose b_j different random values of i and sample H^i and H^{i+1} each $T_j\delta'^{-2}$ times and run D on each of the samples. If the difference in the number of 1-outputs is at least $(2^{(j-1)/2}T_j - \sqrt{T_j/2})\delta'^{-1}$ choose this i and halt. If no i is ever chosen halt with failure. We need to analyze the procedure and determine T_j.

Suppose that at stage j an i is picked such that $\delta_i \geq 2^{(j-1)/2}\delta'$. We claim that the algorithm halts with this i as output with probability at least 1/2. To establish this first consider the following fact, the proof of which we leave to the reader.

Fact. *Let X be a random variable with mean μ and standard deviation σ. Then we have*

$$\Pr[X \leq \mu - \sigma] \leq 1/2.$$

From this, the above claim now follows since the expected difference in the number of 1-outputs when $\delta_i \geq 2^{(j-1)/2}\delta'$ is at least $2^{(j-1)/2}T_j\delta'^{-1}$ and the standard deviation (being the sum of $T_j\delta'^{-2}$ variables each being the difference of two 0/1-valued variables) is at most $\delta'^{-1}\sqrt{T_j/2}$. This implies that the probability that the algorithm halts for an individual iteration during stage j is at least $a_j/2$. The probability that algorithm will fail to output any number is thus bounded by

$$\prod_j (1 - a_j/2)^{b_j} \leq e^{-\sum_j a_j b_j/2} \leq e^{-2},$$

where the last inequality follows from (3) and the definition of b_j.

We must bound the probability that algorithm terminates with an i such that $\delta_i \leq \delta'/2$. Let us analyze the probability that such an i would be output during an individual run of stage j provided that it is chosen as a candidate. The expected difference of the number of 1-outputs in the two experiments is at most $T_j\delta'^{-1}/2$ and we have to estimate the probability that it is at least $(T_j 2^{(j-1)/2} - \sqrt{T_j/2})\delta'^{-1}$. This is, provided

$$T_j(2^{(j-1)/2} - 1/2) - \sqrt{T_j/2} \geq 0, \tag{4}$$

by a simple invocation of Chernoff bounds, at most

$$e^{-\frac{(T_j(2^{(j-1)/2}-1/2)-\sqrt{T_j/2})^2}{2T_j}}.$$

Let us call this probability p_j. The overall probability of ever outputting an i with $\delta_i \leq \delta'/2$ is bounded by

$$\sum_j b_j p_j.$$

We now define T_j to be the smallest number satisfying (4) such that $p_j < 2^{-(j+3)}b_j^{-1}$ and such that $T_j\delta'^{-2}$ is an integer. We get that with this choice the probability of outputting an i with $\delta_i \leq \delta'/2$ is at most 1/4 and hence the probability that we do get a good output is at least $(1 - e^{-2})\frac{3}{4} \geq .64$. The total number of samples of the algorithm is bounded by $c_1\delta'^{-2}$, where

$$c_1 \triangleq 2\sum_j b_j T_j. \tag{5}$$

Note that this sum converges since $T_j \in O(j2^{-j})$ and $b_j \in O(2^{j/2})$. In fact, it can numerically be calculated to be bounded by 5300. Moreover, the sum is completely dominated by the first term which is over 4600, and the sum of all but the first three terms is bounded by 250. Thus, a more careful analysis what to do for small j could lead to considerable improvements in this constant. □

Before we continue let us make some needed definitions. Let $\text{bin}(i)$ be the map that sends the integer i, $0 \leq i < 2^m$ to its binary representation as an m-bit string. In the sequel, we perform some computations in $\mathbb{F}_{2^k}$, the finite field of 2^k elements, represented as $\mathbb{Z}_2[t]/(q(t))$ where $q(t)$ is a polynomial of degree k, irreducible over $\mathbb{Z}_2$. We assume that such q is available to us. If not, it can be found in expected time at most k^4 which is negligible compared to our other running times considered. Viewing $\mathbb{F}_{2^k}$ as a vector space over $\mathbb{F}_2$, for any $\gamma = \sum_{i=0}^{k-1} \gamma_i t^i \in \mathbb{F}_{2^k}$, we let in the natural way $\text{bin}(\gamma)$ denote the vector $(\gamma_0, \ldots, \gamma_{k-1})$ corresponding to γ's representation over the standard polynomial basis. Note also that $\text{bin}(\gamma)$ can be interpreted as a subset of $[0..k-1]$ in the obvious way.

Proof (of Lemma 3). First choose randomly and independently m n-bit strings, $a_0, \ldots, a_{m-1}$ and k strings $b_0, \ldots, b_{k-1}$, each also of length n. The jth matrix, R^j is now defined by $\{a_i\}$, $\{b_l\}$, and an element $\alpha_j \in \mathbb{F}_{2^k}$ as follows. Its ith row, R^j_i, $0 \leq i < m$, is defined by

$$R^j_i \triangleq a_i \oplus \left(\oplus_{l \in \text{bin}(\alpha_j \cdot t^i)} b_l \right),$$

where α_j is the lexicographically jth element of $\mathbb{F}_{2^k}$ (i.e. the lexicographically jth binary string), and the multiplication, $\alpha_j \cdot t^i$, is carried out in $\mathbb{F}_{2^k}$, and $\oplus$ is bitwise addition mod 2.

Clearly the matrices are uniformly distributed, since the a_i are chosen at random. To show pairwise independence it suffices to show that an exclusive-or of any subset of elements from any two matrices is unbiased. Since the columns are independent, it is enough to show that the exclusive-or of any non-empty set of rows from two distinct matrices R^{j_1} and R^{j_2} is unbiased. Take such a set of rows, $S_1 \subset R^{j_1}$, and $S_2 \subset R^{j_2}$. We may actually assume that $S_1 = S_2 = S$, say, since otherwise, the a-vectors makes the result uniformly distributed. In this case the xor can be written as

$$\oplus_{i \in S} \oplus_{l \in \text{bin}((\alpha_{j_1} + \alpha_{j_2}) \cdot t^i)} b_l,$$

but this is the same as

$$\oplus_{l \in \text{bin}((\alpha_{j_1} + \alpha_{j_2}) \cdot (\sum_{i \in S} t^i))} b_l,$$

which is unbiased if, and only if, $\text{bin}((\alpha_{j_1} + \alpha_{j_2}) \cdot (\sum_{i \in S} t^i)) \neq 0$. However, $\sum_{i \in S} t^i \neq 0$, and as $\alpha_{j_1} \neq \alpha_{j_2}$, $\alpha_{j_1} + \alpha_{j_2} \neq 0$ too, so we have two nonzero elements and hence their product is nonzero.

Notice that if we know $\sum_i a_{li} x_i$ and $\sum_i b_{li} x_i$ mod 2 for all a_l, b_l (a total of $m + k$ bits), then by the linearity of the above construction, we also know the matrix-vector products $R^j x$ for all j. To calculate all the matrices we first compute the reduction of t^i for all $i = k+1, \ldots, 2k$ in $GF[2^k]$. Using an iterative procedure this can be done with $3k$ operations on k bit words and since we only care about $k \leq n$ these can be done in unit time. Now generate the vectors a and b in time $m + k$ operations. Then we compute $\oplus_{l \in bin(t^i)} b_l$ for each $i = 0, \ldots, 2k$ using k^2 operations. By using a gray-code construction each row of a matrix can now be generated with two operations and thus the total number of operations is $2m2^k + k^2 + m + 4k$. □

Proof (of Theorem 2). First we apply Lemma 1 to see that there is an i for which we have an algorithm $P^{(i)}$ that when given $f^{(i)}(x)$ runs in time $S(L)+T(L)$ and distinguishes $B_R^m(f^{(i-1)}(x))$ from random bits with advantage at least δ'', where δ'' is $\delta'/2$ or δ' depending on whether we want to find i efficiently, or only show existence (i.e. uniform/non-uniform algorithm). Since δ'' is an average over all x we need to do some work before we can apply Theorem 3.

For each x we have an advantage δ_x. Let a_j be the fraction of x with $\delta_j \geq 2^{(j-1)/2}\delta''$. Since the expected value of δ_x is δ'' we have

$$a_0 + \sum_{j=1}^{\infty} a_j (2^{(j-1)/2} - 2^{(j-2)/2}) \geq 1 - 2^{-1/2}. \tag{6}$$

Now define

$$d_0 \triangleq \frac{1}{2}(1 - 2^{-1/2}) \tag{7}$$

and

$$d_j \triangleq (2j(j+1)2^{(j-1)/2})^{-1} \tag{8}$$

for $j \geq 1$. Since

$$d_0 + \sum_{j=1}^{\infty} d_j (2^{(j-1)/2} - 2^{(j-2)/2}) = 1 - 2^{-1/2}, \tag{9}$$

we must have $a_j \geq d_j$ for some j and this is our choice for j in the existential part. We now apply Theorem 3 with $\epsilon = 2^{(j-1)/2}\delta'$. To eliminate the list we apply f to each element in it to see if it is a correct pre-image in which case it is output. Since whenever $\delta_x \geq \epsilon$ we have a probability $1/2$ of having $f^{(i-1)}(x)$ in the list and hence the probability of being successful for a random x is at least $d_j/2$.

To get a uniform algorithm, we need to sample to find a suitable value of j. Consider the following procedure for parameters d and T_j to be determined.

For $j = -2\log\delta'', -2\log\delta'' - 1, \ldots, 0$ choose $d(j+3)d_j^{-1}$ different random values of x and run $P^{(i)}$, for each x, $T_j\delta''^{-2}$ each on the two distributions given by choosing the m extra bits as $B_R^m(f^{(i-1)}(x))$ or as random bits. If the difference in the number of 1-outputs for the two distributions is at least $(2^{(j-1)/2}T_j - \sqrt{T_j/2})\delta''^{-1}$ for at least $d(j+3)/4$ different values, choose this j and apply the algorithm of Theorem 3 with $\epsilon = 2^{(j-2)/2}\delta'' = 2^{(j-4)/2}\delta'$.

First we analyze the probability that the algorithm outputs j if it ever gets to a stage where $a_j \geq d_j$. For each x chosen, the probability that it will satisfy $\delta_x \geq 2^{(j-1)/2}\delta''$ and yield the desired difference is by the choice of j and Fact A, at least $a_j/2 \geq d_j/2$. Thus, for sufficiently large d, with probability at least $1 - 2^{-(j+3)}$, this desirable distance will be detected $d(j+3)/4$ times and j will be output. Hence, except with this probability the algorithm will produce some output and we have to analyze the probability that a worse j is output at an earlier stage.

We claim that unless $a_{j-1} \geq d_j/8$, the probability of j being output is $2^{-(j+3)}$. Suppose that $a_{j-1} < d_j/8$ and consider an individual execution in stage j. For a suitable choice of T_j we will prove that the probability that we observe a difference greater than $(2^{(j-1)/2}T_j - \sqrt{T_j/2})\delta''^{-1}$ is bounded by $d_j/6$. This is sufficient, for large enough d, to establish the claim.

By assumption $\delta_x \leq 2^{(j-2)/2}\delta''$ except with probability $d_j/8$ and thus we need to prove that given that this inequality is true, the probability to get the desired difference is at most $d_j/24$. By assumption the expected value of the observed difference is $2^{(j-2)/2}T_j\delta''^{-1}$, and by applying Chernoff bounds it is hence sufficient to choose T_j large enough so that

$$e^{-\frac{(T_j(2^{(j-1)/2}-2^{(j-2)/2})-\sqrt{T_j/2})^2}{2T_j}} \leq \frac{d_j}{24}.$$

This can be done with $T_j = O((j+3)2^{-j})$. The expected number of samples computed, given that j_0 is the largest value such that $a_{j_0} \geq d_{j_0}$, is at most

$$\sum_{j=j_0}^{\infty} d(j+3)d_j^{-1}T_j\delta''^{-2} + 2^{-(j_0+3)}\sum_{j=0}^{j_0-1} d(j+3)d_j^{-1}T_j\delta''^{-2},$$

which is $O(j_0^4 2^{-j_0/2}\delta'^{-2})$.

In the case where we efficiently find i and j, the final value of ϵ for which we call upon Theorem 3 is a factor $2^{-3/2}$ smaller than in the existential case, and hence the increase in the running time is increased by a factor $8+o(1)$, where the $o(1)$ comes from the increase in the additive term k. By the above argument the guarantee for the fraction of the inputs for which the procedure has probability at least $1/2$ of finding the inverse image, is at least $1/8$ of that in the existential case. □

Autocorrelation Coefficients and Correlation Immunity of Boolean Functions

Yuriy Tarannikov *, Peter Korolev **, and Anton Botev ***

Mech. & Math. Department
Moscow State University
119899 Moscow, Russia

Abstract. We apply autocorrelation and Walsh coefficients for the investigation of correlation immune and resilient Boolean functions. We prove new lower bound for the absolute indicator of resilient functions that improves significantly (for $m > (n-3)/2$) the bound of Zheng and Zhang [18] on this value. We prove new upper bound for the number of nonlinear variables in high resilient Boolean function. This result supersedes the previous record. We characterize all possible values of resiliency orders for quadratic functions and give a complete description of quadratic Boolean functions that achieve the upper bound on resiliency. We establish new necessary condition that connects the number of variables, the resiliency and the weight of an unbalanced nonconstant correlation immune function and prove that such functions do not exist for $m > 0.75n - 1.25$. For high orders of m this surprising fact supersedes the well-known Bierbrauer–Friedman bound [8], [1] and was not formulated before even as a conjecture. We improve the upper bound of Zheng and Zhang [18] for the nonlinearity of high order correlation immune unbalanced Boolean functions and establish that for high orders of resiliency the maximum possible nonlinearity for unbalanced correlation immune functions is smaller than for balanced.

Keywords: Boolean functions, stream ciphers, correlation immunity, resiliency, nonlinearity, balancedness, Walsh Transform, autocorrelation coefficients, global avalanche characteristics, bounds.

1 Introduction

Different types of ciphers use Boolean functions. So, LFSR based stream ciphers use Boolean functions as a nonlinear combiner or a nonlinear filter, block ciphers use Boolean functions in substitution boxes and so on. Boolean functions used in ciphers must satisfy some specific properties to resist different attacks. One of the most important desired properties of Boolean functions in LFSR based stream ciphers is *correlation immunity* introduced by Siegenthaler [13].

* yutaran@mech.math.msu.su, taran@vertex.inria.msu.ru
** peter-korolev@mtu-net.ru
*** stony_m@mail.ru

C. Boyd (Ed.): ASIACRYPT 2001, LNCS 2248, pp. 460–479, 2001.
© Springer-Verlag Berlin Heidelberg 2001

Another important properties are nonlinearity, algebraic degree and so on. For Boolean functions used in block ciphers the most important properties are nonlinearity and differential (or autocorrelation) characteristics (propagation degree, avalanche criterion, the absolute indicator and so on) based on the autocorrelation coefficients of Boolean functions. Note that in recent research differential characteristics are considered as important for stream ciphers too.

Correlation immunity (or resiliency) is the property important in cryptography not only in stream ciphers. This is an important property if we want that the knowledge of some specified number of input bits does not give a (statistical) information about the output bit. In this respect such functions are considered in [6], [3] and other works.

Many works (see for example [5]) demonstrate that correlation immunity and autocorrelation characteristics are in strong contradiction. Some of results in our paper confirm it. Nevertheless, it appears that autocorrelation coefficients of a Boolean function is a power tool for the investigation of correlation immunity and other properties even without a direct relation to differential characteristics. The results of our paper demonstrate it.

In Section 2 we give preliminary concepts and notions. In Section 3 we prove new lower bound $\Delta_f \geq \left(\frac{2m-n+3}{n+1}\right) 2^n$ for the absolute indicator of resilient functions that improves significantly (for $m > (n-3)/2$) the bound of Zheng and Zhang [18] on this value. In Section 4 we prove that the number of nonlinear variables in n-variable $(n-k)$-resilient Boolean function does not exceed $(k-1)2^{k-2}$. This result supersedes the previous record $n \leq (k-1)4^{k-2}$ of Tarannikov and Kirienko [16]. As a consequence we give the sufficient condition on m and n that the absolute indicator of n-variable m-resilient function is equal to the maximum possible value 2^n. In Section 5 we characterize all possible values of resiliency orders for quadratic functions, i. e. functions with algebraic degree 2 in each variable. In Section 6 we give a complete description of quadratic n-variable m-resilient Boolean functions that achieve the bound $m \leq \frac{n}{2} - 1$. In Section 7 we establish new necessary condition that connects m, n and the weight of an n-variable unbalanced nonconstant mth order correlation immune function and prove that such functions do not exist for $m > 0.75n - 1.25$. For high orders of m this surprising fact supersedes the well-known Bierbrauer–Friedman bound [8], [1] and was not formulated before even as a conjecture. In Section 8 we prove that for $m \geq \frac{1}{2}n + \frac{1}{2}\log_2 n + \frac{1}{2}\log_2\left(\frac{\pi}{2}e^{8/9}\right) - 1$, $n \geq 12$, the nonlinearity of an unbalanced mth order correlation immune function of n variables does not exceed $2^{n-1} - 2^{m+1}$, and for $m \geq \frac{1}{2}n + \frac{3}{2}\log_2 n + \log_2\left(\frac{1}{4} + \frac{1}{n}\right) + \frac{1}{2}\log_2\left(\frac{\pi}{2}e^{8/9}\right) - 2$, $n \geq 24$, this nonlinearity does not exceed $2^{n-1} - 2^{m+2}$. These facts improve significantly correspondent results of Zheng and Zhang [18] and demonstrate that for higher orders of resiliency the maximum possible nonlinearity for balanced functions is greater than for unbalanced.

Along all paper we apply actively autocorrelation and Walsh coefficients for the investigation of correlation immune and resilient Boolean functions. Our new results demonstrate the power of this approach.

2 Preliminary Concepts and Notions

We consider ${F_2}^n$, the vector space of n-tuples of elements from F_2. An n-variable Boolean function is a map from ${F_2}^n$ into F_2. The *weight* of a vector x is the number of ones in x and is denoted by $|x|$. We say that the vector x *precedes* to the vector y and denote it as $x \preceq y$ if $x_i \le y_i$ for each $i = 1, 2, \ldots, n$. The *scalar product* of vectors x and u is defined as $< x, u >= \sum_{i=1}^{n} x_i u_i$.

The *weight* $wt(f)$ of a function f on ${F_2}^n$ is the number of vectors x on ${F_2}^n$ such that $f(x) = 1$. A function f is said to be *balanced* if $wt(f) = wt(f \oplus 1) = 2^{n-1}$. A *subfunction* of the Boolean function f is a function f' obtained by substituting some constants for some variables in f.

It is well known that a function f on ${F_2}^n$ can be uniquely represented by a polynomial on F_2 whose degree in each variable in each term is at most 1. Namely,

$$f(x_1, \ldots, x_n) = \bigoplus_{(a_1,\ldots,a_n) \in {F_2}^n} g(a_1, \ldots, a_n) x_1^{a_1} \ldots x_n^{a_n}$$

where g is also a function on ${F_2}^n$. This polynomial representation of f is called the *algebraic normal form* (briefly, ANF) of the function and each $x_1^{a_1} \ldots x_n^{a_n}$ is called a *term* in ANF of f. The *algebraic degree* of f, denoted by $\deg(f)$, is defined as the number of variables in the longest term of f. The *algebraic degree of variable* x_i in f, denoted by $\deg(f, x_i)$, is the number of variables in the longest term of f that contains x_i. If $\deg(f, x_i) = 1$, we say that f depends on x_i *linearly.* If $\deg(f, x_i) \neq 1$, we say that f depends on x_i *nonlinearly.* A term of length 1 is called a *linear* term. If $\deg(f) \le 1$ then f is called an *affine* function. If f is an affine function and $f(0) = 0$ then f is called a *linear* function.

Definition 1. *We say that the Boolean function f is* quadratic *if an algebraic degree of each variable in f is 2, i. e. if* $\deg(f, x_i) = 2$ *for each* $i = 1, 2, \ldots, n$.

The *Hamming distance* $d(x_1, x_2)$ between two vectors x_1 and x_2 is the number of components where vectors x_1 and x_2 differ. For two Boolean functions f_1 and f_2 on ${F_2}^n$, we define the distance between f_1 and f_2 by $d(f_1, f_2) = \#\{x \in {F_2}^n | f_1(x) \neq f_2(x)\}$. It is easy to see that $d(f_1, f_2) = wt(f_1 \oplus f_2)$. The minimum distance between f and the set of all affine functions is called the *nonlinearity* of f and denoted by $nl(f)$.

Definition 2. *The Walsh Transform of a Boolean function f is an integer-valued function over ${F_2}^n$ that can be defined as*

$$W_f(u) = \sum_{x \in {F_2}^n} (-1)^{f(x) + <u,x>}.$$

Walsh coefficients satisfy *Parseval's equation* $\sum_{u \in {F_2}^n} W_f^2(u) = 2^{2n}$.

Lemma 1. *Let f be an arbitrary Boolean function on ${F_2}^n$. Then*

$$wt(f) = 2^{n-1} - \frac{1}{2}W_f(0).$$

It is well known that $nl(f) = 2^{n-1} - \frac{1}{2}\max_{u \in F_2^n} |W_f(u)|$.

A Boolean function f on ${F_2}^n$ is said to be *correlation immune of order* m, with $1 \leq m \leq n$, if the output of f and any m input variables are statistically independent. This concept was introduced by Siegenthaler [13]. In equivalent non-probabilistic formulation the Boolean function f is called correlation immune of order m if $wt(f') = wt(f)/2^m$ for any its subfunction f' of $n-m$ variables. A balanced mth order correlation immune function is called an m-*resilient* function. In other words the Boolean function f is called m-resilient if $wt(f') = 2^{n-m-1}$ for any its subfunction f' of $n-m$ variables. In [9] a characterization of correlation immune functions by means of Walsh coefficients is given:

Theorem 1. [9] *A Boolean function f on ${F_2}^n$ is correlation-immune of order m if and only if $W_f(u) = 0$ for all vectors $u \in {F_2}^n$ such that $1 \leq |u| \leq m$.*

Theorem 2. [12] *If f is an mth order correlation immune function on ${F_2}^n$, $m \leq n-1$, then $W_f(u) \equiv 0 \pmod{2^{m+1}}$. Moreover, if f is m-resilient, $m \leq n-2$, then $W_f(u) \equiv 0 \pmod{2^{m+2}}$.*

Definition 3. *Let f be a Boolean function on ${F_2}^n$. For each $u \in {F_2}^n$ the autocorrelation coefficient of the function f at the vector u is defined as*

$$\Delta_f(u) = \sum_{x \in {F_2}^n} (-1)^{f(x)+f(x+u)}.$$

Zhang and Zheng [17] proposed the idea of Global Avalanche Characteristics (GAC). One of important indicators of GAC is *the absolute indicator.*

Definition 4. *Let f be a Boolean function on ${F_2}^n$. The absolute indicator of f is defined as*

$$\Delta_f = \max_{x \in {F_2}^n \setminus \{0\}} |\Delta_f(x)|.$$

3 New Lower Bound for the Absolute Indicator of Resilient Functions

In this section we prove new lower bound for the absolute indicator of resilient functions. At first, we establish an important technical formula. Note that this formula can be deduced from the relation $W_f^2(x) = \sum_{u \in {F_2}^n} (-1)^{<x,u>} \Delta_f(u)$ given in [5] and [4] but we prefer to give a direct proof in the Appendix A.

Theorem 3.

$$\Delta_f(u) = -2^n + 2^{1-n} \sum_{\substack{x \in {F_2}^n \\ <x,u> \equiv 0 \pmod 2}} W_f^2(x).$$

We denote by e_i the vector of the length n that has an one in ith component and zeroes in all other components.

Lemma 2. *Let f be an m-resilient Boolean function on ${F_2}^n$. Then*

$$\Delta_f \geq \left(\frac{2m-n+2}{n}\right) 2^n.$$

Proof. We form the matrix B with n column writing in rows of B each binary vector $u \in {F_2}^n$ exactly $W_f^2(u)$ times. By Parseval's equality the matrix B contains exactly 2^{2n} rows. By Xiao Guo-Zhen–Massey spectral characterization [9] each row of the matrix B contains at most $n-m-1$ zeroes. It follows that the total number of zeroes in B is at most $(n-m-1)2^{2n}$. Therefore, there exists some ith column in B that contains at most $\frac{(n-m-1)2^{2n}}{n}$ zeroes. By construction it follows that $\sum\limits_{\substack{x \in {F_2}^n \\ x_i=0}} W_f^2(x) \leq \frac{(n-m-1)2^{2n}}{n}$. Then by Theorem 3 we have

$$\Delta_f(e_i) = -2^n + 2^{1-n} \sum_{\substack{x \in {F_2}^n \\ x_i=0}} W_f^2(x) \leq -2^n + \frac{(n-m-1)}{n} 2^{n+1} \leq \frac{(n-2m-2)}{n} 2^n.$$

It follows that $\Delta_f \geq \left(\frac{2m-n+2}{n}\right) 2^n$. □

In the next theorem we improve the lower bound of Lemma 2.

Theorem 4. *Let f be an m-resilient Boolean function on ${F_2}^n$. Then $\Delta_f \geq \left(\frac{2m-n+3}{n+1}\right) 2^n$.*

Proof. Suppose that in the proof of Lemma 2 the matrix B contains exactly $h2^{2n}$ rows with less than $n-m-1$ zeroes. Then repeating the arguments from the proof of Lemma 2 we have

$$\Delta_f \geq \left(\frac{2m-n+2+2h}{n}\right) 2^n. \tag{1}$$

At the same time it is not hard to see that

$$\Delta_f(1\dots1) = -2^n + 2^{1-n} \sum_{\substack{x \in {F_2}^n \\ |x| \equiv 0 \pmod 2}} W_f^2(x)$$

and

$$\Delta_f \geq |\Delta_f(1\dots1)| \geq (1-2h)2^n. \tag{2}$$

The right part in (1) is increasing on h whereas the right part in (2) is decreasing on h. The right parts in (1) and (2) are equal when $h = \frac{n-m-1}{n+1}$. Therefore, $\Delta_f \geq \left(\frac{2m-n+3}{n+1}\right) 2^n$. □

In [19] Zheng and Zhang proved that for balanced mth order correlation immune function f on ${F_2}^n$ the bound $\Delta_f \geq \frac{2^n}{2^{n-m}-1}$ holds. It follows that $\Delta_f \geq 2^m + 2$. Our Theorem 4 improves significantly this result for $m > (n-3)/2$.

4 Upper Bound for the Number of Nonlinear Variables in High Order Resilient Functions

In this section we prove the new upper bound for the number of nonlinear variables in high order resilient functions.

The next lemma is well-known.

Lemma 3. [11] *Let f be a Boolean function on ${F_2}^n$, $\deg(f) \geq 1$. Then*

$$2^{n-\deg(f)} \leq wt(f) \leq 2^n - 2^{n-\deg(f)}.$$

The next lemma is obvious.

Lemma 4. *Let f be a Boolean function on ${F_2}^n$, $\deg(f) \geq 1$. Then* $\deg(f(x) \oplus f(x+e_i)) \leq \deg(f(x)) - 1$.

Lemma 5. *Let f be a Boolean function on ${F_2}^n$, $\deg(f, x_i) \geq 2$. Then*

$$\sum_{\substack{u \in {F_2}^n \\ u_i = 0}} W_f^2(u) \geq 2^{2n-\deg(f)+1}.$$

Proof. By Theorem 3 using Lemmas 3 and 4 we have

$$-2^n + 2^{1-n} \sum_{\substack{u \in {F_2}^n \\ u_i = 0}} W_f^2(u) = \Delta_f(e_i) = \sum_{x \in {F_2}^n} (-1)^{f(x)+f(x+e_i)} =$$

$$2^n - 2wt(f(x) \oplus f(x+e_i)) \geq 2^n - 2\left(2^n - 2^{n-(\deg(f)-1)}\right) = -2^n + 2^{n-\deg(f)+2}.$$

It follows that $\sum_{\substack{u \in {F_2}^n \\ u_i = 0}} W_f^2(u) \geq 2^{2n-\deg(f)+1}$. □

Theorem 5. *Let f be an $(m = n-k)$-resilient Boolean function on ${F_2}^n$, $k \geq 2$, and $\deg(f, x_i) \geq 2$ for each $i = 1, \ldots, n$. Then $n \leq (k-1)2^{\deg(f)-1}$.*

Proof. We form the matrix B with n column writing in rows of B each binary vector $u \in {F_2}^n$ exactly $W_f^2(u)$ times. By Parseval's equality the matrix B contains exactly 2^{2n} rows. By Xiao Guo-Zhen–Massey spectral characterization [9] each row of the matrix B contains at most $k-1$ zeroes. It follows that the total number of zeroes in B is at most $(k-1)2^{2n}$. By Lemma 5 each column of B contains at least $2^{2n-\deg(f)+1}$ zeroes. Therefore $n \leq \frac{(k-1)2^{2n}}{2^{2n-\deg(f)+1}} = (k-1)2^{\deg(f)-1}$. □

Theorem 6. *Let f be an $(m = n-k)$-resilient Boolean function on ${F_2}^n$, $k \geq 2$, and $\deg(f, x_i) \geq 2$ for each $i = 1, \ldots, n$. Then $n \leq (k-1)2^{k-2}$.*

Proof. By Siegenthaler's Inequality [13] we have $\deg(f) \leq k-1$. This fact together with Theorem 5 follow the result. □

In [16] it is proved that $n \leq (k-1)4^{k-2}$. Our Theorem 6 improves significantly this result. Note that there exists $(n-k)$-resilient function on ${F_2}^n$, $n = 3 \cdot 2^{k-2} - 2$, that depends nonlinearly on all its n variables (see constructions in [14]).

Corollary 1. *Let f be an m-resilient Boolean function on ${F_2}^n$. If $n \geq (n-m-1)2^{n-m-2}$ then $\Delta_f = 2^n$.*

Proof. If $n > (n-m-1)2^{n-m-2}$ then by Theorem 6 the function f depends on some variable linearly, hence, $\Delta_f = 2^n$. If $n = (n-m-1)2^{n-m-2}$ and f depends on all its variables nonlinearly then according to the proofs of Theorems 5 and 6 we have that each row of the matrix B contains exactly $n-m-1$ zeroes. But in this case $|\Delta_f(1\dots1)| = 2^n$, so, $\Delta_f = 2^n$. □

5 Resiliency Orders of Quadratic Functions

In the next two sections we apply the autocorrelation coefficients for the analysis of quadratic Boolean functions, i. e. functions with algebraic degree 2 in each variable.

Lemma 6. *For any Boolean function g on ${F_2}^{n-1}$ the function $f(x_1, x_2, x_3, \dots, x_n) = g(x_1 \oplus x_2, x_3, \dots, x_n) \oplus x_1$ is balanced.*

Proof. We combine all vector from ${F_2}^n$ into pairs (y', y'') such that y' and y'' differ only in first and second components and coincide in all other components. Then $f(y') = f(y'') \oplus 1$ and $wt(f) = \sum\limits_{(y',y'')} (f(y') + f(y'')) = 2^{n-1}$. □

Lemma 7. *For each function $g(y_1, \dots, y_n)$ on ${F_2}^n$ the function $f(x_1, \dots, x_{2n}) = g(x_1 \oplus x_{n+1}, x_2 \oplus x_{n+2}, \dots, x_n \oplus x_{2n}) \oplus x_1 \oplus x_2 \oplus \dots \oplus x_n$ is $(n-1)$-resilient.*

Proof. Consider an arbitrary subfunction f' obtained from f by substitution of $n-1$ constants for some $n-1$ variables. Then there exists j such that both variables x_j and x_{n+j} remain free. Then f' has the form $f' = g'(\dots, x_j \oplus x_{n+j}, \dots) \oplus x_j$ and by Lemma 6 the function f is balanced. Hence, f is $(n-1)$-resilient. □

Theorem 7. *Quadratic m-resilient functions of n variables exist if and only if $m \leq \frac{n}{2} - 1$.*

Proof. Substitutuing to Theorem 5 the value $\deg(f) = 2$ we have $n \leq 2(n-m-1)$. It follows that $m \leq \frac{n}{2} - 1$. Now suppose that $m \leq \frac{n}{2} - 1$. Consider the function $f(x_1, \dots, x_{2(n-m-1)}) = g(x_1 \oplus x_{n-m}, x_2 \oplus x_{n-m+1}, \dots, x_{n-m-1} \oplus x_{2(n-m-1)}) \oplus x_1 \oplus x_2 \oplus \dots \oplus x_{n-m-1}$ where g is some quadratic function. By Lemma 7 the function f is a $(2(n-m-1))$-variable $(n-m-2)$-resilient quadratic function. It is easy to check that if we substitute some constants for the variables $x_{n+1}, \dots, x_{2(n-m-1)}$ in f then we obtain a desired n-variable m-resilient function. □

6 Complete Description of Quadratic Boolean Functions with Maximum Resiliency Order

In this section we give a complete description of quadratic resilient Boolean functions that achieve the bound $m \leq \frac{n}{2} - 1$. It is obvious that for such functions n is even. Therefore in this section we consider for convenience $(N = 2n)$-variable $(m = n-1)$-resilient functions.

Definition 5. *The notation* $f(x_1,\ldots,x_n) \stackrel{\sigma}{=} g(x_1,\ldots,x_n)$ *means that the Boolean functions* f *and* g *are equal up to permutation of indices of variables.*

Theorem 8. *Let* f *be an* $(N = 2n)$*-variable* $(m = n-1)$*-resilient quadratic function. Then* $W_f(u) \neq 0$ *only if* $|u| = n$.

Proof. By Theorem 1 we have that $W_f(u) \neq 0$ only if $|u| \geq m+1 = n$.

We form the matrix B with N columns writing in rows of B each binary vector $u \in {F_2}^N$ exactly $W_f^2(u)$ times. By Parseval's equality B contains exactly $2^{2N} = 2^{4n}$ rows. Each row has at most n zeroes, therefore the matrix B contains at most $n2^{4n}$ zeroes.

On the other hand, by Lemma 5 each column of the matrix B contains at least $2^{2N-\deg(f)+1} = 2^{4n-1}$ zeroes, i. e. the matrix B contains at least $2n2^{4n-1} = n2^{4n}$ zeroes.

Thus the matrix B contains exactly $n2^{4n}$ zeroes and each row of B has exactly n zeroes and n ones. □

Lemma 8. *Let* $e_{pq} = (0,\ldots,0,\underset{p}{1},0,\ldots,0,\underset{q}{1},0,\ldots,0) \in {F_2}^n$, $p \neq q$, *and* f *is a quadratic function on* ${F_2}^n$. *Then* $\Delta_f(e_{pq}) \in \{0, \pm 2^n\}$ *and the next statements hold:*

$$\begin{aligned}\Delta_f(e_{pq}) = 2^n &\iff f(x) = g(\ldots, x_p \oplus x_q, \ldots),\ g \text{ is quadratic},\\ \Delta_f(e_{pq}) = -2^n &\iff f(x) = g(\ldots, x_p \oplus x_q, \ldots) \oplus x_p,\ g \text{ is quadratic}.\end{aligned}$$

Proof.

We write the function f in the form $f(x) = \bigoplus_{1\leq i<j\leq n} a_{ij}x_ix_j \oplus \bigoplus_{1\leq i\leq n} b_ix_i \oplus c$ where $a_{ij} = a_{ji}$ and $a_{ii} = 0$.

Then $\Delta_f(e_{pq}) = \sum_{x\in {F_2}^n} (-1)^{\bigoplus_{i\neq p,q}(a_{pi}\oplus a_{qi})x_i \oplus a_{pq}(x_p\oplus x_q\oplus 1)\oplus b_p\oplus b_q}$.

If the expression $\bigoplus_{i\neq p,q}(a_{pi}\oplus a_{qi})x_i \oplus a_{pq}(x_p\oplus x_q\oplus 1)\oplus b_p\oplus b_q$ contains at least one linear term x_k then we have $\Delta_f(e_{pq}) = 0$. If this expression does not contain linear terms, it means that $a_{pi} = a_{qi}$ for all i. Then $\Delta_f(e_{pq}) = 2^n(-1)^{b_p\oplus b_q}$ and the function f can be represented in the form

$$f(x) = \bigoplus_{\substack{i<j\\ i,j\neq p,q}} a_{ij}x_ix_j \oplus \bigoplus_{i\neq p,q} b_ix_i \oplus c \oplus \left(\bigoplus_{i\neq p,q} a_{pi}x_i \oplus b_q\right)(x_p\oplus x_q) \oplus (b_p\oplus b_q)x_p,$$

that completes the proof. □

Theorem 9. *Let* $f(x_1,\ldots,x_{2n})$ *be a quadratic function on* ${F_2}^{2n}$. *Then*

$$\sum_{1\leq p<q\leq 2n} \Delta_f(e_{pq}) = -n2^{2n} \text{ if and only if}$$

$$f \stackrel{\sigma}{=} g(x_1\oplus x_{n+1},\ldots,x_n\oplus x_{2n}) \oplus x_1 \oplus \ldots \oplus x_n$$

where $g(y_1,\ldots,y_n)$ *is a quadratic function on* ${F_2}^n$.

Proof. Consider an arbitrary quadratic function f on ${F_2}^{2n}$. At the set of vertices $V = \{1, \ldots, 2n\}$ we construct the graph $G = (V, E)$ by the next rule: $(p, q) \in E$ if and only if $\Delta_f(e_{pq}) \neq 0$.

Each connected component $H^t = (V^t, E^t)$ of this graph is a complete graph since by Lemma 8 we have that $(p, q) \in E^t$ if and only if $a_{pi} = a_{qi}$ for all i.

We divide V^t into two subsets $V_0^t \sqcup V_1^t$ such that $i \in V_{b_i}^t$. Let us denote $v_0^t := |V_0^t|$, $v_1^t := |V_1^t|$.

Then for p and q from the same subset of V^t by Lemma 8 we have $\Delta_f(e_{pq}) = 2^{2n}$ and for p and q from different subsets we have $\Delta_f(e_{pq}) = -2^{2n}$.

Let us estimate the sum

$$\sum_{(p,q)\in E^t} \Delta_f(e_{pq}) = 2^{2n}\left(\frac{v_0^t(v_0^t-1)}{2} + \frac{v_1^t(v_1^t-1)}{2}\right) - 2^{2n}v_0^t v_1^t =$$

$$= 2^{2n-1}((v_0^t - v_1^t)^2 - (v_0^t + v_1^t)) \geq -2^{2n-1}|V^t|.$$

The equality is achieved only for $v_0^t = v_1^t = v^t$.

Hence, $\sum\limits_{1\leq p<q\leq 2n} \Delta_f(e_{pq}) \geq -2^{2n-1}\sum\limits_t |V^t| = -n2^{2n}$, moreover, the equality is achieved only if $v_0^t = v_1^t$ for all t.

Thus, if we have the equality $\sum \Delta_f(e_{pq}) = -n2^{2n}$ then it is possible to divide the set of all variables into the pairs (i_k^t, j_k^t) where $i_k^t \in V_0^t$, $j_k^t \in V_1^t$. Then the function will be represented in the form $f(x_1, \ldots, x_{2n}) = g(x_{i_1^1} \oplus x_{j_1^1}, \ldots, x_{i_{v^1}^1} \oplus x_{j_{v^1}^1}, \ldots) \oplus x_{i_1^1} \oplus \ldots \oplus x_{i_{v^1}^1} \oplus \ldots$, i. e. in desired form.

Now suppose that the function has the form $g(x_1 \oplus x_{n+1}, \ldots, x_n \oplus x_{2n}) \oplus x_1 \oplus \ldots \oplus x_n$. Then after the construction of the graph G and the partitioning it into components, we have $i \in V_{b_i}^t$ and $i+n \in V_{b_i\oplus 1}^t$ for all i, $i \leq n$. It follows that $v_0^t = v_1^t$ for all t. □

Theorem 10. *Let $f(x_1, \ldots, x_{2n})$ be an $(2n)$-variable $(n-1)$-resilient quadratic function. Then there exists a quadratic function $g(y_1, \ldots, y_n)$ such that*

$$f(x_1, \ldots, x_{2n}) \stackrel{\sigma}{=} g(x_1 \oplus x_{n+1}, \ldots, x_n \oplus x_{2n}) \oplus x_1 \oplus \ldots \oplus x_n.$$

Proof. Substitute the equation from Theorem 3 into Theorem 9:

$$\sum_{1\leq p<q\leq 2n} \Delta_f(e_{pq}) = \sum_{p<q}\Bigg(-2^{2n} + 2^{1-2n}\sum_{\substack{x\in {F_2}^{2n}\\ <x,e_{pq}>\equiv 0 \pmod 2}} W_f^2(x)\Bigg) =$$

$$= -n(2n-1)2^{2n} + 2^{1-2n}\sum_{p<q}\sum_{x_p=x_q} W_f^2(x).$$

By Theorem 8 for $|x| \neq n$ we have $W_f(x) = 0$, hence

$$\sum_{p<q}\sum_{x_p=x_q} W_f^2(x) = \sum_{p<q}\sum_{\substack{x_p=x_q\\ |x|=n}} W_f^2(x) = \sum_{|x|=n}\left(W_f^2(x)\sum_{p<q\,:\,x_p=x_q} 1\right) =$$

$$= (n^2-n)\sum_{|x|=n} W_f^2(x) = (n^2-n)2^{4n}.$$

Therefore,

$$\sum_{1\le p<q\le 2n} \Delta_f(e_{pq}) = -n(2n-1)2^{2n} + 2^{1-2n}(n^2-n)2^{4n} = -n2^{2n}.$$

It follows by Theorem 9 that all $(2n)$-variable $(n-1)$-resilient quadratic functions have the given form. □

7 Nonexistence of Unbalanced Nonconstant mth Order Correlation Immune Boolean Functions on ${F_2}^n$ for $m > 0.75n - 1.25$

In this section we prove that unbalanced nonconstant mth order correlation immune Boolean functions on ${F_2}^n$ do not exist for $m > 0.75n - 1.25$. Similar statements are known for multioutputs functions (see [2], [10]) but for usual Boolean functions until now statements of such type were not formulated even as conjectures.

Theorem 11. *Let f be an arbitrary Boolean function on ${F_2}^n$. Let $w \in {F_2}^n \setminus \{0\}$. Then*

$$\sum_{\substack{x\in {F_2}^n\\ <x,w>=0}} W_f^2(x) = 2^{n-|w|} \sum_{\substack{u\in {F_2}^n\\ u\preceq w}} \Delta_f(u).$$

Proof. Summing $\Delta_f(u)$ over all u, $u \preceq w$, by Theorem 3 we have

$$\sum_{\substack{u\in {F_2}^n\\ u\preceq w}} \Delta_f(u) = \sum_{\substack{u\in {F_2}^n\\ u\preceq w}} \left(-2^n + 2^{1-n} \sum_{\substack{x\in {F_2}^n\\ <x,u>\equiv 0 \pmod 2}} W_f^2(x) \right) =$$

$$-2^{n+|w|} + 2^{1-n} \sum_{\substack{u\in {F_2}^n\\ u\preceq w}} \sum_{\substack{x\in {F_2}^n\\ <x,u>\equiv 0 \pmod 2}} W_f^2(x) =$$

$$-2^{n+|w|} + 2^{1-n} \left(2^{|w|} \sum_{\substack{x\in {F_2}^n\\ <x,w>=0}} W_f^2(x) + 2^{|w|-1} \sum_{\substack{x\in {F_2}^n\\ <x,w>\,>0}} W_f^2(x) \right) =$$

$$-2^{n+|w|} + 2^{1-n} \left(2^{|w|-1}\cdot 2^{2n} + 2^{|w|-1} \sum_{\substack{x\in {F_2}^n\\ <x,w>=0}} W_f^2(x) \right) = 2^{|w|-n} \sum_{\substack{x\in {F_2}^n\\ <x,w>=0}} W_f^2(x).$$

□

Theorem 12. *Let f be an arbitrary Boolean function on ${F_2}^n$. Then*

$$\sum_{\substack{u\in {F_2}^n\\ u\preceq w}} \Delta_f(u) = \sum_{f'} \left(2^{|w|} - 2wt(f')\right)^2$$

where the last sum is taken over all $2^{n-|w|}$ subfunctions f' of $|w|$ variables obtained from f by substituting constants for all x_i such that $w_i = 0$.

Proof.

$$\sum_{\substack{u\in F_2{}^n\\ u\preceq w}} \Delta_f(u) = \sum_{\substack{u\in F_2{}^n\\ u\preceq w}} \sum_{x\in F_2{}^n} (-1)^{f(x)+f(x+u)} =$$

$$\sum_{x\in F_2{}^n} \sum_{\substack{u\in F_2{}^n\\ u\preceq w}} (-1)^{f(x)+f(x+u)} = \sum_{f'} \sum_{x,y \text{ of } f'} (-1)^{f(x)+f(y)} =$$

$$\sum_{f'} \left(wt^2(f') + (2^{|w|} - wt(f'))^2 - 2wt(f')(2^{|w|} - wt(f'))\right) = \sum_{f'} \left(2^{|w|} - 2wt(f')\right)^2.$$

□

Corollary 2. *Let f be an arbitrary Boolean function on $F_2{}^n$. Then*

$$\sum_{\substack{x\in F_2{}^n\\ <x,w>=0}} W_f^2(x) = 2^{n-|w|+2} \sum_{f'} \left(2^{|w|-1} - wt(f')\right)^2$$

where the last sum is taken over all $2^{n-|w|}$ subfunctions f' of $|w|$ variables obtained from f by substituting constants for all x_i such that $w_i = 0$.

Proof. It follows immediately from Theorems 11 and 12. □

Remark 1. If f is an $(n-k)$th order nonaffine correlation immune Boolean function on $F_2{}^n$ then by (2) we have $W_f(0) \equiv 0 \pmod{2^{n-k+1}}$. Therefore $W_f(0) \equiv 2^{n-i} \pmod{2^{n-i+1}}$ for some i, $i \in \{1, 2, \ldots, k-1\}$.

Theorem 13. *Let f be an unbalanced nonconstant $(n-k)$th order correlation immune Boolean function on $F_2{}^n$. Let $W_f(0) = \pm p \cdot 2^{n-i}$ where p is some odd positive integer, $i \in \{1, 2, \ldots, k-1\}$. Then*

$$\binom{n}{i} \le (2^{2i} - p^2)\binom{k-1}{i}. \tag{3}$$

Proof. By Lemma 1 we have that $\left|2^{n-1} - wt(f)\right| = p \cdot 2^{n-i-1}$. Let $w \in F_2{}^n$ be an arbitrary vector such that $|w| = i$. Then

$$\sum_{f'} \left|2^{i-1} - wt(f')\right| \ge \left|2^{n-1} - wt(f)\right| = p \cdot 2^{n-i-1}$$

where the sum is taken over all 2^{n-i} subfunctions f' of i variables obtained from f by substituting constants for all x_i such that $w_i = 0$. All terms in the sum are integer. It follows that

$$\sum_{f'} (2^{i-1} - wt(f'))^2 \ge \left(\left(\frac{p+1}{2}\right)^2 + \left(\frac{p-1}{2}\right)^2\right) \cdot 2^{n-i-1}.$$

Therefore by Corollary 2 we have

$$\sum_{\substack{x\in F_2{}^n\\ <x,w>=0}} W_f^2(x) \ge 2^{n-i+2} \cdot \left(\frac{p^2+1}{2}\right) \cdot 2^{n-i-1} = (p^2+1) \cdot 2^{2n-2i}.$$

Hence,

$$\sum_{\substack{x \in F_2{}^n \\ <x,w>=0}} W_f^2(x) - W_f^2(0) \geq 2^{2n-2i}. \tag{4}$$

Next, we form the matrix B with n columns writing in rows of B each binary vector $x \in F_2{}^n$ exactly $W_f^2(x)$ times. By Parseval's equality the matrix B contains exactly 2^{2n} rows. The total number of nonzero rows of B is $2^{2n} - p^2 \cdot 2^{2n-2i}$. By Xiao Guo-Zhen–Massey spectral characterization [9] each nonzero row of the matrix B contains at most $k-1$ zeroes. It follows that each nonzero row in B contains at most $\binom{k-1}{i}$ subsets of i zeroes. All nonzero rows in B contain at most $(2^{2n} - p^2 \cdot 2^{2n-2i})\binom{k-1}{i}$ subsets of i zeroes. At the same time by (4) for any i columns in B there exist at least 2^{2n-2i} nonzero rows that contain only zeroes in these i columns. Therefore,

$$\frac{(2^{2n} - p^2 \cdot 2^{2n-2i})\binom{k-1}{i}}{2^{2n-2i}} \geq \binom{n}{i}.$$

□

Corollary 3. *Let f be an mth order correlation immune Boolean function on $F_2{}^n$. Let $wt(f) = u \cdot 2^h$ where u is odd positive integer, h is integer. Then*

$$\binom{n}{h+1} \leq u(2^{n-h} - u)\binom{n-m-1}{h-m}.$$

Proof. It follows immediately from Theorem 13 and Lemma 1. □

Theorem 14. *Let f be an unbalanced nonconstant $(n-k)$th order correlation immune Boolean function on $F_2{}^n$. Then $n \leq 4k-5$.*

Proof. By Remark 1 we can assume that $W_f(0) \equiv 2^{n-i} \pmod{2^{n-i+1}}$ for some i, $i \in \{1, 2, \ldots, k-1\}$. Then by Theorem 13 we have

$$n(n-1)\ldots(n-i+1) \leq (2^{2i}-1)(k-1)(k-2)\ldots(k-i). \tag{5}$$

Suppose that $n \geq 4(k-1)$. Then $n(n-1)\ldots(n-i+1) \geq 2^{2i}(k-1)(k-2)\ldots(k-i)$ that contradicts to (5). □

Corollary 4. *For $m > 0.75n - 1.25$ there do not exist unbalanced nonconstant mth order correlation immune Boolean functions on $F_2{}^n$.*

It is easy to check that the 3-variable function f that takes the value 1 only at two vectors $(0,0,0)$ and $(1,1,1)$ is correlation immune of order 1. Therefore the bound in Corollary 4 is tight.

Remark 2. Until now Bierbrauer–Friedman bound [8], [1]

$$wt(f) \geq 2^n \frac{2(m+1)-n}{2(m+1)} \tag{6}$$

was the best known lower bound for the weight of high order correlation immune nonconstant functions. If we substitute $m > 0.75n - 1.25$ to (6) we obtain $wt(f) > 2^n \frac{n-1}{3n-1}$. In fact, our Corollary 4 follows that in this case $wt(f) = 2^{n-1}$.

8 Tradeoff between Correlation Immunity and Nonlinearity for Unbalanced Boolean Functions

In [12] Sarkar and Maitra proved (this result was obtained independently also in [14] and [18]) that for an n-variable mth order correlation immune Boolean function f, $n-m \geq 1$, the inequality $nl(f) \leq 2^{n-1}-2^m$ holds. Moreover, if f is balanced (i. e. m-resilient), $n-m \geq 2$, then $nl(f) \leq 2^{n-1}-2^{m+1}$. In [18] Zheng and Zhang proved that for unbalanced Boolean functions, $m \geq 0.6n-0.4$, the nonlinearity $2^{n-1}-2^m$ can not be achieved. Therefore for an n-variable mth order correlation immune Boolean function f, $0.6n-0.4 \leq m \leq n-1$, the inequality $nl(f) \leq 2^{n-1}-2^{m+1}$ holds. (Note that by our Corollary 4 for $m > 0.75n-1.25$ unbalanced n-variable mth order correlation immune functions do not exist at all!) At the same time in [15] Tarannikov gives the constructions of n-variable m-resilient Boolean functions with the nonlinearity $2^{n-1}-2^{m+1}$ for $0.6n-1 \leq m \leq n-2$. Thus, although the upper bound in [12] for unbalanced functions is higher than for balanced, nevertheless, at least for $0.6n-0.4 \leq m \leq n-2$ the maximum possible nonlinearity of m-resilient Boolean functions is not less than the maximum possible nonlinearity of mth order correlation immune unbalanced Boolean functions. In this section we continue the investigations in this direction and give new improvements of upper bounds for the nonlinearity of high order correlation immune unbalanced Boolean functions. In our investigation we use the inequality (3) obtained in Theorem 13.

Theorem 15. *Let f be an unbalanced mth order correlation immune function on ${F_2}^n$. Suppose that $W_f(0) \equiv 2^{m+1} \pmod{2^{m+2}}$. Then for $n \geq 12$ the inequality*

$$m < \frac{1}{2}n + \frac{1}{2}\log_2 n + \text{const}$$

holds where $\text{const} = \frac{1}{2}\log_2\left(\frac{\pi}{2}e^{8/9}\right) - 1$.

The proof of Theorem 15 is given in the Appendix B.

Corollary 5. *Let f be an unbalanced mth order correlation immune function on ${F_2}^n$. If $m \geq \frac{1}{2}n + \frac{1}{2}\log_2 n + \frac{1}{2}\log_2\left(\frac{\pi}{2}e^{8/9}\right) - 1$, $n \geq 12$, then $nl(f) \leq 2^{n-1}-2^{m+1}$.*

Proof. By Theorem 15 we have $W_f(0) \not\equiv 2^{m+1} \pmod{2^{m+2}}$. It follows that $|W_f(0)| \geq 2^{m+2}$. Therefore, $nl(f) = 2^{n-1} - \frac{1}{2}\max_{x \in {F_2}^n} |W_f(x)| \leq 2^{n-1} - \frac{1}{2}|W_f(0)| \leq 2^{n-1}-2^{m+1}$. □

Theorem 16. *Let f be an unbalanced mth order correlation immune function on ${F_2}^n$. Suppose that $W_f(0) \equiv 2^{m+2} \pmod{2^{m+3}}$. Then for $n \geq 24$ the inequality*

$$m < \frac{1}{2}n + \frac{3}{2}\log_2 n + \log_2\left(\frac{1}{4} + \frac{1}{n}\right) + \text{const}$$

holds where $\text{const} = \frac{1}{2}\log_2\left(\frac{\pi}{2}e^{8/9}\right) - 2$.

The proof of Theorem 16 is given in the Appendix C.

Corollary 6. *Let f be an unbalanced mth order correlation immune function on ${F_2}^n$. If $m \geq \frac{1}{2}n + \frac{3}{2}\log_2 n + \log_2\left(\frac{1}{4} + \frac{1}{n}\right) + \frac{1}{2}\log_2\left(\frac{\pi}{2}e^{8/9}\right) - 2$, $n \geq 24$, then $nl(f) \leq 2^{n-1}-2^{m+2}$.*

Proof. By Theorems 15 and 16 we have that $|W_f(0)| \geq 2^{m+3}$. Therefore, $nl(f) \leq 2^{n-1} - \frac{1}{2}|W_f(0)| \leq 2^{n-1} - 2^{m+2}$. □

Thus, we see that although the upper bounds in [12] for the nonlinearity of unbalanced functions is higher than for balanced, nevertheless, for higher m balanced functions are "better" than unbalanced in this respect.

The authors are grateful to Oktay Kasim-Zadeh for valuable advices on the analysis of inequality (7).

References

1. J. Bierbrauer, Bounds on orthogonal arrays and resilient functions, Journal of Combinatorial Designs, V. 3, 1995, pp. 179–183.
2. J. Bierbrauer, K. Gopalakrishnan, D. R. Stinson, Orthogonal arrays, resilient functions, error correcting codes and linear programming bounds, SIAM Journal of Discrete Mathematics, V. 9, 1996, pp. 424–452.
3. R. Canetti, Y. Dodis, S. Halevi, E. Kushilevitz, A. Sahai, Exposure-resilient functions and all-or-nothing transforms, In Advanced in Cryptology: Eurocrypt 2000, Proceedings, Lecture Notes in Computer Science, V. 1807, 2000, pp. 453–469.
4. A. Canteaut, C. Carlet, P. Charpin, C. Fontaine, Propagation characteristics and correlation-immunity of highly nonlinear Boolean functions, In Advanced in Cryptology: Eurocrypt 2000, Proceedings, Lecture Notes in Computer Science, V. 1807, 2000, pp. 507–522.
5. C. Carlet, Partially-bent functions, In Advanced in Cryptology: Crypto 1992, Proceedings, Lecture Notes in Computer Science, V. 740, 1992, pp. 280–291.
6. B. Chor, O. Goldreich, J. Hastad, J. Friedman, S. Rudich, R. Smolensky, The bit extraction problem or t-resilient functions, IEEE Symposium on Foundations of Computer Science, V. 26, 1985, pp. 396–407.
7. W. Feller, An introduction to probability theory and its applications, John Wiley & Sons, New York, 3rd edition, 1968.
8. J. Friedman, On the bit extraction problem, Proc. 33rd IEEE Symposium on Foundations of Computer Science, 1992, pp. 314–319.
9. Xiao Guo-Zhen, J. Massey, A spectral characterization of correlation-immune combining functions, IEEE Transactions on Information Theory, V. 34, No 3, May 1988, pp. 569–571.
10. V. Levenshtein, Split orthogonal arrays and maximum independent resilient systems of functions, Designs, Codes and Cryptography, V. 12, 1997, pp. 131–160.
11. F. J. Mac Williams, N. J. A. Sloane, The theory of error correcting codes, North-Holland, Amsterdam, 1977.
12. P. Sarkar, S. Maitra, Nonlinearity bounds and constructions of resilient Boolean functions, In Advanced in Cryptology: Crypto 2000, Proceedings, Lecture Notes in Computer Science, V. 1880, 2000, pp. 515–532.
13. T. Siegenthaler, Correlation-immunity of nonlinear combining functions for cryptographic applications, IEEE Transactions on Information theory, V. IT-30, No 5, 1984, p. 776–780.
14. Yu. Tarannikov, On resilient Boolean functions with maximal possible nonlinearity, Proceedings of Indocrypt 2000, Lecture Notes in Computer Science, V. 1977, pp. 19–30, Springer-Verlag, 2000.
15. Yu. Tarannikov, New constructions of resilient Boolean functions with maximal nonlinearity, Preproceedings of 8th Fast Software Encryption Workshop, Yokohama, Japan, April 2–4, 2001, pp.70-81.

16. Yu. Tarannikov, D. Kirienko, Spectral analysis of high order correlation immune functions, Proceedings of 2001 IEEE International Symposium on Information Theory ISIT2001, Washington, DC, USA, June 2001, p. 69, full version is available at Cryptology ePrint archive (http://eprint.iacr.org/), Report 2000/050, October 2000, 8 pp.
17. X. M. Zhang, Y. Zheng, GAC — the criterion for global avalanche characteristics and nonlinearity of cryptographic functions, Journal of Universal Computer Science, V. 1, 1995, pp. 136–150.
18. Y. Zheng, X. M. Zhang, Improved upper bound on the nonlinearity of high order correlation immune functions, Selected Areas in Cryptography, 7th Annual International Workshop, SAC2000, Lecture Notes in Computer Science, V. 2012, pp. 264–274, Springer-Verlag, 2001.
19. Y. Zheng, X. M. Zhang, New results on correlation immune functions, The 3nd International Conference on Information Security and Cryptology (ICISC 2000), Seoul, Korea, Lecture Notes in Computer Science, V. 2015, pp. 49–63, Springer-Verlag, 2001.

A Proof of Theorem 3

If $u = 0$ then obviously $\Delta_f(u) = 2^n$, and $\sum\limits_{\substack{x \in {F_2}^n \\ <x,u> \equiv 0 \pmod 2}} W_f^2(x) = \sum\limits_{x \in {F_2}^n} W_f^2(x) = 2^{2n}$, therefore, the equality holds. So, we can assume that $u \neq 0$. Next,

$$\sum_{\substack{x \in {F_2}^n \\ <x,u> \equiv 0 \pmod 2}} W_f^2(x) = \sum_{\substack{x \in {F_2}^n \\ <x,u> \equiv 0 \pmod 2}} \left(\sum_{y \in {F_2}^n} (-1)^{f(y)+<x,y>} \right)^2 =$$

$$\sum_{\substack{x \in {F_2}^n \\ <x,u> \equiv 0 \pmod 2}} \left(2^n + \sum_{y' \neq y'' \in {F_2}^n} (-1)^{f(y')+f(y'')+<x,y'+y''>} \right) = 2^{2n-1} +$$

$$\sum_{y' \neq y'' \in {F_2}^n} (-1)^{f(y')+f(y'')} \sum_{x \in {F_2}^n} \left(\frac{1}{2} + \frac{1}{2}(-1)^{<x,u>} \right) (-1)^{<x,y'+y''>} = 2^{2n-1} +$$

$$\frac{1}{2} \sum_{y' \neq y'' \in {F_2}^n} (-1)^{f(y')+f(y'')} \left(\sum_{x \in {F_2}^n} (-1)^{<x,y'+y''>} + \sum_{x \in {F_2}^n} (-1)^{<x,u+y'+y''>} \right) =$$

$$2^{2n-1} + \frac{1}{2} \sum_{\substack{y',y'' \in {F_2}^n \\ y'+y''=u}} (-1)^{f(y')+f(y'')} \left(0 + \sum_{x \in {F_2}^n} 1 \right) =$$

$$2^{2n-1} + 2^{n-1} \sum_{y \in {F_2}^n} (-1)^{f(y)+f(y+u)} = 2^{2n-1} + 2^{n-1} \Delta_f(u).$$

□

B Proof of Theorem 15

For $i = k - 1 = n - m - 1$ in (3) we have

$$\binom{n}{i} < 4^i. \tag{7}$$

For each $i, 0 < i < n$, we have $\binom{n}{i} > (\frac{n}{i})^i$. It follows that if for some i the inequality (7) holds then the inequality $(\frac{n}{i})^i < 4^i$ holds too. Therefore $\frac{n}{i} < 4$ and $\frac{n}{4} < i$. Thus, we obtain the simplest bound on i: $i > \frac{n}{4}$.

By means of the lower and upper bounds for $n!$ (see[7])

$$\sqrt{2\pi}n^{n+1/2}e^{-n}e^{(12n+1)^{-1}} < n! < \sqrt{2\pi}n^{n+1/2}e^{-n}e^{(12n)^{-1}}$$

it is easy to deduce the inequality

$$\binom{n}{i} > \frac{2^{H(\frac{i}{n})n}}{\sqrt{2\pi n\frac{i}{n}(1-\frac{i}{n})}} e^{-\frac{1}{12n\frac{i}{n}(1-\frac{i}{n})}}, \tag{8}$$

that holds for any $0 < i < n$.

(Here $H(x) = -x\log_2 x - (1-x)\log_2(1-x)$ is the *entropy* of x, $0 < x < 1$).

If $\frac{n}{4} < i < \frac{n}{2}$ then $\frac{1}{4} < \frac{i}{n} < \frac{1}{2}$.

Consider the function $x(1-x)$. If $1/4 < x < 1/2$ then $3/16 < x(1-x) < 1/4$. It follows that for $\frac{n}{4} < i < \frac{n}{2}$ we have

$$\frac{3}{16} < \frac{i}{n}\left(1 - \frac{i}{n}\right) < \frac{1}{4}.$$

Then

$$\frac{1}{\frac{i}{n}(1-\frac{i}{n})} > 4 \quad \text{and} \quad \frac{1}{\sqrt{\frac{i}{n}(1-\frac{i}{n})}} > 2,$$

i. e.

$$\frac{1}{\sqrt{2\pi\frac{i}{n}(1-\frac{i}{n})}} > \frac{2}{\sqrt{2\pi}} = \sqrt{\frac{2}{\pi}}. \tag{9}$$

Next,

$$\frac{1}{\frac{i}{n}(1-\frac{i}{n})} < \frac{16}{3}, \quad \text{it follows} \quad \frac{1}{12n\frac{i}{n}(1-\frac{i}{n})} < \frac{16}{12n\cdot 3} = \frac{4}{9n} \leq \frac{4}{9},$$

since $n \geq 1$.

Therefore,

$$e^{-\frac{1}{12n\frac{i}{n}(1-\frac{i}{n})}} > e^{-\frac{4}{9}}. \tag{10}$$

From (8) using (9) and (10) we have for any i, $\frac{n}{4} < i < \frac{n}{2}$, that

$$\binom{n}{i} > \sqrt{\frac{2}{\pi}}e^{-\frac{4}{9}}\frac{2^{H(\frac{i}{n})n}}{\sqrt{n}}. \tag{11}$$

The inequalities (7) and (11) follow the inequality

$$4^i > \frac{\sqrt{\frac{2}{\pi}}e^{-\frac{4}{9}}2^{H\left(\frac{i}{n}\right)n}}{\sqrt{n}}. \tag{12}$$

Taking the logarithm in (12) we have

$$2i > H\left(\frac{i}{n}\right)n - \frac{1}{2}\log_2 n + \alpha$$

where $\alpha = \log_2\left(\sqrt{\frac{2}{\pi}}e^{-4/9}\right)$. Dividing by n we have

$$2\frac{i}{n} > H\left(\frac{i}{n}\right) - \frac{1}{2n}\log_2 n + \frac{\alpha}{n}.$$

Denoting $x = \frac{i}{n}$ we obtain the inequality

$$2x < H(x) - \frac{1}{2n}\log_2 n + \frac{\alpha}{n},$$

or

$$H(x) - 2x < a(n) \tag{13}$$

where $a(n) = \frac{1}{2n}\log_2 n - \frac{\alpha}{n}$.

Thus, the problem is reduced to the obtaining of lower bound for x satisfying (13) under the condition $1/4 < x < 1/2$.

Now put $y = \frac{1}{2} - x$. Then conditions on $x : 1/4 < x < 1/2$ transform into conditions on $y : 0 < y < 1/4$. To find the lower bound for x satisfying (13) is the same as to find the upper bound for y satisfying

$$H\left(\frac{1}{2} - y\right) - 2\left(\frac{1}{2} - y\right) < a(n),$$

or

$$H\left(\frac{1}{2} - y\right) - 1 + 2y < a(n). \tag{14}$$

By Taylor's formula

$$H\left(\frac{1}{2} - y\right) = H\left(\frac{1}{2}\right) - H'\left(\frac{1}{2}\right)y + \frac{1}{2}H''(\xi)y^2 \tag{15}$$

where ξ is some number from the interval $1/2 - y < \xi < 1/2$. Taking into account that $y < 1/4$ we have $1/4 < \xi < 1/2$.

We differentiate and find that $H'(x) = \log_2 \frac{1-x}{x}$, $H''(x) = -\frac{1}{\ln 2}\frac{1}{x(1-x)}$.

It follows $H'(\frac{1}{2}) = 0$, also for $1/4 < \xi < 1/2$ the inequality $H''(\frac{1}{4}) < H''(\xi) < H''(\frac{1}{2})$ holds, in particular, $H''(\xi) > -\frac{16}{3\ln 2}$ (the function $\frac{-1}{x(1-x)}$ increases for $0 < x < 1/2$). Also we take into account that $H(\frac{1}{2}) = 1$.

From (15) we have for any y, $0 < y < 1/4$,

$$H\left(\frac{1}{2} - y\right) > H\left(\frac{1}{2}\right) - H'\left(\frac{1}{2}\right) y + H''\left(\frac{1}{4}\right) y^2 = 1 - \frac{8}{3\ln 2} y^2.$$

Taking into account the last inequality in (14) we have

$$1 - \frac{8}{3\ln 2} y^2 - 1 + 2y < a(n),$$

or

$$0 < \frac{8}{3\ln 2} y^2 - 2y + a(n). \tag{16}$$

The inequality (16) is quadratic with respect to y and depends on the parameter n. The coefficient in quadratic term is positive, therefore y can be determined from the conditions $y < y_1$ or $y > y_2$ where $y_1 < y_2$ are roots of characteristic equation. The second condition is irrelevant and does not correspond to the sense of this problem. A discriminant is equal to

$$1 - \frac{8}{3\ln 2} a(n).$$

Note that $\sqrt{\frac{2}{\pi}} e^{-4/9} < 1$, it follows $\alpha < 0$. Let $\beta = -\alpha > 0$. Then $a(n) = \frac{1}{2n} \log_2 n + \frac{\beta}{n}$ where $\beta > 0$.

Thus, it is sufficient to solve the inequality

$$0 < \gamma y^2 - 2y + b(n) \tag{17}$$

where $\gamma = \frac{8}{3\ln 2}$.

Positiveness of a discriminant means that $1 - \gamma b(n) > 0$ or $b(n) < 1/\gamma$, i. e.

$$\frac{1}{2n} \log_2 n + \frac{\beta}{n} < \frac{1}{\gamma}. \tag{18}$$

The function $\frac{\ln x}{x}$ has the maximum for $x = e$. Let $n \geq 12$, then $\frac{1}{2n} \log_2 n \leq \frac{1}{24} \log_2 12$. Hence, it is sufficient to demonstrate that

$$\frac{1}{24} \log_2 12 + \frac{\log_2(\sqrt{\frac{\pi}{2}} e^{4/9})}{12} < \frac{3\ln 2}{8}$$

or

$$\frac{1}{3}(2 + \log_2 3) + \frac{2}{3}(\log_2 \sqrt{\frac{\pi}{2}} e^{4/9}) < 3\ln 2. \tag{19}$$

The right part of (19) is greater than 2 since $e^2 < 8 = e^{3\ln 2}$. Consider the left part of (19). It is equal to

$$\frac{2}{3} + \frac{\log_2 3}{3} + \frac{2}{3}\left(\log_2 \sqrt{\frac{\pi}{2}} e^{4/9}\right) < \frac{2}{3} + \frac{1}{3}\left(\log_2 \frac{3\pi e}{2}\right).$$

The product $\pi e < 10$, therefore $\frac{3\pi e}{2} < 16$. Hence, the left part of (19) is less than 2. It follows that for $n \geq 12$ a discriminant of the equation (17) is positive and required upper bound for y follows from the inequality

$$y < y_1 = \frac{1 - \sqrt{1 - \frac{8}{3\ln 2}a(n)}}{\frac{8}{3\ln 2}} \leq \frac{1 - 1 + \frac{8}{3\ln 2}a(n)}{\frac{8}{3\ln 2}} = a(n)$$

where y_1 is a root of the equation correspondent to the inequality (16). Pointing in a view that $y = \frac{1}{2} - x = \frac{1}{2} - \frac{i}{n} = \frac{1}{2} - \frac{n-m-1}{n} = \frac{m+1}{n} - \frac{1}{2}$ we have:

$$\frac{m+1}{n} - \frac{1}{2} < \frac{1}{2n}\log_2 n + \frac{1}{2n}\log_2\left(\frac{\pi}{2}e^{8/9}\right).$$

For $n \geq 12$ it follows

$$m < \frac{1}{2}n + \frac{1}{2}\log_2 n + \text{const}$$

where const $= \frac{1}{2}\log_2(\frac{\pi}{2}e^{8/9}) - 1$. □

C Proof of Theorem 16

For $i = k - 2 = n - m - 2$ in (3) we have

$$\binom{n}{i} \leq (4^i - 1)(i + 1). \tag{20}$$

As in the previous proof we use the inequality (8), the bounds (9) and (10) valid for sufficiently high n and the inequality (11). Combining (11) and (20) we have

$$4^i(i+1) > \sqrt{\frac{2}{\pi}}e^{-\frac{4}{9}}\frac{2^{H(\frac{i}{n})n}}{\sqrt{n}}.$$

Taking the logarithm in the last inequality we have:

$$2i + \log_2(i+1) > H\left(\frac{i}{n}\right)n - \frac{1}{2}\log_2 n + \alpha,$$

$\alpha = \log_2\left(\sqrt{\frac{2}{\pi}}e^{-\frac{4}{9}}\right)$.

Introducing new variable $x = \frac{i}{n}$ and dividing by n we obtain

$$2x + \frac{\log_2(x + \frac{1}{n})}{n} > H(x) - \frac{3}{2}\frac{\log_2 n}{n} + \frac{\alpha}{n},$$

Taking into account that $\log_2(x + \frac{1}{n}) \geq \log_2(\frac{1}{4} + \frac{1}{n})$ we have:

$$H(x) - 2x < a(n)$$

where $a(n) = \frac{\log_2(\frac{1}{4}+\frac{1}{n})}{n} + \frac{3}{2}\frac{\log_2 n}{n} - \frac{\alpha}{n}$.

This inequality is analogous to the inequality (13); the only difference is in the function $a(n)$. Using the reasonings completely analogous to the reasoning in the previous proof we deduce the inequality

$$0 < \gamma y^2 - 2y + b(n) \tag{21}$$

where $\gamma = \frac{8}{3\ln 2}$, $y = \frac{1}{2} - x$, $b(n) = \frac{3}{2}\frac{\log_2 n}{n} + \frac{\log_2(\frac{1}{4}+\frac{1}{n})}{n} + \frac{\beta}{n}$, $\beta = -\alpha$.

The solutions of this inequality satisfy $y < a(n)$ (see Appendix B). It means that $\frac{1}{2} - \frac{n-(m+2)}{n} < \frac{\log_2(\frac{1}{4}+\frac{1}{n})}{n} + \frac{3}{2}\frac{\log_2 n}{n} - \frac{\alpha}{n}$ or rewriting

$$m < \frac{1}{2}n + \frac{3}{2}\log_2 n + \log_2\left(\frac{1}{4} + \frac{1}{n}\right) + c,$$

$c = -\alpha - 2$.

Now we need only to know for which n this inequality is satisfied, i. e. beginning with which n a discriminant of inequality (21) is nonnegative, or $b(n) < 1/\gamma$, or

$$\frac{3\log_2 n}{2n} + \frac{\log_2(\frac{1}{4}+\frac{1}{n})}{n} + \frac{\log_2\left(\sqrt{\frac{\pi}{2}}e^{4/9}\right)}{n} < \frac{3\ln 2}{8}.$$

Computer analysis shows that this inequality is true beginning with $n = 24$. It completes the proof. □

An Extension of Kedlaya's Point-Counting Algorithm to Superelliptic Curves

Pierrick Gaudry and Nicolas Gürel

LIX, École polytechnique, 91128 Palaiseau Cedex, France
{gaudry, gurel}@lix.polytechnique.fr

Abstract. We present an algorithm for counting points on superelliptic curves $y^r = f(x)$ over a finite field $\mathbb{F}_q$ of small characteristic different from r. This is an extension of an algorithm for hyperelliptic curves due to Kedlaya. In this extension, the complexity, assuming r and the genus are fixed, is $O(\log^{3+\varepsilon} q)$ in time and space, just like for hyperelliptic curves. We give some numerical examples obtained with our first implementation, thus proving that cryptographic sizes are now reachable.

1 Introduction

In the past few years a lot of candidates have been proposed to enlarge the set of groups that can be used in protocols based on the discrete logarithm problem like Diffie–Hellman or ElGamal. Beside the classical multiplicative groups of finite fields, the most famous are certainly the systems based on elliptic curves [21, 26]. Indeed, for these systems the only general attacks known are variants of the Pollard Rho method which require exponential time computation; in practice it means that the key size is much shorter than in a system that uses finite fields. Thereafter, systems based on hyperelliptic curves were proposed [22]. They seem to have the same advantages as elliptic curve cryptosystems (at least when the genus is less than 4 [1,14]).

More recently, systems based on the discrete logarithm problem in the Jacobians of other curves were designed. Namely, in the literature, we can now find algorithms for working in Jacobians of superelliptic curves [13] and of C_{ab} curves [2]. Several works related to these curves have already been published, concerning security issues [4], efficiency [17,6], building curves with known number of points [3], or possible use in a Weil restriction attack on elliptic curves [5]. The next step for studying the possible cryptographic use of these curves is to conceive an algorithm for counting points of the Jacobian of a *random* curve. Indeed, this is thought to be one of the most secure ways of building a cryptosystem by a large part of the community.

In the case of elliptic curves, this problem of point counting has been a challenge of the past 15 years and nowadays we have satisfactory solutions. When the characteristic of the base field is large the best known method is Schoof's algorithm and all the improvements leading to the so-called Schoof–Elkies–Atkin algorithm. We refer to [7] or [23] for surveys of these techniques and to the references therein. Besides some theoretical results [29] and an attempt to make

C. Boyd (Ed.): ASIACRYPT 2001, LNCS 2248, pp. 480–494, 2001.
© Springer-Verlag Berlin Heidelberg 2001

them practical [15], extending the SEA algorithm to higher genus has not yet proven to be enough for cryptographic sizes. The situation is quite different in small characteristic: two years ago, Satoh [32] showed that p-adic methods using the canonical lift could lead to an algorithm asymptotically faster than SEA. Some work has been done consequently on the subject to extend it to characteristic 2 [33,9], to implement it and obtain new records [9], to use less memory [34], and to combine it with an early-abort strategy for generating secure curves [10]. Mestre, Harley and Gaudry recently proposed a related algorithm, based on the arithmetic-geometric mean, for elliptic curves and hyperelliptic curves of genus 2 in characterstic 2; a nice feature of this technique is that it does not explicitly make use of j-invariants, of modular equations nor of Vélu-type formulae, and these had previously been the main obstructions to generalizing beyond the elliptic case. However, the AGM method does not seem to extend easily to non-hyperelliptic curves. Another approach, also using p-adic methods but not based on canonical lifting, has been proposed by Kedlaya [18]. His method applies to hyperelliptic curves in small odd characteristic. The complexity in time is $O(\log^{3+\varepsilon} q)$, for curves over $\mathbb{F}_q$ of fixed genus, i.e. the same as all the variants of Satoh's method and the complexity in space is $O(\log^3 q)$ which is the same as Satoh's original algorithm, but bad compared to the algorithm of [34] or AGM.

The contribution of this paper is twofold: firstly we show that Kedlaya's algorithm can be extended in a rather straightforward way to superelliptic curves; secondly we report some results obtained with our first implementation written in MAGMA. To our knowledge, these are the first published point counting computations for random hyperelliptic and superelliptic curves of cryptographic sizes.

The paper is organized as follows: after recalling some basics about curves and p-adic numbers, we describe Kedlaya's original algorithm and show how to adapt it for superelliptic curves. Then we give some more details on the way these algorithms can be handled in practice and we estimate the complexity. We conclude by numerical examples and remarks about the use of these curves in cryptography.

2 Background on Algebraic Curves and p-Adic Number Rings

In this section, we recall some basic facts about algebraic curves over finite fields and p-adic numbers. We shall not give precise definitions and we refer the reader to classical books on the subject ([12,20,19,24] for instance).

2.1 Hyperelliptic and Superelliptic Curves

Let $\mathbb{F}_q$ be a finite field with $q = p^n$ elements. We shall consider only two types of curves over $\mathbb{F}_q$, namely hyperelliptic and superelliptic curves.

Definition 1 *A* superelliptic *curve is a plane curve $\mathcal{C}$ which admits an affine equation of the form*

$$y^r = f(x),$$

where r is a prime different from p and f is monic, squarefree of degree d coprime to r.

With such a definition, $\mathcal{C}$ is non-singular in its affine part, and admits a unique place of degree 1 at infinity. Moreover its genus is given by $g = \frac{(d-1)(r-1)}{2}$.

Definition 2 *In characteristic different from 2, a* hyperelliptic *curve is a superelliptic curve whose equation is of the form $y^2 = f(x)$, with $r = 2$ and f of degree $2g + 1$.*

Note that there exists a more general definition of hyperelliptic curves which do not exclude the case of characteristic 2. But the algorithms we will describe work only for this particular case.

Let $\mathcal{C}$ be a superelliptic curve of genus g. Associated to this curve, one can define its *Jacobian*, noted $\mathrm{J}(\mathcal{C})$, which is a finite abelian group. In the past few years, several algorithms were developed to compute explicitly in this group [13, 2,17,6]. The next step is to study the order of $\mathrm{J}(\mathcal{C})$. For this the q-th power Frobenius endomorphism and its characteristic polynomial $\chi(T)$ are key tools. More precisely, $\chi(T)$ can be written as

$$\chi(T) = \sum_{i=0}^{2g} a_i T^i, \quad \text{with} \quad a_{2g} = 1, \quad a_i = q^{g-i} a_{2g-i} \quad \text{for } i = 0, \ldots, g-1,$$

and all its roots have absolute value $\sqrt{q}$. This is essentially the Riemann Hypothesis for zeta functions of curves. For us, the interesting fact is that $\#\mathrm{J}(\mathcal{C}) = \chi(1)$. Our goal in this paper is to compute $\chi(T)$ and to obtain $\#\mathrm{J}(\mathcal{C})$ as a byproduct.

2.2 The Ring $\mathbb{Z}_q$

Let K be the (unique up to isomorphism) unramified extension of degree n of $\mathbb{Q}_p$; its residual field is $\mathbb{F}_q$. We denote by $\mathbb{Z}_q$ the ring of integers of K. In order to construct it, we can start with the polynomial $\overline{P}(t)$ which defines $\mathbb{F}_q$ as an algebraic extension of $\mathbb{F}_p$; we then consider the extension

$$\mathbb{Z}_q := \mathbb{Z}_p[t]/(P(t)),$$

where the polynomial $P(t)$ is obtained from $\overline{P}(t)$ by lifting trivially its coefficients to p-adic integers. In practice, an element z of $\mathbb{Z}_q$ can be represented as a polynomial $z = z_{n-1}t^{n-1} + z_{n-2}t^{n-2} + \cdots + z_1 t + z_0$ taken modulo $P(t)$ and where the z_i are integers modulo a power of p called the *precision* at which the computation is done.

It can be shown that the Galois group of K over $\mathbb{Q}_p$ is cyclic. We will denote by σ the unique generator, also called Frobenius, of this Galois group that reduces modulo p to the p-th power Frobenius in $\mathbb{F}_q$. There is no trivial formula for writing z^σ for an element z in $\mathbb{Z}_q$ expressed on a polynomial basis as above. Later on, we will describe how to precompute t^σ and then z^σ is obtained as follows:

$$z^\sigma = \left(\sum_{i=0}^{n-1} z_i t^i\right)^\sigma = \sum_{i=0}^{n-1} z_i (t^\sigma)^i.$$

3 Kedlaya's Algorithm and Its Extension

3.1 Overview of Kedlaya's Algorithm for Hyperelliptic Curves

Let $\mathcal{C}$ be a hyperelliptic curve of genus g given by its equation $y^2 = \overline{f}(x)$ over $\mathbb{F}_q$. Following the construction of Kedlaya (see also [20], page 72), we consider the curve $\mathcal{C}'$ obtained from $\mathcal{C}$ by removing the point at infinity and the points with vertical tangent (i.e. $y = 0$).

There is a way to lift the coordinate ring of $\mathcal{C}'$ called the weak completion [27], with the nice property that its cohomology verifies a "Lefschetz trace formula" [28] and hence gives information about the cardinalities of the initial curve $\mathcal{C}$.

Taking a lowbrow point of view in which we can forget about the curve $\mathcal{C}'$, we shall work on the vector space generated over the p-adic number field K by the following differential forms:

$$\mathcal{D} = \left\langle \frac{x^i dx}{y}; \quad i \in [0, 2g-1] \right\rangle,$$

in which we have the relations coming from the equation of the curve and $d\varphi(x, y) \equiv 0$ for every rational function φ. On the differential forms one can define a Frobenius action which is compatible with the p-th power Frobenius on $\mathcal{C}$: take $x^\sigma = x^p$, y^σ given by $(y^\sigma)^2 = f(x)^\sigma$ and $(dx)^\sigma = px^{p-1}dx$. Kedlaya shows in a constructive way that the space $\mathcal{D}$ is stable under the action of this σ. Hence σ is an endomorphism of a vector space of dimension $2g$; and everything is done in order for its characteristic polynomial to be closely related to the $\chi(T)$ we are looking for. The heart of Kedlaya's algorithm is then to compute the matrix of σ for the given basis of $\mathcal{D}$.

For each i in $[0, 2g-1]$,

$$\left(\frac{x^i dx}{y}\right)^\sigma = \frac{1}{y^\sigma} px^{ip+p-1} dx,$$

therefore the tricky part is the computation of $\frac{1}{y^\sigma}$. This is not defined in a lifted coordinate ring because it involves a square root and that is a reason why we use the weak completion. From a practical point of view, it means that we shall be able to expand $\frac{1}{y^\sigma}$ as a power series in $\tau = \frac{1}{y^2}$: starting with the definition $(y^\sigma)^2 = f(x)^\sigma$, we have

$$\begin{aligned}\tfrac{1}{y^\sigma} &= (f(x)^\sigma)^{-1/2} \\ &= (f(x)^\sigma - f(x)^p + f(x)^p)^{-1/2} \\ &= (f(x)^p)^{-1/2}\left(1 + \tfrac{f(x)^\sigma - f(x)^p}{f(x)^p}\right)^{-1/2} \\ &= \tfrac{1}{y^p}\left(1 + \tau^p(f(x)^\sigma - f(x)^p)\right)^{-1/2} .\end{aligned}$$

By the usual power series expansion of $(1+X)^{-1/2}$ we get an expression of the form

$$\frac{1}{y^\sigma} = y^{-p}\sum_{k\geq 0} P_k(x)\tau^{pk} = y^{-1}\tau^{(p-1)/2}\sum_{k\geq 0} P_k(x)\tau^{pk}.$$

Note that p divides $(f(x)^\sigma - f(x)^p)$ so that the power of p dividing $P_k(x)$ tends to infinity as k grows (actually this is what is expected due to the theoretical construction of the weak completion). We can now write

$$\left(\frac{x^i dx}{y}\right)^\sigma = \left(\sum_{k\geq 0} Q_k(x)\tau^k\right)\frac{dx}{y},$$

where $Q_k(x)$ are polynomials. The algorithm proceeds as follows: we compute this expression up to some precision in τ, and then we use the relations in $\mathcal{D}$ described above to reduce the expression to a polynomial of degree at most $2g-1$, times $\frac{dx}{y}$. In this way we shall prove that $\mathcal{D}$ is indeed σ-stable and moreover we obtain an explict description of the action of σ on the basis. For this we will use three strategies of reduction:

Red 1. First of all, using the equation of the curve, one can write

$$Q_k(x)\tau^k = (\alpha_k(x)f(x) + \beta_k(x))\tau^k = \alpha_k(x)\tau^{k-1} + \beta_k(x)\tau^k,$$

where α_k and β_k are the quotient and the remainder in the division of Q_k by f. Therefore one can assume that $Q_k(x)$ is of degree at most $2g$ for all k, except for $Q_0(x)$ for which one can show that the degree is at most $2pg-1$.

Red 2. Then we use the relations of cohomology to rewrite the series in the form $Q(x)\frac{dx}{y}$. Fix $k \geq 1$ and consider the term $Q_k(x)\tau^k\frac{dx}{y}$. Let $U(x)$ and $V(x)$ be such that $Q_k(x) = U(x)f(x) + V(x)f'(x)$ (they do exist because f is squarefree). Using

$$d\left(\frac{V(x)}{y^{2k-1}}\right) \equiv 0,$$

one obtains

$$Q_k(x)\,\tau^k\frac{dx}{y} \equiv \left(U(x) + \frac{2}{2k-1}V'(x)\right)\,\tau^{k-1}\frac{dx}{y}.$$

Repeating this for decreasing k's, we can rewrite everything on the constant term of the series.

Red 3. Finally, in the expression $Q(x)\frac{dx}{y}$ that we obtained, one can reduce the degree δ of Q to at most $2g-1$ in the following way. Assume $\delta \geq 2g$: using

$$d(x^{\delta-2g}y) \equiv 0,$$

one gets a polynomial of degree δ that can be subtracted from Q.

At this point, we have computed a $2g \times 2g$ matrix M such that

$$\begin{pmatrix} \frac{dx}{y} \\ \vdots \\ \frac{x^{2g-1}dx}{y} \end{pmatrix}^{\sigma} = M \begin{pmatrix} \frac{dx}{y} \\ \vdots \\ \frac{x^{2g-1}dx}{y} \end{pmatrix}.$$

Most of the operations done during the computation involve elements of $\mathbb{Z}_q$, but at the end we may have to divide by small powers of p. Finally the coefficients of M lie in $p^{-s}\mathbb{Z}_q$ with a small, predictable s, which depends only on p and g.

The final step is then to compute the characteristic polynomial of the matrix

$$MM^{\sigma} \cdots M^{\sigma^{n-1}},$$

which has coefficients in $\mathbb{Z}_2$ and is a p-adic approximation of $\chi(T)$.

3.2 Superelliptic Curves

Let C be a superelliptic curve given by its equation $y^r = \overline{f}(x)$ with $\overline{f}$ of degree d over $\mathbb{F}_q$. The theory is exactly the same as for hyperelliptic curves. In the present case, the space of differential forms we consider is

$$\left\langle \frac{x^i dx}{y^j};\ i \in [0, d-2],\ j \in [1, r-1] \right\rangle.$$

The Frobenius action lifting the p-th power Frobenius on C is defined similarly: take $x^\sigma = x^p$, y^σ given by $(y^\sigma)^r = f(x)^\sigma$ and $(dx)^\sigma = px^{p-1}dx$.

Again, the space of differential forms has been chosen such that it is stable under the action of σ; we will now describe the reduction process which allows us to rewrite $\left(\frac{x^i dx}{y^j}\right)^\sigma$ over the basis. Fix an $i \in [0, d-2]$ and a $j \in [1, r-1]$. We can write $\left(\frac{1}{y^j}\right)^\sigma$ as a power series

$$\left(\frac{1}{y^j}\right)^\sigma = y^{-jp}\left(1 + \frac{f(x)^\sigma - f(x)^p}{y^{rp}}\right)^{-j/r} = y^{-jp}\sum_{k\geq 0} P_k(x)\tau^{pk},$$

where we have set $\tau = y^{-r}$. Hence we can write

$$\left(\frac{x^i dx}{y^j}\right)^\sigma = \left(\sum_{k\geq 0} Q_k(x)\tau^k\right)\frac{dx}{y^{jp \bmod r}}.$$

In the following, we let $\ell = jp \bmod r$. We now proceed with three reduction steps similar to those we had for hyperelliptic curves.

Red 1. First, use the equation of the curve to obtain a series where the $Q_k(x)$ are of degree at most $d-1$, except for the first one.

Red 2. Then, rewrite the term in τ^k as a term in τ^{k-1}. For $k \geq 1$, let $U(x)$ and $V(x)$ be such that $Q_k(x) = U(x)f(x) + V(x)f'(x)$, one has

$$Q_k(x)\,\tau^k \frac{dx}{y^\ell} \equiv \left(U(x) + \frac{r}{r(k-1)+\ell} V'(x)\right) \tau^{k-1} \frac{dx}{y^\ell}\ .$$

Red 3. Finally, we are left with an expression of the form $Q(x)\frac{dx}{y^\ell}$, where $Q(x)$ is a polynomial of degree δ that we can reduce to degree at most $d-2$: assume $\delta \geq d-1$, the exact differential $d(x^{\delta-d+1}y^{r-l}) \equiv 0$ gives a polynomial of degree δ that can be subtracted from $Q(x)$.

We obtain a $2g \times 2g$ matrix M and we conclude as before by taking the characteristic polynomial of its "norm".

Note that the differential forms in $\frac{dx}{y^j}$ are sent by σ to the subspace generated by forms in $\frac{dx}{y^\ell}$ with $\ell = jp \bmod r$. As a consequence, M is a matrix that can be viewed in blocks of size $d-1$, with the property that there is exactly one non-zero block in each row block and each column block.

4 Details and Complexity

4.1 Precision of the Computation

The intermediate result obtained from the algorithm of section 3 is an approximation of the polynomial $\chi(T)$ that we are looking for, and by computing to sufficient precision we can determine it exactly. Two parameters have to be tuned, to ensure that at the end we get enough information to conclude. The first is the p-adic precision p^ν at which we truncate elements of $\mathbb{Z}_p$. The second is the τ-adic precision at which we truncate the series.

Bounds on the coefficients of $\chi(T)$ can be deduced from the bounds on its roots: $|a_i| \leq \binom{2g}{i} q^{i/2}$ for $i \in [1, g]$. We assume that q is large compared to the genus, so that a_g determines the required precision. Hence we need to know $\chi(T)$ modulo $\lceil 2\binom{2g}{g} q^{g/2} \rceil$ to be sure to recover all the coefficients. Therefore the working precision should be at least

$$\nu = \left\lceil \log_p \left(2\binom{2g}{g} q^{g/2} \right) \right\rceil .$$

The precision in τ is more problematic: at first sight it is not clear that we do not need *all* the terms of the series to get a result which makes sense even modulo p. Actually in the power series expansion, one can see that the coefficient in τ^k (which is a polynomial over $\mathbb{Z}_q$) is divisible by a power of p which grows to infinity at the speed of k/p. Hence it appears that the precision μ in τ should be at least p times the p-adic precision ν. Moreover, the reduction process also perturbs things: starting with a term $Q_k(x)\tau^k \frac{dx}{y^\ell}$, with p^m dividing $Q_k(x)$, one reduces to a differential form $Q(x)\frac{dx}{y^\ell}$ and p^m does not divide $Q(x)$ any more. In Lemma 1 of [18], Kedlaya shows that we can bound the loss of precision by $\log_p(rk+\ell)$. Acordingly, it is sufficient to enlarge μ slightly to ensure that at the end we have the required precision. A tedious calculation leads to the following choice for μ: we take the smallest μ such that

$$\mu > p\nu - \frac{p}{r} + p\log_p((r+1)\mu - 1).$$

4.2 Detailed Algorithm

We summarize the algorithm in the following:

Input: A superelliptic curve $y^r = \overline{f}(x)$ over $\mathbb{F}_q$, $q = p^n$, the degree of f is noted d, $g = \frac{(d-1)(r-1)}{2}$.
Output: The characteristic polynomial $\chi(T)$.

1. Set the p-adic working precision $\nu = \lceil \log_p(2\binom{2g}{g}q^{g/2})\rceil$ and set the maximal precision μ for the series to be the smallest value such that $\mu > p\nu - \frac{p}{r} + p\log_p((r+1)\mu - 1)$.
2. Let $S = 1 + (f(x)^\sigma - f(x)^p)\,\tau^p$, where $f(x)$ is the polynomial $\overline{f}(x)$ where the coefficients are lifted arbitrarily from $\mathbb{F}_q$ to $\mathbb{Z}_q$.
3. Compute $S^{-1/r}$ as a truncated series in τ, to precision τ^μ.
   For this, use a Newton iteration $X \leftarrow \frac{1}{r}((r+1)X - SX^{r+1})$, initialized with $X = 1$. At each step in the recursion, use **Red1** to keep the coefficients of the series of degree at most $d-1$.
4. Compute $S^{-j/r}$ for $j \in [2, r-1]$ up to precision τ^μ. This is done by multiplying $S^{-1/r}$ by itself repeatedly; again, use **Red1** after each multiplication.
5. For each $i \in [0, d-2]$ and $j \in [1, r-1]$ do
   a. Compute $\omega_{ij} = \left(\frac{x^i dx}{y^j}\right)^\sigma = p\tau^{jp \operatorname{div} r} x^{ip+p-1} S^{-j/r} \frac{dx}{y^{jp \bmod r}}$.
   b. Use **Red 2** to write ω_{ij} in the form $Q(x)\frac{dx}{y^{jp \bmod r}}$.
   During this reduction it is sometimes necessary to divide by an integer which is divisible by p. In theory, this ought to reduce the precision of the computation. Actually, when this occurs, one adds some arbitrary noise to "force" the precision to remain maximal. This strange way of doing things does not actually affect the final result because this noise will cancel out during the whole process. This is ensured by Lemma 1 in [18], which extends naturally to the superelliptic case.
   c. Use **Red 3** to reduce the degree of $Q(x)$.
6. Compute the matrix M, its norm and its characteristic polynomial $\widetilde{\chi}(T) = \sum_{0\le k\le 2g} \tilde{a}_k T^k$.
7. For $k \in [1, g]$, find the integer a_k in $[-\frac{p^\nu}{2}, \frac{p^\nu}{2}]$ congruent to $\tilde{a}_k$ modulo p^ν. Return the corresponding $\chi(T)$.

4.3 Complexity

For the complexity analysis we shall make the following assumptions:

- The characteristic p is fixed;
- The parameters r and d of the curves are fixed, hence also the genus;
- Each time we have to do a multiplication between two elements of a rather complicated structure (truncated series over polynomials over polynomials over integers), we assume that we pack everything into large integers and that we use Schönhage's fast multiplication algorithm. A multiplication between two objects of bit-size N is then assumed to take time $O(N^{1+\varepsilon})$.

In Step 2 we have to apply Frobenius to some elements of $\mathbb{Z}_q$. For this, note that t being a root of $P(t)$, so is t^σ. Therefore, t^σ can be obtained by a Newton iteration $X \leftarrow X - P(X)/P'(X)$ initialized with t^p. This is just a precomputation and moreover the cost is comparable to the rest of the algorithm. Thereafter, it is possible to obtain the Frobenius of an element in $\mathbb{Z}_q$ in time $O(n^{3+\varepsilon})$.

Step 3 is a Newton lifting. The cost is bounded by a constant times the cost of the last iteration. This last iteration costs a few multiplications between objects which are polynomials of degree μ over polynomials of degree $d-1$ with coefficients in $\mathbb{Z}_q$. An element of $\mathbb{Z}_q$ is of bit-size $n\nu$, therefore the bit-size of the objects is $n\nu\mu d = O(n^{3+\varepsilon})$. Hence the $O(1)$ multiplications we have to do in the final iteration take time in $O(n^{3+\varepsilon})$. Applying **Red 1** to the result has the same asymptotic complexity (we have to visit the whole object and the runtime is linear in its size) but is faster in practice. Finally the overall complexity of Step 3 is in $O(n^{3+\varepsilon})$.

In Step 4 we do a constant number of multiplications (remember r is constant) and then an application of **Red 1** to objects of size in $O(n^{3+\varepsilon})$. Again the complexity is $O(n^{3+\varepsilon})$. Note that in the hyperelliptic case, this step does not exist.

In Step 5 we repeat a reduction process $2g$ times using **Red 2** and **Red 3**. More precisely, Substep 5.a is only reorganizing and applying **Red 1**; this takes negligible time. In Substep 5.b we repeat μ times a process which involves elementary operations over polynomials of degree at most d over $\mathbb{Z}_q$, i.e. a constant number of operations in $\mathbb{Z}_q$. Hence Substep 5.b has a cost in $O(n^{3+\varepsilon})$. The third reduction in Substep 5.c is negligible.

In Step 6 the costly part is to compute the norm of the matrix. By a recursive "divide and conquer" computation, we can save some of the costly Frobenius computations and obtain a runtime again in $O(n^{3+\varepsilon})$. In [31], Satoh proposes another method which can moreover save memory.

Putting everything together, the complexity of the algorithm is $O(n^{3+\varepsilon})$ in time and in space.

5 Numerical Results and Cryptographic Significance

As far as we know, even the original algorithm of Kedlaya has not yet been tested in practice. Therefore, we did our first implementation with the first aim of validating Kedlaya's algorithm and our extension. We used MAGMA, version 2.7, which allowed us to easily manipulate quite complicated objects: it is possible in MAGMA to construct the ring $\mathbb{Z}_q$ and to build the ring of series over polynomials over $\mathbb{Z}_q$ which is required. However, by taking such a high programming level, we can not really hope to do all the optimizations we could dream of; furthermore, there are some small bugs in our version of MAGMA which make us lose precision from time to time and we had to take a (constant) added margin in the precision of the computations. Therefore the results we give here are just meant to show that the algorithm works in practice and that cryptographic sizes are clearly reachable. We are currently working on an optimized implementation in *C* which should reduce the runtime significantly.

All the examples have been run on an Alpha EV6 at 667 MHz. The numbers of points are small enough that it is possible to factor them and prove the results. The space requirement was roughly 150 MB.

5.1 Hyperelliptic Examples

In the hyperelliptic case, we cannot take a field of characteristic 2 for which the algorithm is not designed. We carried out our experiments with finite fields of characteristic 3.

Example 1.

In $\mathbb{F}_{3^{53}}$, we take the generator t given by $t^{53} + 2t^4 + 2t^3 + 2t^2 + 1 = 0$ and consider the genus 2 randomly chosen curve given by

$$\begin{aligned} y^2 = x^5 &+ t^{232112179875500370302098 9}x^4 + t^{8444066873716648223072527}x^3 + \\ &t^{7946343052437940195139141}x^2 + t^{10959512142684015392587300}x + \\ &t^{11366373156356845343093334} . \end{aligned}$$

After about 22 hours of computation we found the coefficients of its characteristic polynomial to be

$$a_1 = 3767947898876,$$
$$a_2 = 16462680188903823501200294,$$

which yields a cardinality of

$$N = 375710212613709295385367112322529717794218564821248.$$

Example 2.

We took a randomly chosen curve over the finite field $\mathbb{F}_q$ with $q = 3^{37}$. Let t with minimal polynomial $t^{37} + t^3 + 2t^2 + 2t + 1$, and consider the genus 3 curve of equation

$$\begin{aligned} y^2 = x^7 &+ t^{1450056053378032 44}x^6 + t^{367106618571281107}x^5 + t^{377813655811225893}x^4 + \\ &t^{472884120990578 87}x^3 + t^{558710154046987 90}x^2 + t^{232037785016055219}x + \\ &t^{286815047052544398} . \end{aligned}$$

After about 30 hours of computation we found the coefficients of its characteristic polynomial to be

$$a_1 = 1128783670,$$
$$a_2 = 1117168429648455309,$$
$$a_3 = 886287268279616285414037148,$$

which yields a cardinality of

$$N = 91297581893980817420223885655399261733128358845689672.$$

5.2 Superelliptic Examples

For superelliptic curves, we concentrate on characteristic 2 which is the most interesting case for practical applications.

Example 1.

In $\mathbb{F}_{2^{53}}$, we take the generator t given by $t^{53}+t^6+t^2+t+1=0$ and consider the randomly chosen curve

$$y^3 = x^4 + t^{2256567407303775}x^3 + t^{7508555791178511}x^2 + t^{1136027055799467}x + t^{4967542575384673}.$$

After about 22 hours of computation we found the coefficients of its characteristic polynomial to be

$$a_1 = 0,$$
$$a_2 = -2299871474212151,$$
$$a_3 = 0,$$

which yields a cardinality of

$$N = 730750818665451438386441787834386121601727865546.$$

The nullity of a_1 and a_2 is not a surprise: it is explained by the absence of third roots of unity in the base field (see below).

Example 2.

In $\mathbb{F}_{2^{58}}$, we take the generator t given by $t^{58}+t^{19}+1=0$ and consider the random curve

$$y^3 = x^4 + t^{184416898722999862}x^3 + t^{138153554162118062}x^2 + t^{90053985362597546}x + t^{159188191651769175}.$$

After about 28 hours of computation we found the coefficients of its characteristic polynomial to be

$$a_1 = 1346491223,$$
$$a_2 = 540650236559852363,$$
$$a_3 = 1067868967585078516467630008,$$

which yields a cardinality of

$$N = 2394524293789162792332288212231614478974438189795 4979.$$

The cardinalities we found were not (almost) prime and these curves should not be used in cryptography. We could have repeated the computations for several curves until we found a good curve. Note that an early-abort strategy cannot be used in this context.

5.3 Some Remarks about Superelliptic Curves in Cryptography

When one wants to build a cryptosystem based upon a curve, there are some security issues that have to be taken into account. Besides the fact that the number of points of the Jacobian should be (almost) prime, the following attacks (or threats) should be avoided:

1. Index-calculus attack for high genus curves [14,4]: the genus of the curve should be at most 3.
2. MOV attack [25,11]: the smallest k such that $\#\mathrm{J}(\mathcal{C}) \mid q^k - 1$ should be large.
3. Rück's attack [30]: the order of the subgroup in which we are working should be coprime to p.
4. The curve should not have "special properties".

Item 1 means that we are left with a small choice of non-elliptic curves useful for cryptography: hyperelliptic curves of genus 2 and 3, and superelliptic curves of the form $y^3 = f(x)$ with f of degree 4.

Items 2 and 3 are almost always fulfilled when we choose random curves and the verification that this is indeed the case for a given curve is straightforward.

The fourth item is less precise but has its importance: nowadays some people do not recommend to use elliptic curves for which the class number of the ring of endomorphism is too small; the base field should be a prime field or a prime extension field due to the threat of an attack by Weil descent [16]; and more generally any special behavior of the curves could be considered as suspect.

Keeping all this in mind, consider now a curve $\mathcal{C}$ of the form $y^3 = f(x)$ with f of degree 4 over a field $\mathbb{F}_q$. Assume that q is congruent to 2 modulo 3. Then every element of $\mathbb{F}_q$ is a cube. Therefore $\#\mathcal{C}/\mathbb{F}_q$ is equal to $q+1$, counting the point at infinity. Furthermore this is the case in every extension of $\mathbb{F}_q$ which does not contain the third roots of unity, namely every odd degree extension. A simple calculation with zeta functions shows that this implies that the coefficients a_1 and a_3 in the characteristic polynomial are zero, therefore $\chi(T)$ is of the form $T^6 + a_2T^4 + qa_2T^2 + q^3$, as we observed in Example 1. It means that this curve is highly "non-random" among all the curves of genus 3. In particular, the 3-torsion part of the Jacobian is partly degenerate which is a first step towards supersingularity. In [35], the reader will find a survey about the gradation from ordinary curves to supersingular curves and the link with the Newton polygon of the characteristic polynomial.

Having noticed this, one is tempted to claim that it is safer to take a base field which includes the third roots of unity. However, we are confronted to another problem, at least in characteristic 2. Indeed, $\mathbb{F}_{2^n}$ will contain the third root of unity if and only if n is even. We could then be subject to a Weil descent attack: if $n = 2m$, by doing a Weil restriction on $\mathrm{J}(\mathcal{C})$, we get an abelian variety of dimension 6 over $\mathbb{F}_{2^m}$. If someone is able to draw a curve of genus 6 on this abelian variety, then the system is broken. As far as we know, nobody is able to find such a curve (if it exists!) but this could be threatening enough to discourage the use $\mathcal{C}$ for cryptography.

This phenomenon is only true when one wants to use a base field of small characteristic. If we use a curve over a prime field, no Weil descent attack is to be feared and one can take a base field with roots of unity. This implies that there are additional automorphisms in the Jacobian and that the key-size should be slightly enlarged accordingly [8].

6 Conclusion

We have presented an extension of Kedlaya's algorithm in order to count points on superelliptic curves over finite fields of small characteristic. The time complexity is the same as the complexity for hyperelliptic curves. This complexity is asymptotically the same as the best known methods for counting points on elliptic curves. Note however that the ε which is involved in the expression $O(\log^{3+\varepsilon} q)$ does not hide the same logarithmic factors. We obtained some numerical examples proving that it is now feasible to count points of random hyperelliptic and superelliptic curves up to genus 3, for cryptographic sizes.

Further research topics are: extend Kedlaya's algorithm for hyperelliptic curves to characteristic 2, reduce the space complexity to $O(\log^2 q)$, extend the algorithm to C_{ab} curves or even to more general varieties (in fact Monsky–Washnitzer cohomology exists for more general varieties).

Acknowledgements. We are grateful to François Morain and to Guillaume Hanrot for their continuous support and many discussions during this work. A special thanks to Robert Harley for his close reading and his helpful comments.

Some of the computations were carried out on the machines of the UMS Medicis at École Polytechnique.

References

1. L. M. Adleman, J. DeMarrais, and M.-D. Huang. A subexponential algorithm for discrete logarithms over the rational subgroup of the jacobians of large genus hyperelliptic curves over finite fields. In L. Adleman and M.-D. Huang, editors, *ANTS-I*, volume 877 of *Lecture Notes in Comput. Sci.*, pages 28–40. Springer-Verlag, 1994.
2. S. Arita. Algorithms for computations in Jacobians of C_{ab} curve and their application to discrete-log-based public key cryptosystems. In *Proceedings of Conference on The Mathematics of Public Key Cryptography, Toronto, June 12–17*, 1999.
3. S. Arita. Construction of secure C_{ab} curves using modular curves. In W. Bosma, editor, *ANTS-IV*, volume 1838 of *Lecture Notes in Comput. Sci.*, pages 113–126. Springer-Verlag, 2000.
4. S. Arita. Gaudry's variant against C_{ab} curve. In H. Imai and Y. Zheng, editors, *Public Key Cryptography*, volume 1751 of *Lecture Notes in Comput. Sci.*, pages 58–67. Springer-Verlag, 2000.
5. S. Arita. Weil descent of elliptic curves over finite fields of characteristic three. In T. Okamoto, editor, *Advances in Cryptolgy – ASIACRYPT 2000*, volume 1976 of *Lecture Notes in Comput. Sci.*, pages 248–258. Springer-Verlag, 2000.

6. A. Basiri, A. Enge, J.-C. Faugère, and N. Gürel. Fast arithmetic on superelliptic cubics. In preparation.
7. I. Blake, G. Seroussi, and N. Smart. *Elliptic curves in cryptography*, volume 265 of *London Math. Soc. Lecture Note Ser.* Cambridge University Press, 1999.
8. I. Duursma, P. Gaudry, and F. Morain. Speeding up the discrete log computation on curves with automorphisms. In K.Y. Lam, E. Okamoto, and C. Xing, editors, *Advances in Cryptology – ASIACRYPT '99*, volume 1716 of *Lecture Notes in Comput. Sci.*, pages 103–121. Springer-Verlag, 1999.
9. M. Fouquet, P. Gaudry, and R. Harley. An extension of Satoh's algorithm and its implementation. *J. Ramanujan Math. Soc.*, 15:281–318, 2000.
10. M. Fouquet, P. Gaudry, and R. Harley. Finding secure curves with the Satoh-FGH algorithm and an early-abort strategy. In B. Pfitzmann, editor, *Advances in Cryptology – EUROCRYPT 2001*, volume 2045 of *Lecture Notes in Comput. Sci.*, pages 14–29. Springer-Verlag, 2001.
11. G. Frey and H.-G. Rück. A remark concerning m-divisibility and the discrete logarithm in the divisor class group of curves. *Math. Comp.*, 62(206):865–874, April 1994.
12. W. Fulton. *Algebraic curves.* Math. Lec. Note Series. W. A. Benjamin Inc, 1969.
13. S. Galbraith, S. Paulus, and N. Smart. Arithmetic on superelliptic curves. To appear Math. Comp.
14. P. Gaudry. An algorithm for solving the discrete log problem on hyperelliptic curves. In B. Preneel, editor, *Advances in Cryptology – EUROCRYPT 2000*, volume 1807 of *Lecture Notes in Comput. Sci.*, pages 19–34. Springer-Verlag, 2000.
15. P. Gaudry and R. Harley. Counting points on hyperelliptic curves over finite fields. In W. Bosma, editor, *ANTS-IV*, volume 1838 of *Lecture Notes in Comput. Sci.*, pages 313–332. Springer-Verlag, 2000.
16. P. Gaudry, F. Hess, and N. Smart. Constructive and destructive facets of Weil descent on elliptic curves. To appear in J. Crypt.
17. R. Harasawa and J. Suzuki. Fast jacobian group arithmetic on C_{ab} curves. In W. Bosma, editor, *ANTS-IV*, volume 1838 of *Lecture Notes in Comput. Sci.*, pages 359–376. Springer-Verlag, 2000.
18. K. Kedlaya. Counting points on hyperelliptic curves using Monsky-Washnitzer cohomology. Preprint, available at `http://arXiv.org/abs/math/0105031`, 2001.
19. N. Koblitz. *p-adic numbers, p-adic analysis and Zeta-functions*, volume 58 of *Graduate Texts in Mathematics.* Springer-Verlag, 1977.
20. N. Koblitz. *p-adic analysis: a short course on recent work*, volume 46 of *London Math. Lec. Note Series.* Cambridge University Press, 1980.
21. N. Koblitz. Elliptic curve cryptosystems. *Math. Comp.*, 48(177):203–209, January 1987.
22. N. Koblitz. Hyperelliptic cryptosystems. *J. of Cryptology*, 1:139–150, 1989.
23. R. Lercier. *Algorithmique des courbes elliptiques dans les corps finis.* Thèse, École polytechnique, June 1997.
24. D. Lorenzini. *An invitation to arithmetic geometry*, volume 106 of *Graduate Studies in Mathematics.* AMS, 1993.
25. A. Menezes, T. Okamoto, and S. A. Vanstone. Reducing elliptic curves logarithms to logarithms in a finite field. In *Proceedings 23rd Annual ACM Symposium on Theory of Computing (STOC)*, pages 80–89. ACM Press, 1991. May 6–8, New Orleans, Louisiana.
26. V. Miller. Use of elliptic curves in cryptography. In A. M. Odlyzko, editor, *Advances in Cryptology – CRYPTO '86*, volume 263 of *Lecture Notes in Comput. Sci.*, pages 417–426. Springer-Verlag, 1987.

27. P. Monsky and G. Washnitzer. Formal cohomology: I. *Ann. of Math. (2)*, 88:181–217, 1968.
28. P. Monsky. Formal cohomology: III. *Ann. of Math. (2)*, 93:315–343, 1971.
29. J. Pila. Frobenius maps of abelian varieties and finding roots of unity in finite fields. *Math. Comp.*, 55(192):745–763, October 1990.
30. H. G. Rück. On the discrete logarithm in the divisor class group of curves. *Math. Comp.*, 68(226):805–806, 1999.
31. T. Satoh. Asymptotically fast algorithm for computing the Frobenius substitution and norms over unramified extension of p-adic number fields. Preprint 2001.
32. T. Satoh. The canonical lift of an ordinary elliptic curve over a finite field and its point counting. *J. Ramanujan Math. Soc.*, 15:247–270, 2000.
33. B. Skjernaa. Satoh's algorithm in characteristic 2. To appear in Math. Comp.
34. F. Vercauteren, B. Preneel, and J. Vandewalle. A memory efficient version of Satoh's algorithm. In B. Pfitzmann, editor, *Advances in Cryptology – EURO-CRYPT 2001*, volume 2045 of *Lecture Notes in Comput. Sci.*, pages 1–13. Springer-Verlag, 2001.
35. N. Yui. On the jacobian varietes of hyperelliptic curves over fields of characteristic $p > 2$. *J. Algebra*, 52:378–410, 1978.

Supersingular Curves in Cryptography

Steven D. Galbraith*

Mathematics Department,
Royal Holloway University of London,
Egham, Surrey TW20 0EX, UK.
Steven.Galbraith@rhul.ac.uk

Abstract. Frey and Rück gave a method to transform the discrete logarithm problem in the divisor class group of a curve over $\mathbb{F}_q$ into a discrete logarithm problem in some finite field extension $\mathbb{F}_{q^k}$. The discrete logarithm problem can therefore be solved using index calculus algorithms as long as k is small.
In the elliptic curve case it was shown by Menezes, Okamoto and Vanstone that for supersingular curves one has $k \leq 6$. In this paper curves of higher genus are studied. Bounds on the possible values for k in the case of supersingular curves are given which imply that supersingular curves are weaker than the general case for cryptography. Ways to ensure that a curve is not supersingular are also discussed.
A constructive application of supersingular curves to cryptography is given, by generalising an identity-based cryptosystem due to Boneh and Franklin. The generalised scheme provides a significant reduction in bandwidth compared with the original scheme.

1 Introduction

Frey and Rück [8] described how the Tate pairing can be used to map the discrete logarithm problem in the divisor class group of a curve C over a finite field $\mathbb{F}_q$ into the multiplicative group $\mathbb{F}_{q^k}^*$ of some extension of the base field. This has significant implications for cryptography as there are well-known subexponential algorithms for solving the discrete logarithm problem in a finite field. Therefore, there is a method for solving the discrete logarithm problem in the divisor class group in those cases where the extension degree k is small.

The extension degree required is the smallest integer k such that the large prime order l of the divisor class group $\mathrm{Pic}_C^0(\mathbb{F}_q)$ is such that $l|(q^k-1)$. In general the value of k depends on both the field and the curve and is very large (i.e., $\log(k) \approx \log(q)$).

Menezes, Okamoto and Vanstone [23] showed that for supersingular elliptic curves the value k above is always less than or equal to 6. This important result

* This research was supported by the Centre for Applied Cryptographic Research at the University of Waterloo, the NRW-Initiative für Wissenschaft und Wirtschaft "Innovationscluster für Neue Medien", cv cryptovision gmbh (Gelsenkirchen) and Hewlett-Packard laboratories, Bristol.

C. Boyd (Ed.): ASIACRYPT 2001, LNCS 2248, pp. 495–513, 2001.
© Springer-Verlag Berlin Heidelberg 2001

implies that supersingular elliptic curves are weaker than the general case for cryptography.

Elliptic curve cryptography was generalised to higher genus curves by Koblitz [16]. Our main result is Theorem 3 which states that for supersingular curves there is an upper bound, which depends only on the genus, on the values of the extension degree k. This bound is sufficiently small (see Table 1) that supersingular curves should be considered weaker than the general case for cryptography.

It is important to be able to detect these weak cases in advance, especially when one is considering curves defined over small fields and using the zeta function to compute the group order over extension fields. Sakai, Sakurai and Ishizuka [27] were unable to find any secure hyperelliptic curves of genus two over $\mathbb{F}_2$. In Section 5 we show why the authors of [27] failed in their search and we explain how to avoid equations for supersingular curves in characteristic two. As an illustration we overcome the problem encountered in [27] and provide examples of secure genus two curves over $\mathbb{F}_2$.

Recently, beginning with the work of Joux [14], the Weil pairing has found positive applications in cryptography. In Section 3 we generalise an identity-based cryptosystem due to Boneh and Franklin [2]. Our scheme provides a significant improvement in bandwidth over the scheme of Boneh and Franklin.

2 The Tate Pairing

In this section we summarise various known results. Throughout the paper C is a non-singular, irreducible curve of genus g over a finite field $\mathbb{F}_q$ where q is a power of a prime p. The Jacobian of the curve C is an abelian variety $\mathrm{Jac}(C)$ of dimension g defined over $\mathbb{F}_q$. The $\mathbb{F}_q$-rational points on the Jacobian correspond to the divisor class group of the curve over $\mathbb{F}_q$, which we denote $\mathrm{Pic}^0_C(\mathbb{F}_q)$ (for background details see [4], [16], [29], [33]).

Those readers only interested in elliptic curves can take C to be an elliptic curve and can think of $\mathrm{Jac}(C)(\mathbb{F}_q) = \mathrm{Pic}^0_C(\mathbb{F}_q) = C(\mathbb{F}_q)$.

2.1 The Tate Pairing

Let l be a positive integer which is coprime to q. In most applications l is a prime and $l | \#\mathrm{Pic}^0_C(\mathbb{F}_q)$. Let k be a positive integer such that the field $\mathbb{F}_{q^k}$ contains the lth roots of unity (in other words, $l | (q^k - 1)$).

Let $G = \mathrm{Pic}^0_C(\mathbb{F}_{q^k})$ and write $G[l]$ for the subgroup of divisors of order l and G/lG for the quotient group The Tate pairing is a mapping

$$\langle \cdot, \cdot \rangle : G[l] \times G/lG \to \mathbb{F}^*_{q^k}/(\mathbb{F}^*_{q^k})^l \quad (1)$$

where the right hand side is the quotient group of elements of $\mathbb{F}^*_{q^k}$ modulo lth powers. Note that all three groups $G[l]$, G/lG and $\mathbb{F}^*_{q^k}/(\mathbb{F}^*_{q^k})^l$ have exponent l. The Tate pairing satisfies the following properties [8]:

1. (Well-defined) $\langle 0, Q\rangle \in (\mathbb{F}_{q^k}^*)^l$ for all $Q \in G$ and $\langle P, Q\rangle \in (\mathbb{F}_{q^k}^*)^l$ for all $P \in G[l]$ and all $Q \in lG$.
2. (Non-degeneracy) For each divisor class $P \in G[l] - \{0\}$ there is some divisor class $Q \in G$ such that $\langle P, Q\rangle \notin (\mathbb{F}_{q^k}^*)^l$.
3. (Bilinearity) For any integer n, $\langle nP, Q\rangle \equiv \langle P, nQ\rangle \equiv \langle P, Q\rangle^n$ modulo lth powers.

The Tate pairing is computed as follows: Let P be a divisor of order l. There is a function f whose divisor, which we write as (f), is equal to lP. Then $\langle P, Q\rangle = f(Q')$ where Q' is a divisor in the same class as Q such that the support of Q' is disjoint with the support of (f). This computation is easily implemented in practice by using the double and add algorithm and evaluating all the intermediate functions at Q' (see [8], [9]).

The value $f(Q')$ lies in $\mathbb{F}_{q^k}^*$. By raising it to the power $(q^k - 1)/l$ we obtain an lth root of unity.

One subtlety when implementing the Tate pairing is finding a divisor Q' with support disjoint from the partial terms in the addition chain for lP. In the elliptic curve case this is done by taking $Q' = (Q + S) - (S)$ where $(Q) - (\infty)$ is the target divisor and where S is an arbitrary point (not necessarily of order l). In the higher genus case general Riemann-Roch algorithms can be used to give an analogous solution. In practice, it is often easier not to choose the class Q first but to just choose two 'random' effective divisors E_1 and E_2 of degree g and set $Q' = E_1 - E_2$. If E_1 and E_2 are chosen randomly over $\mathbb{F}_{q^k}$ then with high probability we expect $\langle P, Q'\rangle \notin (\mathbb{F}_{q^k}^*)^l$.

In the case of elliptic curves one can compare the Tate pairing with the Weil pairing. In general there is no relationship between the Tate pairing and the Weil pairing, as they are defined on different sets. However, when E is an elliptic curve such that $l^2 \| \#E(\mathbb{F}_{q^k})$ and P, Q are independent points in $E(\mathbb{F}_{q^k})[l]$ then we have $e_l(P, Q) = \langle P, Q\rangle / \langle Q, P\rangle$. A consequence of this is that the Tate pairing is not symmetric.

The Weil pairing requires working over the field $\mathbb{F}_q(E[l])$ generated by the coordinates of all the l-division points. In general, one would expect this field to be larger than that used for the Tate pairing, however at ECC '97 Koblitz observed that these fields are usually the same. Finally, the Weil pairing requires roughly twice the computation time as the Tate pairing, although this is partly offset by the added cost of a finite field exponentiation (to the power $(q^k - 1)/l$) in the case of the Tate pairing if a unique value is required.

2.2 The Frey-Rück Attack

We now recall how the Tate pairing is used to attack the discrete logarithm problem in the divisor class group of a curve (this approach is often called the Frey-Rück attack, after [8]). Let $D_1, D_2 \in \mathrm{Pic}_C^0(\mathbb{F}_q)$ be divisors of order l for which we want to solve the discrete logarithm problem $D_2 = \lambda D_1$. Let k be the smallest integer such that the pairing is non-degenerate (hence $l | (q^k - 1)$). The method proceeds as follows:

1. Choose random divisors $Q \in \mathrm{Pic}^0_C(\mathbb{F}_{q^k})$ until $\langle D_1, Q\rangle \notin (\mathbb{F}^*_{q^k})^l$.
2. Compute $\zeta_i = \langle D_i, Q\rangle \in \mathbb{F}^*_{q^k}$.
3. Raise ζ_i to the power $(q^k-1)/l$ (this stage is optional since the linear algebra in the index calculus method below should be performed modulo l).
4. Solve the discrete logarithm problem $\zeta_2 = \zeta_1^\lambda$ in the finite field $\mathbb{F}^*_{q^k}$ using an index calculus method.

This strategy is practical when k is small. This leads to the following important question for cryptography:

Question: Are there certain weak cases of curves for which k is always small?

One of the goals of this paper is to show that, as in the case of elliptic curves, supersingular curves always have small k. Of course, there are lots of non-supersingular curves for which the Frey-Rück attack applies (e.g., elliptic curves over $\mathbb{F}_p$ with $p-1$ points).

2.3 Non-degeneracy of the Tate Pairing

We now discuss the non-degeneracy property a little more closely. Let $P \in G[l]$. We consider the possibilities for $\langle P, P\rangle$. To compute $\langle P, P\rangle$ it is necessary to compute a divisor Q in the same class as P but which has support disjoint from all the intermediate terms in the computation of lP. One can then compute $\langle P, Q\rangle$ to obtain the value of the pairing. If $P \in lG$ then $\langle P, P\rangle \in (\mathbb{F}^*_{q^k})^l$. If $P \in \mathrm{Pic}^0_C(\mathbb{F}_q)$ then $\langle P, P\rangle \in \mathbb{F}^*_q$, but if l is prime and if l does not divide $(q-1)$ then $\langle P, P\rangle \in (\mathbb{F}^*_{q^k})^l$ since every element of $\mathbb{F}^*_{q^k}$ is an lth power in that case. Hence to have $\langle P, P\rangle$ nontrivial it is necessary (but not sufficient) that $l|(q-1)$ and so $k=1$.

The following result originates from the work of [2] and [36]. It provides a very useful technique for finding points where the pairing is non-degenerate.

Lemma 1. *Let E be an elliptic curve. Let $P \in E(\mathbb{F}_q)$ be a point of prime order l. Let $\mathbb{F}_{q^k}$ be the extension over which all points of order l are defined, and write $G = E(\mathbb{F}_{q^k})$. Suppose that $l^2 \| \#G$ (i.e., that $G[l] \cong G/lG$). Let ψ be an endomorphism of E which is not defined over $\mathbb{F}_q$. If $\psi(P) \notin E(\mathbb{F}_q)$ then $\langle P, \psi(P)\rangle^{(q^k-1)/l} \neq 1$.*

For the proof see the full version [11]. We refer to the maps ψ as 'non-$\mathbb{F}_q$-rational endomorphisms' (Verheul [36] calls them 'distortion maps').

In the case of curves of genus greater than one then this result is no longer true. On the other hand, in this setting there are usually many endomorphisms ψ available. Indeed, for supersingular abelian varieties it will generally be true that, for all P, there is some endomorphism ψ such that $\langle P, \psi(P)\rangle^{(q^k-1)/l} \neq 1$.

3 Identity-Based Cryptosystems Using the Tate Pairing

Identity based cryptography was proposed by Shamir [28] as a response to the problem of managing public keys. The basic principle is that it should be possible to derive a user's public data only from their identity. It is therefore necessary to have a trusted dealer who can provide a user with the secret key corresponding to the public key which is derived from their identity. It has turned out to be rather difficult to construct efficient and secure identity-based cryptosystems.

Recently, Boneh and Franklin [2] developed a new identity-based cryptosystem using the Weil pairing on a specific supersingular elliptic curve. In this section we show that the use of other supersingular curves leads to significant efficiency improvements over the original scheme.

3.1 Dealer's System Parameters

The dealer sets up the scheme by choosing a finite field $\mathbb{F}_q$ and a curve C over $\mathbb{F}_q$ of genus g such that:

1. There is a large prime l dividing the order of the group $\mathrm{Pic}^0_C(\mathbb{F}_q)$.
2. The degree k needed for the Tate pairing embedding of the subgroup of order l (i.e., the smallest k such that $l|(q^k-1)$) is relatively small.

One approach is to take C to be a supersingular curve.

The dealer then chooses a divisor $P \in \mathrm{Pic}^0_C(\mathbb{F}_q)$ of order l and a secret integer $1 < s < l$ and computes $P' = sP$. The dealer publishes q, C, l, k, P, P' and keeps the integer s secret. The public data for the scheme also includes two hash functions H_1 and H_2 (these are called G and H in [2]). The function H_1 is used to map identities to bitstrings which are then used to represent divisors in $\mathrm{Pic}^0_C(\mathbb{F}_{q^k})$. The function H_2 maps elements of the subgroup of order l of $\mathbb{F}^*_{q^k}$ to bitstrings of a certain length N. Both hash functions are required to be cryptographically strong and are modelled in the security proofs of [2] as random oracles.

3.2 User's Public and Private Key

We now discuss how a user's identity gives rise to a public key. There must be a procedure to convert the identity of user A (such as their name or email address). to a divisor $Q_A \in G = \mathrm{Pic}^0_C(\mathbb{F}_{q^k})$ such that:

1. $\langle P, Q_A\rangle \notin (\mathbb{F}^*_{q^k})^l$.
2. The process should be one-way, in the sense that it be infeasible to find an identity which gives rise to a given point Q_A.
3. The points Q_A should be distributed uniformly in an appropriate set.

In [2] this process (which Boneh and Franklin call 'MapToPoint') is solved using a cryptographically strong hash function H_1 and a non-$\mathbb{F}_q$-rational endomorphism ψ. We now sketch a generalisation of their method.

The identity bitstring is concatenated with a padding string and then passed through the hash function H_1 (which is constructed to yield a full domain output). This process is repeated using a deterministic sequence of padding strings until the output is the x-coordinate (or $a(x)$-term in the higher genus case) of an element Q of $\mathrm{Pic}^0_C(\mathbb{F}_q)$. It is then easy to find the rest of the representation of Q. One then sets $Q_A = \psi(mQ) \in G$ for a suitable non-$\mathbb{F}_q$-rational endomorphism from the available possibilities where m is the cofactor $\#\mathrm{Pic}^0_C(\mathbb{F}_q)/l$. This process is repeated until $\langle P, Q_A\rangle^{(q^k-1)/l} \neq 1$.

A more general scheme, which does not require non-$\mathbb{F}_q$-rational endomorphisms, is given in [11].

To summarise, every user A has a public key consisting of the divisor Q_A and everyone can obtain this public key just knowing the identity of the user. Each user asks the dealer for a private key $Q'_A = sQ_A$. This must be transmitted to the user using a secure channel.

3.3 Encryption and Decryption

Let the message M be a bitstring of length N and suppose we want to send this to user A. First derive the public key Q_A from the identity of A and obtain the dealer's public keys P and P'. The remaining steps are

1. Choose a random integer $1 \le r \le l$.
2. Compute $R = rP$.
3. Compute $S = M \oplus H_2(\langle P', Q_A\rangle^{r(q^k-1)/l})$. (Recall that $\langle P', Q_A\rangle \in \mathbb{F}^*_{q^k}$.)
4. Send (R, S).

To decrypt, user A simply uses their private key Q'_A to compute $\langle R, Q'_A\rangle$. Recall that $\langle rP, sQ_A\rangle \equiv \langle P, Q_A\rangle^{rs} \equiv \langle P', Q_A\rangle^r$ modulo lth powers. Hence the message is recovered from

$$M = S \oplus H_2(\langle R, Q'_A\rangle^{(q^k-1)/l}).$$

A more versatile encryption process is obtained by using $H_2(\langle P', Q_A\rangle^{r(q^k-1)/l})$ as the key for a fixed symmetric encryption function.

3.4 Security

The security of this system relies on the following variant of the Diffie-Hellman problem:

Definition 1. *The* **Tate-Diffie-Hellman problem (TDH)** *is the following: Let G and l be as above. Given divisors $P, P' = sP, R = rP$ and $Q_A \in G$ of order l such that $\langle P, Q_A\rangle^{(q^k-1)/l} \neq 1$ compute $\zeta = \langle P, Q_A\rangle^{rs(q^k-1)/l}$.*

Let $P \in \mathrm{Pic}^0_C(\mathbb{F}_q)$ be any divisor of large prime order l. We make the assumption that the Tate-Diffie-Hellman problem is hard over random P', R, Q_A, i.e., where $Q_A = \psi(Q)$ (for a suitable non-$\mathbb{F}_q$-rational endomorphism) and where $P', R, Q \in \langle P \rangle$ are chosen uniformly at random.

If one can solve the elliptic curve Diffie-Hellman problem then one can compute rsP and thus $\langle rsP, Q_A \rangle$. Similarly, if one can solve the Diffie-Hellman problem in $\mathbb{F}_{q^k}^*$ then one can solve the TDH.

To produce a cryptosystem with strong security properties (indistinguishability of encryptions under a chosen ciphertext attack) one uses a method of Fujisaki and Okamoto which is discussed thoroughly in [2]. First it is necessary to establish that the basic scheme has the 'one-way encryption' (ID-OWE) security property (see Section 2 of [2]). The security proof for the scheme above is completely analogous to the proof of Theorem 4.1 of [2] and it holds under the assumptions that the hash functions H_1 and H_2 are random oracles and that the TDH problem is hard.

3.5 Parameter Sizes and Performance

For security it is necessary that $q^g \geq 2^{160}$ and $q^k \geq 2^{1024}$. Boneh and Franklin [2] use $g = 1$ and $k = 2$ and so they must take q to be of size at least 512 bits[1]. The whole point of our generalisation is the observation that if k can be taken to be larger than 2 then q may be taken to be smaller. In Section 3.6 we give the details for a curve with $k = 6$. Hence there are the following advantages of the generalised scheme compared with the scheme of [2].

- The bandwidth (number of bits) of an encryption (R, S) can be reduced (see Section 3.6 below).
- For the same reason, the dealer's public keys also require less storage and communication bandwidth with the new scheme.
- The dominant cost in encryption and decryption is the evaluation of the Tate pairing. Since this involves computations in the large field $\mathbb{F}_{q^k}$ the cost of encryption and decryption is roughly comparable for both schemes, although there are some savings available in characteristic two.

As mentioned in [2], the computation of the Weil and Tate pairings can be made much faster by choosing the prime l of size around 160 bits.

3.6 Characteristic Three Example

With elliptic curves one can realise an improvement of k from 2 to 6 by taking the elliptic curves

$$E_1 : y^2 = x^3 - x + 1 \qquad \text{and} \qquad E_2 : y^2 = x^3 - x - 1$$

over $\mathbb{F}_{3^l}$, which have characteristic polynomial of Frobenius $P_{E_1}(X) = X^2 + 3X + 3$ and $P_{E_2}(X) = X^2 - 3X + 3$ respectively. These curves are thoroughly discussed by Koblitz in [18].

[1] Actually, in [2] it is specified that q have 1024 bits, but 512 bits seems to be sufficient.

A convenient non-$\mathbb{F}_3$-rational endomorphism for these curves is

$$\psi : (x, y) \mapsto (-\alpha - x, iy)$$

where $i \in \mathbb{F}_{3^2}$ satisfies $i^2 = -1$ and $\alpha \in \mathbb{F}_{3^3}$ satisfies $\alpha^3 - \alpha + 1 = 0$.

We list some values of m such that the group order of $E_i(\mathbb{F}_{3^m})$ is equal to a small cofactor c times a large prime l.

m	i	# bits in l	c
79	2	125	1
97	1	151	7
149	1	220	$7 \cdot 15199$
163	1	256	7
163	1	259	1
167	1	262	7
167	2	237	$8017 \cdot 44089$
173	2	241	16420688749
193	2	306	1
239	2	379	1

Consider, say, the case $m = 163$ which is a 259 bit field. Since $k = 6$ the size of the field $\mathbb{F}_{q^k}$ is 1551 bits. If messages are of length $N = 160$ bits then an encryption requires $160 + 260 = 420$ bits (259 bits for the x-coordinate of the point and one bit to specify the y-coordinate). For equivalent security using the Boneh-Franklin scheme with $k = 2$ one must take p to be $\lceil 1551/2 \rceil = 776$ bits and so an encryption will require $160 + 776 = 936$ bits (we have 776 as the Boneh-Franklin scheme only requires sending the y-coordinate). Hence our scheme requires less than half the bandwidth of the Boneh-Franklin scheme for the same security level.

3.7 Characteristic Two Example

In characteristic two there are curves available which attain the Frey-Rück embedding degree $k = 4$. In these cases the bandwidth improvement is not as significant as that seen with the characteristic three example above. However, it is easy to get an improvement in performance over the scheme in [2].

Consider the elliptic curves

$$E_1 : y^2 + y = x^3 + x \qquad \text{and} \qquad E_2 : y^2 + y = x^3 + x + 1$$

over $\mathbb{F}_2$. Then E_1 has characteristic polynomial of Frobenius $P_{E_1}(X) = X^2 + 2X + 2$ while E_2 is the quadratic twist of E_1 and has $P_{E_2}(X) = X^2 - 2X + 2$.

We list some values of m such that $\#E_i(\mathbb{F}_{2^m}) = cl$ where l is a large prime and where c is a cofactor.

m	i	# bits in l	c
233	1	210	$5 \cdot 3108221$
239	2	239	1
241	2	241	1
271	1	252	$5 \cdot 97561$
283	1	281	5
283	2	283	1
353	2	353	1
367	2	367	1
397	2	397	1
457	2	457	1

A convenient non-$\mathbb{F}_2$-rational endomorphism for both these curves is given by

$$\psi : (x, y) \mapsto (u^2 x + s^2, y + u^2 s x + s)$$

where $u \in \mathbb{F}_{2^2}$ satisfies $u^2 + u + 1 = 0$ and $s \in \mathbb{F}_{2^4}$ satisfies $s^2 + (u + 1)s + 1 = 0$.

We give a comparison between characteristic 2 and large characteristic p for equivalent sized finite fields. We give the average time (in seconds) for the computation of the Tate pairing and the finite field exponentiation using the Magma computer algebra package. We also give a comparison of the communication bandwidth (number of bits) for the basic scheme (assuming a 160 bit hash function H).

The first case is with 965 bit finite field security (i.e., using E_2 over $\mathbb{F}_{2^{241}}$, which has a prime number of points).

Characteristic	Time	Bandwidth
2	2.4	402
p	4.3	642

Now for 1132 bit finite field security. This time using $E_1(\mathbb{F}_{2^{283}})$ whose number of points is 5 times a prime.

Characteristic	Time	Bandwidth
2	3.4	444
p	6.1	726

Clearly, the elliptic curves used by Boneh and Franklin lead to a scheme which requires about twice the computation time and over one and a half times the bandwidth compared with using curves in characteristic two.

3.8 Open Questions

We have seen that larger values of k help to make a more efficient identity-based cryptosystem. The problem is therefore to find curves C which have suitable large values of k (without being too large). This is very closely related to the question of Section 2.2

For supersingular curves we will show in Section 4.3 that there is an upper bound $k(g)$ (depending only on the genus g) for the values of k. The values of $k(g)$ are large enough to give good performance for the identity-based cryptosystem. However, it seems that one cannot realise these large values for $k(g)$ with suitable Jacobians of curves. It seems that the supersingular elliptic curves with $k = 4$ and $k = 6$ are the optimal choice for the identity-based cryptosystem and other applications using supersingular curves. More research is needed to clarify this.

It is not necessary to insist on using supersingular curves for the identity-based cryptosystem, since there should exist non-supersingular elliptic curves E over certain finite fields $\mathbb{F}_q$ with relatively small values of k. However, for such E it is usually the case that the order of $E(\mathbb{F}_q)$ is not divisible by a large prime (one exception is the case $p = 2l + 1$, but these only have $k = 1$). This phenomenon is indicated by the results of Balasubramanian and Koblitz [1] and is confirmed by computer experiments. It would be extremely interesting to have a construction for non-supersingular curves with relatively small values of k.

4 Supersingular Curves over Finite Fields

In this section we recall some facts about supersingular curves and we give our main result (Theorem 3). More details can be found in the full version of this paper [11].

As before, C is a non-singular, irreducible curve of genus g over a finite field $\mathbb{F}_q$. The Frobenius endomorphism π on $\mathrm{Jac}(C)$ satisfies a characteristic polynomial $P(X)$ of degree $2g$ with integer coefficients. We can factor $P(X)$ over the complex numbers as $P(X) = \prod_{i=1}^{2g}(X - \alpha_i)$. It turns out that the algebraic integers α_i have certain remarkable properties. In particular:

1. The numbers α_i satisfy $|\alpha_i| = \sqrt{q}$ and they can be indexed such that $\alpha_i \alpha_{g+i} = q$.
2. $P(X)$ has the following form
$$X^{2g} + a_1 X^{2g-1} + a_2 X^{2g-2} + \cdots + a_g X^g + q a_{g-1} X^{g-1} + \cdots + q^{g-1} a_1 X + q^g.$$
3. For any integer $r \geq 1$ we have $\#C(\mathbb{F}_{q^r}) = q^r + 1 - \sum_{i=1}^{2g} \alpha_i^r$.
4. For any integer $r \geq 1$ we have $\#\mathrm{Jac}(C)(\mathbb{F}_{q^r}) = \prod_{i=1}^{2g}(1 - \alpha_i^r)$.

The formula of property 4 for $\#\mathrm{Jac}(C)(\mathbb{F}_{q^r})$ gives an efficient method for computing the number of points in the divisor class group of a curve over a large-degree extension of the field $\mathbb{F}_q$ once one has computed $P(X)$ (see Appendix 1 for details about computing $P(X)$). For cryptography one wants a curve such that $\#\mathrm{Jac}(C)(\mathbb{F}_{q^r})$ is divisible by a large prime l and such that the group resists the known attacks ([8], [26]) on the discrete logarithm problem.

A common strategy is to try values of r until one is found for which the large prime l satisfies $\gcd(l, q) = 1$ and $q^{kr} \not\equiv 1 \pmod{l}$ for 'small' k. If the original curve is supersingular then, as we will show, it is futile to try many different values for r since the Frey-Rück attack will always work. Hence, it is important to know that such curves should be discarded right from the start.

4.1 Supersingularity

Recall that an elliptic curve E over $\mathbb{F}_{p^m}$ is supersingular if $E(\overline{\mathbb{F}}_p)$ has no points of order p (see [29]).

Definition 2. *(Oort [24]) An abelian variety A over $\mathbb{F}_q$ is called* **supersingular** *if A is isogenous to a product of supersingular elliptic curves. A curve C over $\mathbb{F}_q$ is called* **supersingular** *if* $\mathrm{Jac}(C)$ *is supersingular.*

The following result follows from the work of Manin and Oort.

Theorem 1. *The following conditions on an abelian variety A over $\mathbb{F}_q$ of dimension g are equivalent.*

1. *A is supersingular.*
2. *A is isogenous (over some finite extension of $\mathbb{F}_q$) to E^g for some supersingular elliptic curve E.*
3. *There is some integer k such that the characteristic polynomial of Frobenius on A over $\mathbb{F}_{q^k}$ is $P(X) = (X \pm q^{k/2})^{2g}$.*
4. *There is some integer k such that $\pi^k = \pm q^{k/2}$.*
5. *For some positive integer k we have $\#A(\mathbb{F}_{q^k}) = (q^{k/2} \pm 1)^{2g}$.*

The fourth property is the one which is most important for our application.

4.2 A Criterion for Supersingularity

The following result follows from Proposition 1 of Stichtenoth and Xing [34]. It gives a simple test for whether or not an abelian variety is supersingular, once $P(X)$ has been computed.

Theorem 2. *Suppose $q = p^n$ and suppose that A is an abelian variety of dimension g over $\mathbb{F}_q$. Let $P(X) = X^{2g} + a_1X^{2g-1} + \cdots + a_gX^g + \cdots + q^g$ be the characteristic polynomial of the Frobenius endomorphism on A. Then A is supersingular if and only if, for all $1 \le j \le g$,*

$$p^{\lceil jn/2 \rceil} \mid a_j.$$

4.3 The Bound on the Extension Degree

The values of k which arise depend on properties of cyclotomic polynomials (i.e., irreducible factors over $\mathbb{Z}$ of $X^m - 1$ for some m). Hence we make the following definitions.

Definition 3. *For each positive integer g let $\mathcal{P}_g = \{p(X) \in \mathbb{Z}[X] : \deg p(X) = 2g, p(X)$ irreducible over $\mathbb{Z}, p(X)|(X^m - 1)$ for some $m\}$. For each $p(X) \in \mathcal{P}_g$ define $m(p(X)) = \min\{m : p(X)|(X^m - 1)\}$. Define $k'(g)$ to be $\max\{m(p(X)) : p(X) \in \mathcal{P}_g\}$. Define $k(g)$ to be*

$$\max\{\mathrm{lcm}(m(p_1(X)), \ldots, m(p_n(X))) : g = \sum_{i=1}^{n} g_i,\ p_i(X) \in \mathcal{P}_{g_i}\}.$$

We now state our main result. We emphasise that the bound $k(g)$ depends only on the genus and not on the abelian variety A.

Theorem 3. *Let A be a supersingular abelian variety of dimension g over a field $\mathbb{F}_q$, then there exists an integer $k \le k(g)$ such that, for all integers $r \ge 1$, the exponent of the group $A(\mathbb{F}_{q^r})$ divides $q^{kr} - 1$.*

Proof. First, take a quadratic extension so that q^r is a square, i.e., consider $q_0 = q^{2r}$. Let $P(X)$ be the characteristic polynomial of the Frobenius endomorphism on A over $\mathbb{F}_{q_0}$ and write α_i for the roots (they are the squares of the values of the roots corresponding to A over $\mathbb{F}_q$).

We follow the proof of Theorem 4.2 of Oort [24] and consider

$$P'(X) = P(\sqrt{q_0}X)/q_0^g = X^{2g} + (a_1/\sqrt{q_0})X^{2g-1} + \cdots + 1$$

which has roots $\alpha_i/\sqrt{q_0}$. By Theorem 2 the coefficients of $P'(X)$ are integers.

The numbers $\alpha_i/\sqrt{q_0}$ are algebraic integers which are units but, by Theorem 4.1 of Manin [21], it follows that they are actually roots of unity. Therefore $P'(X)$ is a product of cyclotomic polynomials.

By definition of $k(g)$ there is some $k \le k(g)$ such that $(\alpha_i/\sqrt{q_0})^k = 1$ for all i. In other words, $\alpha_i^k = q_0^{k/2}$ for all i and so $\pi^k = q_0^{k/2}$. For all points $P \in \mathrm{Pic}_C^0(\mathbb{F}_{q^r})$ we have $P = \pi^r(P) = [q_0^{rk/2}]P$. It follows that the exponent of $A(\mathbb{F}_{q_0^k})$ divides $q_0^{k/2} - 1$ (also see Stichtenoth and Xing [34] Proposition 2). Since $q_0^{k/2} - 1 = q^{rk} - 1$ the result is proven. □

We now consider the values of $k(g)$. Cyclotomic polynomials $X^m - 1$ factor into products of polynomials $\Phi_n(X)$ for each $n|m$ (see Lang [19] VI.3). The polynomials $\Phi_n(X)$ have degree $\varphi(n)$ (this is the Euler φ-function) so the values of $k'(g)$ are related to the problem of finding the largest value of n for which $\varphi(n) = 2g$. The extremal case is when n is the product of the first k primes and so $\varphi(n) = n\frac{1}{2}\frac{2}{3}\cdots\frac{p_k-1}{p_k}$ (e.g., $\varphi(6) = 2, \varphi(30) = 8, \varphi(210) = 48$ etc). The values of $k(g)$ relate to the ways of taking least common multiples of the $m(p(X))$.

Table 1. Values of $k(g)$. The symbol $\star$ indicates the fact that there are no irreducible cyclotomic polynomials of degree 14 (since there are no integers N with $\varphi(N) = 14$).

g	$k'(g)$	$k(g)$	$k(g)/g$
1	6	6	6
2	12	12	6
3	18	30 = lcm(6, 10)	10
4	30	60 = lcm(10, 12)	15
5	22	120 = lcm(8, 10, 6)	24
6	42	210 = lcm(6,10,14)	30
7	$\star$	420 = lcm(5,7,12)	60
8	60	840 = lcm(3,5,7,8)	105

Table 1 gives some values for $k(g)$. We only list values for $g \leq 8$ since there are various algorithms (see [12]) for solving the discrete logarithm problem on high-genus curves. The notation indicates how the maximum value is attained. For example the case $k(3) = 30$ comes from the cyclotomic polynomials $\Phi_6(X) = X^2 - X + 1$ and $\Phi_{10}(X) = X^4 - X^3 + X^2 - X + 1$. It follows that the smallest degree m such that $\Phi_6(X)\Phi_{10}(X)|(X^m - 1)$ is $m = \text{lcm}(6, 10) = 30$. Hence an abelian variety with $P(X) = q^3\Phi_6(X/\sqrt{q})\Phi_{10}(X/\sqrt{q})$ (which must exist by the Honda-Tate theorem [35]) would have embedding degree 30. However, we have not found a curve whose Jacobian is isogenous to such an abelian variety.

The bounds given are sharp, in the sense that there exists an abelian variety over some finite field $\mathbb{F}_q$ for which the bound $k(g)$ is attained (note also that we recover the bound $k = 6$ in the elliptic curve case). However, we are more interested in Jacobian varieties of curves than in general abelian varieties. It is therefore important to determine which values for k can arise as the Jacobian of a curve. We return to this problem in Section 4.4.

What do these results tell us about the security of the discrete logarithm problem in the divisor class group of a curve? Recall that the advantage of the divisor class group of a curve of genus g over $\mathbb{F}_q$ is that, over a field $\mathbb{F}_q$ the group has size approximately q^g. Hence, to determine the applicability of the subexponential algorithms for solving the discrete logarithm problem in finite fields, we really should consider the ratio $k(g)/g$, which is seen in Table 1 to grow rather slowly. This supports the notion that supersingular curves are weaker than the general case for standard discrete logarithm based cryptosystems.

4.4 Are Large Values of k Attained for Curves?

In this section some examples of curves with relatively large values for k are given (see Table 2). When $g > 2$ it is seen that the values are much smaller than the upper bounds given in Table 1. It is an interesting open problem to find the exact largest values of k for each genus, and we hope that this paper motivates further work on the problem.

The fact that the maximum value of k is attained in the case of genus one and two curves is not surprising since every elliptic curve is a Jacobian, and every isogeny class of abelian varieties of dimension two contains a representative which is either a product of elliptic curves or the Jacobian of a hyperelliptic curve (possibly this process requires an extension of the ground field). However, in the case of dimension four or more we would not necessarily expect the bounds to be attained.

The case of dimension three is particularly interesting. Simple abelian varieties of dimension three should be isogenous to a Jacobian of a genus three curve (not necessarily hyperelliptic) over some extension field. However, we have not found any supersingular curves giving large values of k. Further, we have not found any supersingular hyperelliptic curves of genus three in characteristic two.

The reason for only listing curves defined over small fields is that, for elliptic curves, one can only obtain $k > 3$ in characteristic two or three, and we expect analogous results in the higher genus case.

Table 2. Table of curves with large k. Notes:
(1) In the first row p must be an odd prime congruent to 2 modulo 3.
(2) This genus 3 curve is a plane quartic and is not hyperelliptic. It can be written as the affine superelliptic curve $z^3 = x^4 + \theta x^2$.

Field	Curve	Genus	# points	k
$\mathbb{F}_p$ [(1)]	$y^2 = x^3 + a$	1	$p+1$	2
$\mathbb{F}_3$	$y^2 = x^3 + 2x \pm 1$	1	7,1	6
$\mathbb{F}_2$	$y^2 + y = x^5 + x^3$	2	13	12
$\mathbb{F}_3$	$y^2 = x^6 + x + 2$	2	13	3
$\mathbb{F}_5$	$y^2 = x^5 + 2x^4 + x^3 + x + 3$	2	11	5
$\mathbb{F}_{2^2} = \mathbb{F}_2(\theta)$	$x^4 + \theta xy^3 + yz^3$ [(2)]	3	57	9
$\mathbb{F}_3$	$y^2 = x^7 + 1$	3	28	6
$\mathbb{F}_5$	$y^2 = x^8 + 2x^4 + 3x^2 + 2$	3	66	10
$\mathbb{F}_7$	$y^2 = x^8 + x^4 + 5x^3$	3	911	14
$\mathbb{F}_2$	$y^2 + y = x^9 + x^4 + 1$	4	5	12

5 Equations of Supersingular Curves

For applications, especially when using subfield curves, it is very important to know in advance which equations are likely to give rise to supersingular curves. For instance, Sakai, Sakurai and Ishizuka [27] suggested some hyperelliptic curves for use in cryptography. On page 172 they state that they were unable to find any secure genus 2 curves over $\mathbb{F}_2$ and speculated that this was caused by their restriction to the field $\mathbb{F}_2$ (instead of using $\mathbb{F}_{2^n}$). In fact, the reason for this is that they only considered equations of the form $C : y^2 + y = f(x)$ with $f(x) \in \mathbb{F}_2[x]$ monic of degree 5. We will show that all genus two curves of this form over all fields $\mathbb{F}_{2^n}$ are supersingular.

The first observation is that any hyperelliptic curve in characteristic two of the form $y^2 + h(x)y = f(x)$ with $1 \leq \deg(h(x)) \leq g+1$ cannot be supersingular. To see this note that any root x_0 of $h(x)$ gives rise to a point (x_0, y_0) (possibly over a quadratic extension) of order 2, but a supersingular curve in characteristic p has no points (even over algebraic extensions) of order p.

Therefore, curves of the form $y^2 + y = f(x)$ are a poor choice in characteristic two if one wants to avoid supersingular cases. However, the argument sketched above does not imply that all such curves are necessarily supersingular. Indeed, there are curves of this form which are not supersingular when the genus is three or more. Our main result in this section is that all such curves are supersingular in the case of genus two.

Theorem 4. *Let C be a genus 2 curve over $\mathbb{F}_{2^n}$ of the form $y^2 + cy = f(x)$ where $f(x)$ is monic of degree 5 and $c \in \mathbb{F}_{2^n}^*$. Then C is supersingular.*

Before giving the proof of the theorem it is necessary to obtain the following result about the polynomials $P(X)$ for curves of this form.

Lemma 2. *Let C be a genus 2 curve over $\mathbb{F}_{2^n}$ of the form $y^2+cy=f(x)$ where $f(x)$ is monic of degree 5 and $c \in \mathbb{F}_{2^n}^*$. Then the coefficients a_1 and a_2 in the polynomial $P(X)$ are both even.*

Proof. For equations of this form the number of points on the curve over all extensions $\mathbb{F}_{2^{nm}}$ is odd, since apart from the point at infinity, points come in pairs (x_0, y_0) and (x_0, y_0+c). The fact that $\#C(\mathbb{F}_{2^n}) = 2^n+1-a_1$ is odd implies that a_1 is even.

On $C(\mathbb{F}_{2^{2n}})$ there are two points for each possible $x_0 \in \mathbb{F}_{2^n}$ (the corresponding y-coordinates may be in $\mathbb{F}_{2^n}$ or $\mathbb{F}_{2^{2n}}$). For any point with $x_0 \notin \mathbb{F}_{2^n}$ there are the four distinct 'conjugates' $(x_0, y_0), (x_0, y_0+c), (\pi(x_0), \pi(y_0)), (\pi(x_0), \pi(y_0)+c)$ where π is the Frobenius automorphism of $\mathbb{F}_{2^{2n}}/\mathbb{F}_{2^n}$. It follows that $\#C(\mathbb{F}_{2^{2n}}) \equiv 1 \pmod{2^{n+1}}$. Write $t_2 = 2^{2n}+1-\#C(\mathbb{F}_{2^{2n}})$. Then t_2 is divisible by 4 and from $a_1^2 = t_2 + 2a_2$ it follows that a_2 is even. □

If the curve C is actually defined over $\mathbb{F}_2$ then Theorem 2 implies that the curve is supersingular. In the general case we need a further argument.

Proof. (of Theorem 4) Using Lemma 2 we see that $P(X) \equiv X^4 \pmod 2$. By a result of Manin [22] (also see Stichtenoth [32] Satz 1) it follows that $\mathrm{Jac}(C)(\overline{\mathbb{F}}_{2^n})$ has no points of order 2. In the case of dimension 2, this condition is known (see Li and Oort [20] p. 9) to be equivalent to supersingularity. □

An alternative proof of the above result can be given by using the theory of the Newton polygon and some class field theory. One shows that, in genus 2, the only polynomials $P(X)$ which satisfy the condition of Lemma 2 also satisfy the condition of Theorem 2 (see Rück [25] for details of this approach).

Note that both of these arguments rely heavily on the fact that we are in the genus two case. In the case of genus three it is possible to give 'safe' examples. For instance, the curve $C : y^2 + y = x^7$ of [27] has $P(X) = X^6 - 2X^3 + 2^3$ and the fact that a_3 is not divisible by $2^{\lceil 3/2 \rceil}$ means that C is not supersingular.

We note that $\#C(\mathbb{F}_2)$ and $\#C(\mathbb{F}_{2^2})$ being odd does not alone imply that C is supersingular. An example is the genus two curve $y^2 + (x^2+x+1)y = x^5 + 1$ which has 3 points over $\mathbb{F}_2$ and 7 points over $\mathbb{F}_{2^2}$ and so $P(X) = X^4 + X^2 + 4$ and C is not supersingular.

The authors of [27] could have considered curves of the form $y^2 + xy = f(x)$ (with degree five $f(x) \in \mathbb{F}_2[x]$). In these cases it is clear that $\#C(\mathbb{F}_{2^n})$ is always even, in which case a_1 is always odd and, by Theorem 2 the curve cannot be supersingular. Indeed, the same argument shows that curves of the form $y^2+xy = f(x)$ with $f(x) \in \mathbb{F}_{2^n}[x]$ of odd degree are an infinite family of non-supersingular hyperelliptic curves. It is easy to find suitable examples of genus 2 curves of this form, for instance $C : y^2 + xy = x^5 + x^2 + 1$ has $P(X) = X^4 - X^3 - 2X + 4$. One can show that

$$\begin{aligned}
&\#\mathrm{Jac}(C)(\mathbb{F}_{2^{97}}) = 2 \cdot 389 \cdot 1747 \cdot \\
&\quad 18473392463868826910318794676754071940716909907019619 \\
&\#\mathrm{Jac}(C)(\mathbb{F}_{2^{103}}) = 2 \cdot 47381 \cdot \\
&\quad 1085287719049570327739050925845914539948927360923370110769
\end{aligned}$$

where the large numbers are proven primes according to Magma. In both cases the Frey-Rück embedding degree exceeds 10^{50}.

The above arguments suggest that, in characteristic two, only curves of the form $y^2 + h(x)y = f(x)$ where $\deg(h(x)) \geq 1$ should be used in cryptography. However, this is not necessarily the conclusion one wants to draw, since equations of the form $y^2 + y = f(x)$ give some implementation efficiency (see Smart [30] Section 1 and [7] Theorem 14).

Another strategy would be to use genus two curves of the form $y^2 + h(x)y = f(x)$ over $\mathbb{F}_{2^n}$ which always have two points at infinity (i.e., $\deg(h(x)) = 3$ such that $h(x)$ has no root in the ground field). In these cases one also has a_1 odd, and so the curves are not supersingular.

Acknowledgements. It is a pleasure to thank Hans-Georg Rück for indicating both proofs of Theorem 4; Nigel Smart, Dan Boneh and Keith Harrison for discussions on the Boneh and Franklin scheme; Pierrick Gaudry for discussions about hyperelliptic curves in characteristic two; and Alice Silverberg for helpful comments on an earlier version of the paper.

References

1. R. Balasubramanian and N. Koblitz, The improbability that an elliptic curve has subexponential discrete log problem under the Menezes-Okamoto-Vanstone algorithm., *J. Cryptology*, **11** no. 2 (1998) 141–145.
2. D. Boneh and M. Franklin, Identity-based encryption from the Weil pairing, in J. Kilian (ed.), CRYPTO 2001, Springer LNCS 2139 (2001) 213–229.
3. J. Buhler and N. Koblitz, Lattice basis reduction, Jacobi sums and hyperelliptic cryptosystems, *Bull. Aust. Math. Soc.*, **58**, No.1 (1998) 147–154.
4. D. G. Cantor, Computing in the Jacobian of a hyperelliptic curve, *Math. Comp.*, **48** (1987) 95–101.
5. H. Cohen, A course in computational number theory, Springer GTM 138 (1993).
6. I. Duursma, P. Gaudry and F. Morain, Speeding up the discrete log computation on curves with automorphisms, in K. Y. Lam et al (eds.), ASIACRYPT '99, Springer LNCS 1716, (1999) 103–121.
7. A. Enge, The extended Euclidean algorithm on polynomials and the computational efficiency of hyperelliptic cryptosystems, *Designs, Codes and Cryptography*, **23** (2001) 53–74.
8. G. Frey, H.-G. Rück, A remark concerning m-divisibility and the discrete logarithm in the divisor class group of curves, *Math. Comp.*, **62**, No.206 (1994) 865–874.
9. G. Frey, M. Müller and H.-G. Rück, The Tate pairing and the discrete logarithm applied to elliptic curve cryptosystems, *IEEE Trans. Inform. Theory*, **45**, no. 5 (1999) 1717–1719.
10. S. D. Galbraith, S. Paulus and N. P. Smart, Arithmetic on superelliptic curves, To appear in *Math. Comp.*
11. S. D. Galbraith, Supersingular curves in cryptography (full version), available from the author's web pages.
12. P. Gaudry, An algorithm for solving the discrete log problem on hyperelliptic curves, in B. Preneel (ed.), EUROCRYPT 2000, Springer, LNCS 1807 (2000) 19–34.

13. R. Harley, Rump session talk, EUROCRYPT 2001, (2001).
14. A. Joux, A one round protocol for tripartite Diffie-Hellman, in W. Bosma (ed.), ANTS-IV, Springer LNCS 1838 (2000) 385–393.
15. K. S. Kedlaya, Counting points on hyperelliptic curves using Monsky-Washnitzer cohomology, preprint (2001).
16. N. Koblitz, Hyperelliptic cryptosystems, *J. Cryptology*, **1**, no. 3 (1989) 139–150.
17. N. Koblitz, A family of jacobians suitable for discrete log cryptosystems, in S. Goldwasser (ed.), CRYPTO '88, Springer LNCS 403 (1990) 94–99.
18. N. Koblitz, An elliptic curve implementation of the finite field digital signature algorithm, in H. Krawczyk (ed.), CRYPTO '98, Springer LNCS 1462 (1998) 327–337.
19. S. Lang, Algebra, 3rd ed., Addison-Wesley, 1993.
20. K.-Z. Li and F. Oort, Moduli of supersingular abelian varieties, Springer LNM 1680 (1998).
21. Yu. I. Manin, The theory of commutative formal groups over fields of finite characteristic, *Russ. Math. Surv.*, **18**, No. 6 (1963) 1–83.
22. Yu. I. Manin, The Hasse-Witt matrix of an algebraic curve, *Translations, II Ser.*, Am. Math. Soc., **45** (1965) 245–264.
23. A. J. Menezes, T. Okamoto and S. A. Vanstone, Reducing elliptic curve logarithms to logarithms in a finite field, *IEEE Trans. Inf. Theory*, **39**, No. 5 (1993) 1639–1646.
24. F. Oort, Subvarieties of moduli spaces, *Inv. Math.*, **24** (1970) 95–119.
25. H.-G. Rück, Abelsche varietäten niderer dimension über endlichen körpern, Habilitation Thesis, University of Essen (1990).
26. H.-G. Rück, On the discrete logarithm in the divisor class group of curves, *Math. Comp.*, **68**, No.226 (1999) 805–806.
27. Y. Sakai, K. Sakurai and H. Ishizuka, Secure hyperelliptic cryptosystems and their performance, in H. Imai et al. (eds.), PKC '98, Springer LNCS 1431 (1998) 164–181.
28. A. Shamir, Identity-based cryptosystems and signature schemes, In G.R. Blakley and D. Chaum (eds.), CRYPTO '84, Springer LNCS 196 (1985) 47–53.
29. J. H. Silverman, The arithmetic of elliptic curves, Springer GTM 106, (1986).
30. N. Smart, On the performance of hyperelliptic cryptosystems, in J. Stern (ed.), EUROCRYPT '99, Springer LNCS 1592 (1999) 165–175.
31. A. Stein and E. Teske, Explicit bounds and heuristics on class numbers in hyperelliptic function fields, To appear in *Math. Comp.*, University of Waterloo technical report CORR 99-26 (1999).
32. H. Stichtenoth, Die Hasse-Witt-invariante eines kongruenzfunktionenkörpers, *Arch. Math.*, **33**, No. 4 (1980) 357–360.
33. H. Stichtenoth, Algebraic function fields and codes, Springer Universitext (1993).
34. H. Stichtenoth and C. Xing, On the structure of the divisor class group of a class of curves over finite fields, *Arch. Math.*, Vol. **65** (1995) 141–150.
35. J. Tate, Classes d'isogénie de variétés abéliennes sur un corps fini (d'après T. Honda), *Sém. Bourbaki*, Exp. 352, Springer LNM 179 (1971) 95–110.
36. E. R. Verheul, Evidence that XTR is more secure than supersingular elliptic curve cryptosystems, in B. Pfitzmann (ed.), EUROCRYPT 2001, Springer LNCS 2045 (2001) 195–210.

Appendix 1. Methods to Compute $P(X)$

Very recently there have been some breakthroughs [15], [13] in algorithms for counting points and computing $P(X)$ on higher genus curves in the case of small characteristic. Nevertheless there is still interest in using subfield curves. We discuss some methods to compute $P(X)$ for curves C defined over small fields $\mathbb{F}_q$.

First we give the most elementary method. Given a curve $C/\mathbb{F}_q$ of genus $g > 1$ compute $\#C(\mathbb{F}_{q^r})$ for $1 \le r \le g$ by exhaustive search. If the curve is given as a non-singular plane curve $f(x, y) = 0$ with a known number of rational points at infinity then the exhaustive search involves trying all values $x_0 \in \mathbb{F}_{q^r}$ and then calculating the number of roots of $f(x_0, y)$ in $\mathbb{F}_{q^r}$. From the values $t_r = q^r + 1 - \#C(\mathbb{F}_{q^r}) = \sum_{i=1}^{2g} \alpha_i^r$ one can obtain the coefficients of $P(X)$ using Newton's identities $a_m = \frac{1}{m}(-t_m - \sum_{i=1}^{m-1} a_i t_{m-i})$ (see Cohen [5] Proposition 4.3.3). This naive algorithm takes time $O(q^g(\log q^g)^c)$ for some constant c, which can also be written as $O(q^{g+\epsilon})$.

One method to speed this up is to compute $\#C(\mathbb{F}_{q^r})$ for $r = 1, \ldots, g-1$ and then to try all values of $\#C(\mathbb{F}_{q^g}) - (q^g + 1)$ (i.e., all integers in the interval $[-2gq^{g/2}, 2gq^{g/2}]$) and test the correctness of the group order probabilistically by computations on $\mathrm{Jac}(C)$ over $\mathbb{F}_q$ or over some extension $\mathbb{F}_{q^m}$. This produces a method of complexity $O(q^{g-1+\epsilon})$.

A variation on the above strategy is to use the method of Stein and Teske [31] which computes $\#\mathrm{Jac}(C)(\mathbb{F}_q)$ in time proportional to q^d where $d \in \mathbb{Z}$ is a suitable rounding of $(2g-1)/5$. One computes $\#C(\mathbb{F}_{q^r})$ for $r = 1, \ldots, g-1$ and then computes $\#\mathrm{Jac}(C)(\mathbb{F}_q)$ from which it is possible to deduce $P(X)$. This method also has complexity $O(q^{g-1+\epsilon})$.

Similarly, one can compute $\#C(\mathbb{F}_{q^r})$ only up to $r = g-2$ and then compute $\#\mathrm{Jac}(C)(\mathbb{F}_q)$ and $\#\mathrm{Jac}(C)(\mathbb{F}_{q^2})$ using [31]. This method has the superior complexity $O(q^{g-2+\epsilon})$ when $g = 4$ or $g \ge 6$. This trick cannot be extended.

Appendix 2. Superelliptic Curves

The case of hyperelliptic curves has been fairly thoroughly explored in the past [16], [17], [3], [27], [30]. In particular, Buhler and Koblitz [3] mention cases which are guaranteed to be non-supersingular.

A superelliptic curve (see [10]) is a curve given by an affine equation of the form $y^n = f(x)$ over $\mathbb{F}_q$ where $\gcd(n, q) = 1$, $\gcd(n, \deg f(x)) = 1$ and $\gcd(f(x), f'(x)) = 1$. Such curves have only one point at infinity and they have genus $\frac{1}{2}(n-1)(\deg f(x) - 1)$.

Note that the curve $y^3 = f(x)$ over $\mathbb{F}_{2^n}$ has exactly $2^n + 1$ points when n is odd (since in those cases 3 is coprime to the order of $\mathbb{F}_{2^n}^*$). This means that, in the case where the ground field is an odd degree extension of $\mathbb{F}_2$, to compute $P(X)$ it is only necessary to count the number of points over even degree extensions of the ground field. In other words, when g is odd, one can compute $P(X)$ in time $O(q^{g-1+\epsilon})$. On the other hand, such curves do not have full 2-torsion and so they are not fully general among all superelliptic curves.

Table 3 lists some non-supersingular superelliptic curves. In all cases the large numbers l are proven primes according to Magma, and the curves are resistant to the Frey-Rück attack. The symbol α represents a generator of the multiplicative group of the field of definition. As usual, one must be careful about the use of curves such as these due to the large automorphism group [6], [12].

Table 3. Examples of superelliptic curves suitable for cryptography.

$g = 3$ $\quad C : y^3 = x^4 + x^3 + \alpha x^2 + x + \alpha$ over $\mathbb{F}_{2^2}$ $P(X) = X^6 + 3X^4 + 4X^3 + 12X^2 + 2^6$ $\#\mathrm{Jac}(C)(\mathbb{F}_{2^{2\cdot 41}}) = 2^2 \cdot 3 \cdot 7 \cdot 1231 \cdot 12547 \cdot 839353\cdot$ 103838175651664516641765501325467649197030008300761187148661 (197 bit)
$g = 3$ $\quad C : y^3 = x^4 + x^3 + \alpha x + 1$ over $\mathbb{F}_{2^5}$ $P(X) = X^6 + 39X^4 + 1248X^2 + 2^{15}$ $\#\mathrm{Jac}(C)(\mathbb{F}_{2^{5\cdot 23}}) = 2^4 \cdot 3^2 \cdot 5^5 \cdot 7 \cdot 11 \cdot 83\cdot$ 249210979849057649603915759933900855778626741247624026770184646815 709788699839224081758315379 59 (314 bit)
$g = 4$ $\quad C : y^3 = x^5 + 1$ over $\mathbb{F}_2$ $P(X) = X^8 - 2X^4 + 16$ $\#\mathrm{Jac}(C)(\mathbb{F}_{2^{43}}) = 3 \cdot 5 \cdot 4129\cdot$ 966547300638956705087962044300576049126085993 11 (157 bit)
$g = 4$ $\quad C : y^3 = x^5 + x + 1$ over $\mathbb{F}_2$ $P(X) = X^8 + 2X^6 + 6X^4 + 8X^2 + 16$ $\#\mathrm{Jac}(C)(\mathbb{F}_{2^{43}}) = 3 \cdot 11\cdot$ 1814033547426563130808781923043653173548257105356 49 (167 bit) $\#\mathrm{Jac}(C)(\mathbb{F}_{2^{61}}) = 3 \cdot 11 \cdot 12323\cdot$ 6951660491088147396353756902913715826706693781009 0081 343111639513643 (226 bit)

Short Signatures from the Weil Pairing

Dan Boneh*, Ben Lynn, and Hovav Shacham

Computer Science Department, Stanford University
{dabo,blynn,hovav}@cs.stanford.edu

Abstract. We introduce a short signature scheme based on the Computational Diffie-Hellman assumption on certain elliptic and hyper-elliptic curves. The signature length is half the size of a DSA signature for a similar level of security. Our short signature scheme is designed for systems where signatures are typed in by a human or signatures are sent over a low-bandwidth channel.

1 Introduction

Short digital signatures are needed in environments where a human is asked to manually key in the signature. For example, product registration systems often ask users to key in a signature provided on a CD label. More generally, short signatures are needed in low-bandwidth communication environments. For example, short signatures are needed when printing a signature on a postage stamp [21,19]. Currently, the two most frequently used signatures schemes, RSA and DSA, provide relatively long signatures compared to the security they provide. For example, when one uses a 1024-bit modulus, RSA signatures are 1024 bits long. Similarly, when one uses a 1024-bit modulus, standard DSA signatures are 320 bits long. Elliptic curve variants of DSA, such as ECDSA, are also 320 bits long [1]. A 320-bit signature is too long to be keyed in by a human.

We propose a signature scheme whose length is approximately 160 bits and provides a level of security similar to 320-bit DSA signatures. Our signature scheme is secure against existential forgery under a chosen message attack (in the random oracle model) assuming the Computational Diffie-Hellman problem (CDH) is hard on certain elliptic curves over a finite field of characteristic three. Generating a signature is a simple multiplication on the curve. Verifying the signature is done using a bilinear pairing on the curve. Our signature scheme inherently uses properties of elliptic curves. Consequently, there is no equivalent of our scheme in $\mathbb{F}_p^*$.

Due to the properties of the curves we use, currently we can only provide signatures of the lengths given below. The best known algorithm for solving the CDH problem in these groups requires a discrete-log on a finite field of characteristic three. The size of this field is given (in bits) in the rightmost column of the table below.

* Supported by NSF and the Packard Foundation.

C. Boyd (Ed.): ASIACRYPT 2001, LNCS 2248, pp. 514–532, 2001.
© Springer-Verlag Berlin Heidelberg 2001

Signature size (bits)	EC group size (bits)	Discrete-log Security (bits)
126	126	752
154	151	923
237	220	1417
259	256	1551
265	262	1589

The second row shows that we can get a signature of length 154 bits with security comparable to 320-bit DSA or 320-bit ECDSA. The best known algorithm to forge a 154-bit signature requires one to solve a CDH problem in a finite field of size 923 bits or on an elliptic curve group of size 151 bits. In Section 3.5 we outline an approach for generalizing our technique and building signatures of any length.

Constructing short signatures is an old problem. Several proposals show how to shorten the DSA signature scheme while preserving the same level of security. Naccache and Stern [19] propose a variant of DSA where the signature length is approximately 240 bits. Mironov [18] suggests a DSA variant with a similar length and gives a concrete security analysis of the construction (in the random oracle model). Another technique proposed for reducing the DSA signature length is signatures with message recovery [21]. In such systems one encodes a part of the message into the signature thus shortening the total length of the message-signature pair. For long messages, one can then achieve a DSA signature overhead of length 160 bits. However, for very short messages (e.g., 64 bits) the total length is still 320 bits. Using our signature scheme, the signature length is always on the order of 160 bits, no matter how short the message is. Note that when the only transmitted data is the signature (the message is not transmitted) DSA signatures with message recovery are not any shorter than standard DSA signatures.

Our signature scheme uses groups where the CDH problem is hard, but the Decision Diffie-Hellman problem (DDH) is easy. The first example of such groups was given in [12] and was previously used in [11,4]. We call such groups Gap Diffie-Hellman groups, or GDH groups for short. Okamoto and Pointcheval [20] commented that a Gap Diffie-Hellman group gives rise to a signature scheme. However, most Gap Diffie-Hellman groups are relatively long and do not lead to short signatures. We prove the security of signatures schemes derived from GDH groups and show how they lead to very short signatures. We experiment with our proposed signature scheme and give running times in Section 5.

2 Signature Schemes Based on Gap-Diffie-Hellman

We present a signature scheme that works in any Gap Diffie-Hellman group. As mentioned above, this scheme is described implicitly by Okamoto and Pointcheval [20]. The scheme resembles the undeniable signature scheme proposed by Chaum and Pederson [5]. In the next section we show how this signature scheme gives rise to very short signatures.

2.1 Gap Diffie-Hellman Groups (GDH Groups)

Consider a (multiplicative) cyclic group $G = \langle g \rangle$, with $p = |G|$ a prime. We are interested in three problems on G.

Group Action. Given $u, v \in G$, find uv.
Decision Diffie-Hellman. For $a, b, c \in \mathbb{Z}_p^*$, given (g, g^a, g^b, g^c) decide whether $c = ab$.
Computational Diffie-Hellman. For $a, b \in \mathbb{Z}_p^*$, given (g, g^a, g^b), compute g^{ab}.

We define a Gap Diffie-Hellman group, in stages.

Definition 1. *G is a τ-decision group for Diffie-Hellman if the group action can be computed in one time unit, and Decision Diffie-Hellman can be computed on G in time at most τ.*

Definition 2. *The advantage of an algorithm $\mathcal{A}$ in solving the Computational Diffie-Hellman problem in a group G is*

$$\mathsf{Adv\,CDH}_{\mathcal{A}} \stackrel{\text{def}}{=} \Pr\left[\mathcal{A}(g, g^a, g^b) = g^{ab} : a, b \stackrel{\text{R}}{\leftarrow} \mathbb{Z}_p^*\right]$$

Where the probability is over the choice of a and b, and the coin tosses of $\mathcal{A}$. We say that an algorithm $\mathcal{A}$ (t, ϵ)-breaks Computational Diffie-Hellman in G if $\mathcal{A}$ runs in time at most t, and $\mathsf{Adv\,CDH}_{\mathcal{A}} \geq \epsilon$.

Definition 3. *A prime order group G is a (τ, t, ϵ)-GDH group if it is a τ-decision group for Diffie-Hellman and no algorithm (t, ϵ)-breaks Computational Diffie-Hellman on it.*

2.2 The GDH Signature Scheme

The GDH Signature Scheme allows the creation of signatures on arbitrary messages $m \in \{0,1\}^*$. A signature σ is an element of G. The base group G and the generator g are system parameters. We denote by G^* the set $G^* = G \backslash \{1\}$ where 1 is the identity of G.

The signature scheme comprises three algorithms, *KeyGen*, *Sign*, and *Verify*. It makes use of a full-domain hash function $h : \{0,1\}^* \to G^*$. The security analysis views h as a random oracle [3]. In Section 3.3 we weaken the requirement on the full-domain hash.

Key Generation. Pick random $x \stackrel{\text{R}}{\leftarrow} \mathbb{Z}_p^*$, and compute $v \leftarrow g^x$. The public key is v. The secret key is x.
Signing. Given a secret key x, and a message $M \in \{0,1\}^*$, Compute $h \leftarrow h(M)$, and $\sigma \leftarrow h^x$. The signature is $\sigma \in G^*$.
Verification. Given a public key v, a message M, and a signature σ, compute $h \leftarrow h(M)$ and verify that (g, v, h, σ) is a valid Diffie-Hellman tuple.

Note that a GDH signature is a single element of G^*. Hence, to construct short signatures we need a GDH group where elements have a short representation. We construct such groups in Section 3.

2.3 Security

We show the security of the GDH signature scheme against existential forgery, under chosen-message attacks.

Definition 4. *The advantage in existentially forging a signature of a forger algorithm $\mathcal{F}$, given access to a signing oracle S, is*

$$\mathsf{Adv\,Sig}_{\mathcal{F}} \stackrel{\text{def}}{=} \Pr\left[\mathit{Verify}(PK, M, \sigma) = \texttt{valid} \; : \; \begin{array}{l} (PK, SK) \stackrel{\text{R}}{\leftarrow} \mathit{KeyGen}, \\ (M, \sigma) \stackrel{\text{R}}{\leftarrow} \mathcal{F}^{S}(PK) \end{array} \right]$$

The probability is taken over the coin tosses of the key-generation algorithm, and of the forger.

Here the adversary $\mathcal{F}$ is allowed to query the signing oracle adaptively: any of its queries may depend on previous answers, but it may not emit a signature for a message on which it had previously queried the oracle. The adversary also has access to the full-domain hash function, which is treated as a random oracle.

Definition 5. *A forger $\mathcal{F}$ (t, q_H, q_S, ϵ)-breaks a signature scheme if $\mathcal{F}$ runs in time at most t, makes at most q_H queries to the hash function and at most q_S queries to the signing oracle S, and $\mathsf{Adv\,Sig}_{\mathcal{F}} \geq \epsilon$.*

Definition 6. *A signature scheme is (t, q_H, q_S, ϵ)-secure against existential forgery on adaptive chosen-message attacks if no forger (t, q_H, q_S, ϵ)-breaks it.*

The following theorem shows that the GDH signature scheme is secure. The proof of the theorem is given in Section 4.

Theorem. *Let G be a (τ, t', ϵ')-gap group for Diffie-Hellman of order p. Then the Gap Signature Scheme on G is (t, q_H, q_S, ϵ)-secure against existential forgery on adaptive chosen-message attacks, where*

$$t \leq t' - 2c_{\mathcal{A}}(\lg p)(q_H + q_S) \quad \text{and} \quad \epsilon \geq 2e \cdot q_S \epsilon',$$

and $c_{\mathcal{A}}$ is a small constant. Here e is the base of the natural logarithm.

3 Building Gap-Diffie-Hellman Groups with Small Representations

Using the Weil pairing, certain elliptic curves may be used as GDH groups. We recall some necessary facts about elliptic curves (see, e.g., [14,22]), and then show how to use certain curves for GDH signatures. In particular, we describe the curves $y^2 = x^3 + 2x \pm 1$ over $\mathbb{F}_{3^\ell}$.

3.1 Elliptic Curves and the Weil Pairing

An elliptic curve can serve as the basis for a GDH signature scheme if we can use it to construct some group G with large prime order on which Computational Diffie-Hellman is difficult, but Decision Diffie-Hellman is easy. First, we characterize a necessary condition for CDH intractability on a subgroup of E.

Definition 7. *Let p be a prime, l a positive exponent, and E an elliptic curve over $\mathbb{F}_{p^l}$ with m points. Let P in E be a point of prime order q where $q^2 \nmid m$. We say that the subgroup $\langle P \rangle$ has a security multiplier α, for some integer $\alpha > 0$, if the order of p^l in $\mathbb{F}_q^*$ is α. In other words:*

$$q \mid p^{l\alpha} - 1 \quad \textit{and} \quad q \nmid p^{lk} - 1 \quad \textit{for all } k = 1, 2, \ldots, \alpha - 1$$

It is well known (as shown below) that for CDH to be hard in the subgroup $\langle P \rangle$ we must have that the security multiplier, α, for this subgroup is not too small. On the other hand, to get an efficient Decision Diffie-Hellman algorithm in $\langle P \rangle$ we need that α is not too large. Therefore, the problem in constructing short signatures is to find curves for which α is sufficiently large for security, but sufficiently small for efficiency. Using current security parameters, $\alpha = 6$ is sufficient for obtaining short signatures. It is an open problem to build elliptic curves with slightly higher α, say $\alpha = 10$ (see Section 3.5).

Discrete-log on elliptic curves: Let $\langle P \rangle$ be a subgroup of $E/\mathbb{F}_{p^l}$ of order q with security multiplier α. We briefly discuss two standard ways for computing discrete-log in $\langle P \rangle$.

1. **MOV:** Use an efficiently computable homomorphism, as in the Menezes-Okamoto-Vanstone reduction [15], to map the discrete log problem in $\langle P \rangle$ to a discrete log problem in some extension of $\mathbb{F}_{p^l}$, say $\mathbb{F}_{p^{li}}$. We require that the image of $\langle P \rangle$ under this homomorphism is a subgroup of $\mathbb{F}_{p^{li}}^*$ of order q. Thus we have $q | (p^{il} - 1)$, which by the definition of α implies that $i \geq \alpha$. Hence, the MOV method can, at best, reduce the discrete log problem in $\langle P \rangle$ to a discrete log problem in a subgroup of $\mathbb{F}_{p^{l\alpha}}^*$. Therefore, to ensure that discrete log is hard in $\langle P \rangle$ we want curves with large α.
2. **Generic:** Generic discrete log algorithms such as the Baby-Step-Giant-Step and Pollard's Rho method [16] have a running time proportional to $\sqrt{q}$. Therefore, we must ensure that q is sufficiently large.

Decision Diffie-Hellman on elliptic curves: Let $P \in E/\mathbb{F}_{p^l}$ be a point of prime order q. Suppose the subgroup $\langle P \rangle$ has security multiplier α. We assume $q \nmid p^l - 1$. A result of Balasubramanian and Koblitz [2] shows that $E/\mathbb{F}_{p^{l\alpha}}$ contains a point Q that is linearly independent of P. Such a point $Q \in E/\mathbb{F}_{p^{l\alpha}}$ can be efficiently found. Note that linear independence of P and Q can be verified via the Weil pairing described below.

With two linearly independent points $P \in E/\mathbb{F}_{p^l}$ and $Q \in E/\mathbb{F}_{p^{l\alpha}}$, each of order q, we can use the Weil pairing to answer certain questions that will allow us to construct a DDH oracle [12]. Let $E[q]$ denote the subgroup of $E/\mathbb{F}_{p^{l\alpha}}$

generated by P and Q. The Weil pairing is a map $e : E[q] \times E[q] \rightarrow \mathbb{F}^*_{p^{l\alpha}}$ with the following properties:

1. Identity: for all $R \in E[q]$, $e(R, R) = 1$.
2. Bilinear: for all $R_1, R_2 \in E[q]$ and $a, b \in \mathbb{Z}$ we have that $e(aR_1, bR_2) = e(R_1, R_2)^{ab}$.
3. Non-degenerate: if for $R \in E[q]$ we have $e(R, R') = 1$ for all $R' \in E[q]$, then $R = \mathcal{O}$.
4. Computable: for all $R_1, R_2 \in E[q]$, the pairing $e(R_1, R_2)$ can be computed efficiently [17].

Note that $e(R_1, R_2) = 1$ if and only if R_1 and R_2 are linearly dependent.

For the linearly independent points P and Q, both of order q, the Weil pairing allows us to determine whether the tuple (P, aP, Q, bQ) is such that $a = b \bmod q$; indeed,

$$a = b \bmod q \iff e(P, bQ) = e(aP, Q).$$

Suppose we also have a computable isomorphism ϕ from $\langle P \rangle$ to $\langle Q \rangle$. Necessarily, ϕ is such that, for all a, $\phi(aP) = axQ$, where $xQ = \phi(P)$. In this case, the Weil pairing allows us to determine whether the tuple (P, aP, bP, cP) is such that $ab = c \bmod q$:

$$ab = c \bmod q \iff e(P, \phi(cP)) = e(aP, \phi(bP)).$$

With the isomorphism ϕ, the Weil pairing provides an algorithm for Decision Diffie-Hellman. Note that the algorithm for DDH requires two evaluations of the Weil pairing for points over $\mathbb{F}_{p^{l\alpha}}$.

3.2 A Special Supersingular Curve

Using the machinery of Section 3.1, we derive GDH groups with small representation from the supersingular elliptic curves E given by $y^2 = x^3 + 2x \pm 1$ over $\mathbb{F}_{3^l}$. As we will see, these are unique supersingular elliptic curves with security multiplier 6. Hence, the MOV reduction maps the discrete log problem in $E/\mathbb{F}_{3^l}$ to $\mathbb{F}^*_{3^{6l}}$. This means that we can use relatively small values of l to obtain short signatures, but the security is dependent on a discrete log problem in a large finite field. We use two simple lemmas to describe the behavior of these curves (see also [23,13]).

Lemma 1. *The curve E^+ defined by $y^2 = x^3 + 2x + 1$ over $\mathbb{F}_{3^l}$ satisfies*

$$\#E^+/\mathbb{F}^{3^l} = \begin{cases} 3^l + 1 + \sqrt{3 \cdot 3^l} \text{ when } l = \pm 1 \bmod 12, \text{ and} \\ 3^l + 1 - \sqrt{3 \cdot 3^l} \text{ when } l = \pm 5 \bmod 12 \end{cases}$$

The curve E^- defined by $y^2 = x^3 + 2x - 1$ over $\mathbb{F}_{3^l}$ satisfies

$$\#E^-/\mathbb{F}^{3^l} = \begin{cases} 3^l + 1 - \sqrt{3 \cdot 3^l} \text{ when } l = \pm 1 \bmod 12, \text{ and} \\ 3^l + 1 + \sqrt{3 \cdot 3^l} \text{ when } l = \pm 5 \bmod 12 \end{cases}$$

Proof. See [13, section 2]. □

We have thus shown how to construct an elliptic curve with $3^l + 1 \pm \sqrt{3 \cdot 3^l}$ points over $\mathbb{F}_{3^l}$, simply by selecting one of E^- and E^+ as appropriate, whenever $l \bmod 12$ equals ± 1 or ± 5.

Lemma 2. *Let E be an elliptic curve defined by $y^2 = x^3 + 2x \pm 1$ over $\mathbb{F}_{3^l}$, where $l \bmod 12$ equals ± 1 or ± 5. Then $\#(E/\mathbb{F}_{3^l})$ divides $3^{6l} - 1$.*

Proof. We have $x^6 - 1 = (x^3 - 1)(x^3 + 1) = (x-1)(x^2+x+1)(x+1)(x^2-x+1)$, so for any integer x it follows that $(x^2 - x + 1) \mid (x^6 - 1)$. In particular, when $x = 3^l$, we see that $(3^{2l} - 3^l + 1) \mid (3^{6l} - 1)$. Now when E is an elliptic curve as above, we know that $\#(E/\mathbb{F}_{3^l})$ is either $3^l + 1 + \sqrt{3 \cdot 3^l}$ or $3^l + 1 - \sqrt{3 \cdot 3^l}$. But $\left((3^l + 1) + \sqrt{3 \cdot 3^l}\right)\left((3^l + 1) - \sqrt{3 \cdot 3^l}\right) = 3^{2l} - 3^l + 1$. Thus $\#(E/\mathbb{F}_{3^l}) \mid (3^{6l} - 1)$.

Together, Lemmas 1 and 2 show that, for the relevant values of l, the curves $E^+/\mathbb{F}_{3^l}$ and $E^-/\mathbb{F}_{3^l}$ will have security parameters α at most 6 (more specifically: $\alpha \mid 6$). Whether the security parameter actually is 6 for a particular prime subgroup of a curve must be determined by computation.

Automorphism *of $E^+, E^-/\mathbb{F}_{3^{6l}}$:* For l such that $l \bmod 12$ equals ± 1 or ± 5, compute three elements of $\mathbb{F}_{3^{6l}}$, u, r^+, and r^-, satisfying $u^2 = -1$, $(r^+)^3 + 2r^+ + 2 = 0$, and $(r^-)^3 + 2r^- - 2 = 0$. Now consider the following maps over $\mathbb{F}_{3^{6l}}$:

$$\phi^+(x, y) = (-x + r^+, uy) \quad \text{and} \quad \phi^-(x, y) = (-x + r^-, uy)$$

Lemma 3. *Let $l \bmod 12$ equal ± 1 or ± 5. Then ϕ^+ is an automorphism of $E^+/\mathbb{F}_{3^{6l}}$ and ϕ^- is an automorphism of $E^-/\mathbb{F}_{3^{6l}}$. Moreover, if P is a point of order q on $E^+/\mathbb{F}_{3^l}$ (or on $E^-/\mathbb{F}_{3^l}$) then $\phi^+(P)$ (or $\phi^-(P)$) is a point of order q that is linearly independent of P.*

Proof. See Silverman [22, p. 326]. □

For a point P of order q on any of these curves, the appropriate automorphism allows us to solve a Decision Diffie-Hellman question on $G = \langle P \rangle$, as we have shown in the previous section.

3.3 Hashing onto Elliptic Curves

The GDH signature scheme needs a hash function $h : \{0,1\}^* \to G^*$ where G is a GDH group. We are proposing to use a subgroup of an elliptic curve as a GDH group. Since it is difficult to build hash functions that hash directly onto a subgroup of an elliptic curve we slightly relax the hashing requirement.

Let $E/\mathbb{F}_{p^l}$ be an elliptic curve of order m defined by $y^2 = f(x)$. Let $P \in E/\mathbb{F}_{p^l}$ be a point of prime order q, where $q^2 \nmid m$. We wish to use the subgroup $G = \langle P \rangle$ as a GDH group for the GDH signature scheme. Suppose we are given a hash function $h' : \{0,1\}^* \to \mathbb{F}_{p^l} \times \{0,1\}$. Such hash functions h' can be built from standard cryptographic hash functions. The security analysis will view h' as a random oracle. We use the following deterministic algorithm called *MapToGroup* to hash messages in $\{0,1\}^*$ onto G^*. Fix a small parameter $I = \lceil \log_2 \log_2(1/\delta) \rceil$, where δ is some desired bound on the probability of failure.

***MapToGroup$_{h'}$*:** The algorithm defines $h : \{0,1\}^* \to G^*$ as follows:

1. Given $M \in \{0,1\}^*$, set $i \leftarrow 0$;
2. Set $(x,b) \leftarrow h'(i \parallel M) \in \mathbb{F}_{p^l} \times \{0,1\}$;
3. If $f(x)$ is a quadratic residue in $\mathbb{F}_{p^l}$ then do:
   3a. Let $y_0, y_1 \in \mathbb{F}_{p^l}$ be the two square roots of $f(x)$. We use $b \in \{0,1\}$ to choose between these roots. View y_0, y_1 as polynomials of degree $l-1$ over $\mathbb{F}_p$. Then ensure that the constant term of y_0 is not greater than the constant term of y_1 when viewed as integers in $[0,p]$ (swapping y_0 and y_1 if necessary). Set $\tilde{P}_M \in E/\mathbb{F}_{p^l}$ to be the point $\tilde{P}_M = (x, y_b)$.
   3b. Compute $P_M = (m/q)\tilde{P}_M$. Then P_M is in G.
   If P_M is in G^* then output $MapToGroup_{h'}(M) = P_M$ and stop.
4. Otherwise, increment i, and goto Step 2; If i reaches 2^I, report failure.

The failure probability can be made arbitrarily small by picking an appropriately large I. For each i, the probability that $h'(i \parallel M)$ leads to a point on G^* is approximately 1/2 (where the probability is over the choice of the random oracle h'). Hence, the expected number of calls to h' is approximately 2, and the probability that a given message M will be found unhashable is $1/2^{2^I} \leq \delta$.

Lemma 4. *Suppose the GDH signature scheme is (t, q_H, q_S, ϵ)-secure in the subgroup G when using a random hash function $h : \{0,1\}^* \to G^*$. Then it is $(t - 2^I q_H \lg m, q_H, q_S, \epsilon)$-secure when the hash function h is computed with $MapToGroup_{h'}$ where h' is a random hash function $h' : \{0,1\}^* \to \mathbb{F}_{p^l} \times \{0,1\}$.*

Proof Sketch: Suppose a forger algorithm $\mathcal{F}'$ (t, q_H, q_S, ϵ)-breaks the Gap Signature Scheme on the subgroup G when the hash function h is computed using $MapToGroup_{h'}$. We construct an algorithm $\mathcal{F}$ that $(t + 2^I q_H \lg m, q_H, q_S, \epsilon)$-breaks the scheme when h is a random oracle $h : \{0,1\}^* \to G^*$.

Our new forger $\mathcal{F}$ will run $\mathcal{F}'$ as a black box. $\mathcal{F}$ will use its own hash oracle $h : \{0,1\}^* \to G^*$ to simulate for $\mathcal{F}'$ the behavior of $MapToGroup_{h'}$. It uses an array s_{ij}, of elements of $\mathbb{F}_{p^l} \times \{0,1\}$. The array has q_H rows and 2^I columns. On initialization, $\mathcal{F}$ fills s_{ij} with uniformly-selected elements of $\mathbb{F}_{p^l} \times \{0,1\}$.

$\mathcal{F}$ then runs $\mathcal{F}'$, and keeps track (and indexes) all the unique messages M_i for which $\mathcal{F}'$ requests an h' hash. When $\mathcal{F}'$ asks for an h' hash of a message $w \parallel M_i$ whose M_i $\mathcal{F}$ had not previously seen (and whose w is an arbitrary I-bit string), $\mathcal{F}$ scans the row s_{ij}, $0 \leq j < 2^I$. For each $(x,b) = s_{ij}$, $\mathcal{F}$ follows Step 3 of *MapToGroup*, above, seeking points in G^*. For the smallest j for which s_{ij} maps into G^*, $\mathcal{F}$ replaces s_{ij} with a different point (x_i, b_i) defined as follows. Let $Q_i = h(M_i) \in G^*$. Then $\mathcal{F}$ constructs a random $\tilde{Q}_i = (x_i, y_i) \in E/\mathbb{F}_{p^l}$ such that $(m/q)\tilde{Q}_i = Q_i$. It sets $s_{ij} = (x_i, b_i)$ where $b_i \in \{0,1\}$ is set so that (x_i, b_i) maps to $\tilde{Q}_i$ in Step 3a of *MapToGroup*. Then $MapToGroup_{h'}(M_i) = h(M_i)$ as required.

Once this preliminary patching has been completed, $\mathcal{F}$ is able to answer h' hash queries by $\mathcal{F}'$ for strings $w' \parallel M_i$ by simply returning $s_{iw'}$. The simulated h' which $\mathcal{F}'$ sees is statistically indistinguishable from that in the real attack. Thus, if $\mathcal{F}'$ succeeds in breaking the signature scheme using $MapToGroup_{h'}$ then $\mathcal{F}$, in running $\mathcal{F}'$ while consulting h, succeeds with the same likelihood, and suffers only a running-time penalty from maintaining the additional information and running the exponentiation in Step 3 of *MapToGroup*. □

3.4 A Concrete Short Signature Scheme

To summarize things so far, we describe a concrete signature scheme using the GDH group derived from the curve $E/\mathbb{F}_{3^l}$ defined by $y^2 = x^3+2x\pm1$. Some useful instantiations of these curves are presented in Table 1. Note that we restrict these instantiations to those where l is prime, to avoid Weil-descent attacks [9,10]. As explained in Section 3.3, we use $MapToGroup_{h'}$ to map arbitrary bit strings to points of order q on E, using a hash function h' from arbitrary strings to elements of $\mathbb{F}_{p^l}$ and an extra bit.

Table 1. Supersingular elliptic curves for GDH Signatures. Here $m = \#(E/\mathbb{F}_{3^l})$, and q is the largest prime dividing m. The MOV reduction maps the curve onto a field with x elements.

curve	l	Signature Size $\lceil \lg_2 m \rceil$	DLog Security $\lceil \lg_2 q \rceil$	Multiplier α	MOV Security $\lceil \lg_2 x \rceil$
E^-	79	126	126	6	752
E^+	97	154	151	6	923
E^+	149	237	220	6	1417
E^+	163	259	256	6	1551
E^-	163	259	259	6	1551
E^+	167	265	262	6	1589

A concrete signature scheme:

Key generation. Given one of the values l in Table 1, let $E/\mathbb{F}_{3^l}$ be the corresponding curve and let q be the largest prime factor of the order of the curve. Let $P \in E/\mathbb{F}_{3^l}$ be a point of order q. pick a random $x \in \mathbb{Z}_q^*$ and set $R \leftarrow xP$. Then (l, q, P, R) is the public key and x is the private key.

Signing. To sign a message $M \in \{0,1\}^*$ use algorithm $MapToGroup_{h'}$ to map M to a point $P_M \in \langle P \rangle$. Set $S_M \leftarrow xP_M$. The signature σ is the x coordinate of S_M. Therefore, $\sigma \in \mathbb{F}_{3^l}$.

Verification. Given a public key (l, q, P, R), a message M, and a signature σ do:

1. Find a point $S \in E/\mathbb{F}_{3^l}$ of order q whose x-coordinate is σ and whose y-coordinate is y for some $y \in \mathbb{F}_{3^l}$. If no such point exists reject the signature as invalid.
2. Set $u \leftarrow e(P, \phi(S))$ and $v \leftarrow e(R, \phi(h(M)))$, where e is the Weil pairing on the curve $E/\mathbb{F}_{3^{6l}}$ and $\phi : E \to E$ is the automorphism of the curve described in Lemma 3.
3. If either $u = v$ or $u^{-1} = v$, accept the signature. Otherwise, reject.

Note that both (σ, y) and $(\sigma, -y)$ are points on $E/\mathbb{F}_{3^l}$ that have σ as their x-coordinate. Either one of these two points can be the point S_M used to generate the signature in the signing algorithm. Indeed, since $(\sigma, y) = -(\sigma, -y)$ on the curve, we have that $e(P, \phi(-S)) = e(P, \phi(S))^{-1}$. Therefore, $u = v$ tests that $(P, R, h(M), S)$ is a Diffie-Hellman tuple, while $u^{-1} = v$ tests that $(P, R, h(M), -S)$ is a Diffie-Hellman tuple.

The next lemma shows that an attacker capable of existential forgery under a chosen message attack (in the random oracle model) is also capable of solving the Diffie-Hellman problem in $E/\mathbb{F}_{3^l}$.

Lemma 5. *Suppose $E/\mathbb{F}_{3^l}$ is one of the curves given in Table 1, q is the largest prime dividing $\#E$, P is a point of order q on E, and no algorithm (t_0, ϵ_0)-breaks Computational Diffie-Hellman on $G = \langle P \rangle$. Let $h' : \{0,1\}^* \to \mathbb{F}_{3^l} \times \{0,1\}$ be a random oracle. Then the concrete signature scheme described above is (t, q_H, q_S, ϵ)-secure against existential forgery on adaptive chosen-message attacks (in the random oracle model), where*

$$t \leq t_0 - 2c_{\mathcal{A}}(\lg q)(q_H + q_S) - 2^I q_H \lg m - 2\tau \quad \textit{and} \quad \epsilon \geq 2e \cdot q_S \epsilon_0,$$

and $c_{\mathcal{A}}$ is a small constant.

Proof. By assumption, G is a (τ, t_0, ϵ_0)-GDH group, where τ is equal to twice the time necessary to compute the Weil pairing on G. Assuming the existence of a random oracle h from arbitrary bit strings to G^*, the generic GDH signature scheme (given in Section 2.2) on G is $(t_1, q_H, q_S, \epsilon_1)$-secure against existential forgery on adaptive chosen-message attacks by the main theorem (Section 4), where

$$t_1 \leq t_0 - 2c_{\mathcal{A}}(\lg q)(q_H + q_S) \quad \text{and} \quad \epsilon_1 \geq 2e \cdot q_S \epsilon_0, \qquad (*)$$

and $c_{\mathcal{A}}$ is a small constant.

By Section 3.3, we can construct a hash function h onto G^* from the hash function h'. By Lemma 4, the generic GDH signature scheme on G, using algorithm $MapToGroup_{h'}$ is $(t_2, q_H, q_S, \epsilon_2)$-secure against existential forgery on adaptive chosen-message attacks by the main theorem (Section 4), where

$$t_2 = t_1 - 2^I q_H \lg m \quad \text{and} \quad \epsilon_2 = \epsilon_1. \qquad (**)$$

The only difference between the generic GDH signature scheme on G and the concrete scheme on G described above is that signatures in the latter scheme are elements of $\mathbb{F}_{3^l}$, rather than G. Given an adversary $\mathcal{F}$ that breaks the concrete scheme, we can construct an algorithm $\mathcal{A}$ that breaks the generic scheme, as follows. The public key is identical in the two schemes, so $\mathcal{A}$ simply provides $\mathcal{F}$ with the R given to it. Hashes are identical in the two schemes, so $\mathcal{A}$ passes $\mathcal{F}$'s hash requests to its own hash oracle, and provides $\mathcal{F}$ with the answer. When $\mathcal{F}$ requests a signature on a message M, $\mathcal{A}$ obtains the signature $S \in E$ from its signature oracle, and gives $\mathcal{F}$ the x-coordinate σ of S. Finally, when $\mathcal{F}$ outputs a forgery σ^* (for the concrete scheme) on a message M^*, $\mathcal{A}$ finds a point $S^* \in E$ whose x-coordinate is σ^*. By the discussion above, either $(P, R, h(M^*), S^*)$ is a Diffie-Hellman tuple, in which case S^* is a signature on M^* in the concrete scheme, or $(P, R, h(M^*), -S^*)$ is a Diffie-Hellman tuple, in which case $-S^*$ is a signature on M^* in the concrete scheme. $\mathcal{A}$ outputs M^* along with the appropriate one of S^* and $-S^*$.

The additional time required for this simulation is dominated by the two additional signature verifications, each of which takes time τ. Thus if the generic GDH scheme is $(t_2, q_H, q_S, \epsilon_2)$-secure, the concrete GDH scheme is $(t_3, q_H, q_S, \epsilon_3)$-secure, where

$$t_3 = t_2 - 2\tau \quad \text{and} \quad \epsilon_3 = \epsilon_2. \tag{***}$$

Combining (∗), (∗∗), and (∗∗∗) yields the required reduction. □

3.5 An Open Problem: Short Signatures with High Security

In the previous section we proposed using a supersingular curve over $\mathbb{F}_{3^\ell}^*$ to build a short signature scheme as secure as discrete log in $\mathbb{F}_{3^{6\ell}}^*$. However, there is no reason to stick with supersingular curves. Using other elliptic or hyper-elliptic curves it might be possible to achieve even higher security multipliers.

In Section 3.2, we showed that the curves E^+ and E^- over $\mathbb{F}_{3^l}$ have security parameter α at most 6. This is, in fact, the maximum value of α for any supersingular curve [15,23]. Instantiating the GDH signature scheme on (necessarily non-supersingular) elliptic curves with slightly higher values of α would increase the work required for verification, but also increase security against MOV-related attacks at comparable signature bit lengths.

Consider an elliptic curve $E/\mathbb{F}_{p^l}$ with m points, a large prime $q \mid m$, a security parameter α for the subgroup of order q, and two linearly independent points, P and Q, of order q, where $P \in E/\mathbb{F}_{p^l}$, and $Q \in E/\mathbb{F}_{p^{l\alpha}}$. Note that a point $Q \in E/\mathbb{F}_{p^{l\alpha}}$ linearly independent of P must exist by [2] assuming $q \nmid p^l - 1$. For such a curve, there is not necessarily an automorphism that maps between $\langle P \rangle$ and $\langle Q \rangle$. We therefore slightly modify the Gap Signature Scheme to use the two groups together.

It is easy to decide whether a tuple (P, aP, Q, bQ) is such that $a = b$, using the Weil pairing. We call this the co-Decision Diffie-Hellman problem, and it has an obvious computational variant: given the tuple (P, Q, aQ), compute aP. The modified (co-gap) signature scheme is as follows.

Key Generation. Let $P \in E/\mathbb{F}_{p^l}$ and $Q \in E/\mathbb{F}_{p^{l\alpha}}$ be two linearly independent points of prime order q as described above. Pick $x \stackrel{R}{\leftarrow} \mathbb{Z}_q^*$, and compute $R \leftarrow xQ$. The public key is $(E/\mathbb{F}_{p^l}, q, Q, R)$. The secret key is x.

Signing. Given a secret key x, and a message $M \in \{0,1\}^*$ use $\mathit{MapToGroup}_{h'}$ to map M to a point $P_M \in \langle P \rangle$. Set $S_M \leftarrow xP_M$. The signature σ is the x-coordinate of S_M, an element of $\mathbb{F}_{p^l}$.

Verification. Given a public key $(E/\mathbb{F}_{p^l}, q, Q, R)$, a message M, and a purported signature σ, let S be a point on $E/\mathbb{F}_{p^l}$ of order q whose x-coordinate is σ and whose y-coordinate is y for some $y \in \mathbb{F}_{p^l}$ (if no such point exists reject the signature as invalid). Set $u \leftarrow e(Q, S)$ and $v \leftarrow e(R, h(M))$. If either $u = v$ or $u^{-1} = v$, accept the signature. Otherwise, reject.

By reasoning analogous to that in Section 3.4, the tests in the verification phase ensure that either $(Q, R, h(M), S)$ or $(Q, R, h(M), -S)$ is a valid co-Diffie-Hellman tuple. While the public key, R, is an element of $E/F_{p^{l\alpha}}$, and thus long, a signature σ is an element of E/F_{p^l}, and thus relatively short. The security of this scheme follows from the assumption that no adversary (t, ϵ)-breaks the co-Computational Diffie-Hellman problem.

The challenge, therefore, is to construct elliptic curves with larger values of α, say $\alpha = 10$. It is currently an open problem to build a family of elliptic curves with security multiplier $\alpha = 10$.

Galbraith [8] constructs supersingular curves of higher genus with a "large" security multiplier. For example, the supersingular curve $y^2 + y = x^5 + x^3$ has security multiplier 12 over $\mathbb{F}_{2^l}$. Since a point on the Jacobian of this curve of genus two is characterized by two values in $\mathbb{F}_{2^l}$ (the two x-coordinates in a reduced divisor) the length of the signature is $2l$ bits. Hence, we might obtain a signature of length $2l$ with security of computing CDH in the finite field $\mathbb{F}_{2^{12l}}$. This factor of 6 between the length of the signature and the degree of the finite field is the same as in the elliptic curve case. Hence, this genus 2 curve does not improve the security of the signature, but does give more variety in signature lengths beyond those given in Table 1. Since this curve is defined over a field of characteristic two it is better suited for computation than curves defined over of fields of characteristic three. Galbraith shows that Jacobians of genus 2 supersingular curves have a maximum security multiplier of 12. Therefore, genus 2 supersingular curves will not give short signature with higher security. It is an open problem whether one can build a family of hyper-elliptic curves of genus 3 that would give short signatures with higher security.

4 Proof of Security Theorem

We prove, in the random oracle model, that GDH signatures are secure in GDH groups. The proof is similar to that given for full-domain hash RSA signatures by Coron [6], but the presentation is different. The point of this method is that the break-probability ϵ for the signature scheme does not depend on the number of hash queries a forger makes, but only depends on the number of signature queries made by the adversary.

Theorem (Gap Signature Security). *If G is a (τ, t', ϵ')-GDH group, then the Gap Signature Scheme on G is (t, q_H, q_S, ϵ)-secure against existential forgery on adaptive chosen-message attacks, where*

$$t \leq t' - 2\tau c_{\mathcal{A}}(q_H + q_S) \quad \text{and} \quad \epsilon \geq 2e \cdot q_S \epsilon',$$

and $c_{\mathcal{A}}$ is a small constant (in practice, at most 2).

The proof follows, in stages.

4.1 Overview

Assume an algorithm $\mathcal{F}$ (t, q_H, q_S, ϵ)-breaks the Gap Signature Scheme on G. We will use $\mathcal{F}$ to construct an algorithm $\mathcal{A}$ that (τ, t', ϵ') breaks Computational Diffie-Hellman on G, where t' and ϵ' are as above.

Given a forger $\mathcal{F}$ for the GDH group G, we build an algorithm $\mathcal{A}$ that uses $\mathcal{F}$ to break CDH on G. $\mathcal{A}$ is given a challenge (g, g^a, g^b). It uses this challenge to construct a public key that it provides to $\mathcal{F}$. It then allows $\mathcal{F}$ to run. At times, $\mathcal{F}$ makes queries to two oracles, one for message hashes and one for message signatures. These oracles are puppets of $\mathcal{A}$, which it manipulates in constructive ways. Finally, if all goes well, the forgery which $\mathcal{F}$ outputs is transformed by $\mathcal{A}$ into an answer to the CDH challenge.

We assume that $\mathcal{F}$ is well-behaved in the sense that it always requests the hash of a message M before it requests a signature for M, and that it always requests a hash of the message M^* that it outputs as its forgery. It is trivial to modify any forger algorithm $\mathcal{F}$ to have this property.

$\mathcal{A}$ needs to engage in a certain amount of bookkeeping. In particular, it must maintain a list of the messages on which $\mathcal{F}$ requests hashes or signatures. Each message M, as it arrives from $\mathcal{F}$, is assigned an index i; i is obviously bounded above by q_H. The message is stored in M_i, its hash in h_i, and its signature (if available) in σ_i.

4.2 Construction of $\mathcal{A}$

Rather than describe $\mathcal{A}$'s behavior and prove its efficacy *in toto*, we will construct $\mathcal{A}$ in a series of "games," in which increasingly sophisticated $\mathcal{A}$-variants run $\mathcal{F}$; the final variant, $\mathcal{A}_6$, is the $\mathcal{A}$ we seek.

(Each of the $\mathcal{A}$-variants will depend on a probability constant ζ, which will be optimized later, to yield the best possible reduction. Define B_ζ to be the probability distribution over $\{0, 1\}$ where 1 is drawn with probability ζ, and 0 with probability $1 - \zeta$.)

Game 1. $\mathcal{A}_1$ is given a challenge (g, g^a, g^b). In setup, it constructs $PK \leftarrow (g^a)$. Then, for each i, $1 \le i \le q_H$, $\mathcal{A}_1$ picks a random bit $s_i \overset{\text{R}}{\leftarrow} B_\zeta$, and a random number $r_i \overset{\text{R}}{\leftarrow} \mathbb{Z}_p^*$. It then sets $h_i \leftarrow g^{r_i}$, and $\sigma_i \leftarrow (g^a)^{r_i}$. Note that (g, g^a, h_i, σ_i) is a valid Diffie-Hellman tuple, so σ_i is a signature on any message whose hash is h_i. $\mathcal{A}_1$ then runs $\mathcal{F}$ with public key PK.

When $\mathcal{F}$ requests a hash on a message M_i, $\mathcal{A}_1$ responds with h_i; when $\mathcal{F}$ requests a signature on a message M_i, $\mathcal{A}_1$ responds with σ_i.

Finally, $\mathcal{F}$ halts, either conceding failure or returning a a forged signature (M^*, σ^*), where $M^* = M_{i^*}$ for some i^* (on which $\mathcal{F}$ had not requested a signature). If $\mathcal{F}$ succeeds in forging, $\mathcal{A}_1$ outputs "`success`"; otherwise, it outputs "`failure`".

The hashes h_i are uniformly distributed in G, so $\mathcal{A}_1$'s hash oracle is a random oracle. Moreover, the signatures σ_i are all valid. In the random oracle model, therefore, $\mathcal{F}$, when run by $\mathcal{A}_1$, behaves exactly as it would when running on its own. Thus

$$\begin{aligned}\mathsf{Adv}_{\mathcal{A}_1} &= \Pr\left[\mathcal{A}_1^{\mathcal{F}}(g, g^a, g^b) = \mathtt{success} : a, b \overset{\mathrm{R}}{\leftarrow} \mathbb{Z}_p^*\right] \\ &= \Pr\left[\mathit{Verify}(PK, M^*, \sigma^*) = \mathtt{valid} : \begin{array}{l}(PK, SK) \overset{\mathrm{R}}{\leftarrow} \mathit{KeyGen}, \\ (M^*, \sigma^*) \overset{\mathrm{R}}{\leftarrow} \mathcal{F}(PK)\end{array}\right] = \epsilon,\end{aligned}$$

where the first probability is taken over the coin tosses of $\mathcal{A}_1$ and $\mathcal{F}$, and over the choices of a and b. Since a is chosen uniformly from $\mathbb{Z}_p^*$, g^a, the public key $\mathcal{A}_1$ provides $\mathcal{F}$, is uniformly distributed in G.

Game 2. $\mathcal{A}_2$ functions as does $\mathcal{A}_1$, with a single exception. If $\mathcal{F}$ fails, $\mathcal{A}_2$ outputs "`failure`"; if $\mathcal{F}$ succeeds, outputting a forgery (M^*, σ^*), where i^* is the index of M^*, then $\mathcal{A}_2$ outputs "`success`" if $s_{i^*} = 1$, but "`failure`" if $s_{i^*} = 0$.

Clearly, $\mathcal{F}$ can get no information about any s_i, so its behavior cannot depend on their values. Thus the final trip test $\mathcal{A}_2$ performs is independent of the game to that point. Thus we have

$$\mathsf{Adv}_{\mathcal{A}_2} = \mathsf{Adv}_{\mathcal{A}_1} \cdot \Pr[s_{i^*} = 1] = \zeta\epsilon,$$

since each s_i is drawn from B_ζ.

Game 3. $\mathcal{A}_3$ functions as does $\mathcal{A}_2$, but, again, with a modification. If $\mathcal{F}$ fails to create a forgery, $\mathcal{A}_3$ also fails. If $\mathcal{F}$ succeeds in finding a forgery on M_{i^*}, $\mathcal{A}$ claims success only if $s_{i^*} = 1$, and $\mathcal{F}$ asked for signatures only on messages M_i for which $s_i = 0$.

Again, $\mathcal{F}$ can get no information about any s_i. Each of its signature requests can cause $\mathcal{A}$ to declare failure at the game's end, with probability ζ, but it cannot know, during the game, whether any of them did. The s_i's are independent, so each of $\mathcal{F}$'s signature requests is an independent trial insofar as disqualification by s_i is concerned. Moreover, s_{i^*} is independent of any s_i's for which $\mathcal{F}$ requests signatures, so the test that s_{i^*} equals 1 is again an independent trial, and the analysis of Game 2 is not affected.

The probability of $\mathcal{F}$'s not being disqualified because of any particular signature request is $1 - \zeta$. If $\mathcal{F}$ makes k signature oracle queries, where k necessarily is at most q_S, and if, moreover, it makes those queries on the messages with indices $i_1, \ldots, i_k$, then

$$\mathsf{Adv}_{\mathcal{A}_3} = \mathsf{Adv}_{\mathcal{A}_2} \cdot \Pr\left[s_{i_j} = 0,\, j = 1, \ldots, k\right] = \zeta\epsilon \cdot (1-\zeta)^k \geq (1-\zeta)^{q_S}\zeta\epsilon.$$

Game 4. $\mathcal{A}_4$ functions as does $\mathcal{A}_3$, except that, if $\mathcal{F}$ requests a signature on a message M_i for which $s_i = 1$, $\mathcal{A}$ declares failure and halts immediately.

We may fully describe a run of $\mathcal{A}$ by fixing the challenge, $\mathcal{A}$'s random bits, and $\mathcal{F}$'s random bits; these collectively determine the value of each s_i, and the indices on which $\mathcal{F}$ requests signatures. Let us call unlucky any runs in which $\mathcal{F}$ requests a signature on some M_i for which $s_i = 1$. $\mathcal{A}_3$ would already declare failure on any unlucky runs: if $\mathcal{F}$ declares failure, $\mathcal{A}_3$ does also; if $\mathcal{F}$ finds a forgery, $\mathcal{A}_3$ fails anyway because of the unlucky signature query. Thus $\mathcal{A}_3$ and $\mathcal{A}_4$ will agree (with output "`failure`") on all unlucky runs; they will also agree on all lucky runs, since the modification of $\mathcal{A}_4$ relative to $\mathcal{A}_3$ is not invoked on those runs. Thus we have

$$\mathsf{Adv}_{\mathcal{A}_4} = \mathsf{Adv}_{\mathcal{A}_3} \geq (1-\zeta)^{q_S}\zeta\epsilon.$$

The immediate halt in unlucky runs is a shortcut and does not affect the outcome distribution.

Game 5. $\mathcal{A}_5$ is based on $\mathcal{A}_4$. In the setup phase, for each i, if $s_i = 1$, $\mathcal{A}_5$ sets $h_i \leftarrow g^b \cdot g^{r_i}$ and $\sigma_i \leftarrow \star$, a placeholder value; if $s_i = 0$, it sets $h_i \leftarrow g^{r_i}$ and $\sigma_i \leftarrow (g^b)^{r_i}$, as before.

G is a cyclic group of prime order, so multiplication by any element of G, and g^b in particular, induces a permutation on G. Thus if r is uniformly distributed in $\mathbb{Z}_p^*$, g^r and $g^b \cdot g^r$ have identical, uniform distributions in G. $\mathcal{F}$ cannot learn any information about the s_i's from examining the h_i's it is given. $\mathcal{A}_5$ is unable to provide signatures on messages for which $s_i = 1$, but that is unimportant, since any runs in which $\mathcal{F}$ asks for such a signature are failed immediately. Therefore, $\mathcal{F}$ will behave under $\mathcal{A}_5$ exactly as it does under $\mathcal{A}_4$, and

$$\mathsf{Adv}_{\mathcal{A}_5} = \mathsf{Adv}_{\mathcal{A}_4} \geq (1-\zeta)^{q_S}\zeta\epsilon.$$

Game 6. $\mathcal{A}_6$ behaves as does $\mathcal{A}_5$. In those games where $\mathcal{A}_5$ outputs "`success`", however, $\mathcal{A}_6$ outputs "`success`" and, in addition, outputs $\sigma^*/(g^a)^{r_{i^*}}$, where i^* is the index of the message M^* for which $\mathcal{F}$ output a forged signature σ^*. ($\mathcal{A}_6$, like the $\mathcal{A}$'s before it, only succeeds when $\mathcal{F}$ succeeds.)

Clearly, $\mathcal{A}_6$ succeeds with precisely the same probability as $\mathcal{A}_5$, so

$$\mathsf{Adv}_{\mathcal{A}_6} = \mathsf{Adv}_{\mathcal{A}_5} \geq (1-\zeta)^{q_S}\zeta\epsilon.$$

Moreover, $\mathcal{A}_6$ only succeeds if $s_{i^*} = 1$, which means that $h_{i^*} = g^b \cdot g^{r_{i^*}}$. If σ^* is a valid signature on $M^* = M_{i^*}$, then $(g, g^a, h_{i^*}, \sigma^*)$ must be a valid Diffie-Hellman tuple, so σ^* must equal $h_{i^*}^a = g^{ab} \cdot (g^{r_{i^*}})^a$. Thus, in every instance on which $\mathcal{A}_6$ claims to succeed, it also outputs $\sigma^*/(g^a)^{r_{i^*}} = g^{ab}$, which is indeed the answer to the Diffie-Hellman challenge posed to it.

4.3 Optimization and Conclusion

The algorithm $\mathcal{A}_6$ thus uses the GDH-signature forger $\mathcal{F}$ to solve CDH challenges. What remains is to optimize the parameter ζ to achieve a maximal probability of success. The function $(1-\zeta)^{q_S}\zeta\epsilon$ is maximized at $\zeta = 1/(q_S+1)$, where it has the value

$$\frac{1}{q_S+1} \cdot \left(1 - \frac{1}{q_S+1}\right)^{q_S} \cdot \epsilon = \frac{1}{q_S} \cdot \left(1 - \frac{1}{q_S+1}\right)^{q_S+1} \cdot \epsilon.$$

(The latter equality follows from taking partial fractions.) Now $\mathcal{A}$'s success probability ϵ' is at least as great as this. For large q_S, $(1 - 1/(q_S + 1))^{q_S+1} \approx 1/e$.

$\mathcal{A}$'s running time includes the running time of $\mathcal{F}$. The additional overhead imposed by $\mathcal{A}$ is dominated by the need to evaluate group exponentiation for each signature and hash request from $\mathcal{F}$. Any one such exponentiation may be computed by using at most $2 \lg p$ group actions, and thus at most $2 \lg p$ time units, on G (see [16]). $\mathcal{A}$ may need to answer as many as $q_H + q_S$ such requests, so its overall running time is $t' \leq t + 2c_{\mathcal{A}}(\lg p)(q_H + q_S)$, Where $c_{\mathcal{A}}$ is a small constant that accounts for the remainder of $\mathcal{A}$'s administrative overhead; in practice, $c_{\mathcal{A}}$ should be at most 2.

To summarize: if there exists a forger algorithm $\mathcal{F}$ that (t, q_H, g_S, ϵ)-breaks the GDH signature scheme on G, then there exists an algorithm $\mathcal{A}$ that (t', ϵ')-breaks CDH on G, where

$$t' = t + 2c_{\mathcal{A}}(\lg p)(q_H + q_S) \quad \text{and} \quad \epsilon' = \frac{1}{q_S} \cdot \left(1 - \frac{1}{q_S + 1}\right)^{q_S+1} \cdot \epsilon .$$

Conversely, if G is a (τ, t', ϵ')-GDH group, then there can exist no algorithm $\mathcal{F}$ that (t, q_H, q_S, ϵ)-breaks the GDH signature scheme, where

$$t = t' - 2c_{\mathcal{A}}(\lg p)(q_H + q_S) \quad \text{and} \quad \epsilon = q_S \epsilon' \Big/ \left(1 - \frac{1}{q_S + 1}\right)^{q_S+1} .$$

For all positive q_S, the radicand in the latter equation is greater than $1/2e$, so the equation may be rewritten as $\epsilon \leq q_S \epsilon' \cdot 2e$. This completes the proof.

5 Experimental Results

5.1 Implementation Details

We experimented with the scheme of Section 3.4. Recall that signing is a single multiplication on the curve $y^2 = x^3 + 2x \pm 1$ over $\mathbb{F}_{3^l}$. Verifying a signature requires two Weil pairing computations over $\mathbb{F}_{3^{6l}}$. Hence, verifying takes more time than signing.

For efficiency, rather than working in $\mathbb{F}_{3^{6l}}$ directly (which involves manipulating polynomials of degree $6l$), we work with extensions of $\mathbb{F}_{3^6}$ of degree l. To speed up arithmetic in $\mathbb{F}_{3^6}$ we construct lookup tables (of size 3^6) for quickly multiplying two elements. Elements of $\mathbb{F}_{3^6}$ are represented by their exponent relative to a chosen generator, so that multiplication and division corresponds to addition modulo $3^6 - 1$. Addition is done using a multiplication, division and table lookup via the identity $a + b = a(1 + a^{-1}b)$. The constants r^+, r^-, u used in the automorphism ϕ also lie in $\mathbb{F}_{3^6}$ and can be quickly found by a brute force search.

We map an element a in $\mathbb{F}_{3^l}$ to an element of $\mathbb{F}_{3^{6l}}$ using the obvious injection: a is represented by a polynomial of degree l with coefficients in $\mathbb{F}_3$, and we simply view it as a polynomial with coefficients in $\mathbb{F}_{3^{6l}}$.

We use the Tate pairing [7] instead of the Weil pairing, since it has similar properties and is easier to compute: the Weil pairing requires two iterations of Miller's algorithm [17] and one division while the Tate pairing needs only one call to Miller's algorithm and an additional exponentiation.

Because Miller's algorithm involves the computation of various quotients, several divisions can be avoided since we may scale the numerator and denominator by arbitrary constants. We used sliding windows for every exponentiation-like operation, that is, exponentiation in $\mathbb{F}_{3^{6l}}$, Miller's algorithm, and multiplication of a point on the curve. Point multiplication can be sped up further by using signed sliding windows, converting to weighted projective coordinates (though this may not help; it depends on the implementation of the field operations), and taking advantage of the fact that some points are fixed for the whole system.

Recall that the output of the Tate pairing is a coset representative in $\mathbb{F}^*_{3^{6l}}$. Signature verification then consists of checking that the output of two Tate pairings lie in the same coset. This could be done by finding the quotient of the outputs, and raising it to the appropriate power (and comparing with the identity element). However, we can replace the division with a multiplication by exploiting the bilinearity of the Tate pairing: dividing by $e(A, B)$ is equivalent to multiplying by $e(A, -B) = 1/e(A, B)$ ($-B$ can be easily computed from B by negating the y-coordinate).

The x-coordinate is an element of $\mathbb{F}_{3^l}$ and is represented as a polynomial of degree at most $l-1$ with coefficients in $\mathbb{F}_3$. For output, it is viewed as a number in base 3, and then encoded in base-64. For $l = 97$, which has 923-bit discrete-log security, an example signature looks as follows: "`KrpIcV0O9CJ8iyBS8MyVkNrMyE`". This is under half the size of the standard 320-bit DSS signature (with 1024-bit discrete security).

5.2 Running Times

The following table shows the time required to verify a signature. Recall that a verification is much more expensive than signature generation because it requires computing two pairings. The program was run on a 1 GHz Pentium III computer running GNU/Linux.

l	sig-length (bits)	Dlog Security $\lceil \log_2 x \rceil$	curve	Running Time (seconds)
79	126	752	E^-	1.6
97	154	923	E^+	2.9
149	237	1417	E^+	9.6
163	259	1551	E^+	13.3
163	259	1551	E^-	13.4
167	265	1589	E^+	14.0

When using elliptic curves to get short GDH signatures we are forced to use a curve over a field of characteristic three. This slows down arithmetic on the curve. It is possible that the running times above can be improved using higher genus curves over fields of characteristic two as discussed at the end of Section 3.5. Similarly, the techniques of [13] for computing on the curves E^+ and E^- over $\mathbb{F}_{3^l}$ may slightly improve these numbers.

6 Conclusions

We presented a short signature based on the Weil pairing. The length of a signature is one element of a finite field. Standard signatures based on discrete log such as DSA require two elements. When working with the curve $y^2 = x^3 + 2x \pm 1$ over $\mathbb{F}_{3^l}$ the MOV attack maps the CDH problem in this curve to a CDH problem in $\mathbb{F}_{3^{6l}}$. Hence, we can use small values of l to obtain short signatures with security comparable to the security of 320-bit DSA. For example, we obtain a signature of length 154 bits where breaking the scheme reduces to solving the Diffie-Hellman problem in a finite field of size approximately 2^{923}. In Section 3.5 we outlined an open problem that would enable us to get even better security while maintaining the same length signatures. We hope future work on constructing elliptic curves or higher genus curves will help in solving this problem.

Acknowledgments. The authors thank Steven Galbraith, Alice Silverberg, and Moni Naor for helpful discussions about this work.

References

1. ANSI X9.62 and FIPS 186-2. Elliptic Curve Digital Signature Algorithm, 1998.
2. R. Balasubramanian and N. Koblitz. The Improbability That an Elliptic Curve Has Subexponential Discrete Log Problem under the Menezes-Okamoto-Vanstone Algorithm. *Journal of Cryptology*, 11(2):141–145, 1998.
3. M. Bellare and P. Rogaway. The Exact Security of Digital Signatures: How to Sign with RSA and Rabin. In U. Maurer, editor, *Proceedings of Eurocrypt '96*, volume 1070 of *LNCS*, pages 399–416. Springer-Verlag, 1996.
4. D. Boneh and M. Franklin. Identity-Based Encryption from the Weil Pairing. In J. Kilian, editor, *Proceedings of Crypto '2001*, volume 2139 of *LNCS*, pages 213–229. Springer-Verlag, 2001.
5. D. Chaum and T. Pederson. Wallet Databases with Observers. In E. Brickell, editor, *Proceedings of Crypto '92*, volume 740 of *LNCS*, pages 89–105. Springer-Verlag, 1992.
6. J.-S. Coron. On the Exact Security of Full Domain Hash. In M. Bellare, editor, *Proceedings of Crypto '2000*, volume 1880 of *LNCS*, pages 229–235. Springer-Verlag, 2000.
7. G. Frey, M. Muller, and H. Ruck. The Tate Pairing and the Discrete Logarithm Applied to Elliptic Curve Cryptosystems. *IEEE Tran. on Info. Th.*, 45(5):1717–1719, 1999.
8. S. Galbraith. Supersingular curves in cryptography. In *Proceedings of Asiacrypt '2001*, LNCS. Springer-Verlag, 2001.
9. S. Galbraith and N. P. Smart. A Cryptographic Application of Weil Descent. In M. Walker, editor, *Cryptology and Coding*, volume 1746 of *LNCS*, pages 191–200. Springer-Verlag, 1999.

10. P. Gaudry, F. Hess, and N. P. Smart. Constructive and Destructive Facets of Weil Descent on Elliptic Curves. Technical Report CSTR-00-016, Department of Computer Science, University of Bristol, 2000.
11. A. Joux. A One Round Protocol for Tripartite Diffie-Hellman. In W. Bosma, editor, *Proceedings of ANTS IV*, volume 1838 of *LNCS*, pages 385–394. Springer-Verlag, 2000.
12. A. Joux and K. Nguyen. Separating Decision Diffie-Hellman from Diffie-Hellman in Cryptographic Groups. Cryptology ePrint Archive, Report 2001/003, 2001. `http://eprint.iacr.org/`.
13. N. Koblitz. An Elliptic Curve Implementation of the Finite Field Digital Signature Algorithm. In H. Krawczyk, editor, *Proceedings of Crypto '98*, volume 1462 of *LNCS*, pages 327–333. Springer-Verlag, 1998.
14. S. Lang. *Elliptic Functions.* Addison-Wesley, Reading, MA, 1973.
15. A. Menezes, T. Okamoto, and P. Vanstone. Reducing Elliptic Curve Logarithms to Logarithms in a Finite Field. *IEEE Transactions on Information Theory*, 39(5):1639–1646, 1993.
16. A. J. Menezes, P. C. Van Oorschot, and S. A. Vanstone. *Handbook of Applied Cryptography.* CRC Press, 1997.
17. V. Miller. Short Programs for Functions on Curves. unpublished manuscript, 1986.
18. I. Mironov. A Short Signature as Secure as DSA. Preprint, 2001.
19. D. Naccache and J. Stern. Signing on a Postcard. In *Proceedings of Financial Cryptography '00*, 2000.
20. T. Okamoto and D. Pointcheval. The Gap Problems: A New Class of Problems for the Security of Cryptographic Primitives. In K. Kim, editor, *Public Key Cryptography, PKC 2001*, volume 1992 of *LNCS*, pages 104–118. Springer-Verlag, 2001.
21. L. Pintsov and S. Vanstone. Postal Revenue Collection in the Digital Age. In *Proceedings of Financial Cryptography '00*, 2000.
22. J. H. Silverman. *The Arithmetic of Elliptic Curves*, volume 106 of *Graduate Texts in Mathematics.* Springer-Verlag, 1986.
23. W. C. Waterhouse. Abelian Varieties over Finite Fields. *Ann. Sci. École Norm. Sup.*, 2:521–60, 1969.

Self-Blindable Credential Certificates from the Weil Pairing

Eric R. Verheul

PricewaterhouseCoopers, GRMS Crypto group, P.O. Box 85096, 3508 AB Utrecht, The Netherlands, eric.verheul@[nl.pwcglobal.com, pobox.com]

Abstract. We describe two simple, efficient and effective credential pseudonymous certificate systems, which also support anonymity without the need for a trusted third party. The second system provides cryptographic protection against the forgery and transfer of credentials. Both systems are based on a new paradigm, called self-blindable certificates. Such certificates can be constructed using the Weil pairing in supersingular elliptic curves.

1 Introduction

Credential pseudonymous certificates (CPCs) were introduced by David Chaum [7] in 1985 to counter some of the privacy problems related to identity certificates. One such problem is that service providers know exactly who they are servicing when a user employs an identity certificate, which for some applications is not required, acceptable or even permissible. Moreover, by combing their logs, service providers can piece together a record of all the user's activities.

A *pseudonym* is a unique identifier (string) by which a user is known by a certain party; typically each party knows the same user by a different pseudonym. These pseudonyms can be references to a user's identity known only by designated parties, or can be completely anonymous, (i.e., known only to the user). Unlike Chaum [7], we do not limit a 'physical' user to only one pseudonym with a given provider. We believe that for some types of providers, e.g., on-line, subscription based, information providers, the use of many different pseudonyms for one physical user, without the provider knowing, can be considered an important feature. However, we do discuss how, if necessary, such unique pseudonyms can be supported by our systems.

A *pseudonymous certificate* binds a user's pseudonym to their public key, the private key to which the user possesses. Such certificates are issued by a trust provider. Identities, pseudonyms and public keys should be unique. A *credential* is a trust provider's statement about the user which is relied upon by other parties, who we simply call service providers. Examples of such statements are properties such as "lives in Amsterdam", qualifications such as "has a PhD in math", or rights such as "can access this secure room". A credential can be *single-use*, such as a prescription, or *multiple-use* such as a driver's license. In this paper we focus on the latter type of credentials.

C. Boyd (Ed.): ASIACRYPT 2001, LNCS 2248, pp. 533–551, 2001.
© Springer-Verlag Berlin Heidelberg 2001

Finally, *credential pseudonymous certificates* (CPCs) are digital certificates that bind credentials to users, known by a pseudonym. Proof of credential possession is given by proving possession of the private key related to the public key referenced in the certificate. Several credentials may be bound to a single pseudonymous certificate and, thus, pseudonym.

In Chaum's model, pseudonyms are *unlinkable*: parties that know a user by different pseudonyms must not have the ability to combine their logs to assemble a dossier on the user.[1] Another requirement in Chaum's model is that CPCs must be *translatable*: a CPC issued under pseudonym A must be usable under pseudonym B. For example, a user may be given a credential asserting his good health from a doctor under pseudonym A, and show this to its insurance company who knows it by pseudonym B. In addition to these two requirements, the system should fulfill the following three basic security requirements:

Protection against pseudonym/credential forgery. It should not be possible for outsiders, malicious users, or other parties involved to generate (credential) pseudonymous certificates without the consent of the relevant trust providers.

Protection against pseudonym/credential sharing. A user could be tempted to share its credentials (e.g., a season pass for public transport) with another user. It should therefore be very difficult or awkward for a user to do so.[2] One potential solution to this problem would be to store credentials on tamper resistant devices that are valuable to the user (e.g., smartcard based passports). A better solution would be an *all-or-nothing* concept for credentials: sharing a credential effectively implies sharing a credential that is highly valuable to the user, most notably one enabling him to take over the user's identity and digitally sign contracts that legally binds the user (cf., [6], [5]).

Revocation of pseudonymous certificates and credentials. Under certain circumstances, it should be possible for the user and trust providers to revoke pseudonymous certificates as well as credentials bound to them. This could be case, for instance, if a user lost secret (key) information or changes jobs.

CPCs such as those described above, counter the privacy problems of identity certificates to some extent, but not completely. Indeed, in that setting, all user's activities with a provider are related to a pseudonym, so that the provider can link the user's activities with the fixed pseudonym. If the user's identity

[1] Unlinkability and pseudonymity of credentials are sometimes difficult to enforce simultaneously in practice. Indeed, even if they are anonymous, credentials implicitly narrow down the number of possible users possessing them. To illustrate, how many people have both a degree in cryptography (credential number one) and Swedish citizenship (credential number two)?

[2] Perhaps complete eradication of credential sharing would be impossible in the virtual world, as the end user might give away everything he knows (passwords) or has (smartcards), leaving only the identification factor "what the user is"(e.g., biometrics) to counter credential sharing.

is compromised, then so are its activities. To prevent this potential problem, a CPC system should preferably support that users can easily and regularly change pseudonyms. A CPC system should also ensure that the translation of credentials includes as few (trusted) parties as possible. In our CPC system, users themselves can both change pseudonyms and translate their credentials.

We remark that, in the above text, we implicitly define the parties as *users*, *trust providers* (providing credentials and pseudonyms to users) and *service providers* (relying on credentials and pseudonyms), which we use in the remainder of this paper without further explanation.

The goal of this paper is to describe a very simple, effective and efficient CPC system that meets the basic requirements of a CPC system and that is based on the new paradigm of *self-blindable* certificates. With this type of certificates the user can, e.g.:

- generate its own new pseudonymous certificates itself (to which it possesses the private key) based on a valid pseudonymous certificate; and
- translate and combine CPCs issued under one pseudonym to another pseudonym, including a one-time-use pseudonym.

1.1 Related Work

As we could probably write an entire paper just discussing and comparing all of the CPC schemes that have been published, we will be brief. The first scheme was introduced by Chaum and Evertse [10] and is based on having a semi-trusted third party involved in all credential translations. Both from an efficiency and a security point of view, this is undesirable. Chen's scheme [12], envisions a trusted party who, amongst other things, should be trusted to refrain from transferring credentials between different users. Damgård's scheme [13], is based on general complexity-theoretic primitives and is therefore not applicable for practical use. The scheme developed by Lysyanskaya, Rivest, Sahai and Wolf [19] is based on one-way functions and general zero-knowledge proofs which also makes it inappropriate for practical use. Our CPC system can be considered as the opposite of the credential scheme [6] constructed by Camenisch and Lysyanskaya, which in effect issues one secret CPC for each trust provider; the scheme's properties of anonymity and untraceablity arise from the zero-knowledge protocols that confirm that a user indeed has such a certificate without revealing it. Although the scheme [6] appears to be of practical use, it is based on rather complex (zero-knowledge) protocols. Our scheme and the required proofs of knowledge are basic (Schnorr and Okamoto). Finally, we mention the work of Brands [5], which deals with the related subject of privacy protecting attribute certificates. In our system, the user itself can translate or combine credentials received from different trust providers without the interaction of any trusted party, generating a new certificate. This is an important distinction from Brands' scheme [5] when applied to the special case of a credential certificate system. As a final note, we remark that the privacy of our scheme can be further improved by the use of "Wallet with Observer" techniques, cf., [5], [11].

Outline of the Paper

- In Section 2, we describe a variant of the Chaum-Pedersen digital signature scheme which is of crucial importance for our constructions of self-blindable certificates.
- In Section 3, we provide a functional description of our model for CPSs.
- In Section 4.1, we present the first technical construction of our model, which assumes that secret key information is stored on tamper-proof devices to provide resistance to credential transfer.
- In Section 4.2, we present the second technical construction of our model, which is more resistant to the transfer of credentials, without requiring the use of tamper-proof devices. The transfer of any credential in this construction to another person will actually result in the transfer of a very valuable signing key, e.g., one enabling the holder to sign legally binding contracts in the user's name.
- In Section 5, we summarize our results.

2 A Proofless Variant of the Chaum-Pedersen Signature Scheme

A digital signature s formed by an entity is a data string, based on a private key under control of the entity, that associates a message m (in digital form) to enable a proof that it originates from the entity and that it has not been changed. If the actual message comprises a public key plus some optional additional attributes, then (m, s) is called a certificate and the entity issuing it is called a Certification Authority (CA).

In this section, we describe a digital signature scheme that enables a CA to issue certificates that are "self-blindable". This will be explained further in Section 3. The digital signature scheme is based on the Chaum-Pedersen signature scheme (cf., [11]). The setting of our scheme is not standard but is based on a group, G, of prime order q, with generator g, in which the Decision Diffie-Hellman problem is simple, while the discrete logarithm and the Diffie-Hellman problems are practically intractable. In the section below, we further explain these notions and indicate how such groups can be constructed. In Section 2.2, we describe our digital signature scheme and its properties.

2.1 Groups in which the DDH Problem Is Simple and DH, DL Are Hard

Recall, that the *Diffie-Hellman* (DH) problem with respect to a generator g of a group G of (prime) order q, is the problem of computing the values of the function $DH_g(g^x, g^y) = g^{xy}$. Two other problems are related to the DH problem. The first one is the *Decision Diffie-Hellman* (DDH) problem with respect to g: given $a, b, c \in G$ decide whether $c = DH_g(a, b)$ or not. An alternative formulation of the Decision Diffie-Hellman problem is: given a quadruple g, g^x, h, h^y in the group G decide whether $x = y$. The second problem related to the DH problem, is the *discrete logarithm* (DL) problem in G with respect to g: given $a = g^x \in G$, with $0 \le x < q$, find $x = DL(\alpha)$. The DL problem is at least as difficult as the DH problem. It is widely assumed that if the DL problem G is hard, then so is the DH problem. Currently, cf. [16], [27], [17], a large class of groups has been discovered in which the DDH problem is simple, while the Diffie-Hellman and discrete logarithm problems are presumably not. This class consists of certain groups of points on supersingular elliptic curves in which setting the DDH problem can be efficiently computed (in polynomial time i.e., in polynomial time and space in length of input) by using the so-called Weil pairing.

As an illustration of such groups and techniques, consider the curve C_a : $y^2 = x^3 + a$ with $p = 2 \bmod 3$ and a any non-zero element in GF(p). Then, the Frobenius trace over GF(p) is equal to 0 (hence the curve is supersingular) and the number of points on the curve in GF(p) is equal to $p + 1$. Moreover, as $p = 2 \bmod 3$, the equation $x^3 = 1$ only has solutions in GF(p^2) other than $x = 1$; let ω be such a solution. Now, if $\langle P \rangle$ is a group of points of (prime) order q on the curve in GF(p) (i.e., q divides $p + 1$) and A, B, C is an instance of the DDH problem with respect to P. Then $C = DH_P(A, B)$ if and only if $e_q(A, D(B)) = e_q(P, D(C))$, where $D(.)$ is the endomorphism (called a *distortion map* in [27]) on C_a that maps a point (x, y) on the curve to the point $(\omega \cdot x, y)$ also on the curve (over GF(p^2)) and where $e_q(.,.)$ is the so-called Weil pairing. See [1], [20] or [26]. As the Weil pairing is efficiently computable, the DDH problem is also efficiently computable in this situation. It is well-known that the DL problem in the group of points on the curve in GF(p) reduces to the DL problem in a subgroup of order q in GF(p^2)* (cf. [21]). That is, to make the DH and DL problems practically intractable against attacks known today, the length of the prime number q should be at least 160 bit and the length of the prime number should be at least 512 bits.

A practical construction of a group in which the DDH problem is efficiently computable and the DH and DL problems are presumably not, is as follows. Choose a 512 bit prime number p of type $p = 6q - 1$ where q is also a prime number and consider the curve $C_1 : y^2 = x^3 + 1$. Let P be any GF(p)-rational point on the curve of order q. This construction is used in [2] in the setting of an identity-based encryption scheme that is also based on the Weil pairing. This paper also analyzes the work needed to solve the DDH problem in the group $\langle P \rangle$, which amounts to a small number of multiplications on the curve.

These techniques generalize to groups of points on supersingular elliptic curves over a finite field, say F, and the work required to compute the DDH problem is asymptotically bounded by $O(k^3 \log(\|F\|)$ bit operations, i.e., the complexity of calculating a Weil pairing. The parameter k is the so-called MOV degree (cf. [21]) and is equal to either 1, 2, 3, 4 or 6 in the setting of supersingular curves.

We end this section with two remarks for later reference. A group of points, G, on a supersingular elliptic curves has the property that there exists an efficiently computable embedding, i.e., an injective homomorphism, of the group in a second group G' where all three of the DDH, DH and the DL problems are believed to be hard. Indeed, this embedding is given by the MOV embedding (cf. [21]) and the second group, G', is a subgroup of the multiplicative group of a finite field. It is shown in [27] that inverting such embeddings is hard; in fact, as hard as the DH problem in the group G. Note that by using a specific choice of G, the group G' could be the XTR group. Compare [18] and [27]. A group of points on a (supersingular) elliptic curve over a finite field used in cryptography is typically chosen in such a way that its order is a prime number times a small number (e.g., 6 in the example above). This means that choosing provable random elements in the subgroup without knowledge of relative discrete logarithms is very simple, e.g., by mapping a hash value into a point on the curve and then mapping it to a point in the subgroup. See also [3].

2.2 The 'Proofless' Variant of the Chaum-Pedersen Scheme

As explained in the previous section, we consider a group, G, of prime order q, with generator g, in which the DDH problem is simple, while the discrete logarithm and the Diffie-Hellman problems are practically intractable. The public key of a participant in the Chaum-Pedersen scheme takes the form $y = g^x$ where $0 \leq x < q$ is the participant's randomly chosen private key. A signature on a message $m \in G$ in the original Chaum-Pedersen scheme, consists of $z = m^x$ plus a proof that $\log_g(y) = \log_m(z)$. Resolving the latter problem is just an instance of the *Decision Diffie-Hellman* (DDH) problem with respect to g. Indeed, one can easily verify that $\log_g(y) = \log_m(z)$ if and only if $z = DH_g(m, y)$. That is, if one applies the Chaum-Pedersen scheme to the group G, one is not required to send along an explicit proof that $\log_g(y) = \log_m(z)$, as anyone can validate that themselves. Or, in other words, the signature on a message $m \in G$ only consists of an element $z = m^x$ of the group G, without the additional proof of knowledge. This is the variant of the Chaum-Pedersen scheme that we use in our schemes. It follows that by choosing a group of points on a supersingular elliptic curve of MOV degree 6 (cf. [21] and the previous section), the representation of the element z requires only $1024/6 \approx 171$ bits to obtain a security level comparable with 1024 bit RSA (with respect to attacks known today). See [3], where it is also shown that the above digital signature scheme is secure in the random oracle model.

An interesting property of this variant is that it is *self-blindable*: it enables easy randomization without losing the verification property and without requiring knowledge of the signing key z. Indeed, given the signed message m, m^z, then by choosing a randomizing factor, k, it can be transformed into m^k, m^{kz}. This property becomes useful when the message m has a property that is inherited by m^k, e.g., knowledge of a certain discrete logarithm, and is explored in the following sections. Another interesting property of this variant (as pointed out to us by Stefan Brands), is its easy blinding property, cf. [7]. When a party wants to obtain a blind signature on a message (typically a hash), M, from a signing party with public key g^x in our variant of the Chaum-Pedersen, it asks the signing party to sign M^r, for a random $0 \leq r < q$, resulting in M^{rx}. The user can deduce M^x from this using r and verify that it is a correct signature on M, which is publicly verifiable. We will delve no further into this property in this paper.

In the terminology we introduced above, we formulate the security assumption that we require for our variant of the Chaum-Pedersen scheme (cf. [8], [9]).

Assumption 21 *If the Diffie-Hellman problem with respect to g is hard, then without knowledge of the private signing key z, the only forged message an attacker can make on the basis of signed messages $(m_1, m_1^z), (m_2, m_2^z), \ldots, (m_n, m_n^z)$ with respect to the public key g^z is of the form $(g^{i_0} \prod_{j=1}^{n} m_n^{i_n}, (g^{i_0} \prod_{j=1}^{n} m_n^{i_n})^z)$, for any integers $i_0, i_1, \ldots i_n$, i.e., a power product of the signed messages.*

3 Our Functional Model for CPCs

In this section, we describe our functional model for CPCs. To this end, we first formulate the requirements for self-blindable pseudonymous certificates and credentials based upon them. Then we explain how these elements can be used to build a CPC system.

3.1 Self-Blindable Certificates

In this section we introduce the notion of *self-blindable* certificates, which is of crucial importance for our schemes. Our introduction is somewhat informal, but can be made formal without much effort.

We assume that one public key crypto system is employed by all users and we denote the collection of all possible user public keys by $\mathcal{U}$. We also assume that one signing public key crypto system is employed by all trust providers for certificate issuance. For simplicity's sake, we also assume that certificate signing is deterministic, i.e., there is only one possible valid certificate on a fixed public key, plus optional fields. We let $\mathcal{T}$ denote the collection of possible verification public keys of trust providers. Our description of a credential on a user public key $P_U \in \mathcal{U}$ from a trust provider with public verification key P_T takes the form

$$\{P_U, Sig(P_U, S_T)\},$$

where S_T stands for the private signing key of the trust provider relating to P_T. This certificate is typically accompanied by a higher-level certificate

$$Cert(P_U, \text{"Trust statement"})$$

on the public verification key P_T. We do not further elaborate on this, but this certificate can be thought of as a standard X.509 certificate with the "Trust statement" in one of its extension fields. We denote the collection of all possible certificates by $\mathcal{C}$.

The certificates are called *self-blindable*, provided there exists a set called *transformation factor space* F and an efficiently computable *transformation map* $D : \mathcal{C} \times F \to \mathcal{C}$ with the following properties:

1. For any certificate $C \in \mathcal{C}$ and $f \in F$ the certificate $D(C, f)$ is signed with the same trust provider public key as C.
2. Let C_1, C_2 be certificates and $f \in F$ known. If $C_2 = D(C_1, f)$ then one can efficiently compute a transformation factor $f' \in F$ such that $C_1 = D(C_2, f')$.
3. If $C_1, C_2 \in \mathcal{C}$ are two different certificates on the same user public key, then so are $D(C_1, f)$ and $D(C_2, f)$. That is, the mapping $D(.,.)$ induces a mapping $\mathcal{U} \times F \to \mathcal{U}$ and although abusive, we also use the notation $D(P_U, f)$ for any user public key P_U and transformation factor f.
4. Let P_U be a user public key and let $f \in F$ be a known transformation factor. Then, a user possesses the private key relating to P_U if and only if it possesses the private key relating to $D(P_U, f)$.
5. If the user's public key $P_U \in \mathcal{U}$ is fixed and if $f \in F$ is a uniformly random element in F, then $D(P_U, f)$ is a uniformly random element in $\mathcal{U}$.

We briefly explain the rationale behind these properties. The first property enables one to transform a user certificate into another one from the same certificate authority; the fourth property ensures that the user still has possession of the private key referenced in the transformed certificate provided he knows the transformation factor. The fifth property states that all user public keys are equally possible in the transformed certificate. As we will explain below, a user typically collects credentials on different certificates formed as transformations of one fixed certificate. Now, the second property enables to invert transformations, allowing to translate all credentials to the fixed certificate and then to other certificates. Finally, the third property is technical and in fact emerged from our constructions. We have chosen it as part of our formal definition, as it enables simple proofs and formulation of other properties, e.g., properties four and five. More complicated requirements are possible to arrive at a more general notion of self-blindable certificates, but we will not explore this.

3.2 A CPC System Based on the Building Blocks

We use the terminology introduced above and we assume that the certificates are self-blindable. Our notion of a pseudonymous credential is the simplest possible and takes the form

$$\{P_U, [Sig(P_U, S_N), Cert(P_N, \text{"PP statement"})]\},$$

where P_U stands for the public key of the user (with related private key S_U). Moreover, $Sig(P_U, S_N)$ is a signature on the user's public key with a signing key of the *pseudonym provider* (PP) and $Cert(P_N, \text{"PP statement"})$ is a (conventional) certificate on the public verification key of the pseudonym provider, with a statement on its applicability included among the usual fields (e.g., expiration date). For evident reasons, this PP certificate must be used by the pseudonym provider for many users to prevent linkage of the issued pseudonymous certificate. Also note that the pseudonym of a user is in fact the user's public key in its certificate, which is reminiscent of the SPKI (Simple Public Key Infrastructure) approach, cf. [24].

Note that the self-blinding properties of the certificates enable the users themselves to generate a new pseudonymous certificate validly signed by the same PP, by choosing a (random) factor and transforming an initially issued pseudonymous certificate.

Our description of a CPC is based upon that of a pseudonymous credential, say $\{P_U, [Sig(P_U, S_N), Cert(P_N, \text{“PP statement”})]\}$ and its simplest form is:

$$\{P_U, [Sig(P_U, S_N), Cert(P_N, \text{“PP statement”})],\\ [Sig(P_U, S_C), Cert(P_C, \text{“CP statement”})]\}.$$

Here, $[Sig(P_U, S_C), Cert(P_C, \text{“CP statement”})]$ is called the *credential field.* In this, $Sig(P_U, S_U)$ is a signature on the public key of the owner with a signing key S_C of the *credential provider* (CP). Also, $Cert(P_C, \text{“CP statement”})$ is a (conventional) certificate on the related credential provider's public verification key, that has a statement on its credential applicability, e.g., “the person having possession of the private key is over 18 years old” included among the usual fields (e.g., expiration date). In a natural fashion one can have several credential fields attached to a pseudonymous credential in the above way, which is in fact the general form of a CPC.

Based on the building blocks explained above, one can now construct a wide variety of types of CPC systems. We provide a high-level description of one such system on which many variations are possible (cf. Figure 1).

System description 31

Initial Registration. The user registers, typically in a non-anonymous fashion, with a pseudonym provider. After registration a First Pseudonymous Certificate (FPC) issuing protocol between the user and the pseudonym provider is started. This protocol is system specific. The pseudonym provider puts the FPC in a public directory. When unique pseudonyms are required, the provider has the option to maintain a private list of physical persons that were issued a pseudonymous certificate; this ensures that at most one such certificate is issued to a physical person.

Credential Issuance. By using a random transformation factor, the user transforms its FPC into a random pseudonymous certificates (RPC). The user securely stores the used transformation factor. Then the user registers with a credential provider using this RPC which includes a proof of possession of the private key referenced in the RPC. This registration need not be anonymous. The user does what is required to obtain a credential (e.g., takes a driver's exam, shows other credentials) and up-on succeeding, is issued a credential on the RPC, that is the CPC. The pseudonym provider has the option to put the CPC in a public directory.

Credential Use. The user registers (typically anonymously) with a service provider using a new RPC, which includes a proof of possession of the private key referenced in the new RPC. The user combines all of the CPCs relating to credentials required by the service provider into one CPC under the registered pseudonym. This is possible by using the second property of self-blinding certificates on the transformation factors related with the individual, original CPCs. That is, a CPC is first translated to the First Pseudonym and then translated to the registered pseudonym (in our constructions these two steps can be performed in one operation). This certificate is presented to the service provider, together with a proof of possession of the private key referenced in this CPC. Once the user is successful in doing so, he will be serviced.
If the service provider wants to be certain that the user has not already been issued another pseudonym, the service provider has the option to require that the user contact a specific trust provider which we refer to as "unicity" provider. The user sends this trust provider the transformation factor(s), transforming the new RPC to the first issued pseudonymous certificate stored in the pseudonym provider's directory (i.e., the FPC). This trust provider then validates that these factor(s) transform the RPC into a FPC on the PP's directory, and that this FPC was not registered before. The trusted party then reports to the service provider that the user has not registered before. Note that the PP directory does not specify user identities, only FPCs, also note that the specific trust provider need not be the user's pseudonym provider.

In the system description above, we have used the FPC list of the pseudonym provider as the reference data for all trust providers that need to verify that a 'physical' user cannot register twice (under different pseudonyms) with a service provider. This means that if two such trust providers conspire, they can link together the different pseudonyms of a user. One can prevent this linkage with a flexible secret sharing technique as follows. During registration, the pseudonym provider and the user, say U, exchange a secret, S. If a trust provider, say T, wants to provide assurance on unique pseudonyms, then provider A is provided a list consisting of transformed FPCs, in such a way that:

- user U's FPC is transformed using a transformation factor based on a secure hash of the name of the provider T and the secret S; and
- the order of the FPCs is randomly permuted.

If user U wants to assure the trust provider T that it is not registering twice (under different pseudonyms) with a service provider via T, then it provides the provider with the transform factor transforming the RPC (see above) into the transformed FPC stored at the provider T. This technique can be iterated: user U can (after proving possession of a transformed FPC at T) be issued another secret by T, and the transformed FPC can be re-transformed by T and stored at another trust provider T_2, etc.. By combining transformation factors, user U can employ provider T_2's service without any interference from provider T.

Moreover, in such a setting, linkage requires that all such trust providers and the pseudonym provider conspire.

In Figure 1 we have depicted the (five) steps from pseudonym issuance to CPC application in a sample voting application. The communication between the "unicity" provider and the service provider (the voting application) is not depicted.

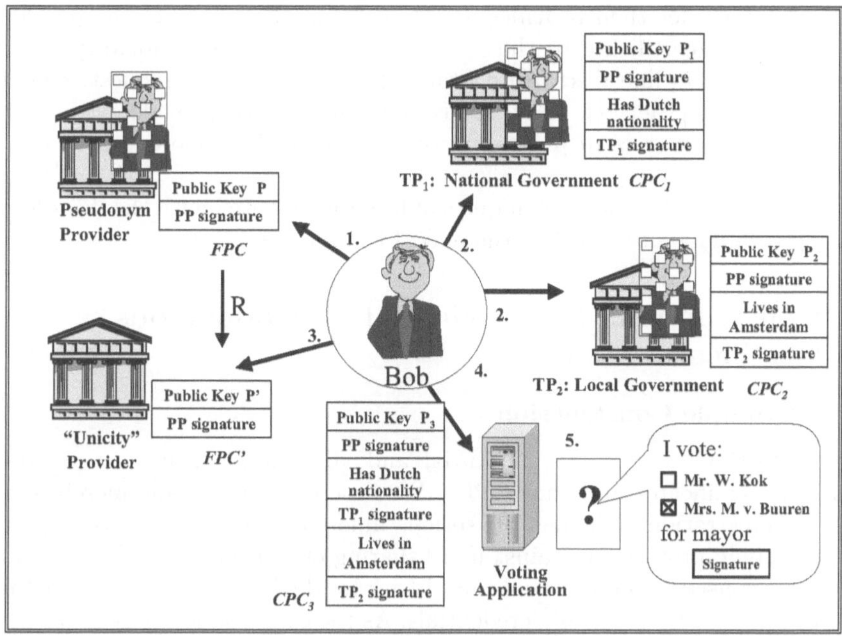

Fig. 1. Overview of system description 31

3.3 Revocation of Certificate Bases

As users typically will not present the originally issued certificates to service providers, certificates cannot be revoked in the conventional way. A primary concern is that the revocation process should not make it possible to link credential use, except, possibly, by certain trusted parties.

There are several methods to address revocation in our model, but we outline only two. The first method is pro-active, and consists of letting the trust providers employ signing keys with a short expiration time (e.g., a week). If a pseudonymous certificate or a credential relating to such a certificate has not been revoked, then the trust provider automatically updates the certificates or credentials in its directory with newly signed ones. A user can collect the updated pseudonymous certificates and credentials, preferably via an anonymous channel to reduce the chances of linkage. To achieve this, the user can, for example, collect many certificates, including the required ones. By revoking its FPC, the user can effectively revoke all credentials based on it.

The second method for revocation we outline consists of sending along specific transformation factors with a (credential) pseudonymous certificate, to a specific trust provider. This trust provider can then retrieve the original issued (credential) pseudonymous certificates and find out if they have been revoked. The trust provider then provides a statement on the status of the (credential) pseudonymous certificate to the service provider. This functionality resembles the use of an On-line Certificate Status Protocol (OCSP) request, commonly used on the Internet (cf. [23]). Of course, the service provider still needs to verify that the user is in possession of the private key referenced in the used randomized CPC.

The second revocation technique can be supplemented with the flexible secret sharing technique described at the end of the previous section.

4 Constructions for Credential Pseudonymous Certificates

4.1 A Simple Construction

In this section we describe an initial and very simple construction for self-blindable certificates and thus CPCs. We describe this scheme merely for purposes of illustration, as it has the serious inherent draw-back of not supporting cryptographic protection against users sharing credentials. Therefore, to implement this construction one would need to trust devices resistant to user tamper to prevent users from sharing credentials. As the construction in Section 4.2 provides cryptographic protection against users sharing credentials this construction is favorable to the scheme presented in this section.

Let $G = \langle g \rangle$ be a group of prime order q in which the DDH problem is efficiently computable, while the discrete logarithm and the Diffie-Hellman problems are practically intractable. We also assume that the (provable) random generation of elements in G without knowing any relative discrete logarithms is also possible (see the end of Section 2.1). The description of the group G, including the g, q are considered as system parameters.

The set $\mathcal{T}$ of all trust provider's public keys takes the form j, j^s where $0 \leq s < q$ is the related private key and where $j \in G \setminus \{1\}$. We assume that each trust provider's public generator j is (provably) randomly chosen, e.g., it could be based on the output of a secure hash algorithm with a fixed input. The set of users public keys $\mathcal{U}$ consists of elements of the form g^x where $0 < x < q$ are all possible user's private keys. There is a subtle reason why $x = 0$ is principally not allowed, see below. Note that a user can prove possession of x in a zero-knowledge fashion with the Schnorr identification protocol [25]. Moreover, several digital signature systems can be based on the user public, private key pairs mentioned above, e.g., DSA [15], ElGamal [14] and Schnorr [25]. Finally, a certificate issued by a trust provider with public key h, h^z on a user public key g^x takes the form:

$$\{g^x, g^{xz}\}.$$

Note that the above certificate is based on the variant of the Chaum-Pedersen signature (as outlined in Section 2) on g^x with respect to the public key h, h^z, i.e., g^{xz}. An important feature of this variant is that it is not required to add an interactive proof that the second component indeed has the form g^{xz} as the DDH problem is assumed to be simple. Due to the restrictions on the first element in the certificate, it cannot be equal to the unity element. If this condition is not also checked by applications, then certificate forgery becomes simple.

The certificates $\mathcal{C}$ constructed in this way are self-blindable. To this end, choose the transformation factor space F equal to $\mathrm{GF}(q)^*$ and define the transformation $D : \mathcal{C} \times F \to \mathcal{C}$ as

$$(\{X, Y\}, f) \to \{X^f, Y^f\}.$$

That is, the certificate $\{g^x, g^{xz}\}$ is transformed to the certificate $\{g^{xf}, g^{xfz}\}$ under factor f. It is a simple verification that $D(.,.)$ satisfies the five properties of a transformation and, thus, that the certificates constructed in this way are self-blindable.

Notice that the transfer of credentials is simple in this construction, if the user is able to retrieve (and transfer) the private key related to the public key of a (transformed) pseudonymous certificate. This problem can be controlled by ensuring that all security operations with respect to credentials take place on a tamper resistant signing device in such a way that private key information of (transformed) certificates can be used ('addressed') but not retrieved. The use of such devices needs to be addressed in the FPC issuing protocol for these certificates, for instance as follows.

1. The user registers, typically in a non-anonymous fashion, with a pseudonym provider.
2. The pseudonym provider generates a random $0 < x < q$, and forms the user public key g^x and the certificate $\{g^x, g^{xz}\}$. All information is put on a tamper resistant signing device, in such a way that private key information of (transformed) certificates can be used but not retrieved.
3. The secure signing device is handed over to the user in a secure fashion.

Having filled in this issuing protocol, our CPC scheme now follows system description 31. Protection against pseudonym/credential linking and pseudonym/credential translation are obvious consequences of the properties of self-blindable certificates. For the other two security properties (protection against forgery and transfer), one needs to trust devices resistant to user tampering.

4.2 A More Robust Construction

This construction is based on the technique in Brands' e-cash scheme to trace double spenders (cf. from [4]). Just as in the previous section our construction is based on the variant of the Chaum-Pedersen signature scheme as introduced in Section 2. So, again, let $G = \langle g \rangle$ be a group of prime order q in which the DDH problem is efficiently computable, while the discrete logarithm and the Diffie-Hellman problems are practically intractable. We also assume that the (provable) random generation of elements in G without knowing any relative discrete logarithms is also possible. In addition to this, we assume that there exists an efficiently computable embedding $E(.)$ from G into a group G' where all three problems DDH, DH and DL are practically intractable. All these requirements are met by suitable groups of points on supersingular elliptic curves, cf. the end of Section 2.1. The description of the groups G, including the g, q, the group G' and the embedding are considered to be system parameters.

As before, the set of all trust providers' public keys, $\mathcal{T}$ takes the form j, j^s where $0 \leq s < q$ is the related private key and where $j \in G \setminus \{1\}$. We assume that each trust provider's public generator j is (provably) randomly chosen, e.g., it could be based on the output of a secure hash algorithm with a fixed input. In addition we assume that the pseudonym provider publishes a certified pair $(r, s) = (r, r^f)$ where $r, s \in G$ and for some $0 < f < q$ which is unknown by all parties. Generation of such a pair consists of choosing two (provable) random r, s which determines f. Alternatively, the pseudonym provider can choose the element r in a provable random fashion and generate a random element $0 < f < q$ and form $s = r^f$. We prefer the first construction, for two reasons. First, it is difficult for the pseudonym provider to convince others that f has been chosen randomly and, second, it is good practice to have as few secret keys in a system as possible.

The set of users public keys $\mathcal{U}$ consists of elements of the form $g_1, g_2, g_1^{x_1} g_2^{x_2}$. Here $0 \leq x_1, x_2 < q$ is the related private key, g_1 is a random generator and $\log_{g_1}(g_2) = f$. As in the previous scheme, we require that $g_1^{x_1} g_2^{x_2}$ be unequal to the unity element. Note that a participant can prove possession of x_1, x_2 in a zero-knowledge fashion with the Okamoto variant of Schnorr's identification protocol [22]. In the same paper, a variant of Schnorr's signature scheme is described based on the user public, private key pairs mentioned above. Finally, a certificate issued by a trust provider with public key h, h^z on a user's public key $g_1, g_2, g_1^{x_1} g_2^{x_2}$ takes the form:

$$\{g_1, g_2, g_1^{x_1} g_2^{x_2}, (g_1^{x_1} g_2^{x_2})^z\}.$$

Again, this is precisely the variant of the Chaum-Pedersen signature (as outlined in Section 2) on the user's public key with respect to the public key h, h^z. As the DDH problem is simple, on basis of the certified pair (r, r^f), anyone can and should verify that the first two parameters in the certificate are indeed correctly formed, i.e., the second one is an f-th power of the first one (cf. the alternative description of the DDH problem in Section 2.1). Due to the restrictions on the three elements in the certificate, none of them can be equal to the unity element. If this condition is not also checked by applications, then certificate forgery becomes simple.

The certificates $\mathcal{C}$ constructed in this way are self-blindable. To this end, define the transformation factor space by $F = \mathrm{GF}(q)^* \times \mathrm{GF}(q)^*$ and the transformation $D : \mathcal{C} \times F \to \mathcal{C}$ as:

$$(\{X, Y, W, Z\}, (k, l)) \rightarrow \{X^l, Y^l, W^{kl}, Z^{kl}\}.$$

That is, the certificate $\{g_1, g_2, g_1^{x_1} g_2^{x_2}, (g_1^{x_1} g_2^{x_2})^z\}$ is transformed into the certificate

$$\{g_1^l, g_2^l, g_1^{x_1 kl} g_2^{x_2 kl}, (g_1^{x_1 kl} g_2^{x_2 kl})^z\}$$

under the transformation factor (k, l). It is a simple verification that $D(.,.)$ satisfies the five properties of a transformation. Notice that two transformation factors (k, l) are used to ensure that a randomly transformed public key is indeed a random element in the user's public key space.

The FPC issuing protocol for these certificates can be filled in as follows, but many variations are possible; the pseudonym provider's public key is denoted as h, h^z, where $h \in G \setminus \{1\}$ is (provably) randomly chosen.

1. The user registers, typically in a non-anonymous fashion, with a pseudonym provider.
2. The pseudonym provider generates a random pair (g_1, g_2) such that $g_2 = g_1^f$, by choosing a (provably) random power of the elements r, s. The pair (g_1, g_2) is sent to the user, or to a party acting on its behalf (e.g., a smart card issuer).
3. The user (or a party acting on its behalf), generates a random private key $0 \leq x < q$ and forms g_2^x. The user sends g_2^x and proves possession of the private key x (i.e., the discrete logarithm with respect to g_2 of the first sent public key), e.g., by using Schnorr's protocol.
4. Based on the elements g_1, g_2 and g_2^x, the pseudonym provider forms the public key $g_1, g_2, g_1 g_2^x$, checks to ensure that the last element is unequal to the unity element and places a Chaum-Pedersen signature on it, i.e., $(g_1 g_2^x)^z$. Moreover, the provider employs the embedding $E : G \rightarrow G'$ and determines the elements $E(g_2), E(g_2^x)$ of the group G' (in which the DDH, DH and DL problems are hard). Next the provider determines a random power r of these elements, i.e., $E(g_2)^r, E(g_2^x)^r$. The provider then forms a conventional non-repudiation certificate (e.g., based on the US Digital Signature Algorithm) on $(E(g_2)^r, E(g_2^x)^r)$. The first pseudonymous certificate and the non-repudiation certificate are issued to the user. Both are also stored in separate directories.

Using the terminology of the above protocol; as the embedding $E(.)$ is a homomorphism it directly follows that the private non-repudiation signing key is equal to x. We have used a non-repudiation signing key only as an example of a private key that is highly important to a user. Many more examples exist (e.g., the user's signing key for financial transactions).

There are two reasons why the user's non-repudiation key is embedded in the group G' in the specified way. First of all, using a group where all three of the DDH, DH and DL problems are hard, seems appropriate for a conventional signature scheme. Second, embedding the non-repudiation key in the specified way, prevents linkage between the first pseudonymous certificate and the non-repudiation certificate. Should a party have access to g_2^r, g_2^{xr} (whose $E(.)$ images appear in the non-repudiation key) then this party would be able to link this to the pair g_2, g_2^x as the DDH problem in G is simple. However, inverting the embedding $E(.)$ is hard (cf. the remarks at the end of Section 2.1), so inverting the values $E(g_2)^r, E(g_2^x)^r$ (deducible from the non-repudiation certificate) is not a practical possibility. Moreover, as the DDH problem is presumed to be hard in G' it would be impossible to relate $E(g_2), E(g_2^x)$ (deducible from the first pseudonymous certificate) to $E(g_2)^r, E(g_2^x)^r$ (deducible from the non-repudiation certificate). Strictly speaking, such a linkage might not be an issue, as users will typically employ transformed pseudonymous credentials. However (cf. the generic description 31), this might become an issue should a service provider want to be certain that the user has not already been issued another pseudonym. Indeed, the user would then need to provide a trust provider with the transformation factor from its registered pseudonymous certificate to the First Pseudonymous Certificate. We finally note that, in the issuing protocol, the pseudonym can alternatively first calculate random r-powers of the elements g_2, g_2^x in the group G and then utilize the embedding $E(.)$. For the same r, this would give the same result as with the method described above.

Having filled in this issuing protocol, our CPC scheme now follows from the system description 31. Protection against pseudonym/credential linking and pseudonym/credential translation are obvious consequences of the properties of self-blindable certificates. We discuss the two other security properties.

Protection against pseudonym/credential forgery
This protection is based on an all-or-nothing concept (see the introduction). The private key in a transformed credential takes the form $(k, k \cdot x \bmod q)$ for some $0 < k < q$. Note that dividing the second part by the first part yields the user's non-repudiation key x. Hence, if the user transfers a credential, then it also transfers a copy of its non-repudiation signing key. We think that this is a sufficient deterrent to transferring credentials (which can be supplemented with the physical security of a signing device).

Protection against pseudonym/credential forgery [*Indication*]
Under Assumption 21, we provide a sketched proof in the appendix that an efficient pseudonym/credential forgery algorithm based on all issued certificates and private keys, will in fact provide an algorithm determining hard discrete logarithms with non-negligible probability.

5 Conclusion

We have described two simple, efficient and effective credential pseudonymous certificate systems, which also support anonymity without the need for a trusted third party. Both systems are based on a new paradigm, called self-blindable certificates. Such certificates were constructed using the Weil pairing in supersingular elliptic curves. The second system provides cryptographic protection against the forgery and transfer of credentials.

Acknowledgments. We want to thank Stefan Brands and Berry Schoenmakers for stimulating discussions. Berry is specifically thanked for pointing us to the double spending preventing technique from E-cash based on the Okamoto identification protocol and Stefan is specifically thanked for providing us with the term "self-blinding signatures and certificates".

References

1. I.F. Blake, G. Seroussi, N.P. Smart, *Elliptic Curves in Cryptography*, Cambridge University Press, 1999.
2. D. Boneh, M. Franklin, *Identity-Based Encryption from the Weil Pairing*, Proceedings of Crypto 2001, LNCS 2139, Springer-Verlag 2001, 213-229.
3. D. Boneh, B. Lynn, H. Shacham *Short Signatures from the Weil Pairing*, these proceedings.
4. S. Brands, *Untraceable Off-line Cash in Wallet with Observers*, Proceedings of Crypto '93, LNCS 911, Springer-Verlag 1994, 302-318.
5. S. Brands, *Rethinking Public Key Infrastructures and Digital Signatures; Building in Privacy*, PhD Thesis, Eindhoven University of Technology, the Netherlands, 1999.
6. J. Camenisch, A. Lysyanskaya, *An Efficient System for Non-transferable Anonymous Credentials with Optional Anonymity Revocation*, Proceedings of Eurocrypt 2001, LNCS 2045, Springer-Verlag 2001, 93-118.
7. D. Chaum, *Security Without Identification: Transaction Systems to Make Big Brother Obsolete*, Communications of the ACM, 1985, 28(10), 1035-1044. See also *Security Without Identification: Card Computers to Make Big Brother Obsolete*, available from www.chaum.com.
8. D. Chaum, *Zero-knowledge Undeniable Signatures*, Proceedings of Eurocrypt'90, LNCS 473, Springer-Verlag 1991, 458-464.
9. D. Chaum, H. van Antwerpen, *Undeniable Signatures*, Proceedings of Crypto'89, LNCS 435, Springer-Verlag 1990, 212-216.
10. D. Chaum, J.-H. Evertse, *A Secure and Privacy-protecting Protocol for Transmitting Personal Information between Organizations*, Proceedings of Crypto '86, LNCS 263, Springer-Verlag 1987, 118-167.
11. D. Chaum, T.P. Pedersen, *Wallet Databases with Observers*, Proceedings of Crypto'92, LNCS 740, Springer-Verlag 1993, 89-105.
12. L. Chen, *Access with Pseudonyms*, In Cryptography: Policy and Algorithms, LNCS 1029,Springer-Verlag 1995, 232-243.
13. I. Damgård, *Efficient Concurrent Zero-knowledge in the Auxiliary String Model*, Proceedings of Eurocrypt 2000, LNCS 1807, Springer-Verlag 2000, 431-444.
14. T. ElGamal *A Public Key Cryptosystem and Signature System Based on Discrete Logarithms*, Proceedings of Crypto '84, LNCS 196, Springer-Verlag 1985, 10-18.
15. FIPS 186, *Digital Signature Standard*, Federal Information Processing Standards publication 186, U.S. Department of Commerce/NIST, 1994.
16. A. Joux, *A One Round Protocol for Tripartite Diffie-Hellman*, 4th International Symposium, Proceedings of ANTS, LNCS 1838, Springer-Verlag, 2000, 385-394.
17. A. Joux, K. Nguyen, *Seperating Decision Diffie-Hellman from Diffie-Hellman in Cryptographic Groups*, in preparation. Available from eprint.iacr.org.
18. A.K. Lenstra, E.R. Verheul, *The XTR Public Key System*, Proceedings of Crypto 2000, LNCS 1880, Springer-Verlag, 2000, 1-19; available from www.ecstr.com.

19. A. Lysyanskaya. R. Rivest, A. Sahai, S. Wolf, *Pseudonym Systems*, In Selected Areas in Cryptography, LNCS 1758, Springer-Verlag 1999.
20. A. Menezes, *Elliptic Curve Public Key Cryptosystems*, Kluwer Academic Publishers, Boston 1993.
21. A. Menezes, T. Okamoto, S.A. Vanstone *Reducing Elliptic Curve Logarithms to a Finite Field*, IEEE Trans. Info. Theory, 39, 1639-1646, 1993.
22. T. Okamoto, *Provable Secure and Practical Identifications and Corresponding Signature Schemes*, Proceedings of Crypto'92, LNCS 740, Springer-Verlag 1993, 31-53.
23. RFC 2560, *Online Certificate Status Protocol (OCSP)*, available from www.ietf.org.
24. RFC 2693, *SPKI Certificate Theory*, available from www.ietf.org.
25. C.P. Schnorr, *Efficient Identification and Signatures for Smart Cards*, Proceedings of Crypto'89, LNCS 435, Springer-Verlag 1990, 239-252.
26. J. Silverman, *The Arithmetic on Elliptic Curves*, Springer-Verlag, New York, 1986.
27. E. Verheul, *Evidence that XTR is More Secure than Supersingular Elliptic Curve Cryptosystems*, Proceedings of Eurocrypt 2001, LNCS 2045, Springer-Verlag 2001, 195-210.

A Appendix: Forgery Protection in the Robust Construction

Suppose that a total of n-number of certificates under one trust provider are issued, e.g., of type:

$$\{g_{1,i}, g_{2,i}, g_{1,i}^{x_{1,i}} g_{2,i}^{x_{2,i}}, (g_{1,i}^{x_{1,i}} g_2^{x_{2,i}})^z\},$$

where the trust providers public key is of the form h, h^z as usual. Also suppose that a forger has access to all private keys $x_{1,i}$ and $x_{2,i}$ and is able to produce a forged certificate, say

$$\{h_1, h_2, h_1^{y_1} h_2^{y_2}, (h_1^{y_1} h_2^{y_2})^z\},$$

where $0 \leq y_1, y_2 < q$ is known to the forger. Notice that (y_1, y_2) should not be equal to $(0, 0)$ as then the certificate contains the unity element. As the h_2 should be an f-th power of h_1, it follows from Assumption 21 that h_1 (resp. h_2) is a power product of the $\{g_{1,i}\}$ and r (resp. $\{g_{2,i}\}$ and s). Likewise, $h_1^{y_1} h_2^{y_2}$ is a power product of all $g_{1,i}^{x_{1,i}} g_{2,i}^{x_{2,i}}$ and h. By choosing the right transformation factors, we may assume without loss of generality that $h_1 = r^b \prod_{i \in I} g_{1,i}$, $h_2 = s^b \prod_{i \in I} g_{2,i}$ and

$$h_1^{y_1} h_2^{y_2} = h^c \prod_{j \in J} g_{1,j}^{x_{1,j}} g_{2,j}^{x_{2,j}}, \tag{1}$$

for some subsets I, J of $\{1, 2, \ldots, n\}$ and $b, c \in \{0, 1\}$.

We now sketch that we can rule out the possibility that either b, c is equal to 1. To this end, suppose the probability that the event that $c = 1$ to be non-negligible. Now, if one simulates f, then one can use the forgery algorithm to determine $\log_r(h)$. Indeed, by feeding the algorithm $g_{1,i}$ (resp. $g_{2,i}$) that are of form r^{t_i} (resp. s^{t_i}), where $0 \leq t_i < q$ known and random and by choosing the $0 \leq x_{1,i}, x_{2,i} < q$ in a random way ($i = 1, 2, \ldots n$). As this is 'correct' input, it will lead to equalities of type (1). Now, if $c = 1$ in any of these equalities then the algorithm has produced $\log_r(h)$, which is assumed to be a hard problem. Likewise, if the probability that $b = 1$ is non-negligible, then simulation of f the forgery algorithm will also enable to determine discrete logarithms with respect to r, by basing all $g_{1,i}, g_{2,i}$ on random powers of an element z for which $\log_r(z)$ is required. Thus we conclude that $b = c = 0$ with overwhelming probability and that actually the equations (1) are of type

$$h_1^{y_1} h_2^{y_2} = \prod_{j \in J} g_{1,j}^{x_{1,j}} g_{2,j}^{x_{2,j}}, \tag{2}$$

where $h_1 = \prod_{i \in I} g_{1,i}$, $h_2 = \prod_{i \in I} g_{2,i}$. Note that the sets I, J cannot be empty as the unity element would then occur in the certificate. Moreover, if the set $I \cup J$ does not contain at least two elements, then $I = J$ is a singleton, and the forgery algorithm has in fact produced a transformed user certificate, which is not considered a forgery. Now, suppose that $\log_a(b)$ is required for some $a, b \in G$, then this can be determined with high probability, by basing 'half' the $g_{i,i}, g_{2,i}$ on random powers of a and the other half on random powers of b. With non-negligible probability, the set $I \cup J$ will contain both a $g_{i,i}, g_{2,i}$ based on a and b, and will hence give a relation providing $\log_a(b)$.

How to Leak a Secret

Ronald L. Rivest[1], Adi Shamir[2], and Yael Tauman[2]

[1] Laboratory for Computer Science, Massachusetts Institute of Technology, Cambridge, MA 02139, rivest@mit.edu

[2] Computer Science department, The Weizmann Institute, Rehovot 76100, Israel. {shamir,tauman}@wisdom.weizmann.ac.il

Abstract. In this paper we formalize the notion of a *ring signature*, which makes it possible to specify a set of possible signers without revealing which member actually produced the signature. Unlike group signatures, ring signatures have no group managers, no setup procedures, no revocation procedures, and no coordination: any user can choose any set of possible signers that includes himself, and sign any message by using his secret key and the others' public keys, without getting their approval or assistance. Ring signatures provide an elegant way to leak authoritative secrets in an anonymous way, to sign casual email in a way which can only be verified by its intended recipient, and to solve other problems in multiparty computations. The main contribution of this paper is a new construction of such signatures which is unconditionally signer-ambiguous, provably secure in the random oracle model, and exceptionally efficient: adding each ring member increases the cost of signing or verifying by a single modular multiplication and a single symmetric encryption.

Keywords: signature scheme, ring signature scheme, signer-ambiguous signature scheme, group signature scheme, designated verifier signature scheme.

1 Introduction

The general notion of a *group signature scheme* was introduced in 1991 by Chaum and van Heyst [2]. In such a scheme, a trusted group manager predefines certain groups of users and distributes specially designed keys to their members. Individual members can then use these keys to anonymously sign messages on behalf of their group. The signatures produced by different group members look indistinguishable to their verifiers, but not to the group manager who can revoke the anonymity of misbehaving signers.

In this paper we formalize the related notion of *ring signature schemes*. These are simplified group signature schemes which have only users and no managers (we call such signatures "ring signatures" instead of "group signatures" since rings are geometric regions with uniform periphery and no center). Group signatures are useful when the members want to cooperate, while ring signatures are useful when the members do not want to cooperate. Both group signatures and

C. Boyd (Ed.): ASIACRYPT 2001, LNCS 2248, pp. 552–565, 2001.
© Springer-Verlag Berlin Heidelberg 2001

ring signatures are *signer-ambiguous*, but in a ring signature scheme there are no prearranged groups of users, there are no procedures for setting, changing, or deleting groups, there is no way to distribute specialized keys, and there is no way to revoke the anonymity of the actual signer (unless he decides to expose himself). Our only assumption is that each member is already associated with the public key of some standard signature scheme such as RSA. To produce a ring signature, the *actual signer* declares an arbitrary set of *possible signers* that includes himself, and computes the signature entirely by himself using only his secret key and the others' public keys. In particular, the other possible signers could have chosen their RSA keys only in order to conduct e-commerce over the internet, and may be completely unaware that their public keys are used by a stranger to produce such a ring signature on a message they have never seen and would not wish to sign.

The notion of ring signatures is not completely new, but previous references do not crisply formalize the notion, and propose constructions that are less efficient and/or that have different, albeit related, objectives. They tend to describe this notion in the context of general group signatures or multiparty constructions, which are quite inefficient. For example, Chaum et al. [2]'s schemes three and four, and the two signature schemes in Definitions 2 and 3 of Camenisch's paper [1] can be viewed as ring signature schemes. However the former schemes require zero-knowledge proofs with each signature, and the latter schemes require as many modular exponentiations as there are members in the ring. Cramer et al. [3] shows how to produce witness-indistinguishable interactive proofs. Such proofs could be combined with the Fiat-Shamir technique to produce ring signature schemes. Similarly, DeSantis et al. [10] show that interactive SZK for random self-reducible languages are closed under monotone boolean operations, and show the applicability of this result to the construction of a ring signature scheme (although they don't use this terminology).

The direct construction of ring signatures proposed in this paper is based on a completely different idea, and is exceptionally efficient for large rings (adding only one modular multiplication and one symmetric encryption per ring member both to generate and to verify such signatures). The resultant signatures are unconditionally signer-ambiguous and provably secure in the random oracle model.

2 Definitions and Applications

2.1 Ring Signatures

Terminology: We call a set of *possible signers* a *ring*. We call the ring member who produces the actual signature the *signer* and each of the other ring members a *non-signer*.

We assume that each possible signer is associated (via a PKI directory or certificate) with a public key P_k that defines his signature scheme and specifies his verification key. The corresponding secret key (which is used to generate regular signatures) is denoted by S_k. The general notion of a ring signature scheme does not require any special properties of these individual signing schemes, but our simplest construction assumes that they use trapdoor one-way permutations (such as the RSA functions) to generate and verify signatures.

A ring signature scheme is defined by two procedures:

- **ring-sign**$(m, P_1, P_2, \ldots, P_r, s, S_s)$ which produces a ring signature σ for the message m, given the public keys P_1, P_2, ..., P_r of the r ring members, together with the secret key S_s of the s-th member (who is the actual signer).
- **ring-verify**(m, σ) which accepts a message m and a signature σ (which includes the public keys of all the possible signers), and outputs either *true* or *false*.

A ring signature scheme is *set-up free*: The signer does not need the knowledge, consent, or assistance of the other ring members to put them in the ring - all he needs is knowledge of their regular public keys. Different members can use different independent public key signature schemes, with different key and signature sizes. Verification must satisfy the usual soundness and completeness conditions, but in addition we want the signatures to be *signer-ambiguous* in the sense that the verifier should be unable to determine the identity of the actual signer in a ring of size r with probability greater than $1/r$. This limited anonymity can be either *computational* or *unconditional*. Our main construction provides unconditional anonymity in the sense that even an infinitely powerful adversary with access to an unbounded number of chosen-message signatures produced by the same ring member cannot guess his identity with any advantage, and cannot link additional signatures to the same signer.

2.2 Leaking Secrets

To motivate the title for this paper, suppose that Bob (also known as "Deep Throat") is a member of the cabinet of Lower Kryptonia, and that Bob wishes to leak a juicy fact to a journalist about the escapades of the Prime Minister, in such a way that Bob remains anonymous, yet such that the journalist is convinced that the leak was indeed from a cabinet member.

Bob cannot send to the journalist a standard digitally signed message, since such a message, although it convinces the journalist that it came from a cabinet member, does so by directly revealing Bob's identity.

It also doesn't work for Bob to send the journalist a message through a standard anonymizer, since the anonymizer strips off all source identification and authentication: the journalist would have no reason to believe that the message really came from a cabinet member at all.

A standard group signature scheme does not solve the problem, since it requires the prior cooperation of the other group members to set up, and leaves Bob vulnerable to later identification by the group manager, who may be controlled by the Prime Minister.

The correct approach is for Bob to send the story to the journalist through an anonymizer, signed with a ring signature scheme that names each cabinet member (including himself) as a ring member. The journalist can verify the ring signature on the message, and learn that it definitely came from a cabinet member. He can even post the ring signature in his paper or web page, to prove to his readers that the juicy story came from a reputable source. However, neither he nor his readers can determine the actual source of the leak, and thus the whistleblower has perfect protection even if the journalist is later forced by a judge to reveal his "source" (the signed document).

2.3 Designated Verifier Signature Schemes

A designated verifier signature scheme is a signature scheme in which signatures can only be verified by a single "designated verifier" chosen by the signer. This concept was first introduced by Jakobsson Sako and Impagliazzo at Eurocrypt 96 [6]. A typical application is to enable users to authenticate casual emails without being legally bound to their contents. For example, two companies may exchange drafts of proposed contracts. They wish to add to each email an authenticator, but not a real signature which can be shown to a third party (immediately or years later) as proof that a particular draft was proposed by the other company. A designated verifier scheme can thus be viewed as a "light signature scheme" which can authenticate messages to their intended recipients without having the nonrepudiation property.

One approach would be to use zero knowledge interactive proofs, which can only convince their verifiers. However, this requires interaction and is difficult to integrate with standard email systems and anonymizers. We can use non-interactive zero knowledge proofs, but then the authenticators become signatures which can be shown to third parties. Another approach is to agree on a shared secret symmetric key k, and to authenticate each contract draft by appending a message authentication code (MAC) for the draft computed with key k. A third party would have to be shown the secret key to validate a MAC, and even then he wouldn't know which of the two companies computed the MAC. However, this requires an initial set-up procedure, in which we still face the problem of authenticating the emailed choice of k without actually signing it.

A designated verifier scheme provides a simple solution to this problem: company A can sign each draft it sends, naming company B as the designated verifier. This can be easily achieved by using a ring signature scheme with companies A and B as the ring members. Just as with a MAC, company B knows that the message came from company A (since no third party could have produced this ring signature), but company B cannot prove to anyone else that the draft of the contract was signed by company A, since company B could have produced this draft by itself. Unlike the case of MAC's, this scheme uses public key cryptography, and thus A can send unsolicited email to B signed with the ring signature without any preparations, interactions, or secret key exchanges. By using our proposed ring signature scheme, we can turn standard signature schemes into designated verifier schemes which can be added at almost no cost as an extra option to any email system.

2.4 Efficiency of Our Ring Signature Scheme

When based on Rabin or RSA signatures, our ring signature scheme is particularly efficient:

- signing requires one modular exponentiation, plus one or two modular multiplications for each non-signer.
- verification requires one or two modular multiplications for each ring member.

In essence, generating or verifying a ring signature costs the same as generating or verifying a regular signature plus an extra multiplication or two for each non-signer, and thus the scheme is truly practical even when the ring contains hundreds of members. It is two to three orders of magnitude faster than Camenisch's scheme, whose claimed efficiency is based on the fact that it is 4 times faster than earlier known schemes (see bottom of page 476 in his paper [1]). In addition, a Camenisch-like scheme uses linear algebra in the exponents, and thus requires all the members to use the same prime modulus p in their individual signature schemes. One of our design criteria is that the signer should be able to assemble an arbitrary ring without any coordination with the other ring members. In reality, if one wants to use other users' public keys, they are much more likely to be RSA keys, and even if they are based on discrete logs, different users are likely to have different moduli p. The only realistic way to arrange a Camenisch-like signature scheme is thus to have a group of consenting parties.

Note that the size of any ring signature must grow linearly with the size of the ring, since it must list the ring members; this is an inherent disadvantage of ring signatures as compared to group signatures that use predefined groups.

3 The Proposed Ring Signature Scheme (RSA Version)

Suppose that Alice wishes to sign a message m with a ring signature for the ring of r individuals $A_1, A_2, \ldots, A_r$, where the signer Alice is A_s, for some value of s, $1 \leq s \leq r$. To simplify the presentation and proof, we first describe a ring signature scheme in which all the ring members use RSA [9] as their individual signature schemes. The same construction can be used for any other trapdoor one way permutation, but we have to modify it slightly in order to use trapdoor one way functions (as in, for example, Rabin's signature scheme [8]).

3.1 RSA Trap-Door Permutations

Each ring member A_i has an RSA public key $P_i = (n_i, e_i)$ which specifies the trapdoor one-way permutation f_i of $\mathbf{Z}_{n_i}$:

$$f_i(x) = x^{e_i} \pmod{n_i} .$$

We assume that only A_i knows how to compute the inverse permutation f_i^{-1} efficiently, using trap-door information; this is the original Diffie-Hellman model [4] for public-key cryptography.

Extending trap-door permutations to a common domain

The trap-door RSA permutations of the various ring members will have domains of different sizes (even if all the moduli n_i have the same number of bits). This makes it awkward to combine the individual signatures, and thus we extend all the trap-door permutations to have as their common domain the same set $\{0,1\}^b$, where 2^b is some power of two which is larger than all the moduli n_i's.

For each trap-door permutation f_i over $\mathbf{Z}_{n_i}$, we define the extended trap-door permutation g_i over $\{0,1\}^b$ in the following way. For any b-bit input m define nonnegative integers q_i and r_i so that $m = q_i n_i + r_i$ and $0 \le r_i < n_i$. Then

$$g_i(m) = \begin{cases} q_i n_i + f_i(r_i) & \text{if } (q_i+1)n_i \le 2^b \\ m & \text{else.} \end{cases}$$

Intuitively, g_i is defined by using f_i to operate on the low-order digit of the n_i-ary representation of m, leaving the higher order digits unchanged. The exception is when this might cause a result larger than $2^b - 1$, in which case m is unchanged. If we choose a sufficiently large b (e.g. 160 bits larger than any of the n_i), the chance that a randomly chosen m is unchanged by the extended g_i becomes negligible. (A stonger but more expensive approach, which we don't need, would use instead of $g_i(m)$ the function $g_i'(m) = g_i((2^b - 1) - g_i(m))$ which can modify all its inputs). The function g_i is clearly a permutation over $\{0,1\}^b$, and it is a one-way trap-door permutation since only someone who knows how to invert f_i can invert g_i efficiently on more than a negligible fraction of the possible inputs.

3.2 Symmetric Encryption

We assume the existence of a publicly defined symmetric encryption algorithm E such that for any key k of length l, the function E_k is a permutation over b-bit strings. Here we use the random (permutation) oracle model which assumes that all the parties have access to an oracle that provides truly random answers to new queries of the form $E_k(x)$ and $E_k^{-1}(y)$, provided only that they are consistent with previous answers and with the requirement that E_k be a permutation (e.g. see [7]).

3.3 Hash Functions

We assume the existence of a publicly defined collision-resistant hash function h that maps arbitrary inputs to strings of length l, which are used as keys for E. We model h as a random oracle. (Since h need not be a permutation, different queries may have the same answer, and we will disallow "h^{-1}" queries.)

3.4 Combining Functions

We define a family of keyed "combining functions" $C_{k,v}(y_1, y_2, \ldots, y_r)$ which take as input a key k, an initialization value v, and arbitrary values y_1, y_2, ..., y_r in $\{0,1\}^b$. Each such combining function uses E_k as a sub-procedure, and produces as output a value z in $\{0,1\}^b$ such that given any fixed values for k and v, we have the following properties.

1. **Permutation on each input:** For each s, $1 \le s \le r$, and for any fixed values of all the other inputs y_i, $i \ne s$, the function $C_{k,v}$ is a one-to-one mapping from y_s to the output z.
2. **Efficiently solvable for any single input:** For each s, $1 \le s \le r$, given a b-bit value z and values for all inputs y_i except y_s, it is possible to efficiently find a b-bit value for y_s such that $C_{k,v}(y_1, y_2, \ldots, y_r) = z$.
3. **Infeasible to solve verification equation for all inputs without trap-doors:** Given k, v, and z, it is infeasible for an adversary to solve the equation
$$C_{k,v}(g_1(x_1), g_2(x_2), \ldots, g_r(x_r)) = z \quad (1)$$
for x_1, x_2, $\ldots$, x_r, (given access to each g_i, and to E_k) if the adversary can't invert any of the trap-door functions g_1, g_2, $\ldots$, g_r.

For example, the function

$$C_{k,v}(y_1, y_2, \ldots, y_r) = y_1 \oplus y_2 \oplus \cdots \oplus y_r$$

(where $\oplus$ is the exclusive-or operation on b-bit words) satisfies the first two of the above conditions, and can be kept in mind as a candidate combining function. Indeed, it was the first one we tried. But it fails the third condition since for any choice of trapdoor one-way permutations g_i, it is possible to use linear algebra when r is large enough to find a solution for $x_1, x_2, \ldots, x_r$ without inverting any of the g_i's. The basic idea of the attack is to choose a random value for each x_i, and to compute each $y_i = g_i(x_i)$ in the easy forward direction. If the number of values r exceeds the number of bits b, we can find with high probability a subset of the y_i bit strings whose XOR is any desired b-bit target z. However, our goal is to represent z as the XOR of all the values $y_1, y_2, \ldots, y_r$ rather than as a XOR of a random subset of these values. To overcome this problem, we choose for each i *two* random values x'_i and x''_i, and compute their corresponding $y'_i = g_i(x'_i)$ and $y''_i = g_i(x''_i)$. We then define for each i $y'''_i = y'_i \oplus y''_i$, and modify the target value to $z' = z \oplus y'_1 \oplus y'_2, \ldots \oplus y'_r$. We use the previous algorithm to represent z' as a XOR of a random subset of y'''_i values. After simplification, we get a representation of the original z as the XOR of a set of r values, with exactly one value chosen from each pair (y'_i, y''_i). By choosing the corresponding value of either x'_i or x''_i, we can solve the verification equation without inverting any of the trapdoor one-way permutations g_i. (One approach to countering this attack, which we don't explore further here, is to let b grow with r.)

Even worse problems can be shown to exist in other natural combining functions such as addition mod 2^b. Assume that we use the RSA trapdoor functions $g_i(x_i) = x_i^3 (\text{mod } n_i)$ where all the moduli n_i have the same size b. It is known [5] that any nonnegative integer z can be efficiently represented as the sum of exactly nine nonnegative integer cubes $x_1^3 + x_2^3 + \ldots + x_9^3$. If z is a b-bit target value, we can expect each one of the x_i^3 to be slightly shorter than z, and thus their values are not likely to be affected by reducing each x_i^3 modulo the corresponding b-bit n_i. Consequently, we can solve the verification equation $(x_1^3 \bmod n_1) + (x_2^3 \bmod n_2) \ldots + (x_9^3 \bmod n_9) = z (\bmod 2^b)$ with nine RSA permutations without inverting any one of them.

Our proposed combining function utilizes the symmetric encryption function E_k as follows:

$$C_{k,v}(y_1, y_2, \ldots, y_r) = E_k(y_r \oplus E_k(y_{r-1} \oplus E_k(y_{r-2} \oplus E_k(\ldots \oplus E_k(y_1 \oplus v) \ldots)))) .$$

This function is applied to the sequence $(y_1, y_2, \ldots, y_r)$, where $y_i = g_i(x_i)$, as shown in Figure 1; the resulting function is provably secure in the random oracle model.

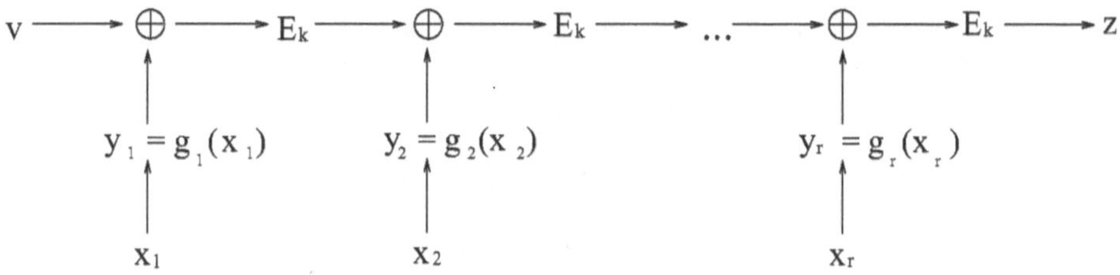

Fig. 1. An illustration of the proposed combining function

It is clearly a permutation on each input, since the XOR, g_i, and E_k functions are permutations. In addition, it is efficiently solvable for any single input since knowledge of k makes it possible to run the evaluation forwards from the initial v and backwards from the final z in order to uniquely compute any missing value y_i. This function can be used to verify signatures by using a hashed version of m to choose the symmetric key k, and forcing the output z to be equal to the input v. This consistency condition $C_{k,v}(y_1, y_2, \ldots, y_r) = v$ bends the line into the ring shape shown in Fig. 2.

A slightly more compact ring signature variant can be obtained by always selecting 0 as the "glue value" v. This variant is also secure, but we prefer the total ring symmetry of our main proposal.

We now formally describe the signature generation and verification procedures:

Generating a ring signature:

Given the message m to be signed, his secret key S_s, and the sequence of public keys $P_1, P_2, \ldots, P_r$ of all the ring members, the signer computes a ring signature as follows.

1. **Choose a key:** The signer first computes the symmetric key k as the hash of the message m to be signed:

$$k = h(m)$$

(a more complicated variant computes k as $h(m, P_1, \ldots, P_r)$; however, the simpler construction is also secure.)

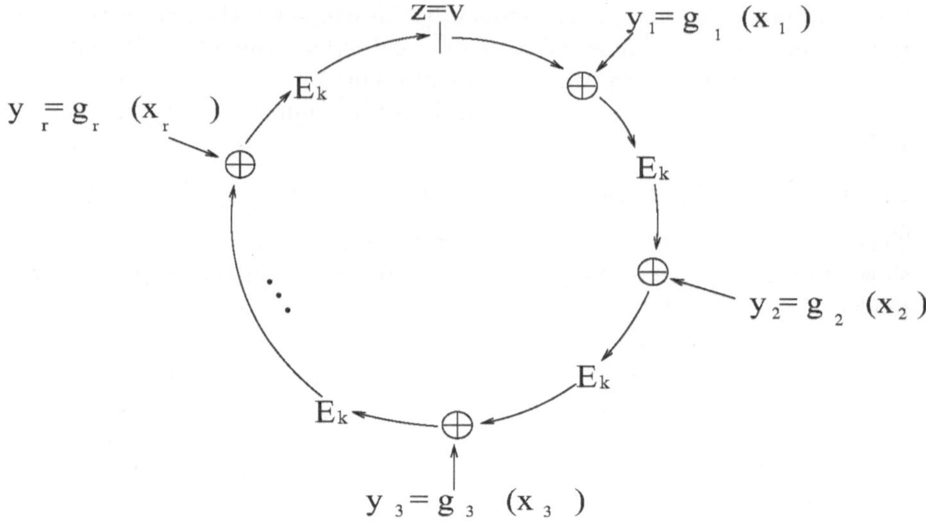

Fig. 2. Ring signatures

2. **Pick a random glue value:** Second, the signer picks an initialization (or "glue") value v uniformly at random from $\{0,1\}^b$.
3. **Pick random x_i's:** Third, the signer picks random x_i for all the other ring members $1 \leq i \leq r$, $i \neq s$ uniformly and independently from $\{0,1\}^b$, and computes
$$y_i = g_i(x_i) \ .$$
4. **Solve for y_s:** Fourth, the signer solves the following ring equation for y_s:
$$C_{k,v}(y_1, y_2, \ldots, y_r) = v \ .$$
By assumption, given arbitrary values for the other inputs, there is a unique value for y_s satisfying the equation, which can be computed efficiently.
5. **Invert the signer's trap-door permutation:** Fifth, the signer uses his knowledge of his trapdoor in order to invert g_s on y_s to obtain x_s:
$$x_s = g_s^{-1}(y_s) \ .$$
6. **Output the ring signature:** The signature on the message m is defined to be the $(2r+1)$-tuple:
$$(P_1, P_2, \ldots, P_r; v; x_1, x_2, \ldots, x_r) \ .$$

Verifying a ring signature:

A verifier can verify an alleged signature
$$(P_1, P_2, \ldots, P_r; v; x_1, x_2, \ldots, x_r) \ .$$
on the message m as follows.

1. **Apply the trap-door permutations:** First, for $i = 1, 2, \ldots, r$ the verifier computes
$$y_i = g_i(x_i) \ .$$
2. **Obtain k:** Second, the verifier hashes the message to compute the encryption key k:
$$k = h(m) \ .$$
3. **Verify the ring equation:** Finally, the verifier checks that the y_i's satisfy the fundamental equation:
$$C_{k,v}(y_1, y_2, \ldots, y_r) = v \ . \quad (2)$$
If the ring equation (2) is satisfied, the verifier accepts the signature as valid. Otherwise the verifier rejects.

3.5 Security

The identity of the signer is unconditionally protected with our ring signature scheme. To see this, note that for each k and v the ring equation has exactly $(2^b)^{(r-1)}$ solutions, and all of them can be chosen by the signature generation procedure with equal probability, regardless of the signer's identity. This argument does not depend on any complexity-theoretic assumptions or on the randomness of the oracle.

The soundness of the ring signature scheme must be computational, since ring signatures cannot be stronger than the individual signature scheme used by the possible signers. Our goal now is to show that in the random oracle model, any forging algorithm A which can generate with non-negligible probability a new ring signature for m by analysing polynomially many ring signatures for other chosen messages $m_j \neq m$, can be turned into an algorithm B which inverts one of the trapdoor one-way functions g_i on random inputs y with non-negligible probability.

Algorithm A accepts the public keys P_1, P_2, $\ldots$, P_r (but not any of the corresponding secret keys) and is given oracle access to h, E, E^{-1}, and to a ring signing oracle. It can work adaptively, querying the oracles at arguments that may depend on previous answers. Eventually, it must produce a valid ring signature on a new message that was not presented to the signing oracle, with a non-negligible probability (over the random answers of the oracles and its own random tape).

Algorithm B uses algorithm A as a black box, but has full control over its oracles. A must query the oracle about all the symmetric encryptions along the forged ring signature of m (otherwise the probability of satisfying the ring equation becomes negligible). Without loss of generality, we can assume that each one of these r symmetric encryptions is queried once either in the "clockwise" E_k direction or in the "counterclockwise" E_k^{-1} direction, but not in both directions since this is redundant. When A makes its polynomially many querries of E_k and E_k^{-1} with various keys $k = h(m)$, B can guess which k will be involved in the actual forgery with non-negligible probability, but it cannot guess which subset of r queries will be used in the final forgery and in which order they will occur along the satisfied ring equation since there are too many possibilities.

Algorithm B can easily simulate the ring signing oracle for all the other m_j by providing random vectors $(v, x_1, x_2, \ldots, x_r)$ as their ring signatures, and adjusting the random answers for queries of the form $E_{h(m_j)}$ and $E^{-1}_{h(m_j)}$ to support the correctness of the ring equation for these messages. Note that A cannot ask relevant oracle questions which will limit B's freedom of choice before providing m_j to the signing oracle since all the values along the actual ring signature (including v) are chosen randomly by B when it provides the requested signature, and cannot be guessed in advance by A. In addition, we use the assumption that h is collision resistant to show that E and E^{-1} queries with key $k_j = h(m_j)$ will not constrain the answers to E and E^{-1} queries with key $k = h(m)$ which will be used in the final forgery, since they use different keys.

The goal of algorithm B is to compute for some i $x_i = g_i^{-1}(y)$ for random inputs y's with non-negligible probability. This will reduce the security of the ring signature to the security of the individual signature schemes. The basic idea of the reduction is to slip this random y as the "gap" between the output and input values of two cyclically consecutive E's along the ring equation of the final forgery, which forces A to close the gap by providing the corresponding x_i in the generated signature. Note that y is a random value which is known to B but not to A, and thus A cannot "recognize the trap" and refuse to sign the corresponding messages.

The main difficuly is that A can close gaps between E values not only by inverting trapdoor one-way functions, but also by evaluating these functions in the easy forward direction (as done by the real signer in the generation of ring signatures). To overcome this difficulty, we note that in any valid ring signature produced by A, there must be a gap somewhere between two cyclically consecutive occurences of E in which the queries were computed in one of the following three ways:

- The oracle for the i-th E was queried in the "clockwise" direction and the oracle for the $i+1$-st E was queried in the "counterclockwise" direction.
- Both E's were queried in the "clockwise" direction, but the i-th E was queried after the $i+1$-st E.
- Both E's were queried in the "counterclockwise" direction, but the i-th E was queried before the $i+1$-st E.

In all these cases, B can provide a random answer to the later query which is based on his knowledge of input and output of the earlier query in such a way that the XOR of the values acros the gap is the desired y. This will force A to compute the corresponding $g_i^{-1}(y)$ in order to fill in this gap in its final ring signature.

B does not know which queries will be these cyclically consecutive queries in the forged ring signature, and thus he has to guess their identity. However, he has to make only two guesses and thus the probability of guessing correctly is $1/Q^2$

where Q is the total number of queries made by the forger A. Consequently, B will manage to compute $g_i^{-1}(y)$ for a random y and some i with non-negligible probability.

When the trapdoor one-way functions g_i are RSA functions, we can slightly strengthen the result. Since RSA is homomorphic, we can randomize y by computing $y' = y * t^{e_i} (\text{mod } n_i)$ for a randomly chosen t. By using y' instead of y, we can show that successful forgeries of ring signatures can be used to extract modular roots from particular numbers such as $y = 2$, and not just from random inputs y. This is not necessarily true for other trapdoor functions, since the forger A can intentionally decide not to produce any forgeries in which one of the gaps between cyclically consecutive E functions happens to be 2.

4 Our Ring Signature Scheme (Rabin Version)

Rabin's public-key cryptosystem [8] has more efficient signature verification than RSA, since verification involves squaring rather than cubing, which reduces the number of modular multiplications from 2 to 1. However, we need to deal with the fact that the Rabin mapping $f_i(x_i) = x_i^2 \ (\text{mod } n_i)$ is not a permutation over $\mathbf{Z}_{n_i}^*$, and thus only one quarter of the messages can be signed, and those which can be signed have multiple signatures.

The operational fix is the natural one: when signing, change your last random choice of x_{s-1} if $g_s^{-1}(y_s)$ is undefined. Since only one trapdoor one-way function has to be inverted, the signer should expect on average to try four times before succeeding in producing a ring signature. The complexity of this search is essentially the same as in the case of regular Rabin signatures, regardless of the size of the ring.

A more important difference is in the proof of unconditional anonymity, which relied on the fact that all the mappings were permutations. When the g_i are not permutations, there can be noticable differences between the distribution of randomly chosen and computed x_i values in given ring signatures. This could lead to the identification of the real signer among all the possible signers, and can be demonstrated to be a real problem in many concrete types of trapdoor one-way functions.

We overcome this difficulty in the case of Rabin signatures with the following simple observation:

Theorem 1. *Let S be a given finite set of "marbles" and let B_1, B_2, ..., B_n be disjoint subsets of S (called "buckets") such that all non-empty buckets have the same number of marbles, and every marble in S is in exactly one bucket. Consider the following sampling procedure: pick a bucket at random until you find a non-empty bucket, and then pick a marble at random from that bucket. Then this procedure picks marbles from S with uniform probability distribution.*

Proof. Trivial. □

Rabin's functions $f_i(x_i) = x_i^2 \ (\text{mod } n_i)$ are extended to functions $g_i(x_i)$ over $\{0,1\}^b$ in the usual way. Both the marbles and the buckets are all the b-bit numbers $u = q_i n_i + r_i$ in which $r_i \in \mathbf{Z}^*_{n_i}$ and $(q_i + 1)n_i \leq 2^b$ Each marble is placed in the bucket to which it is mapped by the extended Rabin mapping g_i. We know that each bucket contains either zero or four marbles, and the lemma inplies that the sampled distribution of the marbles x_i is exactly the same regardless of whether they were chosen at random or picked at random among the computed inverses in a randomly chosen bucket. Consequently, even an infinitely powerful adversary cannot distinguish between signers and nonsigners by analysing actual ring signatures produced by one of the possible signers.

5 Generalizations and Special Cases

The notion of ring signatures has many interesting extensions and special cases. In particular, ring signatures with $r = 1$ can be viewed as a randomized version of Rabin's signature scheme: As shown in Fig. 3, the verification condition can be written as $(x^2 \bmod n) = v \oplus E^{-1}_{h(m)}(v)$. The right hand side is essentially a hash of the message m, randomized by the choice of v.

Ring signatures with $r = 2$ have the ring equation:

$$E_{h(m)}(x_2^2 \oplus E_{h(m)}(x_1^2 \oplus v)) = v$$

(see Fig. 3). A simpler ring equation (which is not equivalent but has the same security properties) is:

$$(x_1^2 \bmod n_1) = E_{h(m)}(x_2^2 \bmod n_2)$$

where the modular squares are extended to $\{0,1\}^b$ in the usual way. This is our recommended method for implementing designated verifier signatures in email systems, where n_1 is the public key of the sender and n_2 is the public key of the recipient.

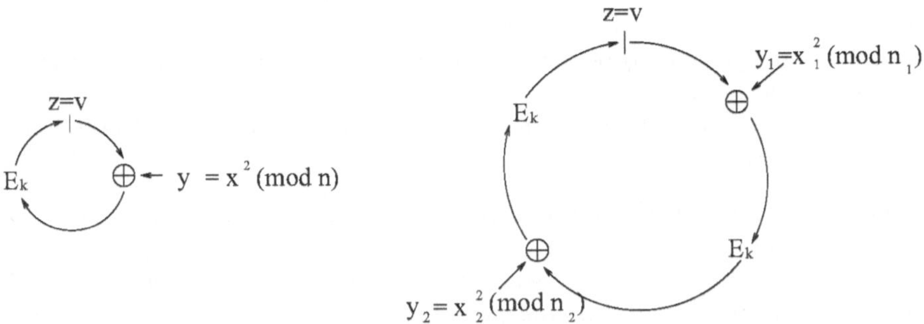

Fig. 3. Rabin-based Ring Signatures with $r = 1, 2$

In regular ring signatures it is provably impossible for an adversary to expose the signer's identity. However, there may be cases in which the signer himself wants to have the option of later proving his authorship of the anonymized email (e.g., if he is successful in toppling the disgraced Prime Minister). Yet another possibility is that the signer A wants to initially use {A,B,C} as the list of possible signers, but later prove that C is *not* the real signer. There is a simple way to implement these options, by choosing the x_i values for the nonsigners in a pseudorandom rather than truly random way. To show that C is *not* the author, A publishes the seed which pseudorandomly generated the part of the signature associated with C. To prove that A *is* the signer, A can reveal a single seed which was used to generate all the nonsigners' parts of the signature. The signer A cannot misuse this technique to prove that he is not the signer since his part is computed rather than generated, and is extremely unlikely to have a corresponding seed. Note that these modified versions can guarantee only computational anonymity, since a powerful adversary can search for such proofs of nonauthorship and use them to expose the signer.

References

1. Jan Camenisch. Efficient and generalized group signatures. In Walter Fumy, editor, *Advances in Cryptology - Eurocrypt '97*, pages 465–479, Berlin, 1997. Springer. Lecture Notes in Computer Science 1233.
2. David Chaum and Eugène Van Heyst. Group signatures. In D.W. Davies, editor, *Advances in Cryptology — Eurocrypt '91*, pages 257–265, Berlin, 1991. Springer-Verlag. Lecture Notes in Computer Science No. 547.
3. Ronald Cramer, Ivan Damgård, and Berry Schoenmakers. Proofs of partial knowledge and simplified design of witness hiding protocols. In Yvo Desmedt, editor, *Advances in Cryptology - CRYPTO '94*, pages 174–187, Berlin, 1994. Springer-Verlag. Lecture Notes in Computer Science Volume 839.
4. W. Diffie and M. E. Hellman. New directions in cryptography. *IEEE Trans. Inform. Theory*, IT-22:644–654, November 1976.
5. G. H. Hardy and E. M. Wright. *An Introduction to the Theory of Numbers.* Oxford, fifth edition, 1979.
6. M. Jakobsson, K. Sako, and R. Impagliazzo. Designated verifier proofs and their applications. In Ueli Maurer, editor, *Advances in Cryptology - EuroCrypt '96*, pages 143–154, Berlin, 1996. Springer-Verlag. Lecture Notes in Computer Science Volume 1070.
7. M. Luby and C. Rackoff. How to construct pseudorandom permutations from pseudorandom functions. *SIAM J. Computing*, 17(2):373–386, April 1988.
8. M. Rabin. Digitalized signatures as intractable as factorization. Technical Report MIT/LCS/TR-212, MIT Laboratory for Computer Science, January 1979.
9. Ronald L. Rivest, Adi Shamir, and Leonard M. Adleman. A method for obtaining digital signatures and public-key cryptosystems. *Communications of the ACM*, 21(2):120–126, 1978.
10. Alfredo De Santis, Giovanni Di Crescenzo, Giuseppe Persiano, and Moti Yung. On monotone formula closure of SZK. In *Proc. 35th FOCS*, pages 454–465. IEEE, 1994.

Key-Privacy in Public-Key Encryption

Mihir Bellare[1], Alexandra Boldyreva[1], Anand Desai[2], and David Pointcheval[3]

[1] Dept of Computer Science & Engineering, University of California, San Diego
La Jolla, California 92093, USA.
{mihir,aboldyre}@cs.ucsd.edu
http://www-cse.ucsd.edu/users/{mihir,aboldyre}

[2] NTT Multimedia Communications Laboratories
Palo Alto, California 94306, USA.
desai@nttmcl.com
http://www-cse.ucsd.edu/users/adesai

[3] Dépt d'Informatique, ENS - CNRS
45 rue d'Ulm, 75230 Paris Cedex 05, France.
David.Pointcheval@ens.fr
http://www.di.ens.fr/users/pointche

Abstract. We consider a novel security requirement of encryption schemes that we call "key-privacy" or "anonymity". It asks that an eavesdropper in possession of a ciphertext not be able to tell which specific key, out of a set of known public keys, is the one under which the ciphertext was created, meaning the receiver is anonymous from the point of view of the adversary. We investigate the anonymity of known encryption schemes. We prove that the El Gamal scheme provides anonymity under chosen-plaintext attack assuming the Decision Diffie-Hellman problem is hard and that the Cramer-Shoup scheme provides anonymity under chosen-ciphertext attack under the same assumption. We also consider anonymity for trapdoor permutations. Known attacks indicate that the RSA trapdoor permutation is not anonymous and neither are the standard encryption schemes based on it. We provide a variant of RSA-OAEP that provides anonymity in the random oracle model assuming RSA is one-way. We also give constructions of anonymous trapdoor permutations, assuming RSA is one-way, which yield anonymous encryption schemes in the standard model.

1 Introduction

The classical security requirement of an encryption scheme is that it provide privacy of the encrypted data. Popular formalizations— such as indistinguishability (semantic security) [22] or non-malleability [15], under either chosen-plaintext or various kinds of chosen-ciphertext attacks [27,29]— are directed at capturing various data-privacy requirements. (See [5] for a comprehensive treatment).

In this paper we consider a different (additional) security requirement of an encryption scheme which we call *key-privacy* or *anonymity*. It asks that the

C. Boyd (Ed.): ASIACRYPT 2001, LNCS 2248, pp. 566–582, 2001.
© Springer-Verlag Berlin Heidelberg 2001

encryption provide (in addition to privacy of the data being encrypted) privacy of the key under which the encryption was performed.

This might sound odd, especially in the public-key setting which is our main focus: here the key under which encryption is performed is the public key of the receiver and being public there might not seem to be anything to keep private about it. The privacy refers to the information conveyed to the adversary regarding which specific key, out of a set of known public keys, is the one under which a given ciphertext was created. We call this anonymity because it means that the receiver is anonymous from the point of view of the adversary.

Anonymity of encryption has surfaced in various different places in the past, and found several applications, as we detail later. However, it lacks a comprehensive treatment. Our goal is to provide definitions, and then systematically study popular asymmetric encryption schemes with regard to their meeting these definitions. Below we discuss our contributions and then discuss related work.

1.1 Definitions

We suggest a notion we call "indistinguishability of keys" to formalize the property of key-privacy. In the formalization, the adversary knows two public keys pk_0, pk_1, corresponding to two different entities, and gets a ciphertext C formed by encrypting some data under one of these keys. Possession of C should not give the adversary an advantage in determining under which of the two keys C was created. This can be considered under either chosen-plaintext attack or chosen-ciphertext attack, yielding two notions of security, IK-CPA and IK-CCA.

We also introduce the notion of an anonymous trapdoor permutation, which will serve as tool in some of the designs.

1.2 The Search for Anonymous Asymmetric Encryption Schemes

In a heterogenous public-key environment, encryption will probably fail to be anonymous for trivial reasons. For example, different users might be using different cryptosytems, or, if the same cryptosystem, have keys of different lengths. (If one possible recipient has a RSA public key with a 1024 bit modulus and the other a RSA public key with a 512 bit modulus, the length of the RSA ciphertext will immediately enable an eavesdropper to know for which recipient the ciphertext is intended.) We can however hope for anonymity in a context where all users use the same security parameter or global parameters. We will look at specific systems with this restriction in mind.

Ideally, we would like to be able to prove that popular, existing and practical encryption schemes have the anonymity property (rather than having to design new schemes.) This would be convenient because then existing encryption-using protocols or software would not have to be altered in order for them to have the anonymity guarantees conferred by those of the encryption scheme. Accordingly, we begin by examining existing schemes. We will consider discrete log based schemes such as El Gamal and Cramer-Shoup, and also RSA-based schemes such as RSA-OAEP.

It is easy to see that an encryption scheme could meet even the strongest notion of data-privacy— namely indistinguishability under chosen-ciphertext attack— yet not provide key-privacy. (The ciphertext could contain the public key.) Accordingly, existing results about data-privacy of asymmetric encryption schemes are not directly applicable. Existing schemes must be re-analyzed with regard to key-privacy.

In approaching this problem, we had no a priori way to predict whether or not a given asymmetric scheme would have the key-privacy property, and, if it did, whether the proof would be a simple modification of the known data privacy proof, or require new techniques. It is only by doing the work that one can tell what is involved.

We found that the above-mentioned discrete log based schemes did have the key-privacy property, and, moreover, that it was possible to prove this, under the same assumptions as used to prove data-privacy, by following the outline of the proofs of data-privacy with appropriate modifications. This perhaps unexpected strength of the discrete log based world (meaning not only the presence of the added security property in the popular schemes, but the fact that the existing techniques are strong enough to lead to a proof) seems important to highlight. In contrast, folklore attacks already rule out key-privacy for standard RSA-based schemes. Accordingly, we provide variants that have the property. Let us now look at these results in more detail.

1.3 Discrete Log Based Schemes

The El Gamal cryptosystem over a group of prime order provably provides data-privacy under chosen-plaintext attack assuming the DDH (Decision Diffie-Hellman) problem is hard in the group [25,12,33,3]. Let us now consider a system of users all of which work over the same group. (To be concrete, let q be a prime such that $2q+1$ is also prime, let G_q be the order q subgroup of quadratic residues of Z^*_{2q+1} and let $g \in G_q$ be a generator of G_q. Then q, g are system wide parameters based on which all users choose keys.) In this setting we prove that the El Gamal scheme meets the notion of IK-CPA under the same assumption used to establish data-privacy, namely the hardness of the DDH problem in the group. Thus the El Gamal scheme provably provides anonymity. Our proof exploits self-reducibility properties of the DDH problem together with ideas from the proof of data-privacy.

The Cramer-Shoup scheme [12] is proven to provide data-privacy under chosen-ciphertext attack, under the assumption that the DDH problem is hard in the group underlying the scheme. Let us again consider a system of users, all of which work over the same group, and for concreteness let it be the group G_q that we considered above. In this setting we prove that the Cramer-Shoup scheme meets the notion of IK-CCA assuming the DDH problem is hard in G_q. Our proof exploits ideas in [12,3].

1.4 RSA-Based Schemes

A simple observation that seems to be folkore is that standard RSA encryption does not provide anonymity, even when all modulii in the system have the same length. In all popular schemes, the ciphertext is (or contains) an element $y = x^e \bmod N$ where x is a random member of Z_N^*. Suppose an adversary knows that the ciphertext is created under one of two keys N_0, e_0 or N_1, e_1, and suppose $N_0 \leq N_1$. If $y \geq N_0$ then the adversary bets it was created under N_1, e_1, else it bets it was created under N_0, e_0. It is not hard to see that this attack has non-negligible advantage.

One approach to anonymizing RSA, suggested by Desmedt [14], is to add random multiples of the modulus N to the ciphertext. This seems to overcome the above attack, at least when the data encrypted is random, but results in a doubling of the length of the ciphertext. We look at a few other approaches.

We consider an RSA-based encryption scheme popular in current practice, namely RSA-OAEP [8]. (It is the PKCS v2.0 standard [28], proved secure against chosen-ciphertext attack in the random oracle model [18].) We suggest a variant which we can prove is anonymous. Recall that OAEP is a randomized (invertible) transform that on input a message M picks a random string r and, using some public hash functions, produces a point $x = \mathsf{OAEP}(r, M) \in Z_N^*$ where N, e is the public key of the receiver. The ciphertext is then $y = x^e \bmod N$. Our variant simply repeats the ciphertext computation, each time using new coins, until the ciphertext y satisfies $1 \leq y \leq 2^{k-2}$, where k is the length of N. We prove that this scheme meets the notion of IK-CCA in the random oracle model assuming RSA is a one-way function. (Data-privacy under chosen-ciphertext attack must be re-proved, but this can be done, under the same assumption, following [18].) The expected number of exponentiations for encryption being two, encryption in our variant is about twice as expensive as for RSA-OAEP itself, but this may be tolerable when the encryption exponent is small. The cost of decryption is the same as for RSA-OAEP itself, namely one exponentiation with the decryption exponent. As compared to Desmedt's scheme, the size of the ciphertext increases by only one bit rather than doubling. Our proof exploits the framework and techniques of [18,8].

1.5 Trapdoor Permutation Based Schemes

We then ask a more theoretical, or foundational, question, namely whether there exists an encryption scheme that can be proven to provide key-privacy based only on the assumption that RSA is one-way, meaning without making use of the random oracle model. To answer this we return to the classical techniques based on hardcore bits. We define a notion of anonymity for trapdoor permutations. We note that the above attack implies that RSA is not an anonymous trapdoor permutation, but we then design some trapdoor permutations which are anonymous and one-way as long as RSA is one-way. Appealing to known results about hardcore bits then yields an encryption scheme whose anonymity is proven based solely on the one-wayness of RSA. The computational costs of this approach, however, prohibit its being useful in practice.

1.6 Applications and Related Work

In recent years, anonymous encryption has arisen in the context of mobile communications. Consider a mobile user A, communicating over a wireless network with some entity B. The latter is sending A ciphertexts encrypted under A's public key. A common case is that B is a base station. A wants to keep her identity private from an eavesdropping adversary. In this case A will be a member of some set of users whose identities and public keys are possibly known to the adversary. The adversary will also be able to see the ciphertexts sent by B to A. If the scheme is anonymous, however, the adversary will be unable to determine A's identity. A particular case of this is anonymous authenticated key exchange, where the communication between roaming user A and base station B is for the purpose of authentication and distribution of a session key based on the parties public keys, but the identity of A should remain unknown to an eavesdropper. Anonymity is targeted in authenticated key exchange protocols such as SKEME [23]. The author notes that a requirement for SKEME to provide anonymous authenticated key exchange is that the public-key encryption scheme used to encrypt under A's public key must have the key-privacy property.

In independent and concurrent work, Camenisch and Lysyanskaya [10] consider anonymous credential systems. Such a sytem enables users to control the dissemination of information about themselves. It is required that it be infeasible to correlate transactions carried out by the same user. The solution to this given in [10] makes use of a *verifiable circular* encryption scheme that needs to have the key-privacy property. They provide a notion similar to ours, but in the context of verifiable encryption. They observe that their variant of the El Gamal scheme is anonymous under chosen-plaintext attack.

Sako [30] considers the problem of achieving bid secrecy and verifiability in auction protocols. Their approach is to express each bid as an encryption of a known message, with the *key* to encrypt it corresponding to the value of the bid. Thus, what needs to be hidden is not the message that is encrypted, but the key used to encrypt it. The bid itself can be identified by finding the corresponding decrypting key that successfully decrypts to the given message. Unlike the previous examples, where the key-privacy property was needed to protect identities, this application shows how that property can be exploited to satisfy a secrecy requirement. Sako also considered a notion similar to ours and gave a variant of the El Gamal scheme that was expected to be secure in that sense.

Formal notions of key-privacy have appeared in the context of symmetric encryption [1,13,17]. Abadi and Rogaway [1] show that popular modes of operation of block ciphers, such as CBC, provide key-privacy if the block cipher is a pseudorandom permutation.

The notion given by Desai [13], like ours, is concerned with the privacy of keys. However, the goal, model and setting in which it is considered differs from ours— the goal there is to capture a security property for block cipher based encryption schemes that implies that exhaustive key-search on them is slowed down proportional to the size of the ciphertext. There is, however, a similarity between our definitions (suitably adapted to the symmetric setting) and those of Abadi and Rogaway [1] and Fischlin [17]. Although the exact formalizations differ, it is not hard to see that there is an equivalence between the three for chosen-plaintext attack.

Chosen-ciphertext attacks do not seem to have been considered before in the context of key-privacy. In fact, Fischlin [17] observes that giving decryption oracles to the adversary in their setting makes its task trivial. However, in our formalization chosen-ciphertext attacks can be modeled by giving decryption oracles and then putting an appropriate restriction on their use. The restriction is the most natural and is anyway in effect for modeling semantic security against chosen-ciphertext attack. This allows us to make a distinction between those encryption schemes that are anonymous under chosen-ciphertext attack, such as Cramer-Shoup, and those that are not, such as El Gamal— just as there are schemes that are semantically secure under chosen-plaintext attack but not under chosen-ciphertext attack.

2 Notions of Key-Privacy

The notions of security typically considered for encryption schemes are "indistinguishability of encryptions under chosen-plaintext attack" [22] and "indistinguishability of encryptions under adaptive chosen-ciphertext attack" [29]. The former is usually denoted IND-CPA, but is denoted IE-CCA in this paper to emphasize that it is about encryptions, not keys. Similarly, the latter notion is usually denoted IND-CCA (or IND-CCA2), but is denoted IE-CCA in this paper. It is well-known that these capture strong data-privacy properties. However, they do not guarantee that some partial information about the underlying *key* is not leaked. Indeed, in a public-key encryption scheme, the entire public-key could be made an explicit part of the ciphertext and yet the scheme could meet the above-mentioned data-privacy notions. We want to make a distinction between such schemes and those that do not leak information about the underlying key. As noted earlier, schemes of the latter kind are necessary if the anonymity of receivers is a concern.

We are interested in formalizing the inability of an adversary, given a challenge ciphertext, to learn any information about the underlying plaintext or key. It is not hard to see that the goals of data-privacy and key-privacy are orthogonal. We recognize that existing encryption schemes are likely to have already been investigated with respect to their data-privacy security properties. Hence it is useful, from a practical point of view, to isolate the key-privacy requirements from the data-privacy ones. We do this in the form of two notions: "indistinguishability of keys under chosen-plaintext attack" (IK-CPA) and "indistinguishability of keys under adaptive chosen-ciphertext attack" (IK-CCA). We begin with a syntax for public-key encryption schemes, divorcing syntax from formal notions of security.

2.1 Syntax

The syntax of an encryption scheme specifies what algorithms make it up. We augment the usual formalization in order to better model practice, where users may share some fixed "global" information.

A *public-key encryption scheme* $\mathcal{PE} = (\mathcal{G}, \mathcal{K}, \mathcal{E}, \mathcal{D})$ consists of four algorithms. The *common-key generation* algorithm $\mathcal{G}$ takes as input some security parameter k and returns some common key I. (Here I may be just a security parameter k, or include some additional information. For example in a Diffie-Hellman based scheme, I might include, in addition to k, a global prime number and generator of a group which all parties use to create their keys.) The *key generation* algorithm $\mathcal{K}$ is a randomized algorithm that takes as input the common key I and returns a pair (pk, sk) of keys, the public key and a matching secret key, respectively; we write $(pk, sk) \stackrel{R}{\leftarrow} \mathcal{K}(I)$. The *encryption* algorithm $\mathcal{E}$ is a randomized algorithm that takes the public key pk and a *plaintext* x to return a *ciphertext* y; we write $y \leftarrow \mathcal{E}_{pk}(x)$. The *decryption* algorithm $\mathcal{D}$ is a deterministic algorithm that takes the secret key sk and a ciphertext y to return the corresponding plaintext x or a special symbol $\perp$ to indicate that the ciphertext was invalid; we write $x \leftarrow \mathcal{D}_{sk}(y)$ when y is valid and $\perp \leftarrow \mathcal{D}_{sk}(y)$ otherwise. Associated to each public key pk is a *message space* $\mathrm{MsgSp}(pk)$ from which x is allowed to be drawn. We require that $\mathcal{D}_{sk}(\mathcal{E}_{pk}(x)) = x$ for all $x \in \mathrm{MsgSp}(pk)$.

2.2 Indistinguishability of Keys

We give notions of key-privacy under chosen-plaintext and chosen-ciphertext attacks. We think of an adversary running in two stages. In the find stage it takes two public keys pk_0 and pk_1 (corresponding to secret keys sk_0 and sk_1, respectively) and outputs a message x together with some state information s. In the guess stage it gets a challenge ciphertext y formed by encrypting at random the messages under one of the two keys, and must say which key was chosen. In the case of a chosen-ciphertext attack the adversary gets oracles for $\mathcal{D}_{sk_0}(\cdot)$ and $\mathcal{D}_{sk_1}(\cdot)$ and is allowed to invoke them on any point with the restriction (on both oracles) of not querying y during the guess stage.

Definition 1. [IK-CPA, IK-CCA] *Let $\mathcal{PE} = (\mathcal{G}, \mathcal{K}, \mathcal{E}, \mathcal{D})$ be an encryption scheme. Let $b \in \{0, 1\}$ and $k \in \mathsf{N}$. Let $A_{\text{cpa}}, A_{\text{cca}}$ be adversaries that run in two stages and where A_{cca} has access to the oracles $\mathcal{D}_{sk_0}(\cdot)$ and $\mathcal{D}_{sk_1}(\cdot)$. Now, we consider the following experiments:*

Experiment $\mathbf{Exp}^{\text{ik-cpa-}b}_{\mathcal{PE},A_{\text{cpa}}}(k)$	*Experiment* $\mathbf{Exp}^{\text{ik-cca-}b}_{\mathcal{PE},A_{\text{cca}}}(k)$
$I \stackrel{R}{\leftarrow} \mathcal{G}(k)$	$I \stackrel{R}{\leftarrow} \mathcal{G}(k)$
$(pk_0, sk_0) \stackrel{R}{\leftarrow} \mathcal{K}(I)$; $(pk_1, sk_1) \stackrel{R}{\leftarrow} \mathcal{K}(I)$	$(pk_0, sk_0) \stackrel{R}{\leftarrow} \mathcal{K}(I)$; $(pk_1, sk_1) \stackrel{R}{\leftarrow} \mathcal{K}(I)$
$(x, s) \leftarrow A_{\text{cpa}}(\mathsf{find}, pk_0, pk_1)$	$(x, s) \leftarrow A_{\text{cca}}^{\mathcal{D}_{sk_0}(\cdot),\mathcal{D}_{sk_1}(\cdot)}(\mathsf{find}, pk_0, pk_1)$
$y \leftarrow \mathcal{E}_{pk_b}(x)$	$y \leftarrow \mathcal{E}_{pk_b}(x)$
$d \leftarrow A_{\text{cpa}}(\mathsf{guess}, y, s)$	$d \leftarrow A_{\text{cca}}^{\mathcal{D}_{sk_0}(\cdot),\mathcal{D}_{sk_1}(\cdot)}(\mathsf{guess}, y, s)$
Return d	*Return* d

Above it is mandated that $A_{\rm cca}$ never queries $\mathcal{D}_{sk_0}(\cdot)$ or $\mathcal{D}_{sk_1}(\cdot)$ on the challenge ciphertext y. For atk $\in$ {cpa, cca} *we define the advantages of the adversaries via*

$$\mathbf{Adv}^{\text{ik-atk}}_{\mathcal{PE},A_{\rm atk}}(k) = \Pr[\,\mathbf{Exp}^{\text{ik-atk-1}}_{\mathcal{PE},A_{\rm atk}}(k) = 1\,] - \Pr[\,\mathbf{Exp}^{\text{ik-atk-0}}_{\mathcal{PE},A_{\rm atk}}(k) = 1\,]\,.$$

The scheme $\mathcal{PE}$ is said to be IK-CPA secure (respectively IK-CCA secure) if the function $\mathbf{Adv}^{\text{ik-cpa}}_{\mathcal{PE},A}(\cdot)$ (resp. $\mathbf{Adv}^{\text{ik-cca}}_{\mathcal{PE},A}(\cdot)$) is negligible for any adversary A whose time complexity is polynomial in k. ∎

The "time-complexity" is the worst case execution time of the experiment plus the size of the code of the adversary, in some fixed RAM model of computation. (Note that the execution time refers to the entire experiment, not just the adversary. In particular, it includes the time for key generation, challenge generation, and computation of responses to oracle queries if any.) The same convention is used for all other definitions in this paper and will not be explicitly mentioned again.

2.3 Anonymous One-Way Functions

A *family of functions* $F = (K, S, E)$ is specified by three algorithms. The randomized *key-generation* algorithm K takes input the security parameter $k \in \mathsf{N}$ and returns a pair (pk, sk) where pk is a public key, and sk is an associated secret key. (In cases where the family is not trapdoor, the secret key is simply the empty string.) The randomized *sampling* algorithm S takes input pk and returns a random point in a set that we call the domain of pk and denote $\mathrm{Dom}_F(pk)$. We usually omit explicit mention of the sampling algorithm and just write $x \stackrel{R}{\leftarrow} \mathrm{Dom}_F(pk)$. The deterministic *evaluation* algorithm E takes input pk and a point $x \in \mathrm{Dom}_F(pk)$ and returns an output we denote by $E_{pk}(x)$. We let $\mathrm{Rng}_F(pk) = \{\, E_{pk}(x) \;:\; x \in \mathrm{Dom}_F(pk)\,\}$ denote the range of the function $E_{pk}(\cdot)$. We say that F is a family of *trapdoor* functions if there exists a deterministic *inversion* algorithm I that takes input sk and a point $y \in \mathrm{Rng}_F(pk)$ and returns a point $x \in \mathrm{Dom}_F(pk)$ such that $E_{pk}(x) = y$. We say that F is a family of *permutations* if $\mathrm{Dom}_F(pk) = \mathrm{Rng}_F(pk)$ and E_{pk} is a permutation on this set.

Definition 2. *Let $F = (K, S, E)$ be a family of functions. Let $b \in \{0, 1\}$ and $k \in \mathsf{N}$ be a security parameter. Let $0 < \theta \le 1$ be a constant. Let A, B be adversaries. Now, we consider the following experiments:*

Experiment $\mathbf{Exp}^{\theta\text{-pow-fnc}}_{F,B}(k)$	*Experiment* $\mathbf{Exp}^{\text{ik-fnc-}b}_{F,A}(k)$
$(pk, sk) \stackrel{R}{\leftarrow} K(k)$	$(pk_0, sk_0) \stackrel{R}{\leftarrow} K(k)$
$x_1\|x_2 \stackrel{R}{\leftarrow} \mathrm{Dom}_F(pk)$ *where* $\lvert x_1\rvert = \lceil \theta \cdot \lvert(x_1\|x_2)\rvert\rceil$	$(pk_1, sk_1) \stackrel{R}{\leftarrow} K(k)$
$y \leftarrow E_{pk}(x_1\|x_2)$	$x \stackrel{R}{\leftarrow} \mathrm{Dom}_F(pk_b)$
$x_1' \leftarrow B(pk, y)$ *where* $\lvert x_1'\rvert = \lvert x_1\rvert$	$y \leftarrow E_{pk_b}(x)$
For any x_2' *if* $E_{pk}(x_1'\|x_2') = y$ *then return 1*	$d \leftarrow A(pk_0, pk_1, y)$
Else return 0	*Return* d

We define the advantages of the adversaries via

$$\mathbf{Adv}^{\theta\text{-pow-fnc}}_{F,B}(k) = \Pr[\,\mathbf{Exp}^{\theta\text{-pow-fnc}}_{F,B}(k) = 1\,]$$
$$\mathbf{Adv}^{\text{ik-fnc}}_{F,A}(k) = \Pr[\,\mathbf{Exp}^{\text{ik-fnc-1}}_{F,A}(k) = 1\,] - \Pr[\,\mathbf{Exp}^{\text{ik-fnc-0}}_{F,A}(k) = 1\,]\,.$$

The family F is said to be θ-partial one-way if the function $\mathbf{Adv}^{\theta\text{-pow-fnc}}_{F,B}(\cdot)$ is negligible for any adversary B whose time complexity is polynomial in k. The family F is said to be anonymous if the function $\mathbf{Adv}^{\text{ik-fnc}}_{F,A}(\cdot)$ is negligible for any adversary A whose time complexity is polynomial in k. The family F is said to be perfectly anonymous if $\mathbf{Adv}^{\text{ik-fnc}}_{F,A}(k) = 0$ for every k and every adversary A. ▮

Note that when $\theta = 1$ the notion of θ-partial one-wayness coincides with the standard notion of one-wayness. As the above indicates, we expect that information-theoretic anonymity is possible for one-way functions, even though not for encryption schemes.

3 Anonymity of DDH-Based Schemes

The DDH-based schemes we consider work over a group of prime order. This could be a subgroup of order q of Z_p^* where p, q are primes such that q divides $p-1$. It could also be an elliptic curve group of prime order. For concreteness our description is for the first case. Specifically if q is a prime such that $2q+1$ is also prime we let G_q be the subgroup of quadratic residues of Z_p^*. It has order q. A *prime-order-group generator* is a probabilistic algorithm that on input the security parameter k returns a pair (q, g) satisfying the following conditions: q is a prime with $2^{k-1} < q < 2^k$; $2q+1$ is a prime; and g is a generator of G_q. (There are numerous possible specific prime-order-group generators.) We will relate the anonymity of the El Gamal and Cramer-Shoup schemes to the hardness of the DDH problem for appropriate prime-order-group generators. Accordingly we next summarize definitions for the latter.

Definition 3. [DDH] *Let $\mathcal{G}$ be a prime-order-group generator. Let D be an adversary that on input q, g and three elements $X, Y, T \in G_q$ returns a bit. We consider the following experiments*

Experiment $\mathbf{Exp}^{\text{ddh-real}}_{\mathcal{G},D}(k)$	*Experiment* $\mathbf{Exp}^{\text{ddh-rand}}_{\mathcal{G},D}(k)$
$(q,g) \xleftarrow{R} \mathcal{G}(k)$	$(q,g) \xleftarrow{R} \mathcal{G}(k)$
$x \xleftarrow{R} Z_q$; $X \leftarrow g^x$	$x \xleftarrow{R} Z_q$; $X \leftarrow g^x$
$y \xleftarrow{R} Z_q$; $Y \leftarrow g^y$	$y \xleftarrow{R} Z_q$; $Y \leftarrow g^y$
$T \leftarrow g^{xy}$	$T \xleftarrow{R} G_q$
$d \leftarrow D(q,g,X,Y,T)$	$d \leftarrow D(q,g,X,Y,T)$
Return d	*Return* d

The advantage of D in solving the Decisional Diffie-Hellman (DDH) problem for $\mathcal{G}$ is the function of the security parameter defined by

$$\mathbf{Adv}^{\text{ddh}}_{\mathcal{G},D}(k) \;=\; \Pr[\,\mathbf{Exp}^{\text{ddh-real}}_{\mathcal{G},D}(k) = 1\,] - \Pr[\,\mathbf{Exp}^{\text{ddh-rand}}_{\mathcal{G},D}(k) = 1\,]\,.$$

We say that the DDH problem is hard for $\mathcal{G}$ if the function $\mathbf{Adv}^{\text{ddh}}_{\mathcal{G},D}(\cdot)$ is negligible for every algorithm D whose time-complexity is polynomial in k. ▮

3.1 El Gamal

The El Gamal scheme in a group of prime order is known to meet the notion of indistinguishability under chosen-plaintext attack under the assumption that the decision Diffie-Hellman (DDH) problem is hard. (This is noted in [25,12] and fully treated in [33]). We want to look at the anonymity of the El Gamal encryption scheme under chosen-plaintext attack.

Let $\mathcal{G}$ be a prime-order-group generator. This is the common key generation algorithm of the associated scheme $\mathcal{EG} = (\mathcal{G}, \mathcal{K}, \mathcal{E}, \mathcal{D})$, the rest of whose algorithms are as follows:

Algorithm $\mathcal{K}(q,g)$	Algorithm $\mathcal{E}_{pk}(M)$	Algorithm $\mathcal{D}_{sk}(Y,W)$
$x \stackrel{R}{\leftarrow} Z_q$	$y \stackrel{R}{\leftarrow} Z_q$	$T \leftarrow Y^x$
$X \leftarrow g^x$	$Y \leftarrow g^y$	$M \leftarrow WT^{-1}$
$pk \leftarrow (q,g,X)$	$T \leftarrow X^y$	Return M
$sk \leftarrow (q,g,x)$	$W \leftarrow TM$	
Return (pk, sk)	Return (Y, W)	

The message space associated to a public key (q, g, X) is the group G_q itself, with the understanding that all messages from G_q are properly encoded as strings of some common length whenever appropriate. Note that a generator g is the output of the common key generation algorithm, which means we fix g for all keys. We do it only for a simplicity reason and will show that all our results hold also for a case when each key uses a random generator g.

We now analyze the anonymity of the El Gamal scheme under chosen-plaintext attack.

Theorem 1. *Let $\mathcal{G}$ be a prime-order-group generator. If the DDH problem is hard for $\mathcal{G}$ then the associated El Gamal scheme $\mathcal{EG}$ is IK-CPA secure. Concretely, for any adversary A there exists a distinguisher D such that for any k*

$$\mathbf{Adv}^{\text{ik-cpa}}_{\mathcal{EG},A}(k) \le 2\mathbf{Adv}^{\text{ddh}}_{\mathcal{G},D}(k) + \frac{1}{2^{k-2}}$$

and the running time of D is that of A plus $O(k^3)$. ∎

The proof of the above is in the full version of this paper [2].

3.2 Cramer-Shoup

The El Gamal scheme provides data privacy and anonymity against chosen-plaintext attack. We now consider the Cramer-Shoup scheme [12] in order to obtain the same security properties under chosen-ciphertext attack. We will use collision-resistant hash functions so we begin by recalling what we need.

A family of hash functions $\mathcal{H} = (\mathcal{GH}, \mathcal{EH})$ is defined by a probabilistic generator algorithm $\mathcal{GH}$ —which takes as input the security parameter k and returns a key K— and a deterministic evaluation algorithm $\mathcal{EH}$ —which takes as input the key K and a string $M \in \{0,1\}^*$ and returns a string $\mathcal{EH}_K(M) \in \{0,1\}^{k-1}$.

Definition 4. *Let $\mathcal{H} = (\mathcal{GH}, \mathcal{EH})$ be a family of hash functions and let C be an adversary that on input a key K returns two strings. Now, we consider the following experiment:*

Experiment $\mathbf{Exp}^{\mathrm{cr}}_{\mathcal{H},C}(k)$
$K \stackrel{R}{\leftarrow} \mathcal{GH}(k)$; $(x_0, x_1) \leftarrow C(K)$
If $(x_0 \neq x_1)$ *and* $\mathcal{EH}_K(x_0) = \mathcal{EH}_K(x_1)$ *then return 1 else return 0*

We define the advantage of adversary C via

$$\mathbf{Adv}^{\mathrm{cr}}_{\mathcal{H},C}(k) = \Pr[\,\mathbf{Exp}^{\mathrm{cr}}_{\mathcal{H},C}(k) = 1\,]\,.$$

We say that the family of hash functions $\mathcal{H}$ is collision-resistant if $\mathbf{Adv}^{\mathrm{cr}}_{\mathcal{H},C}(\cdot)$ is negligible for every algorithm C whose time-complexity is polynomial in k. ▮

Let $\overline{\mathcal{G}}$ be a prime-order-group generator. The common key generation algorithm of the associated Cramer-Shoup scheme $\mathcal{CS} = (\mathcal{G}, \mathcal{K}, \mathcal{E}, \mathcal{D})$ is:

$$\text{Algorithm } \mathcal{G}(k):\ (q, g_1) \stackrel{R}{\leftarrow} \overline{\mathcal{G}};\ g_2 \stackrel{R}{\leftarrow} G_q;\ K \stackrel{R}{\leftarrow} \mathcal{GH}(k);\ \text{ Return } (q, g_1, g_2, K).$$

The rest of algorithms are specified as follows:

Algorithm $\mathcal{K}(q, g_1, g_2, K)$	Algorithm $\mathcal{E}_{pk}(M)$	Algorithm $\mathcal{D}_{sk}(u_1, u_2, e, v)$
$g_1 \leftarrow g$	$r \stackrel{R}{\leftarrow} Z_q$	$\alpha \leftarrow \mathcal{EH}_K(u_1, u_2, e)$
$x_1, x_2, y_1, y_2, z \stackrel{R}{\leftarrow} Z_q$	$u_1 \leftarrow g_1^r$; $u_2 \leftarrow g_2^r$	If ${u_1}^{x_1+y_1\alpha}{u_2}^{x_2+y_2\alpha} = v$
$c \leftarrow g_1^{x_1} g_2^{x_2}$; $d \leftarrow g_1^{y_1} g_2^{y_2}$	$e \leftarrow h^r M$	then $M \leftarrow e/{u_1}^z$
$h \leftarrow g_1^z$	$\alpha \leftarrow \mathcal{EH}_K(u_1, u_2, e)$	else $M \leftarrow \perp$
$pk \leftarrow (g_1, g_2, c, d, h, K)$	$v \leftarrow c^r d^{r\alpha}$	Return M
$sk \leftarrow (x_1, x_2, y_1, y_2, z)$	Return (u_1, u_2, e, v)	
Return (pk, sk)		

The message space is the group G_q. Note that the range of the hash function $\mathcal{EH}_K$ is $\{0,1\}^{k-1}$ which we identify with $\{0, \ldots, 2^{k-1}\}$. Since $q > 2^{k-1}$ this is a subset of Z_q. Again for simplicity we assume that g_1, g_2 are fixed for all keys but we will show that our results hold even if g_1, g_2 are chosen at random for all keys.

We now analyze the anonymity of $\mathcal{CS}$ under chosen-ciphertext attack.

Theorem 2. *Let $\overline{\mathcal{G}}$ be a prime-order-group generator and let $\mathcal{CS}$ be the associated Cramer-Shoup scheme. If the DDH problem is hard for $\overline{\mathcal{G}}$ then $\mathcal{CS}$ is anonymous in the sense of IK-CCA. Concretely, for any adversary A attacking the anonymity of $\mathcal{CS}$ under a chosen-ciphertext attack and making in total $q_{\mathrm{dec}}(\cdot)$ decryption oracle queries, there exists a distinguisher D for DDH and an adversary C attacking the collision-resistance of $\mathcal{H}$ such that*

$$\mathbf{Adv}^{\text{ik-cca}}_{\mathcal{CS},A}(k) \;\leq\; 2\mathbf{Adv}^{\mathrm{ddh}}_{\overline{\mathcal{G}},D}(k) + 2\mathbf{Adv}^{\mathrm{cr}}_{\mathcal{H},C}(k) + \frac{q_{\mathrm{dec}}(k)+2}{2^{k-3}}.$$

and the running time of D and C is that of A plus $O(k^3)$. ▮

The proof of the above is in the full version of this paper [2]. Note that security of the Cramer-Shoup scheme in the IE-CCA sense has been proven in [12] using a weaker assumption on the hash function $\mathcal{H}$ than the one we have here. They do not require that $\mathcal{H}$ be collision-resistant, as we do, but only that it be a universal one-way family of hash functions (UOWHF) [26]. We have at this time not determined if the scheme can also be proven secure in the IK-CCA sense assuming $\mathcal{H}$ to be a UOWHF.

4 Anonymity of RSA-Based Schemes

The attack on RSA mentioned in Section 1 implies that the RSA family of trapdoor permutations is not anonymous. This means that all traditional RSA-based encryption schemes are not anonymous. We provide several ways to implement anonymous RSA-based encryption. First we take a direct approach, specifying an anonymous RSA-OAEP variant based on repetition and proving it secure in the random oracle model. Then we show how to construct anonymous trapdoor permutation families based on RSA and derive anonymous RSA-based encryption schemes from them. In particular, the latter leads to anonymous encryption schemes whose proofs of security are in the standard rather than the random oracle model. We begin with a description of the RSA family of trapdoor permutations we will use in this section. See Section 2 for notions of security for families of trapdoor permutations.

Example 1. The specifications of the *standard RSA family* of trapdoor permutations $\mathsf{RSA} = (K, S, E)$ are as follows. The key generation algorithm takes as input a security parameter k and picks random, distinct primes p, q in the range $2^{k/2-1} < p, q < 2^{k/2}$. (If k is odd, increment it by 1 before picking the primes.) It sets $N = pq$. It picks $e, d \in Z^*_{\varphi(N)}$ such that $ed \equiv 1 \pmod{\varphi(N)}$ where $\varphi(N) = (p-1)(q-1)$. The public key is N, e and the secret key is N, d. The sets $\mathrm{Dom}_{\mathsf{RSA}}(N, e)$ and $\mathrm{Rng}_{\mathsf{RSA}}(N, e)$ are both equal to Z^*_N. The evaluation algorithm is $E_{N,e}(x) = x^e \bmod N$ and the inversion algorithm is $I_{N,d}(y) = y^d \bmod N$. The sampling algorithm returns a random point in Z^*_N. ▮

The anonymity attack on RSA carries over to most encryption schemes based on it, including the most popular one, RSA-OAEP. We next describe a variant of RSA-OAEP that preserves its data-privacy properties but is in addition anonymous.

4.1 Anonymous Variant of RSA-OAEP

The original scheme and our variant are described in the random-oracle (RO) model [7]. All the notions of security, defined earlier, can be "lifted" to the RO setting in a straightforward manner. To modify the definitions, begin the experiment defining advantage by choosing random functions G and H, each from the set of all functions from some appropriate domain to appropriate range. Then provide a G-oracle and H-oracle to the adversaries, and allow that $\mathcal{E}_{pk}$ and $\mathcal{D}_{sk}$ may depend on G and H (which we write as $\mathcal{E}^{G,H}_{pk}$ and $\mathcal{D}^{G,H}_{sk}$).

The idea behind our variant is to repeat the standard encryption procedure under RSA-OAEP, until the ciphertext falls in some "safe" range. We refer to our scheme as RSA-RAEP (for *repeated* asymmetric encryption with padding). More concretely, for $\mathsf{RSA} = (K, S, E)$, our scheme RSA-RAEP $= (\mathcal{G}, \mathcal{K}, \mathcal{E}, \mathcal{D})$ is as follows. The common key generator algorithm $\mathcal{G}$ takes a security parameter k and returns parameters k, k_0 and k_1 such that $k_0(k) + k_1(k) < k$ for all $k > 1$. This defines an associated plaintext-length function $n(k) = k - k_0(k) - k_1(k)$. The key generation algorithm $\mathcal{K}$ takes k, k_0, k_1 and runs the key-generation algorithm of the RSA family, namely K on k to get a public key (N, e) and secret key (N, d) (see Example 1). The public key for the scheme *pk* is $(N, e), k, k_0, k_1$ and the secret key *sk* is $(N, d), k, k_0, k_1$. The other algorithms are depicted below. The oracles G and H which $\mathcal{E}_{pk}$ and $\mathcal{D}_{sk}$ reference below map bit strings as follows: $G: \{0,1\}^{k_0} \mapsto \{0,1\}^{n+k_1}$ and $H: \{0,1\}^{n+k_1} \mapsto \{0,1\}^{k_0}$.

Algorithm $\mathcal{E}_{pk}^{G,H}(x)$
- $ctr = -1$
- Repeat
  - $ctr \leftarrow ctr + 1$
  - $r \stackrel{R}{\leftarrow} \{0,1\}^{k_0}$
  - $s \leftarrow (x \| 0^{k_1}) \oplus G(r)$
  - $t \leftarrow r \oplus H(s)$
  - $v \leftarrow (s \| t)^e \bmod N$
- Until $(v < 2^{k-2}) \vee (ctr = k_1)$
- If $ctr = k_1$ then $y \leftarrow 1 \| 0^{k_0+k_1} \| x$
- Else $y \leftarrow 0 \| v$
- Return y

Algorithm $\mathcal{D}_{sk}^{G,H}(y)$
- Parse y as $b \| v$ where b is a bit
- If $b = 1$ then parse v as $w \| x$ where $|x| = n$
  - If $w = 0^{k_0+k_1}$ then $z \leftarrow x$
  - Else (if $w \neq 0^{k_0+k_1}$) $z \leftarrow \perp$
- Else (if $b = 0$)
  - $(s \| t) \leftarrow v^d \bmod N$ where:
    - $|s| = k_1 + n$ and $|t| = k_0$
  - $r \leftarrow t \oplus H(s)$
  - $(x \| p) \leftarrow s \oplus G(r)$ where:
    - $|x| = n$ and $|p| = k_1$
  - If $p = 0^{k_1}$ then $z \leftarrow x$
  - Else $z \leftarrow \perp$
- Return z

Note that the valid ciphertexts under RSA-OAEP are (uniformly) distributed in $\mathrm{Rng}_{\mathsf{RSA}}(N, e)$, which is Z_N^*. Under RSA-RAEP, valid ciphertexts take the form $0 \| v$ where $v \in (Z_N^* \cap [1, 2^{k-2}])$. The expected running time of this scheme is approximately twice that of RSA-OAEP (and k_1 times more, in the worst case). The ciphertext is longer by one bit. However, unlike RSA-OAEP, this scheme turns out to be IK-CCA secure. The (data-privacy) security of RSA-OAEP under CCA has already been established [18]. It is not hard to see that this result holds for RSA-RAEP as well. We omit the (simple) proof of this, noting only that the security (relative to RSA-OAEP) degrades roughly by the probability that after k_1 repetitions, the ciphertext was still not in the desired range (and consequently, the plaintext had to be sent in the clear). Given this, we turn to determining its security in the IK-CCA sense. We show that if the RSA family of trapdoor permutations is *partial* one-way then RSA-RAEP is anonymous.

Theorem 3. *If the* RSA *family of trapdoor permutations is partial one-way then Π = RSA-RAEP is anonymous. Concretely, for any adversary A attacking the anonymity of Π under a chosen-ciphertext attack, and making at most q_{dec} decryption oracle queries, q_{gen} G-oracle queries and q_{hash} H-oracle queries, there exists a θ-partial inverting adversary M_A for the* RSA *family, such that for any $k, k_0(k), k_1(k)$ and $\theta = \frac{k - k_0(k)}{k}$,*

$$\mathbf{Adv}^{\text{ik-cca}}_{\Pi,A}(k) \leq 32 q_{\text{hash}} \cdot ((1-\epsilon_1)\cdot(1-\epsilon_2)\cdot(1-\epsilon_3))^{-1} \cdot \mathbf{Adv}^{\theta\text{-pow-fnc}}_{\mathsf{RSA},M_A}(k) + q_{\text{gen}} \cdot (1-\epsilon_3)^{-1} \cdot 2^{-k+2}$$

where

$$\epsilon_1 = 4 \cdot \left(\frac{3}{4}\right)^{k/2-1} \; ; \qquad \epsilon_2 = \frac{1}{2^{k/2-3}-1} \; ;$$

$$\epsilon_3 = \frac{2q_{\text{gen}} + q_{\text{dec}} + 2q_{\text{gen}}q_{\text{dec}}}{2^{k_0}} + \frac{2q_{\text{dec}}}{2^{k_1}} + \frac{2q_{\text{hash}}}{2^{k-k_0}} ,$$

and the running time of M_A is that of A plus $q_{\text{gen}} \cdot q_{\text{hash}} \cdot O(k^3)$. ∎

The proof of the above is in the full version of this paper [2]. Note that for typical parameters $k_0(k), k_1(k)$, and number of allowed queries $q_{\text{gen}}, q_{\text{hash}}$ and q_{dec}, the values of ϵ_1, ϵ_2 and ϵ_3 are very small. This means that if there exists an adversary that is successful in breaking RSA-RAEP in the IK-CCA sense, then there exists a partial inverting adversary for the RSA family of trapdoor permutations that has a comparable advantage and running time.

The θ-partial one-wayness of RSA has been shown to be equivalent to the one-wayness of RSA, for $\theta > 0.5$ [18]. In RSA-RAEP (as also in RSA-OAEP) this is usually the case. (In general, the equivalence holds if any constant fraction of the most significant bits of the pre-image can be recovered, but the reduction is proportionately weaker [18].) Using this and Theorem 3 we are able to prove the security of RSA-RAEP in the IK-CCA sense assuming RSA to be one-way. A theorem to this effect, with concrete bounds, can be found in the full version of this paper [2].

4.2 Encryption with Anonymous Trapdoor Permutations

Given that the standard RSA family is not anonymous, we seek families that are. We describe some simple RSA-derived anonymous families.

Construction 1 We define a family $F = (K, S, E)$ as follows. The key generation algorithm is the same as in the standard RSA family of Example 1. Let (N, e) be a public key and k the corresponding security parameter. We set $\text{Dom}_F(N, e) = \text{Rng}_F(N, e) = \{0,1\}^k$. Viewing Z_N^* as a subset of $\{0,1\}^k$ we define

$$E_{N,e}(x) = \begin{cases} x^e \bmod N & \text{if } x \in Z_N^* \\ x & \text{otherwise} \end{cases}$$

for any $x \in \{0,1\}^k$. This is a permutation on $\{0,1\}^k$. The sampling algorithm S on input N, e simply returns a random k-bit string. It is easy to see that this family is trapdoor. ∎

As we will see, the family F is perfectly anonymous. But it is not one-way. However, it is weakly one-way. (Meaning, for every polynomial-time adversary

B, there is a polynomial $\beta(\cdot)$ such that $\mathbf{Adv}_{F,B}^{\text{1-pow-fnc}}(k) \le 1 - 1/\beta(k)$ for all sufficiently large k.) Thus, standard transformations of weak to strong one-way functions (cf. [19, Section 2.3]) can be applied. Most of these preserve anonymity. To be concrete, let us use one.

Construction 2 Let $\overline{F} = (K, \overline{S}, \overline{E})$ be obtained from F of Construction 1 by Yao's cross-product construction [34]. In detail, the key-generation algorithm is unchanged and for any key N, e we set $\mathrm{Dom}_{\overline{F}}(N,e) = \mathrm{Rng}_{\overline{F}}(N,e) = \{0,1\}^{k^2}$. Parsing a point from this domain as a sequence of k-bit strings we set $\overline{E}_{N,e}(x_1,\ldots,x_k) = (E_{N,e}(x_1),\ldots,E_{N,e}(x_k))$. The sampling algorithm is obvious and it is easy to see the family is trapdoor. ▌

Proposition 1. *The family $\overline{F}$ of Construction 2 is a perfectly anonymous family of trapdoor, one-way permutations, under the assumption that the standard RSA family is one-way.* ▌

The proof of one-wayness is a direct consequence of the known results on the security of the cross-product construction. (A proof of Yao's result can be found in [19, Section 2.3].) The anonymity is easy to see. Regardless of the key, the adversary simply gets a random string of length k^2, and can have no advantage in determining the key based on it.

The drawback of the construction is that the cross product construction is costly, increasing both the computational and the space requirements. There are alternative amplification methods that are better and in particular do not increase space requirements, but we know of none that do not increase the computational cost.

Standard methods of trapdoor permutation based encryption yield anonymous schemes provided the underlying trapdoor permutation is anonymous. This means any encryption method based on hardcore bits [21].

These methods lead to appreciable losses of concrete security, which is why we do not state concrete security versions of the results.

Acknowledgements. The UCSD authors are supported in part by Bellare's 1996 Packard Foundation Fellowship in Science and Engineering.

References

1. M. Abadi and P. Rogaway, "Reconciling two views of cryptography (The computational soundness of formal encryption)," *Proceedings of the First IFIP International Conference on Theoretical Computer Science*, LNCS Vol. 1872, Springer-Verlag, 2000.
2. M. Bellare, A. Boldyreva, A. Desai and D. Pointcheval, "Key-privacy in public-key encryption," Full version of this paper, available via `http://www-cse.ucsd.edu/users/mihir/`.
3. M. Bellare, A. Boldyreva and S. Micali, "Public-key encryption in a multi-user setting: security proofs and improvements," *Advances in Cryptology – EUROCRYPT '00*, LNCS Vol. 1807, B. Preneel ed., Springer-Verlag, 2000.

4. M. Bellare, A. Desai, E. Jokipii and P. Rogaway, "A concrete security treatment of symmetric encryption ," *Proceedings of the 38th Symposium on Foundations of Computer Science*, IEEE, 1997.
5. M. Bellare, A. Desai, D. Pointcheval and P. Rogaway, "Relations among notions of security for public-key encryption schemes," *Advances in Cryptology – CRYPTO '98*, LNCS Vol. 1462, H. Krawczyk ed., Springer-Verlag, 1998.
6. M. Bellare, J. Kilian and P. Rogaway, "The security of the cipher block chaining message authentication code," *Advances in Cryptology – CRYPTO '94*, LNCS Vol. 839, Y. Desmedt ed., Springer-Verlag, 1994.
7. M. Bellare and P. Rogaway, Random oracles are practical: a paradigm for designing efficient protocols. *First ACM Conference on Computer and Communications Security*, ACM, 1993.
8. M. Bellare and P. Rogaway, "Optimal asymmetric encryption – How to encrypt with RSA," *Advances in Cryptology – EUROCRYPT '95*, LNCS Vol. 921, L. Guillou and J. Quisquater ed., Springer-Verlag, 1995.
9. M. Blum and S. Goldwasser, "An efficient probabilistic public-key encryption scheme which hides all partial information," *Advances in Cryptology – CRYPTO '84*, LNCS Vol. 196, R. Blakely ed., Springer-Verlag, 1984.
10. J. Camenisch and A. Lysyanskaya, "Efficient non-transferable anonymous multi-show credential system with optional anonymity revocation," *Advances in Cryptology – EUROCRYPT '01*, LNCS Vol. 2045, B. Pfitzmann ed., Springer-Verlag, 2001.
11. D. Coppersmith, "Finding a small root of a bivariate integer equation; factoring with high bits known," *Advances in Cryptology – EUROCRYPT '96*, LNCS Vol. 1070, U. Maurer ed., Springer-Verlag, 1996.
12. R. Cramer and V. Shoup, "A practical public key cryptosystem provably secure against adaptive chosen ciphertext attack," *Advances in Cryptology – CRYPTO '98*, LNCS Vol. 1462, H. Krawczyk ed., Springer-Verlag, 1998.
13. A. Desai, "The security of all-or-nothing encryption: protecting against exhaustive key search," *Advances in Cryptology – CRYPTO '00*, LNCS Vol. 1880, M. Bellare ed., Springer-Verlag, 2000.
14. Y. Desmedt, "Securing traceability of ciphertexts: Towards a secure software escrow scheme," *Advances in Cryptology – EUROCRYPT '95*, LNCS Vol. 921, L. Guillou and J. Quisquater ed., Springer-Verlag, 1995.
15. D. Dolev, C. Dwork and M. Naor, "Non-malleable cryptography," *SIAM J. on Computing*, Vol. 30, No. 2, 2000, pp. 391–437.
16. T. ElGamal, "A public key cryptosystem and signature scheme based on discrete logarithms," *IEEE Transactions on Information Theory*, vol 31, 1985, pp. 469–472.
17. M. Fischlin, "Pseudorandom Function Tribe Ensembles based on one-way permutations: Improvements and applications," *Advances in Cryptology – EUROCRYPT '99*, LNCS Vol. 1592, J. Stern ed., Springer-Verlag, 1999.
18. E. Fujisaki, T. Okamoto, D. Pointcheval and J. Stern, "RSA-OAEP is Secure under the RSA Assumption," *Advances in Cryptology – CRYPTO '01*, LNCS Vol. 2139, J. Kilian ed., Springer-Verlag, 2001.
19. O. Goldreich, "Foundations of Cryptography, Basic Tools," Cambridge University Press, 2001.
20. O. Goldreich, S. Goldwasser and S. Micali, "How to construct random functions," *Journal of the ACM,* Vol. 33, No. 4, 1986, pp. 210–217.

21. O. Goldreich and L. Levin, "A hard-core predicate for all one-way functions," *Proceedings of the 21st Annual Symposium on the Theory of Computing*, ACM, 1989.
22. S. Goldwasser and S. Micali, "Probabilistic encryption," *J. of Computer and System Sciences*, Vol. 28, April 1984, pp. 270–299.
23. H. Krawczyk, "SKEME: A Versatile Secure Key Exchange Mechanism for Internet," *Proceedings of the 1996 Internet Society Symposium on Network and Distributed System Security*, 1996.
24. National Bureau of Standards, NBS FIPS PUB 81, "DES modes of operation," U.S Department of Commerce, 1980.
25. M. Naor and O. Reingold, "Number-theoretic constructions of efficient pseudo-random functions," *Proceedings of the 38th Symposium on Foundations of Computer Science*, IEEE, 1997.
26. M. Naor and M. Yung, "Universal one-way hash functions and their cryptographic applications," *Proceedings of the 21st Annual Symposium on the Theory of Computing*, ACM, 1989.
27. M. Naor and M. Yung, "Public-key cryptosystems provably secure against chosen ciphertext attacks," *Proceedings of the 22nd Annual Symposium on the Theory of Computing*, ACM, 1990.
28. RSA Labs, "PKCS-1," `http://www.rsasecurity.com/rsalabs/pkcs/pkcs-1/`.
29. C. Rackoff and D. Simon, "Non-interactive zero-knowledge proof of knowledge and chosen-ciphertext attack," *Advances in Cryptology – CRYPTO '91*, LNCS Vol. 576, J. Feigenbaum ed., Springer-Verlag, 1991.
30. K. Sako, "An auction protocol which hides bids of losers," *Proceedings of the Third International workshop on practice and theory in Public Key Cryptography (PKC 2000)*, LNCS Vol. 1751, H. Imai and Y. Zheng eds., Springer-Verlag, 2000.
31. V. Shoup, "On formal models for secure key exchange, " Technical report. Theory of Cryptography Library: 1999 Records.
32. M. Stadler, "Publicly verifiable secret sharing," *Advances in Cryptology – EUROCRYPT '96*, LNCS Vol. 1070, U. Maurer ed., Springer-Verlag, 1996.
33. Y. Tsiounis and M. Yung, "On the security of El Gamal based encryption," *Proceedings of the First International workshop on practice and theory in Public Key Cryptography (PKC'98)*, LNCS Vol. 1431, H. Imai and Y. Zheng eds., Springer-Verlag, 1998.
34. A. Yao, "Theory and applications of trapdoor functions, " *Proceedings of the 23rd Symposium on Foundations of Computer Science*, IEEE, 1982.

Provably Secure Fair Blind Signatures with Tight Revocation

Masayuki Abe[1] and Miyako Ohkubo[2]

[1] NTT Information Sharing Platform Laboratories. 1-1 Hikari-no-oka, Yokosuka-shi, 239-0847 JAPAN
abe@isl.ntt.co.jp
[2] NTT East. A-15F Shinagawa InterCity, 2-15-1 Kounan, Minato-ku, Tokyo, 108-6015 JAPAN
ookubo.miyako@east.ntt.co.jp

Abstract. A fair blind signature scheme allows the trustee to revoke blindness so that it provides authenticity and anonymity to honest users while preventing malicious users from abusing the anonymity to conduct blackmail etc. Although plausible constructions that offer efficient tricks for anonymity revocation have been published, security, especially one-more unforgeability and revocability against adaptive and parallel attacks, has not been studied well. We point out a concrete vulnerability of some of the previous schemes and present an efficient fair blind signature scheme with a security proof against most general attacks. Our scheme offers tight revocation where each signature and issuing session can be linked by the trustee.

1 Introduction

Fair blind signature schemes are a variant of blind signature schemes; they allow a trustee to revoke the blindness in such ways that

- given a view of a signature issuing session conducted with an authenticated user, the trustee can identify the resulting signature (Signature Tracing), or
- given a signature, the trustee can identify the issuing session that yielded the signature, which eventually identifies the user who conducted the session (Session Tracing).

Such schemes will play an important role in applications that must offer both privacy and authenticity while preventing users from abusing anonymity. See [25] for a concrete example.

The notion of fair blind signatures was introduced independently in [6,9] for the construction of anonymous electronic payment schemes. Since then, some efficient constructions have been shown [23,7] and several different approaches to the same goal have been taken [12,16]. These previous schemes provide efficient

C. Boyd (Ed.): ASIACRYPT 2001, LNCS 2248, pp. 583–601, 2001.
© Springer-Verlag Berlin Heidelberg 2001

revocation mechanisms but their security, especially in terms of revocability and unforgeability against adaptive and parallel attacks, has not been rigorously studied. Indeed, even the security of ordinary blind signatures against parallel attacks has been studied formally only in recent works [20,17,22,2,1].

In some schemes, revocation is limited to linking a signature to its owner. There are some other schemes that allow a signature to be linked to a particular issuing session. Such a fine revocation, for instance, allows one to know the issuing time of the target signature from the session log. Typically, revocation in this type of schemes reveals the randomness generated by the user during the issuing session. Accordingly, if a malicious user broadcasts a value via the Internet and encourages all other users to use it as the random parameter in issuing sessions, revocation becomes useless. Some known schemes, e.g. [16,7, 15], are vulnerable against this attack, or they implicitly resort to on-the-fly freshness checking, which is expensive in practice.

Our contribution is an efficient fair-blind signature scheme that is secure against adaptive and parallel attacks. Assuming the existence of ideal hash functions [5], its blindness is proven under the decision Diffie-Hellman assumption, and revocability and one-more unforgeability against adaptive and parallel attacks are proven under the discrete logarithm assumption. Another advantage of our scheme is that it offers tight revocation. That is, given a signature, revocation identifies the issuing session that uniquely produced the signature, and, given a session view, revocation identifies the unique signature created in the session. Naturally, once such tight revocability is achieved, the scheme also provides one-more unforgeability since tight and bi-directional revocability guarantees one-to-one mapping between issuing sessions and resulting signatures.

The rest of this paper is organized as follows. Section 2 defines the security of fair blind signatures. Section 3 reviews underlying ideas and building blocks. Section 4 presents our scheme in detail. A security analysis is given in Section 5. Section 6 gives several remarks. It includes weakness of our scheme, modifications, and open problems.

2 Definitions

Let $(\mathcal{G}_S, \mathcal{S}, \mathcal{U}, \mathcal{V})$ be a blind signature scheme where $\mathcal{G}_S$ is a signing key generation algorithm, $\mathcal{S}$ and $\mathcal{U}$ are interactive Turing machines called signer and user, and $\mathcal{V}$ is a signature verification algorithm. (Please refer to [17,22] for a formal functional definition of blind signature schemes.) Informally, a fair blind signature scheme with off-line trustee is a blind signature scheme with five additional probabilistic polynomial-time algorithms, $\mathcal{G}_T$, $\mathcal{R}_{sig}$, $\mathcal{R}_{sid}$, $\mathcal{M}_{sig}$, and $\mathcal{M}_{sid}$ as follows.

$\mathcal{G}_T$ is a revocation key generation algorithm that takes a public key of a signer, say *pk*, and outputs a private and public revocation key pair. The keys can be independent of the public key of the signer (thus only one revocation key pair for all signers); $(rsk, rpk) \leftarrow \mathcal{G}_T(1^n, pk)$.

$\mathcal{R}_{sig}$ is a revocation algorithm that generates signature identifier I_{sig} that identifies the signature yielded from the target session. It takes the view of the signer during the target session and revocation key; $I_{sig} \leftarrow \mathcal{R}_{sig}(view_i, rsk)$.

$\mathcal{R}_{sid}$ is a revocation algorithm that generates session identifier I_{sid} that identifies the session that has produced target signature-message pair Σ_m. $I_{sid} \leftarrow \mathcal{R}_{sid}(\Sigma_m, rsk)$.

$\mathcal{M}_{sig}$ is a matching algorithm that examines whether I_{sig} matches to signature-message pair Σ_m or not. It outputs 1 if they match, 0 otherwise; $0/1 \leftarrow \mathcal{M}_{sig}(I_{sig}, \Sigma_m)$.

$\mathcal{M}_{sid}$ is a matching algorithm that examines whether I_{sid} matches to $view_i$ or not. It outputs 1 if they match, 0 otherwise; $0/1 \leftarrow \mathcal{M}_{sid}(I_{sid}, view_i)$.

These algorithms also take public data such as pk and rpk if needed. Although $view_i$ include everything that the signer can see during the session, which includes his own private key, what is really necessary to complete revocation differs $\mathcal{M}_{ss}$ differ depending on the specific revocation mechanism used.

We start the security definitions with traceability. Intuition states that a scheme is session traceable if no adversary can output a signature that can not be associated with the corresponding session, or can be associated with more than two sessions by revocation. Accordingly, it assures that each valid signature should be linked to a single session. Similarly, a scheme is signature traceable if no adversary can output two signatures that will be associated to the same session. Hence, it assures that every session should be linked to a single valid signature. If a scheme provides both types of traceability, shown below, we say that the scheme offers tight revocation.

Definition 1. *(Signature Traceability) A fair blind signature scheme is signature traceable if, for any probabilistic polynomial-time algorithm $\mathcal{U}^*$ that, after interacting with legitimate signer $\mathcal{S}$ at most ℓ times in an adaptive and arbitrarily interleaving manner, outputs*

- *a valid signature-message pair, say Σ_m, such that, for $I_{sig} = \mathcal{R}_{sig}(view_i, rsk)$, $\mathcal{M}_{sig}(I_{sig}, \Sigma_M) = 0$ holds for all $i = 1, \ldots, \ell$, or*
- *two valid and different signature-message pairs, say Σ_{m0}, Σ_{m1}, such that, there exists i in $1, \ldots, \ell$ such that $\mathcal{M}_{sig}(I_{sig}, \Sigma_{m0}) = \mathcal{M}_{sig}(I_{sig}, \Sigma_{m1}) = 1$ where $I_{sig} = \mathcal{R}_{sig}(view_i, rsk)$,*

with probability at most $1/n^c$ for sufficiently large n and some constant c. The probability is taken over the coin flips of $\mathcal{G}_S$, $\mathcal{G}_T$, $\mathcal{S}$, and $\mathcal{U}^*$.

Definition 2. *(Session Traceability) A fair blind signature scheme is session traceable if, for any probabilistic polynomial-time algorithm $\mathcal{U}^*$ that, after interacting with legitimate signer $\mathcal{S}$ at most ℓ times in an adaptive and arbitrarily interleaving manner, outputs a valid signature-message pair Σ_m such that*

- *for $I_{sid} = \mathcal{R}_{sid}(\Sigma_m, rsk)$, $\mathcal{M}_{sid}(I_{sid}, view_i) = 0$ holds for all $i = 1, \ldots, \ell$, or*
- *there exists $i, j, i \neq j$ such that $\mathcal{M}_{sid}(I_{sid}, view_i) = \mathcal{M}_{sid}(I_{sid}, view_j) = 1$,*

with probability at most $1/n^c$ *for sufficiently large* n *and some constant* c. *The probability is taken over the coin flips of* $\mathcal{G}_S$, $\mathcal{G}_T$, $\mathcal{S}$, *and* $\mathcal{U}^*$.

Note that, in the random oracle model, these success probabilities also depend on the choice of random oracles.

Next is blindness, which informally means that any adversary that colludes with the signer can distinguish two session views only with negligible advantage when one of the views results in a given signature.

Definition 3. *(Blindness) Let* $\mathcal{S}^*$ *and* $\mathcal{D}^*$ *be probabilistic poly-time algorithms that play the following game with honest user* $\mathcal{U}_0$ *and* $\mathcal{U}_1$.

1. $(pk, sk) \leftarrow \mathcal{G}_S(1^n)$, $(rsk, rpk) \leftarrow \mathcal{G}_T(1^n, pk)$
2. $(\mathsf{msg}_0, \mathsf{msg}_1) \leftarrow \mathcal{S}^*(sk, rpk)$
3. *For* $b \in_U \{0,1\}$, msg_b *is given to* $\mathcal{U}_0$, *and* msg_{1-b} *is given to* $\mathcal{U}_1$.
4. $\mathcal{S}^*$ *engages in the signature issuing protocol with* $\mathcal{U}_0$, $\mathcal{U}_1$ *in arbitrary order.*
5. *Resulting signature* Σ_0 *for* msg_0 *is given to* $\mathcal{D}^*$. $\mathcal{D}^*$ *also allowed to take any information from* $\mathcal{S}^*$.
6. $\mathcal{D}^*$ *outputs* $b' \in \{0,1\}$.

The signature scheme is blind if, for all polynomial-time $\mathcal{S}^*$ *and* $\mathcal{D}^*$, $b' = b$ *happens with probability at most* $1/2 + 1/n^c$ *for sufficiently large* n *and some constant* c. *The probability is taken over the coin flips of* $\mathcal{G}_T$, $\mathcal{G}_S$, $\mathcal{S}^*$, $\mathcal{D}^*$ *and* $\mathcal{U}_0$, $\mathcal{U}_1$ *and* b.

Finally, we define one-more unforgeability in such a sense that it is infeasible to output $\ell + 1$ valid signatures after interacting with the signer ℓ times.

Definition 4. *(One-more unforgeability) A blind signature scheme is* $(\ell, \ell+1)$ *unforgeable if, for any probabilistic polynomial-time algorithm* $\mathcal{U}^*$, $\mathcal{U}^*$ *outputs* $\ell + 1$ *valid signatures with probability at most* $1/n^c$ *for sufficiently large* n *and some constant* c *after interacting with legitimate signer* $\mathcal{S}$ *at most* ℓ *times. The interaction can be done in an adaptive and arbitrarily interleaving manner. The probability is taken over the coin flips of* $\mathcal{G}$, $\mathcal{S}$, *and* $\mathcal{U}^*$.

It is important to see that if a scheme provides tight revocability, the scheme is one-more unforgeable since tight revocability assures that there is one-to-one correspondence between successful sessions and valid signatures. Accordingly, it suffice to prove blindness and tight revocability for our scheme.

The above definitions are weak since the adversaries have no access to the trustee. Thus it is important for the trustee not to show the tracing information to anybody to prevent the adversaries from using the trustee as an oracle. When revocation is done only for private purposes such as criminal investigation, such weak definitions may suffice. Although our scheme provides security only in a weak sense, one can define a stronger notion of security by modifying the above definitions. Informally, the scheme provides *strong* signature/session traceability if traceability is retained even if the private revocation key *rsk* is given to $\mathcal{U}^*$ in Definition 1 and 2. Similarly, we say a scheme provides strong blindness if blindness is retained even if $\mathcal{S}^*$ and $\mathcal{D}^*$ are allowed to ask the trustee for revocation except for the sessions and the signature in question.

3 Underlying Idea and Building Blocks

3.1 Efficient Revocation Mechanism

We take an approach similar to that introduced in [24,7]. Let $x_t, y_t(= g^{x_t})$ be the revocation key pair. Let z be a part of the signer's public key. To ask a signature, the user sends $(z^{1/\gamma}, g^{\gamma})$ to the signer where γ is a blinding factor that will be used later in blinding. The signer then blindly issues a signature bringing a pair $(z^{1/\gamma}, y_t)$ into the issuing protocol in such a way that a valid signature can be obtained only if the pair is blinded into $(z, {y_t}^{\gamma})$. The user can get a valid signature as he can do the conversion by taking the γ-th power. The signer is left blind since z is common to all signatures and $(y_t, g^{\gamma}, {y_t}^{\gamma})$ is assumed to be indistinguishable from $(y_t, g^{\gamma}, {y_t}^{\gamma'})$ with random γ' used for another signature.

Given a signature that contains ${y_t}^{\gamma}$, the trustee can trace the session that contains g^{γ} by computing $({y_t}^{\gamma})^{1/x_t}(= g^{\gamma})$. Similarly, given a session log that contains g^{γ}, the trustee can trace the resulting signature that must contain ${y_t}^{\gamma}$ by computing $(g^{\gamma})^{x_t}(= {y_t}^{\gamma})$.

For the above revocation mechanism to function, we must be sure that blinding by exponentiation, $(z^{1/\gamma}, y_t) \to (z, y_t^{\gamma})$, is the only way to get a valid signature. A blind signature scheme from [1] suits this purpose. As well as its security against adaptive and parallel attacks, one good property we can exploit is the restrictive blinding property. That is, when the signer issues a signature based on $(\mathsf{z}, \mathsf{z}_1)$ a user has to blind it into $(\mathsf{z}^{\gamma}, \mathsf{z}_1^{\gamma})$ to have the signature correctly blinded. So if we set $(\mathsf{z}, \mathsf{z}_1) = (z^{1/\gamma}, y_t)$, it must be transformed into $(z, {y_t}^{\gamma})$.

This trick, however, offers tight revocation only if all users are honest in choosing a unique γ in each session. Our idea for tight revocation is to add extra randomness v to the blinding factor from the signer's side so that ${y_t}^{\gamma v}$ is involved in the signature. With this adaptation, the signer can randomize blinding factor γ chosen by the user into γv so that it is unique in every session.

3.2 Verifiable Encryption of DL

For the reduction in our security proof to work, we need the trustee (simulator) to be able to extract not only ${y_t}^{\gamma}$ but also γ itself. For this purpose, a user encrypts γ with the public encryption key of the trustee and proves that γ can certainly be recovered from the ciphertext. Generally speaking, an encryption scheme accompanied by a non-interactive proof that assures the receiver that the embedded plaintext satisfies some poly-time computable predicate is often called a verifiable encryption scheme. Concrete examples can be seen in the literature, e.g. [4,3,8].

Let $C = (z_u, \xi) = (z^{1/\gamma}, g^{\gamma})$ be a commitment of witness γ. Let $(\mathcal{G}_E, \mathcal{E}, \mathcal{D})$ be a public-key encryption scheme. Let $(ek, dk) \leftarrow \mathcal{G}_E(1^k)$ and $E \leftarrow \mathcal{E}_{ek}(\gamma; \omega)$ where ω is a random tape. Let $\mathcal{R}$ be a relation between C and E such that

$$(C, E) \in \mathcal{R} \Leftrightarrow \log_{z_u} z \equiv \log_g \xi \equiv \mathcal{D}_{dk}(E) \bmod q.$$

Let $(\mathcal{P}, \mathcal{V})$ be a non-interactive zero-knowledge proof (argument) system for relation $\mathcal{R}$ such that $P \leftarrow \mathcal{P}(C, E, \gamma, \omega, ek)$ and $0/1 \leftarrow \mathcal{V}(C, E, P)$. We assume that it provides correctness, soundness, and computational zero-knowledge. Note that when it is zero-knowledge argument the soundness is conditionally achieved under some intractability assumptions.

On top of this standard security, we need it to be *simulatable* in such a sense that, for $C = (z^{1/\gamma}, g^{\gamma})$, there exists a poly-time simulator which, without being given γ and dk, outputs $(\tilde{E}, \tilde{P})$ such that $(C, \tilde{E}) \notin R$ and $(\tilde{E}, \tilde{P})$ is computationally indistinguishable from correct (E, P) that satisfies $(C, E) \in \mathcal{R}$ and $\mathcal{V}(C, E, P) = 1$. We say that a verifiable encryption scheme is secure and simulatable if it provides all these properties. Note that we only consider passive adversaries who have no access to the decryption oracle. When the encryption scheme is semantically secure against chosen plaintext attacks and the proof system is a public-coin honest verifier zero-knowledge proof made non-interactive with the Fiat-Shamir technique [11], simulatability is provided under the embedded assumption for the semantic security of the encryption scheme and the random oracle assumption.

Appendix A and B show two examples of verifiable encryption that provide all of the security properties we need in our construction. These schemes have different flavors. The scheme in Appendix A is taken from [3] and is based on Okamoto-Uchiyama encryption [19] combined with the statistical zero-knowledge argument of [14]. In this scheme, it is assumed that the decryption key is not given to the adversary in order to assure soundness. Accordingly, if this scheme is integrated in our construction, one has to assume that the trustee and the users are not colluding. The second scheme in Appendix B is newly constructed based on ElGamal encryption and a log-round perfect zero-knowledge proof. Though its efficiency is worse than that of the first one, this scheme provides a stronger property in that soundness holds even if the decryption key of the trustee is given to the adversary.

4 Our Scheme

[Signing Key Generation]

Let $\mathcal{G}$ be a probabilistic polynomial-time algorithm that generates a group parameter, (p, q, g, h) where p, q are primes and g, h are generators of subgroup of order q in $\mathbb{Z}_p{}^*$. A signer selects three hash functions $\mathcal{H}_1 : \{0,1\}^* \to \langle g \rangle$, $\mathcal{H}_{2,3} : \{0,1\}^* \to \{0,1\}^{|q|}$ and generates public-key $pk = (p, q, g, h, y, z)$ and private-key $sk = (x)$ as follows;

$$
\begin{aligned}
&(p, q, g, h) \leftarrow \mathcal{G}(1^n), \\
&x \in_U \mathbb{Z}_q, \\
&y = g^x \bmod p, \\
&z = \mathcal{H}_1(p, q, g, h, y).
\end{aligned}
$$

All arithmetic operations are done in $\langle g \rangle$ hereafter unless otherwise noted.

[Revocation Key Generation]

Given the public key of a signer, the trustee generates secret-key $rsk = (x_t, dk)$ and public-key $rpk = (y_t, ek)$ where $x_t \in_U \mathbb{Z}_q^*$, $y_t = g^{x_t}$, and ek, dk are the key pair for verifiable encryption scheme described in Section 3.2.

Depending on the encryption algorithm $\mathcal{E}$ used for verifiable encryption, (ek, dk) can be common for all signers. Similarly, if (p, q, g) are common as system-wide parameters, x_t, y_t can be common, too.

[Signature Generation]

Here, we describe the signature issuing protocol in a higher level. Details can be found in Figure 1.

1. The user chooses blinding factor γ and computes $z_u = z^{1/\gamma}$ and $\xi = g^{\gamma}$. He then executes verifiable encryption where γ is encrypted into E and the relation among z_u, ξ, E is proven by providing P.
2. The signer verifies (E, P). He generates v randomly, and computes $z_1 = {y_t}^v$ and $z_2 = z_u/z_1$. He then proves to the user that z_1 is made as it should be by providing Schnorr zero-knowledge proof $P_s = (\sigma_s, c_s)$ where $c_s = \mathcal{H}_3(z_1 \| {y_t}^{r_s})$ and $\sigma_s = r_s - c_s v \bmod q$ for $r_s \in_U \mathbb{Z}_q$. The proof will be verified by the user as $c_s \stackrel{?}{=} \mathcal{H}_3(z_1 \| {y_t}^{\sigma_s} {z_1}^{c_s})$.
3. Based on y, z_1, z_2, the signer and the user engages in an interactive proof protocol. For the signer, the protocol is a witness indistinguishable proof of knowledge of
$$\log_g y \vee (\log_g z_1 \wedge \log_h(z_u/z_1)).$$
The signer converts the proof into the one for
$$\log_g y \vee (\log_g \zeta_1 \wedge \log_h(z/\zeta_1))$$
by exponentiating $(z_1, z_u) \xrightarrow{\gamma} (\zeta_1, z)$ and blinding it with the standard diversion technique [18]. The converted proof is eventually transformed to a signature with Fiat-Shamir technique.
4. The signer stores ξ^v as the identity of this session.
5. The user outputs a signature, $\Sigma = (\zeta_1, \rho, \varpi, \sigma_1, \sigma_2, \delta)$ for message m.

Note that ξ^v can be published, though it is not necessary to the user. The signer may provide extra Schnorr zero-knowledge proof that proves $\log_\xi(\xi^v) = \log_{y_t} z_1$.

[Verification]

A signature-message pair, (Σ, m), is valid if it satisfies

$$\varpi + \delta \stackrel{?}{=} \mathcal{H}_2(\zeta_1 \| g^\rho y^\varpi \| g^{\sigma_1} {\zeta_1}^\delta \| h^{\sigma_2} (z/\zeta_1)^\delta \| \mathsf{m}) \bmod q. \tag{1}$$

[Revocation]

Signature Tracing: Given valid (z_u, ξ, E, P) and ξ^v, the trustee computes $I_{sig} = (\xi^v)^{x_t}$. Observe that

$$I_{sig} = (\xi^v)^{x_t} = g^{\gamma v x_t} = {y_t}^{\gamma v} = \zeta_1. \tag{2}$$

Thus, I_{sig} identifies the resulting signature.

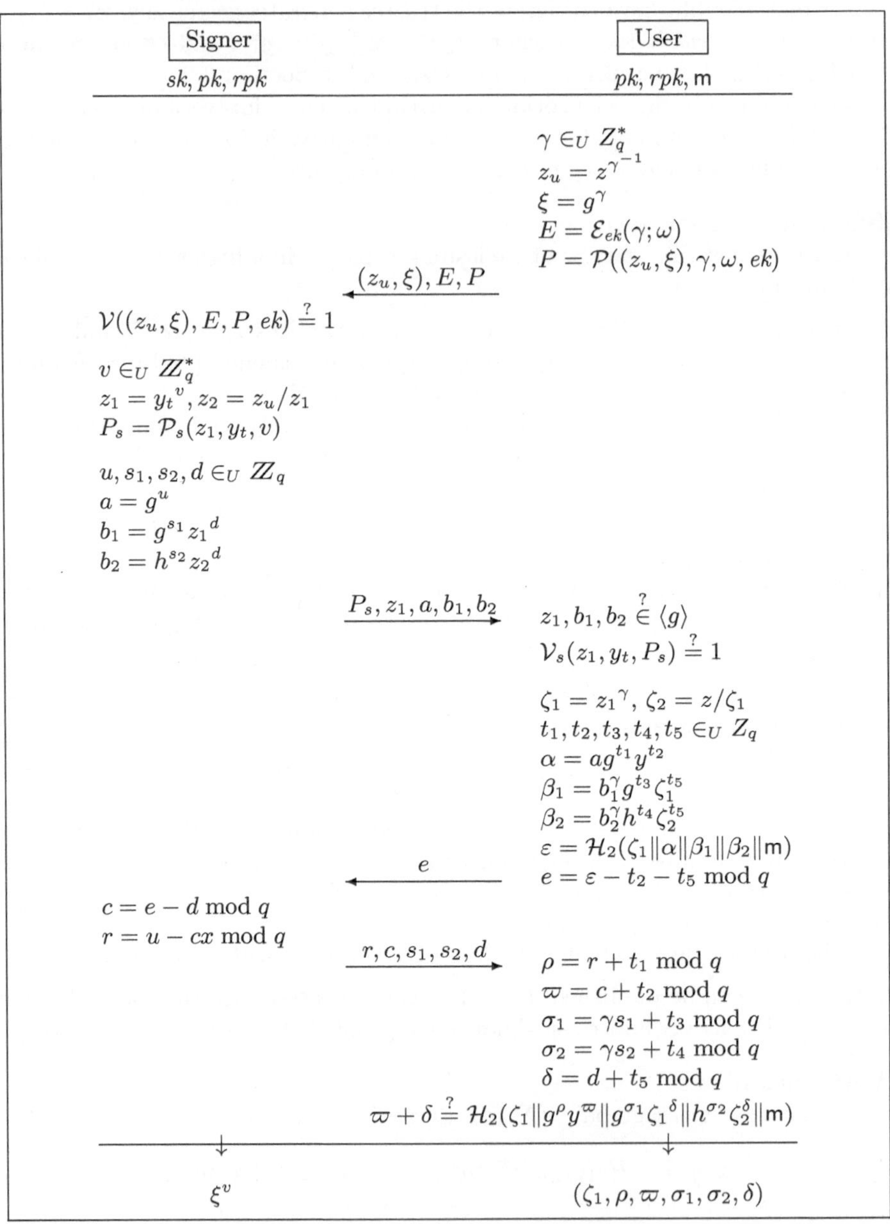

Fig. 1. The signature issuing protocol. The session aborts if any of the checks ($\stackrel{?}{=}, \stackrel{?}{\in}$) fails. $\mathcal{E}, \mathcal{P}$ are from the underlying verifiable encryption scheme of Section 3.2. ($\mathcal{P}_s, \mathcal{V}_s$) is a Schnorr-type proof of knowledge of v w.r.t y_t and z_1. The trustee is off-line, i.e., not involved in the issuing protocol.

Session Tracing: Given a valid signature, the trustee computes $I_{ss} = \zeta_1^{1/x_t}$. Observe that

$$I_{ss} = {\zeta_1}^{1/x_t} = {z_1}^{\gamma/x_t} = {y_t}^{v\gamma/x_t} = g^{v\gamma} = \xi^v. \quad (3)$$

Since ξ^v is stored or published by the signer, I_{ss} identifies the session that issued the signature.

5 Security Proofs

5.1 Correctness

Theorem 1. *If the signer and the user follow the issuing protocol, the protocol completes with a valid signature with probability 1.*

Proof. There are four verifications denoted by $\stackrel{?}{=}$ in the issuing protocol. The verification for P and P_s in each side will accept the proof with probability 1 due to the correctness of these proof systems. It is clear that z_1, b_1, b_2 are in $\langle g \rangle$. For the last one, which is equivalent to the verification predicate, observe that the following holds.

$$\begin{aligned}
\varpi + \delta &= c + t_2 + d + t_5 = e + t_2 + t_5 = \varepsilon \pmod q\\
g^\rho y^\varpi &= g^{r+t_1} y^{c+t_2} = g^{r+cx} g^{t_1} y^{t_2} = a g^{t_1} y^{t_2} = \alpha\\
g^{\sigma_1} {\zeta_1}^\delta &= g^{\gamma s_1 + t_3} {z_1}^{\gamma\delta} = (g^{s_1} {z_1}^d)^\gamma g^{t_3} {z_1}^{\gamma t_5} = {b_1}^\gamma g^{t_3} {\zeta_1}^{t_5} = \beta_1\\
h^{\sigma_2} {\zeta_2}^\delta &= h^{\gamma s_2 + t_4} {\zeta_2}^{d+t_5} = h^{\gamma s_2 + t_4} (z_u/z_1)^{\gamma d} (z/\zeta_1)^{t_5} = {b_2}^\gamma h^{t_4} {\zeta_2}^{t_5} = \beta_2
\end{aligned}$$

Thus, the protocol always stops with a valid signature if both parties follow the protocol. □

5.2 Blindness

Theorem 2. *The proposed scheme is blind if all hash functions are random oracles, the decision Diffie-Hellman problem is intractable, and the underlying verifiable encryption scheme is secure and simulatable in the random oracle model.*

Proof. Suppose that $(\mathcal{S}^*, \mathcal{D}^*)$ is successful in breaking blindness with probability $1/2+\epsilon$ where ϵ is not negligible. We show that $\mathcal{S}^*$ and $\mathcal{D}^*$ can be used to solve the DDH problem. Define $DH = \{(X_1, X_2, X_3) \in \langle g \rangle^3 | \log_g X_1 \log_g X_2 = \log_g X_3\}$ and $RND = \{(X_1, X_2, X_3) \in \langle g \rangle^3\}$. Let $(A, B, C) \in \langle g \rangle^3$ be a DDH instance, i.e., taken from DH or RND with equal probability. Let $(A, B, C) = (g^{\mathbf{a}}, g^{\mathbf{b}}, g^{\mathbf{c}})$. If any of $\mathbf{a}, \mathbf{b}, \mathbf{c}$ is zero, we can immediately determine whether the instance is in DH or not. So we assume that none of them are zero hereafter.

Simulation proceeds as follows. We simulate hash function $\mathcal{H}_1$ so that it outputs $B^{r_{1i}}$ by selecting $r_{1i} \in_U \mathbb{Z}_q^*$ for each fresh query. Suppose that r_1 is selected for $z = \mathcal{H}_1(p, q, ...) = B^{r_1}$. Next choose $r_2 \in_u \mathbb{Z}_q^*$ and set the revocation public key as $y_t = A^{r_2}$. Select $\mathsf{d} \in_U \{0, 1\}$ and execute the issuing protocol with

$\mathcal{S}^*$ twice. Label the executions run_0 and run_1. In $\mathsf{run}_{1-\mathsf{d}}$, we simply follow the protocol. In $\mathsf{run}_{\mathsf{d}}$, we first set $z_u = g^{r_1}$ and $\xi = B$. Observe that z, z_u, and ξ are perfectly simulated no matter whether (A, B, C) is from *DH* or *RND* since z, z_u, ξ satisfies $\log_{z_u} z = \log_g \xi = (\log_g B)$. We then simulate E by encrypting $r_3 \in_U \mathbb{Z}_q^*$. Since $r_3 \neq \log_g B$ in general, $((z_u, \xi), E) \notin \mathcal{R}$. However, the simulator can produce P in such a way that (E, P) is computationally indistinguishable from the real ones since we assume that the underlying verifiable encryption is simulatable in such sense. Now send z_u, ξ, E, P and receive P_s, z_1 and etc from S^*. At this point, we rewind S^* to extract v from P_s by applying the Forking Lemma [21]. We then continue and complete the issuing session.

For message m_0 is given by $\mathcal{S}^*$ at the beginning, the simulator generates a signature-message pair, say (Σ, m_0), with regard to $\zeta_1 = C^{r_2 v}$. Other variables except for ζ_1 in Σ are generated by using the standard zero knowledge simulation technique; randomly choose $\rho, \varpi, \sigma_1, \sigma_2, \delta$, and then freely define $\mathcal{H}_2$ so that they look consistent. Given (Σ, m) and views from $\mathcal{S}^*$, distinguisher $\mathcal{D}^*$ outputs d'. If $\mathsf{d}' = \mathsf{d}$, we conclude that (A, B, C) is in *DH*. It is in *RND*, otherwise.

We now claim that if $(A, B, C) \in DH$, Σ is a valid signature that could have been produced in $\mathsf{run}_{\mathsf{d}}$. Observe that, for z, z_1 used in run_b,

$$(z_u, z, z_1, \zeta_1) = (g^{r_1}, g^{\mathbf{b} r_1}, g^{\mathbf{a} r_2 v}, g^{\mathbf{c} r_2 v}).$$

So if $\mathbf{ab} = \mathbf{c}$, we have a consistent blinding factor, $\gamma = \mathbf{b}$ which satisfies $\gamma = \log_{z_u} z = \log_{z_1} \zeta_1$ for z_u and z_1 used in $\mathsf{run}_{\mathsf{d}}$. Furthermore, there are blinding factors t_1, t_2, t_3, t_4, t_5 that convert the view of $\mathsf{run}_{\mathsf{d}}$ into the remaining elements in Σ. On the other hand, Σ could have been produced by $\mathsf{run}_{1-\mathsf{d}}$ only with negligible probability as z_u, z_1 should differ in $\mathsf{run}_{1-\mathsf{d}}$. Accordingly, given Σ, $\mathcal{D}^*$ outputs $\mathsf{d}' = \mathsf{d}$ with probability $1/2 + \epsilon$.

Next, we claim that if $(A, B, C) \in RND$, Σ is statistically independent of the views of the signer in run_0 and run_1 since $\log_{z_u} z \neq \log_{z_1} \zeta_1$ holds for (z_u, z_1) in both runs except with negligible probability. Hence, d is also statistically independent of the view of the signer, and $\mathsf{d}' = \mathsf{d}$ happens with probability close to $1/2$ except for a negligible fraction.

In total, the success probability is $1/2(1/2+\epsilon)+1/2(1/2) = 1/2+\epsilon/2$, which contradicts the DDH assumption when ϵ is not negligible.

□

5.3 Tight Revocability

Theorem 3. *The proposed scheme is session traceable if all hash functions are random oracles, the discrete logarithm is intractable, and the soundness condition of the underlying verifiable encryption scheme holds.*

Proof. Here we must show two properties. We first show that it is infeasible for a user to produce a signature $\Sigma^\star = (\zeta_1, \rho, \varpi, \sigma_1, \sigma_2, \delta)$ such that $\log_{z_u} z \neq \log_{z_1} \zeta_1$ for all (z_u, z_1) used in issuing sessions. We then show that a valid signature cannot be linked to more than one session.

Assume that having at most q_h accesses to $\mathcal{H}_2$ and asking at most ℓ signatures to $\mathcal{S}$, $\mathcal{U}_0^*$ outputs signature $\Sigma^\star = (\zeta_1, \rho, \varpi, \sigma_1, \sigma_2, \delta)$ that satisfies $\log_{z_u} z \neq \log_{z_1} \zeta_1$ for (z_u, z_1) used in any session. Here, q_h and ℓ are bound by a polynomial of security parameter n. Let ϵ_0 be the success probability of $\mathcal{U}_0^*$, which is not negligible in n. We randomly fix an index $Q \in \{1, \ldots, q_h\}$ and regard $\mathcal{U}_0^*$ as successful only if the resulting signature corresponds to the Q-th query to $\mathcal{H}_2$. (If it does not correspond to any query, $\mathcal{U}_0^*$ is successful only with negligible probability due to the randomness of $\mathcal{H}_2$.) Accordingly, it is equivalent to assuming an adversary, say $\mathcal{U}_1^*$, that asks $\mathcal{H}_2$ only once and succeeds with probability $\epsilon_1 \geq \epsilon_0/q_h$. By using $\mathcal{U}_1^*$, we construct machine $\mathcal{M}_1$ that solves the discrete-log problem. Let $(\mathbf{p}, \mathbf{q}, \mathbf{g}, \mathbf{Y})$ be an instance of the discrete-log problem to solve $\mathbf{X} = \log_{\mathbf{g}} \mathbf{Y}$ in $\mathbb{Z}_{\mathbf{q}}$.

Reduction Algorithm: $\mathcal{M}_1$ first sets $(p, q, g) := (\mathbf{p}, \mathbf{q}, \mathbf{g})$. It also generates key pair (dk, ek) for the underlying verifiable encryption scheme. It then flips a coin $\chi \in_U \{0, 1\}$ to select either $y := \mathbf{Y}$ (case $\chi = 0$) , or $h := \mathbf{Y}$ (case $\chi = 1$).

Case $\chi = 0$:

Intuition: We set $y = \mathbf{Y}$ and attempt to extract the y-side witness by simulating the signing oracle with z-side witness, which is $\log_g z_1$ and $\log_h z_2$. We run $\mathcal{U}_1^*$ twice with a different answer from $\mathcal{H}_2$ and apply the Forking Lemma. It should cause a change of either δ or ϖ in the resulting signatures. If we are lucky, we have different ϖ's and can extract the y-side witness.

1. $\mathcal{M}_1$ sets $y = \mathbf{Y}$.
2. $\mathcal{M}_1$ selects $w, w_0, w_1 \in_U \mathbb{Z}_q^*$ and sets $h := g^w$, $z := \mathcal{H}_1(p\|q\|g\|y) = g^{w_0}$, and $y_t = g^{w_1}$.
3. $\mathcal{M}_1$ runs $\mathcal{U}_1^*$ and simulates $\mathcal{S}$ for i-th query in the following way.
   a) Given $(z_{ui}, \xi_i, E_i, P_i)$ from the user, check P_i and reject if incorrect. Otherwise, decrypt $E_i \to \gamma_i$.
   b) Compute $a_i := g^{r_i} y^{c_i}$ for $c_i, r_i \in_U \mathbb{Z}_q$.
   c) Compute $w_{1i} = w_1 v_i \bmod q$ and $w_{2i} = (w_0/\gamma_i - w_{1i})/w \bmod q$ for $v_i \in_U \mathbb{Z}_q^*$. Then set $z_{1i} = g^{w_{1i}}$ and $z_{2i} = h^{w_{2i}}$.
   d) Compute P_{si} by using legitimate witness v_i.
   e) Compute $b_{1i} := g^{u_{1i}}$ and $b_{2i} := h^{u_{2i}}$ with $u_{1i}, u_{2i} \in_U \mathbb{Z}_q$.
   f) Send $P_{si}, a_i, b_{1i}, b_{2i}$ to $\mathcal{U}_1^*$.
   g) Given e_i from $\mathcal{U}_1^*$, compute $d_i := e_i - c_i \bmod q$, $s_{1i} := u_{1i} - d_i w_{1i} \bmod q$, and $s_{2i} := u_{2i} - d_i w_{2i} \bmod q$.
   h) Send $r_i, c_i, s_{1i}, s_{2i}, d_i$ to $\mathcal{U}_1^*$.

   $\mathcal{M}_1$ simulates $\mathcal{H}_2$ by returning $\varepsilon \in_U \mathbb{Z}_q$.
4. $\mathcal{U}_1^*$ outputs a signature, say $(\zeta_1, \rho, \varpi, \sigma_1, \sigma_2, \delta)$, that corresponds to ε.
5. Reset and restart $\mathcal{U}_1^*$ with the same setting. $\mathcal{M}_1$ simulates $\mathcal{H}_2$ with $\varepsilon' \in_U \mathbb{Z}_q$. In this second run, $\mathcal{M}_1$ also uses the same random tape.
6. $\mathcal{U}_1^*$ outputs a signature, say $(\zeta_1, \rho', \varpi', \sigma_1', \sigma_2', \delta')$, that corresponds to ε'.
7. If $\varpi \neq \varpi'$, $\mathcal{M}_1$ outputs $X := (\rho - \rho')/(\varpi' - \varpi) \bmod q$. The simulation fails, otherwise.

Case $\chi = 1$:

Intuition: We set $h = \mathbf{Y}$, $z = g^{w_1} h^{w_2}$ with random w_1, w_2, and attempt to extract different representation of z, that leads $\log_g h$. The signing oracle is simulated with y-side witness except for one query. For the one randomly chosen J-th query, we use y-side witness and z-side witness, i.e., (w_1, w_2), together. We rewind $\mathcal{U}_1^*$ to apply the Forking Lemma. But this time, we fork the process by changing d in the J-th issuing session, which is used as a challenge to the z-side proof. We can answer to two different d's in the J-th session since the z-side witness in this session is (w_1, w_2). Now if δ is sensitive to the change of d, we have different δ's and can extract the z-side witness which is different from (w_1, w_2).

1. $\mathcal{M}_1$ sets $h = \mathbf{Y}$.
2. $\mathcal{M}_1$ selects $x \in_U \mathbb{Z}_q$ and sets $y := g^x$. It also selects $w_1, w_2 \in_U \mathbb{Z}_q$ and sets $z := \mathcal{H}_1(p \| q \| g \| y) = g^{w_1} h^{w_2}$.
3. $\mathcal{M}_1$ selects $J \in_U \{1, \ldots, \ell\}$. It also selects v_J and set $y_t = g^{w_1 / v_J}$.
4. $\mathcal{M}_1$ runs $\mathcal{U}_1^*$ and simulates the signing oracle for the i-th query in the following way.
   a) For $i \neq J$, $\mathcal{M}_1$ follows the protocol with y-side witness, x. H_2 is simulated by returning random choices from $\langle g \rangle$.
   b) For $i = J$, $\mathcal{M}_1$ engages in the issuing protocol using x and (w_1, w_2) as follows.
      i. Given $(z_{ui}, \xi_i, E_i, P_i)$ from the user, check P_i and reject if incorrect. Otherwise, decrypt $E_i \to \gamma_i$.
      ii. Set $z_{1J} = {y_t}^{v_J}$. (Accordingly, $z_{1J} = g^{w_1}$ and $z_{2J} = h^{w_2}$.)
      iii. Compute $a_J = g^{u_J}, b_{1J} = g^{u_{1J}}, b_{2J} = h^{u_{2J}}$ with $u_J, u_{1J}, u_{2J} \in_U \mathbb{Z}_q$.
      iv. Send $(v_J, a_J, b_{1J}, b_{2J})$ to $\mathcal{U}_1^*$.
      v. Given e_J from $\mathcal{U}_1^*$, choose $d_J \in_U \mathbb{Z}_q$ and compute $c_J := e_J - d_J \bmod q$, $r_J := u_J - c_J x \bmod q$, $s_{1J} := u_{1J} - d_J w_1 \bmod q$, and $s_{2J} := u_{2J} - d_J w_2 \bmod q$.
      vi. Send $(r_J, c_J, s_{1J}, s_{2J}, d_J)$ to $\mathcal{U}_1^*$.

   $\mathcal{M}_1$ simulates $\mathcal{H}_2$ by returning $\varepsilon \in_U \mathbb{Z}_q$.
5. $\mathcal{U}_1^*$ outputs a signature, say $(\zeta_1, \rho, \varpi, \sigma_1, \sigma_2, \delta)$, that corresponds to ε.
6. Rewind and restart $\mathcal{U}_1^*$ with the same setting. Then choose $I \in_U \{0, \ldots, \ell\}$.
   - If $I = 0$, $\mathcal{M}_1$ simulates $\mathcal{H}_2$ by returning $\varepsilon' \in_U \mathbb{Z}_q$. Otherwise, set $\varepsilon' = \varepsilon$.
   - If $I \neq 0$ and run_J have not yet been completed before the query to $\mathcal{H}_2$ is sent, $\mathcal{M}_1$ simulates the execution by using both y-side and z-side witnesses as above choosing $d'_J \in_U \mathbb{Z}_q$. Otherwise, $\mathcal{M}_1$ simulates only with y-side witness choosing $d'_J = d_J$.
7. $\mathcal{U}_1^*$ outputs a signature, say $(\zeta_1, \rho', \varpi', \sigma'_1, \sigma'_2, \delta')$, that corresponds to ε'.
8. If $\delta = \delta'$, simulation fails. Otherwise, $\mathcal{M}_1$ computes $w'_1 = (\sigma_1 - \sigma'_1)/(\delta' - \delta) \bmod q$, $w'_2 = (\sigma_2 - \sigma'_2)/(\delta - \delta') \bmod q$, and outputs $\mathbf{X} = (w_1 - w'_1)/(w'_2 - w_2) \bmod q$.

Sketch of success probability evaluation:

Suppose that all random variables chosen by the simulating signer are determined purely from the random tape so that they are fixed before the simulation starts. We consider how δ in $\Sigma^\star$ is sensitive to the alteration of ε and $\{d_{i_{k+1}}, \ldots, d_{i_\ell}\}$ which are given after ε is given to $\mathcal{U}_1^*$. Observe that independent variables given to $\mathcal{U}_1^*$ are p, q, g, h, y, $\mathcal{H}_1$, $\mathcal{H}_2$, $\mathsf{sid}_i, a_i, b_{1i}, b_{2i}$, d_i for all i, and ε and the random tape of $\mathcal{U}_1^*$. All other variables are uniquely determined by these independent variables and outputs of $\mathcal{U}_1^*$. We wrap all these independent variables into Λ, except for $\{\varepsilon, d_{i_{k+1}}, \ldots, d_{i_\ell}\}$, which is denoted by D_ε hereafter. Let D denote $D_\varepsilon \setminus \{\varepsilon\}$.

Let S be the set of all (Λ, D_ε) that leads $\mathcal{U}_1^*$ to a success, i.e., $\Pr_{\Lambda,D_\varepsilon}[(\Lambda, D_\varepsilon) \in S] \geq \epsilon_1$. According to the Splitting Lemma [11,22], with probability at least $\epsilon_1/2$, randomly selected Λ satisfies $\Pr_{D_\varepsilon}[(\Lambda, D_\varepsilon) \in S] \geq \epsilon_1/2$. Once Λ is fixed, δ is uniquely determined by D_ε. By $\delta \leftarrow D_\varepsilon$, we denote the map from $(\Lambda, D_\varepsilon) \in S$ to δ. If $(\Lambda, D_\varepsilon) \notin S$, we denote $\perp \leftarrow D_\varepsilon$.

Define function ψ as

$$\psi(\delta) = \Pr_{D_\varepsilon}[\delta \leftarrow D_\varepsilon].$$

Let δ_{max} be the value of δ that maximizes $\psi(\delta)$. That is, δ_{max} is the value of δ that is most likely to appear in $\Sigma^\star$. Let $\psi_{max} = \psi(\delta_{max})$. We consider two cases.

Case 1 (ψ_{max} is not negligible) :

In this case, for randomly chosen D_ε and D'_ε, the adversary is likely to output signatures that contain δ_{max} with sufficiently large probability. When δ is the same for different ε from $\mathcal{H}_2$, ϖ must differ as $\delta + \varpi = \varepsilon$. Consequently, with sufficient probability, we obtain $\varpi \neq \varpi'$ with which y-side witness can be extracted as written in Step-7 of Case $\chi = 0$. For more details, we refer to the proof of Lemma 3 of [1].

Case 2 (ψ_{max} is negligible) :

In this case, δ tends to change if D_ε is altered. Due to [1], randomly chosen D_ε and D'_ε that differ only at one position lead $\mathcal{U}_1^*$ to output two corresponding signatures $(\zeta_1, \rho, \varpi, \sigma_1, \sigma_2, \delta)$ and $(\zeta_1, \rho', \varpi', \sigma'_1, \sigma'_2, \delta')$ with sufficiently large probability. From these signatures, we can extract w'_1, w'_2 that satisfy $\zeta_1 = g^{w'_1}$ and $\zeta/\zeta_1 = h^{w'_2}$. By assumption, $\log_{z_u} z \neq \log_{z_1} \zeta_1$. So $w_1 \neq w'_1$ and $w_2 \neq w'_2$ holds. Accordingly $\mathbf{X} = \log_g h = (w_1 - w'_1)/(w'_2 - w_2) \bmod q$ is computable.

The probability distribution over these cases depends on Λ and the strategy of $\mathcal{U}_1^*$. Note that the distribution of Λ does not depend on the choice of χ as the protocol is witness indistinguishable and the public key is generated so that it distributes uniformly. Accordingly, the coin flip of χ turns the simulation to the proper case with probability 1/2.

In the above, we proved that for ζ_1 in a valid signature, there exists at least one session that includes (z_u, z_1) that satisfies $\log_{z_u} z = \log_{z_1} \zeta_1$. Since $z_1 (= {z_u}^v)$ depends on random v chosen by the honest signer and z_u is in $\langle g \rangle$ when P is valid, z_1 is unique among all sessions with overwhelming probability if only polynomially many sessions are executed.

We also need to prove that a signature cannot be produced without interacting with the legitimate signer. This can be done by a standard argument that uses the Forking Lemma and so is omitted here.

Finally, we need to show that a session that includes target (z_u, z_1) can be identified from ζ_1^{1/x_t}. For this, observe that the rightmost equality in Equation 3 holds because $\xi = g^\gamma$ for $\gamma = \log_{z_u} z = \log_{z_1} \zeta_1$ with overwhelming probability due to the soundness of P. □

Theorem 4. *The proposed scheme is signature traceable if all hash functions are random oracles, the discrete logarithm is intractable, and the soundness condition of the verifiable encryption scheme holds.*

Proof. We need to show that no adversary can generate a signature containing ζ_1 such that $\zeta_1 \neq (\xi^v)^{x_t}$ for any (ξ^v) stored by the signer. This can be done in the same way as done in the proof of Theorem 3.

In the following, we show that it is infeasible for the user to output two valid signatures that contain the same ζ_1 regardless of the user's behaviour.

The proof is done by contradiction. Suppose that there exists an adversary $\mathcal{U}_2^*$ that outputs two valid signatures that result in the same session by revocation with success probability ϵ_2. Here, ϵ_2 is not negligible in n and $\mathcal{U}_2^*$ is allowed to interact with $\mathcal{S}$ at most ℓ times in an arbitrary fashion. Let $\ell \geq 1$. ($\ell = 0$ was considered in Theorem 3.)

Now there exist two queries to $\mathcal{H}_2$ that correspond to those two signatures. In a similar way as used in the proof of Theorem 3, we guess the indexes of these queries and regard $\mathcal{U}_2^*$ as being successful only if the guess is correct. Accordingly, this is equivalent to an adversary, say $\mathcal{U}_3^*$, that asks $\mathcal{H}_2$ only twice and succeeds with probability $\epsilon_3 = \epsilon_2 / \binom{q_h}{2}$ in producing two signatures in the expected relation.

We construct a machine $\mathcal{M}_2$ that, given $(\mathbf{p}, \mathbf{q}, \mathbf{g}, \mathbf{Y})$, solves $\mathbf{X} = \log_g \mathbf{Y}$ in $\mathbb{Z}_q$ by using $\mathcal{U}_3^*$.

Reduction algorithm:

1. $\mathcal{M}_2$ sets $(p, q, g) := (\mathbf{p}, \mathbf{q}, \mathbf{g})$.
2. $\mathcal{M}_2$ sets either $y = \mathbf{Y}$ or $y = g^x$ for $x \in_U \mathbb{Z}_q^*$ by flipping coin χ.
3. $\mathcal{M}_2$ selects $w, w_0, w_1 \in_U \mathbb{Z}_q$ and sets $h := g^w$ and $z := g^{w_0}$, $y_t := g^{w_1}$.
4. $\mathcal{M}_2$ selects $I \in_U \{1, \ldots, \ell\}$.
5. $\mathcal{M}_2$ runs $\mathcal{U}_3^*$ simulating $\mathcal{S}$ as follows.
   - For run_i $(i \neq I)$, $\mathcal{M}_2$ simulates with z-side witness in the same way as shown in Step-3 of Case $\chi = 0$ in the proof of Theorem 3.
   - For run_I,
     - if $y = \mathbf{Y}$, $\mathcal{M}_2$ simulates with z-side witness as above, otherwise
     - it sets $z_{1I} = \mathbf{Y}$ and simulate P_s in the standard way by setting $\mathcal{H}_3$ conveniently. Then follow the rest of the protocol using x. Save γ_I by decrypting E_I.

   $\mathcal{M}_2$ simulates $\mathcal{H}_2$ by returning random values, say ε_1 and ε_2.

6. $\mathcal{U}_3^*$ outputs two signatures.
7. $\mathcal{M}_2$ rewinds and restarts $\mathcal{U}_3^*$ with the same setting. It selects $J \in_U \{1, 2\}$ and answers to J-th query to $\mathcal{H}_2$ with $\varepsilon'_J \in_U \mathbb{Z}_q$.
8. $\mathcal{U}_3^*$ outputs two signatures.
9. Let $(\zeta_1, \rho, \varpi, \sigma_1, \sigma_2, \delta)$ and $(\zeta_1, \rho', \varpi', \sigma'_1, \sigma'_2, \delta')$ be the resulting signatures that correspond to ε_J and ε'_J respectively. (If any of the resulting signatures does not correspond to the hash value, $\mathcal{M}_2$ fails.) If $\chi = 0$ and $\varpi \neq \varpi'$, $\mathcal{M}_2$ outputs $\log_g y = \log_g \mathbf{Y} = (\rho - \rho')/(\varpi' - \varpi) \bmod q$. If $\chi = 1$ and $\delta \neq \delta'$, it outputs $\log_g z_{1I} = \log_g \mathbf{Y} = (\sigma_1 - \sigma'_1)/\gamma_I(\delta - \delta') \bmod q$. $\mathcal{M}_2$ fails, otherwise.

We omit the evaluation of success probability as it can be done in the same way as shown in the proof of Theorem 3 of [1]. □

Due to Theorem 4 and 3, the mapping between each session and valid signature is bijective with overwhelming probability. Accordingly, we have the following corollary.

Corollary 1. *The proposed scheme is $(\ell, \ell + 1)$-unforgeable for polynomially bound ℓ if the discrete logarithm is intractable, all hash functions are random oracles, and the verifiable encryption is secure and simulatable.*

6 Remarks and Open Problems

- When each user uses a unique (z_u, ξ, E, P) repeatedly in all issuing sessions, i.e. as a public-key of the user, the scheme provides blindness (and unlinkability) in a weak sense. That is, signatures are computationally independent of each other unless the signer cooperates with the attacker. Such low-level privacy may be acceptable in applications as it offers less computation and communication complexity instead.
- As briefly mentioned in Section 2, the security definitions and the proofs confirm the security under the assumption that the trustee will never be abused as an oracle. Accordingly, the trustee must not show the tracing information to anybody. To provide stronger security in blindness where the trustee can publish the tracing information, we need the following properties. First, the verifiable encryption must be non-malleable against adaptive chosen message attacks. It also has to provide public verifiability. Second, the signature scheme must be unforgeable even for the signer in such a sense that for target signature Σ produced from a session identified by ξ^v the signer should not be able to produce valid signature $\Sigma'(\neq \Sigma)$ that results in tracing information that is relative to ξ^v. This property is not achieved in our construction even if we restrict Σ' to be different from Σ in the part necessary for revocation, which is ζ_1 in our case. A particular attack on the strong blindness is as follows. The signer transforms ζ_1 in challenge signature Σ into $\zeta' = \zeta_1^a$ with random a and creates signature Σ' that includes ζ' by using real signing key x. Session tracing information computed from ζ' will be $(\xi^v)^a$ and the signer can obtain target session identifier ξ^v. This

particular attack can be prevented but we leave a provably secure solution for this issue an open problem.

- It is important to point out that, since the trustee can recover γ from E, he can produce signature Σ' that results in the same tracing information ξ^v linked from signature Σ legitimately produced by the user. Such a threat can be eliminated by encrypting γ with a encryption key whose decryption key is not known to anybody. (Remember that the decryption-key is not necessary for the trustee to complete revocation.) But for the sake of security proof, the simulator must be able to decrypt it. This is possible, for instance, with the verifiable encryption scheme in Appendix B. By generating encryption key y as $y = \mathcal{H}(\mathsf{str})$ where str is a fixed public string and $\mathcal{H}$ is a hash function $\mathcal{H} : \{0,1\}^* \to \langle g \rangle$. In this way, any party can be convinced that no one knows the decryption key corresponding to y, but a simulator that simulates the hash function as a random oracle in the proof of revocability can assign arbitrary g^x as $\mathcal{H}(\mathsf{str})$ so that x is known only to the simulator.
- Since revocation only identifies a specific randomness appearing in a issuing session, it would be necessary to assure that the session is really done by the user. An easy solution would be to have the transcript signed by the user. Although the signer may flame the user by creating Σ' from Σ so that they result in the same session tracing information in the similar way shown in the second remark, one can see that it is not the user who created the second signature due to Theorem 3.

Acknowledgements. The authors thank David Pointcheval for helpful comments. Contribution from Eiichiro Fujisaki about verifiable encryption schemes is appreciated very much.

References

1. M. Abe. A three-move blind signature scheme secure for pollynomially many signatures. In B. Pfitzmann, editor, *Advances in Cryptology — EUROCRYPT '01*, volume 2045 of *Lecture Notes in Computer Science*, pages 136–151. Springer-Verlag, 2001.
2. M. Abe and T. Okamoto. Provably secure partially blind signatures. In M. Bellare, editor, *Advances in Cryptology — CRYPTO 2000*, volume 1880 of *Lecture Notes in Computer Science*, pages 271–286. Springer-Verlag, 2000.
3. G. Ateniese. Efficient verifiable encryption (and fair exchange) of digital signatures. In *ACM CCS'99*, pages 138–146. Association for Computing Machinery, 1999.
4. F. Bao. An efficient verifiable encryption scheme for encryption of discrete logarithms. In *CARDIS'98*, 1998.
5. M. Bellare and P. Rogaway. Random oracles are practical: a paradigm for designing efficient protocols. In *First ACM Conference on Computer and Communication Security*, pages 62–73. Association for Computing Machinery, 1993.
6. E. Brickell, P. Gemmell, and D. Kravitz. Trustee-based tracking extensions to anonymous cash and the making of anonymous change. In *Proceedings of Sixth Annual ACM-SIAM Symposium on Discrete Algorithms*, pages 457–466. ACM, 1995.

7. J. Camenisch. *Group Signature Schemes and Payment Systems Based on the Discrete Logarithm Problem.* PhD thesis, ETH Zürich, 1998.
8. J. Camenisch and I. Damgård. Verifiable encryption, group encryption, and their applications to separable group signatures and signature sharing schemes. In T. Okamoto, editor, *Advances in Cryptology – Asiacrypt 2000*, volume 1976 of *Lecture Notes in Computer Science*, pages 331–345. Springer-Verlag, 2000.
9. J. Camenisch, J.-M. Piveteau, and M. Stadler. Fair blind signatures. In L. C. Guillou and J.-J. Quisquater, editors, *Advances in Cryptology — EUROCRYPT '95*, volume 921 of *Lecture Notes in Computer Science*, pages 209–219. Springer-Verlag, 1995.
10. D. L. Chaum and T. P. Pedersen. Wallet databases with observers. In E. F. Brickell, editor, *Advances in Cryptology — CRYPTO '92*, volume 740 of *Lecture Notes in Computer Science*, pages 89–105. Springer-Verlag, 1993.
11. A. Fiat and A. Shamir. How to prove yourself: Practical solutions to identification and signature problems. In A. M. Odlyzko, editor, *Advances in Cryptology — CRYPTO '86*, volume 263 of *Lecture Notes in Computer Science*, pages 186–199. Springer-Verlag, 1987.
12. Y. Frankel, Y. Tsiounis, and M. Yung. "Indirect discourse proofs": Achieving efficient fair off-line e-cash. In K. Kim and T. Matsumoto, editors, *Advances in Cryptology — ASIACRYPT '96*, volume 1163 of *Lecture Notes in Computer Science*, pages 286–300. Springer-Verlag, 1996.
13. E. Fujisaki. A simple approach to secretly sharing a factoring witness in publicly-verifiable manner. (unpublished manuscript), 2001.
14. E. Fujisaki and T. Okamoto. Statistical zero knowledge protocols to prove modular polynomial relations. In B. S. Kaliski Jr., editor, *Advances in Cryptology — CRYPTO '97*, volume 1294 of *Lecture Notes in Computer Science*, pages 16–30. Springer-Verlag, 1997.
15. M. Jakobsson and J. Müller. Improved magic ink signatures using hints. In *Financial Cryptography'99*, 1999.
16. M. Jakobsson and M. Yung. Distributed "Magic Ink" signatures. In W. Fumy, editor, *Advances in Cryptology — EUROCRYPT '97*, volume 1233 of *Lecture Notes in Computer Science*, pages 450–464. Springer-Verlag, 1997.
17. A. Juels, M. Luby, and R. Ostrovsky. Security of blind digital signatures. In B. S. Kaliski Jr., editor, *Advances in Cryptology — CRYPTO '97*, volume 1294 of *Lecture Notes in Computer Science*, pages 150–164. Springer-Verlag, 1997.
18. T. Okamoto and K. Ohta. Divertible zero knowledge interactive proofs and commutative random self-reducibility. In J.-J. Quisquater and J. Vandewalle, editors, *Advances in Cryptology - EUROCRYPT '89*, volume 434 of *Lecture Notes in Computer Science*, pages 134–149. Springer-Verlag, 1990.
19. T. Okamoto and S. Uchiyama. A new public-key cryptosystem as secure as factoring. In K. Nyberg, editor, *Advances in Cryptology — EUROCRYPT '98*, volume 1403 of *Lecture Notes in Computer Science*, pages 308–318. Springer-Verlag, 1998.
20. D. Pointcheval and J. Stern. Provably secure blind signature schemes. In K. Kim and T. Matsumoto, editors, *Advances in Cryptology - ASIACRYPT '96*, volume 1163 of *Lecture Notes in Computer Science*, pages 252–265. Springer-Verlag, 1996.
21. D. Pointcheval and J. Stern. Security proofs for signature schemes. In U. Maurer, editor, *Advances in Cryptology — EUROCRYPT '96*, volume 1070 of *Lecture Notes in Computer Science*, pages 387–398. Springer-Verlag, 1996.
22. D. Pointcheval and J. Stern. Security arguments for digital signatures and blind signatures. *Journal of Cryptology*, 2000.

23. M. Stadler. *Cryptographic Protocols for Revocable Privacy.* PhD thesis, Swiss Federal Institute of Technology Zürich, 1996.
24. M. Stadler. Publicly verifiable secret sharing. In U. Maurer, editor, *Advances in Cryptology — EUROCRYPT '96*, volume 1070 of *Lecture Notes in Computer Science*, pages 190–199. Springer-Verlag, 1996.
25. S. von Solms and D. Naccache. On blind signatures and perfect crime. *Computer & Security*, 11:581–583, 1992.
26. A. Young and M. Yung. Finding length-3 positive cunningham chains and their cryptographic significance. In *ANTS '98*, Lecture Notes in Computer Science. Springer-Verlag, 1998.

Appendix A

The following verifiable encryption scheme is taken from [13]. Let $(\mathsf{n}, \mathsf{g}, \mathsf{h}, \ell_\mathsf{g})$ be the public key and (p, q) be the secret key of the Okamoto-Uchiyama encryption scheme. Here, $\mathsf{n} = \mathsf{p}^2\mathsf{q}$, and g is in $\mathbb{Z}_\mathsf{n}$ that satisfies $\mathrm{ord}(\mathsf{g} \bmod \mathsf{p}^2) = \mathsf{p}(\mathsf{p}-1)$, $\mathsf{h} = {\mathsf{h}_0}^\mathsf{n} \bmod \mathsf{n}$ for randomly chosen $\mathsf{h}_0 \in \mathbb{Z}_n$, and ℓ_g is the bit length of the order of g. We assume that $\ell_g > 2\ell_q$ where ℓ_q is the bit length of q. Let $\mathcal{H}_4 : \{0,1\}^* \to \{0,1\}^{\ell_q}$ be a hash function.

Now, γ is encrypted by Okamoto-Uchiyama encryption as $E = \mathsf{g}^\gamma \mathsf{h}^{t_u} \bmod \mathsf{n}$ where $t_u \in_U \mathbb{Z}_n$. For $C = (z_u, \xi) = (z^{1/\gamma}, g^\gamma)$, $(C, E) \in \mathcal{R}$ is proven by providing $P = (c_u, s_{1u}, s_{2u})$ computed by the prover as follows.

1. Choose $k_1 \in_U \{0,1\}^{\epsilon_s \ell_g}$ and $k_2 \in_U \{0,1\}^{\epsilon_s(\ell_g + \ell_q)}$.
2. Compute $c_u = \mathcal{H}_4(z_u, \xi, E, {z_u}^{k_1} \bmod p, {y_t}^{k_1} \bmod p, \mathsf{g}^{k_1}\mathsf{h}^{k_2} \bmod \mathsf{n})$.
3. Compute $s_1 = k_1 - c_u\gamma$ and $s_2 = k_2 - c_u t_u$ in $\mathbb{Z}$.

Here ϵ_s is a security parameter larger than 1. P is valid if it satisfies

$$c_u \in \{0,1\}^{\ell_q},$$
$$s_{1u} \in \{0,1\}^{\epsilon_s \ell_\mathsf{g}}, \text{ and}$$
$$c_u = \mathcal{H}_4(z_u, \xi, E, {z_u}^{s_{1u}} g^{c_u} \bmod p, {y_{t_u}}^{s_{2u}} \xi^{c_u} \bmod p, \mathsf{g}^{s_{1u}} \mathsf{h}^{s_{2u}} E^{c_u} \bmod \mathsf{n}).$$

The above protocol is a statistical zero-knowledge argument for relation R. Soundness is due to the strong RSA assumption over n. The detailed security proof can be found in [13].

Appendix B

In this section, we require that $p = 2q+1$ and $q = 2s+1$ for prime s. (See [26] for generating such Cunningham Chains.) Let h be a generator of a prime subgroup in $\mathbb{Z}_q$ where $\mathrm{ord}(\mathsf{h}) = s$. Let $(\mathsf{x}, \mathsf{y}) \in \mathbb{Z}_s \times \langle \mathsf{h} \rangle$ be a key pair of ElGamal encryption defined over $\langle \mathsf{h} \rangle$. That is, $\mathsf{y} = \mathsf{h}^\mathsf{x} \bmod q$.

For $\gamma \in \mathbb{Z}_q$ and $C = (z_u, \xi) = (z^{1/\gamma}, g^\gamma)$, (E, P) is computed as follows. We first transform γ into $\gamma^\star \in \langle \mathsf{h} \rangle$ by

$$\gamma^\star = J_q(\gamma) \cdot \gamma \bmod q.$$

Here, $J_q(\gamma)$ is the Jacobi symbol, $(\frac{\gamma}{q})$. $\gamma^\star$ is then encrypted into $E = (C_1, C_2)$ using ElGamal encryption as

$$C_1 = \gamma^\star \cdot \mathsf{y}^\omega \bmod q,$$
$$C_2 = \mathsf{h}^\omega \bmod q,$$

where $\omega \in_U \mathbb{Z}_s$. When E is decrypted into $\gamma^\star$ and $g^{\gamma^\star} \bmod p \neq \xi$, γ is obtained by $\gamma = -1 \cdot \gamma^\star \bmod q$. Otherwise, $\gamma = \gamma^\star$.

The proof is done in two steps. In the first step, the prover proves relation $\log_{z_u} z = \log_g \xi$ by the Chaum-Pedersen protocol [10]. In the second step, we prove in zero-knowledge manner that $\mathcal{D}(E) = J_q(\log_g \xi) \cdot \log_g \xi \bmod q$ by repeating the following protocol sufficiently many times.

1. The prover selects $a \in_U \mathbb{Z}_q^*$ and $b \in_U \mathbb{Z}_s$ and sends

$$T_0 = \xi^a \bmod p,$$
$$T_1 = C_1 \cdot J_q(a) \cdot a \cdot \mathsf{y}^b \bmod q, \text{ and}$$
$$T_2 = C_2 \cdot \mathsf{h}^b \bmod q$$

   to the verifier.
2. The verifier sends $c \in_U \{0, 1\}$ to the prover.
3. The prover sends (α, β) where $(\alpha, \beta) = (a, b)$ when $c = 0$, and $(\alpha, \beta) = (a\gamma \bmod q, b + \omega \bmod q)$ when $c = 1$.
4. The verifier accepts if, for $c = 0$,

$$T_0 = \xi^\alpha \bmod p,$$
$$T_1 = C_1 \cdot J_q(\alpha) \cdot \alpha \cdot \mathsf{y}^\beta \bmod q, \text{ and}$$
$$T_2 = C_2 \cdot \mathsf{h}^\beta \bmod q,$$

   and for $c = 1$,

$$T_0 = g^\alpha \bmod p,$$
$$T_1 = J_q(\alpha) \cdot \alpha \cdot \mathsf{y}^\beta \bmod q, \text{ and}$$
$$T_2 = \mathsf{h}^\beta \bmod q.$$

It is not hard to see that the above is correct, sound, and perfectly zero-knowledge for any verifier. As usual, this method can be made non-interactive by executing all repetitions in parallel and creating the challenge c by hashing all data before the second step.